Graduate Texts in Mathematics 138

Springer
New York
Berlin
Heidelberg
Barcelona
Hong Kong
London
Milan
Paris
Singapore
Tokyo

Graduate Texts in Mathematics

(continued after index)

Henri Cohen

A Course in Computational Algebraic Number Theory

Springer

Henri Cohen
U.F.R. de Mathématiques et Informatique
Université Bordeaux I
351 Cours de la Libération
F-33405 Talence Cedex, France

Fourth Printing 2000

With 1 Figure

Mathematics Subject Classification (1991): 11Y05, 11Y11, 11Y16,
11Y40, 11A51, 11C08, 11C20, 11R09, 11R11, 11R29

ISSN 0072-5285
ISBN 978-3-642-08142-2

Cataloging-In-Publication Data applied for

Die Deutsche Bibliothek - CIP-Einheitsaufnahme

Cohen, Henri:
A course in computational algebraic number theory / Henri
Cohen. - 3., corr. print. - Berlin ; Heidelberg ; New York :
Springer, 1996
(Graduate texts in mathematics ; 138)

NE: GT

© Springer-Verlag Berlin Heidelberg 2010
Printed in Germany

Acknowledgments

This book grew from notes prepared for graduate courses in computational number theory given at the University of Bordeaux I. When preparing this book, it seemed natural to include both more details and more advanced subjects than could be given in such a course. By doing this, I hope that the book can serve two audiences: the mathematician who might only need the details of certain algorithms as well as the mathematician wanting to go further with algorithmic number theory.

In 1991, we started a graduate program in computational number theory in Bordeaux, and this book was also meant to provide a framework for future courses in this area.

In roughly chronological order I need to thank, Horst Zimmer, whose Springer Lecture Notes on the subject [Zim] was both a source of inspiration and of excellent references for many people at the time when it was published.

Then, certainly, thanks must go to Donald Knuth, whose (unfortunately unfinished) series on the Art of Computer Programming ([Knu1], [Knu2] and [Knu3]) contains many marvels for a mathematician. In particular, the second edition of his second volume. Parts of the contents of Chapters 1 and 3 of this book are taken with little or no modifications from Knuth's book. In the (very rare) cases where Knuth goes wrong, this is explicitly mentioned.

My thesis advisor and now colleague Jacques Martinet, has been very influential, both in developing the subject in Bordeaux and more generally in the rest of France–several of his former students are now professors. He also helped to make me aware of the beauty of the subject, since my personal inclination was more towards analytic aspects of number theory, like modular forms or L-functions. Even during the strenuous period (for him!) when he was Chairman of our department, he always took the time to listen or enthusiastically explain.

I also want to thank Hendrik Lenstra, with whom I have had the pleasure of writing a few joint papers in this area. Also Arjen Lenstra, who took the trouble of debugging and improving a big Pascal program which I wrote, which is still, in practice, one of the fastest primality proving programs. Together and separately they have contributed many extremely important algorithms, in particular LLL and its applications (see Section 2.6). My only regret is that they both are now in the U.S.A., so collaboration is more difficult.

Although he is not strictly speaking in the algorithmic field, I must also thank Don Zagier, first for his personal and mathematical friendship and also for his continuing invitations first to Maryland, then at the Max Planck Institute in Bonn, but also because he is a mathematician who takes both real pleasure and real interest in creating or using algorithmic tools in number theory. In fact, we are currently finishing a large algorithmic project, jointly with Nils Skoruppa.

Daniel Shanks, both as an author and as editor of Mathematics of Computation, has also had a great influence on the development of algorithmic algebraic number theory. I have had the pleasure of collaborating with him during my 1982 stay at the University of Maryland, and then in a few subsequent meetings.

My colleagues Christian Batut, Dominique Bernardi and Michel Olivier need to be especially thanked for the enormous amount of unrewarding work that they put in the writing of the PARI system under my supervision. This system is now completely operational (even though a few unavoidable bugs crop up from time to time), and is extremely useful for us in Bordeaux, and for the (many) people who have a copy of it elsewhere. It has been and continues to be a great pleasure to work with them.

I also thank my colleague Francois Dress for having collaborated with me to write our first multi-precision interpreter ISABELLE, which, although considerably less ambitious than PARI, was a useful first step.

I met Johannes Buchmann several years ago at an international meeting. Thanks to the administrative work of Jacques Martinet on the French side, we now have a bilateral agreement between Bordeaux and Saarbrücken. This has allowed several visits, and a medium term joint research plan has been informally decided upon. Special thanks are also due to Johannes Buchmann and Horst Zimmer for this. I need to thank Johannes Buchmann for the many algorithms and techniques which I have learned from him both in published work and in his preprints. A large part of this book could not have been what it is without his direct or indirect help. Of course, I take complete responsibility for the errors that may have appeared!

Although I have met Michael Pohst and Hans Zassenhaus[1] only in meetings and did not have the opportunity to work with them directly, they have greatly influenced the development of modern methods in algorithmic number theory. They have written a book [Poh-Zas] which is a landmark in the subject. I recommend it heartily for further reading, since it goes into subjects which could not be covered in this book.

I have benefited from discussions with many other people on computational number theory, which in alphabetical order are, Oliver Atkin, Anne-Marie Bergé, Bryan Birch, Francisco Diaz y Diaz, Philippe Flajolet, Guy Henniart, Kevin McCurley, Jean-Francois Mestre, Francois Morain, Jean-Louis

---

[1]Hans Zassenhaus died on November 21, 1991.

Nicolas, Andrew Odlyzko, Joseph Oesterlé, Johannes Graf von Schmettow, Claus-Peter Schnorr, Rene Schoof, Jean-Pierre Serre, Bob Silverman, Harold Stark, Nelson Stephens, Larry Washington. There are many others that could not be listed here. I have taken the liberty of borrowing some of their algorithms, and I hope that I will be forgiven if their names are not always mentioned.

The theoretical as well as practical developments in Computational Number Theory which have taken place in the last few years in Bordeaux would probably not have been possible without a large amount of paperwork and financial support. Hence, special thanks go to the people who made this possible, and in particular to Jean-Marc Deshouillers, Francois Dress and Jacques Martinet as well as the relevant local and national funding committees and agencies.

I must thank a number of persons without whose help we would have been essentially incapable of using our workstations, in particular "Achille" Braquelaire, Laurent Fallot, Patrick Henry, Viviane Sauquet-Deletage, Robert Strandh and Bernard Vauquelin.

Although I do not know anybody there, I would also like to thank the GNU project and its creator Richard Stallman, for the excellent software they produce, which is not only free (as in "freedom", but also as in "freeware"), but is generally superior to commercial products. Most of the software that we use comes from GNU.

Finally, I thank all the people, too numerous to mention, who have helped me in some way or another to improve the quality of this book, and in particular to Dominique Bernardi and Don Zagier who very carefully read drafts of this book. But special thanks go to Gary Cornell who suggested improvements to my English style and grammar in almost every line.

In addition, several people contributed directly or helped me write specific sections of the book. In alphabetical order they are D. Bernardi (algorithms on elliptic curves), J. Buchmann (Hermite normal forms and sub-exponential algorithms), J.-M. Couveignes (number field sieve), H. W. Lenstra (in several sections and exercises), C. Pomerance (factoring and primality testing), B. Vallée (LLL algorithms), P. Zimmermann (Appendix A).

Preface

With the advent of powerful computing tools and numerous advances in mathematics, computer science and cryptography, algorithmic number theory has become an important subject in its own right. Both external and internal pressures gave a powerful impetus to the development of more powerful algorithms. These in turn led to a large number of spectacular breakthroughs. To mention but a few, the LLL algorithm which has a wide range of applications, including real world applications to integer programming, primality testing and factoring algorithms, sub-exponential class group and regulator algorithms, etc . . .

Several books exist which treat parts of this subject. (It is essentially impossible for an author to keep up with the rapid pace of progress in all areas of this subject.) Each book emphasizes a different area, corresponding to the author's tastes and interests. The most famous, but unfortunately the oldest, is Knuth's *Art of Computer Programming*, especially Chapter 4.

The present book has two goals. First, to give a reasonably comprehensive **introductory** course in computational number theory. In particular, although we study some subjects in great detail, others are only mentioned, but with suitable pointers to the literature. Hence, we hope that this book can serve as a first course on the subject. A natural sequel would be to study more specialized subjects in the existing literature.

The prerequisites for reading this book are contained in introductory texts in number theory such as Hardy and Wright [H-W] and Borevitch and Shafarevitch [Bo-Sh]. The reader also needs some feeling or taste for algorithms and their implementation. To make the book as self-contained as possible, the main definitions are given when necessary. However, it would be more reasonable for the reader to first acquire some basic knowledge of the subject before studying the algorithmic part. On the other hand, algorithms often give natural proofs of important results, and this nicely complements the more theoretical proofs which may be given in other books.

The second goal of this course is **practicality**. The author's primary intentions were not only to give fundamental and interesting algorithms, but also to concentrate on practical aspects of the implementation of these algorithms. Indeed, the theory of algorithms being not only fascinating but rich, can be (somewhat arbitrarily) split up into four closely related parts. The first is the discovery of new algorithms to solve particular problems. The second is the detailed mathematical analysis of these algorithms. This is usually quite mathematical in nature, and quite often intractable, although the algorithms seem to perform rather well in practice. The third task is to study the complexity of the problem. This is where notions of fundamental importance in complexity theory such as NP-completeness come in. The last task, which some may consider the least noble of the four, is to actually implement the algorithms. But this task is of course as essential as the others for the actual resolution of the problem.

In this book we give the algorithms, the mathematical analysis and in some cases the complexity, without proofs in some cases, especially when it suffices to look at the existing literature such as Knuth's book. On the other hand, we have usually tried as carefully as we could, to give the algorithms in a ready to program form–in as optimized a form as possible. This has the drawback that some algorithms are unnecessarily clumsy (this is unavoidable if one optimizes), but has the great advantage that a casual user of these algorithms can simply take them as written and program them in his/her favorite programming language. In fact, the author himself has implemented almost all the algorithms of this book in the number theory package PARI (see Appendix A).

The approach used here as well as the style of presentation of the algorithms is similar to that of Knuth (analysis of algorithms excepted), and is also similar in spirit to the book of Press et al [PFTV] *Numerical Recipes (in Fortran, Pascal or C)*, although the subject matter is completely different.

For the practicality criterion to be compatible with a book of reasonable size, some compromises had to be made. In particular, on the mathematical side, many proofs are not given, especially when they can easily be found in the literature. From the computer science side, essentially no complexity results are proved, although the important ones are stated.

The book is organized as follows. The first chapter gives the fundamental algorithms that are constantly used in number theory, in particular algorithms connected with powering modulo N and with the Euclidean algorithm.

Many number-theoretic problems require algorithms from linear algebra over a field or over $\mathbb{Z}$. This is the subject matter of Chapter 2. The highlights of this chapter are the Hermite and Smith normal forms, and the fundamental LLL algorithm.

In Chapter 3 we explain in great detail the Berlekamp-Cantor-Zassenhaus methods used to factor polynomials over finite fields and over $\mathbb{Q}$, and we also give an algorithm for finding all the complex roots of a polynomial.

Chapter 4 gives an introduction to the algorithmic techniques used in number fields, and the basic definitions and results about algebraic numbers and number fields. The highlights of these chapters are the use of the Hermite Normal Form representation of modules and ideals, an algorithm due to Diaz y Diaz and the author for finding "simple" polynomials defining a number field, and the subfield and field isomorphism problems.

Quadratic fields provide an excellent testing and training ground for the techniques of algorithmic number theory (and for algebraic number theory in general). This is because although they can easily be generated, many non-trivial problems exist, most of which are unsolved (are there infinitely many real quadratic fields with class number 1?). They are studied in great detail in Chapter 5. In particular, this chapter includes recent advances on the efficient computation in class groups of quadratic fields (Shanks's NUCOMP as modified by Atkin), and sub-exponential algorithms for computing class groups and regulators of quadratic fields (McCurley-Hafner, Buchmann).

Chapter 6 studies more advanced topics in computational algebraic number theory. We first give an efficient algorithm for computing integral bases in number fields (Zassenhaus's round 2 algorithm), and a related algorithm which allows us to compute explicitly prime decompositions in field extensions as well as valuations of elements and ideals at prime ideals. Then, for number fields of degree less than or equal to 7 we give detailed algorithms for computing the Galois group of the Galois closure. We also study in some detail certain classes of cubic fields. This chapter concludes with a general algorithm for computing class groups and units in general number fields. This is a generalization of the sub-exponential algorithms of Chapter 5, and works quite well. For other approaches, I refer to [Poh-Zas] and to a forthcoming paper of J. Buchmann. This subject is quite involved so, unlike most other situations in this book, I have not attempted to give an efficient algorithm, just one which works reasonably well in practice.

Chapters 1 to 6 may be thought of as one unit and describe many of the most interesting aspects of the theory. These chapters are suitable for a two semester graduate (or even a senior undergraduate) level course in number theory. Chapter 6, and in particular the class group and unit algorithm, can certainly be considered as a climax of the first part of this book.

A number theorist, especially in the algorithmic field, must have a minimum knowledge of elliptic curves. This is the subject of chapter 7. Excellent books exist about elliptic curves (for example [Sil] and [Sil3]), but our aim is a little different since we are primarily concerned with applications of elliptic curves. But a minimum amount of culture is also necessary, and so the flavor of this chapter is quite different from the others chapters. In the first three sections, we give the essential definitions, and we give the basic and most striking results of the theory, with no pretense to completeness and no algorithms.

The theory of elliptic curves is one of the most marvelous mathematical theories of the twentieth century, and abounds with important conjectures. They are also mentioned in these sections. The last sections of Chapter 7, give a number of useful algorithms for working on elliptic curves, with little or no proofs.

The reader is warned that, apart from the material necessary for later chapters, Chapter 7 needs a much higher mathematical background than the other chapters. It can be skipped if necessary without impairing the understanding of the subsequent chapters.

Chapter 8 (whose title is borrowed from a talk of Hendrik Lenstra) considers the techniques used for primality testing and factoring prior to the 1970's, with the exception of the continued fraction method of Brillhart-Morrison which belongs in Chapter 10.

Chapter 9 explains the theory and practice of the two modern primality testing algorithms, the Adleman-Pomerance-Rumely test as modified by H. W. Lenstra and the author, which uses Fermat's (little) theorem in cyclotomic fields, and Atkin's test which uses elliptic curves with complex multiplication.

Chapter 10 is devoted to modern factoring methods, i.e. those which run in sub-exponential time, and in particular to the Elliptic Curve Method of Lenstra, the Multiple Polynomial Quadratic Sieve of Pomerance and the Number Field Sieve of Pollard. Since many of the methods described in Chapters 9 and 10 are quite complex, it is not reasonable to give ready-to-program algorithms as in the preceding chapters, and the implementation of any one of these complex methods can form the subject of a three month student project.

In Appendix A, we describe what a serious user should know about computer packages for number theory. The reader should keep in mind that the author of this book is biased since he has written such a package himself (this package being available without cost by anonymous ftp).

Appendix B has a number of tables which we think may useful to the reader. For example, they can be used to check the correctness of the implementation of certain algorithms.

What I have tried to cover in this book is so large a subject that, necessarily, it cannot be treated in as much detail as I would have liked. For further reading, I suggest the following books.

For Chapters 1 and 3, [Knu1] and [Knu2]. This is the bible for algorithm analysis. Note that the sections on primality testing and factoring are outdated. Also, algorithms like the LLL algorithm which did not exist at the time he wrote are, obviously, not mentioned. The recent book [GCL] contains essentially all of our Chapter 3, as well as many more polynomial algorithms which we have not covered in this book such as Gröbner bases computation.

For Chapters 4 and 5, [Bo-Sh], [Mar] and [Ire-Ros]. In particular, [Mar] and [Ire-Ros] contain a large number of practical exercises, which are not far from the spirit of the present book, [Ire-Ros] being more advanced.

For Chapter 6, [Poh-Zas] contains a large number of algorithms, and treats in great detail the question of computing units and class groups in general number fields. Unfortunately the presentation is sometimes obscured by quite complicated notations, and a lot of work is often needed to implement the algorithms given there.

For Chapter 7, [Sil] and [Sil3] are excellent books, and contain numerous exercises. Another good reference is [Hus], as well as [Ire-Ros] for material on zeta-functions of varieties. The algorithmic aspect of elliptic curves is beautifully treated in [Cre], which I also heartily recommend.

For Chapters 8 to 10, the best reference to date, in addition to [Knu2], is [Rie]. In addition, Riesel has several chapters on prime number theory.

Note on the exercises. The exercises have a wide range of difficulty, from extremely easy to unsolved research problems. Many are actually implementation problems, and hence not mathematical in nature. No attempt has been made to grade the level of difficulty of the exercises as in Knuth, except of course that unsolved problems are mentioned as such. The ordering follows roughly the corresponding material in the text.

WARNING. Almost all of the algorithms given in this book have been programmed by the author and colleagues, in particular as a part of the Pari package. The programming has not however, always been synchronized with the writing of this book, so it may be that some algorithms are incorrect, and others may contain slight typographical errors which of course also invalidate them. Hence, the author and Springer-Verlag do not assume any responsibility for consequences which may directly or indirectly occur from the use of the algorithms given in this book. Apart from the preceding legalese, the author would appreciate corrections, improvements and so forth to the algorithms given, so that this book may improve if further editions are printed. The simplest is to send an e-mail message to

cohen@math.u-bordeaux.fr

or else to write to the author's address. In addition, a regularly updated errata file is available by anonymous ftp from megrez.math.u-bordeaux.fr (147.210.16.17), directory pub/cohenbook.

Contents

Chapter 2 Algorithms for Linear Algebra and Lattices 46

Chapter 3 Algorithms on Polynomials 109

Chapter 1

Fundamental Number-Theoretic Algorithms

1.1 Introduction

This book describes in detail a number of algorithms used in algebraic number theory and the theory of elliptic curves. It also gives applications to problems such as factoring and primality testing. Although the algorithms and the theory behind them are sufficiently interesting in themselves, I strongly advise the reader to take the time to implement them on her/his favorite machine. Indeed, one gets a feel for an algorithm mainly after executing it several times. (This book does help by providing many tricks that will be useful for doing this.)

We give the necessary background on number fields and classical algebraic number theory in Chapter 4, and the necessary prerequisites on elliptic curves in Chapter 7. This chapter shows you some basic algorithms used almost constantly in number theory. The best reference here is [Knu2].

1.1.1 Algorithms

Before we can describe even the simplest algorithms, it is necessary to precisely define a few notions. However, we will do this without entering into the sometimes excessively detailed descriptions used in Computer Science. For us, an *algorithm* will be a method which, given certain types of inputs, gives an answer after a finite amount of time.

Several things must be considered when one describes an algorithm. The first is to prove that it is correct, i.e. that it gives the desired result when it stops. Then, since we are interested in practical implementations, we must give an estimate of the algorithm's running time, if possible both in the worst case, and on average. Here, one must be careful: the running time will always be measured in *bit operations*, i.e. logical or arithmetic operations on zeros and ones. This is the most realistic model, if one assumes that one is using real computers, and not idealized ones. Third, the space requirement (measured in bits) must also be considered. In many algorithms, this is negligible, and then we will not bother mentioning it. In certain algorithms however, it becomes an important issue which has to be addressed.

First, some useful terminology: The size of the inputs for an algorithm will usually be measured by the number of bits that they require. For example, the size of a positive integer N is $\lfloor \lg N \rfloor + 1$ (see below for notations). We will say that an algorithm is *linear*, *quadratic* or *polynomial time* if it requires time $O(\ln N)$, $O(\ln^2 N)$, $O(P(\ln N))$ respectively, where P is a polynomial. If the time required is $O(N^\alpha)$, we say that the algorithm is exponential time. Finally, many algorithms have some intermediate running time, for example

$$e^{C\sqrt{\ln N \ln \ln N}} \, ,$$

which is the approximate expected running time of many factoring algorithms and of recent algorithms for computing class groups. In this case we say that the algorithm is *sub-exponential*.

The definition of algorithm which we have given above, although a little vague, is often still too strict for practical use. We need also *probabilistic algorithms*, which depend on a source of random numbers. These "algorithms" should in principle not be called algorithms since there is a possibility (of probability zero) that they do not terminate. Experience shows, however, that probabilistic algorithms are usually more efficient than non-probabilistic ones; in many cases they are even the only ones available.

Probabilistic algorithms should not be mistaken with methods (which I refuse to call algorithms), which produce a result which has a high probability of being correct. It is essential that an algorithm produces correct results (discounting human or computer errors), even if this happens after a very long time. A typical example of a non-algorithmic method is the following: suppose N is large and you suspect that it is prime (because it is not divisible by small numbers). Then you can compute

$$2^{N-1} \bmod N$$

using the powering Algorithm 1.2.1 below. If it is not 1 mod N, then this proves that N is not prime by Fermat's theorem. On the other hand, if it is equal to 1 mod N, there is a very good chance that N is indeed a prime. But this is not a proof, hence not an algorithm for primality testing (the smallest counterexample is $N = 341$).

Another point to keep in mind for probabilistic algorithms is that the idea of absolute running time no longer makes much sense. This is replaced by the notion of expected running time, which is self-explanatory.

1.1.2 Multi-precision

Since the numbers involved in our algorithms will almost always become quite large, a prerequisite to any implementation is some sort of multi-precision package. This package should be able to handle numbers having up to 1000 decimal digits. Such a package is easy to write, and one is described in detail in Riesel's book ([Rie]). One can also use existing packages or languages, such as Axiom, Bignum, Derive, Gmp, Lisp, Macsyma, Magma, Maple, Mathematica, Pari, Reduce, or Ubasic (see Appendix A). Even without a multi-precision package, some algorithms can be nicely tested, but their scope becomes more limited.

The pencil and paper method for doing the usual operations can be implemented without difficulty. One should not use a base-10 representation, but rather a base suited to the computer's hardware.

Such a bare-bones multi-precision package must include at the very least:

• Addition and subtraction of two n-bit numbers (time linear in n).

• Multiplication and Euclidean division of two n-bit numbers (time linear in n^2).

• Multiplication and division of an n-bit number by a short integer (time linear in n). Here the meaning of short integer depends on the machine. Usually this means a number of absolute value less than 2^{15}, 2^{31}, 2^{35} or 2^{63}.

• Left and right shifts of an n bit number by small integers (time linear in n).

• Input and output of an n-bit number (time linear in n or in n^2 depending whether the base is a power of 10 or not).

Remark. Contrary to the choice made by some systems such as Maple, I strongly advise using a power of 2 as a base, since usually the time needed for input/output is only a very small part of the total time, and it is also often dominated by the time needed for physical printing or displaying the results.

There exist algorithms for multiplication and division which as n gets large are much faster than $O(n^2)$, the best, due to Schönhage and Strassen, running in $O(n \ln n \ln \ln n)$ bit operations. Since we will be working mostly with numbers of up to roughly 100 decimal digits, it is not worthwhile to implement these more sophisticated algorithms. (These algorithms become practical only for numbers having more than several hundred decimal digits.) On the other hand, simpler schemes such as the method of Karatsuba (see [Knu2] and Exercise 2) can be useful for much smaller numbers.

The times given above for the basic operations should constantly be kept in mind.

Implementation advice. For people who want to write their own bare-bones multi-precision package as described above, by far the best reference is [Knu2] (see also [Rie]). A few words of advice are however necessary. A priori, one can write the package in one's favorite high level language. As will be immediately seen, this limits the multi-precision base to roughly the square root of the word size. For example, on a typical 32 bit machine, a high level language will be able to multiply two 16-bit numbers, but not two 32-bit ones since the result would not fit. Since the multiplication algorithm used is quadratic, this immediately implies a loss of a factor 4, which in fact usually becomes a factor of 8 or 10 compared to what could be done with the machine's central processor. This is intolerable. Another alternative is to write everything in assembly language. This is extremely long and painful, usually bug-ridden, and in addition not portable, but at least it is fast. This is the solution used in systems such as Pari and Ubasic, which are much faster than their competitors when it comes to pure number crunching.

There is a third possibility which is a reasonable compromise. Declare global variables (known to all the files, including the assembly language files if any) which we will call remainder and overflow say.

Then write in any way you like (in assembly language or as high level language macros) nine functions that do the following. Assume a,b,c are unsigned word-sized variables, and let M be the chosen multi-precision base, so all variables will be less than M (for example $M = 2^{32}$). Then we need the following functions, where $0 \le c < M$ and overflow is equal to 0 or 1:

c=add(a,b) corresponding to the formula a+b=overflow·M+c.

c=addx(a,b) corresponding to the formula a+b+overflow=overflow·M+c.

c=sub(a,b) corresponding to the formula a-b=c-overflow·M.

c=subx(a,b) corresponding to the formula a-b-overflow=c-overflow·M.

c=mul(a,b) corresponding to the formula a·b=remainder·M+c,

in other words c contains the low order part of the product, and remainder the high order part.

c=div(a,b) corresponding to the formula remainder·M+a=b·c+remainder,

where we may assume that remainder<b.

For the last three functions we assume that M is equal to a power of 2, say $M = 2^m$.

c=shiftl(a,k) corresponding to the formula $2^k a$=remainder·M+c.

c=shiftr(a,k) corresponding to the formula a·M/2^k=c·M+remainder,

where we assume for these last two functions that $0 \le k < m$.

k=bfffo(a) corresponding to the formula $M/2 \le 2^k a < M$, i.e. $k = \lceil \lg(M/(2a)) \rceil$ when $a \ne 0$, $k = m$ when $a = 0$.

The advantage of this scheme is that the rest of the multi-precision package can be written in a high level language without much sacrifice of speed, and that the black boxes described above are short and easy to write in assembly language. The portability problem also disappears since these functions can easily be rewritten for another machine.

Knowledgeable readers may have noticed that the functions above correspond to a simulation of a few machine language instructions of the 68020/68030/68040 processors. It may be worthwhile to work at a higher level, for example by implementing in assembly language a few of the multi-precision functions mentioned at the beginning of this section. By doing this to a limited extent one can avoid many debugging problems. This also avoids much function call overhead, and allows easier optimizing. As usual, the price paid is portability and robustness.

Remark. One of the most common operations used in number theory is *modular multiplication*, i.e. the computation of $a \cdot b$ modulo some number N, where a and b are non-negative integers less than N. This can, of course, be trivially done using the formula div(mul(a,b),N), the result being the value of remainder. When many such operations are needed using the *same* modulus N (this happens for example in most factoring methods, see Chapters 8, 9 an 10), there is a more clever way of doing this, due to P. Montgomery which can save 10 to 20 percent of the running time, and this is not a negligible saving since it is an absolutely basic operation. We refer to his paper [Mon1] for the description of this method.

1.1.3 Base Fields and Rings

Many of the algorithms that we give (for example the linear algebra algorithms of Chapter 2 or some of the algorithms for working with polynomials in Chapter 3) are valid over any base ring or field R where we know how to compute. We must emphasize however that the behavior of these algorithms will be quite different depending on the base ring. Let us look at the most important examples.

The simplest rings are the rings $R = \mathbb{Z}/N\mathbb{Z}$, especially when N is small. Operations in R are simply operations "modulo N" and the elements of R can always be represented by an integer less than N, hence of bounded size. Using the standard algorithms mentioned in the preceding section, and a suitable version of Euclid's extended algorithm to perform division (see Section 1.3.2), all operations need only $O(\ln^2 N)$ bit operations (in fact $O(1)$ since N is considered as fixed!). An important special case of these rings R is when $N = p$ is a prime, and then $R = \mathbb{F}_p$ the finite field with p elements. More generally, it is easy to see that operations on any finite field $\mathbb{F}_q$ with $q = p^k$ can be done quickly.

The next example is that of $R = \mathbb{Z}$. In many algorithms, it is possible to give an upper bound N on the size of the numbers to be handled. In this case we are back in the preceding situation, except that the bound N is no longer fixed, hence the running time of the basic operations is really $O(\ln^2 N)$ bit operations and not $O(1)$. Unfortunately, in most algorithms some divisions are needed, hence we are no longer working in $\mathbb{Z}$ but rather in $\mathbb{Q}$. It is possible to rewrite some of these algorithms so that non-integral rational numbers never occur (see for example the Gauss-Bareiss Algorithm 2.2.6, the integral LLL Algorithm 2.6.7, the sub-resultant Algorithms 3.3.1 and 3.3.7). These versions are then preferable.

The third example is when $R = \mathbb{Q}$. The main phenomenon which occurs in practically all algorithms here is "coefficient explosion". This means that in the course of the algorithm the numerator and denominators of the rational numbers which occur become very large; their size is almost impossible to control. The main reason for this is that the numerator and denominator of the sum or difference of two rational numbers is usually of the same order of magnitude as those of their product. Consequently it is not easy to give running times in bit operations for algorithms using rational numbers.

The fourth example is that of $R = \mathbb{R}$ (or $R = \mathbb{C}$). A new phenomenon occurs here. How can we represent a real number? The truthful answer is that it is in practice impossible, not only because the set $\mathbb{R}$ is uncountable, but also because it will always be impossible for an algorithm to tell whether two real numbers are equal, since this requires in general an infinite amount of time (on the other hand if two real numbers are different, it is possible to prove it by computing them to sufficient accuracy). So we must be content with approximations (or with interval arithmetic, i.e. we give for each real number involved in an algorithm a rational lower and upper bound), increasing the closeness of the approximation to suit our needs. A nasty specter is waiting for us in the dark, which has haunted generations of numerical analysts: numerical instability. We will see an example of this in the case of the LLL algorithm (see Remark (4) after Algorithm 2.6.3). Since this is not a book on numerical analysis, we do not dwell on this problem, but it should be kept in mind.

As far as the bit complexity of the basic operations are concerned, since we must work with limited accuracy the situation is analogous to that of $\mathbb{Z}$ when an upper bound N is known. If the accuracy used for the real number is of the order of $1/N$, the number of bit operations for performing the basic operations is $O(\ln^2 N)$.

Although not much used in this book, a last example I would like to mention is that of $R = \mathbb{Q}_p$, the field of p-adic numbers. This is similar to the case of real numbers in that we must work with a limited precision, hence the running times are of the same order of magnitude. Since the p-adic valuation is non-Archimedean, i.e. the accuracy of the sum or product of p-adic numbers with a given accuracy is at least of the same accuracy, the phenomenon of numerical instability essentially disappears.

1.1.4 Notations

We will use Knuth's notations, which have become a *de facto* standard in the theory of algorithms. Also, some algorithms are directly adapted from Knuth (why change a well written algorithm?). However the algorithmic style of writing used by Knuth is not well suited to structured programming. The reader may therefore find it completely straightforward to write the corresponding programs in assembly language, Basic or Fortran, say, but may find it slightly less so to write them in Pascal or in C.

A warning: presenting an algorithms as a series of steps as is done in this book is only one of the ways in which an algorithm can be described. The presentation may look old-fashioned to some readers, but in the author's opinion it is the best way to explain all the details of an algorithm. In particular it is perhaps better than using some pseudo-Pascal language (pseudo-code). Of course, this is debatable, but this is the choice that has been made in this book. Note however that, as a consequence, the reader should read as carefully as possible the exact phrasing of the algorithm, as well as the accompanying explanations, to avoid any possible ambiguity. This is particularly true in if (conditional) expressions. Some additional explanation is sometimes added to diminish the possibility of ambiguity. For example, if the if condition is not satisfied, the usual word used is otherwise. If if expressions are nested, one of them will use otherwise, and the other will usually use else. I admit that this is not a very elegant solution.

A typical example is step 7 in Algorithm 6.2.9. The initial statement If $c = 0$ do the following: implies that the whole step will be executed only if $c = 0$, and must be skipped if $c \neq 0$. Then there is the expression if $j = i$ followed by an otherwise, and nested inside the otherwise clause is another if dim(...) $< n$, and the else go to step 7 which follows refers to this last if, i.e. we go to step 7 if dim(...) $\geq n$.

I apologize to the reader if this causes any confusion, but I believe that this style of presentation is a good compromise.

$\lfloor x \rfloor$ denotes the floor of x, i.e. the largest integer less than or equal to x. Thus $\lfloor 3.4 \rfloor = 3$, $\lfloor -3.4 \rfloor = -4$.

$\lceil x \rceil$ denotes the ceiling of x, i.e. the smallest integer greater than or equal to x. We have $\lceil x \rceil = -\lfloor -x \rfloor$.

$\lfloor x \rceil$ denotes an integer nearest to x, i.e. $\lfloor x \rceil = \lfloor x + 1/2 \rfloor$.

$[a, b[$ denotes the real interval from a to b including a but excluding b. Similarly $]a, b]$ includes b and excludes a, and $]a, b[$ is the open interval excluding a and b. (This differs from the American notations $[a, b)$, $(a, b]$ and (a, b) which in my opinion are terrible. In particular, in this book (a, b) will usually mean the GCD of a and b, and sometimes the ordered pair (a, b).)

lg x denotes the base 2 logarithm of x.

If E is a finite set, $|E|$ denotes the cardinality of E.

If A is a matrix, A^t denotes the transpose of the matrix A. A $1 \times n$ (resp. $n \times 1$) matrix is called a row (resp. column) vector. The reader is warned that many authors use a different notation where the transpose sign is put on the left of the matrix.

If a and b are integers with $b \neq 0$, then except when explicitly mentioned otherwise, a mod b denotes the *non-negative* remainder in the Euclidean division of a by b, i.e. the unique number r such that $a \equiv r \pmod{b}$ and $0 \leq r < |b|$.

The notation $d \mid n$ means that d divides n, while $d\|n$ will mean that $d \mid n$ and $(d, n/d) = 1$. Furthermore, the notations $p \mid n$ and $p^\alpha\|n$ are always taken to imply that p is prime, so for example $p^\alpha\|n$ means that p^α is the highest power of p dividing n.

Finally, if a and b are elements in a Euclidean ring (typically $\mathbb{Z}$ or the ring of polynomials over a field), we will denote the greatest common divisor (abbreviated GCD in the text) of a and b by $\gcd(a, b)$, or simply by (a, b) when there is no risk of confusion.

1.2 The Powering Algorithms

In almost every non-trivial algorithm in number theory, it is necessary at some point to compute the n-th power of an element in a group, where n may be some very large integer (i.e. for instance greater than 10^{100}). That this is actually possible and very easy is fundamental and one of the first things that one must understand in algorithmic number theory. These algorithms are general and can be used in any group. In fact, when the exponent is non-negative, they can be used in any monoid with unit. We give an abstract version, which can be trivially adapted for any specific situation.

Let $(G, \times)$ be a group. We want to compute g^n for $g \in G$ and $n \in \mathbb{Z}$ in an efficient manner. Assume for example that $n > 0$. The naïve method requires $n - 1$ group multiplications. We can however do much better (A note: although Gauss was very proficient in hand calculations, he seems to have missed this method.) The idea is as follows. If $n = \sum_i \epsilon_i 2^i$ is the base 2 expansion of n with $\epsilon_i = 0$ or 1, then

$$g^n = \prod_{\epsilon_i = 1} \left(g^{2^i} \right) ,$$

hence if we keep track in an auxiliary variable of the quantities g^{2^i} which we compute by successive squarings, we obtain the following algorithm.

Algorithm 1.2.1 (Right-Left Binary). Given $g \in G$ and $n \in \mathbb{Z}$, this algorithm computes g^n in G. We write 1 for the unit element of G.

1. [Initialize] Set $y \leftarrow 1$. If $n = 0$, output y and terminate. If $n < 0$ let $N \leftarrow -n$ and $z \leftarrow g^{-1}$. Otherwise, set $N \leftarrow n$ and $z \leftarrow g$.

2. [Multiply?] If N is odd set $y \leftarrow z \cdot y$.

3. [Halve N] Set $N \leftarrow \lfloor N/2 \rfloor$. If $N = 0$, output y as the answer and terminate the algorithm. Otherwise, set $z \leftarrow z \cdot z$ and go to step 2.

Examining this algorithm shows that the number of multiplication steps is equal to the number of binary digits of $|n|$ plus the number of ones in the binary representation of $|n|$ minus 1. So, it is at most equal to $2\lfloor \lg |n| \rfloor + 1$, and on average approximately equal to $1.5 \lg |n|$. Hence, if one can compute rapidly in G, it is not unreasonable to have exponents with several million decimal digits. For example, if $G = (\mathbb{Z}/m\mathbb{Z})^*$, the time of the powering algorithm is $O(\ln^2 m \ln |n|)$, since one multiplication in G takes time $O(\ln^2 m)$.

The validity of Algorithm 1.2.1 can be checked immediately by noticing that at the start of step 2 one has $g^n = y \cdot z^N$. This corresponds to a right-to-left scan of the binary digits of $|n|$.

We can make several changes to this basic algorithm. First, we can write a similar algorithm based on a left to right scan of the binary digits of $|n|$. In other words, we use the formula $g^n = (g^{n/2})^2$ if n is even and $g^n = g \cdot (g^{(n-1)/2})^2$ if n is odd.

This assumes however that we know the position of the leftmost bit of $|n|$ (or that we have taken the time to look for it beforehand), i.e. that we know the integer e such that $2^e \leq |n| < 2^{e+1}$. Such an integer can be found using a standard binary search on the binary digits of n, hence the time taken to find it is $O(\lg \lg |n|)$, and this is completely negligible with respect to the other operations. This leads to the following algorithm.

Algorithm 1.2.2 (Left-Right Binary). Given $g \in G$ and $n \in \mathbb{Z}$, this algorithm computes g^n in G. If $n \neq 0$, we assume also given the unique integer e such that $2^e \leq |n| < 2^{e+1}$. We write 1 for the unit element of G.

1. [Initialize] If $n = 0$, output 1 and terminate. If $n < 0$ set $N \leftarrow -n$ and $z \leftarrow g^{-1}$. Otherwise, set $N \leftarrow n$ and $z \leftarrow g$. Finally, set $y \leftarrow z$, $E \leftarrow 2^e$, $N \leftarrow N - E$.

2. [Finished?] If $E = 1$, output y and terminate the algorithm. Otherwise, set $E \leftarrow E/2$.

3. [Multiply?] Set $y \leftarrow y \cdot y$ and if $N \geq E$, set $N \leftarrow N - E$ and $y \leftarrow y \cdot z$. Go to step 2.

Note that E takes as values the decreasing powers of 2 from 2^e down to 1, hence when implementing this algorithm, all operations using E must be thought of as bit operations. For example, instead of keeping explicitly the (large) number E, one can just keep its exponent (which will go from e down to 0). Similarly, one does not really subtract E from N or compare N with E, but simply look whether a particular bit of N is 0 or not. To be specific, assume that we have written a little program bit(N, f) which outputs bit number f of N, bit 0 being, by definition, the least significant bit. Then we can rewrite Algorithm 1.2.2 as follows.

Algorithm 1.2.3 (Left-Right Binary, Using Bits). Given $g \in G$ and $n \in \mathbb{Z}$, this algorithm computes g^n in G. If $n \neq 0$, we assume also that we are given the unique integer e such that $2^e \leq |n| < 2^{e+1}$. We write 1 for the unit element of G.

1. [Initialize] If $n = 0$, output 1 and terminate. If $n < 0$ set $N \leftarrow -n$ and $z \leftarrow g^{-1}$. Otherwise, set $N \leftarrow n$ and $z \leftarrow g$. Finally, set $y \leftarrow z$, $f \leftarrow e$.

2. [Finished?] If $f = 0$, output y and terminate the algorithm. Otherwise, set $f \leftarrow f - 1$.

3. [Multiply?] Set $y \leftarrow y \cdot y$ and if bit(N, f) = 1, set $y \leftarrow y \cdot z$. Go to step 2.

The main advantage of this algorithm over Algorithm 1.2.1 is that in step 3 above, z is always the initial g (or its inverse if $n < 0$). Hence, if g is represented by a small integer, this may mean a linear time multiplication instead of a quadratic time one. For example, if $G = (\mathbb{Z}/m\mathbb{Z})^*$ and if g (or g^{-1} if $n < 0$) is represented by the class of a single precision integer, the running time of Algorithms 1.2.2 and 1.2.3 will be in average up to 1.5 times faster than Algorithm 1.2.1.

Algorithm 1.2.3 can be improved by making use of the representation of $|n|$ in a base equal to a power of 2, instead of base 2 itself. In this case, only the left-right version exists.

This is done as follows (we may assume $n > 0$). Choose a suitable positive integer k (we will see in the analysis how to choose it optimally). Precompute g^2 and by induction the odd powers g^3, g^5, $\ldots$, g^{2^k-1}, and initialize y to g as in Algorithm 1.2.3. Now if we scan the 2^k-representation of $|n|$ from left to right (i.e. k bits at a time of the binary representation), we will encounter digits a in base 2^k, hence such that $0 \le a < 2^k$. If $a=0$, we square k times our current y. If $a \neq 0$, we can write $a = 2^t b$ with b odd and less than 2^k, and $0 \le t < k$. We must set $y \leftarrow y^{2^k} \cdot g^{2^t b}$, and this is done by computing first $y^{2^{k-t}} \cdot g^b$ (which involves $k - t$ squarings plus one multiplication since g^b has been precomputed), then squaring t times the result. This leads to the following algorithm. Here we assume that we have an algorithm digit(k, N, f) which gives digit number f of N expressed in base 2^k.

Algorithm 1.2.4 (Left-Right Base 2^k).   Given $g \in G$ and $n \in \mathbb{Z}$, this algorithm computes g^n in G. If $n \neq 0$, we assume also given the unique integer e such that $2^{ke} \le |n| < 2^{k(e+1)}$. We write 1 for the unit element of G.

1. [Initialize] If $n = 0$, output 1 and terminate. If $n < 0$ set $N \leftarrow -n$ and $z \leftarrow g^{-1}$. Otherwise, set $N \leftarrow n$ and $z \leftarrow g$. Finally set $f \leftarrow e$.

2. [Precomputations] Compute and store z^3, z^5, $\ldots$, z^{2^k-1}.

3. [Multiply] Set $a \leftarrow$ digit(k, N, f). If $a = 0$, repeat k times $y \leftarrow y \cdot y$. Otherwise, write $a = 2^t b$ with b odd, and if $f \neq e$ repeat $k - t$ times $y \leftarrow y \cdot y$ and set $y \leftarrow y \cdot z^b$, while if $f = e$ set $y \leftarrow z^b$ (using the precomputed value of z^b), and finally (still if $a \neq 0$) repeat t times $y \leftarrow y \cdot y$.

4. [Finished?] If $f = 0$, output y and terminate the algorithm. Otherwise, set $f \leftarrow f - 1$ and go to step 3.

Implementation Remark. Although the splitting of a in the form $2^t b$ takes very little time compared to the rest of the algorithm, it is a nuisance to have to repeat it all the time. Hence, we suggest precomputing all pairs (t, b) for a given k (including $(k, 0)$ for $a = 0$) so that t and b can be found simply by table lookup. Note that this precomputation depends only on the value of k chosen for Algorithm 1.2.4, and not on the actual value of the exponent n.

Let us now analyze the average behavior of Algorithm 1.2.4 so that we can choose k optimally. As we have already explained, we will regard as negligible the time spent in computing e or in extracting bits or digits in base 2^k.

The precomputations require 2^{k-1} multiplications. The total number of squarings is exactly the same as in the binary algorithm, i.e. $\lfloor \lg |n| \rfloor$, and the number of multiplications is equal to the number of non-zero digits of $|n|$ in base 2^k, i.e. on average

$$\left(1 - \frac{1}{2^k}\right)\left(\left\lfloor\frac{\lg|n|}{k}\right\rfloor + 1\right) ,$$

so the total number of multiplications which are not squarings is on average approximately equal to

$$m(k) = 2^{k-1} + \left(\frac{2^k - 1}{k2^k}\right)\lg|n| .$$

Now, if we compute $m(k+1) - m(k)$, we see that it is non-negative as long as

$$\lg|n| \le \frac{k(k+1)2^{2k}}{2^{k+1} - k - 2} .$$

Hence, for the highest efficiency, one should choose k equal to the smallest integer satisfying the above inequality, and this gives $k = 1$ for $|n| \le 256$, $k = 2$ for $|n| \le 2^{24}$, etc .... For example, if $|n|$ has between 60 and 162 decimal digits, the optimal value of k is $k = 5$. For a more specific example, assume that n has 100 decimal digits (i.e. $\lg n$ approximately equal to 332) and that the time for squaring is about 3/4 of the time for multiplication (this is quite a reasonable assumption). Then, counting multiplication steps, the ordinary binary algorithm takes on average $(3/4)332 + 332/2 = 415$ steps. On the other hand, the base 2^5 algorithm takes on average $(3/4)332 + 16 + (31/160)332 \approx 329$ multiplication steps, an improvement of more than 20%.

There is however another point to take into account. When, for instance $G = (\mathbb{Z}/m\mathbb{Z})^*$ and g (or g^{-1} when $n < 0$) is represented by the (residue) class of a single precision integer, replacing multiplication by g by multiplication by its small odd powers may have the disadvantage compared to Algorithm 1.2.3 that these powers may not be single precision. Hence, in this case, it may be preferable, either to use Algorithm 1.2.3, or to use the highest power of k less than or equal to the optimal one which keeps all the z^b with b odd and $1 \le b \le 2^k - 1$ represented by single precision integers.

Quite a different way to improve on Algorithm 1.2.1 is to try to find a near optimal "addition chain" for $|n|$, and this also can lead to improvements, especially when the same exponent is used repeatedly (see [BCS]. For a detailed discussion of addition chains, see [Knu2].) In practice, we suggest using the 2^k-algorithm for a suitable value of k.

The powering algorithm is used very often with the ring $\mathbb{Z}/m\mathbb{Z}$. In this case multiplication does not give a group law, but the algorithm is valid nonetheless if either n is non-negative or if g is an invertible element. Furthermore, the group multiplication is "multiplication followed by reduction modulo m". Depending on the size of m, it may be worthwhile to not do the reductions each time, but to do them only when necessary to avoid overflow or loss of time.

We will use the powering algorithm in many other contexts in this book, in particular when computing in class groups of number fields, or when working with elliptic curves over finite fields.

Note that for many groups it is possible (and desirable) to write a squaring routine which is faster than the general-purpose multiplication routine. In situations where the powering algorithm is used intensively, it is essential to use this squaring routine when multiplications of the type $y \leftarrow y \cdot y$ are encountered.

1.3 Euclid's Algorithms

We now consider the problem of computing the GCD of two integers a and b. The naïve answer to this problem would be to factor a and b, and then multiply together the common prime factors raised to suitable powers. Indeed, this method works well when a and b are *very* small, say less than 100, or when a or b is known to be prime (then a single division is sufficient). In general this is not feasible, because one of the important facts of life in number theory is that factorization is difficult and slow. We will have many occasions to come back to this. Hence, we must use better methods to compute GCD's. This is done using Euclid's algorithm, probably the oldest and most important algorithm in number theory.

Although very simple, this algorithm has several variants, and, because of its usefulness, we are going to study it in detail. We shall write (a, b) for the GCD of a and b when there is no risk of confusion with the pair (a, b). By definition, (a, b) is the unique non-negative generator of the additive subgroup of $\mathbb{Z}$ generated by a and b. In particular, $(a, 0) = (0, a) = |a|$ and $(a, b) = (|a|, |b|)$. Hence we can always assume that a and b are non-negative.

1.3.1 Euclid's and Lehmer's Algorithms

Euclid's algorithm is as follows:

Algorithm 1.3.1 (Euclid).   Given two non-negative integers a and b, this algorithm finds their GCD.

1. [Finished?] If $b = 0$ then output a as the answer and terminate the algorithm.

2. [Euclidean step] Set $r \leftarrow a \bmod b$, $a \leftarrow b$, $b \leftarrow r$ and go to step 1.

If either a or b is less than a given number N, the number of Euclidean steps in this algorithm is bounded by a constant times $\ln N$, in both the worst case and on average. More precisely we have the following theorem (see [Knu2]):

Theorem 1.3.2. *Assume that a and b are randomly distributed between 1 and N. Then*

(1) *The number of Euclidean steps is at most equal to*

$$\left\lceil \frac{\ln(\sqrt{5}N)}{\ln((1+\sqrt{5})/2)} \right\rceil - 2 \approx 2.078 \ln N + 1.672 \; .$$

(2) *The average number of Euclidean steps is approximately equal to*

$$\frac{12 \ln 2}{\pi^2} \ln N + 0.14 \approx 0.843 \ln N + 0.14 \; .$$

However, Algorithm 1.3.1 is far from being the whole story. First, it is not well suited to handling large numbers (in our sense, say numbers with 50 or 100 decimal digits). This is because each Euclidean step requires a long division, which takes time $O(\ln^2 N)$. When carelessly programmed, the algorithm takes time $O(\ln^3 N)$. If, however, at each step the precision is decreased as a function of a and b, and if one also notices that the time to compute a Euclidean step $a = bq + r$ is $O((\ln a)(\ln q + 1))$, then the total time is bounded by $O((\ln N)((\sum \ln q) + O(\ln N)))$. But $\sum \ln q = \ln \prod q \le \ln a \le \ln N$, hence if programmed carefully, the running time is only $O(\ln^2 N)$. There is a useful variant due to Lehmer which also brings down the running time to $O(\ln^2 N)$. The idea is that the Euclidean quotient depends generally only on the first few digits of the numbers. Therefore it can usually be obtained using a single precision calculation. The following algorithm is taken directly from Knuth. Let $M = m^p$ be the base used for multi-precision numbers. Typical choices are $m = 2$, $p = 15, 16, 31$, or 32, or $m = 10$, $p = 4$ or 9.

Algorithm 1.3.3 (Lehmer). Let a and b be non-negative multi-precision integers, and assume that $a \ge b$. This algorithm computes (a, b), using the following auxiliary variables. $\hat{a}$, $\hat{b}$, A, B, C, D, T and q are single precision (i.e. less than M), and t and r are multi-precision variables.

1. [Initialize] If $b < M$, i.e. is single precision, compute (a, b) using Algorithm 1.3.1 and terminate. Otherwise, let $\hat{a}$ (resp. $\hat{b}$) be the single precision number formed by the highest non-zero base M digit of a (resp. b). Set $A \leftarrow 1$, $B \leftarrow 0$, $C \leftarrow 0$, $D \leftarrow 1$.

2. [Test quotient] If $\hat{b} + C = 0$ or $\hat{b} + D = 0$ go to step 4. Otherwise, set $q \leftarrow \lfloor (\hat{a} + A)/(\hat{b} + C) \rfloor$. If $q \ne \lfloor (\hat{a} + B)/(\hat{b} + D) \rfloor$, go to step 4. Note that one always has the conditions

$$0 \le \hat{a} + A \le M \; , \quad 0 \le \hat{b} + C < M \; ,$$

$$0 \le \hat{a} + B < M \; , \quad 0 \le \hat{b} + D \le M \; .$$

Notice that one can have a single precision overflow in this step, which must be taken into account. (This can occur only if $\hat{a} = M - 1$ and $A = 1$ or if $\hat{b} = M - 1$ and $D = 1$.)

3. [Euclidean step] Set $T \leftarrow A - qC$, $A \leftarrow C$, $C \leftarrow T$, $T \leftarrow B - qD$, $B \leftarrow D$, $D \leftarrow T$, $T \leftarrow \hat{a} - q\hat{b}$, $\hat{a} \leftarrow \hat{b}$, $\hat{b} \leftarrow T$ and go to step 2 (all these operations are single precision operations).

4. [Multi-precision step] If $B = 0$, set $t \leftarrow a \bmod b$, $a \leftarrow b$, $b \leftarrow t$, using multi-precision division (this happens with a very small probability, on the order of $1.4/M$) and go to step 1. Otherwise, set $t \leftarrow Aa$, $t \leftarrow t + Bb$, $r \leftarrow Ca$, $r \leftarrow r + Db$, $a \leftarrow t$, $b \leftarrow r$, using linear-time multi-precision operations, and go to step 1.

Note that the number of steps in this algorithm will be the same as in Algorithm 1.3.1, i.e. $O(\ln N)$ if a and b are less than N, but each loop now consists only of linear time operations (except for the case $B = 0$ in step 4 which is so rare as not to matter in practice). Therefore, even without using variable precision, the running time is now only of order $O(\ln^2 N)$ and not $O(\ln^3 N)$. Of course, there is much more bookkeeping involved, so it is not clear how large N must be before a particular implementation of this algorithm becomes faster than a crude implementation of Algorithm 1.3.1. Or, even whether a careful implementation of Algorithm 1.3.1 will not compete favorably in practice. Testing needs to be done before choosing which of these algorithms to use.

Another variant of Euclid's algorithm which is also useful in practice is the so-called binary algorithm. Here, no long division steps are used, except at the beginning, instead only subtraction steps and divisions by 2, which are simply integer shifts. The number of steps needed is greater, but the operations used are much faster, and so there is a net gain, which can be quite large for multi-precision numbers. Furthermore, using subtractions instead of divisions is quite reasonable in any case, since most Euclidean quotients are small. More precisely, we can state:

Theorem 1.3.4. *In a suitable sense, the probability $P(q)$ that a Euclidean quotient be equal to q is*

$$P(q) = \lg((q + 1)^2/((q + 1)^2 - 1)) \ .$$

For example, $P(1) = 0.41504\ldots$, $P(2) = 0.16992\ldots$, $P(3) = 0.09311\ldots$, $P(4) = 0.05890\ldots$.

For example, from this theorem, one can see that the probability of occurrence of $B = 0$ in step 4 of Algorithm 1.3.3 is $\lg(1 + 1/M)$, and this is negligible in practice.

One version of the binary algorithm is as follows.

Algorithm 1.3.5 (Binary GCD). Given two non-negative integers a and b, this algorithm finds their GCD.

1. [Reduce size once] If $a < b$ exchange a and b. Now if $b = 0$, output a and terminate the algorithm. Otherwise, set $r \leftarrow a \bmod b$, $a \leftarrow b$ and $b \leftarrow r$.

2. [Compute power of 2] If $b = 0$ output a and terminate the algorithm. Otherwise, set $k \leftarrow 0$, and then while a and b are both even, set $k \leftarrow k + 1$, $a \leftarrow a/2$, $b \leftarrow b/2$.

3. [Remove initial powers of 2] If a is even, repeat $a \leftarrow a/2$ until a is odd. Otherwise, if b is even, repeat $b \leftarrow b/2$ until b is odd.

4. [Subtract] (Here a and b are both odd.) Set $t \leftarrow (a - b)/2$. If $t = 0$, output $2^k a$ and terminate the algorithm.

5. [Loop] While t is even, set $t \leftarrow t/2$. Then if $t > 0$ set $a \leftarrow t$, else set $b \leftarrow -t$ and go to step 4.

Remarks.

(1) The binary algorithm is especially well suited for computing the GCD of multi-precision numbers. This is because no divisions are performed, except on the first step. Hence we suggest using it systematically in this case.

(2) All the divisions by 2 performed in this algorithm must be done using shifts or Boolean operations, otherwise the algorithm loses much of its attractiveness. In particular, it may be worthwhile to program it in a low-level language, and even in assembly language, if it is going to be used extensively. Note that some applications, such as computing in class groups, use GCD as a basic operation, hence it is essential to optimize the speed of the algorithm for these applications.

(3) One could directly start the binary algorithm in step 2, avoiding division altogether. We feel however that this is not such a good idea, since a and b may have widely differing magnitudes, and step 1 ensures that we will work on numbers at most the size of the *smallest* of the two numbers a and b, and not of the largest, as would be the case if we avoided step 1. In addition, it is quite common for b to divide a when starting the algorithm. In this case, of course, the algorithm immediately terminates after step 1.

(4) Note that the sign of t in step 4 of the algorithm enables the algorithm to keep track of the larger of a and b, so that we can replace the larger of the two by $|t|$ in step 5. We can also keep track of this data in a separate variable and thereby work only with non-negative numbers.

(5) Finally, note that the binary algorithm can use the ideas of Algorithm 1.3.3 for multi-precision numbers. The resulting algorithm is complex and its efficiency is implementation dependent. For more details, see [Knu2 p.599].

The proof of the validity of the binary algorithm is easy and left to the reader. On the other hand, a detailed analysis of the average running time of the binary algorithm is a challenging mathematical problem (see [Knu2] once again). Evidently, as was the case for Euclid's algorithm, the running time will be $O(\ln^2 N)$ bit operations when suitably implemented, where N is an upper bound on the size of the inputs a and b. The mathematical problem is to find an asymptotic estimate for the number of steps and the number of shifts performed in Algorithm 1.3.5, but this has an influence only on the O constant, not on the qualitative behavior.   □

1.3.2 Euclid's Extended Algorithms

The information given by Euclid's algorithm is not always sufficient for many problems. In particular, by definition of the GCD, if $d = (a, b)$ there exists integers u and v such that $au + bv = d$. It is often necessary to extend Euclid's algorithm so as to be able to compute u and v. While u and v are not unique, u is defined modulo b/d, and v is defined modulo a/d.

There are two ways of doing this. One is by storing the Euclidean quotients as they come along, and then, once d is found, backtracking to the initial values. This method is the most efficient, but can require a lot of storage. In some situations where this information is used extensively (such as Shanks's and Atkin's NUCOMP in Section 5.4.2), any little gain should be taken, and so one should do it this way.

The other method requires very little storage and is only slightly slower. This requires using a few auxiliary variables so as to do the computations as we go along. We first give a version which does not take into account multiprecision numbers.

Algorithm 1.3.6 (Euclid Extended). Given non-negative integers a and b, this algorithm determines (u, v, d) such that $au + bv = d$ and $d = (a, b)$. We use auxiliary variables v_1, v_3, t_1, t_3.

1. [Initialize] Set $u \leftarrow 1$, $d \leftarrow a$. If $b = 0$, set $v \leftarrow 0$ and terminate the algorithm, otherwise set $v_1 \leftarrow 0$ and $v_3 \leftarrow b$.

2. [Finished?] If $v_3 = 0$ then set $v \leftarrow (d - au)/b$ and terminate the algorithm.

3. [Euclidean step] Let $q \leftarrow \lfloor d/v_3 \rfloor$ and simultaneously $t_3 \leftarrow d \bmod v_3$. Then set $t_1 \leftarrow u - qv_1$, $u \leftarrow v_1$, $d \leftarrow v_3$, $v_1 \leftarrow t_1$, $v_3 \leftarrow t_3$ and go to step 2.

"Simultaneously" in step 3 means that if this algorithm is implemented in assembly language, then, since the division instruction usually gives both the quotient and remainder, this should of course be used. Even if this algorithm is not programmed in assembly language, but a and b are multi-precision numbers, the division routine in the multi-precision library should also return both quotient and remainder. Note also that in step 2, the division of $d - au$ by b is exact.

Proof of the Algorithm. Introduce three more variables v_2, t_2 and v. We want the following relations to hold each time one begins step 2:

$$at_1 + bt_2 = t_3 \ , \quad au + bv = d \ , \quad av_1 + bv_2 = v_3 \ .$$

For this to be true after the initialization step, it suffices to set $v \leftarrow 0$, $v_2 \leftarrow 1$. (It is not necessary to initialize the t variables.) Then, it is easy to check that step 3 preserves these relations if we update suitably the three auxiliary variables (by $(v_2, t_2, v) \leftarrow (t_2, v - qv_2, v_2)$). Therefore, at the end of the algorithm, d contains the GCD (since we have simply added some extra work to the initial Euclidean algorithm), and we also have $au + bv = d$. $\square$

As an exercise, the reader can show that at the end of the algorithm, we have $v_1 = \pm b/d$ (and $v_2 = \mp a/d$ in the proof), and that throughout the algorithm, $|v_1|$, $|u|$, $|t_1|$ stay less than or equal to b/d (and $|v_2|$, $|v|$, $|t_2|$ stay less than or equal to a/d).

This algorithm can be improved for multi-precision numbers exactly as in Lehmer's Algorithm 1.3.3. Since it is a simple blend of Algorithms 1.3.3 and 1.3.5, we do not give a detailed proof. (Notice however that the variables d and v_3 have become a and b.)

Algorithm 1.3.7 (Lehmer Extended).   Let a and b be non-negative multi-precision integers, and assume that $a \geq b$. This algorithm computes (u, v, d) such that $au + bv = d = (a, b)$, using the following auxiliary variables. $\hat{a}$, $\hat{b}$, A, B, C, D, T and q are single precision (i.e. less than M), and t, r, v_1, v_3 are multi-precision variables.

1. [Initialize] Set $A \leftarrow 1$, $B \leftarrow 0$, $C \leftarrow 0$, $D \leftarrow 1$, $u \leftarrow 1$, $v_1 \leftarrow 0$.

2. [Finished?] If $b < M$, i.e. is single precision, compute (u, v, d) using Algorithm 1.3.6 and terminate. Otherwise, let $\hat{a}$ (resp. $\hat{b}$) be the single precision number formed by the p most significant digits of a (resp. b).

3. [Test quotient] If $\hat{b} + C = 0$ or $\hat{b} + D = 0$ go to step 5. Otherwise, set $q \leftarrow \lfloor (\hat{a} + A)/(\hat{b} + C) \rfloor$. If $q \neq \lfloor (\hat{a} + B)/(\hat{b} + D) \rfloor$, go to step 5.

4. [Euclidean step] Set $T \leftarrow A - qC$, $A \leftarrow C$, $C \leftarrow T$, $T \leftarrow B - qD$, $B \leftarrow D$, $D \leftarrow T$, $T \leftarrow \hat{a} - q\hat{b}$, $\hat{a} \leftarrow \hat{b}$, $\hat{b} \leftarrow T$ and go to step 3 (all these operations are single precision operations).

5. [Multi-precision step] If $B = 0$, set $q \leftarrow \lfloor a/b \rfloor$ and simultaneously $t \leftarrow a \bmod b$ using multi-precision division, then $a \leftarrow b$, $b \leftarrow t$, $t \leftarrow u - qv_1$, $u \leftarrow v_1$, $v_1 \leftarrow t$ and go to step 2.

   Otherwise, set $t \leftarrow Aa$, $t \leftarrow t + Bb$, $r \leftarrow Ca$, $r \leftarrow r + Db$, $a \leftarrow t$, $b \leftarrow r$, $t \leftarrow Au$, $t \leftarrow t + Bv_1$, $r \leftarrow Cu$, $r \leftarrow r + Dv_1$, $u \leftarrow t$, $v_1 \leftarrow r$ using linear-time multi-precision operations, and go to step 2.

In a similar way, the binary algorithm can be extended to find u and v. The algorithm is as follows.

Algorithm 1.3.8 (Binary Extended).   Given non-negative integers a and b, this algorithm determines (u, v, d) such that $au + bv = d$ and $d = (a, b)$. We use auxiliary variables v_1, v_3, t_1, t_3, and two Boolean flags f_1 and f_2.

1. [Reduce size once] If $a < b$ exchange a and b and set $f_1 \leftarrow 1$, otherwise set $f_1 \leftarrow 0$. Now if $b = 0$, output $(1,0,a)$ if $f_1 = 0$, $(0,1,a)$ if $f_1 = 1$ and terminate the algorithm. Otherwise, let $a = bq + r$ be the Euclidean division of a by b, where $0 \leq r < b$, and set $a \leftarrow b$ and $b \leftarrow r$.

2. [Compute power of 2] If $b = 0$, output $(0,1,a)$ if $f_1 = 0$, $(1,0,a)$ if $f_1 = 1$ and terminate the algorithm. Otherwise, set $k \leftarrow 0$, and while a and b are both even, set $k \leftarrow k+1$, $a \leftarrow a/2$, $b \leftarrow b/2$.

3. [Initialize] If b is even, exchange a and b and set $f_2 \leftarrow 1$, otherwise set $f_2 \leftarrow 0$. Then set $u \leftarrow 1$, $d \leftarrow a$, $v_1 \leftarrow b$, $v_3 \leftarrow b$. If a is odd, set $t_1 \leftarrow 0$, $t_3 \leftarrow -b$ and go to step 5, else set $t_1 \leftarrow (1+b)/2$, $t_3 \leftarrow a/2$.

4. [Remove powers of 2] If t_3 is even do as follows. Set $t_3 \leftarrow t_3/2$, $t_1 \leftarrow t_1/2$ if t_1 is even and $t_1 \leftarrow (t_1 + b)/2$ if t_1 is odd, and repeat step 4.

5. [Loop] If $t_3 > 0$, set $u \leftarrow t_1$ and $d \leftarrow t_3$, otherwise, set $v_1 \leftarrow b - t_1$, $v_3 \leftarrow -t_3$.

6. [Subtract] Set $t_1 \leftarrow u - v_1$, $t_3 \leftarrow d - v_3$. If $t_1 < 0$, set $t_1 \leftarrow t_1 + b$. Finally, if $t_3 \neq 0$, go to step 4.

7. [Terminate] Set $v \leftarrow (d - au)/b$ and $d \leftarrow 2^k d$. If $f_2 = 1$ exchange u and v. Then set $u \leftarrow u - vq$. Finally, output (u,v,d) if $f_1 = 1$, (v,u,d) if $f_1 = 0$, and terminate the algorithm.

Proof. The proof is similar to that of Algorithm 1.3.6. We introduce three more variables v_2, t_2 and v and we require that at the start of step 4 we always have

$$At_1 + Bt_2 = t_3 , \quad Au + Bv = d , \quad Av_1 + Bv_2 = v_3 ,$$

where A and B are the values of a and b after step 3. For this to be true, we must initialize them by setting (in step 3) $v \leftarrow 0$, $v_2 \leftarrow 1 - a$ and $t_2 \leftarrow -1$ if a is odd, $t_2 \leftarrow -a/2$ if a is even. After this, the three relations will continue to be true provided we suitably update v_2, t_2 and v. Since, when the algorithm terminates d will be the GCD of A and B, it suffices to backtrack from both the division step and the exchanges done in the first few steps in order to obtain the correct values of u and v (as is done in step 7). We leave the details to the reader. $\square$

Euclid's "extended" algorithm, i.e. the algorithm used to compute (u,v,d) and not d alone, is useful in many different contexts. For example, one frequent use is to compute an inverse (or more generally a division) modulo m. Assume one wants to compute the inverse of a number b modulo m. Then, using Algorithm 1.3.6, 1.3.7 or 1.3.8, compute (u,v,d) such that $bu + mv = d = (b,m)$. If $d > 1$ send an error message stating that b is not invertible, otherwise the inverse of b is u. Notice that in this case, we can avoid computing v in step 2 of Algorithm 1.3.6 and in the analogous steps in the other algorithms.

There are other methods to compute $b^{-1} \bmod m$ when the factorization of m is known, for example when m is a prime. By Euler-Fermat's Theorem

1.4.2, we know that, if $(b, m) = 1$ (which can be tested very quickly since the factorization of m is known), then

$$b^{\phi(m)} \equiv 1 \qquad (\mathrm{mod}\ m)\ ,$$

where $\phi(m)$ is Euler's ϕ function (see [H-W]). Hence, the inverse of b modulo m can be obtained by computing

$$b^{-1} = b^{\phi(m)-1} \qquad (\mathrm{mod}\ m)\ ,$$

using the powering Algorithm 1.2.1.

Note however that the powering algorithms are $O(\ln^3 m)$ algorithms, which is worse than the time for Euclid's extended algorithm. Nonetheless they can be useful in certain cases. A practical comparison of these methods is done in [Bre1].

1.3.3 The Chinese Remainder Theorem

We recall the following theorem:

Theorem 1.3.9 (Chinese Remainder Theorem). *Let $m_1, \ldots, m_k$ and $x_1, \ldots, x_k$ be integers. Assume that for every pair (i, j) we have*

$$x_i \equiv x_j \quad (\mathrm{mod}\ \gcd(m_i, m_j))\ .$$

There exists an integer x such that

$$x \equiv x_i \quad (\mathrm{mod}\ m_i) \qquad for\ 1 \le i \le k\ .$$

Furthermore, x is unique modulo the least common multiple of $m_1, \ldots, m_k$.

Corollary 1.3.10. *Let $m_1, \ldots, m_k$ be pairwise coprime integers, i.e. such that*

$$\gcd(m_i, m_j) = 1 \qquad when\ i \ne j\ .$$

Then, for any integers x_i, there exists an integer x, unique modulo $\prod m_i$, such that

$$x \equiv x_i \quad (\mathrm{mod}\ m_i) \qquad for\ 1 \le i \le k\ .$$

We need an algorithm to compute x. We will consider only the case where the m_i are pairwise coprime, since this is by far the most useful situation. Set $M = \prod_{1 \le i \le k} m_i$ and $M_i = M/m_i$. Since the m_i are coprime in pairs, $\gcd(M_i, m_i) = 1$ hence by Euclid's extended algorithm we can find a_i such that $a_i M_i \equiv 1 \pmod{m_i}$. If we set

$$x = \sum_{1 \leq i \leq k} a_i M_i x_i \ ,$$

it is clear that x satisfies the required conditions. Therefore, we can output $x \bmod M$ as the result.

This method could be written explicitly as a formal algorithm. However we want to make one improvement before doing so. Notice that the necessary constants a_i are small (less than m_i), but the M_i or the $a_i M_i$ which are also needed can be very large. There is an ingenious way to avoid using such large numbers, and this leads to the following algorithm. Its verification is left to the reader.

Algorithm 1.3.11 (Chinese). Given pairwise coprime integers m_i $(1 \leq i \leq k)$ and integers x_i, this algorithm finds an integer x such that $x \equiv x_i \pmod{m_i}$ for all i. Note that steps 1 and 2 are a precomputation which needs to be done only once when the m_i are fixed and the x_i vary.

1. [Initialize] Set $j \leftarrow 2$, $C_1 \leftarrow 1$. In addition, if it is not too costly, reorder the m_i (and hence the x_i) so that they are in increasing order.

2. [Precomputations] Set $p \leftarrow m_1 m_2 \cdots m_{j-1} \pmod{m_j}$. Compute (u, v, d) such that $up + vm_j = d = \gcd(p, m_j)$ using a suitable version of Euclid's extended algorithm. If $d > 1$ output an error message (the m_i are not pairwise coprime). Otherwise, set $C_j \leftarrow u$, $j \leftarrow j + 1$, and go to step 2 if $j \leq k$.

3. [Compute auxiliary constants] Set $y_1 \leftarrow x_1 \bmod m_1$, and for $j = 2, \ldots, k$ compute (as written)

$$y_j \leftarrow (x_j - (y_1 + m_1(y_2 + m_2(y_3 + \cdots + m_{j-2}y_{j-1}) \cdots)))C_j \bmod m_j \ .$$

4. [Terminate] Output

$$x \leftarrow y_1 + m_1(y_2 + m_2(y_3 + \cdots + m_{k-1}y_k) \cdots)) \ ,$$

and terminate the algorithm.

Note that we will have $0 \leq x < M = \prod m_i$.

As an exercise, the reader can give an algorithm which finds x in the more general case of Theorem 1.3.9 where the m_i are not assumed to be pairwise coprime. It is enough to write an algorithm such as the one described before Algorithm 1.3.11, since it will not be used very often (Exercise 9).

Since this algorithm is more complex than the algorithm mentioned previously, it should only be used when the m_i are fixed moduli, and not just for a one shot problem. In this last case is it preferable to use the formula for two numbers inductively as follows. We want $x \equiv x_i \pmod{m_i}$ for $i = 1, 2$. Since the m_i are relatively prime, using Euclid's extended algorithm we can find u and v such that

$$um_1 + vm_2 = 1 \ .$$

It is clear that

$$x = um_1x_2 + vm_2x_1 \bmod m_1m_2$$

is a solution to our problem. This leads to the following.

Algorithm 1.3.12 (Inductive Chinese). Given pairwise coprime integers m_i ($1 \le i \le k$) and integers x_i, this algorithm finds an integer x such that $x \equiv x_i$ (mod m_i) for all i.

1. [Initialize] Set $i \leftarrow 1$, $m \leftarrow m_1$, $x \leftarrow x_1$.
2. [Finished?] If $i = k$ output x and terminate the algorithm. Otherwise, set $i \leftarrow i+1$, and by a suitable version of Euclid's extended algorithm compute u and v such that $um + vm_i = 1$.
3. [Compute next x] Set $x \leftarrow umx_i + vm_ix$, $m \leftarrow mm_i$, $x \leftarrow x \bmod m$ and go to step 2.

   Note that the results and algorithms of this section remain true if we replace $\mathbb{Z}$ by any Euclidean domain, for example the polynomial ring $K[X]$ where K is a field.

1.3.4 Continued Fraction Expansions of Real Numbers

We now come to a subject which though closely linked to Euclid's algorithm, has a different flavor. Consider first the following apparently simple problem. Let $x \in \mathbb{R}$ be given by an approximation (for example a decimal or binary one). Decide if x is a rational number or not. Of course, this question as posed does not really make sense, since an approximation is usually itself a rational number. In practice however the question does make a lot of sense in many different contexts, and we can make it algorithmically more precise. For example, assume that one has an algorithm which allows us to compute x to as many decimal places as one likes (this is usually the case). Then, if one claims that x is (approximately) equal to a rational number p/q, this means that p/q should still be extremely close to x whatever the number of decimals asked for, p and q being fixed. This is still not completely rigorous, but it comes quite close to actual practice, so we shall be content with this notion.

   Now how does one find p and q if x is indeed a rational number? The standard (and algorithmically excellent) answer is to compute the *continued fraction expansion* of x, i.e. find integers a_i such that $a_i \ge 1$ for $i \ge 1$ and

$$x = a_0 + \cfrac{1}{a_1 + \cfrac{1}{a_2 + \cfrac{1}{a_3 + \cdots}}},$$

which we shall write as $x = [a_0, a_1, a_2, a_3, \dots]$. If a/b is the given (rational) approximation to x, then the a_i are obtained by simply using Euclid's algorithm on the pair (a, b), the a_i being the successive partial quotients. The number x is rational if and only if its continued fraction expansion is finite, i.e. if and only if one of the a_i is infinite. Since x is only given with the finite precision a/b, x will be considered rational if x has a very large partial quotient a_i in its continued fraction expansion. Of course this is subjective, but should be put to the stringent test mentioned above. For example, if one uses the approximation $\pi \approx 3.1415926$ one finds that the continued fraction for π should start with $[3, 7, 15, 1, 243, \dots]$ and 243 does seem a suspiciously large partial quotient, so we suspect that $\pi = 355/113$, which is the rational number whose continued fraction is exactly $[3, 7, 15, 1]$. If we compute a few more decimals of π however, we see that this equality is not true. Nonetheless, $355/113$ is still an excellent approximation to π (the continued fraction expansion of π starts in fact $[3, 7, 15, 1, 292, 1, \dots]$).

To implement a method for computing continued fractions of real numbers, I suggest using the following algorithm, which says exactly when to stop.

Algorithm 1.3.13 (Lehmer).　Given a real number x by two rational numbers a/b and a'/b' such that $a/b \le x \le a'/b'$, this algorithm computes the continued fraction expansion of x and stops exactly when it is not possible to determine the next partial quotient from the given approximants a/b and a'/b', and it gives lower and upper bounds for this next partial quotient.

1. [Initialize] Set $i \leftarrow 0$.

2. [Euclidean step] Let $a = bq + r$ the Euclidean division of a by b, and set $r' \leftarrow a' - b'q$. If $r' < 0$ or $r' \ge b'$ set $q' \leftarrow \lfloor a'/b' \rfloor$ and go to step 4.

3. [Output quotient] Set $a_i \leftarrow q$ and output a_i, then set $i \leftarrow i+1$, $a \leftarrow b$, $b \leftarrow r$, $a' \leftarrow b'$ and $b' \leftarrow r'$. If b and b' are non-zero, go to step 2. If $b = b' = 0$, terminate the algorithm. Finally, if $b = 0$ set $q \leftarrow \infty$ and $q' \leftarrow \lfloor a'/b' \rfloor$ while if $b' = 0$ set $q \leftarrow \lfloor a/b \rfloor$ and $q' \leftarrow \infty$.

4. [Terminate] If $q > q'$ output the inequality $q' \le a_i \le q$, otherwise output $q \le a_i \le q'$. Terminate the algorithm.

Note that the ∞ mentioned in step 3 is only a mathematical abstraction needed to make step 4 make sense, but it does not need to be represented in a machine by anything more than some special code.

This algorithm runs in at most twice the time needed for the Euclidean algorithm on a and b alone, since, in addition to doing one Euclidean division at each step, we also multiply q by b'.

We can now solve the following problem: given two complex numbers z_1 and z_2, are they $\mathbb{Q}$-linearly dependent? This is equivalent to z_1/z_2 being rational, so the solution is this: compute $z \leftarrow z_1/z_2$. If the imaginary part of z is non-zero (to the degree of approximation that one has), then z_1 and z_2 are not even $\mathbb{R}$-linearly dependent. If it is zero, then compute the continued fraction expansion of the real part of z using algorithm 1.3.13, and look for large partial quotients as explained above.

We will see in Section 2.7.2 that the LLL algorithms allow us to determine in a satisfactory way the problem of Q-linear dependence of more than two complex or real numbers.

Another closely related problem is the following: given two vectors $\mathbf{a}$ and $\mathbf{b}$ in a Euclidean vector space, determine the shortest non-zero vector which is a Z-linear combination of $\mathbf{a}$ and $\mathbf{b}$ (we will see in Chapter 2 that the set of such Z-linear combinations is called a *lattice*, here of dimension 2). One solution, called Gaussian reduction, is again a form of Euclid's algorithm, and is as follows.

Algorithm 1.3.14 (Gauss). Given two linearly independent vectors $\mathbf{a}$ and $\mathbf{b}$ in a Euclidean vector space, this algorithm determines one of the shortest non-zero vectors which is a Z-linear combination of $\mathbf{a}$ and $\mathbf{b}$. We denote by $\cdot$ the Euclidean inner product and write $|\mathbf{a}|^2 = \mathbf{a} \cdot \mathbf{a}$. We use a temporary scalar variable T, and a temporary vector variable $\mathbf{t}$.

1. [Initialize] Set $A \leftarrow |\mathbf{a}|^2$, $B \leftarrow |\mathbf{b}|^2$. If $A < B$ then exchange $\mathbf{a}$ and $\mathbf{b}$ and exchange A and B.

2. [Euclidean step] Set $n \leftarrow \mathbf{a} \cdot \mathbf{b}$, $r \leftarrow \lfloor n/B \rceil$, where $\lfloor x \rceil = \lfloor x + 1/2 \rfloor$ is the nearest integer to x, and $T \leftarrow A - 2rn + r^2 B$.

3. [Finished?] If $T \geq B$ then output $\mathbf{b}$ and terminate the algorithm. Otherwise, set $\mathbf{t} \leftarrow \mathbf{a} - r\mathbf{b}$, $\mathbf{a} \leftarrow \mathbf{b}$, $\mathbf{b} \leftarrow \mathbf{t}$, $A \leftarrow B$, $B \leftarrow T$ and go to step 2.

Proof. Note that A and B are always equal to $|\mathbf{a}|^2$ and $|\mathbf{b}|^2$ respectively. I first claim that an integer r such that $|\mathbf{a} - r\mathbf{b}|$ has minimal length is given by the formula of step 2. Indeed, we have

$$|\mathbf{a} - x\mathbf{b}|^2 = Bx^2 - 2\mathbf{a} \cdot \mathbf{b}x + A \ ,$$

and this is minimum for *real* x for $x = \mathbf{a} \cdot \mathbf{b}/B$. Hence, since a parabola is symmetrical at its minimum, the minimum for integral x is the nearest integer (or one of the two nearest integers) to the minimum, and this is the formula given in step 2.

Thus, at the end of the algorithm we know that $|\mathbf{a} - m\mathbf{b}| \geq |\mathbf{b}|$ for all integers m. It is clear that the transformation which sends the pair $(\mathbf{a}, \mathbf{b})$ to the pair $(\mathbf{b}, \mathbf{a} - r\mathbf{b})$ has determinant -1, hence the Z-module L generated by $\mathbf{a}$ and $\mathbf{b}$ stays the same during the algorithm. Therefore, let $\mathbf{x} = u\mathbf{a} + v\mathbf{b}$ be a non-zero element of L. If $u = 0$, we must have $v \neq 0$ hence trivially $|\mathbf{x}| \geq |\mathbf{b}|$. Otherwise, let $v = uq + r$ be the Euclidean division of v by u, where $0 \leq r < |u|$. Then we have

$$|\mathbf{x}| = |u(\mathbf{a} + q\mathbf{b}) + r\mathbf{b}| \geq |u||\mathbf{a} + q\mathbf{b}| - |r||\mathbf{b}| \geq (|u| - |r|)|\mathbf{b}| \geq |\mathbf{b}|$$

since by our above claim $|\mathbf{a} + q\mathbf{b}| \geq |\mathbf{b}|$ for any integer q, hence $\mathbf{b}$ is indeed one of the shortest vectors of L, proving the validity of the algorithm.

Note that the algorithm must terminate since there are only a finite number of vectors of L with norm less than or equal to a given constant (compact+discrete=finite!). In fact the number of steps can easily be seen to be comparable to that of the Euclidean algorithm, hence this algorithm is very efficient. □

We will see in Section 2.6 that the LLL algorithm allows us to determine efficiently small $\mathbb{Z}$-linear combinations for more than two linearly independent vectors in a Euclidean space. It does not always give an optimal solution, but, in most situations, the results are sufficiently good to be very useful.

1.4 The Legendre Symbol

1.4.1 The Groups $(\mathbb{Z}/n\mathbb{Z})^*$

By definition, when A is a commutative ring with unit, we will denote by A^* the group of *units* of A, i.e. of invertible elements of A. It is clear that A^* is a group, and also that $A^* = A \setminus \{0\}$ if and only if A is a field. Now we have the following fundamental theorem which gives the structure of $(\mathbb{Z}/n\mathbb{Z})^*$ (see [Ser] and Exercise 13).

Theorem 1.4.1. *We have*

$$|(\mathbb{Z}/n\mathbb{Z})^*| = \phi(n) = n \prod_{p \mid n} \left(1 - \frac{1}{p}\right) ,$$

and more precisely

$$(\mathbb{Z}/n\mathbb{Z})^* \simeq \prod_{p^\alpha \| n} (\mathbb{Z}/p^\alpha\mathbb{Z})^* ,$$

where

$$(\mathbb{Z}/p^\alpha\mathbb{Z})^* \simeq \mathbb{Z}/(p-1)p^{\alpha-1}\mathbb{Z}$$

(i.e. is cyclic) when $p \geq 3$ or $p = 2$ and $\alpha \leq 2$, and

$$(\mathbb{Z}/2^\alpha\mathbb{Z})^* \simeq \mathbb{Z}/2\mathbb{Z} \times \mathbb{Z}/2^{\alpha-2}\mathbb{Z}$$

when $p = 2$ and $\alpha \geq 3$.

Now when $(\mathbb{Z}/n\mathbb{Z})^*$ is cyclic, i.e. by the above theorem when n is equal either to p^α, $2p^\alpha$ with p an odd prime, or $n = 2$ or 4, an integer g such that the class of g generates $(\mathbb{Z}/n\mathbb{Z})^*$ will be called a *primitive root* modulo n. Recall that the *order* of an element g in a group is the least positive integer n such that g^n is equal to the identity element of the group. When the group is finite, the order of any element divides the order of the group. Furthermore, g is a primitive root of $(\mathbb{Z}/n\mathbb{Z})^*$ if and only if its order is exactly equal to $\phi(n)$. As a corollary of the above results, we obtain the following:

Proposition 1.4.2.

(1) (Fermat). *If p is a prime and a is not divisible by p, then we have*

$$a^{p-1} \equiv 1 \pmod{p} .$$

(2) (Euler). *More generally, if n is a positive integer, then for any integer a coprime to n we have*

$$a^{\phi(n)} \equiv 1 \pmod{n} ,$$

and even

$$a^{\phi(n)/2} \equiv 1 \pmod{n}$$

if n is not equal to 2, 4, p^α or $2p^\alpha$ with p an odd prime.

To compute the order of an element in a finite group G, we use the following straightforward algorithm.

Algorithm 1.4.3 (Order of an Element). Given a finite group G of cardinality $h = |G|$, and an element $g \in G$, this algorithm computes the order of g in G. We denote by 1 the unit element of G.

1. [Initialize] Compute the prime factorization of h, say $h = p_1^{v_1} p_2^{v_2} \cdots p_k^{v_k}$, and set $e \leftarrow h$, $i \leftarrow 0$.

2. [Next p_i] Set $i \leftarrow i + 1$. If $i > k$, output e and terminate the algorithm. Otherwise, set $e \leftarrow e/p_i^{v_i}$, $g_1 \leftarrow g^e$.

3. [Compute local order] While $g_1 \neq 1$, set $g_1 \leftarrow g_1^{p_i}$ and $e \leftarrow e \cdot p_i$. Go to step 2.

Note that we need the complete factorization of h for this algorithm to work. This may be difficult when the group is very large.

Let p be a prime. To find a primitive root modulo p there seems to be no better way than to proceed as follows. Try $g = 2$, $g = 3$, etc ... until g is a primitive root. One should avoid perfect powers since if $g = g_0^k$, then if g is a primitive root, so is g_0 which has already been tested.

To see whether g is a primitive root, we could compute the order of g using the above algorithm. But it is more efficient to proceed as follows.

Algorithm 1.4.4 (Primitive Root). Given an odd prime p, this algorithm finds a primitive root modulo p.

1. [Initialize a] Set $a \leftarrow 1$ and let $p - 1 = p_1^{v_1} p_2^{v_2} \cdots p_k^{v_k}$ be the complete factorization of $p - 1$.

2. [Initialize check] Set $a \leftarrow a + 1$ and $i \leftarrow 1$.

3. [Check p_i] Compute $e \leftarrow a^{(p-1)/p_i}$. If $e = 1$ go to step 2. Otherwise, set $i \leftarrow i + 1$.

4. [finished?] If $i > k$ output a and terminate the algorithm, otherwise go to step 3.

Note that we do not avoid testing prime powers, hence this simple algorithm can still be improved if desired. In addition, the test for $p_i = 2$ can be replaced by the more efficient check that the Legendre symbol $\left(\frac{a}{p}\right)$ is equal to -1 (see Algorithm 1.4.10 below).

If n is not a prime, but is such that there exists a primitive root modulo n, we could, of course, use the above two algorithms by modifying them suitably. It is more efficient to proceed as follows.

First, if $n = 2$ or $n = 4$, $g = n - 1$ is a primitive root. When $n = 2^a$ is a power of 2 with $a \geq 3$, $(\mathbb{Z}/n\mathbb{Z})^*$ is not cyclic any more, but is isomorphic to the product of $\mathbb{Z}/2\mathbb{Z}$ with a cyclic group of order 2^{a-2}. Then $g = 5$ is always a generator of this cyclic subgroup (see Exercise 14), and can serve as a substitute in this case if needed.

When $n = p^a$ is a power of an odd prime, with $a \geq 2$, then we use the following lemma.

Lemma 1.4.5. *Let p be an odd prime, and let g be a primitive root modulo p. Then either g or $g + p$ is a primitive root modulo every power of p.*

Proof. For any m we have $m^p \equiv m \pmod{p}$, hence it follows that for every prime l dividing $p - 1$, $g^{p^{a-1}(p-1)/l} \equiv g^{(p-1)/l} \not\equiv 1 \pmod{p}$. So for g to be a primitive root, we need only that $g^{p^{a-2}(p-1)} \not\equiv 1 \pmod{p^a}$. But one checks immediately by induction that $x^p \equiv 1 \pmod{p^a}$ implies that $x \equiv 1 \pmod{p^b}$ for every $b \leq a - 1$. Applying this to $x = g^{p^{a-2}(p-1)}$ we see that our condition on g is equivalent to the same condition with a replaced by $a - 1$, hence by induction to the condition $g^{p-1} \not\equiv 1 \pmod{p^2}$. But if $g^{p-1} \equiv 1 \pmod{p^2}$, then by the binomial theorem $(g + p)^{p-1} \equiv 1 - pg^{p-2} \not\equiv 1 \pmod{p^2}$, thus proving the lemma. $\square$

Therefore to find a primitive root modulo p^a for p an odd prime and $a \geq 2$, proceed as follows: first compute g a primitive root modulo p using Algorithm 1.4.4, then compute $g_1 = g^{p-1} \bmod p^2$. If $g_1 \neq 1$, g is a primitive root modulo p^a for every a, otherwise $g + p$ is.

Finally, note that when p is an odd prime, if g is a primitive root modulo p^a then g or $g + p^a$ (whichever is odd) is a primitive root modulo $2p^a$.

1.4.2 The Legendre-Jacobi-Kronecker Symbol

Let p be an odd prime. Then it is easy to see that for a given integer a, the congruence

$$x^2 \equiv a \pmod{p}$$

can have either no solution (we say in this case that a is a quadratic non-residue mod p), one solution if $a \equiv 0 \pmod{p}$, or two solutions (we then say that a is a quadratic residue mod p). Define the Legendre symbol $\left(\frac{a}{p}\right)$ as being -1 if a is a quadratic non-residue, 0 if $a = 0$, and 1 if a is a quadratic residue. Then the number of solutions modulo p of the above congruence is $1 + \left(\frac{a}{p}\right)$. Furthermore, one can easily show that this symbol has the following properties (see e.g. [H-W]):

Proposition 1.4.6.

(1) *The Legendre symbol is multiplicative, i.e.*

$$\left(\frac{a}{p}\right)\left(\frac{b}{p}\right) = \left(\frac{ab}{p}\right) .$$

In particular, the product of two quadratic non-residues is a quadratic residue.

(2) *We have the congruence*

$$a^{(p-1)/2} \equiv \left(\frac{a}{p}\right) \pmod{p} .$$

(3) *There are as many quadratic residues as non-residues mod p, i.e. $(p-1)/2$.*

We will see that the Legendre symbol is fundamental in many problems. Thus, we need a way to compute it. One idea is to use the congruence $a^{(p-1)/2} \equiv \left(\frac{a}{p}\right) \pmod{p}$. Using the powering Algorithm 1.2.1, this enables us to compute the Legendre symbol in time $O(\ln^3 p)$. We can improve on this by using the Legendre-Gauss quadratic reciprocity law, which is itself a result of fundamental importance:

Theorem 1.4.7. *Let p be an odd prime. Then:*

(1)

$$\left(\frac{-1}{p}\right) = (-1)^{(p-1)/2} , \qquad \left(\frac{2}{p}\right) = (-1)^{(p^2-1)/8} .$$

(2) *If q is an odd prime different from p, then we have the reciprocity law:*

$$\left(\frac{p}{q}\right)\left(\frac{q}{p}\right) = (-1)^{(p-1)(q-1)/4} .$$

For a proof, see Exercises 16 and 18 and standard textbooks (e.g. [H-W], [Ire-Ros]).

This theorem can certainly help us to compute Legendre symbols since $\left(\frac{a}{p}\right)$ is multiplicative in a and depends only on a modulo p. A direct use of Theorem 1.4.7 would require factoring all the numbers into primes, and this is very slow. Luckily, there is an extension of this theorem which takes care of this problem. We first need to extend the definition of the Legendre symbol.

Definition 1.4.8. *We define the Kronecker (or Kronecker-Jacobi) symbol $\left(\frac{a}{b}\right)$ for any a and b in $\mathbb{Z}$ in the following way.*

(1) *If $b = 0$, then $\left(\frac{a}{0}\right) = 1$ if $a = \pm 1$, and is equal to 0 otherwise.*
(2) *For $b \neq 0$, write $b = \prod p$, where the p are not necessarily distinct primes (including $p = 2$), or $p = -1$ to take care of the sign. Then we set*

$$\left(\frac{a}{b}\right) = \prod \left(\frac{a}{p}\right) ,$$

where $\left(\frac{a}{p}\right)$ is the Legendre symbol defined above for $p > 2$, and where we define

$$\left(\frac{a}{2}\right) = \begin{cases} 0, & \text{if } a \text{ is even} \\ (-1)^{(a^2-1)/8}, & \text{if } a \text{ is odd.} \end{cases}$$

and also

$$\left(\frac{a}{-1}\right) = \begin{cases} 1, & \text{if } a \geq 0 \\ -1, & \text{if } a < 0. \end{cases}$$

Then, from the properties of the Legendre symbol, and in particular from the reciprocity law 1.4.7, one can prove that the Kronecker symbol has the following properties:

Theorem 1.4.9.

(1) $\left(\frac{a}{b}\right) = 0$ *if and only if $(a, b) \neq 1$*
(2) *For all a, b and c we have*

$$\left(\frac{ab}{c}\right) = \left(\frac{a}{c}\right)\left(\frac{b}{c}\right) , \quad \left(\frac{a}{bc}\right) = \left(\frac{a}{b}\right)\left(\frac{a}{c}\right) \quad \text{if } bc \neq 0$$

(3) $b > 0$ *being fixed, the symbol $\left(\frac{a}{b}\right)$ is periodic in a of period b if $b \not\equiv 2$ (mod 4), otherwise it is periodic of period $4b$.*
(4) $a \neq 0$ *being fixed (positive or negative), the symbol $\left(\frac{a}{b}\right)$ is periodic in b of period $|a|$ if $a \equiv 0$ or 1 (mod 4), otherwise it is periodic of period $4|a|$.*
(5) *The formulas of Theorem 1.4.7 are still true if p and q are only supposed to be positive odd integers, not necessarily prime.*

Note that in this theorem (as in the rest of this book), when we say that a function $f(x)$ is periodic of period b, this means that for all x, $f(x+b) = f(x)$, but b need not be the smallest possible period.

Theorem 1.4.9 is a necessary prerequisite for any study of quadratic fields, and the reader is urged to prove it by himself (Exercise 17).

As has been mentioned, a consequence of this theorem is that it is easy to design a fast algorithm to compute Legendre symbols, and more generally Kronecker symbols if desired.

Algorithm 1.4.10 (Kronecker). Given $a, b \in \mathbb{Z}$, this algorithm computes the Kronecker symbol $\left(\frac{a}{b}\right)$ (hence the Legendre symbol when b is an odd prime).

1. [Test b equal to 0] If $b = 0$ then output 0 if $|a| \neq 1$, 1 if $|a| = 1$ and terminate the algorithm.

2. [Remove 2's from b] If a and b are both even, output 0 and terminate the algorithm. Otherwise, set $v \leftarrow 0$ and while b is even set $v \leftarrow v + 1$ and $b \leftarrow b/2$. Then if v is even set $k \leftarrow 1$, otherwise set $k \leftarrow (-1)^{(a^2-1)/8}$ (by table lookup, *not* by computing $(a^2 - 1)/8$). Finally if $b < 0$ set $b \leftarrow -b$, and if in addition $a < 0$ set $k \leftarrow -k$.

3. [Finished?] (Here b is odd and $b > 0$.) If $a = 0$ then output 0 if $b > 1$, k if $b = 1$, and terminate the algorithm. Otherwise, set $v \leftarrow 0$ and while a is even do $v \leftarrow v + 1$ and $a \leftarrow a/2$. If v is odd set $k \leftarrow (-1)^{(b^2-1)/8}k$.

4. [Apply reciprocity] Set

$$k \leftarrow (-1)^{(a-1)(b-1)/4}k ,$$

(using if statements and no multiplications), and then $r \leftarrow |a|$, $a \leftarrow b \bmod r$, $b \leftarrow r$ and go to step 3.

Remarks.

(1) As mentioned, the expressions $(-1)^{(a^2-1)/8}$ and $(-1)^{(a-1)(b-1)/4}$ should not be computed as powers, even though they are written this way. For example, to compute the first expression, set up and save a table tab2 containing

$$\{0, 1, 0, -1, 0, -1, 0, 1\} ,$$

and then the formula $(-1)^{(a^2-1)/8} =$ tab2[a&7], the & symbol denoting bitwise and, which is a very fast operation compared to multiplication (note that a&7 is equivalent to a mod 8). The instruction $k \leftarrow (-1)^{(a-1)(b-1)/4}k$ is very efficiently translated in C by

```
if(a&b&2) k= -k;
```

(2) We need to prove that the algorithm is valid! It terminates since, because except possibly the first time, at the beginning of step 3 we have $0 < b < a$ and the value of b is strictly decreasing. It gives the correct result because of the following lemma which is an immediate corollary of Theorem 1.4.9:

Lemma 1.4.11. *If a and b are odd integers with $b > 0$ (but not necessarily $a > 0$), then we have*

$$\left(\frac{a}{b}\right) = (-1)^{(a-1)(b-1)/4}\left(\frac{b}{|a|}\right).$$

(3) We may want to avoid cleaning out the powers of 2 in step 3 at each pass through the loop. We can do this by slightly changing step 4 so as to always end up with an odd value of a. This however may have disastrous effects on the running time, which may become exponential instead of polynomial time (see [Bac-Sha] and Exercise 24).

Note that Algorithm 1.4.10 can be slightly improved (by a small constant factor) by adding the following statement at the end of the assignments of step 4, before going back to step 3: If $a > r/2$, then $a = a - r$. This simply means that we ask, not for the residue of $a \bmod r$ which is between 0 and $r - 1$, but for the one which is least in absolute value, i.e. between $-r/2$ and $r/2$. This modification could also be used in Euclid's algorithms if desired, if tests suggest that it is faster in practice.

One can also use the binary version of Euclid's algorithm to compute Kronecker symbols. Since, in any case, the prime 2 plays a special role, this does not really increase the complexity, and gives the following algorithm.

Algorithm 1.4.12 (Kronecker-Binary). Given $a, b \in \mathbb{Z}$, this algorithm computes the Kronecker symbol $\left(\frac{a}{b}\right)$ (hence the Legendre symbol when b is an odd prime).

1. [Test $b = 0$] If $b = 0$ then output 0 if $|a| \neq 1$, 1 if $|a| = 1$ and terminate the algorithm.

2. [Remove 2's from b] If a and b are both even, output 0 and terminate the algorithm. Otherwise, set $v \leftarrow 0$ and while b is even set $v \leftarrow v + 1$ and $b \leftarrow b/2$. Then if v is even set $k \leftarrow 1$, otherwise set $k \leftarrow (-1)^{(a^2-1)/8}$ (by table lookup, *not* by computing $(a^2 - 1)/8$). Finally, if $b < 0$ set $b \leftarrow -b$, and if in addition $a < 0$ set $k \leftarrow -k$.

3. [Reduce size once] (Here b is odd and $b > 0$.) Set $a \leftarrow a \bmod b$.

4. [Finished?] If $a = 0$, output 0 if $b > 1$, k if $b = 1$, and terminate the algorithm.

5. [Remove powers of 2] Set $v \leftarrow 0$ and, while a is even, set $v \leftarrow v + 1$ and $a \leftarrow a/2$. If v is odd, set $k \leftarrow (-1)^{(b^2-1)/8}k$.

6. [Subtract and apply reciprocity] (Here a and b are odd.) Set $r \leftarrow b - a$. If $r > 0$, then set $k \leftarrow (-1)^{(a-1)(b-1)/4}k$ (using if statements), $b \leftarrow a$ and $a \leftarrow r$, else set $a \leftarrow -r$. Go to step 4.

Note that we cannot immediately reduce a modulo b at the beginning of the algorithm. This is because when b is even the Kronecker symbol $\left(\frac{a}{b}\right)$ is not periodic of period b in general, but only of period $4b$. Apart from this remark, the proof of the validity of this algorithm follows immediately from Theorem 1.4.10 and the validity of the binary algorithm.   □

The running time of all of these Legendre symbol algorithms has the same order of magnitude as Euclid's algorithm, i.e. $O(\ln^2 N)$ when carefully programmed, where N is an upper bound on the size of the inputs a and b. Note however that the constants will be different because of the special treatment of even numbers.

1.5 Computing Square Roots Modulo p

We now come to a slightly more specialized question. Let p be an odd prime number, and suppose that we have just checked that $\left(\frac{a}{p}\right) = 1$ using one of the algorithms given above. Then by definition, there exists an x such that $x^2 \equiv a$ (mod p). How do we find x? Of course, a brute force search would take time $O(p)$ and, even for p moderately large, is out of the question. We need a faster algorithm to do this. At this point the reader might want to try and find one himself before reading further. This would give a feel for the difficulty of the problem. (Note that we will be considering much more difficult and general problems later on, so it is better to start with a simple one.)

There is an easy solution which comes to mind that works for half of the primes p, i.e. primes $p \equiv 3$ (mod 4). I claim that in this case a solution is given by

$$x = a^{(p+1)/4} \quad (\text{mod } p) \ ,$$

the computation being done using the powering Algorithm 1.2.1. Indeed, since a is a quadratic residue, we have $a^{(p-1)/2} \equiv 1$ (mod p) hence

$$x^2 \equiv a^{(p+1)/2} \equiv a \cdot a^{(p-1)/2} \equiv a \quad (\text{mod } p)$$

as claimed.

A less trivial solution works for half of the remaining primes, i.e. primes $p \equiv 5$ (mod 8). Since we have $a^{(p-1)/2} \equiv 1$ (mod p) and since $\mathbb{F}_p = \mathbb{Z}/p\mathbb{Z}$ is a field, we must have

$$a^{(p-1)/4} \equiv \pm 1 \quad (\text{mod } p) \ .$$

Now, if the sign is $+$, then the reader can easily check as above that

$$x = a^{(p+3)/8} \quad (\text{mod } p)$$

is a solution. Otherwise, using $p \equiv 5$ (mod 8) and Theorem 1.4.7, we know that $2^{(p-1)/2} \equiv -1$ (mod p). Then one can check that

$$x = 2a \cdot (4a)^{(p-5)/8} \quad (\text{mod } p)$$

is a solution.

Thus the only remaining case is $p \equiv 1 \pmod 8$. Unfortunately, this is the hardest case. Although, by methods similar to the one given above, one could give an infinite number of families of solutions, this would not be practical in any sense.

1.5.1 The Algorithm of Tonelli and Shanks

There are essentially three algorithms for solving the above problem. One is a special case of a general method for factoring polynomials modulo p, which we will study in Chapter 3. Another is due to Schoof and it is the only non-probabilistic polynomial time algorithm known for this problem. It is quite complex since it involves the use of elliptic curves (see Chapter 7), and its practicality is not clear, although quite a lot of progress has been achieved by Atkin. Therefore, we will not discuss it here. The third and last algorithm is due to Tonelli and Shanks, and although probabilistic, it is quite efficient. It is the most natural generalization of the special cases studied above. We describe this algorithm here.

We can always write

$$p - 1 = 2^e \cdot q, \quad \text{with } q \text{ odd.}$$

The multiplicative group $(\mathbb{Z}/p\mathbb{Z})^*$ is isomorphic to the (additive) group $\mathbb{Z}/(p-1)\mathbb{Z}$, hence its 2-Sylow subgroup G is a cyclic group of order 2^e. Assume that one can find a generator z of G. The squares in G are the elements of order dividing 2^{e-1}, and are also the even powers of z. Hence, if a is a quadratic residue mod p, then, since

$$a^{(p-1)/2} = (a^q)^{(2^{e-1})} \equiv 1 \pmod p ,$$

$b = a^q \bmod p$ is a square in G, so there exists an even integer k with $0 \le k < 2^e$ such that

$$a^q z^k = 1 \quad \text{in } G .$$

If one sets

$$x = a^{(q+1)/2} z^{k/2} ,$$

it is clear that $x^2 \equiv a \pmod p$, hence x is the answer. To obtain an algorithm, we need to solve two problems: finding a generator z of G, and computing the exponent k. Although very simple to solve in practice, the first problem is the probabilistic part of the algorithm. The best way to find z is as follows: choose at random an integer n, and compute $z = n^q \bmod p$. Then it is clear that z is a generator of G (i.e. $z^{2^{e-1}} = -1$ in G) if and only if n is a quadratic non-residue mod p, and this occurs with probability close to $1/2$ (exactly $(p-1)/(2p)$). Therefore, in practice, we will find a non-residue very quickly. For example, the probability that one does not find one after 20 trials is lower than 10^{-6}.

Finding the exponent k is slightly more difficult, and in fact is not needed explicitly (only $a^{(q+1)/2}z^{k/2}$ is needed). The method is explained in the following complete algorithm, which in this form is due to Shanks.

Algorithm 1.5.1 (Square Root Mod p). Let p be an odd prime, and $a \in \mathbb{Z}$. Write $p - 1 = 2^e \cdot q$ with q odd. This algorithm, either outputs an x such that $x^2 \equiv a \pmod{p}$, or says that such an x does not exist (i.e. that a is a quadratic non-residue mod p).

1. [Find generator] Choose numbers n at random until $\left(\frac{n}{p}\right) = -1$. Then set $z \leftarrow n^q \pmod{p}$.

2. [Initialize] Set $y \leftarrow z$, $r \leftarrow e$, $x \leftarrow a^{(q-1)/2} \pmod{p}$, $b \leftarrow ax^2 \pmod{p}$, $x \leftarrow ax \pmod{p}$.

3. [Find exponent] If $b \equiv 1 \pmod{p}$, output x and terminate the algorithm. Otherwise, find the smallest $m \geq 1$ such that $b^{2^m} \equiv 1 \pmod{p}$. If $m = r$, output a message saying that a is not a quadratic residue mod p.

4. [Reduce exponent] Set $t \leftarrow y^{2^{r-m-1}}$, $y \leftarrow t^2$, $r \leftarrow m$, $x \leftarrow xt$, $b \leftarrow by$ (all operations done modulo p), and go to step 3.

Note that at the beginning of step 3 we always have the congruences modulo p:

$$ab \equiv x^2 \ , \qquad y^{2^{r-1}} \equiv -1 \ , \qquad b^{2^{r-1}} \equiv 1 \ .$$

If G_r is the subgroup of G whose elements have an order dividing 2^r, then this says that y is a generator of G_r and that b is in G_{r-1}, in other words that b is a square in G_r. Since r is strictly decreasing at each loop of the algorithm, the number of loops is at most e. When $r \leq 1$ we have $b = 1$ hence the algorithm terminates, and the above congruence shows that x is one of the square roots of a mod p.

It is easy to show that, on average, steps 3 and 4 will require $e^2/4$ multiplications mod p, and at most e^2. Hence the expected running time of this algorithm is $O(\ln^4 p)$. □

Remarks.

(1) In the algorithm above, we have not explicitly computed the value of the exponent k such that $a^q z^k = 1$ but it is easy to do so if needed (see Exercise 25).

(2) As already mentioned, Shanks's algorithm is probabilistic, although the only non-deterministic part is finding a quadratic non-residue mod p, which seems quite a harmless task. One could try making it completely deterministic by successively trying $n = 2, 3 \ldots$ in step 1 until a non-residue is found. This is a reasonable method, but unfortunately the most powerful analytical tools only allow us to prove that the smallest quadratic non-residue is $O(p^\alpha)$ for a non-zero α. Thus, this deterministic algorithm,

although correct, may have, as far as we know, an exponential running time.

If one assumes the Generalized Riemann Hypothesis (GRH), then one can prove much more, i.e. that the smallest quadratic non-residue is $O(\ln^2 p)$, hence this gives a polynomial running time (in $O(\ln^4 p)$ since computing a Legendre symbol is in $O(\ln^2 p)$). In fact, Bach [Bach] has proved that for $p > 1000$ the smallest non-residue is less than $2 \ln^2 p$. In any case, in practice the probabilistic method and the sequential method (i.e. choosing $n = 2, 3 \cdots$) give essentially equivalent running times.

(3) If m is an integer whose factorization into a product of prime powers is completely known, it is easy to write an algorithm to solve the more general problem $x^2 \equiv a \pmod{m}$ (see Exercise 30).

1.5.2 The Algorithm of Cornacchia

A well known theorem of Fermat (see [H-W]) says that an odd prime p is a sum of two squares if and only if $p \equiv 1 \bmod 4$, i.e. if and only if -1 is a quadratic residue mod p. Furthermore, up to sign and exchange, the representation of p as a sum of two squares is unique. Thus, it is natural to ask for an algorithm to compute x and y such that $x^2 + y^2 = p$ when $p \equiv 1 \bmod 4$. More generally, given a positive integer d and an odd prime p, one can ask whether the equation

$$x^2 + dy^2 = p$$

has a solution, and for an algorithm to find x and y when they exist. There is a pretty algorithm due to Cornacchia which solves both problems simultaneously. For the beautiful and deep *theory* concerning the first problem, which is closely related to complex multiplication (see Section 7.2) see [Cox].

First, note that a necessary condition for the existence of a solution is that $-d$ be a quadratic residue modulo p. Indeed, we clearly must have $y \not\equiv 0 \bmod p$ hence

$$\left(xy^{-1}\right)^2 \equiv -d \bmod p \; ,$$

where y^{-1} denotes the inverse of y modulo p. We therefore assume that this condition is satisfied. By using Algorithm 1.5.1 we can find an integer x_0 such that

$$x_0^2 \equiv -d \bmod p$$

and we may assume that $p/2 < x_0 < p$. Cornacchia's algorithm tells us that we should simply apply Euclid's Algorithm 1.3.1 to the pair $(a, b) = (p, x_0)$ until we obtain a number b such that $b < \sqrt{p}$. Then we set $c \leftarrow (p - b^2)/d$, and if c is the square of an integer s, the equation $x^2 + dy^2 = p$ has $(x, y) = (b, s)$ as (essentially unique) solution, otherwise it has no solution. This leads to the following algorithm.

Algorithm 1.5.2 (Cornacchia). Let p be a prime number and d be an integer such that $0 < d < p$. This algorithm either outputs an integer solution (x, y) to the Diophantine equation $x^2 + dy^2 = p$, or says that such a solution does not exist.

1. [Test if residue] Using Algorithm 1.4.12 compute $k \leftarrow \left(\frac{-d}{p}\right)$. If $k = -1$, say that the equation has no solution and terminate the algorithm.

2. [Compute square root] Using Shanks's Algorithm 1.5.1, compute an integer x_0 such that $x_0^2 \equiv -d \bmod p$, and change x_0 into $\pm x_0 + kp$ so that $p/2 < x_0 < p$. Then set $a \leftarrow p$, $b \leftarrow x_0$ and $l \leftarrow \lfloor \sqrt{p} \rfloor$.

3. [Euclidean algorithm] If $b > l$, set $r \leftarrow a \bmod b$, $a \leftarrow b$, $b \leftarrow r$ and go to step 3.

4. [Test solution] If d does not divide $p - b^2$ or if $c = (p - b^2)/d$ is not the square of an integer (see Algorithm 1.7.3), say that the equation has no solution and terminate the algorithm. Otherwise, output $(x, y) = (b, \sqrt{c})$ and terminate the algorithm.

Let us give a numerical example. Assume that we want to solve $x^2 + 2y^2 = 97$. In step 1, we first compute $\left(\frac{-2}{97}\right)$ by Algorithm 1.4.12 (or directly since here it is easy), and find that -2 is a quadratic residue mod 97. Thus the equation may have a solution (and in fact it must have one since the class number of the ring of integers of $\mathbb{Q}(\sqrt{2})$ is equal to 1, see Chapter 5). In step 2, we compute x_0 such that $x_0^2 \equiv -2 \bmod 97$ using Algorithm 1.5.1. Using $n = 5$ hence $z = 28$, we readily find $x_0 = 17$. Then the Euclidean algorithm in step 3 gives $97 = 5 \cdot 17 + 12$, $17 = 1 \cdot 12 + 5$ and hence $b = 5$ is the first number obtained in the Euclidean stage, which is less than or equal to the square root of 97. Now $c = (97 - 5^2)/2 = 36$ is a square, hence a solution (unique) to our equation is $(x, y) = (5, 6)$. Of course, this could have been found much more quickly by inspection, but for larger numbers we need to use the algorithm as written.

The proof of this algorithm is not really difficult, but is a little painful so we refer to [Mor-Nic]. A nice proof due to H. W. Lenstra can be found in [Scho2]. Note also that Algorithm 1.3.14 above can also be used to solve the problem, and the proof that we gave of the validity of that algorithm is similar, but simpler.

When working in complex quadratic orders of discriminant $D < 0$ congruent to 0 or 1 modulo 4 (see Chapter 5), it is more natural to solve the equation

$$x^2 + |D|y^2 = 4p$$

where p is an odd prime (we will for example need this in Chapter 9).

If $4 \mid D$, we must have $2 \mid x$, hence the equation is equivalent to $x'^2 + dy^2 = p$ with $x' = x/2$ and $d = |D|/4$, which we can solve by using Algorithm 1.5.2.

If $D \equiv 1 \pmod 8$, we must have $x^2 - y^2 \equiv 4 \pmod 8$ and this is possible only when x and y are even, hence our equation is equivalent to $x'^2 + dy'^2 = p$ with $x' = x/2$, $y' = y/2$ and $d = |D|$, which is again solved by Algorithm 1.5.2

Finally, if $D \equiv 5 \pmod{8}$, the parity of x and y is not a priori determined. Therefore Algorithm 1.5.2 cannot be applied as written. There is however a modification of Algorithm 1.5.2 which enables us to treat this problem.

For this compute x_0 such that $x_0^2 \equiv D \pmod{p}$ using Algorithm 1.5.1, and if necessary change x_0 into $p - x_0$ so that in fact $x_0^2 \equiv D \pmod{4p}$. Then apply the algorithm as written, starting with $(a, b) = (2p, x_0)$, and stopping as soon as $b < l$, where $l = \lfloor 2\sqrt{p} \rfloor$. Then, as in [Mor-Nic] one can show that this gives the (essentially unique) solution to $x^2 + |D|y^2 = 4p$. This gives the following algorithm.

Algorithm 1.5.3 (Modified Cornacchia). Let p be a prime number and D be a negative integer such that $D \equiv 0$ or 1 modulo 4 and $|D| < 4p$. This algorithm either outputs an integer solution (x, y) to the Diophantine equation $x^2 + |D|y^2 = 4p$, or says that such a solution does not exist.

1. [Case $p = 2$] If $p = 2$ do as follows. If $D + 8$ is the square of an integer, output $(\sqrt{D + 8}, 1)$, otherwise say that the equation has no solution. Then terminate the algorithm.

2. [Test if residue] Using Algorithm 1.4.12 compute $k \leftarrow \left(\frac{D}{p}\right)$. If $k = -1$, say that the equation has no solution and terminate the algorithm.

3. [Compute square root] Using Shanks's Algorithm 1.5.1, compute an integer x_0 such that $x_0^2 \equiv D \bmod p$ and $0 \le x_0 < p$, and if $x_0 \not\equiv D \pmod{2}$, set $x_0 \leftarrow p - x_0$. Finally, set $a \leftarrow 2p$, $b \leftarrow x_0$ and $l \leftarrow \lfloor 2\sqrt{p} \rfloor$.

4. [Euclidean algorithm] If $b > l$, set $r \leftarrow a \bmod b$, $a \leftarrow b$, $b \leftarrow r$ and go to step 4.

5. [Test solution] If $|D|$ does not divide $4p - b^2$ or if $c = (4p - b^2)/|D|$ is not the square of an integer (see Algorithm 1.7.3), say that the equation has no solution and terminate the algorithm. Otherwise, output $(x, y) = (b, \sqrt{c})$ and terminate the algorithm.

1.6 Solving Polynomial Equations Modulo p

We will consider more generally in Chapter 3 the problem of factoring polynomials mod p. If one wants only to find the linear factors, i.e. the roots mod p, then for small degrees one can use the standard formulas. To avoid writing congruences all the time, we implicitly assume that we work in $\mathbb{F}_p = \mathbb{Z}/p\mathbb{Z}$.

In degree one, the solution of the equation $ax + b = 0$ is $x = -b \cdot a^{-1}$, where a^{-1} is computed using Euclid's extended algorithm.

In degree two, the solutions of the equation $ax^2 + bx + c = 0$ where $a \ne 0$ and $p \ne 2$, are given as follows. Set $D = b^2 - 4ac$. If $\left(\frac{D}{p}\right) = -1$, then there are no solutions in $\mathbb{F}_p$. If $\left(\frac{D}{p}\right) = 0$, i.e. if $p \mid D$, then there is a unique (double) solution given by $x = -b \cdot (2a)^{-1}$. Finally, if $\left(\frac{D}{p}\right) = 1$, there are two solutions, obtained in the following way: compute an s such that $s^2 = D$ using one of the algorithms of the preceding section. Then the solutions are as usual

$$(-b \pm s) \cdot (2a)^{-1} .$$

In degree three, Cardano's formulas can be used (see Exercise 28 of Chapter 3). There are however two difficulties which must be taken care of. The first is that we must find an algorithm to compute cube roots. This can be done in a manner similar to the case of square roots. The second difficulty lies in the handling of square roots when these square roots are not in $\mathbb{F}_p$ (they are then in $\mathbb{F}_{p^2}$). This is completely analogous to handling complex numbers when a real cubic equation has three real roots. The reader will find it an amusing exercise to try and iron out all these problems (see Exercise 28). Otherwise, see [Wil-Zar] and [Mor1], who also gives the analogous recipes for degree four equations (note that for computing fourth roots one can simply compute two square roots).

In degree 5 and higher, the general equations have a non-solvable Galois group, hence as in the complex case, no special-purpose algorithms are known, and one must rely on general methods, which are slower. These methods will be seen in Section 3.4, to which we refer for notations and definitions, but in the special case of root finding, the algorithm is much simpler. We assume $p > 2$ since for $p = 2$ there are just two values to try.

Algorithm 1.6.1 (Roots Mod p). Given a prime number $p \geq 3$ and a polynomial $P \in \mathbb{F}_p[X]$, this algorithm outputs the roots of P in $\mathbb{F}_p$. This algorithm will be called recursively, and it is understood that all the operations are done in $\mathbb{F}_p$.

1. [Isolate roots in $\mathbb{F}_p$] Compute $A(X) \leftarrow (X^p - X, P(X))$ as explained below. If $A(0) = 0$, output 0 and set $A(X) \leftarrow A(X)/X$.

2. [Small degree?] If $\deg(A) = 0$, terminate the algorithm. If $\deg(A) = 1$, and $A(X) = a_1 X + a_0$, output $-a_0/a_1$ and terminate the algorithm. If $\deg(A) = 2$ and $A(X) = a_2 X^2 + a_1 X + a_0$, set $d \leftarrow a_1^2 - 4a_0 a_2$, compute $e \leftarrow \sqrt{d}$ using Algorithm 1.5.1, output $(-a_1 + e)/(2a_2)$ and $(-a_1 - e)/(2a_2)$, and terminate the algorithm. (Note that e will exist.)

3. [Random splitting] Choose a random $a \in \mathbb{F}_p$, and compute $B(X) \leftarrow ((X + a)^{(p-1)/2} - 1, A(X))$ as explained below. If $\deg(B) = 0$ or $\deg(B) = \deg(A)$, go to step 3.

4. [Recurse] Output the roots of B and A/B using the present algorithm recursively (skipping step 1), and terminate the algorithm.

Proof. The elements of $\mathbb{F}_p$ are the elements x of an algebraic closure which satisfy $x^p = x$. Hence, the polynomial A computed in step 1 is, up to a constant factor, equal to the product of the $X - x$ where the x are the roots of P in $\mathbb{F}_p$. Step 3 then splits the roots x in two parts: the roots such that $x + a$ is a quadratic residue mod p, and the others. Since a is random, this has approximately one chance in $2^{\deg(A)-1}$ of not splitting the polynomial A into smaller pieces, and this shows that the algorithm is valid. $\qquad\qquad\square$

Implementation Remarks.

(1) step 2 can be simplified by not taking into account the case of degree 2, but this gives a slightly less efficient algorithm. Also, if step 2 is kept as it is, it may be worthwhile to compute once and for all the quadratic non-residue mod p which is needed in Algorithm 1.5.1.

(2) When we are asked to compute a GCD of the form $\gcd(u^n - b, c)$, we must *not* compute $u^n - b$, but instead we compute $d \leftarrow u^n \bmod c$ using the powering algorithm. Then we have $\gcd(u^n - b, c) = \gcd(d - b, c)$. In addition, since $u = X + a$ is a very simple polynomial, the left-right versions of the powering algorithm (Algorithms 1.2.3 and 1.2.4) are more advantageous here.

(3) When p is small, and in particular when p is smaller than the degree of $A(X)$, it may be faster to simply test all values $X = 0, \ldots, p - 1$. Thus, the above algorithm is really useful when p is not too small. In that case, it may be faster to compute $\gcd(X^{(p-1)/2} - 1, A(X - a))$ than $\gcd((X + a)^{(p-1)/2} - 1, A(X))$.

1.7 Power Detection

In many algorithms, it is necessary to detect whether a number is a square or more generally a perfect power, and if it is, to compute the root. We consider here the three most frequent problems of this sort and give simple *arithmetic* algorithms to solve them. Of course, to test whether $n = m^k$, you can always compute the nearest integer to $e^{\ln n/k}$ by transcendental means, and see if the kth power of that integer is equal to n. This needs to be tried only for $k \leq \lg n$. This is clearly quite inefficient, and also requires the use of transcendental functions, so we turn to better methods.

1.7.1 Integer Square Roots

We start by giving an algorithm which computes the integer part of the square root of any positive integer n. It uses a variant of Newton's method, but works entirely with integers. The algorithm is as follows.

Algorithm 1.7.1 (Integer Square Root). Given a positive integer n, this algorithm computes the integer part of the square root of n, i.e. the number m such that $m^2 \leq n < (m + 1)^2$.

1. [Initialize] Set $x \leftarrow n$ (see discussion).

2. [Newtonian step] Set $y \leftarrow \lfloor (x + \lfloor n/x \rfloor)/2 \rfloor$ using integer divides and shifts.

3. [Finished?] If $y < x$ set $x \leftarrow y$ and go to step 2. Otherwise, output x and terminate the algorithm.

Proof. By step 3, the value of x is strictly decreasing, hence the algorithm terminates. We must show that the output is correct. Let us set $q = \lfloor \sqrt{n} \rfloor$.

Since $(t + n/t)/2 \geq \sqrt{n}$ for any positive real value of t, it is clear that the inequality $x \geq q$ is satisfied throughout the algorithm (note that it is also satisfied also after the initialization step). Now assume that the termination condition in step 3 is satisfied, i.e. that $y = \lfloor (x + n/x)/2 \rfloor \geq x$. We must show that $x = q$. Assume the contrary, i.e. that $x \geq q + 1$. Then,

$$y - x = \left\lfloor \frac{x + n/x}{2} \right\rfloor - x = \left\lfloor \frac{n/x - x}{2} \right\rfloor = \left\lfloor \frac{n - x^2}{2x} \right\rfloor .$$

Since $x \geq q + 1 > \sqrt{n}$, we have $n - x^2 < 0$, hence $y - x < 0$ contradiction. This shows the validity of the algorithm. □

Remarks.

(1) We have written the formula in step 2 using the integer part function twice to emphasize that every operation must be done using integer arithmetic, but of course mathematically speaking, the outermost one would be enough.

(2) When actually implementing this algorithm, the initialization step must be modified. As can be seen from the proof, the only condition which must be satisfied in the initialization step is that x be greater or equal to the integer part of $\sqrt{n}$. One should try to initialize x as close as possible to this number. For example, after a $O(\ln \ln n)$ search, as in the left-right binary powering Algorithm 1.2.2, one can find e such that $2^e \leq n < 2^{e+1}$. Then, one can take $x \leftarrow 2^{\lfloor (e+2)/2 \rfloor}$. Another option is to compute a single precision floating point approximation to the square root of n and to take the ceiling of that. The choice between these options is machine dependent.

(3) Let us estimate the running time of the algorithm. As written, we will spend a lot of time essentially dividing x by 2 until we are in the right ball-park, and this requires $O(\ln n)$ steps, hence $O(\ln^3 n)$ running time. However, if care is taken in the initialization step as mentioned above, we can reduce this to the usual number of steps for a quadratically convergent algorithm, i.e. $O(\ln \ln n)$. In addition, if the precision is decreased at each iteration, it is not difficult to see that one can obtain an algorithm which runs in $O(\ln^2 n)$ bit operations, hence only a constant times slower than multiplication/division.

1.7.2 Square Detection

Given a positive integer n, we want to determine whether n is a square or not. One method of course would be to compute the integer square root of

n using Algorithm 1.7.1, and to check whether n is equal to the square of the result. This is far from being the most efficient method. We could also use Exercise 22 which says that a number is a square if and only if it is a quadratic residue modulo every prime not dividing it, and compute a few Legendre symbols using the algorithms of Section 1.4.2. We will use a variant of this method which replaces Legendre symbol computation by table lookup. One possibility is to use the following algorithm.

Precomputations 1.7.2. This is to be done and stored once and for all.

1. [Fill 11] For $k = 0$ to 10 set $q11[k] \leftarrow 0$. Then for $k = 0$ to 5 set $q11[k^2 \bmod 11] \leftarrow 1$.

2. [Fill 63] For $k = 0$ to 62 set $q63[k] \leftarrow 0$. Then for $k = 0$ to 31 set $q63[k^2 \bmod 63] \leftarrow 1$.

3. [Fill 64] For $k = 0$ to 63 set $q64[k] \leftarrow 0$. Then for $k = 0$ to 31 set $q64[k^2 \bmod 64] \leftarrow 1$.

4. [Fill 65] For $k = 0$ to 64 set $q65[k] \leftarrow 0$. Then for $k = 0$ to 32 set $q65[k^2 \bmod 65] \leftarrow 1$.

Once the precomputations are made, the algorithm is simply as follows.

Algorithm 1.7.3 (Square Test). Given a positive integer n, this algorithm determines whether n is a square or not, and if it is, outputs the square root of n. We assume that the precomputations 1.7.2 have been made.

1. [Test 64] Set $t \leftarrow n \bmod 64$ (using if possible only an and statement). If $q64[t] = 0$, n is not a square and terminate the algorithm. Otherwise, set $r \leftarrow n \bmod 45045$.

2. [Test 63] If $q63[r \bmod 63] = 0$, n is not a square and terminate the algorithm.

3. [Test 65] If $q65[r \bmod 65] = 0$, n is not a square and terminate the algorithm.

4. [Test 11] If $q11[r \bmod 11] = 0$, n is not a square and terminate the algorithm.

5. [Compute square root] Compute $q \leftarrow \lfloor \sqrt{n} \rfloor$ using Algorithm 1.7.1. If $n \neq q^2$, n is not a square and terminate the algorithm. Otherwise n is a square, output q and terminate the algorithm.

The validity of this algorithm is clear since if n is a square, it must be a square modulo k for any k. Let us explain the choice of the moduli. Note first that the number of squares modulo 64,63,65,11 is 12,16,21,6 respectively (see Exercise 23). Thus, if n is not a square, the probability that this will not have been detected in the four table lookups is equal to

$$\frac{12}{64} \frac{16}{63} \frac{21}{65} \frac{6}{11} = \frac{6}{715}$$

and this is less than one percent. Therefore, the actual computation of the integer square root in step 5 will rarely be done when n is not a square. This is the reason for the choice of the moduli. The order in which the tests are done comes from the inequalities

$$\frac{12}{64} < \frac{16}{63} < \frac{21}{65} < \frac{6}{11} .$$

If one is not afraid to spend memory, one can also store the squares modulo $45045 = 63 \cdot 65 \cdot 11$, and then only one test is necessary instead of three, in addition to the modulo 64 test.

Of course, other choices of moduli are possible (see [Nic]), but in practice the above choice works well.

1.7.3 Prime Power Detection

The last problem we will consider in this section is that of determining whether n is a prime power or not. This is a test which is sometimes needed, for example in some of the modern factoring algorithms (see Chapter 10). We will not consider the problem of testing whether n is a power of a general number, since it is rarely needed.

The idea is to use the following proposition.

Proposition 1.7.4. *Let $n = p^k$ be a prime power. Then*

(1) *For any a we have $p \mid (a^n - a, n)$.*
(2) *If $k \geq 2$ and $p > 2$, let a be a witness to the compositeness of n given by the Rabin-Miller test 8.2.2, i.e. such that $(a, n) = 1$, and if $n - 1 = 2^t q$ with q odd, then $a^q \not\equiv 1 \pmod{n}$ and for all e such that $0 \leq e \leq t-1$ then $a^{2^e q} \not\equiv -1 \pmod{n}$. Then $(a^n - a, n)$ is a non-trivial divisor of n (i.e. is different from 1 and n).*

Proof. By Fermat's theorem, we have $a^n \equiv a \pmod{p}$, hence (1) is clear. Let us prove (2). Let a be a witness to the compositeness of n as defined above. By (1), we already know that $(a^n - a, n) > 1$. Assume that $(a^n - a, n) = n$, i.e. that $a^n \equiv a \pmod{n}$. Since $(a, n) = 1$ this is equivalent to $a^{n-1} \equiv 1 \pmod{n}$, i.e. $a^{2^t q} \equiv 1 \pmod{n}$. Let f be the smallest non-negative integer such that $a^{2^f q} \equiv 1 \pmod{n}$. Thus f exists and $f \leq t$. If we had $f = 0$, this would contradict the definition of a witness ($a^q \not\equiv 1 \pmod{n}$). So $f > 0$. But then we can write

$$p^k \mid (a^{2^{f-1}q} - 1)(a^{2^{f-1}q} + 1)$$

and since p is an odd prime, this implies that p^k divides one of the two factors. But $p^k \mid (a^{2^{f-1}q} - 1)$ contradicts the minimality of f, and $p^k \mid (a^{2^{f-1}q} + 1)$ contradicts the fact that a is a witness (we cannot have $a^{2^e q} \equiv -1 \pmod{n}$ for $e < t$), hence we have a contradiction in every case thus proving the proposition. $\square$

This leads to the following algorithm.

Algorithm 1.7.5 (Prime Power Test). Given a positive integer $n > 1$, this algorithm tests whether or not n is of the form p^k with p prime, and if it is, outputs the prime p.

1. [Case n even] If n is even, set $p \leftarrow 2$ and go to step 4. Otherwise, set $q \leftarrow n$.

2. [Apply Rabin-Miller] By using Algorithm 8.2.2 show that either q is a probable prime or exhibit a witness a to the compositeness of q. If q is a probable prime, set $p \leftarrow q$ and go to step 4.

3. [Compute GCD] Set $d \leftarrow (a^q - a, q)$. If $d = 1$ or $d = q$, then n is not a prime power and terminate the algorithm. Otherwise set $q \leftarrow d$ and go to step 2.

4. [Final test] (Here p is a divisor of n which is almost certainly prime.) Using a primality test (see Chapters 8 and 9) prove that p is prime. If it is not (an exceedingly rare occurence), set $q \leftarrow p$ and go to step 2. Otherwise, by dividing n by p repeatedly, check whether n is a power of p or not. If it is not, n is not a prime power, otherwise output p. Terminate the algorithm.

We have been a little sloppy in this algorithm. For example in step 4, instead of repeatedly dividing by p we could use a binary search analogous to the binary powering algorithm. We leave this as an exercise for the reader (Exercise 4).

1.8 Exercises for Chapter 1

1. Write a bare-bones multi-precision package as explained in Section 1.1.2.

2. Improve your package by adding a squaring operation which operates faster than multiplication, and based on the identity $(aX + b)^2 = a^2 X^2 + b^2 + ((a + b)^2 - a^2 - b^2)X$, where X is a power of the base. Test when a similar method applied to multiplication (see Section 3.1.2) becomes faster than the straightforward method.

3. Given a 32-bit non-negative integer x, assume that we want to compute quickly the highest power of 2 dividing x (32 if $x = 0$). Denoting by $e(x)$ the exponent of this power of 2, show that this can be done using the formula

$$e(x) = t[(x\char`\^(x - 1)) \bmod 37]$$

where t is a suitable table of 37 values indexed from 0 to 36, and $a\char`\^ b$ denotes bitwise exclusive or (addition modulo 2 on bits). Show also that 37 is the least integer having this property, and find an analogous formula for 64-bit numbers.

4. Given two integers n and p, give an algorithm which uses ideas similar to the binary powering algorithm, to check whether n is a power of p. Also, if p is known to be prime, show that one can use only repeated squarings followed by a final divisibility test.

5. Write a version of the binary GCD algorithm which uses ideas of Lehmer's algorithm, in particular keeping information about the low order words and the high order words. Try also to write an extended version.

6. Write an algorithm which computes (u, v, d) as in Algorithm 1.3.6, by storing the partial quotients and climbing back. Compare the speed with the algorithms of the text.

7. Prove that at the end of Algorithm 1.3.6, one has $v_1 = \pm b/d$ and $v_2 = \mp a/d$, and determine the sign as a function of the number of Euclidean steps.

8. Write an algorithm for finding a solution to the system of congruences $x \equiv x_1$ (mod m_1) and $x \equiv x_2$ (mod m_2) assuming that $x_1 \equiv x_2$ (mod $\gcd(m_1, m_2)$).

9. Generalizing Exercise 8 and Algorithm 1.3.12, write a general algorithm for finding an x satisfying Theorem 1.3.9.

10. Show that the use of Gauss's Algorithm 1.3.14 leads to a slightly different algorithm than Cornacchia's Algorithm 1.5.2 for solving the equation $x^2 + dy^2 = p$ (consider $a = (p, 0)$ and $b = (x_0, \sqrt{d})$).

11. Show how to modify Lehmer's Algorithm 1.3.13 for finding the continued fraction expansion of a real number, using the ideas of Algorithm 1.3.3, so as to avoid almost all multi-precision operations.

12. Using Algorithm 1.3.13, compute at least 30 partial quotients of the continued fraction expansions of the numbers e, e^2, e^3, $e^{2/3}$ (you will need some kind of multi-precision to do this). What do you observe? Experiment with number of the form $e^{a/b}$, and try to see for which a/b one sees a pattern. Then try and prove it (this is difficult. It is advised to start by doing a good bibliographic search).

13. Prove that if $n = n_1 n_2$ with n_1 and n_2 coprime, then $(\mathbb{Z}/n\mathbb{Z})^* \simeq (\mathbb{Z}/n_1\mathbb{Z})^* \times (\mathbb{Z}/n_2\mathbb{Z})^*$. Then prove Theorem 1.4.1.

14. Show that when $a > 2$, $g = 5$ is always a generator of the cyclic subgroup of order 2^{a-2} of $(\mathbb{Z}/2^a\mathbb{Z})^*$.

15. Prove Proposition 1.4.6.

16. Give a proof of Theorem 1.4.7 (2) along the following lines (read Chapter 4 first if you are not familiar with number fields). Let p and q be distinct odd primes. Set $\zeta = e^{2i\pi/p}$, $R = \mathbb{Z}[\zeta]$ and

$$\tau(p) = \sum_{a \bmod p} \left(\frac{a}{p}\right) \zeta^a .$$

a) Show that $\tau(p)^2 = (-1)^{(p-1)/2} p$ and that $\tau(p)$ is invertible in R/qR.
b) Show that $\tau(p)^q \equiv \left(\frac{q}{p}\right) \tau(p)$ (mod qR).
c) Prove Theorem 1.4.7 (2), and modify the above arguments so as to prove Theorem 1.4.7 (1).

17. Prove Theorem 1.4.9 and Lemma 1.4.11.

18. Let p be an odd prime and n and integer prime to p. Then multiplication by n induces a permutation γ_n of the finite set $(\mathbb{Z}/p\mathbb{Z})^*$. Show that the signature of this permutation is equal to the Legendre symbol $\left(\frac{n}{p}\right)$. Deduce from this another proof of the quadratic reciprocity law (Theorem 1.4.7).

19. Generalizing Lemma 1.4.11, show the following general reciprocity law: if a and b are non-zero and $a = 2^\alpha a_1$ (resp. $b = 2^\beta b_1$) with a_1 and b_1 odd, then

$$\left(\frac{a}{b}\right) = (-1)^{(a_1-1)(b_1-1)/4+(\text{sign}(a_1)-1)(\text{sign}(b_1)-1)/4}\left(\frac{b}{a}\right).$$

20. Implement the modification suggested after Algorithm 1.4.10 (i.e. taking the smallest residue in absolute value instead of the smallest non-negative one) and compare its speed with that of the unmodified algorithm.

21. Using the quadratic reciprocity law, find the number of solutions of the congruence $x^3 \equiv 1 \pmod{p}$. Deduce from this the number of $cubic$ residues mod p, i.e. numbers a not divisible by p such that the congruence $x^3 \equiv a \pmod{p}$ has a solution.

22. Show that an integer n is a square if and only if $\left(\frac{n}{p}\right) = 1$ for every prime p not dividing n.

23. Given a modulus m, give an exact formula for $s(m)$, the number of squares modulo m, in other words the cardinality of the image of the squaring map from $\mathbb{Z}/m\mathbb{Z}$ into itself. Apply your formula to the special case $m = 64, 63, 65, 11$.

24. Show that the running time of Algorithm 1.4.10 modified by keeping b odd, may be exponential time for some inputs.

25. Modify Algorithm 1.5.1 so that in addition to computing x, it also computes the (even) exponent k such that $a^q z^k = 1$ in G, using the notations of the text.

26. Give an algorithm analogous to Shanks's Algorithm 1.5.1, to find the cube roots of a mod p when a is a cubic residue. It may be useful to consider separately the cases $p \equiv 2 \pmod{3}$ and $p \equiv 1 \pmod{3}$.

27. Given a prime number p and a quadratic non-residue a mod p, we can consider $K = \mathbb{F}_{p^2} = \mathbb{F}_p(\sqrt{a})$. Explain how to do the usual arithmetic operations in K. Give an algorithm for computing square roots in K, assuming that the result is in K.

28. Generalizing Exercise 27, give an algorithm for computing cube roots in $\mathbb{F}_{p^2}$, and give also an algorithm for computing roots of equations of degree 3 by Cardano's formulas (see Exercise 28 of Chapter 3).

29. Show that, as claimed in the proof of Algorithm 1.5.1, steps 3 and 4 will require in average $e^2/4$ and at most e^2 multiplications modulo p.

30. Let $m = \prod_p p^{e_p}$ be any positive integer for which we know the complete factorization into primes, and let $a \in \mathbb{Z}$.

   a) Give a necessary and sufficient condition for a to be congruent to a square modulo m, using several Legendre symbols.

   b) Give a closed formula for the number of solutions of the congruence $x^2 \equiv a \pmod{m}$.

   c) Using Shanks's Algorithm 1.5.1 as a sub-algorithm, write an algorithm for computing a solution to $x^2 \equiv a \pmod{m}$ if a solution exists (you should take care to handle separately the power of 2 dividing m).

31. Implement Algorithm 1.6.1 with and without the variant explained in Remark (3) following the algorithm, as well as the systematic trial of $X = 0, \dots, p-1$,

and compare the speed of these three algorithms for different values of p and $\deg(P)$ or $\deg(A)$.

32. By imitating Newton's method once again, design an algorithm for computing integer cube roots which works only with integers.

Chapter 2

Algorithms for Linear Algebra and Lattices

2.1 Introduction

In many algorithms, and in particular in number-theoretic ones, it is necessary to use algorithms to solve common problems of linear algebra. For example, solving a linear system of equations is such a problem. Apart from stability considerations, such problems and algorithms can be solved by a single algorithm independently of the base field (or more generally of the base ring if we work with modules). Those algorithms will naturally be called *linear algebra* algorithms.

On the other hand, many algorithms of the same general kind specifically deal with problems based on specific properties of the base ring. For example, if the base ring is $\mathbb{Z}$ (or more generally any Euclidean domain), and if L is a submodule of rank n of $\mathbb{Z}^n$, then $\mathbb{Z}^n/L$ is a finite Abelian group, and we may want to know its structure once a generating system of elements of L is known. This kind of problem can loosely be called an arithmetic linear algebra problem. Such problems are trivial if $\mathbb{Z}$ is replaced by a field K. (In our example we would have $L = K^n$ hence the quotient group would always be trivial.) In fact we will see that a submodule of $\mathbb{Z}^n$ is called a *lattice*, and that essentially all arithmetic linear algebra problems deal with lattices, so we will use the term *lattice algorithms* to describe the kind of algorithms that are used for solving arithmetic linear algebra problems.

This chapter is therefore divided into two parts. In the first part, we give algorithms for solving the most common linear algebra problems. It must be emphasized that the goal will be to give general algorithms valid over any field, but that in the case of *imprecise* fields such as the field of real numbers, care must be taken to insure stability. This becomes an important problem of numerical analysis, and we refer the reader to the many excellent books on the subject ([Gol-Van], [PFTV]). Apart from mentioning the difficulties, given the spirit of this book we will not dwell on this aspect of linear algebra.

In the second part, we recall the definitions and properties of lattices. We will assume that the base ring is $\mathbb{Z}$, but essentially everything carries over to the case where the base ring is a principal ideal domain (PID), for example $K[X]$, where K is a field. Then we describe algorithms for lattices. In particular we discuss in great detail the LLL algorithm which is of fundamental importance, and give a number of applications.

2.2 Linear Algebra Algorithms on Square Matrices

2.2.1 Generalities on Linear Algebra Algorithms

Let K be a field. Linear algebra over K is the study of K-vector spaces and K-linear maps between them. We will always assume that the vector spaces that we use are finite-dimensional. Of course, infinite-dimensional vector spaces arise naturally, for example the space $K[X]$ of polynomials in one variable over K. Usually, however when one needs to perform linear algebra on these spaces it is almost always on finite-dimensional subspaces.

A K-vector space V is an abstract object, but in practice, we will assume that V is given by a basis of n linearly independent vectors $v_1, \ldots v_n$ in some K^m (where m is greater or equal, but not necessarily equal to n). This is of course highly non-canonical, but we can always reduce to that situation.

Since K^m has by definition a canonical basis, we can consider V as being given by an $m \times n$ matrix $M(V)$ (i.e. a matrix with m rows and n columns) such that the *columns* of $M(V)$ represent the coordinates in the canonical basis of K^m of the vectors v_i. If $n = m$, the linear independence of the v_i means, of course, that $M(V)$ is an invertible matrix. (The notation $M(V)$ is slightly improper since $M(V)$ is attached, not to the vector space V, but to the chosen basis v_i.)

Note that changing bases in V is equivalent to multiplying $M(V)$ on the right by an invertible $n \times n$ matrix. In particular, we may want the matrix $M(V)$ to satisfy certain properties, for example being in upper triangular form. We will see below (Algorithm 2.3.11) how to do this.

A linear map f between two vector spaces V and W of respective dimensions n and m will in practice be represented by an $m \times n$ matrix $M(f)$, $M(f)$ being the matrix of the map f with respect to the bases $M(V)$ and $M(W)$ of V and W respectively. In other words, the j-th column of $M(f)$ represents the coordinates of $f(v_j)$ in the basis w_i, where the v_j correspond to the columns of $M(V)$, and the w_i to the columns of $M(W)$.

Note that in the above we use column-representation of vectors and not row-representation; this is quite arbitrary, but corresponds to traditional usage. Once a choice is made however, one must consistently stick with it.

Thus, the objects with which we will have to work with in performing linear algebra operations are matrices and (row or column) vectors. This is only for practical purposes, but keep in mind that it rarely corresponds to anything canonical. The internal representation of vectors is completely straightforward (i.e. as a linear array).

For matrices, essentially three equivalent kinds of representation are possible. The particular one which should be chosen depends on the language in which the algorithms will be implemented. For example, it will not be the same in Fortran and in C.

One representation is to consider matrices as (row) vectors of (column) vectors. (We could also consider them as column vectors of row vectors but the former is preferable since we have chosen to represent vectors mainly in column-representation.) A second method is to represent matrices as two-dimensional arrays. Finally, we can also represent matrices as one-dimensional arrays, by adding suitable macro-definitions so as to be able to access individual elements by row and column indices.

Whatever representation is chosen, we must also choose the index numbering for rows and columns. Although many languages such as C take 0 as the starting index, for consistency with usual mathematical notation we will assume that the first index for vectors or for rows and columns of matrices is always taken to be equal to 1. This is *not* meant to suggest that one should use this in a particular implementation, it is simply for elegance of exposition. In any given implementation, it may be preferable to make the necessary trivial changes so as to use 0 as the starting index. Again, this is a language-dependent issue.

2.2.2 Gaussian Elimination and Solving Linear Systems

The basic operation which is used in linear algebra algorithms is that of *Gaussian elimination*, sometimes also known as *Gaussian pivoting*. This consists in replacing a column (resp. a row) C by some linear combination of all the columns (resp. rows) where the coefficient of C must be non-zero, so that (for example) some coefficient becomes equal to zero. Another operation is that of exchanging two columns (resp. rows). Together, these two basic types of operations (which we will call *elementary operations* on columns or rows) will allow us to perform all the tasks that we will need in linear algebra. Note that they do not change the vector space spanned by the columns (resp. rows). Also, in matrix terms, performing a series of elementary operations on columns (resp. rows) is equivalent to right (resp. left) multiplication by an invertible square matrix of the appropriate size. Conversely, one can show (see Exercise 1) that an invertible square matrix is equal to a product of matrices corresponding to elementary operations.

The linear algebra algorithms that we give are simply adaptations of these basic principles to the specific problems that we must solve, but the underlying strategy is always the same, i.e. reduce a matrix to some simpler form (i.e. with many zeros at suitable places) so that the problem can be solved very simply. The proofs of the algorithms are usually completely straightforward, hence will be given only when really necessary. We will systematically use the following notation: if M is a matrix, M_j denotes its j-th *column*, M_i' its i-th row, and $m_{i,j}$ the entry at row i and column j. If B is a (column or row) vector, b_i will denote its i-th coordinate.

Perhaps the best way to see Gaussian elimination in action is in solving square linear systems of equations.

Algorithm 2.2.1 (Square Linear System). Let M be an $n \times n$ matrix and B a column vector. This algorithm either outputs a message saying that M is not invertible, or outputs a column vector X such that $MX = B$. We use an auxiliary column vector C.

1. [Initialize] Set $j \leftarrow 0$.
2. [Finished?] Let $j \leftarrow j + 1$. If $j > n$ go to step 6.
3. [Find non-zero entry] If $m_{i,j} = 0$ for all $i \geq j$, output a message saying that M is not invertible and terminate the algorithm. Otherwise, let $i \geq j$ be some index such that $m_{i,j} \neq 0$.
4. [Swap?] If $i > j$, for $l = j, \ldots, n$ exchange $m_{i,l}$ and $m_{j,l}$, and exchange b_i and b_j.
5. [Eliminate] (Here $m_{j,j} \neq 0$.) Set $d \leftarrow m_{j,j}^{-1}$ and for all $k > j$ set $c_k \leftarrow dm_{k,j}$. Then, for all $k > j$ and $l > j$ set $m_{k,l} \leftarrow m_{k,l} - c_k m_{j,l}$. (Note that we do not need to compute this for $l = j$ since it is equal to zero.) Finally, for $k > j$ set $b_k \leftarrow b_k - c_k b_j$ and go to step 2.
6. [Solve triangular system] (Here M is an upper triangular matrix.) For $i = n, n-1, \ldots, 1$ (in that order) set $x_i \leftarrow (b_i - \sum_{i<j\leq n} m_{i,j} x_j)/m_{i,i}$, output $X = (x_i)_{1 \leq i \leq n}$ and terminate the algorithm.

Note that steps 4 and 5 (the swap and elimination operations) are really row operations, but we have written them as working on entries since it is not necessary to take into account the first $j - 1$ columns.

Note also in step 5 that we start by computing the inverse of $m_{j,j}$ since in fields like $\mathbb{F}_p$ division is usually much more time-consuming than multiplication.

The number of necessary multiplications/divisions in this algorithm is clearly asymptotic to $n^3/3$ in the general case. Note however that this does not represent the true complexity of the algorithm, which should be counted in bit operations. This of course depends on the base field (see Section 1.1.3). This remark also applies to all the other linear algebra algorithms given in this chapter.

Inverting a square matrix M means solving the linear systems $MX = E_i$, where the E_i are the canonical basis vectors of K^n, hence one can achieve this by successive applications of Algorithm 2.2.1. Clearly, it is a waste of time to use Gaussian elimination on the matrix for each linear system. (More generally, this is true when we must solve several linear systems with the same matrix M but different right hand sides B.) We should compute the inverse of M, and then the solution of a linear system requires only a simple matrix times vector multiplication requiring n^2 field multiplications.

To obtain the inverse of M, only a slight modification of Algorithm 2.2.1 is necessary.

Algorithm 2.2.2 (Inverse of a Matrix). Let M be an $n \times n$ matrix. This algorithm either outputs a message saying that M is not invertible, or outputs the inverse of M. We use an auxiliary column vector C and we recall that B_i' (resp. X_i') denotes the i-th row of B (resp. X).

1. [Initialize] Set $j \leftarrow 0$, $B \leftarrow I_n$, where I_n is the $n \times n$ identity matrix.

2. [Finished?] Let $j \leftarrow j + 1$. If $j > n$, go to step 6.

3. [Find non-zero entry] If $m_{i,j} = 0$ for all $i \geq j$, output a message saying that M is not invertible and terminate the algorithm. Otherwise, let $i \geq j$ be some index such that $m_{i,j} \neq 0$.

4. [Swap?] If $i > j$, for $l = j, \ldots, n$ exchange $m_{i,l}$ and $m_{j,l}$, and exchange the rows B'_i and B'_j.

5. [Eliminate] (Here $m_{j,j} \neq 0$.) Set $d \leftarrow m_{j,j}^{-1}$ and for all $k > j$ set $c_k \leftarrow dm_{k,j}$. Then for all $k > j$ and $l > j$ set $m_{k,l} \leftarrow m_{k,l} - c_k m_{j,l}$. (Note that we do not need to compute this for $l = j$ since it is equal to zero.) Finally, for all $k > j$ set $B'_k \leftarrow B'_k - c_k B'_j$ and go to step 2.

6. [Solve triangular system] (Here M is an upper triangular matrix.) For $i = n, n - 1, \ldots, 1$ (in that order) set $X'_i \leftarrow (B'_i - \sum_{i < j \leq n} m_{i,j} X'_j)/m_{i,i}$, output the matrix X and terminate the algorithm.

It is easy to check that the number of multiplications/divisions needed is asymptotic to $4n^3/3$ in the general case. This is only four times longer than the number required for solving a single linear system. Thus as soon as more than four linear systems with the same matrix need to be solved, it is worthwhile to compute the inverse matrix.

Remarks.

(1) In step 1 of the algorithm, the matrix B is initialized to I_n. If instead, we initialize B to be any $n \times m$ matrix N for any m, the result is the matrix $M^{-1}N$, and this is of course faster than computing M^{-1} and then the matrix product. The case $m = 1$ is exactly Algorithm 2.2.1.

(2) Instead of explicitly computing the inverse of M, it is worthwhile for many applications to put M in LUP *form* , i.e. to find a lower triangular matrix L and an upper triangular matrix U such that $M = LUP$ for some permutation matrix P. (Recall that a *permutation matrix* is a square matrix whose elements are only 0 or 1 such that each row and column has exactly one 1.) Exercise 3 shows how this can be done. Once M is in this form, solving linear systems, inverting M, computing $\det(M)$, etc $\ldots$ is much simpler (see [AHU] and [PFTV]).

2.2.3 Computing Determinants

To compute determinants, we can simply use Gaussian elimination as in Algorithm 2.2.1. Since the final matrix is triangular, the determinant is trivial to compute. This gives the following algorithm.

Algorithm 2.2.3 (Determinant, Using Ordinary Elimination). Let M be an $n \times n$ matrix. This algorithm outputs the determinant of M. We use an auxiliary column vector C.

1. [Initialize] Set $j \leftarrow 0$, $x \leftarrow 1$.

2. [Finished?] Let $j \leftarrow j + 1$. If $j > n$ output x and terminate the algorithm.

3. [Find non-zero entry] If $m_{i,j} = 0$ for all $i \geq j$, output 0 and terminate the algorithm. Otherwise, let $i \geq j$ be some index such that $m_{i,j} \neq 0$.

4. [Swap?] If $i > j$, for $l = j, \ldots, n$ exchange $m_{i,l}$ and $m_{j,l}$, and set $x \leftarrow -x$.

5. [Eliminate] (Here $m_{j,j} \neq 0$.) Set $d \leftarrow m_{j,j}^{-1}$ and for all $k > j$ set $c_k \leftarrow dm_{k,j}$. Then for all $k > j$ and $l > j$ set $m_{k,l} \leftarrow m_{k,l} - c_k m_{j,l}$. (Note that we do not need to compute this for $l = j$ since it is equal to zero.) Finally, set $x \leftarrow x \cdot m_{j,j}$ and go to step 2.

The number of multiplications/divisions needed in this algorithm is clearly of the same order as Algorithm 2.2.1, i.e. asymptotic to $n^3/3$ in general.

Very often, this algorithm will be used in the case where the matrix M has entries in $\mathbb{Z}$ or some polynomial ring. In this case, the elimination step will introduce denominators, and these have a tendency to get very large. Furthermore, the coefficients of the intermediate matrices will be in $\mathbb{Q}$ (or some rational function field), and hence large GCD computations will be necessary which will slow down the algorithm even more. All this is of course valid for the other straightforward elimination algorithms that we have seen.

On the other hand, if the base field is a finite field $\mathbb{F}_q$, we do not have such problems. If the base field is inexact, like the real or complex numbers or the p-adic numbers, care must be taken for numerical stability. For example, numerical analysis books advise taking the largest non-zero entry (in absolute value) and not the first non-zero one found. We refer to [Gol-Van], [PFTV] for more details on these stability problems.

To overcome the problems that we encounter when the matrix M has integer coefficients, several methods can be used (and similarly when M has coefficients in a polynomial ring). The first method is to compute $\det(M)$ modulo sufficiently many primes (using Algorithm 2.2.3 which is efficient here), and then use the Chinese remainder Theorem 1.3.9 to obtain the exact value of $\det(M)$. This can be done as soon as we know an a priori upper bound for $|\det(M)|$. (We then simply choose sufficiently many primes p_i so that the product of the p_i is greater than twice the upper bound.) Such an upper bound is given by Hadamard's inequality which we will prove below (Corollary 2.5.5; note that this corollary is proved in the context of real matrices, i.e. Euclidean vector spaces, but its proof is identical for Hermitian vector spaces).

Proposition 2.2.4 (Hadamard's Inequality). *If $M = (m_{ij})_{1 \leq i,j \leq n}$ is a square matrix with complex coefficients, then*

$$|\det(M)| \leq \prod_{1 \leq i \leq n} \left(\sum_{1 \leq j \leq n} |m_{ij}|^2 \right)^{1/2}.$$

This method for computing determinants can be much faster than a direct computation using Algorithm 2.2.3, but will be slower when the number of primes needed for the Chinese remainder theorem is large. This happens because the size of the Hadamard bound is often far from ideal.

Another method is based on the following easily proved proposition due to Dodgson (alias Lewis Caroll), which is a special case of a general theorem due to Bareiss [Bar].

Proposition 2.2.5. *Let $M_0 = (a_{i,j}^0)_{1 \le i,j \le n}$ be an $n \times n$ matrix where the coefficients are considered as independent variables. Set $c_0 = 1$ and for $1 \le k < n$, define recursively*

$$a_{i,j}^{(k)} = \frac{1}{c_{k-1}} \begin{vmatrix} a_{k,k}^{(k-1)} & a_{k,j}^{(k-1)} \\ a_{i,k}^{(k-1)} & a_{i,j}^{(k-1)} \end{vmatrix} \quad, \quad M_k = (a_{i,j}^{(k)})_{k+1 \le i,j \le n} \quad \text{and} \quad c_k = a_{k,k}^{(k-1)} \quad .$$

Finally, let $c_n = a_{n,n}^{(n-1)}$. Then all the divisions by c_{k-1} are exact; we have $\det(M_k) = c_k^{n-k-1} \det(M_0)$, and in particular $\det(M_0) = c_n$.

Proof (Sketch). Going from M_{k-1} to M_k is essentially Gaussian elimination, except that the denominators are removed. This shows that

$$\det(M_k) = \frac{c_k^{n-k-1}}{c_{k-1}^{n-k}} \det(M_{k-1})$$

thus proving the formula for $\det(M_k)$ by induction.

That all the divisions by c_{k-1} are exact comes from the easily checked fact that we can explicitly write the coefficients $a_{i,j}^{(k)}$ as $(k+1) \times (k+1)$ minors of the matrix M_0 (see Exercise 5).   □

We have stated this proposition with matrices having coefficients considered as independent variables. For more special rings, some c_k may vanish, in which case one must exchange rows or columns, as in Algorithm 2.2.3, and keep track of the sign changes. This leads to the following method for computing determinants.

Algorithm 2.2.6 (Determinant Using Gauss-Bareiss). Given an $n \times n$ matrix M with coefficients in an integral domain $\mathcal{R}$, this algorithm computes the determinant of M. All the intermediate results are in $\mathcal{R}$.

1. [Initialize] Set $k \leftarrow 0$, $c \leftarrow 1$, $s \leftarrow 1$.

2. [Increase k] Set $k \leftarrow k+1$. If $k = n$ output $sm_{n,n}$ and terminate the algorithm. Otherwise, set $p \leftarrow m_{k,k}$.

3. [Is $p = 0$?] If $p \ne 0$ go to step 4. Otherwise, look for the first non-zero coefficient $m_{i,k}$ in the k-th column. If no such coefficient exists, output 0 and

terminate the algorithm. If it does, for $j = k, \ldots, n$ exchange $m_{i,j}$ and $m_{k,j}$, then set $s \leftarrow -s$ and $p \leftarrow m_{k,k}$.

4. [Main step] (p is now non-zero.) For $i = k+1, \ldots, n$ and $j = k+1, \ldots, n$ set $t \leftarrow pm_{i,j} - m_{i,k}m_{k,j}$, then $m_{i,j} \leftarrow t/c$ where the division is exact. Then set $c \leftarrow p$ and go to step 2.

Although this algorithm is particularly well suited to the computation of determinants when the matrix M has integer (or similar type) entries, it can of course, be used in general. There is however a subtlety which must be taken into account when dealing with inexact entries.

Assume for example that the coefficients of M are polynomials with real coefficients. These in general will be imprecise. Then in step 4, the division t/c will, in general, not give a polynomial, but rather a rational function. This is because when we perform the Euclidean division of t by c, there may be a very small but non-zero remainder. In this case, when implementing the algorithm, it is essential to compute t/c using Euclidean division, and *discard* the remainder, if any.

The number of necessary multiplications/divisions in this modified algorithm is asymptotic to n^3 instead of $n^3/3$ in Algorithm 2.2.3, but using Gauss-Bareiss considerably improves on the time needed for the basic multiplications and divisions and this usually more than compensates for the factor of 3.

Finally, note that although we have explained the Gauss-Bareiss method for computing determinants, it can usually be applied to any other algorithmic problem using Gaussian elimination, where the coefficients are integers (see Exercise 6).

2.2.4 Computing the Characteristic Polynomial

Recall that if M is an $n \times n$ square matrix, the *characteristic polynomial* of M is the monic polynomial of degree n defined by

$$P(X) = \det(XI_n - M) ,$$

where as usual I_n is the $n \times n$ identity matrix. We want to compute the coefficients of $P(X)$. Note that the constant term of $P(X)$ is equal to $(-1)^n \det(M)$, and more generally the coefficients of $P(X)$ can be expressed as the sum of the so-called *principal minors* of M which are sub-determinants of M. To compute the coefficients of $P(X)$ in this manner is usually not the best way to proceed. (In fact the number of such minors grows exponentially with n.) In addition to the method which I have just mentioned, there are essentially four methods for computing $P(X)$.

The first method is to apply the definition directly, and to use the Gauss-Bareiss algorithm for computing $\det(XI_n - M)$, this matrix considered as having coefficients in the ring $K[X]$. Although computing in $K[X]$ is more expensive than computing in K, this method can be quite fast in some cases.

The second method is to apply Lagrange interpolation. In our special case, this gives the following formula.

$$\det(XI_n - M) = \sum_{k=0}^{n} \det(kI_n - M) \prod_{0 \le j \le n, j \ne k} \frac{(X - j)}{(k - j)} .$$

This formula is easily checked since both sides are polynomials of degree less than or equal to n which agree on the $n + 1$ points $X = i$ for $0 \le i \le n$.

Hence, to compute the characteristic polynomial of M, it is enough to compute $n + 1$ determinants, and this is usually faster than the first method. Since multiplication and division by small constants can be neglected in timing estimates, this method requires asymptotically $n^4/3$ multiplications/divisions when we use ordinary Gaussian elimination.

The third method is based on the computation of the *adjoint matrix* or *comatrix* of M, i.e. the matrix M^{adj} whose coefficient of row i and column j is equal to $(-1)^{i+j}$ times the sub-determinant of M obtained by removing row j and column i (note that i and j are reversed). From the expansion rule of determinants along rows or columns, it is clear that this matrix satisfies the identity

$$MM^{\text{adj}} = M^{\text{adj}}M = \det(M)I_n .$$

We give the method as an algorithm.

Algorithm 2.2.7 (Characteristic Polynomial and Adjoint Matrix). Given an $n \times n$ matrix M, this algorithm computes the characteristic polynomial $P(X) = \det(XI_n - M)$ of M and the adjoint matrix M^{adj} of M. We use an auxiliary matrix C and auxiliary elements a_i.

1. [Initialize] Set $i \leftarrow 0$, $C \leftarrow I_n$, $a_0 \leftarrow 1$.

2. [Finished?] Set $i \leftarrow i + 1$. If $i = n$ set $a_n \leftarrow -\operatorname{Tr}(MC)/n$, output $P(X) \leftarrow \sum_{0 \le i \le n} a_i X^{n-i}$, $M^{\text{adj}} \leftarrow (-1)^{n-1}C$ and terminate the algorithm.

3. [Compute next a_i and C] Set $C \leftarrow MC$, $a_i \leftarrow -\operatorname{Tr}(C)/i$, $C \leftarrow C + a_i I_n$ and go to step 2.

Before proving the validity of this algorithm, we prove a lemma.

Lemma 2.2.8. *Let M be an $n \times n$ matrix, $A(X)$ be the adjoint matrix of $XI_n - M$, and $P(X)$ the characteristic polynomial of M. We have the identity*

$$\operatorname{Tr}(A(X)) = P'(X) .$$

Proof. Recall that the determinant is multilinear, hence the derivative of an $n \times n$ determinant is equal to the sum of the n determinants obtained by replacing the j-th column by its derivative, for $1 \le j \le n$. In our case, calling

E_j the columns of the identity matrix (i.e. the canonical basis of K^n), we have, after expanding the determinants along the j-th column

$$P'(X) = (\det(XI - M))' = \sum_{1 \leq j \leq n} A_{j,j}(X)$$

where $A_{j,j}(X)$ is the $n - 1 \times n - 1$ sub-determinant of $XI - M$ obtaining by removing row and column j, i.e. $A_{j,j}$ is the coefficient of row and column j of the adjoint matrix $A(X)$, and this proves the lemma. □

Proof of the Algorithm. Call $A(X)$ the adjoint matrix of $XI_n - M$. We can write $A(X) = \sum_{0 \leq i \leq n-1} C_i X^{n-i-1}$ with constant matrices C_i. From the lemma, it follows that if $P(X) = \sum_{0 \leq i \leq n} a_i X^{n-i}$ we have

$$(n - i)a_i = \text{Tr}(C_i) \ .$$

On the other hand, since $P(X)I_n = (XI_n - M)A(X)$, we obtain by comparing coefficients $C_0 = I_n$ and for $i \geq 1$

$$C_i = MC_{i-1} + a_i I_n \ .$$

Taking traces, this gives $(n-i)a_i = \text{Tr}(MC_{i-1}) + na_i$, i.e. $a_i = -\text{Tr}(MC_{i-1})/i$. Finally, it is clear that $A(0) = C_{n-1}$ is the adjoint matrix of $-M$, hence $(-1)^{n-1}C_{n-1}$ is the adjoint matrix of M, thus showing the validity of the algorithm. □

The total number of operations is easily seen to be asymptotic to n^4 multiplications, and this may seem slower (by a factor of 3) than the method based on Lagrange interpolation. However, since no divisions are required the basic multiplication/division time is reduced considerably—especially when the matrix M has integral entries, and hence this algorithm is in fact *faster*. In addition, it gives for free the adjoint matrix of M (and even of $XI_n - M$ if we want it).

The fourth and last method is based on the notion of *Hessenberg form* of a matrix. We first compute a matrix H which is similar to M (i.e. is of the form PMP^{-1}), and in particular has the same characteristic polynomial as M, and which has the following form (Hessenberg form)

$$H = \begin{pmatrix} h_{1,1} & h_{1,2} & h_{1,3} & \cdots & h_{1,n} \\ k_2 & h_{2,2} & h_{2,3} & \cdots & h_{2,n} \\ 0 & k_3 & h_{3,3} & \cdots & h_{3,n} \\ \vdots & \ddots & \ddots & \ddots & \vdots \\ 0 & \cdots & 0 & k_n & h_{n,n} \end{pmatrix}$$

In this form, since we have a big triangle of zeros on the bottom left, it is not difficult to obtain a recursive relation for the characteristic polynomial of H, hence of M. More precisely, if $p_m(X)$ is the characteristic polynomial of the sub-matrix of H formed by the first m rows and columns, we have $p_0(X) = 1$ and the recursion:

$$p_m(X) = (X - h_{m,m})p_{m-1}(X) - \sum_{i=1}^{m-1}\left(h_{i,m}\left(\prod_{j=i+1}^{m}k_j\right)p_{i-1}(X)\right) .$$

This leads to the following algorithm.

Algorithm 2.2.9 (Hessenberg). Given an $n \times n$ matrix $M = (m_{i,j})$ with coefficients in a field, this algorithm computes the characteristic polynomial of M by first transforming M into a Hessenberg matrix as above.

1. [Initialize] Set $H \leftarrow M$, $m \leftarrow 2$.

2. [Search for non-zero] If all the $h_{i,m-1}$ with $i > m$ are equal to 0, go to step 4. Otherwise, let $i \geq m$ be the smallest index such that $h_{i,m-1} \neq 0$. Set $t \leftarrow h_{i,m-1}$. Then if $i > m$, for all $j \geq m - 1$ exchange $h_{i,j}$ and $h_{m,j}$ and exchange column H_i with column H_m.

3. [Eliminate] For $i = m+1, \ldots, n$ do the following if $h_{i,m-1} \neq 0$: $u \leftarrow h_{i,m-1}/t$, for all $j \geq m$ set $h_{i,j} \leftarrow h_{i,j} - uh_{m,j}$, set $h_{i,m-1} \leftarrow 0$, and finally set column $H_m \leftarrow H_m + uH_i$.

4. [Hessenberg finished?] If $m < n - 1$, set $m \leftarrow m + 1$ and go to step 2.

5. [Initialize characteristic polynomial] Set $p_0(X) \leftarrow 1$ and $m \leftarrow 1$.

6. [Initialize computation] Set $p_m(X) \leftarrow (X - h_{m,m})p_{m-1}(X)$ and $t \leftarrow 1$.

7. [Compute p_m] For $i = 1, \ldots, m - 1$ do the following: set $t \leftarrow th_{m-i+1,m-i}$, $p_m(X) \leftarrow p_m(X) - th_{m-i,m}p_{m-i-1}(X)$.

8. [Finished?] If $m < n$ set $m \leftarrow m + 1$ and go to step 6. Otherwise, output $p_n(X)$ and terminate the algorithm.

This algorithm requires asymptotically only n^3 multiplications/divisions in the general case, and this is much better than the preceding algorithms when n is large. If M has integer coefficients however, the Hessenberg form as well as the intermediate results will usually be non-integral rational numbers, hence we lose all the advantage of the reduced operation count, since the time needed for the basic multiplications/divisions will be large. In that case, one should not use the Hessenberg algorithm directly. Instead, one should apply it to compute the characteristic polynomial modulo sufficiently many primes and use the Chinese remainder theorem, exactly as we did for the determinant. For this, we need bounds for the coefficients of the characteristic polynomial, analogous to the Hadamard bound. The following result, although not optimal, is easy to prove and gives a reasonably good estimate.

Proposition 2.2.10. Let $M = (m_{i,j})$ be an $n \times n$ matrix, and write $\det(XI_n - M) = \sum_{0 \leq k \leq n} a_k X^{n-k}$ with $a_0 = 1$. Let B be an upper bound for the moduli of all the $m_{i,j}$. Then the coefficients a_k satisfy the inequality

$$|a_k| \leq \binom{n}{k} k^{k/2} B^k \ .$$

Proof. As already mentioned, the coefficient a_k is up to sign equal to the sum of the $\binom{n}{k}$ principal $k \times k$ minors. By Hadamard's inequality (Proposition 2.2.4), each of these minors is bounded by $\prod(\sum |m_{ij}|^2)^{1/2}$ where the product and the sums have k terms. Hence the minors are bounded by $(kB^2)^{k/2} = k^{k/2} B^k$, and this gives the proposition. $\square$

Remarks.

(1) The optimal form for computing the characteristic polynomial of a matrix would be triangular. This is however not possible if the eigenvalues of the matrix are not in the base field, hence the Hessenberg form can be considered as the second best choice.

(2) A problem related to computing the characteristic polynomial, is to compute the eigenvalues (and eigenvectors) of a matrix, say with real or complex coefficients. These are by definition the roots of the characteristic polynomial $P(X)$. Therefore, we could compute $P(X)$ using one of the above methods, then find the roots of $P(X)$ using algorithm 3.6.6 which we will see later, and finally apply algorithm 2.2.1 to get the eigenvectors. This is however *not* the way to proceed in general since much better methods based on iterative processes are available from numerical analysis (see [Gol-Van], [PFTV]), and we will not study this subject here.

2.3 Linear Algebra on General Matrices

2.3.1 Kernel and Image

We now come to linear algebra problems which deal with arbitrary $m \times n$ matrices M with coefficients in a field K. Recall from above that M can be viewed as giving a generating set for the subspace of K^m generated by the columns of M, or as the matrix of a linear map from an n-dimensional space to an m-dimensional space with respect to some bases. (Beware of the order of m and n.) It is usually conceptually easier to think of M in this way.

The first basic algorithm that we will need is for computing the kernel of M, i.e. a basis for the space of column vectors X such that $MX = 0$. The following algorithm is adapted from [Knu2].

Algorithm 2.3.1 (Kernel of a Matrix).   Given an $m \times n$ matrix $M = (m_{i,j})$ with $1 \leq i \leq m$ and $1 \leq j \leq n$ having coefficients in a field K, this algorithm outputs a basis of the kernel of M, i.e. of column vectors X such that $MX = 0$. We use auxiliary constants c_i $(1 \le i \le m)$ and d_i $(1 \le i \le n)$.

1. [Initialize] Set $r \leftarrow 0$, $k \leftarrow 1$ and for $i = 1, \ldots, m$, set $c_i \leftarrow 0$ (there is no need to initialize d_i).

2. [Scan column] If there does not exist a j such that $1 \le j \le m$ with $m_{j,k} \ne 0$ and $c_j = 0$ then set $r \leftarrow r + 1$, $d_k \leftarrow 0$ and go to step 4.

3. [Eliminate] Set $d \leftarrow -m_{j,k}^{-1}$, $m_{j,k} \leftarrow -1$ and for $s = k + 1, \ldots, n$ set $m_{j,s} \leftarrow dm_{j,s}$. Then for all i such that $1 \le i \le m$ and $i \ne j$ set $d \leftarrow m_{i,k}$, $m_{i,k} \leftarrow 0$ and for $s = k + 1, \ldots, n$ set $m_{i,s} \leftarrow m_{i,s} + dm_{j,s}$. Finally, set $c_j \leftarrow k$ and $d_k \leftarrow j$.

4. [Finished?] If $k < n$ set $k \leftarrow k + 1$ and go to step 2.

5. [Output kernel] (Here r is the dimension of the kernel.) For every k such that $1 \le k \le n$ and $d_k = 0$ (there will be exactly r such k), output the column vector $X = (x_i)_{1 \le i \le n}$ defined by

$$
x_i = \begin{cases} m_{d_i,k}, & \text{if } d_i > 0 \\ 1, & \text{if } i = k \\ 0, & \text{otherwise.} \end{cases}
$$

These r vectors form a basis for the kernel of M. Terminate the algorithm.

The proof of the validity of this algorithm is not difficult and is left as an exercise for the reader (see Exercise 8). In fact, the main point is that $c_j > 0$ if and only if $m_{j,c_j} = -1$ and all other entries in column c_j are equal to zero.

Note also that step 3 looks complicated because I wanted to give as efficient an algorithm as possible, but in fact it corresponds to elementary row operations.

Only a slight modification of this algorithm gives the image of M, i.e. a basis for the vector space spanned by the columns of M. In fact, apart from the need to make a copy of the initial matrix M, only step 5 needs to be changed.

Algorithm 2.3.2 (Image of a Matrix).   Given an $m \times n$ matrix $M = (m_{i,j})$ with $1 \le i \le m$ and $1 \le j \le n$ having coefficients in a field K, this algorithm outputs a basis of the image of M, i.e. the vector space spanned by the columns of M. We use auxiliary constants c_i $(1 \le i \le m)$.

1. [Initialize] Set $r \leftarrow 0$, $k \leftarrow 1$ and for $i = 1, \ldots, m$, set $c_i \leftarrow 0$, and let $N \leftarrow M$ (we need to keep a copy of the initial matrix M).

2. [Scan column] If there does not exists a j such that $1 \le j \le m$ with $m_{j,k} \ne 0$ and $c_j = 0$ then set $r \leftarrow r + 1$, $d_k \leftarrow 0$ and go to step 4.

3. [Eliminate] Set $d \leftarrow -m_{j,k}^{-1}$, $m_{j,k} \leftarrow -1$ and for $s = k + 1, \ldots, n$ set $m_{j,s} \leftarrow dm_{j,s}$. Then for all i such that $1 \le i \le m$ and $i \ne j$ set $d \leftarrow m_{i,k}$, $m_{i,k} \leftarrow 0$

and for $s = k+1, \ldots, n$ set $m_{i,s} \leftarrow m_{i,s} + dm_{j,s}$. Finally, set $c_j \leftarrow k$ and $d_k \leftarrow j$.

4. [Finished?] If $k < n$ set $k \leftarrow k+1$ and go to step 2.

5. [Output image] (Here $n - r$ is the dimension of the image, i.e. the rank of the matrix M.) For every j such that $1 \leq j \leq m$ and $c_j \neq 0$ (there will be exactly $n - r$ such j), output the column vector N_{c_j} (where N_k is the k-th column of the initial matrix M). These $n - r$ vectors form a basis for the image of M. Terminate the algorithm.

One checks easily that both the kernel and image algorithms require asymptotically $n^2 m/2$ multiplications/divisions in general.

There are many possible variations on this algorithm for determining the image. For example if only the rank of the matrix M is needed and not an actual basis of the image, simply output the number $n - r$ in step 5. If one needs to also know the precise rows and columns that must be extracted from the matrix M to obtain a non-zero $(n-r) \times (n-r)$ determinant, we output the pairs (j, c_j) for each $j \leq m$ such that $c_j \neq 0$, where j gives the row number, and c_j the column number.

Finally, if the columns of M represent a generating set for a subspace of K^m, the image algorithm enables us to extract a basis for this subspace.

Remark. We recall the following definition.

Definition 2.3.3. *We will say that an $m \times n$ matrix M is in* column echelon form *if there exists $r \leq n$ and a strictly increasing map f from $[r+1, n]$ to $[1, m]$ satisfying the following properties.*

(1) *For $r+1 \leq j \leq n$, $m_{f(j),j} = 1$, $m_{i,j} = 0$ if $i > f(j)$ and $m_{f(k),j} = 0$ if $k < j$.*

(2) *The first r columns of M are equal to 0.*

It is clear that the definition implies that the last $n - r$ columns (i.e. the non-zero columns) of M are linearly independent.

It can be seen that Algorithm 2.3.1 gives the basis of the kernel in column echelon form. This property can be useful in other contexts, and hence, if necessary, we may assume that the basis which is output has this property. In fact we will see later that any subspace can be represented by a matrix in column echelon form (Algorithm 2.3.11).

For the image, the basis is simply extracted from the columns of M, no linear combination being taken.

2.3.2 Inverse Image and Supplement

A common problem is to solve linear systems whose matrix is either not square or not invertible. In other words, we want to generalize algorithm 2.2.1 for solving $MX = B$ where M is an $m \times n$ matrix. If X_0 is a particular solution of this system, the general solution is given by $X = X_0 + Y$ where $Y \in \ker(M)$, and $\ker(M)$ can be computed using Algorithm 2.3.1, so the only problem is to find one particular solution to our system (or to show that none exist). We will naturally call this the *inverse image* problem.

If we want the complete inverse image and not just a single solution, the best way is probably to use the kernel Algorithm 2.3.1. Indeed, consider the augmented $m \times (n + 1)$ matrix M_1 obtained by adding B as an $n + 1$-st column to the matrix M. If X is a solution to $MX = B$, and if X_1 is the $n + 1$-vector obtained from X by adding -1 as $n + 1$-st component, we clearly have $M_1 X_1 = 0$. Conversely, if X_1 is any solution of $M_1 X_1 = 0$, then either the $n + 1$-st component of X_1 is equal to 0 (corresponding to elements of the kernel of M), or it is non-zero, and by a suitable normalization we may assume that it is equal to -1, and then the first n components give a solution to $MX = B$. This leads to the following algorithm.

Algorithm 2.3.4 (Inverse Image). Given an $m \times n$ matrix M and an m-dimensional column vector B, this algorithm outputs a solution to $MX = B$ or outputs a message saying that none exist. (The algorithm can be trivially modified to output the complete inverse image if desired.)

1. [Compute kernel] Let M_1 be the $m \times (n + 1)$ matrix whose first n columns are those of M and whose $n + 1$-st column is equal to B. Using Algorithm 2.3.1, compute a matrix V whose columns form a basis for the kernel of M_1. Let r be the number of columns of V.

2. [Solution exists?] If $v_{n+1,j} = 0$ for all j such that $1 \leq j \leq r$, output a message saying that the equation $MX = B$ has no solution. Otherwise, let $j \leq r$ be such that $v_{n+1,j} \neq 0$ and set $d \leftarrow -1/v_{n+1,j}$.

3. [Output solution] Let $X = (x_i)_{1 \leq i \leq n}$ be the column vector obtained by setting $x_i \leftarrow d v_{i,j}$. Output X and terminate the algorithm.

Note that as for the kernel algorithm, this requires asymptotically $n^2 m/2$ multiplications/divisions, hence is roughly three times slower than algorithm 2.2.1 when $n = m$.

If we want only one solution, or if we want several inverse images corresponding to the same matrix but different vectors, it is more efficient to directly use Gaussian elimination once again. A simple modification of Algorithm 2.2.2 does this as follows.

Algorithm 2.3.5 (Inverse Image Matrix). Let M be an $m \times n$ matrix and V be an $m \times r$ matrix, where $n \leq m$. This algorithm either outputs a message saying that some column vector of V is not in the image of M, or outputs an

$n \times r$ matrix X such that $V = MX$. We assume that the columns of M are linearly independent. We use an auxiliary column vector C and we recall that B'_i (resp. M'_i, X'_i) denotes the i-th row of B (resp. M, X).

1. [Initialize] Set $j \leftarrow 0$ and $B \leftarrow V$.

2. [Finished?] Let $j \leftarrow j + 1$. If $j > n$ go to step 6.

3. [Find non-zero entry] If $m_{i,j} = 0$ for all i such that $m \geq i \geq j$, output a message saying that the columns of M are not linearly independent and terminate the algorithm. Otherwise, let i be some index such that $m \geq i \geq j$ and $m_{i,j} \neq 0$.

4. [Swap?] If $i > j$, for $l = j, \ldots, n$ exchange $m_{i,l}$ and $m_{j,l}$, and exchange the rows B'_i and B'_j.

5. [Eliminate] (Here $m_{j,j} \neq 0$.) Set $d \leftarrow m_{j,j}^{-1}$ and for all k such that $m \geq k > j$ set $c_k \leftarrow dm_{k,j}$. Then for all k and l such that $m \geq k > j$ and $n \geq l > j$ set $m_{k,l} \leftarrow m_{k,l} - c_k m_{j,l}$. Finally, for all k such that $m \geq k > j$ set $B'_k \leftarrow B'_k - c_k B'_j$ and go to step 2.

6. [Solve triangular system] (Here the first n rows of M form an upper triangular matrix.) For $i = n, n - 1, \ldots, 1$ (in that order) set $X'_i \leftarrow (B'_i - \sum_{i < j \leq n} m_{i,j} X'_j)/m_{i,i}$.

7. [Check rest of matrix] Check whether for each k such that $m \geq k > n$ we have $B'_k = M'_k X$. If this is not the case, output a message that some column vector of V is not in the image of M. Otherwise, output the matrix X and terminate the algorithm.

Note that in practice the columns of M represent a basis of some vector space hence are linearly independent. However, it is not difficult to modify this algorithm to work without the assumption that the columns of M are linearly independent.

Another problem which often arises is to find a *supplement* to a subspace in a vector space. The subspace can be considered as given by the coordinates of a basis on some basis of the full space, hence as an $n \times k$ matrix M with $k \leq n$ of rank equal to k. The problem is to supplement this basis, i.e. to find an *invertible* $n \times n$ matrix B such that the first k columns of B form the matrix M. A basis for a supplement of our subspace is then given by the last $n - k$ columns of B.

This can be done using the following algorithm.

Algorithm 2.3.6 (Supplement a Basis). Given an $n \times k$ matrix M with $k \leq n$ having coefficients in a field K, this algorithm either outputs a message saying that M is of rank less than k, or outputs an invertible $n \times n$ matrix B such that the first k columns of B form the matrix M. Recall that we denote by B_j the columns of B.

1. [Initialize] Set $s \leftarrow 0$ and $B \leftarrow I_n$.

2. [Finished?] If $s = k$, then output B and terminate the algorithm.

3. [Search for non-zero] Set $s \leftarrow s + 1$. Let t be the smallest $j \geq s$ such that $m_{t,s} \neq 0$, and set $d \leftarrow m_{t,s}^{-1}$. If such a $t \leq n$ does not exist, output a message saying that the matrix M is of rank less than k and terminate the algorithm.

4. [Modify basis and eliminate] Set $B_t \leftarrow B_s$ (if $t \neq s$), then set $B_s \leftarrow M_s$. Then for $j = s + 1, \ldots, k$, do as follows. Exchange $m_{s,j}$ and $m_{t,j}$ (if $t \neq s$). Set $m_{s,j} \leftarrow dm_{s,j}$. Then, for all $i \neq s$ and $i \neq t$, set $m_{i,j} \leftarrow m_{i,j} - m_{i,s}m_{s,j}$. Finally, go to step 2.

Proof. This is an easy exercise in linear algebra and is left to the reader (Exercise 9). Note that the elimination part of step 4 ensures that the matrix BM stays constant throughout the algorithm, and at the end of the algorithm the first k rows of the matrix M form the identity matrix I_k, and the last $n - k$ rows are equal to 0.   □

Often one needs to find the supplement of a subspace in another subspace and not in the whole space. In this case, the simplest solution is to use a combination of Algorithms 2.3.5 and 2.3.6 as follows.

Algorithm 2.3.7 (Supplement a Subspace in Another). Let V (resp. M) be an $m \times r$ (resp. $m \times n$) matrix whose columns form a basis of some subspace F (resp. E) of K^m with $r \leq n \leq m$. This algorithm either finds a basis for a supplement of F in E or outputs a message saying that F is not a subspace of E.

1. [Find new coordinates] Using Algorithm 2.3.5, find an $n \times r$ inverse image matrix X such that $V = MX$. If such a matrix does not exist, output a message saying that F is not a subspace of E and terminate the algorithm.

2. [Supplement X] Apply Algorithm 2.3.6 to the matrix X, thus giving an $n \times n$ matrix B whose first r columns form the matrix X.

3. [Supplement F in E] Let C be the $n \times n - r$ matrix formed by the last $n - r$ columns of B. Output MC and terminate the algorithm (the columns of MC will form a basis for a supplement of F in E).

Note that in addition to the error message of step 1, Algorithms 2.3.5 and 2.3.6 will also output error messages if the columns of V or M are not linearly independent.

2.3.3 Operations on Subspaces

The final algorithms that we will study concern the sum and intersection of two subspaces. If M and M' are $m \times n$ and $m \times n'$ matrices respectively, the columns of M (resp. M') span subspaces V (resp. V') of K^m. To obtain a basis for the sum $V + V'$ is very easy.

Algorithm 2.3.8 (Sum of Subspaces). Given an $m \times n$ (resp. $m \times n'$) matrix M (resp. M') whose columns span a subspace V (resp. V') of K^m, this algorithm finds a matrix N whose columns form a basis for $V + V'$.

1. [Concatenate] Let M_1 be the $m \times (n + n')$ matrix obtained by concatenating side by side the matrices M and M'. (Hence the first n columns of M_1 are those of M, the last n' those of M'.)

2. Using Algorithm 2.3.2 output a basis of the image of M_1 and terminate the algorithm.

Obtaining a basis for the intersection $V \cap V'$ is not much more difficult.

Algorithm 2.3.9 (Intersection of Subspaces). Given an $m \times n$ (resp. $m \times n'$) matrix M (resp. M') whose columns span a subspace V (resp. V') of K^m, this algorithm finds a matrix N whose columns form a basis for $V \cap V'$.

1. [Compute kernel] Let M_1 be the $m \times (n+n')$ matrix obtained by concatenating side by side the matrices M and M'. (Hence the first n columns of M_1 are those of M, the last n' those of M'.) Using Algorithm 2.3.1 compute a basis of the kernel of M_1, given by an $(n + n') \times p$ matrix N for some p.

2. [Compute intersection] Let N_1 be the $n \times p$ matrix obtained by extracting from N the first n *rows*. Set $M_2 \leftarrow MN_1$, output the matrix obtained by applying Algorithm 2.3.2 to M_2 and terminate the algorithm. (Note that if we know beforehand that the columns of M (resp. M') are also linearly independent, i.e. form a basis of V (resp. V'), we can simply output the matrix M_2 without applying Algorithm 2.3.2.)

Proof. We will constantly use the trivial fact that a column vector B is in the span of the columns of a matrix M if and only if there exists a column vector X such that $B = MX$.

Let N_1' be the $n' \times p$ matrix obtained by extracting from N the last n' rows. By block matrix multiplication, we have $MN_1 + M'N_1' = 0$. If B_i is the i-th column of $M_2 = MN_1$ then $B_i \in V$, but B_i is also equal to the opposite of the i-th column of $M'N_1'$, hence $B_i \in V'$. Conversely, let $B \in V \cap V'$. Then we can write $B = MX = M'X'$ for some column vectors X and X'. If Y is the $n + n'$-dimensional column vector whose first n (resp. last n') components are X (resp. $-X'$), we clearly have $M_1Y = 0$, hence $Y = NC$ for some column vector C. In particular, $X = N_1C$ hence $B = MN_1C = M_2C$, so B belongs to the space spanned by the columns of M_2. It follows that this space is equal to $V \cap V'$, and the image algorithm gives us a basis.

If the columns of M (resp. M') are linearly independent, then it is left as an easy exercise for the reader to check that the columns of M_2 are also linearly independent (Exercise 12), thus proving the validity of the algorithm. $\square$

As mentioned earlier, a subspace V of K^m can be represented as an $m \times n$ matrix $M = M(V)$ whose columns are the coordinates of a basis of V on the canonical basis of K^m. This representation depends entirely on the basis, so we may hope to find a more canonical representation. For example, how do we decide whether two subspaces V and W of K^m are equal? One method is of course to check whether every basis element of W is in the image of the matrix V and conversely, using Algorithm 2.3.4.

A better method is to represent V by a matrix having a special form, in the present case in column echelon form (see Definition 2.3.3).

Proposition 2.3.10. *If V is a subspace of K^m, there exists a unique basis of V such that the corresponding matrix $M(V)$ is in column echelon form.*

Proof. This will follow immediately from the following algorithm.   □

Algorithm 2.3.11 (Column Echelon Form). Given an $m \times n$ matrix M this algorithm outputs a matrix N in column echelon form whose image is equal to the image of M (i.e. $N = MP$ for some invertible $n \times n$ matrix P).

1. [Initialize] Set $i \leftarrow m$ and $k \leftarrow n$.

2. [Search for non-zero] Search for the largest integer $j \leq k$ such that $m_{i,j} \neq 0$. If such a j does not exist, go to step 4. Otherwise, set $d \leftarrow 1/m_{i,j}$, then for $l = 1, \ldots, i$ set $t \leftarrow dm_{l,j}$, $m_{l,j} \leftarrow m_{l,k}$ (if $j \neq k$) and $m_{l,k} \leftarrow t$.

3. [Eliminate] For all j such that $1 \leq j \leq n$ and $j \neq k$ and for all l such that $1 \leq l \leq i$ set $m_{l,j} \leftarrow m_{l,j} - m_{l,k}m_{i,j}$. Finally, set $k \leftarrow k - 1$.

4. [Next row] If $i = 1$ output M and terminate the algorithm. Otherwise, set $i \leftarrow i - 1$ and go to step 2.

The proof of the validity of this algorithm is easy and left to the reader (see Exercise 11). The number of required multiplications/divisions is asymptotically $n^2(2m - n)/2$ if $n \leq m$ and $nm^2/2$ if $n > m$.

Since the non-zero columns of a matrix which is in column echelon form are linearly independent, this algorithm gives us an alternate way to compute the image of a matrix. Instead of obtaining a basis of the image as a subset of the columns, we obtain a matrix in column echelon form. This is preferable in many situations. Comparing the number of multiplications/divisions needed, this algorithm is slower than Algorithm 2.3.2 for $n \leq m$, but faster when $n > m$.

2.3.4 Remarks on Modules

We can study most of the above linear algebra problems in the context of modules over a commutative ring with unit R instead of vector spaces over a field. If the ring R is an integral domain, we can work over its field of fractions K. (This is what we did in the algorithms given above when we assumed that the matrices had integral entries.) However, this is not completely satisfactory, since the answer that we want may be different. For example, to compute the kernel of a map defined between two free modules of finite rank (given as usual by a matrix), finding the kernel as a K-vector space is not sufficient, since we want it as an R-module. In fact, this kernel will usually not be a free module, hence cannot be represented by a matrix whose columns form a basis. One important special case where it will be free is when R is a principal ideal domain (PID, see Chapter 4). In this case all submodules of a free module of finite rank are free of finite rank. This happens when $R = \mathbb{Z}$ or $R = k[X]$ for a field k. In this case, asking for a basis of the kernel makes perfectly good sense, and the algorithm that we have given is not sufficient. We will see later (Algorithm 2.4.10) how to solve this problem.

A second difficulty arises when R is not an integral domain, because of the presence of zero-divisors. Since almost all linear algebra algorithms involve elimination, i.e. division by an element of R, we are bound at some point to get a non-zero non-invertible entry as divisor. In this case, we are in more trouble. Sometimes however, we can work around this difficulty. Let us consider for example the problem of solving a square linear system over $\mathbb{Z}/r\mathbb{Z}$, where r is not necessarily a prime. If we know the factorization of r into prime powers, we can use the Chinese remainder Theorem 1.3.9 to reduce to the case where r is a prime power. If r is prime, Algorithm 2.2.1 solves the problem, and if r is a higher power of a prime, we can still use Algorithm 2.2.1 applied to the field $K = \mathbb{Q}_p$ of p-adic numbers (see Exercise 2).

But what are we to do if we do not know the complete factorization of r? This is quite common, since as we will see in Chapters 8, 9 and 10 large numbers (say more than 80 decimal digits) are quite hard to factor. Fortunately, we do not really care. After extracting the known factors of r, we are left with a linear system modulo a new r for which we know (or expect) that it does not have any small factors (say none less than 10^6). We then simply apply Algorithm 2.2.1. Two things may happen. Either the algorithm goes through with no problem, and this will happen as long as all the elements which are used to perform the elimination (which we will call the pivots) are coprime to r. This will almost always be the case since r has no small factors. We then get the solution to the system. Note that this solution must be unique since the determinant of M, which is essentially equal to the product of the pivots, is coprime to r.

The other possibility is that we obtain a pivot p which is not coprime to r. Since the pivot is non-zero (modulo r), this means that the GCD (p, r) gives a non-trivial factor of r, hence we split r as a product of smaller (coprime) numbers and apply Algorithm 2.2.1 once again. The idea of working "as if" r was a prime can be applied to many number-theoretic algorithms where the basic assumption is that $\mathbb{Z}/r\mathbb{Z}$ is a field, and usually the same procedure can be made to work. H. W. Lenstra calls the case where working this way we find a non-trivial factor of r a *side exit*. In fact, this is sometimes the main purpose of an algorithm. For example, the elliptic curve factoring algorithm (Algorithm 10.3.3) uses exactly this kind of side exit to factor r.

2.4 Z-Modules and the Hermite and Smith Normal Forms

2.4.1 Introduction to Z-Modules

The most common kinds of modules that one encounters in number theory, apart from vector spaces, are evidently Z-modules, i.e. Abelian groups. The Z-modules V that we consider will be assumed to be *finitely generated*, in other words there exists a finite set $(v_i)_{1 \le i \le k}$ of elements of V such that any element of V can be expressed as a linear combination of the v_i with integral coefficients. The basic results about such Z-modules are summarized in the following theorem, whose proof can be found in any standard text (see for example [Lang]).

Theorem 2.4.1. *Let V be a finitely generated Z-module (i.e. Abelian group).*

(1) *If V_{tors} is the torsion subgroup of V, i.e. the set of elements $v \in V$ such that there exists $m \in \mathbb{Z} \setminus \{0\}$ with $mv = 0$, then V_{tors} is a finite group, and there exists a non-negative integer n and an isomorphism*

$$V \simeq V_{\text{tors}} \times \mathbb{Z}^n$$

(the number n is called the rank *of V).*
(2) *If V is a free Z-module (i.e. if $V \simeq \mathbb{Z}^n$, or equivalently by (1) if $V_{\text{tors}} = \{0\}$), then any submodule of V is free of rank less than or equal to that of V.*
(3) *If V is a finite Z-module (i.e. by (1) if V is of zero rank), there exists n and a submodule L of $\mathbb{Z}^n$ (which is free by (2)) such that $V \simeq \mathbb{Z}^n/L$.*

Note that (2) and (3) are easy consequences of (1) (see Exercise 13).

This theorem shows that the study of finitely generated Z-modules splits naturally into, on the one hand the study of finite Z-modules (which we will usually denote by the letter G for (finite Abelian) group), and on the other hand the study of free Z-modules of finite rank (which we will usually denote by the letter L for lattice (see Section 2.5)). Furthermore, (3) shows that these notions are in some sense dual to each other, so that we can in fact study only free Z-modules, finite Z-modules being considered as quotients of free modules.

Studying free modules L puts us in almost the same situation as studying vector spaces. In particular, we will usually consider L to be a submodule of some $\mathbb{Z}^m$, and we will represent L as an $m \times n$ matrix M whose columns give the coordinates of a basis of L on the canonical basis of $\mathbb{Z}^m$. Such a representation is of course not unique, since it depends on the choice of a basis for L. In the case of vector spaces, one of the ways to obtain a more canonical representation was to transform the matrix M into column echelon form. Since this involves elimination, this is not possible anymore over $\mathbb{Z}$. Nonetheless, there exists an analogous notion which is just as useful, called the *Hermite normal form* (abbreviated HNF). Another notion, called the *Smith normal form* (abbreviated SNF) allows us to represent finite $\mathbb{Z}$-modules.

2.4.2 The Hermite Normal Form

The following definition is the analog of Definition 2.3.3 for $\mathbb{Z}$-modules.

Definition 2.4.2. *We will say that an $m \times n$ matrix $M = (m_{i,j})$ with integer coefficients is in* Hermite normal form *(abbreviated HNF) if there exists $r \leq n$ and a strictly increasing map f from $[r+1, n]$ to $[1, m]$ satisfying the following properties.*

(1) *For $r + 1 \leq j \leq n$, $m_{f(j),j} \geq 1$, $m_{i,j} = 0$ if $i > f(j)$ and $0 \leq m_{f(k),j} < m_{f(k),k}$ if $k < j$.*
(2) *The first r columns of M are equal to 0.*

Remark. In the important special case where $m = n$ and $f(k) = k$ (or equivalently $\det(M) \neq 0$), M is in HNF if it satisfies the following conditions.

(1) M is an upper triangular matrix, i.e. $m_{i,j} = 0$ if $i > j$.
(2) For every i, we have $m_{i,i} > 0$.
(3) For every $j > i$ we have $0 \leq m_{i,j} < m_{i,i}$.

More generally, if $n \geq m$, a matrix M in HNF has the following shape

$$\begin{pmatrix} 0 & 0 & \cdots & 0 & * & * & \cdots & * \\ 0 & 0 & \cdots & 0 & 0 & * & \cdots & * \\ \vdots & \vdots & \ddots & \vdots & \vdots & & \ddots & \vdots \\ 0 & 0 & \cdots & 0 & 0 & \cdots & 0 & * \end{pmatrix}$$

where the last m columns form a matrix in HNF.

Theorem 2.4.3. *Let A be an $m \times n$ matrix with coefficients in $\mathbb{Z}$. Then there exists a unique $m \times n$ matrix $B = (b_{i,j})$ in HNF of the form $B = AU$ with $U \in \mathrm{GL}_n(\mathbb{Z})$, where $\mathrm{GL}_n(\mathbb{Z})$ is the group of matrices with integer coefficients which are invertible, i.e. whose determinant is equal to ± 1.*

Note that although B is unique, the matrix U will *not* be unique.

The matrix W formed by the non-zero columns of B will be called the Hermite normal form of the matrix A. Note that if A is the matrix of any *generating set* of a sub-$\mathbb{Z}$-module L of $\mathbb{Z}^m$, and not only of a basis, the columns of W give the unique basis of L whose matrix is in HNF. This basis will be called the HNF basis of the $\mathbb{Z}$-module L, and the matrix W the HNF of L.

In the special case where the $\mathbb{Z}$-module L is of rank equal to m, the matrix W will be upper triangular, and will sometimes be called the upper triangular HNF of L.

We give the proof of Theorem 2.4.3 as an algorithm.

Algorithm 2.4.4 (Hermite Normal Form). Given an $m \times n$ matrix A with integer coefficients $(a_{i,j})$ this algorithm finds the Hermite normal form W of A. As usual, we write $w_{i,j}$ for the coefficients of W, A_i (resp. W_i) for the columns of A (resp. W).

1. [Initialize] Set $i \leftarrow m$, $k \leftarrow n$, $l \leftarrow 1$ if $m \leq n$, $l \leftarrow m - n + 1$ if $m > n$.

2. [Row finished?] If all the $a_{i,j}$ with $j < k$ are zero, then if $a_{i,k} < 0$ replace column A_k by $-A_k$ and go to step 5.

3. [Choose non-zero entry] Pick among the non-zero $a_{i,j}$ for $j \leq k$ one with the smallest absolute value, say a_{i,j_0}. Then if $j_0 < k$, exchange column A_k with column A_{j_0}. In addition, if $a_{i,k} < 0$ replace column A_k by $-A_k$. Set $b \leftarrow a_{i,k}$.

4. [Reduce] For $j = 1, \ldots, k - 1$ do the following: set $q \leftarrow \lfloor a_{i,j}/b \rfloor$, and $A_j \leftarrow A_j - qA_k$. Then go to step 2.

5. [Final reductions] Set $b \leftarrow a_{i,k}$. If $b = 0$, set $k \leftarrow k + 1$ and go to step 6. Otherwise, for $j > k$ do the following: set $q \leftarrow \lfloor a_{i,j}/b \rfloor$, and $A_j \leftarrow A_j - qA_k$.

6. [Finished?] If $i = l$ then for $j = 1, \ldots, n - k + 1$ set $W_j \leftarrow A_{j+k-1}$ and terminate the algorithm. Otherwise, set $i \leftarrow i - 1$, $k \leftarrow k - 1$ and go to step 2.

This algorithm terminates since one can easily prove that $|a_{i,k}|$ is strictly decreasing each time we return to step 2 from step 4. Upon termination, it is clear that W is in Hermite normal form, and since it has been obtained from A by elementary column operations of determinant ± 1, W is the HNF of A. We leave the uniqueness statement of Theorem 2.4.3 as an exercise for the reader (Exercise 14). □

Remarks.

(1) It is easy to modify the above algorithm (as well as the subsequent ones) so as to give the lower triangular HNF of A in the case where A is of rank equal to m.

(2) If we also want the matrix $U \in \mathrm{GL}_n(\mathbb{Z})$, it is easy to add the corresponding statements (see for example Algorithm 2.4.10).

Consider the very special case $m = 1$, $n = 2$ of this algorithm. The result will be (usually) a 1×1 matrix whose unique element is equal to the GCD $(a_{1,1}, a_{1,2})$. Hence, it is conceptually easier, and usually faster, to replace in the above algorithm divisions by (extended) GCD's. We can then choose among several available methods for computing these GCD's. This gives the following algorithm.

Algorithm 2.4.5 (Hermite Normal Form).   Given an $m \times n$ matrix A with integer coefficients $(a_{i,j})$ this algorithm finds the Hermite normal form W of A. We use an auxiliary column vector B.

1. [Initialize] Set $i \leftarrow m$, $j \leftarrow n$, $k \leftarrow n$, $l = 1$ if $m \leq n$, $l = m - n + 1$ if $m > n$.

2. [Check zero] If $j = 1$ go to step 4. Otherwise, set $j \leftarrow j - 1$, and if $a_{i,j} = 0$ go to step 2.

3. [Euclidean step] Using Euclid's extended algorithm, compute (u, v, d) such that $ua_{i,k} + va_{i,j} = d = \gcd(a_{i,k}, a_{i,j})$, with $|u|$ and $|v|$ minimal (see below). Then set $B \leftarrow uA_k + vA_j$, $A_j \leftarrow (a_{i,k}/d)A_j - (a_{i,j}/d)A_k$, $A_k \leftarrow B$, and go to step 2.

4. [Final reductions] Set $b \leftarrow a_{i,k}$. If $b < 0$ set $A_k \leftarrow -A_k$ and $b \leftarrow -b$. Now if $b = 0$, set $k \leftarrow k + 1$ and go to step 5, otherwise for $j > k$ do the following: set $q \leftarrow \lfloor a_{i,j}/b \rfloor$, and $A_j \leftarrow A_j - qA_k$.

5. [Finished?] If $i = l$ then for $j = 1, \ldots, n - k + 1$ set $W_j \leftarrow A_{j+k-1}$ and terminate the algorithm. Otherwise, set $i \leftarrow i - 1$, $k \leftarrow k - 1$, $j \leftarrow k$ and go to step 2.

Important Remark. In step 3, we are asked to compute (u, v, d) with $|u|$ and $|v|$ minimal. The meaning of this is as follows. We must choose among all possible (u, v), the unique pair such that

$$-\frac{|a|}{d} < v \operatorname{sign}(b) \leq 0 \quad \text{and} \quad 1 \leq u \operatorname{sign}(a) \leq \frac{|b|}{d}.$$

In fact, the condition on u is equivalent to the condition on v and that such a pair exists and is unique is an exercise left to the reader (Exercise 15). The sign conditions are not important, they could be reversed if desired, but it is essential that when $d = |a|$, i.e. when $a \mid b$, we take $v = 0$. If this condition is not obeyed, the algorithm may enter into an infinite loop. This remark applies also to all the Hermite and Smith normal form algorithms that we shall see below.

Algorithms 2.4.4 and 2.4.5 work entirely with integers, and there are no divisions except for Euclidean divisions, hence one could expect that it behaves reasonably well with respect to the size of the integers involved. Unfortunately, this is absolutely not the case, and the coefficient explosion phenomenon occurs here also, even in very reasonable situations. For example, Hafner-McCurley ([Haf-McCur2]) give an example of a 20×20 integer matrix whose coefficients are less than or equal to 10, but which needs integers of up to 1500 decimal digits in the computations of Algorithm 2.4.4 or Algorithm 2.4.5 leading to its HNF. Hence, it is necessary to improve these algorithms.

One modification of Algorithm 2.4.5 would be for a fixed row i, instead of setting equal to zero the successive $a_{i,j}$ for $j = k - 1, k - 2, \ldots, 1$ by doing column operations between columns i and j, to set these $a_{i,j}$ equal to zero in the same order, but now doing operations between columns k and $k - 1$, then $k - 1$ and $k - 2$, and so on until columns 2 and 1, and then exchanging columns 1 and k. This idea is due to Bradley [Bra].

Still another modification is the following. In Algorithm 2.4.5, we perform the column operations as follows: $(k, k - 1)$, $(k, k - 2)$, $\ldots$, $(k, 1)$. In the modified version just mentioned, the order is $(k, k - 1)$, $(k - 1, k - 2)$, $\ldots$, $(2, 1)$, $(1, k)$. One can also for row i do as follows. Work with the pair of columns (j_1, j_2) where a_{i,j_1} and a_{i,j_2} are the largest and second largest non-zero elements of row i with $j \leq k$. Then experiments show that the coefficient explosion is considerably reduced, and actual computational experience shows that it is faster than the preceding versions. However this is still insufficient for our needs.

When $m \leq n$ and A is of rank m (in which case W is an upper triangular matrix with non-zero determinant D), an important improvement suggested by several authors (see for example [Kan-Bac]) is to work modulo a multiple of the determinant of W, or even modulo a multiple of the *exponent* of $\mathbb{Z}^m/W$. (Note that D is equal to the order of the finite $\mathbb{Z}$-module $\mathbb{Z}^m/W$; the exponent is by definition the smallest positive integer e such that $e\mathbb{Z}^m \subset W$. It divides the determinant.)

In the case where $m = n$, we have $\det(W) = \pm \det(A)$ hence the determinant can be computed before doing the reduction if needed. In the general case however one does not know $\det(W)$ in advance, but in practice, the HNF is often used for obtaining a HNF-basis for a $\mathbb{Z}$-module L in a number field (see Chapter 4), and in that case one usually knows a multiple of the determinant of L. One can modify all of the above mentioned algorithms in this way.

These modifications are based on the following additional algorithm, essentially due to Hafner and McCurley (see [Haf-McCur2]):

Algorithm 2.4.6 (HNF Modulo D). Let A be an $m \times n$ integer matrix of rank m. Let $L = (l_{i,j})_{1 \leq i,j \leq m}$ be the $m \times m$ upper triangular matrix obtained from A by doing all operations modulo D in any of the above mentioned algorithms, where D is a positive multiple of the determinant of the module generated by the columns of A (or equivalently of the determinant of the HNF of A). This algorithm outputs the true upper triangular Hermite normal form $W = (w_{i,j})_{1 \leq i,j \leq m}$ of A. We write W_i and L_i for the i-th columns of W and L respectively.

1. [Initialize] Set $b \leftarrow D$, $i \leftarrow m$.

2. [Euclidean step] Using a form of Euclid's extended algorithm, compute (u, v, d) such that $ul_{i,i} + vb = d = \gcd(l_{i,i}, b)$. Then set $W_i \leftarrow (uL_i \bmod b)$ (recall that $a \bmod b$ is the least non-negative residue of a modulo b). If $d = b$ (i.e. if $b \mid l_{i,i}$) set in addition $w_{i,i} \leftarrow d$ (if $d \neq b$, this will already be true, but if $d = b$ we would have $w_{i,i} = 0$ if we do not include this additional assignment).

3. [Finished?] If $i > 1$, set $b \leftarrow b/d$, $i \leftarrow i - 1$ and go to step 2. Otherwise, for $i = m - 1, m - 2, \ldots, 1$, and for $j = i + 1, \ldots, m$ set $q \leftarrow \lfloor w_{i,j}/w_{i,i} \rfloor$, $W_j \leftarrow W_j - qW_i$. Output the matrix $W = (w_{i,j})_{1 \leq i,j \leq m}$ and terminate the algorithm.

We must prove that this algorithm is valid. Since step 2 is executed exactly m times, the algorithm terminates, so what we need to prove is that the matrix W that the algorithm produces is indeed the HNF of A. For any $m \times n$ matrix M of rank m, denote by $\gamma_i(M)$ the GCD of all the $i \times i$ sub-determinants obtained from the last i rows of M for $1 \leq i \leq m$. It is clear that elementary column operations like those of Algorithms 2.4.4 or 2.4.5 leave these quantities unchanged. Furthermore, reduction modulo D changes these $i \times i$ sub-determinants by multiples of D, hence does not change the GCD of $\gamma_i(M)$ with D. It is clear that $\gamma_{m-i+1}(W) = w_{i,i} \cdots w_{m,m}$ divides $\det(W)$, hence divides D. Therefore we have:

$$w_{i,i} \cdots w_{m,m} = \gcd(D, \gamma_{m-i+1}(W))$$
$$= \gcd(D, \gamma_{m-i+1}(A))$$
$$= \gcd(D, \gamma_{m-i+1}(L))$$
$$= \gcd(D, l_{i,i} \cdots l_{m,m}). \tag{1_i}$$

hence the value given by Algorithm 2.4.6 for $w_{m,m}$ is correct. Call D_i the value of b for the value i, and set $P_i = w_{i+1,i+1} \cdots w_{m,m}$. Then if we assume that the diagonal elements $w_{j,j}$ are correct for $j > i$, we have by definition $D_i = D/P_i$. Hence, if we divide equation (1_{i+1}) by P_i we obtain

$$1 = \gcd(D_i, (l_{i+1,i+1} \cdots l_{m,m})/P_i)$$

for $1 \leq i < m$. Now if we divide equation (1_i) by P_i we obtain

$$w_{i,i} = \gcd(D_i, (l_{i,i} \cdots l_{m,m})/P_i) = \gcd(D_i, l_{i,i})$$

by the preceding formula, hence the diagonal elements of the matrix W which are output by Algorithm 2.4.6 are correct. Since W is an upper triangular matrix, it follows that its determinant is equal to the determinant of the HNF of A.

To finish the proof that Algorithm 2.4.6 is valid, we will show that the columns $W_i = (uL_i \bmod D_i)$ output by the algorithm are in the Z-module L generated by the columns of A. By the remark just made, this will show that, in fact, the W_i are a basis of L, hence that W is obtained from A by elementary transformations. Since step 3 of the algorithm finishes to transform W into a Hermite normal form, W must be equal to the HNF of A. Since

$$W_i = \sum_{1 \leq j \leq m} c_{i,j} A_j + D_i B_i$$

where the A_j are the columns of A, B_i is a (column) vector in $\mathbb{Z}^m$ whose components of index greater than i are zero, and the $c_{i,j}$ are integers, the claim concerning the W_i follows immediately from the following lemma:

Lemma 2.4.7. *With the above notations, for every i with $1 \leq i \leq m$ and any vector B whose components of index greater than i are zero, we have $D_i B \in L$.*

Proof. Consider the $i \times i$ matrix N_i formed by the first i rows and columns of the true HNF of A. We already have proved that the diagonal elements are $w_{j,j}$ as output by the algorithm. Now if one considers $\mathbb{Z}^i$ as a submodule of $\mathbb{Z}^m$ by considering the last $m - i$ components to be equal to 0, then we see that the columns of N_i (extended by $m - i$ zeros) are $\mathbb{Z}$-linear combinations of the columns A_i of A, i.e. are in L. Now $\det(N_i) = w_{1,1} \cdots w_{i,i}$ and by definition D_i is a multiple of $w_{1,1} \cdots w_{i,i}$. Hence, if L_i is the submodule of $\mathbb{Z}^i$ generated by the columns of N_i, we have on the one hand $L_i \subset \mathbb{Z}^i \cap L$, and on the other hand, since $\det(N_i) = [\mathbb{Z}^i : L_i]$, we have $\det(N_i)\mathbb{Z}^i \subset L_i$ which implies $D_i\mathbb{Z}^i \subset L$, and this is equivalent to the statement of the lemma. This concludes the proof of the validity of Algorithm 2.4.6.   □

Note that if we work modulo D in Algorithm 2.4.5, the order in which the columns are treated, which is what distinguishes Algorithm 2.4.5 from its variants, is not really important. Furthermore, the proof of Algorithm 2.4.6 shows that it is not necessary to work modulo the full multiple of the determinant D in Algorithm 2.4.5, but that at row i one can work modulo D_i, which can be much smaller. Finally, note that in step 2 of Algorithm 2.4.5, if we have worked modulo D (or D_i), it may happen that $a_{i,k} = 0$. In that case, it is necessary to set $a_{i,k} \leftarrow D_i$ (or any non-zero multiple of D_i). Combining these observations leads to the following algorithm, essentially due to Domich et al. [DKT].

It should be emphasized that all reductions modulo R should be taken in the interval $] - R/2, R/2]$, and not in the interval $[0, R[$. Otherwise, small negative coefficients will become large positive ones, and this may lead to infinite loops.

Algorithm 2.4.8 (HNF Modulo D). Given an $m \times n$ matrix A with integer coefficients $(a_{i,j})$ of rank m (hence such that $n \geq m$), and a positive integer D which is known to be a multiple of the determinant of the $\mathbb{Z}$-module generated by the columns of A, this algorithm finds the Hermite normal form W of A. We use an auxiliary column vector B.

1. [Initialize] Set $i \leftarrow m$, $j \leftarrow n$, $k \leftarrow n$, $R \leftarrow D$.

2. [Check zero] If $j = 1$ go to step 4. Otherwise, set $j \leftarrow j - 1$, and if $a_{i,j} = 0$ go to step 2.

3. [Euclidean step] Using Euclid's extended algorithm, compute (u, v, d) such that $ua_{i,k} + va_{i,j} = d = \gcd(a_{i,k}, a_{i,j})$, with $|u|$ and $|v|$ minimal. Then set $B \leftarrow uA_k + vA_j$, $A_j \leftarrow ((a_{i,k}/d)A_j - (a_{i,j}/d)A_k) \bmod R$, $A_k \leftarrow B \bmod R$, and go to step 2.

4. [Next row] Using Euclid's extended algorithm, find (u, v, d) such that $ua_{i,k} + vR = d = \gcd(a_{i,k}, R)$. Set $W_i \leftarrow uA_k \bmod R$ (here taken in the interval $[0, R-1]$). If $w_{i,i} = 0$ set $w_{i,i} \leftarrow R$. For $j = i+1, \ldots, m$ set $q \leftarrow \lfloor w_{i,j}/w_{i,i} \rfloor$ and $W_j \leftarrow W_j - qW_i$. If $i = 1$, output the matrix $W = (w_{i,j})_{1 \leq i,j \leq m}$ and terminate the algorithm. Otherwise, set $R \leftarrow R/d$, $i \leftarrow i - 1$, $k \leftarrow k - 1$, $j \leftarrow k$, and if $a_{i,k} = 0$ set $a_{i,k} \leftarrow R$. Go to step 2.

This will be our algorithm of choice for HNF reduction, at least when some D is known and A is of rank m.

Remark. It has been noted (see Remark (2) after Algorithm 2.4.4) that it is easy to add statements so as to obtain the matrix U such that $B = AU$ where B is the $n \times m$ matrix in Hermite normal form whose non-zero columns form the HNF of A. In the case of modulo D algorithms such as the one above, it seems more difficult to do so.

2.4.3 Applications of the Hermite Normal Form

In this section, we will see a few basic applications of the HNF form of a matrix representing a free $\mathbb{Z}$-module. Further applications will be seen in the context of number fields (Chapter 4).

Image of an Integer Matrix. First note that finding the HNF of a matrix using Algorithm 2.4.5 is essentially analogous to finding the column eche-lon form in the case of vector spaces (Algorithm 2.3.11). In particular, if the columns of the matrix represents a generating set for a free module L, Algo-rithm 2.4.5 allows us to find a basis (in fact of quite a special form), hence it also performs the same role as Algorithm 2.3.2. Contrary to the case of vector spaces, however, it is not possible in general to extract a basis from a generating set (this would mean that $(a, b) = |a|$ or $(a, b) = |b|$ in the case $m = 1, n = 2$), hence an analog of Algorithm 2.3.2 cannot exist.

Kernel of an Integer Matrix. We can also use Algorithm 2.4.5 to find the kernel of an $m \times n$ integer matrix A, i.e. a $\mathbb{Z}$-basis for the free sub-$\mathbb{Z}$-module of $\mathbb{Z}^n$ which is the set of column vectors X such that $AX = 0$. Note that this *cannot* be done (at least not without considerable extra work) by using Algorithm 2.3.1 which gives only a $\mathbb{Q}$-basis. What we must do is simply keep track of the matrix $U \in \mathrm{GL}_n(\mathbb{Z})$ such that $B = AU$ is in HNF. Indeed, we have the following proposition.

Proposition 2.4.9. *Let A be an $m \times n$ matrix, $B = AU$ its HNF with $U \in \mathrm{GL}_n(\mathbb{Z})$, and let r be such that the first r columns of B are equal to 0. Then a $\mathbb{Z}$-basis for the kernel of A is given by the first r columns of U.*

Proof. If U_i is the i-th column of U, then AU_i is the i-th column of B so is equal to 0 if $i \leq r$. Conversely, let X be a column vector such that $AX = 0$ or equivalently $BY = 0$ with $Y = U^{-1}X$. Solving the system $BY = 0$ from bottom up, $b_{f(k),k} > 0$ for $k > r$ (with the notation of Definition 2.4.2) implies that the last $n - r + 1$ coordinates of Y are equal to 0, and the first r are arbitrary, hence the first r canonical basis elements of $\mathbb{Z}^n$ form a $\mathbb{Z}$-basis for the kernel of B, and upon left multiplication by U we obtain the proposition. $\square$

This gives the following algorithm.

Algorithm 2.4.10 (Kernel over $\mathbb{Z}$). Given an $m \times n$ matrix A with integer coefficients $(a_{i,j})$, this algorithm finds a $\mathbb{Z}$-basis for the kernel of A. We use an auxiliary column vector B and an auxiliary $n \times n$ matrix U.

1. [Initialize] Set $i \leftarrow m$, $j \leftarrow n$, $k \leftarrow n$, $U \leftarrow I_n$, $l \leftarrow 1$ if $m \leq n$, $l \leftarrow m-n+1$ if $m > n$.

2. [Check zero] If $j = 1$ go to step 4. Otherwise, set $j \leftarrow j - 1$, and if $a_{i,j} = 0$ go to step 2.

3. [Euclidean step] Using Euclid's extended algorithm, compute (u, v, d) such that $ua_{i,k} + va_{i,j} = d = \gcd(a_{i,k}, a_{i,j})$, with $|u|$ and $|v|$ minimal. Then set $B \leftarrow uA_k + vA_j$, $A_j \leftarrow (a_{i,k}/d)A_j - (a_{i,j}/d)A_k$, $A_k \leftarrow B$; similarly set $B \leftarrow uU_k + vU_j$, $U_j \leftarrow (a_{i,k}/d)U_j - (a_{i,j}/d)U_k$, $U_k \leftarrow B$, then go to step 2.

4. [Final reductions] Set $b \leftarrow a_{i,k}$. If $b < 0$ set $A_k \leftarrow -A_k$, $U_k \leftarrow -U_k$ and $b \leftarrow -b$. Now if $b = 0$, set $k \leftarrow k + 1$ and go to step 5, otherwise for $j > k$ do the following: set $q \leftarrow \lfloor a_{i,j}/b \rfloor$, $A_j \leftarrow A_j - qA_k$ and $U_j \leftarrow U_j - qU_k$.

5. [Finished?] If $i = l$ then for $j = 1, \ldots, k - 1$ set $M_j \leftarrow U_j$, output the matrix M and terminate the algorithm. Otherwise, set $i \leftarrow i - 1$, $k \leftarrow k - 1$, $j \leftarrow k$ and go to step 2.

Remark. Although this algorithm correctly gives a $\mathbb{Z}$-basis for the kernel of A, the coefficients that are obtained are usually large. To obtain a really useful algorithm, it is necessary to *reduce* the basis that is obtained, for example using one of the variants of the LLL algorithm that we will see below (see Section 2.6). However, it is desirable to obtain directly a basis of good quality that avoids introducing large coefficients. This can be done using the MLLL algorithm (see Algorithm 2.7.2), and gives an algorithm which is usually preferable.

In view of the applications to number fields, limiting ourselves to free submodules of some $\mathbb{Z}^m$ is a little too restrictive. In what follows we will simply say that L is a *module* if it is a free sub-$\mathbb{Z}$-module of rank m of $\mathbb{Q}^m$. Considering basis elements of L, it is clear that there exists a minimal positive integer d such that $dL \subset \mathbb{Z}^m$. We will call d the *denominator* of L with respect to $\mathbb{Z}^m$. Then the HNF of L will be by definition the pair (W, d), where W is the HNF of dL, and d is the denominator of L.

Test for Equality. Since the HNF representation of a free module L is unique, it is clear that one can trivially test equality of modules: their denominator and their HNF must be the same.

Sum of Modules. Given two modules L and L' by their HNF, we can compute their sum $L + L' = \{x + x', x \in L, x' \in L'\}$ in the following way. Let (W, d) and (W', d') be their HNF representation. Let $D = dd'/(d, d')$ be the least common multiple of d and d'. Denoting as usual by A_i the i-th column of a matrix A, consider the $m \times 2m$ matrix A such that $A_i = (D/d)W_i$ and $A_{m+i} = (D/d')W_i'$ for $1 \leq i \leq m$, then it is clear that the columns of A generate $D(L + L')$, hence if we compute the HNF H of A and divide D and H by the greatest common divisor of D and of all the coefficients of H, we obtain the HNF normal form of $L + L'$. Apart from the treatment of denominators, this is similar to Algorithm 2.3.8.

Test for Inclusion. To test whether $L' \subset L$, where L and L' are given by their HNF, the most efficient way is probably to compute $N = L + L'$ as above, and then test the equality $N = L$. Note that if d and d' are the denominators of L and L' respectively, a necessary condition for $L' \subset L$ is that $d' \mid d$, hence the LCM D must be equal to d.

Product by a Constant. This is especially easy: if $c = p/q \in \mathbb{Q}$ with $(p, q) = 1$ and $q > 0$, the HNF of cL is obtained as follows. Let d_1 be the GCD of all the coefficients of the HNF of L. Then the denominator of cL is $qd/((p, d)(q, d_1))$, and the HNF matrix is equal to $p/((p, d)(q, d_1))$ times the HNF matrix of L.

We will see that the HNF is quite practical for other problems also, but the above list is, I hope, sufficiently convincing.

2.4.4 The Smith Normal Form and Applications

We have seen that the Hermite normal form permits us to handle free $\mathbb{Z}$-modules of finite rank quite nicely. We would now like a similar notion which would allow us to handle finite $\mathbb{Z}$-modules G. Recall from Theorem 2.4.1 (3) that such a module is isomorphic (in many ways of course) to a quotient $\mathbb{Z}^n/L$ where L is a (necessarily free) submodule of $\mathbb{Z}^n$ of rank equal to n. More elegantly perhaps, we can say that G is isomorphic to a quotient L'/L of free $\mathbb{Z}$-modules of the same (finite) rank n. Thus we can represent G (still non-canonically) by an $n \times n$ matrix A giving the coordinates of some $\mathbb{Z}$-basis of L on some $\mathbb{Z}$-basis of L'. In particular, A will have non-zero determinant, and in fact the absolute value of the determinant of A is equal to the cardinality of G, i.e. to the index $[L' : L]$ (see Exercise 18).

The freedom we now have is as follows. Changing the $\mathbb{Z}$-basis of L is equivalent to right multiplication of A by a matrix $U \in \mathrm{GL}_n(\mathbb{Z})$, as in the HNF case. Changing the $\mathbb{Z}$-basis of L' is on the other hand equivalent to *left* multiplication of A by a matrix $V \in \mathrm{GL}_n(\mathbb{Z})$. In other words, we are allowed to perform elementary column *and* row operations on the matrix A without changing (the isomorphism class of) G. This leads to the notion of *Smith normal form* of A.

Definition 2.4.11. *We say that an $n \times n$ matrix B is in Smith normal form (abbreviated SNF) if B is a diagonal matrix with nonnegative integer coefficients such that $b_{i+1,i+1} \mid b_{i,i}$ for all $i < n$.*

Then the basic theorem which explains the use of this definition is as follows.

Theorem 2.4.12. *Let A be an $n \times n$ matrix with coefficients in $\mathbb{Z}$ and non-zero determinant. Then there exists a unique matrix in Smith normal form B such that $B = VAU$ with U and V elements of $\mathrm{GL}_n(\mathbb{Z})$.*

If we set $d_i = b_{i,i}$, the d_i are called the *elementary divisors* of the matrix A, and the theorem can be written

$$A = V^{-1} \begin{pmatrix} d_1 & 0 & \cdots & 0 \\ 0 & d_2 & \ddots & \vdots \\ \vdots & \ddots & \ddots & 0 \\ 0 & \cdots & 0 & d_n \end{pmatrix} U^{-1}$$

with $d_{i+1} \mid d_i$ for $1 \leq i < n$.

This theorem, stated for matrices, is equivalent to the following theorem for $\mathbb{Z}$-modules.

Theorem 2.4.13 (Elementary Divisor Theorem). *Let L be a $\mathbb{Z}$-submodule of a free module L' and of the same rank. Then there exist positive integers $d_1, \ldots, d_n$ (called the elementary divisors of L in L') satisfying the following conditions:*

(1) *For every i such that $1 \leq i < n$ we have $d_{i+1} \mid d_i$.*
(2) *As $\mathbb{Z}$-modules, we have the isomorphism*

$$L'/L \simeq \bigoplus_{1 \leq i \leq n} (\mathbb{Z}/d_i\mathbb{Z}) \ ,$$

and in particular $[L' : L] = d_1 \cdots d_n$ and d_1 is the exponent of L'/L.
(3) *There exists a $\mathbb{Z}$-basis $(v_1, \ldots, v_n)$ of L' such that $(d_1v_1, \ldots, d_nv_n)$ is a $\mathbb{Z}$-basis of L.*

Furthermore, the d_i are uniquely determined by L and L'.

Remarks.

(1) This fundamental theorem is valid more generally. It holds for finitely generated (torsion) free modules over a principal ideal domain (PID, see Chapter 4). It is *false* if the base ring R is not a PID: applying the theorem to $n = 1$, $L' = R$ and L any integral ideal of R, it is clear that the truth of this theorem is equivalent to the PID condition.
(2) We have stated Theorem 2.4.12 only for square matrices of non-zero determinant. As in the Hermite case, it would be easy to state a generalization valid for general matrices (including non-square ones). In practice, this is not really needed since we can always first perform a Hermite reduction.

The proof of these two theorems can be found in any standard textbook but it follows immediately from the algorithm below.

Since we are going to deal with square matrices, as with the case of the HNF, it is worthwhile to work modulo the determinant (or a multiple). In most cases this determinant (or a multiple of it) is known in advance. It should also be emphasized again that all reductions modulo R should be taken in the interval $] - R/2, R/2]$, and not in the interval $[0, R[$.

The following algorithm is essentially due to Hafner and McCurley (see [Haf-McCur2]).

Algorithm 2.4.14 (Smith Normal Form). Given an $n \times n$ non-singular integral matrix $A = (a_{i,j})$, this algorithm finds the Smith normal form of A, i.e. outputs the diagonal elements d_i such that $d_{i+1} \mid d_i$. Recall that we denote by A_i (resp. A'_i) the columns (resp. the rows) of the matrix A. We use a temporary (column or row) vector variable B.

1. [Initialize i] Set $i \leftarrow n$, $R \leftarrow |\det(A)|$. If $n = 1$, output $d_1 \leftarrow R$ and terminate the algorithm.

2. [Initialize j for row reduction] Set $j \leftarrow i$, $c \leftarrow 0$.

3. [Check zero] If $j = 1$ go to step 5. Otherwise, set $j \leftarrow j - 1$. If $a_{i,j} = 0$ go to step 3.

4. [Euclidean step] Using Euclid's extended algorithm, compute (u, v, d) such that $ua_{i,i} + va_{i,j} = d = \gcd(a_{i,i}, a_{i,j})$, with u and v minimal (see remark after Algorithm 2.4.5). Then set $B \leftarrow uA_i + vA_j$, $A_j \leftarrow ((a_{i,i}/d)A_j - (a_{i,j}/d)A_i)$ mod R, $A_i \leftarrow B$ mod R and go to step 3.

5. [Initialize j for column reduction] Set $j \leftarrow i$.

6. [Check zero] If $j = 1$ go to step 8. Otherwise, set $j \leftarrow j - 1$, and if $a_{j,i} = 0$ go to step 6.

7. [Euclidean step] Using Euclid's extended algorithm, compute (u, v, d) such that $ua_{i,i} + va_{j,i} = d = \gcd(a_{i,i}, a_{j,i})$, with u and v minimal (see remark after Algorithm 2.4.5). Then set $B \leftarrow uA'_i + vA'_j$, $A'_j \leftarrow ((a_{i,i}/d)A'_j - (a_{j,i}/d)A'_i)$ mod R, $A'_i \leftarrow B$ mod R, $c \leftarrow c + 1$ and go to step 6.

8. [Repeat stage i?] If $c > 0$ go to step 2.

9. [Check the rest of the matrix] Set $b \leftarrow a_{i,i}$. For $1 \leq k, l < i$ check whether $b \mid a_{k,l}$. As soon as some coefficient $a_{k,l}$ is not divisible by b, set $A'_i \leftarrow A'_i + A'_k$ and go to step 2.

10. [Next stage] (Here all the $a_{k,l}$ for $1 \leq k, l < i$ are divisible by b). Output $d_i = \gcd(a_{i,i}, R)$ and set $R \leftarrow R/d_i$. If $i = 2$, output $d_1 = \gcd(a_{1,1}, R)$ and terminate the algorithm. Otherwise, set $i \leftarrow i - 1$ and go to step 2.

This algorithm seems complicated at first, but one can see that it is actually quite straightforward, using elementary row and column operations of determinant ± 1 to reduce the matrix A.

This algorithm terminates (and does not take too many steps!) since each time one returns to step 2 from step 9, the coefficient $a_{i,i}$ has been reduced at least by a factor of 2.

The proof that this algorithm is valid, i.e. that the result is correct, follows exactly the proof of the validity of Algorithm 2.4.6. If we never reduced modulo R in Algorithm 2.4.14, it is clear that the result would be correct (however the coefficients would explode). Incidentally, this gives a proof of Theorems 2.4.12 and 2.4.13.

Hence, we must simply show that the transformations done in step 10 correctly restore the values of d_i. Denote by $\delta_i(A)$ the GCD of the determinants of *all* $i \times i$ sub-matrices of A, and not only from the first i rows as in the proof of Algorithm 2.4.6. Then, in a similar manner, these δ_i are invariant under elementary row and column operations of determinant ± 1. Hence, denoting by Δ the diagonal SNF of A, by D the determinant of A, and by $S = (a_{i,j})$ the final form of the matrix A at the end of Algorithm 2.4.14, we have:

$$\begin{aligned} d_i \cdots d_n &= \gcd(D, \delta_{n-i+1}(\Delta)) \\ &= \gcd(D, \delta_{n-i+1}(A)) \\ &= \gcd(D, \delta_{n-i+1}(S)) \\ &= \gcd(D, a_{i,i} \cdots a_{n,n}). \end{aligned} \tag{2_i}$$

Hence, if we set $P_i = d_{i+1} \cdots d_n$, exactly as in the proof of Algorithm 2.4.6 we obtain

$$1 = (D/P_i, (a_{i+1,i+1} \cdots a_{n,n})/P_i)$$

(divide formula (2_{i+1}) by P_i), then

$$d_i = (D/P_i, (a_{i,i}a_{i+1,i+1} \cdots a_{n,n})/P_i)$$

(divide (2_i) by P_i), and hence

$$d_i = (D/P_i, a_{i,i}) \ .$$

But clearly in stage i of the algorithm, $R = D/P_i$, thus proving the validity of the algorithm. $\square$

Note that we have chosen an order for the d_i which is consistent with our choice for Hermite normal forms, but which is the reverse of the one which is found in most texts. The modifications to Algorithm 2.4.14 so that the order is reversed are trivial (essentially make i and j go up instead of down) and are left to the reader.

The Smith normal form will mainly be used as follows. Let G be a finite $\mathbb{Z}$-module (i.e. a finite Abelian group). We want to determine the structure of G, and in particular its cardinality. Note that a corollary of Theorem 2.4.13 is the structure theorem for finite Abelian groups: such a group is isomorphic to a unique direct sum of cyclic groups $\mathbb{Z}/d_i\mathbb{Z}$ with $d_{i+1} \mid d_i$.

We can then proceed as follows. By theoretical means, we find some integer n and a free module L' of rank n such that G is isomorphic to a quotient L'/L, where L is also of rank n but unknown. We then determine as many elements of L as possible (how to do this depends, of course, entirely on the specific problem) so as to have at least n elements which are $\mathbb{Q}$-linearly independent. Using the Hermite normal form Algorithm 2.4.5, we can then find the HNF basis for the submodule L_1 of L generated by the elements that we have found. Computing the determinant of this basis (which is trivial since the basis is in triangular form) already gives us the cardinality of L'/L_1. If we know bounds for the order of G (for example, if we know the order of G up to a factor of $\sqrt{2}$ from above and below), we can check whether $L_1 = L$. If not, we continue finding new elements of L until the cardinality check shows that $L_1 = L$. We can then compute the SNF of the HNF basis (note that the determinant is now known), and this gives us the complete structure of G.

We will see a concrete application of the process just described in the sub-exponential computations of class groups (see Chapter 5).

Remark. The diagonal elements which are obtained after a Hermite Normal Form computation are usually *not* equal to the Smith invariants. For example, the matrix $\begin{pmatrix} 2 & 1 \\ 0 & 2 \end{pmatrix}$ is in HNF, but its Smith normal form has as diagonal elements $(4, 1)$.

2.5 Generalities on Lattices

2.5.1 Lattices and Quadratic Forms

We are now going to add some extra structure to free $\mathbb{Z}$-modules of finite rank. Recall the following definition.

Definition 2.5.1. *Let K be a field of characteristic different from 2, and let V be a K-vector space. We say that a map q from V to K is a quadratic form if the following two conditions are satisfied:*

(1) *For every $\lambda \in K$ and $x \in V$ we have*

$$q(\lambda \cdot x) = \lambda^2 q(x) \ .$$

(2) *If we set $b(x, y) = \frac{1}{2}(q(x+y) - q(x) - q(y))$ then b is a (symmetric) bilinear form, i.e. $b(x + x', y) = b(x, y) + b(x', y)$ and $b(\lambda \cdot x, y) = \lambda b(x, y)$ for all $\lambda \in K$, x, x' and y in V (the similar conditions on the second variable follow from the fact that $b(y, x) = b(x, y)$).*

The identity $b(x, x) = q(x)$ allows us to recover q from b.

In the case where $K = \mathbb{R}$, we say that q is *positive definite* if for all non-zero $x \in V$ we have $q(x) > 0$.

Definition 2.5.2. *A lattice L is a free $\mathbb{Z}$-module of finite rank together with a positive definite quadratic form q on $L \otimes \mathbb{R}$.*

Let $(\mathbf{b}_i)_{1 \le i \le n}$ be a $\mathbb{Z}$-basis of L. If $\mathbf{x} = \sum_{1 \le i \le n} x_i \mathbf{b}_i \in L$ with $x_i \in \mathbb{Z}$, the definition of a quadratic form implies that

$$q(\mathbf{x}) = \sum_{1 \le i,j \le n} q_{i,j} x_i x_j \quad \text{with } q_{i,j} = b(\mathbf{b}_i, \mathbf{b}_j)$$

where as above, b denotes the symmetric bilinear form associated to q.

The matrix $Q = (q_{i,j})_{1 \le i,j \le n}$ is then a symmetric matrix which is positive definite when q is positive definite. We have $b(\mathbf{x}, \mathbf{y}) = Y^t Q X$ and in particular $q(\mathbf{x}) = X^t Q X$ where X and Y are the column vectors giving the coordinates of $\mathbf{x}$ and $\mathbf{y}$ respectively in the basis $(\mathbf{b}_i)$.

We will say that two lattices (L, q) and (L', q') are equivalent if there exists a $\mathbb{Z}$-module isomorphism between L and L' sending q to q'. We will identify equivalent lattices. Also, when the quadratic form is understood, we will write L instead of (L, q).

A lattice (L, q) can be represented in several ways all of which are useful. First, one can choose a $\mathbb{Z}$-basis $(\mathbf{b}_i)_{1 \le i \le n}$ of the lattice. Then an element of $\mathbf{x} \in L$ will be considered as a (column) vector X giving the (integral) coordinates of $\mathbf{x}$ on the basis. The quadratic form q is then represented by the positive definite symmetric matrix Q as we have seen above.

Changing the $\mathbb{Z}$-basis amounts to replacing X by PX for some $P \in \mathrm{GL}_n(\mathbb{Z})$, hence $q(x) = (PX)^t Q (PX) = X^t Q' X$ with $Q' = P^t Q P$. Hence, equivalence classes of lattices correspond to equivalence classes of positive definite symmetric matrices under the equivalence relation $Q' \sim Q$ if and only if there exists $P \in \mathrm{GL}_n(\mathbb{Z})$ such that $Q' = P^t Q P$. Note that $\det(P) = \pm 1$, hence the determinant of Q is independent of the choice of the basis. Since Q is positive definite, $\det(Q) > 0$ and we will set $d(L) = \det(Q)^{1/2}$ and call it the *determinant of the lattice*.

A second way to represent a lattice (L, q) is to consider L as a discrete subgroup of rank n of the Euclidean vector space $E = L \otimes \mathbb{R}$. Then if $(\mathbf{b}_i)_{1 \le i \le n}$ is a $\mathbb{Z}$-basis of L, it is also by definition of the tensor product an $\mathbb{R}$-basis of E. The matrix of scalar products $Q = (\mathbf{b}_i \cdot \mathbf{b}_j)_{1 \le i,j \le n}$ (where $\mathbf{b}_i \cdot \mathbf{b}_j = b(\mathbf{b}_i, \mathbf{b}_j)$) is then called the *Gram matrix* of the $\mathbf{b}_i$. If we choose some orthonormal basis of E, we can then identify E with the Euclidean space $\mathbb{R}^n$ with the usual Euclidean structure coming from the quadratic form $q(\mathbf{x}) = x_1^2 + \cdots + x_n^2$.

If B is the $n \times n$ matrix whose columns give the coordinates of the $\mathbf{b}_i$ on the chosen orthonormal basis of E, it is clear that $Q = B^t B$. In particular, $d(L) = |\det(B)|$. Furthermore, if another choice of orthonormal basis is made, the new matrix B' will be of the form $B' = KB$ where K is an *orthogonal matrix*, i.e. a matrix such that $K^t K = KK^t = I_n$. Thus we have proved the following proposition.

Proposition 2.5.3.

(1) *If Q is the matrix of a positive definite quadratic form, then Q is the Gram matrix of some lattice basis, i.e. there exists a matrix $B \in \mathrm{GL}_n(\mathbb{R})$ such that $Q = B^t B$*

(2) *The Gram matrix of a lattice basis $\mathbf{b}_i$ determines this basis uniquely up to isometry. In other words, if the $\mathbf{b}_i$ and the $\mathbf{b}'_i$ have the same Gram matrix, then the $\mathbf{b}'_i$ can be obtained from the $\mathbf{b}_i$ by an orthogonal transformation. In matrix terms, $B' = KB$ where K is an orthogonal matrix.*

It is not difficult to give a completely matrix-theoretic proof of this proposition (see Exercise 20).

It follows from the above results that when dealing with lattices, it is not necessary to give the coordinates of the $\mathbf{b}_i$ on some orthonormal basis. We can simply give a positive definite matrix which we can then think of as being the Gram matrix of the $\mathbf{b}_i$.

We see from the above discussion that there are natural bijections between the following three sets.

$$\{\text{Isomorphism classes of lattices of rank } n\} ,$$

$$\{\text{Classes of positive definite symmetric matrices } Q\}/ \sim ,$$

where $Q' \sim Q$ if and only if $Q' = P^t Q P$ for some $P \in \mathrm{GL}_n(\mathbb{Z})$, and

$$\mathrm{GL}_n(\mathbb{R})/ \sim ,$$

where $B' \sim B$ if and only if $B' = KBP$ for some $P \in \mathrm{GL}_n(\mathbb{Z})$ and some orthogonal matrix K.

Remarks.

(1) We have considered L in particular as a free discrete sub-$\mathbb{Z}$-module of the n-dimensional Euclidean space $L \otimes \mathbb{R}$. In many situations, it is desirable to consider L as a free discrete sub-$\mathbb{Z}$-module of some Euclidean space E of dimension m larger than n. The matrix B of coordinates of a basis of L on some orthonormal basis of E will then be an $m \times n$ matrix, but the Gram matrix $Q = B^t B$ will still be an $n \times n$ symmetric matrix.

(2) By abuse of language, we will frequently say that a free $\mathbb{Z}$-module of finite rank is a lattice even if there is no implicit quadratic form.

2.5.2 The Gram-Schmidt Orthogonalization Procedure

The existence of an orthonormal basis in a Euclidean vector space is often proved by using Gram-Schmidt orthonormalization (see any standard textbook). Doing this requires taking square roots, since the final vectors must be of length equal to 1.

For our purposes, we will need only an *orthogonal basis*, i.e. a set of mutually orthogonal vectors which are not necessarily of length 1. The same procedure works, except we do not normalize the length, and we will also call this the Gram-Schmidt orthogonalization procedure. It is summarized in the following proposition.

Proposition 2.5.4 (Gram-Schmidt). *Let* $\mathbf{b}_i$ *be a basis of a Euclidean vector space* E. *Define by induction:*

$$\mathbf{b}_i^* = \mathbf{b}_i - \sum_{j=1}^{i-1} \mu_{i,j} \mathbf{b}_j^* \qquad (1 \le i \le n) ,$$

where

$$\mu_{i,j} = \mathbf{b}_i \cdot \mathbf{b}_j^* / \mathbf{b}_j^* \cdot \mathbf{b}_j^* \qquad (1 \le j < i \le n) ,$$

then the $\mathbf{b}_i^*$ *form an orthogonal (but not necessarily orthonormal) basis of* E, $\mathbf{b}_i^*$ *is the projection of* $\mathbf{b}_i$ *on the orthogonal complement of* $\sum_{j=1}^{i-1} \mathbb{R}\mathbf{b}_j = \sum_{j=1}^{i-1} \mathbb{R}\mathbf{b}_j^*$, *and the matrix* M *whose columns gives the coordinates of the* $\mathbf{b}_i^*$ *in terms of the* $\mathbf{b}_i$ *is an upper triangular matrix with diagonal terms equal to 1. In particular, if* $d(L)$ *is the determinant of the lattice* L, *we have* $d(L)^2 = \prod_{1 \le i \le n} \|\mathbf{b}_i^*\|^2$.

The proof is trivial using induction.   □

We will now give a number of corollaries of this construction.

Corollary 2.5.5 (Hadamard's Inequality). *Let* (L, q) *be a lattice of determinant* $d(L)$, $(\mathbf{b}_i)_{1 \le i \le n}$ *a* $\mathbb{Z}$-*basis of* L, *and for* $x \in L$ *write* $|\mathbf{x}|$ *for* $q(\mathbf{x})^{1/2}$. *Then*

$$d(L) \le \prod_{i=1}^{n} |\mathbf{b}_i| .$$

Equivalently, if B *is an* $n \times n$ *matrix then*

$$|\det(B)| \le \prod_{1 \le i \le n} \left(\sum_{1 \le j \le n} |b_{i,j}|^2 \right)^{1/2} .$$

Proof. If we set $B_i = |\mathbf{b}_i^*|^2$, the orthogonality of the $\mathbf{b}_i^*$ implies that

$$q(\mathbf{b}_i) = |\mathbf{b}_i|^2 = B_i + \sum_{1 \le j < i} \mu_{i,j}^2 B_j$$

hence $d(L)^2 = \prod_{1 \le i \le n} B_i \le \prod_{1 \le i \le n} |\mathbf{b}_i|^2$. □

Corollary 2.5.6. *Let B be an invertible matrix with coefficients in $\mathbb{R}$. Then there exist unique matrices K, A and N such that:*

(1) $B = KAN$.
(2) K *is an orthogonal matrix, in other words $K^t = K^{-1}$.*
(3) A *is a diagonal matrix with positive diagonal coefficients.*
(4) N *is an upper triangular matrix with diagonal terms equal to 1.*

Note that this Corollary is sometimes called the Iwasawa decomposition of B since it is in fact true in a much more general setting than that of the group $\mathrm{GL}_n(\mathbb{R})$.

Proof. Let B' be the matrix obtained by applying the Gram-Schmidt process to the vectors whose coordinates are the columns of B on the standard basis of $\mathbb{R}^n$. Then, by the proposition we have $B' = BN$ where N is an upper triangular matrix with diagonal terms equal to 1. Now the Gram-Schmidt process gives an orthogonal basis, in other words the Gram matrix of the $\mathbf{b}_i^*$ is a diagonal matrix D with positive entries. Let A be the diagonal matrix obtained from D by taking the positive square root of each coefficient (we will call A the square root of D). Then the equality $B'^t B' = D$ is equivalent to $B' = KA$ for an orthogonal matrix K, hence $BN = KA$ which is equivalent to the existence statement of the corollary.

The uniqueness statement also follows since the equality $B' = BN = KA$ means that the $\mathbf{b}_i^*$ form an orthogonal basis which can be expressed on the $\mathbf{b}_i$ via an upper triangular matrix with diagonal terms equal to 1, and the procedure for obtaining this basis (i.e. the Gram-Schmidt coefficients) is clearly unique. □

Remarks.

(1) The requirement that the diagonal coefficients of A be positive is not essential, and is given only to insure uniqueness.
(2) By considering the inverse matrix and/or the transpose matrix of B, one has the same result with N lower triangular, or with $B = NAK$ instead of KAN.
(3) $T = AN$ is an upper triangular matrix with positive diagonal coefficients, and clearly any such upper triangular matrix T can be written uniquely in the form AN where A and N are as in the corollary. Hence we can use interchangeably both notations.

Another result is as follows.

Proposition 2.5.7. *If Q is the matrix of a positive definite quadratic form, then there exists a unique upper triangular matrix T with positive diagonal coefficients such that $Q = T^t T$ (or equivalently $Q = N^t D N$ where N is an upper triangular matrix with diagonal terms equal to 1 and D is a diagonal matrix with positive diagonal coefficients).*

Proof. By Proposition 2.5.3, we know that there exists $B \in \mathrm{GL}_n(\mathbb{R})$ such that $Q = B^t B$. On the other hand, by the Iwasawa decomposition we know that there exists matrices K and T such that $B = KT$ with K orthogonal and T upper triangular with positive diagonal coefficients ($T = AN$ in the notation of Proposition 2.5.6). Hence $Q = B^t B = T^t T$ thus showing the existence of T.

For the uniqueness, note that if $T^t T = T'^t T'$ with T and T' upper triangular, then

$$T'^{t-1} T^t = T' T^{-1} ,$$

where taking inverses is justified since Q is a positive definite matrix. But the left hand side of this equality is a lower triangular matrix, while the right hand side is an upper triangular one, hence both sides must be equal to some diagonal matrix D, and plugging back in the initial equality and using again the invertibility of T, we obtain that D^2 is equal to the identity matrix. Now since the diagonal coefficients of $D = T' T^{-1}$ must be positive, we deduce that D itself is equal to the identity matrix, thus proving the proposition. □

We will give later an algorithm to find the matrix T (Algorithm 2.7.6).

2.6 Lattice Reduction Algorithms

2.6.1 The LLL Algorithm

Among all the $\mathbb{Z}$ bases of a lattice L, some are better than others. The ones whose elements are the shortest (for the corresponding norm associated to the quadratic form q) are called *reduced*. Since the bases all have the same determinant, to be reduced implies also that a basis is not too far from being orthogonal.

The notion of reduced basis is quite old, and in fact in some sense one can even define an optimal notion of reduced basis. The problem with this is that no really satisfactory algorithm is known to find such a basis in a reasonable time, except in dimension 2 (Algorithm 1.3.14), and quite recently in dimension 3 from the work of B. Vallée [Val].

A real breakthrough came in 1982 when A. K. Lenstra, H. W. Lenstra and L. Lovász succeeded in giving a new notion of reduction (what is now called

LLL-reduction) and simultaneously a reduction algorithm which is deterministic and polynomial time (see [LLL]). This has proved invaluable.

The LLL notion of reduction is as follows. Let $b_1, b_2, \ldots, b_n$ be a basis of L. Using the Gram-Schmidt orthogonalization process, we can find an orthogonal (not orthonormal) basis $b_1^*, b_2^*, \ldots, b_n^*$ as explained in Proposition 2.5.4.

Definition 2.6.1. *With the above notations, the basis* $b_1, b_2, \ldots, b_n$ *is called LLL-reduced if*

$$|\mu_{i,j}| \leq \frac{1}{2} \quad \text{for } 1 \leq j < i \leq n$$

and

$$|b_i^* + \mu_{i,i-1} b_{i-1}^*|^2 \geq \frac{3}{4} |b_{i-1}^*|^2 \quad \text{for } 1 < i \leq n ,$$

or equivalently

$$|b_i^*|^2 \geq \left(\frac{3}{4} - \mu_{i,i-1}^2 \right) |b_{i-1}^*|^2 .$$

Note that the vectors $b_i^* + \mu_{i,i-1} b_{i-1}^*$ and b_{i-1}^* are the projections of b_i and b_{i-1} on the orthogonal complement of $\sum_{j=1}^{i-2} \mathbb{R} b_j$.

Then we have the following theorem:

Theorem 2.6.2. *Let* $b_1, b_2, \ldots, b_n$ *be an LLL-reduced basis of a lattice L. Then*

(1)

$$d(L) \leq \prod_{i=1}^{n} |b_i| \leq 2^{n(n-1)/4} d(L) ,$$

(2)

$$|b_j| \leq 2^{(i-1)/2} |b_i^*|, \quad \text{if } 1 \leq j \leq i \leq n ,$$

(3)

$$|b_1| \leq 2^{(n-1)/4} d(L)^{1/n} ,$$

(4) *For every* $x \in L$ *with* $x \neq 0$ *we have*

$$|b_1| \leq 2^{(n-1)/2} |x| ,$$

(5) *More generally, for any linearly independent vectors* $x_1, \ldots, x_t \in L$ *we have*

$$|b_j| \leq 2^{(n-1)/2} \max(|x_1|, \ldots, |x_t|) \quad \text{for } 1 \leq j \leq t .$$

We see that the vector $\mathbf{b}_1$ in a reduced basis is, in a very precise sense, not too far from being the shortest non-zero vector of L. In fact, it often is the shortest, and when it is not, one can, most of the time, work with $\mathbf{b}_1$ instead of the actual shortest vector.

Notation. In the rest of this chapter, we will use the notation $\mathbf{x} \cdot \mathbf{y}$ instead of $b(\mathbf{x}, \mathbf{y})$ where b is the bilinear form associated to q, and write $\mathbf{x}^2$ instead of $\mathbf{x} \cdot \mathbf{x} = q(\mathbf{x})$.

Proof. As in Corollary 2.5.5, we set $B_i = |\mathbf{b}_i^*|^2$. The first inequality of (1) is Corollary 2.5.5, Since the $\mathbf{b}_i$ are LLL-reduced, we have $B_i \geq (3/4 - \mu_{i,i-1}^2)B_{i-1} \geq B_{i-1}/2$ since $|\mu_{i,i-1}| \leq 1/2$. By induction, this shows that $B_j \leq 2^{i-j}B_i$ for $i \geq j$, hence

$$\mathbf{b}_i^2 \leq \frac{2^{i-1}+1}{2}B_i ,$$

and this trivially implies Theorem 2.6.2 (1), in fact with a slightly better exponent of 2. Combining the two inequalities which we just obtained, we get for all $j \leq i$, $\mathbf{b}_j^2 \leq (2^{i-2} + 2^{i-j-1})B_i$ which implies (2). If we set $j = 1$ in (2) and take the product of (2) for $i = 1$ to $i = n$, we obtain $(\mathbf{b}_1^2)^n \leq 2^{n(n-1)/2} \prod_{1 \leq i \leq n} B_i = 2^{n(n-1)/2}d(L)^2$, proving (3). For (4), there exists an i such that $\mathbf{x} = \sum_{1 \leq j \leq i} r_j\mathbf{b}_j = \sum_{1 \leq j \leq i} s_j\mathbf{b}_j^*$ and $r_i \neq 0$, where $r_j \in \mathbb{Z}$ and $s_j \in \mathbb{R}$. It is clear from the definition of the $\mathbf{b}_j^*$ that $r_i = s_i$, hence

$$|\mathbf{x}|^2 \geq s_i^2 B_i = r_i^2 B_i \geq B_i$$

since r_i is a non-zero integer, and since by (2) we know that $B_i \geq 2^{1-i}|\mathbf{b}_1|^2 \geq 2^{1-n}|\mathbf{b}_1|^2$, (4) is proved. (5) is proved by a generalization of the present argument and is left to the reader. □

Remark. Although we have lost a little in the exponent of 2 in Theorem 2.6.2 (1), the proof shows that even using the optimal value given in our proof would not improve the estimate in (4). On the other hand, we have not completely used the full LLL-reduction inequalities. In particular, the inequalities on the $\mu_{i,j}$ can be weakened to $\mu_{i,j}^2 \leq 1/2$ for all $j < i - 1$ and $|\mu_{i,i-1}| \leq 1/2$. This can be used to speed up the reduction algorithm which follows.

As has already been mentioned, what makes all these notions and theorems so valuable is that there is a very simple and efficient algorithm to find a reduced basis in a lattice. We now describe this algorithm in its simplest form. The idea is as follows. Assume that the vectors $\mathbf{b}_1, \dots, \mathbf{b}_{k-1}$ are already LLL-reduced (i.e. form an LLL-reduced basis of the lattice they generate). This will be initially the case for $k = 2$. The vector $\mathbf{b}_k$ first needs to be reduced so that $|\mu_{k,j}| \leq 1/2$ for all $j < k$ (some authors call this *size reduction*). This is done by replacing $\mathbf{b}_k$ by $\mathbf{b}_k - \sum_{j<k} a_j\mathbf{b}_j$ for some $a_j \in \mathbb{Z}$ in the following way. Assume that $|\mu_{k,j}| \leq 1/2$ for $l < j < k$ (initially with $l = k$). Then, if

$q = \lfloor \mu_{k,l} \rceil$ is the nearest integer to $\mu_{k,l}$, and, if we replace $\mathbf{b}_k$ by $\mathbf{b}_k - q\mathbf{b}_l$, then $\mu_{k,j}$ is not modified for $j > l$ (since $\mathbf{b}_j^*$ is orthogonal to $\mathbf{b}_l$ for $l < j$), and $\mu_{k,l}$ is replaced by $\mu_{k,l} - q$ (since $\mathbf{b}_l \cdot \mathbf{b}_l^* = \mathbf{b}_l^* \cdot \mathbf{b}_l^*$) and $|\mu_{k,l} - q| \le 1/2$ hence the modified $\mu_{k,j}$ satisfy $|\mu_{k,j}| \le 1/2$ for $l - 1 < j < k$.

Now that size reduction is done for the vector $\mathbf{b}_k$, we also need to satisfy the so-called *Lovász condition*, i.e. $B_k \ge (3/4 - \mu_{k,k-1}^2)B_{k-1}$. If this condition is satisfied, we increase k by 1 and start on the next vector $\mathbf{b}_k$ (if there is one). If it is not satisfied, we exchange the vectors $\mathbf{b}_k$ and $\mathbf{b}_{k-1}$, but then we must decrease k by 1 since we only know that $\mathbf{b}_1, \ldots, \mathbf{b}_{k-2}$ is LLL-reduced. A priori it is not clear that this succession of increments and decrements of k will ever terminate, but we will prove that this is indeed the case (and that the number of steps is not large) after giving the algorithm.

We could compute all the Gram-Schmidt coefficients $\mu_{k,j}$ and B_k at the beginning of the algorithm, and then update them during the algorithm. After each exchange step however, the coefficients $\mu_{i,k}$ and $\mu_{i,k-1}$ for $i > k$ must be updated, and this is usually a waste of time since they will probably change before they are used. Hence, it is a better idea to compute the Gram-Schmidt coefficients as needed, keeping in a variable k_{max} the maximal value of k that has been attained.

Another improvement on the basic idea is to reduce only the coefficient $\mu_{k,k-1}$ and not all the $\mu_{k,l}$ for $l < k$ during size-reduction, since this is the only coefficient which must be less than $1/2$ in absolute value before testing the Lovász condition. All this leads to the following algorithm.

Algorithm 2.6.3 (LLL Algorithm). Given a basis $\mathbf{b}_1, \mathbf{b}_2, \ldots, \mathbf{b}_n$ of a lattice (L, q) (either by coordinates on the canonical basis of $\mathbb{R}^m$ for some $m \ge n$ or by its Gram matrix), this algorithm transforms the vectors $\mathbf{b}_i$ so that when the algorithm terminates, the $\mathbf{b}_i$ form an LLL-reduced basis. In addition, the algorithm outputs a matrix H giving the coordinates of the LLL-reduced basis in terms of the initial basis. As usual we will denote by H_i the columns of H.

1. [Initialize] Set $k \leftarrow 2$, $k_{max} \leftarrow 1$, $\mathbf{b}_1^* \leftarrow \mathbf{b}_1$, $B_1 \leftarrow \mathbf{b}_1 \cdot \mathbf{b}_1$ and $H \leftarrow I_n$.

2. [Incremental Gram-Schmidt] If $k \le k_{max}$ go to step 3. Otherwise, set $k_{max} \leftarrow k$, $\mathbf{b}_k^* \leftarrow \mathbf{b}_k$, then for $j = 1, \ldots, k-1$ set $\mu_{k,j} \leftarrow \mathbf{b}_k \cdot \mathbf{b}_j^*/B_j$ and $\mathbf{b}_k^* \leftarrow \mathbf{b}_k^* - \mu_{k,j}\mathbf{b}_j^*$. Finally, set $B_k \leftarrow \mathbf{b}_k^* \cdot \mathbf{b}_k^*$ (see Remark (2) below for the corresponding step if only the Gram matrix of the $\mathbf{b}_i$ is given). If $B_k = 0$ output an error message saying that the $\mathbf{b}_i$ did not form a basis and terminate the algorithm.

3. [Test LLL condition] Execute Sub-algorithm RED($k, k-1$) below. If $B_k < (0.75 - \mu_{k,k-1}^2)B_{k-1}$, execute Sub-algorithm SWAP(k) below, set $k \leftarrow \max(2, k-1)$ and go to step 3. Otherwise, for $l = k-2, k-3, \ldots, 1$ execute Sub-algorithm RED(k, l), then set $k \leftarrow k+1$.

4. [Finished?] If $k \le n$, then go to step 2. Otherwise, output the LLL reduced basis $\mathbf{b}_i$, the transformation matrix $H \in GL_n(\mathbb{Z})$ and terminate the algorithm.

Sub-algorithm RED(k, l). If $|\mu_{k,l}| \le 0.5$ terminate the sub-algorithm. Otherwise, let q be the integer nearest to $\mu_{k,l}$, i.e.

$$q \leftarrow \lfloor \mu_{k,l} \rceil = \lfloor 0.5 + \mu_{k,l} \rfloor \ .$$

Set $\mathbf{b}_k \leftarrow \mathbf{b}_k - q\mathbf{b}_l$, $H_k \leftarrow H_k - qH_l$, $\mu_{k,l} \leftarrow \mu_{k,l} - q$, for all i such that $1 \leq i \leq l - 1$, set $\mu_{k,i} \leftarrow \mu_{k,i} - q\mu_{l,i}$ and terminate the sub-algorithm.

Sub-algorithm SWAP(k). Exchange the vectors $\mathbf{b}_k$ and $\mathbf{b}_{k-1}$, H_k and H_{k-1}, and if $k > 2$, for all j such that $1 \leq j \leq k - 2$ exchange $\mu_{k,j}$ with $\mu_{k-1,j}$. Then set (in this order) $\mu \leftarrow \mu_{k,k-1}$, $B \leftarrow B_k + \mu^2 B_{k-1}$, $\mu_{k,k-1} \leftarrow \mu B_{k-1}/B$, $\mathbf{b} \leftarrow \mathbf{b}_{k-1}^*$, $\mathbf{b}_{k-1}^* \leftarrow \mathbf{b}_k^* + \mu\mathbf{b}$, $\mathbf{b}_k^* \leftarrow -\mu_{k,k-1}\mathbf{b}_{k-1}^* + (B_k/B)\mathbf{b}$, $B_k \leftarrow B_{k-1}B_k/B$ and $B_{k-1} \leftarrow B$. Finally, for $i = k+1, k+2, \ldots, k_{max}$ set (in this order) $t \leftarrow \mu_{i,k}$, $\mu_{i,k} \leftarrow \mu_{i,k-1} - \mu t$, $\mu_{i,k-1} \leftarrow t + \mu_{k,k-1}\mu_{i,k}$ and terminate the sub-algorithm.

Proof. It is easy to show that at the beginning of step 4, the LLL conditions of Definition 2.6.1 are valid for $i \leq k - 1$. Hence, if $k > n$, we have indeed obtained an LLL-reduced family, and since it is clear that the operations which are performed on the $\mathbf{b}_i$ are of determinant ± 1, this family is a basis of L, hence the output of the algorithm is correct. What we must show is that the algorithm does in fact terminate.

If we set for $0 \leq i \leq n$

$$d_i = \det \left((\mathbf{b}_r \cdot \mathbf{b}_s)_{1 \leq r,s \leq i} \right) \ ,$$

we easily check that

$$d_i = \prod_{1 \leq j \leq i} B_j \ ,$$

where as usual $B_i = |\mathbf{b}_i^*|^2$, and in particular $d_i > 0$, and it is clear from this that $d_0 = 1$ and $d_n = d(L)^2$. Set

$$D = \prod_{1 \leq i \leq n-1} d_i \ .$$

This can change only if some B_i changes, and this can occur only in Sub-algorithm SWAP. In that sub-algorithm the d_i are unchanged for $i < k - 1$ and for $i \geq k$, and by the condition of step 3, d_{k-1} is multiplied by a factor at most equal to $3/4$. Hence D is also reduced by a factor at most equal to $3/4$. Let L_i be the lattice of dimension i generated by the $\mathbf{b}_j$ for $j \leq i$, and let s_i be the smallest non-zero value of the quadratic form q in L_i. Using Proposition 6.4.1 which we will give in Chapter 6, we obtain

$$d_i \geq s_i^i \gamma_i^{-i} \geq s_n^i \gamma_i^{-i} \ ,$$

and since s_n is the smallest non-zero value of $q(x)$ on L, this last expression depends only on i but not on the $\mathbf{b}_j$. It follows that d_i is bounded from below by a positive constant depending only on i and L. Hence D is bounded from below by a positive constant depending only on L, and this shows that the number of times that Sub-algorithm SWAP is executed must be finite.

Since this is the only place where k can decrease (after execution of the sub-algorithm) the algorithm must terminate, and this finishes the proof of its validity. $\square$

A more careful analysis shows that the running time of the LLL algorithm is at most $O(n^6 \ln^3 B)$, if $|b_i|^2 \leq B$ for all i. In practice however, this upper bound is quite pessimistic.

Remarks.

(1) If the matrix transformation H is not desired, one can suppress from the algorithm all the statements concerning it, since it does not play any real role.

(2) On the other hand if the b_i are given only by their Gram matrix, the b_i and b_i^* exist only abstractly. Hence, the only output of the algorithm is the matrix H, and the updating of the vectors b_i done in Sub-algorithms RED and SWAP must be done directly on the Gram matrix.

In particular, step 2 must then be replaced as follows (see Exercise 21).

2. [Incremental Gram-Schmidt] If $k \leq k_{\max}$ go to step 3. Otherwise, set $k_{\max} \leftarrow k$ then for $j = 1, \ldots, k-1$ set $a_{k,j} \leftarrow b_k \cdot b_j - \sum_{i=1}^{j-1} \mu_{j,i} a_{k,i}$ and $\mu_{k,j} \leftarrow a_{k,j}/B_j$, then set $B_k \leftarrow b_k \cdot b_k - \sum_{i=1}^{k-1} \mu_{k,i} a_{k,i}$. If $B_k = 0$ output an error message saying that the b_i did not form a basis and terminate the algorithm.

The auxiliary array $a_{k,j}$ is used to minimize the number of operations, otherwise we could of course write the formulas directly with $\mu_{k,j}$.

Asymptotically, this requires $n^3/6$ multiplications/divisions, and this is much faster than the $n^2 m/2$ required by Gram-Schmidt when only the coordinates of the b_i are known. Since the computation of the Gram matrix from the coordinates of the b_i also requires asymptotically $n^2 m/2$ multiplications, one should use directly the formulas of Algorithm 2.6.3 when the Gram matrix is not given.

(3) The constant 0.75 in step 3 of the algorithm can be replaced by any constant c such that $1/4 < c < 1$. Of course, this changes the estimates given by Theorem 2.6.2. (In the results and proof of the theorem, replace 2 by $\alpha = 1/(c - 1/4)$, and use the weaker inequality $\mu_{k,l}^2 \leq (\alpha - 1)/\alpha$.) The speed of the algorithm and the "quality" of the final basis which one obtains, are relatively insensitive to the value of the constant. In practice, one should perhaps use $c = 0.99$. The ideal value would be $c = 1$, but in this case one does not know whether the LLL algorithm runs in polynomial time, although in practice this seems to be the case.

(4) Another possibility, suggested by LaMacchia in [LaM] is to *vary* the constant c in the course of the algorithm, starting the reduction with a constant c slightly larger than 1/4 (so that the reduction is as fast as possible), and increasing it so as to reach $c = 0.99$ at the end of the reduction, so

that the quality of the reduced basis is a good as possible. We refer to [LaM] for details.

(5) We can also replace the LLL condition $B_k \geq (3/4 - \mu_{k,k-1}^2)B_{k-1}$ by the so-called Siegel condition $B_k \geq B_{k-1}/2$. Indeed, since $|\mu_{k,k-1}| \leq 1/2$, the LLL condition with the constant $c = 3/4$ implies the Siegel condition, and conversely the Siegel condition implies the LLL condition for the constant $c = 1/2$. In that case the preliminary reduction RED$(k, k-1)$ should be performed after the test, together with the other RED(k, l).

(6) If the Gram matrix does not necessarily have rational coefficients, the $\mu_{i,j}$ and B_i must be represented approximately using floating point arithmetic. Even if the Gram matrix is rational or even integral, it is often worthwhile to work using floating point arithmetic. The main problem with this approach is that roundoff errors may prevent the final basis from being LLL reduced. In many cases, this is not really important since the basis is not far from being LLL reduced. It may happen however that the roundoff errors cause catastrophic divergence from the LLL algorithm, and consequently give a basis which is very far from being reduced in any sense. Hence we must be careful. Let r be the number of relative precision bits.

First, during step 2 it is possible to replace the computation of the products $\mathbf{b}_i \cdot \mathbf{b}_j$ by floating point approximations (of course only in the case where the $\mathbf{b}_i$ are given by coordinates, otherwise there is nothing to compute). This should not be done if $\mathbf{b}_i$ and $\mathbf{b}_j$ are nearly orthogonal, i.e. if $\mathbf{b}_i \cdot \mathbf{b}_j/|\mathbf{b}_i||\mathbf{b}_j|$ is smaller than $2^{-r/2}$ say. In that case, $\mathbf{b}_i \cdot \mathbf{b}_j$ should be computed as exactly as possible using the given data.

Second, at the beginning of Sub-algorithm RED, the nearest integer q to $\mu_{k,l}$ is computed. If q is too large, say $q > 2^{r/2}$, then $\mu_{k,l} - q$ will have a small relative precision and the values of the $\mu_{k,l}$ will soon become incorrect. In that case, we should recompute the $\mu_{k,l}$, $\mu_{k-1,l}$, B_{k-1} and B_k directly from the Gram-Schmidt formulas, set $k \leftarrow \max(k-1, 2)$ and start again at step 3.

These modifications (and many more) are explained in a rigorous theoretical setting in [Schn], and for practical uses in [Schn-Euch] to which we refer.

(7) The algorithm assumes that the $\mathbf{b}_i$ are linearly independent. If they are not, we will get an error message in the Gram-Schmidt stage of the algorithm. It is possible to modify the algorithm so that it will not only work in this case, but in fact output a true basis and a set of linearly independent relations for the initial set of vectors (see Algorithm 2.6.8).

2.6.2 The LLL Algorithm with Deep Insertions

A modification of the LLL algorithm due to Schnorr and Euchner ([Schn-Euc]) is the following. It may be argued that the Lovász condition $B_k \geq (0.75 - \mu_{k,k-1}^2)B_{k-1}$ (in addition to the requirement $\mu_{k,j} \leq 1/2$) should be strengthened, taking into account the earlier B_j. If this is done rashly however, it leads to a non-polynomial time algorithm, both in theory and in practice. This is, of course, one of the reasons for the choice of a weaker condition. Schnorr and Euchner (loc. cit.) have observed however that one can strengthen the above condition without losing much on the practical speed of the algorithm, although in the worst case the resulting algorithm is no longer polynomial time. They report that in many cases, this leads to considerably shorter lattice vectors than the basic LLL algorithm.

The idea is as follows. If $\mathbf{b}_k$ is inserted between $\mathbf{b}_{i-1}$ and $\mathbf{b}_i$ for some $i < k$, then (Exercise 22) the new B_i will become

$$\mathbf{b}_k \cdot \mathbf{b}_k - \sum_{1 \le j < i} \mu_{k,j}^2 B_j = B_k + \sum_{i \le j < k} \mu_{k,j}^2 B_j .$$

If this is significantly smaller than the old B_i (say at most $\frac{3}{4} B_i$ as in our initial version of LLL), then it is reasonable to do this insertion. Note that the case $i = k - 1$ of this test is exactly the original LLL condition. For these tests to make sense, Algorithm RED(k, l) must be executed before the test for all $l < k$ and not only for $l = k - 1$ as in Algorithm 2.6.3.

Inserting $\mathbf{b}_k$ just after $\mathbf{b}_{i-1}$ for some $i < k$ will be called a *deep insertion*. After such an insertion, k must be set back to $\max(i - 1, 2)$, and the $\mu_{j,l}$ and B_j must be updated. When $i < k - 1$ however, the formulas become complicated and it is probably best to recompute the new Gram-Schmidt coefficients instead of updating them. One consequence of this is that we do not need to keep track of the largest value $k_{\max}$ that k has attained.

This leads to the following algorithm, due in essence to Schnorr and Euchner ([Schn-Euc]).

Algorithm 2.6.4 (LLL Algorithm with Deep Insertions). Given a basis $\mathbf{b}_1$, $\mathbf{b}_2, \ldots, \mathbf{b}_n$ of a lattice (L, q) (either by coordinates in the canonical basis of $\mathbb{R}^m$ for some $m \ge n$ or by its Gram matrix), this algorithm transforms the vectors $\mathbf{b}_i$ so that when the algorithm terminates, the $\mathbf{b}_i$ form an LLL-reduced basis. In addition, the algorithm outputs a matrix H giving the coordinates of the LLL-reduced basis in terms of the initial basis. As usual we will denote by H_i the columns H.

1. [Initialize] Set $k \leftarrow 1$ and $H \leftarrow I_n$.

2. [Incremental Gram-Schmidt] Set $\mathbf{b}_k^* \leftarrow \mathbf{b}_k$, then for $j = 1, \ldots, k - 1$ set $\mu_{k,j} \leftarrow \mathbf{b}_k \cdot \mathbf{b}_j^* / B_j$ and $\mathbf{b}_k^* \leftarrow \mathbf{b}_k^* - \mu_{k,j}\mathbf{b}_j^*$. Then set $B_k \leftarrow \mathbf{b}_k^* \cdot \mathbf{b}_k^*$. If $B_k = 0$ output an error message saying that the $\mathbf{b}_i$ did not form a basis and terminate the algorithm. Finally, if $k = 1$, set $k \leftarrow 2$ and go to step 5.

3. [Initialize test] For $l = k - 1, k - 2, \ldots, 1$ execute Sub-algorithm RED(k, l) above. Set $B \leftarrow \mathbf{b}_k \cdot \mathbf{b}_k$ and $i \leftarrow 1$.

4. [Deep LLL test] If $i = k$, set $k \leftarrow k + 1$ and go to step 5. Otherwise, do as follows. If $\frac{3}{4} B_i \le B$ set $B \leftarrow B - \mu_{k,i}^2 B_i$, $i \leftarrow i + 1$ and go to step 4.

Otherwise, execute Algorithm INSERT(k, i) below. If $i \geq 2$ set $k \leftarrow i - 1$, $B \leftarrow \mathbf{b}_k \cdot \mathbf{b}_k$, $i \leftarrow 1$ and go to step 4. If $i = 1$, set $k \leftarrow 1$ and go to step 2.

5. [Finished?] If $k \leq n$, then go to step 2. Otherwise, output the LLL reduced basis $\mathbf{b}_i$, the transformation matrix $H \in GL_n(\mathbb{Z})$ and terminate the algorithm.

Sub-algorithm INSERT(k, i). Set $\mathbf{b} \leftarrow \mathbf{b}_k$, $V \leftarrow H_k$, for $j = k, k-1, \ldots, i+1$ set $\mathbf{b}_j \leftarrow \mathbf{b}_{j-1}$ and $H_j \leftarrow H_{j-1}$, and finally set $\mathbf{b}_i \leftarrow \mathbf{b}$ and $H_i \leftarrow V$. Terminate the sub-algorithm.

2.6.3 The Integral LLL Algorithm

If the Gram matrix of the $\mathbf{b}_i$ has integral coefficients, the $\mu_{i,j}$ and the B_k will be rational and it may be tempting to do all the computation with rational numbers. Unfortunately, the repeated GCD computations necessary for performing rational arithmetic during the algorithm slows it down considerably. There are essentially two ways to overcome this problem. The first is to do only approximate computations of the $\mu_{i,j}$ and the B_i as mentioned above.

The second is as follows. In the proof of Algorithm 2.6.3 we have introduced quantities d_i which are clearly integral in our case, since they are equal to subdeterminants of our Gram matrix. We have the following integrality results.

Proposition 2.6.5. *Assume that the Gram matrix* $(\mathbf{b}_i \cdot \mathbf{b}_j)$ *is integral, and set*

$$d_i = \det((\mathbf{b}_r \cdot \mathbf{b}_s)_{1 \leq r, s \leq i}) = \prod_{1 \leq j \leq i} B_j .$$

Then for all i and for all $j < i$

(1)
$$d_{i-1} B_i \in \mathbb{Z} \qquad and \qquad d_j \mu_{i,j} \in \mathbb{Z} .$$

(2) *for all m such that $j < m \leq i$*

$$d_j \sum_{1 \leq k \leq j} \mu_{i,k} \mu_{m,k} B_k \in \mathbb{Z} .$$

Proof. We have seen above that $d_i = \prod_{1 \leq k \leq i} B_k$ hence $d_{i-1} B_i = d_i \in \mathbb{Z}$. For the second statement of (1), let $j < i$ and consider the vector

$$\mathbf{v} = \mathbf{b}_i - \sum_{1 \leq k \leq j} \mu_{i,k} \mathbf{b}_k^* = \mathbf{b}_i^* + \sum_{j < k < i} \mu_{i,k} \mathbf{b}_k^* .$$

From the second expression it is clear that $\mathbf{b}_k^* \cdot \mathbf{v} = 0$ for all k such that $1 \leq k \leq j$, or equivalently since the $\mathbb{R}$-span of the $\mathbf{b}_k^*$ $(1 \leq k \leq j)$ is equal to the $\mathbb{R}$-span of the $\mathbf{b}_k$,

$$\mathbf{b}_k \cdot \mathbf{v} = 0 \quad \text{for} \quad 1 \leq k \leq j .$$

For the same reason, we can write

$$\mathbf{v} = \mathbf{b}_i - \sum_{1 \leq k \leq j} x_k \mathbf{b}_k$$

for some $x_k \in \mathbb{R}$. Then the above equations can be written in matrix form

$$\begin{pmatrix} \mathbf{b}_1 \cdot \mathbf{b}_1 & \cdots & \mathbf{b}_1 \cdot \mathbf{b}_j \\ \vdots & \ddots & \vdots \\ \mathbf{b}_j \cdot \mathbf{b}_1 & \cdots & \mathbf{b}_j \cdot \mathbf{b}_j \end{pmatrix} \begin{pmatrix} x_1 \\ \vdots \\ x_j \end{pmatrix} = \begin{pmatrix} \mathbf{b}_i \cdot \mathbf{b}_1 \\ \vdots \\ \mathbf{b}_i \cdot \mathbf{b}_j \end{pmatrix} .$$

In particular, since the determinant of the matrix is equal by definition to d_j, by inverting the matrix we see that the x_k are of the form m_k/d_j for some $m_k \in \mathbb{Z}$ (since the Gram matrix is integral). Furthermore, the equality $\sum_{1 \leq k \leq j} x_k \mathbf{b}_k = \sum_{1 \leq k \leq j} \mu_{i,k} \mathbf{b}_k^*$ shows by projection on $\mathbf{b}_j^*$ that $x_j = \mu_{i,j}$, thus proving (1).

For (2) we note that by what we have proved, $d_j \mathbf{v}$ is an integral linear combination of the $\mathbf{b}_k$ (in other words it belongs to the lattice), hence in particular $d_j \mathbf{v} \cdot \mathbf{b}_m \in \mathbb{Z}$ for all m such that $1 \leq m \leq n$. Since $\mathbf{v} = \mathbf{b}_i - \sum_{1 \leq k \leq j} \mu_{i,k} \mathbf{b}_k^*$, we obtain (2). $\qquad\square$

Corollary 2.6.6. *With the same hypotheses and notations as the proposition, set $\lambda_{i,j} = d_j \mu_{i,j}$ for $j < i$ (so $\lambda_{i,j} \in \mathbb{Z}$) and $\lambda_{i,i} = d_i$. Then for $j \leq i$ fixed, if we define the sequence u_k by $u_0 = \mathbf{b}_i \cdot \mathbf{b}_j$ and for $1 \leq k < j$*

$$u_k = \frac{d_k u_{k-1} - \lambda_{i,k} \lambda_{j,k}}{d_{k-1}} ,$$

then $u_k \in \mathbb{Z}$ and $u_{j-1} = \lambda_{i,j}$.

Proof. It is easy to check by induction on k that

$$u_k = d_k \left(\mathbf{b}_i \cdot \mathbf{b}_j - \sum_{1 \leq l \leq k} \frac{\lambda_{i,l} \lambda_{j,l}}{d_l d_{l-1}} \right) = d_k \left(\mathbf{b}_i \cdot \mathbf{b}_j - \sum_{1 \leq l \leq k} \mu_{i,l} \mu_{j,l} B_l \right)$$

and the proposition shows that this last expression is integral. We also have $u_{j-1} = B_j d_{j-1} \mu_{i,j} = d_j \mu_{i,j} = \lambda_{i,j}$ thus proving the corollary. $\qquad\square$

Using these results, it is easy to modify Algorithm 2.6.3 so as to work entirely with integers. This leads to the following algorithm, where it is assumed that the basis is given by its Gram-Schmidt matrix. (Hence, if the basis is given in terms of coordinates, compute first the Gram-Schmidt matrix before applying the algorithm, or modify appropriately the formulas of step 1.) Essentially the same algorithm is given in [de Weg].

Algorithm 2.6.7 (Integral LLL Algorithm).   Given a basis $b_1, b_2, \ldots, b_n$ of a lattice (L, q) by its Gram matrix which is assumed to have integral coefficients, this algorithm transforms the vectors b_i so that when the algorithm terminates, the b_i form an LLL-reduced basis. The algorithm outputs a matrix H giving the coordinates of the LLL-reduced basis in terms of the initial basis. We will denote by H_i the column vectors of H. All computations are done using integers only.

1. [Initialize] Set $k \leftarrow 2$, $k_{max} \leftarrow 1$, $d_0 \leftarrow 1$, $d_1 \leftarrow b_1 \cdot b_1$ and $H \leftarrow I_n$.

2. [Incremental Gram-Schmidt] If $k \leq k_{max}$ go to step 3. Otherwise, set $k_{max} \leftarrow k$ and for $j = 1, \ldots, k$ (in that order) do as follows: set $u \leftarrow b_k \cdot b_j$ and for $i = 1, \ldots, j - 1$ set

$$u \leftarrow \frac{d_i u - \lambda_{k,i} \lambda_{j,i}}{d_{i-1}}$$

   (the result is in $\mathbb{Z}$), then if $j < k$ set $\lambda_{k,j} \leftarrow u$ and if $j = k$ set $d_k \leftarrow u$. If $d_k = 0$, the b_i did not form a basis, hence output an error message and terminate the algorithm (but see also Algorithm 2.6.8).

3. [Test LLL condition] Execute Sub-algorithm REDI($k, k-1$) below. If $d_k d_{k-2} < \frac{3}{4} d_{k-1}^2 - \lambda_{k,k-1}^2$, execute algorithm SWAPI(k) below, set $k \leftarrow \max(2, k - 1)$ and go to step 3. Otherwise, for $l = k - 2, k - 3, \ldots, 1$ execute Sub-algorithm REDI(k, l), then set $k \leftarrow k + 1$.

4. [Finished?] If $k \leq n$ go to step 2. Otherwise, output the transformation matrix $H \in GL_n(\mathbb{Z})$ and terminate the algorithm.

Sub-algorithm REDI(k, l).   If $|2\lambda_{k,l}| \leq d_l$ terminate the sub-algorithm. Otherwise, let q be the integer nearest to $\lambda_{k,l}/d_l$, i.e. the quotient of the Euclidean division of $2\lambda_{k,l} + d_l$ by $2d_l$. Set $H_k \leftarrow H_k - qH_l$, $b_k \leftarrow b_k - qb_l$, $\lambda_{k,l} \leftarrow \lambda_{k,l} - qd_l$, for all i such that $1 \leq i \leq l - 1$ set $\lambda_{k,i} \leftarrow \lambda_{k,i} - q\lambda_{l,i}$ and terminate the sub-algorithm.

Sub-algorithm SWAPI(k).   Exchange the vectors H_k and H_{k-1}, exchange b_k and b_{k-1}, and if $k > 2$, for all j such that $1 \leq j \leq k - 2$ exchange $\lambda_{k,j}$ with $\lambda_{k-1,j}$. Then set $\lambda \leftarrow \lambda_{k,k-1}$, $B \leftarrow (d_{k-2}d_k + \lambda^2)/d_{k-1}$, then for $i = k + 1, k + 2, \ldots k_{max}$ set (in this order) $t \leftarrow \lambda_{i,k}$, $\lambda_{i,k} \leftarrow (d_k\lambda_{i,k-1} - \lambda t)/d_{k-1}$ and $\lambda_{i,k-1} \leftarrow (Bt + \lambda\lambda_{i,k})/d_k$. Finally, set $d_{k-1} \leftarrow B$ and terminate the sub-algorithm.

It is an easy exercise (Exercise 24) to check that these formulas correspond exactly to the formulas of Algorithm 2.6.3.

Remark. In step 3, the fundamental LLL comparison $d_k d_{k-2} < \frac{3}{4} d_{k-1}^2 - \lambda_{k,k-1}^2$ involves the non-integral number $\frac{3}{4}$ (it could also be 0.99). This is not really a problem since this comparison can be done any way one likes (by multiplying by 4, or using floating point arithmetic), since a roundoff error at that point is totally unimportant.

2.6.4 LLL Algorithms for Linearly Dependent Vectors

As has been said above, the LLL algorithm cannot be applied directly to a system of linearly dependent vectors b_i. It can however be modified so as to work in this case, and to output a basis and a system of relations. The problem is that in the Gram-Schmidt orthogonalization procedure we will have at some point $B_i = b_i^* \cdot b_i^* = 0$. This means of course that b_i is equal to a linear combination of the b_j for $j < i$. Since Gram-Schmidt performs projections of the successive vectors on the subspace generated by the preceding ones, this means that we can forget the index i in the rest of the orthogonalization (although not the vector b_i itself). This leads to the following algorithm which is very close to Algorithm 2.6.3 and whose proof is left to the reader.

Algorithm 2.6.8 (LLL Algorithm on Not Necessarily Independent Vectors). Given n non-zero vectors b_1, b_2, ..., b_n generating a lattice (L, q) (either by coordinates or by their Gram matrix), this algorithm transforms the vectors b_i and computes the rank p of the lattice L so that when the algorithm terminates $b_i = 0$ for $1 \le i \le n - p$ and the b_i for $n - p < i \le n$ form an LLL-reduced basis of L. In addition, the algorithm outputs a matrix H giving the coordinates of the new b_i in terms of the initial ones. In particular, the first $n - p$ columns H_i of H will be a basis of relation vectors for the b_i, i.e. of vectors r such that $\sum_{1 \le i \le n} r_i b_i = 0$.

1. [Initialize] Set $k \leftarrow 2$, $k_{max} \leftarrow 1$, $b_1^* \leftarrow b_1$, $B_1 \leftarrow b_1 \cdot b_1$ and $H \leftarrow I_n$.

2. [Incremental Gram-Schmidt] If $k \le k_{max}$ go to step 3. Otherwise, set $k_{max} \leftarrow k$ and for $j = 1, \ldots, k - 1$ set $\mu_{k,j} \leftarrow b_k \cdot b_j^*/B_j$ if $B_j \neq 0$ and $\mu_{k,j} \leftarrow 0$ if $B_j = 0$, then set $b_k^* \leftarrow b_k - \sum_{j=1}^{k-1} \mu_{k,j} b_j^*$ and $B_k \leftarrow b_k^* \cdot b_k^*$ (use the formulas given in Remark (2) above if the b_i are given by their Gram matrix).

3. [Test LLL condition] Execute Sub-algorithm RED$(k, k - 1)$ above. If $B_k < (0.75 - \mu_{k,k-1}^2)B_{k-1}$, execute Sub-algorithm SWAPG(k) below, set $k \leftarrow \max(2, k - 1)$ and go to step 3. Otherwise, for $l = k - 2, k - 3, \ldots, 1$ execute Sub-algorithm RED(k, l), then set $k \leftarrow k + 1$.

4. [Finished?] If $k \le n$ go to step 2. Otherwise, let r be the number of initial vectors b_i which are equal to zero, output $p \leftarrow n - r$, the vectors b_i for $r + 1 \le i \le n$ (which form an LLL-reduced basis of L), the transformation matrix $H \in GL_n(\mathbb{Z})$ and terminate the algorithm.

Sub-algorithm SWAPG(k). Exchange the vectors b_k and b_{k-1}, H_k and H_{k-1}, and if $k > 2$, for all j such that $1 \le j \le k - 2$ exchange $\mu_{k,j}$ with $\mu_{k-1,j}$. Then set $\mu \leftarrow \mu_{k,k-1}$ and $B \leftarrow B_k + \mu^2 B_{k-1}$. Now, in the case $B = 0$ (i.e. $B_k = \mu = 0$), exchange B_k and B_{k-1}, exchange b_k^* and b_{k-1}^* and for $i = k + 1, k + 2, \ldots k_{max}$ exchange $\mu_{i,k}$ and $\mu_{i,k-1}$.

In the case $B_k = 0$ and $\mu \neq 0$, set $B_{k-1} \leftarrow B$, $b_{k-1}^* \leftarrow \mu b_{k-1}^*$, $\mu_{k,k-1} \leftarrow 1/\mu$ and for $i = k + 1, k + 2, \ldots, k_{max}$ set $\mu_{i,k-1} \leftarrow \mu_{i,k-1}/\mu$.

Finally, in the case $B_k \neq 0$, set (in this order) $t \leftarrow B_{k-1}/B$, $\mu_{k,k-1} \leftarrow \mu t$,

$\mathbf{b} \leftarrow \mathbf{b}_{k-1}^*$, $\mathbf{b}_{k-1}^* \leftarrow \mathbf{b}_k^* + \mu\mathbf{b}$, $\mathbf{b}_k^* \leftarrow -\mu_{k,k-1}\mathbf{b}_k^* + (B_k/B)\mathbf{b}$, $B_k \leftarrow B_k t$, $B_{k-1} \leftarrow B$, then for $i = k+1, k+2, \ldots, k_{\max}$ set (in this order) $t \leftarrow \mu_{i,k}$, $\mu_{i,k} \leftarrow \mu_{i,k-1} - \mu t$, $\mu_{i,k-1} \leftarrow t + \mu_{k,k-1}\mu_{i,k}$. Terminate the sub-algorithm.

Note that in this sub-algorithm, in the case $B = 0$, we have $B_k = 0$ and hence $\mu_{i,k} = 0$ for $i > k$, so the exchanges are equivalent to setting $B_k \leftarrow B_{k-1}$, $B_{k-1} \leftarrow 0$ and for $i \geq k+1$, $\mu_{i,k} \leftarrow \mu_{i,k-1}$ and $\mu_{i,k-1} \leftarrow 0$.

An important point must be made concerning this algorithm. Since several steps of the algorithm test whether some quantity is equal to zero or not, it can be applied only to vectors with exact (i.e. rational) entries. Indeed, for vectors with non-exact entries, the notion of relation vector is itself not completely precise since some degree of approximation must be given in advance. Thus the reader is advised to use caution when using LLL algorithms for linearly dependent vectors when they are non-exact. (For instance, we could replace a test $B_k = 0$ by $B_k \leq \varepsilon$ for a suitable ε.)

We must prove that this algorithm is valid. To show that it terminates, we use a similar quantity to the one used in the proof of the validity of Algorithm 2.6.3. We set

$$d_k = \prod_{i \leq k, B_i \neq 0} B_i \quad \text{and} \quad D = \prod_{k \leq n, B_k \neq 0} d_k \prod_{k \leq n, B_k = 0} 2^k \; .$$

This quantity is modified only in Sub-algorithm SWAPG(k). If $B = B_k + \mu^2 B_{k-1} \neq 0$, then d_{k-1} is multiplied by a factor which is smaller than $3/4$ and the others are unchanged, hence D decreases by a factor at least $3/4$ as in the usual LLL algorithm. If $B = 0$, then B_{k-1} becomes 0 and B_k becomes equal to B_{k-1}, hence d_{k-1} becomes equal to d_{k-2}, d_k stays the same (since $B_{k-1}d_{k-2} = d_{k-1} = d_k$ when $B_k = 0$) as well as the others, so D is multiplied by $2^{k-1}/2^k = 1/2$ hence decreases multiplicatively again, thus showing that the algorithm terminates since D is bounded from below.

When the algorithm terminates, we have for all i, j and k the conditions $B_k \geq (3/4 - \mu_{k,k-1}^2)B_{k-1}$ and $|\mu_{i,j}| \leq 1/2$. If p is the rank of the lattice L, it follows that $n - p$ of the B_i must be equal to zero, and these inequalities show that it must be the *first* $n - p$ B_i, since $B_i = 0$ implies $B_j = 0$ for $j < i$. Since the vector space generated by the $\mathbf{b}_i^*$ for $i \leq n - p$ is the same as the space generated by the $\mathbf{b}_i$ for $i \leq n - p$, it follows that $\mathbf{b}_i = 0$ for $i \leq n - p$. Since the $\mathbf{b}_i$ form a generating set for L over $\mathbb{Z}$ throughout the algorithm, the $\mathbf{b}_i$ for $i > n - p$ also generate L, hence they form a basis since there are exactly p of them, and this basis is LLL reduced by construction. It also follows from the vanishing of the $\mathbf{b}_i$ for $i \leq n - p$ that the first $n - p$ columns H_i of H are relation vectors for our initial $\mathbf{b}_i$. Since H is an integer matrix with determinant ± 1, it is an easy exercise to see that these columns form a basis of the space of relation vectors for the initial $\mathbf{b}_i$ (Exercise 25). $\square$

This algorithm is essentially due to M. Pohst and called by him the MLLL algorithm (for Modified LLL, see [Poh2]).

We leave as an excellent exercise for the reader to write an all-integer version of Algorithm 2.6.8 when the Gram matrix is integral (see Exercise 26).

Summary. We have seen a number of modifications and variations on the basic LLL Algorithm 2.6.3. Most of these can be combined. We summarize them here.

(1) The Gram-Schmidt formulas of step 2 can be modified to use only the Gram matrix of the b_i (see Remark (2) after Algorithm 2.6.3).
(2) If the Gram-Schmidt matrix is integral, the computation can be done entirely with integers (see Algorithm 2.6.7).
(3) If floating point computations are used, care must be taken during the computation of the $b_i \cdot b_j$ and when the nearest integer to a $\mu_{k,l}$ is computed (see Remark (4) after Algorithm 2.6.3).
(4) If we want better quality vectors than those output by the LLL algorithm, we can use deep insertion to improve the output (see Algorithm 2.6.4).
(5) If the vectors b_i are not linearly independent, we must use Algorithm 2.6.8, combined if desired with any of the preceding variations.

2.7 Applications of the LLL Algorithm

2.7.1 Computing the Integer Kernel and Image of a Matrix

In Section 2.4.3 we have seen how to apply the Hermite normal form algorithms to the computation of the image and kernel of an integer matrix A. It is clear that this can also be done using the MLLL algorithm (in fact its integer version, see Exercise 26). Indeed if we set b_j to be the columns of A, the vectors b_i output by Algorithm 2.6.8 form an LLL-reduced basis of the image of A and the relation vectors H_i for $i \le r = n - p$ form a basis of the integer kernel of A. If desired, the result given by Algorithm 2.6.8 can be improved in two ways. First, the relation vectors H_i for $i \le r$ are not LLL-reduced, so it is useful to LLL-reduce them to obtain small relations. This means that we will multiply the first r column of H on the right by an $r \times r$ invertible matrix over $\mathbb{Z}$, and this of course leaves H unimodular.

Second, although the basis b_i for $r < i \le n$ is already an LLL-reduced basis for the image of A hence cannot be improved much, the last p columns of H (which express the LLL-reduced b_i in terms of the initial b_i) can be large and in many situations it is desirable to reduce their size. Here we must not LLL-reduce these columns since the corresponding image vectors b_i would not be anymore LLL-reduced in general. (This is of course a special case of the important but difficult problem of simultaneously reducing a lattice basis and its dual, see [Sey2].) We still have some freedom however since we can replace any column H_i for $i > r$ by

$$H_i - \sum_{j \leq r} m_j H_j$$

for any $m_j \in \mathbb{Z}$ since this will not change the b_i and will preserve the relation $\det(H) = \pm 1$. To choose the m_j close to optimally we proceed as follows. Let C be the Gram matrix of the vectors H_j for $j \leq r$. Using Algorithm 2.2.1 compute $X = (x_1, \ldots, x_r)^t$ solution to the linear system $CX = V_i$, where V_i is the column vector whose j-th element is equal to $H_i \cdot H_j$ (here the scalar product is the usual one). Then by elementary geometric arguments it is clear that the vector $\sum_{j \leq r} x_j H_j$ is the projection of H_i on the real vector space generated by the H_j for $j \leq r$, hence a close to optimal choice of the m_j is to choose $m_j = \lfloor x_j \rceil$. Since we have several linear systems to solve using the same matrix, it is preferable to invert the matrix using Algorithm 2.2.2 and this gives the following algorithm.

Algorithm 2.7.1 (Kernel and Image of a Matrix Using LLL). Given an $m \times n$ matrix A with integral entries, this algorithm computes an $n \times n$ matrix H and a number p with the following properties. The matrix H has integral entries and is of determinant equal to ± 1 (i.e. $H \in \mathrm{GL}_n(\mathbb{Z})$). The first $n - p$ columns of H form an LLL-reduced basis of the integer kernel of A. The product of A with the last p columns of H give an LLL-reduced basis of the image of A, and the coefficients of these last p columns are small.

1. [Apply MLLL] Perform Algorithm 2.6.8 on the vectors b_i equal to the columns of A, the Euclidean scalar product being the usual scalar product on vectors. We thus obtain p and a matrix $H \in \mathrm{GL}_n(\mathbb{Z})$. Set $r \leftarrow n - p$.

2. [LLL-reduce the kernel] Using the integral LLL-Algorithm 2.6.7, replace the first r vectors of H by an LLL-reduced basis of the lattice that they generate.

3. [Compute inverse of Gram matrix] Let C be the Gram matrix of the H_j for $j \leq r$ (i.e. $C_{j,k} = H_j \cdot H_k$ for $1 \leq j, k \leq r$), set $D \leftarrow C^{-1}$ computed using Algorithm 2.2.2, and set $i \leftarrow r$.

4. [Finished?] Set $i \leftarrow i + 1$. If $i > n$, output the matrix H and the number p and terminate the algorithm.

5. [Modify H_i] Let V be the r-dimensional column vector whose j-th coordinate is $H_i \cdot H_j$. Set $X \leftarrow DV$, and for $j \leq r$ set $m_j \leftarrow \lfloor x_j \rceil$, where x_j is the j-th component of X. Finally, set $H_i \leftarrow H_i - \sum_{1 \leq j \leq r} m_j H_j$ and go to step 4.

A practical implementation of this algorithm should use only an all-integer version of Algorithm 2.6.8 (see Exercise 26), and the other steps can be similarly modified so that all the computations are done with integers only.

If only the integer kernel of A is wanted, we may replace the test $B_k < (0.75 - \mu_{k,k-1}^2) B_{k-1}$ by $B_k = 0$, which avoids most of the swaps and gives a much faster algorithm. Since this algorithm is very useful, we give explicitly the complete integer version.

Algorithm 2.7.2 (Kernel over $\mathbb{Z}$ Using LLL).   Given an $m \times n$ matrix A with integral entries, this algorithm finds an LLL-reduced $\mathbb{Z}$-basis for the kernel of A. We use an auxiliary $n \times n$ integral matrix H. We denote by H_j the j-th column of H and (to keep notations similar to the other LLL algorithms) by $\mathbf{b}_j$ the j-th column of A. All computations are done using integers only. We use an auxiliary set of flags $f_1, \ldots, f_n$ (which will be such that $f_k = 0$ if and only if $B_k = 0$).

1. [Initialize] Set $k \leftarrow 2$, $k_{\max} \leftarrow 1$, $d_0 \leftarrow 1$, $t \leftarrow \mathbf{b}_1 \cdot \mathbf{b}_1$ and $H \leftarrow I_n$. If $t \neq 0$ set $d_1 \leftarrow t$ and $f_1 \leftarrow 1$, otherwise set $d_1 \leftarrow 1$ and $f_1 \leftarrow 0$.

2. [Incremental Gram-Schmidt] If $k \leq k_{\max}$ go to step 3. Otherwise, set $k_{\max} \leftarrow k$ and for $j = 1, \ldots, k$ (in that order) do as follows. If $f_j = 0$ and $j < k$, set $\lambda_{k,j} \leftarrow 0$. Otherwise, set $u \leftarrow \mathbf{b}_k \cdot \mathbf{b}_j$ and for each $i = 1, \ldots, j-1$ (in that order) such that $f_i \neq 0$ set

$$u \leftarrow \frac{d_i u - \lambda_{k,i}\lambda_{j,i}}{d_{i-1}}$$

(the result is in $\mathbb{Z}$), then, if $j < k$ set $\lambda_{k,j} \leftarrow u$ and if $j = k$ set $d_k \leftarrow u$ and $f_k \leftarrow 1$ if $u \neq 0$, $d_k \leftarrow d_{k-1}$ and $f_k \leftarrow 0$ if $u = 0$.

3. [Test $f_k = 0$ and $f_{k-1} \neq 0$] If $f_{k-1} \neq 0$, execute Sub-algorithm REDI$(k, k-1)$ above. If $f_{k-1} \neq 0$ and $f_k = 0$, execute Sub-algorithm SWAPK(k) below, set $k \leftarrow \max(2, k-1)$ and go to step 3. Otherwise, for each $l = k-2, k-3, \ldots, 1$ (in this order) such that $f_l \neq 0$, execute Sub-algorithm REDI(k, l) above, then set $k \leftarrow k + 1$.

4. [Finished?] If $k \leq n$ go to step 2. Otherwise, let $r + 1$ be the least index such that $f_i \neq 0$ ($r = n$ if all f_i are equal to 0). Using Algorithm 2.6.7, output an LLL-reduced basis of the lattice generated by the linearly independent vectors $H_1, \ldots, H_r$ and terminate the algorithm.

Sub-algorithm SWAPK(k).   Exchange the vectors H_k and H_{k-1}, and if $k > 2$, for all j such that $1 \leq j \leq k - 2$ exchange $\lambda_{k,j}$ with $\lambda_{k-1,j}$. Set $\lambda \leftarrow \lambda_{k,k-1}$. If $\lambda = 0$, set $d_{k-1} \leftarrow d_{k-2}$, exchange f_{k-1} and f_k (i.e. set $f_{k-1} \leftarrow 0$ and $f_k \leftarrow 1$), set $\lambda_{k,k-1} \leftarrow 0$ and for $i = k + 1, \ldots, k_{\max}$ set $\lambda_{i,k} \leftarrow \lambda_{i,k-1}$ and $\lambda_{i,k-1} \leftarrow 0$.

   If $\lambda \neq 0$, for $i = k + 1, \ldots, k_{\max}$ set $\lambda_{i,k-1} \leftarrow \lambda\lambda_{i,k-1}/d_{k-1}$, then set $t \leftarrow d_k$, $d_{k-1} \leftarrow \lambda^2/d_{k-1}$, $d_k \leftarrow d_{k-1}$ then for $j = k + 1, \ldots, k_{\max} - 1$ and for $i = j + 1, \ldots, k_{\max}$ set $\lambda_{i,j} \leftarrow \lambda_{i,j}d_{k-1}/t$ and finally for $j = k + 1, \ldots, k_{\max}$ set $d_j \leftarrow d_j d_{k-1}/t$. Terminate the sub-algorithm.

Remarks.

(1) Since $f_i = 0$ implies $\lambda_{k,i} = 0$, time can be saved in a few places by first testing whether f_i vanishes. The proof of the validity of this algorithm is left as an exercise (Exercise 24).

(2) It is an easy exercise to show that in this algorithm

$$d_k = \det\left((\mathbf{b}_i \cdot \mathbf{b}_j)_{1 \leq i, j \leq k, B_i B_j \neq 0}\right)$$

and that $d_j \mu_{i,j} \in \mathbb{Z}$ (see Exercise 29).

(3) An annoying aspect of Algorithm SWAPK is that when $\lambda \neq 0$, in addition to the usual updating, we must also update the quantities d_j and $\lambda_{i,j}$ for all i and j such that $k+1 \leq j < i \leq k_{max}$. This comes from the single fact that the new value of d_k is different from the old one, and suggests that a suitable modification of the definition of d_k can suppress this additional updating. This is indeed the case (see Exercise 30). Unfortunately, with this modification, it is the reduction algorithm REDI which needs much additional updating. I do not see how to suppress the extra updating in SWAPK and in REDI simultaneously.

2.7.2 Linear and Algebraic Dependence Using LLL

Now let us see how to apply the LLL algorithm to the problem of $\mathbb{Z}$-linear independence. Let $z_1, z_2, \ldots, z_n$ be n complex numbers, and the problem is to find a $\mathbb{Z}$-dependence relation between them, if one exists. Assume first that the z_i are real. For a large number N, consider the positive definite quadratic form in the a_i:

$$Q(\mathbf{a}) = a_2^2 + a_3^2 + \cdots + a_n^2 + N(z_1 a_1 + z_2 a_2 + \cdots + z_n a_n)^2 \ .$$

This form is represented as a sum of n squares of linearly independent linear forms in the a_i, hence defines a Euclidean scalar product on $\mathbb{R}^n$, as long as $z_1 \neq 0$, which we can of course assume. If N is large, a "short" vector of $\mathbb{Z}^n$ for this form will necessarily be such that $|z_1 a_1 + \cdots + z_n a_n|$ is small, and also the a_i for $i > 1$ not too large. Hence, if the z_i are really $\mathbb{Z}$-linearly dependent, by choosing a suitable constant N the dependence relation (which will make $z_1 a_1 + \cdots + z_n a_n$ equal to 0 up to roundoff errors) will be discovered. The choice of the constant N is subtle, and depends in part on what one knows about the problem. If the $|z_i|$ are not too far from 1 (meaning between 10^{-6} and 10^6, say), and are known with an absolute (or relative) precision ϵ, then one should take N between $1/\epsilon$ and $1/\epsilon^2$, but ϵ should also be taken quite small: if one expects the coefficients a_i to be of the order of a, then one might take $\epsilon = a^{-1.5n}$, but in any case $\epsilon < a^{-n}$.

Hence, we will start with the $\mathbf{b}_i$ being the standard basis of $\mathbb{Z}^n$, and use LLL with the quadratic form above. One nice thing is that step 2 of the LLL algorithm can be avoided completely. Indeed, one has the following lemma.

Lemma 2.7.3. *With the above notations, if we execute the complete Gram-Schmidt orthogonalization procedure on the standard basis of $\mathbb{Z}^n$ and the quadratic form*

$$Q(\mathbf{a}) = a_2^2 + a_3^2 + \cdots + a_n^2 + N(z_1 a_1 + z_2 a_2 + \cdots + z_n a_n)^2$$

we have $\mu_{i,1} = z_i/z_1$ *for* $2 \leq i \leq n$, $\mu_{i,j} = 0$ *if* $2 \leq j < i \leq n$, $\mathbf{b}_i^* = \mathbf{b}_i - (z_i/z_1)\mathbf{b}_1$, $B_1 = N z_1^2$, *and* $B_k = 1$ *for* $2 \leq k \leq n$.

The proof is trivial by induction.

It is easy to modify these ideas to obtain an algorithm which also works for complex numbers z_i. In this case, the quadratic form that we can take is

$$Q(\mathbf{a}) = a_3^2 + \cdots + a_n^2 + N|z_1 a_1 + z_2 a_2 + \cdots + z_n a_n|^2 ,$$

since the expression which multiplies N is now a sum of *two* squares of linear forms, and these forms will be independent if and only if z_1/z_2 is not real. We can however always satisfy this condition by a suitable reordering: if there exists i and j such that $z_i/z_j \notin \mathbb{R}$, then by applying a suitable permutation of the z_i, we may assume that $z_1/z_2 \notin \mathbb{R}$. On the other hand, if $z_i/z_j \in \mathbb{R}$ for all i and j, then we can apply the algorithm to the real numbers $1, z_2/z_1, \ldots, z_n/z_1$.

All this leads to the following algorithm.

Algorithm 2.7.4 (Linear Dependence). Given n complex numbers $z_1, \ldots, z_n$, (as approximations), a large number N chosen as explained above, this algorithm finds $\mathbb{Z}$-linear combinations of small modulus between the z_i. We assume that all the z_i are non-zero, and that if one of the ratios z_i/z_j is not real, the z_i are reordered so that the ratio z_2/z_1 is not real.

1. [Initialize] Set $\mathbf{b}_i \leftarrow [0, \ldots, 1, \ldots, 0]^t$, i.e. as a column vector the ith element of the standard basis of $\mathbb{Z}^n$. Then, set $\mu_{i,j} \leftarrow 0$ for all i and j with $3 \leq j < i \leq n$, $B_1 \leftarrow |z_1|^2$, $B_2 \leftarrow \text{Im}(z_1 \bar{z}_2)$, $B_k \leftarrow 1$ for $3 \leq k \leq n$, $\mu_{i,1} \leftarrow \text{Re}(z_1 \bar{z}_i)/B_1$ for $2 \leq i \leq n$.

   Now if $B_2 \neq 0$ (i.e. if we are in the complex case), do the following: set $\mu_{i,2} \leftarrow \text{Im}(z_1 \bar{z}_i)/B_2$ for $3 \leq i \leq n$, $B_2 \leftarrow N \cdot B_2^2/B_1$. Otherwise (in the real case), set $\mu_{i,2} \leftarrow 0$ for $3 \leq i \leq n$, $B_2 \leftarrow 1$.

2. [Execute LLL] Set $B_1 \leftarrow N B_1$, $k \leftarrow 2$, $k_{\max} \leftarrow n$, $H \leftarrow I_n$ and go to step 3 of the LLL Algorithm 2.6.3.

3. [Terminate] Output the coefficients $\mathbf{b}_i$ as coefficients of linear combinations of the z_i with small modulus, the best one being probably $\mathbf{b}_1$.

Implementation advice. Algorithm 2.7.4 performs slightly better if z_1 is the number with the largest modulus. Hence one should try to reorder the z_i so that this is the case. (Note that it may not be possible to do so, since if the z_i are not all real, one must have z_2/z_1 non-real.)

Remarks.

(1) The reason why the first component plays a special role comes from the choice of the quadratic form. To be more symmetrical, one could choose instead

$$Q(\mathbf{a}) = a_1^2 + a_2^2 + a_3^2 + \cdots + a_n^2 + N|z_1 a_1 + z_2 a_2 + \cdots + z_n a_n|^2$$

both in the real and complex case. The result would be more symmetrical in the variables a_i, but then we cannot avoid executing step 2 of the LLL algorithm, i.e. the Gram-Schmidt reduction procedure, which in practice can take a non-negligible proportion of the running time. Hence the above non-symmetric version (due to W. Neumann) is probably better.

(2) We can express the linear dependence algorithm in terms of matrices instead of quadratic forms as follows (for simplicity we use the symmetrical version and we assume the z_i real). Set $S = \sqrt{N}$. We must then find the LLL reduction of the following $(n+1) \times n$ matrix:

$$\begin{pmatrix} 1 & 0 & \cdots & 0 \\ 0 & \ddots & \ddots & \vdots \\ \vdots & \ddots & \ddots & 0 \\ 0 & \cdots & 0 & 1 \\ Sz_1 & Sz_2 & \cdots & Sz_n \end{pmatrix} .$$

(3) We have not used at all the multiplicative structure of the field $\mathbb{C}$. This means that essentially the same algorithm can be used to find linear dependencies between elements of a k-dimensional vector space over $\mathbb{R}$ for any k. This essentially reduces to the MLLL algorithm, except that thanks to the number N we can better handle imprecise vectors.

(4) A different method for finding linear dependence relations based on an algorithm which is a little different from the LLL algorithm, is explained and analyzed in detail in [HJLS]. It is not clear which should be preferred.

A special case of Algorithm 2.7.4 is when $z_i = \alpha^{i-1}$, where α is a given complex number. Then finding a $\mathbb{Z}$-linear relation between the z_i is equivalent to finding a polynomial $A \in \mathbb{Z}[X]$ such that $A(\alpha) = 0$, i.e. an algebraic relation for α. This is very useful in practice. (From the implementation advice given above we should choose $z_i = \alpha^{n-i}$ instead if $\alpha > 1$.)

In this case however, some modifications may be useful. First note that Lemma 2.7.3 stays essentially the same if we replace the quadratic form $Q(\mathbf{a})$ by

$$Q(\mathbf{a}) = \lambda_2 a_2^2 + \lambda_3 a_3^2 + \cdots + \lambda_n a_n^2 + N|z_1 a_1 + z_2 a_2 + \cdots + z_n a_n|^2$$

where the λ_i are arbitrary positive real numbers (see Exercise 32). Now when testing for algebraic relations, we may or may not know in advance the degree of the relation. Assume that we do. (For example, if $\alpha = \sqrt{2} + \sqrt{3} + \sqrt{5}$ we know that the relation will be of degree 8.) Then (choosing $z_i = \alpha^{n-i}$) we would like to have small coefficients for α^{n-i} with i small, and allow larger ones for i large. This amounts to choosing λ_i large for small i, and small for large i. One choice could be $\lambda_i = A^{n-i}$ for some reasonable constant $A > 1$ (at least such that A^n is much smaller than N). In other words, we look for an algebraic relation for z_i/A.

In other situations, we do not know in advance the degree of the relation, or even if the number is algebraic or not. In this case, it is probably not necessary to modify Algorithm 2.7.4, i.e. we simply choose $\lambda_i = 1$ for all i.

2.7.3 Finding Small Vectors in Lattices

For many applications, even though the LLL algorithm does not always give us the smallest vector in a lattice, the vectors which are obtained are sufficiently reasonable to give good results. We have seen one such example in the preceding section, where LLL was used to find linear dependence relations between real or complex numbers. In some cases, however, it is absolutely necessary to find one of the smallest vectors in a lattice, or more generally all vectors having norm less than or equal to some constant. This problem is hard, and in a slightly modified form is known to be NP-complete, i.e. equivalent to the most difficult reasonable problems in computer science for which no polynomial time algorithm is known. (For a thorough discussion of NP-completeness and related matters, see for example [AHU].) Nonetheless, we must give an algorithm to solve it, keeping in mind that any algorithm will probably be exponential time with respect to the dimension.

Using well known linear algebra algorithms (over $\mathbb{R}$ and not over $\mathbb{Z}$), we can assume that the matrix defining the Euclidean inner product on $\mathbb{R}^n$ is diagonal with respect to the canonical basis, say $Q(\mathbf{x}) = q_{1,1}x_1^2 + q_{2,2}x_2^2 + \cdots + q_{n,n}x_n^2$. If we want $Q(\mathbf{x}) \leq C$, say, then we must choose $|x_1| \leq \sqrt{C/q_{1,1}}$. Once x_1 is chosen, we choose $|x_2| \leq \sqrt{(C - q_{1,1}x_1^2)/q_{2,2}}$, and so on. This leads to n nested loops, and in addition it is desirable to have n variable and not fixed. Hence it is not as straightforward to implement as it may seem. The idea is to use implicitly a lexicographic ordering of the vectors $\mathbf{x}$. If we generalize this to non-diagonal quadratic forms, this leads to the following algorithm.

Algorithm 2.7.5 (Short Vectors). If Q is a positive definite quadratic form given by

$$Q(\mathbf{x}) = \sum_{i=1}^{n} q_{i,i} \left(x_i + \sum_{j=i+1}^{n} q_{i,j}x_j \right)^2$$

and a positive constant C, this algorithm outputs all the non-zero vectors $\mathbf{x} \in \mathbb{Z}^n$ such that $Q(\mathbf{x}) \leq C$, as well as the value of $Q(\mathbf{x})$. Only one of the two vectors in the pair $(\mathbf{x}, -\mathbf{x})$ is actually given.

1. [Initialize] Set $i \leftarrow n$, $T_i \leftarrow C$, $U_i \leftarrow 0$.

2. [Compute bounds] Set $Z \leftarrow \sqrt{T_i/q_{i,i}}$, $L_i \leftarrow \lfloor Z - U_i \rfloor$, $x_i \leftarrow \lceil -Z - U_i \rceil - 1$.

3. [Main loop] Set $x_i \leftarrow x_i + 1$. If $x_i > L_i$, set $i \leftarrow i+1$ and go to step 3. Otherwise, if $i > 1$, set $T_{i-1} \leftarrow T_i - q_{i,i}(x_i + U_i)^2$, $i \leftarrow i - 1$, $U_i \leftarrow \sum_{j=i+1}^{n} q_{i,j}x_j$, and go to step 2.

4. [Solution found] If $\mathbf{x} = 0$, terminate the algorithm, otherwise output $\mathbf{x}$, $Q(\mathbf{x}) = C - T_1 + q_{1,1}(x_1 + U_1)^2$ and go to step 3.

Now, although this algorithm (due in this form to Fincke and Pohst) is quite efficient in small dimensions, it is far from being the whole story. Since we have at our disposal the LLL algorithm which is efficient for finding short vectors in a lattice, we can use it to modify our quadratic form so as to shorten the length of the search. More precisely, let $R = (r_{i,j})$ be the upper triangular matrix defined by $r_{i,i} = \sqrt{q_{i,i}}$, $r_{i,j} = r_{i,i}q_{i,j}$ for $1 \le i < j \le n$, $r_{i,j} = 0$ for $1 \le j < i \le n$. Then

$$Q(\mathbf{x}) = \mathbf{x}^t R^t R \mathbf{x} .$$

Now call $\mathbf{r}_i$ the columns of R and $\mathbf{r}'_i$ the rows of R^{-1}. Then from the identity $R^{-1}R\mathbf{x} = \mathbf{x}$ we obtain $x_i = \mathbf{r}'_i R\mathbf{x}$, hence by the Cauchy-Schwarz inequality,

$$x_i^2 \le \|\mathbf{r}'_i\|^2(\mathbf{x}^t R^t R\mathbf{x}) \le \|\mathbf{r}'_i\|^2 C .$$

This bound is quite sharp since for example when the quadratic form is diagonal, we have $\|\mathbf{r}'_i\|^2 = 1/q_{i,i}$ and the bound that we obtain for x_1, say, is as usual $\sqrt{C/q_{1,1}}$. Using the LLL algorithm on the rows of R^{-1}, however, will in general drastically reduce the norms of these rows, and hence improve correspondingly the search for short vectors.

As a final improvement, we note that the implicit lexicographic ordering on the vectors $\mathbf{x}$ used in Algorithm 2.7.5 is not unique, and in particular we can permute the coordinates as we like. This adds some more freedom on our reduction of the matrix R. Before giving the final algorithm, due to Fincke and Pohst, we give the standard method to obtain the so-called Cholesky decomposition of a positive definite quadratic form, i.e. to obtain Q in the form used in Algorithm 2.7.5.

Algorithm 2.7.6 (Cholesky Decomposition). Let A be a real symmetric matrix of order n defining a positive definite quadratic form Q. This algorithm computes constants $q_{i,j}$ and a matrix R such that

$$Q(\mathbf{x}) = \sum_{i=1}^{n} q_{i,i} \left(x_i + \sum_{j=i+1}^{n} q_{i,j} x_j \right)^2$$

or equivalently in matrix form $A = R^t R$.

1. [Initialize] For all i and j such that $1 \le i \le j \le n$ set $q_{i,j} \leftarrow a_{i,j}$, then set $i \leftarrow 0$.

2. [Loop on i] Set $i \leftarrow i+1$. If $i = n$, go to step 4. Otherwise, for $j = i+1, \ldots, n$ set $q_{j,i} \leftarrow q_{i,j}$ and $q_{i,j} \leftarrow q_{i,j}/q_{i,i}$.

3. [Main loop] For all k and l such that $i + 1 \le k \le l \le n$ set

$$q_{k,l} \leftarrow q_{k,l} - q_{k,i}q_{i,l}$$

and go to step 2.

4. [Find matrix R] For $i = 1, \ldots, n$ set $r_{i,i} \leftarrow \sqrt{q_{i,i}}$, then set $r_{i,j} = 0$ if $1 \le j < i \le n$ and $r_{i,j} = r_{i,i}q_{i,j}$ if $1 \le i < j \le n$ and terminate the algorithm.

Note that this algorithm is essentially a reformulation of the Gram-Schmidt orthogonalization procedure in the case where only the Gram matrix is known. (See Proposition 2.5.7 and Remark (2) after Algorithm 2.6.3.)

We can now give the algorithm of Fincke-Pohst for finding vectors of small norm ([Fin-Poh])

Algorithm 2.7.7 (Fincke-Pohst). Let A be a real symmetric matrix of order n defining a positive definite quadratic form Q, and C be a positive constant. This algorithm outputs all non-zero vectors $\mathbf{x} \in \mathbb{Z}^n$ such that $Q(\mathbf{x}) \leq C$ and the corresponding values of $Q(\mathbf{x})$. As in Algorithm 2.7.5, only one of the two vectors $(\mathbf{x}, -\mathbf{x})$ is actually given.

1. [Cholesky] Apply the Cholesky decomposition Algorithm 2.7.6 to the matrix A, thus obtaining an upper triangular matrix R. Compute also R^{-1} (note that this is easy since R is triangular).

2. [LLL reduction] Apply the LLL algorithm to the n vectors formed by the *rows* of R^{-1}, thus obtaining a unimodular matrix U and a matrix S^{-1} such that $S^{-1} = U^{-1}R^{-1}$. Compute also $S = RU$. (Note that U will simply be the inverse transpose of the matrix H obtained in Algorithm 2.6.3, and this can be directly obtained instead of H in that algorithm, in other words it is not necessary to compute a matrix inverse).

3. [Reorder the columns of S] Call $\mathbf{s}_i$ the columns of S and $\mathbf{s}'_i$ the rows of S^{-1}. Find a permutation σ on $[1, \ldots, n]$ such that

$$\|\mathbf{s}'_{\sigma(1)}\| \geq \|\mathbf{s}'_{\sigma(2)}\| \geq \cdots \geq \|\mathbf{s}'_{\sigma(n)}\| \ .$$

Then permute the *columns* of S using the same permutation σ, i.e. replace S by the matrix whose i^{th} column is $\mathbf{s}_{\sigma(i)}$ for $1 \leq i \leq n$.

4. Compute $A_1 \leftarrow S^t S$, and find the coefficients $q_{i,j}$ of the Cholesky decomposition of A_1 using the first three steps of Algorithm 2.7.6 (it is not necessary to compute the new matrix R).

5. Using Algorithm 2.7.5 on the quadratic form Q_1 defined by the symmetric matrix A_1, compute all the non-zero vectors $\mathbf{y}$ such that $Q_1(\mathbf{y}) \leq C$, and for each such vector output $\mathbf{x} = U(y_{\sigma^{-1}(1)}, \ldots, y_{\sigma^{-1}(n)})^t$ and $Q(\mathbf{x}) = Q_1(\mathbf{y})$.

Although this algorithm is still exponential time, and is more complex than Algorithm 2.7.5, in theory and in practice it is much better and should be used systematically except if n is very small (less than 5, say).

Remark. If we want not only small vectors but *minimal* non-zero vectors, the Fincke-Pohst algorithm should be used as follows. First, use the LLL algorithm on the lattice $(\mathbb{Z}^n, Q)$. This will give small vectors in this lattice, and then choose as constant C the smallest norm among the vectors found by LLL, then apply Algorithm 2.7.7.

2.8 Exercises for Chapter 2

1. Prove that if K is a field, any invertible matrix over K is equal to a product of matrices corresponding to elementary column operations. Is this still true if K is not a field, for example for $\mathbb{Z}$?

2. Let $MX = B$ be a square linear system with coefficients in the ring $\mathbb{Z}/p^r\mathbb{Z}$ for some prime number p and some integer $r > 1$. Show how to use Algorithm 2.2.1 over the field $\mathbb{Q}_p$ to obtain at least one solution to the system, if such a solution exists. Compute in particular the necessary p-adic precision.

3. Write an algorithm which decomposes a square matrix M in the form $M = LUP$ as mentioned in the text, where P is a permutation matrix, and L and U are lower and upper triangular matrices respectively (see [AHU] or [PFTV] if you need help).

4. Give a detailed proof of Proposition 2.2.5.

5. Using the notation of Proposition 2.2.5, show that for $k + 1 \le i, j \le n$, the coefficient $a_{i,j}^{(k)}$ is equal to the $(k + 1) \times (k + 1)$ minor of M_0 obtained by taking the first k rows and the i-th row, and the first k columns and the j-th column of M_0.

6. Generalize the Gauss-Bareiss method for computing determinants, to the computation of the inverse of a matrix with integer coefficients, and more generally to the other algorithms of this chapter which use elimination.

7. Is it possible to modify the Hessenberg Algorithm 2.2.9 so that when the matrix M has coefficients in $\mathbb{Z}$ all (or most) operations are done on integers and not on rational numbers? (I do not know the answer to this question.)

8. Prove the validity of Algorithm 2.3.1.

9. Prove the validity of Algorithm 2.3.6.

10. Write an algorithm for computing one element of the inverse image, analogous to Algorithm 2.3.4 but using elimination directly instead of using Algorithm 2.3.1, and compare the asymptotic speed with that of Algorithm 2.3.4.

11. Prove the validity of Algorithm 2.3.11 and the uniqueness statement of Proposition 2.3.10.

12. In Algorithm 2.3.9, show that if the columns of M and M' are linearly independent then so are the columns of M_2.

13. Assuming Theorem 2.4.1 (1), prove parts (2) and (3). Also, try and prove (1).

14. Prove the uniqueness part of Theorem 2.4.3.

15. Show that among all possible pairs (u, v) such that $au + bv = d = \gcd(a, b)$, there exists exactly one such that $-|a|/d < v\,\text{sign}(b) \le 0$, and that in addition we will also have $1 \le u\,\text{sign}(a) \le |b|/d$.

16. Generalize Algorithm 2.4.14 to the case where the $n \times n$ square matrix A is not assumed to be non-singular.

17. Let $A = \begin{pmatrix} a & b \\ c & d \end{pmatrix}$ be a 2×2 matrix with integral coefficients such that $ad - bc \ne 0$. If we set $d_2 = \gcd(a, b, c, d)$ and $d_1 = (ad - bc)/d_2$ show directly that there

exists two matrices U and V in $\mathrm{GL}_2(\mathbb{Z})$ such that $A = V \begin{pmatrix} d_1 & 0 \\ 0 & d_2 \end{pmatrix} U$ (this is the special case $n = 2$ of Theorem 2.4.12).

18. Let G be a finite $\mathbb{Z}$-module, hence isomorphic to a quotient L'/L, and let A be a matrix giving the coordinates of some $\mathbb{Z}$-basis of L on some $\mathbb{Z}$-basis of L'. Show that the absolute value of $\det(A)$ is equal to the cardinality of G.

19. Let B be an invertible matrix with real coefficients. Show that there exist matrices K_1, K_2 and A such that $B = K_1 A K_2$, where A is a diagonal matrix with positive diagonal coefficients, and K_1 and K_2 are orthogonal matrices (this is called the *Cartan decomposition* of B). What extra condition can be added so that the decomposition is unique?

20. Prove Proposition 2.5.3 using only matrix-theoretical tools (hint: the matrix Q is diagonalizable since it is real symmetric).

21. Give recursive formulas for the computation of the Gram-Schmidt coefficients $\mu_{i,j}$ and B_i when only the Gram matrix $(\mathbf{b}_i \cdot \mathbf{b}_j)$ is known.

22. Assume that the vector $\mathbf{b}_i$ is replaced by some other vector $\mathbf{b}_k$ in the Gram-Schmidt process. Compute the new value of $B_i = \mathbf{b}_i^* \cdot \mathbf{b}_i^*$ in terms of the $\mu_{k,j}$ and B_j for $j < i$.

23. Prove Theorem 2.6.2 (5) and the validity of the LLL Algorithm 2.6.3.

24. Prove that the formulas of Algorithm 2.6.3 become those of Algorithm 2.6.7 when we set $\lambda_{i,j} \leftarrow d_j \mu_{i,j}$ and $d_i \leftarrow d_{i-1} B_i$.

25. Show that at the end of Algorithm 2.6.8 the first $n - p$ columns H_i of the matrix H form a basis of the space of relation vectors for the initial $\mathbf{b}_i$.

26. Write an all integer version of Algorithm 2.6.8, generalizing Algorithm 2.6.7 to not necessarily independent vectors. The case corresponding to $B_k = 0$ but $\mu_{k,k-1} \neq 0$ must be treated with special care.

27. (This is not really an exercise, just food for thought). Generalize to modules over principal ideal domains R the results and algorithms given about lattices. For example, generalize the LLL algorithm to the case where R is either the ring of integers of a number field (see Chapter 4) assumed to be principal, or is the ring $K[X]$ where $K = \mathbb{Q}$, $K = \mathbb{R}$ or $K = \mathbb{C}$. What can be said when $K = \mathbb{F}_p$? Give applications to the problem of linear or algebraic dependence of power series.

28. Compare the performance of Algorithms 2.7.2 and 2.4.10 (in the author's implementations, Algorithm 2.7.2 is by far superior).

29. Prove that the quantities that occur in Algorithm 2.7.2 are indeed all integral. In particular, show that $d_k = \det(\mathbf{b}_i \cdot \mathbf{b}_j)_{1 \leq i,j \leq k, B_i B_j \neq 0}$ and that $d_j \mu_{i,j} \in \mathbb{Z}$.

30. Set by convention $\mu_{k,0} = 1$, $\mu_{k,k} = B_k$, $j(k) = \max\{j, 0 \leq j \leq k, \mu_{k,j} \neq 0\}$, $d_k = \prod_{1 \leq i \leq k} \mu_{i,j(i)}$ and $\lambda_{k,j} = d_j \mu_{k,j}$ for $k > j$.

    a) Modify Sub-algorithm SWAPK so that it uses this new definition of d_k and $\lambda_{k,j}$. In other words, find the formulas giving the new values of the d_j, f_j and $\lambda_{k,j}$ in terms of the old ones after exchanging $\mathbf{b}_k$ and $\mathbf{b}_{k-1}$. In particular show that, contrary to Sub-algorithm SWAPK, d_k is always unchanged.

    b) Modify also Sub-algorithm REDI accordingly. (Warning: d_k may be modified, hence all d_j and $\lambda_{i,j}$ for $i > j > k$.)

    c) Show that we still have $d_j \in \mathbb{Z}$ and $\lambda_{k,j} \in \mathbb{Z}$ (this is much more difficult and is analogous to the integrality property of the Gauss-Bareiss Algorithm 2.2.6 and the sub-resultant Algorithm 3.3.1 that we will study in Chapter 3).

31. It can be proved that $s_k = \sum_{n \geq 1}(n(n+1) \cdots (n+k-1))^{-3}$ is of the form $a\pi^2 + b$ where a and b are rational numbers when k is even, and also when k is odd if the middle coefficient $(n + (k-1)/2)$ is only raised to the power -2 instead of -3. Compute s_k for $k \leq 4$ using Algorithm 2.7.4.

32. Prove Lemma 2.7.3 and its generalization mentioned after Algorithm 2.7.4. Write the corresponding algebraic dependence algorithm.

33. Let U be a non-singular real square matrix of order n, and let Q be the positive definite quadratic form defined by the real symmetric matrix $U^t U$. Using explicitly the inverse matrix V of U, generalize Algorithm 2.7.5 to find small values of Q on $\mathbf{Z}^n$ (Algorithm 2.7.5 corresponds to the case where U is a triangular matrix). Hint: if you have trouble, see [Knu2] Section 3.3.4.C.

Chapter 3

Algorithms on Polynomials

Excellent book references on this subject are [Knu2] and [GCL].

3.1 Basic Algorithms

3.1.1 Representation of Polynomials

Before studying algorithms on polynomials, we need to decide how they will be represented in an actual program. The straightforward way is to represent a polynomial

$$P(X) = a_n X^n + a_{n-1} X^{n-1} + \cdots a_1 X + a_0$$

by an array $a[0]$, $a[1]$, $\ldots a[n]$. The only difference between different implementations is that the array of coefficients can also be written in reverse order, with $a[0]$ being the coefficient of X^n. We will always use the first representation. Note that the leading coefficient a_n may be equal to 0, although usually this will not be the case.

The true degree of the polynomial P will be denoted by $\deg(P)$, and the coefficient of $X^{\deg(P)}$, called the leading coefficient of P, will be denoted by $\ell(P)$. In the example above, if, as is usually the case, $a_n \neq 0$, then $\deg(P) = n$ and $\ell(P) = a_n$.

The coefficients a_i may belong to any commutative ring with unit, but for many algorithms it will be necessary to specify the base ring. If this base ring is itself a ring of polynomials, we are then dealing with polynomials in several variables, and the representation given above (called the dense representation) is very inefficient, since multivariate polynomials usually have very few non-zero coefficients. In this situation, it is better to use the so-called *sparse* representation, where only the exponents and coefficients of the non-zero monomials are stored. The study of algorithms based on this kind of representation would however carry us too far afield, and will not be considered here. In any case, practically all the algorithms that we will need use only polynomials in one variable.

The operations of addition, subtraction and multiplication by a scalar, i.e. the vector space operations, are completely straightforward and need not be discussed. On the other hand, it is necessary to be more specific concerning multiplication and division.

3.1.2 Multiplication of Polynomials

As far as multiplication is concerned, one can of course use the straightforward method based on the formula:

$$\left(\sum_{i=0}^{m} a_i X^i\right)\left(\sum_{j=0}^{n} b_j X^j\right) = \sum_{k=0}^{n+m} c_k X^k ,$$

where

$$c_k = \sum_{i=0}^{k} a_i b_{k-i} ,$$

where it is understood that $a_i = 0$ if $i > m$ and $b_j = 0$ if $j > n$. This method requires $(m + 1)(n + 1)$ multiplications and mn additions. Since in general multiplications are much slower than additions, especially if the coefficients are multi-precision numbers, it is reasonable to count only the multiplication time. If $T(M)$ is the time for multiplication of elements in the base ring, the running time is thus $O(mnT(M))$. It is possible to multiply polynomials faster than this, however. We will not study this in detail, but will give an example. Assume we want to multiply two polynomials of degree 1. The straightforward method above gives:

$$(a_1 X + a_0)(b_1 X + b_0) = c_2 X^2 + c_1 X + c_0 ,$$

with

$$c_0 = a_0 b_0 , \quad c_1 = a_0 b_1 + a_1 b_0 , \quad c_2 = a_1 b_1 .$$

As mentioned, this requires 4 multiplications and 1 addition. Consider instead the following alternate method for computing the c_k:

$$c_0 = a_0 b_0 , \qquad c_2 = a_1 b_1 ,$$

$$d = (a_1 - a_0)(b_1 - b_0) , \qquad c_1 = c_0 + (c_2 - d) .$$

This requires only 3 multiplications, but 4 additions (subtraction and addition times are considered identical). Hence it is faster if one multiplication in the base ring is slower than 3 additions. This is almost always the case, especially if the base ring is not too simple or involves large integers. Furthermore, this method can be used for any degree, by recursively splitting the polynomials in two pieces of approximately equal degrees.

There is a generalization of the above method which is based on Lagrange's interpolation formula. To compute $A(X)B(X)$, which is a polynomial of degree $m+n$, compute its value at $m+n+1$ suitably chosen points. This involves only $m+n+1$ multiplications. One can then recover the coefficients of $A(X)B(X)$ (at least if the ring has characteristic zero) by using a suitable algorithmic form of Lagrange's interpolation formula. The overhead which this implies is unfortunately quite large, and for practical implementations, the reader is advised either to stick to the straightforward method, or to use the recursive splitting procedure mentioned above.

3.1.3 Division of Polynomials

We assume here that the polynomials involved have coefficients in a field K, (or at least that all the divisions which occur make sense. Note that if the coefficients belong to an integral domain, one can extend the scalars and assume that they in fact belong to the quotient field). The ring $K[X]$ is then a Euclidean domain, and this means that given two polynomials A and B with $B \neq 0$, there exist unique polynomials Q and R such that

$$A = BQ + R, \quad \text{with } \deg(R) < \deg(B)$$

(where as usual we set $\deg(0) = -\infty$). As we will see in the next section, this means that most of the algorithms described in Chapter 1 for the Euclidean domain $\mathbb{Z}$ can be applied here as well.

First however we must describe algorithms for computing Q and R. The straightforward method can easily be implemented as follows. For a non-zero polynomial Z, recall that $\ell(Z)$ is the leading coefficient of Z. Then:

Algorithm 3.1.1 (Euclidean Division). Given two polynomials A and B in $K[X]$ with $B \neq 0$, this algorithm finds Q and R such that $A = BQ + R$ and $\deg(R) < \deg(B)$.

1. [Initialize] Set $R \leftarrow A$, $Q \leftarrow 0$.

2. [Finished?] If $\deg(R) < \deg(B)$ then terminate the algorithm.

3. [Find coefficient] Set

$$S \leftarrow \frac{\ell(R)}{\ell(B)} X^{\deg(R)-\deg(B)},$$

then $Q \leftarrow Q + S$, $R \leftarrow R - S \cdot B$ and go to step 2.

Note that the multiplication $S \cdot B$ in step 3 is not really a polynomial multiplication, but simply a scalar multiplication followed by a shift of coefficients. Also, if division is much slower than multiplication, it is worthwhile to compute only once the inverse of $\ell(B)$, so as to have only multiplications in step 3. The running time of this algorithm is hence

$$O(\deg(B)(\deg(Q) + 1)T(M)) ,$$

(of course, $\deg(Q) = \deg(A) - \deg(B)$ if $\deg(A) \geq \deg(B)$).

Remark. The subtraction $R \leftarrow R - S \cdot B$ in step 3 of the algorithm must be carefully written: by definition of S, the coefficient of $X^{\deg R}$ must become exactly zero, so that the degree of R decreases. If however the base field is for example $\mathbb{R}$ or $\mathbb{C}$, the elements of K will only be represented with finite precision, and in general the operation $\ell(R) - \ell(B)(\ell(R)/\ell(B))$ will not give exactly zero but a very small number. Hence it is absolutely necessary to set it exactly equal to zero when implementing the algorithm.

Note that the assumption that K is a field is not strictly necessary. Since the only divisions which take place in the algorithm are divisions by the leading coefficient of B, it is sufficient to assume that this coefficient is invertible in K, as for example is the case if B is monic. We will see an example of this in Algorithm 3.5.5 below (see also Exercise 3).

The abstract value $T(M)$ does not reflect correctly the computational complexity of the situation. In the case of multiplication, the abstract $T(M)$ used made reasonable sense. For example, if the base ring K was $\mathbb{Z}$, then $T(M)$ would be the time needed to multiply two integers whose size was bounded by the coefficients of the polynomials A and B. On the contrary, in Algorithm 3.1.1 the coefficients explode, as can easily be seen, hence this abstract measure of complexity $T(M)$ does not make sense, at least in $\mathbb{Z}$ or $\mathbb{Q}$. On the other hand, in a field like $\mathbb{F}_p$, $T(M)$ does make sense.

Now these theoretical considerations are in fact very important in practice: Among the most used base fields (or rings), there can be no coefficient explosion in $\mathbb{F}_p$ (or more generally any finite field), or in $\mathbb{R}$ or $\mathbb{C}$ (since in that case the coefficients are represented as limited precision quantities). On the other hand, in the most important case of $\mathbb{Q}$ or $\mathbb{Z}$, such an explosion does take place, and one must be ready to deal with it.

There is however one other important special case where no explosion takes place, that is when B is a monic polynomial ($\ell(B) = 1$), and A and B are in $\mathbb{Z}[X]$. In this case, there is no division in step 3 of the algorithm.

In the general case, one can avoid divisions by multiplying the polynomial A by $\ell(B)^{\deg(A)-\deg(B)+1}$. This gives an algorithm which is not really more efficient than Algorithm 3.1.1, but which is neater and will be used in the next section. Knuth calls it "pseudo-division" of polynomials. It is as follows:

Algorithm 3.1.2 (Pseudo-Division).   Let K be a ring, A and B be two polynomials in $K[X]$ with $B \neq 0$, and set $m \leftarrow \deg(A)$, $n \leftarrow \deg(B)$, $d \leftarrow \ell(B)$. Assume that $m \geq n$. This algorithm finds Q and R such that $d^{m-n+1}A = BQ+R$ and $\deg(R) < \deg(B)$.

1. [Initialize] Set $R \leftarrow A$, $Q \leftarrow 0$, $e \leftarrow m - n + 1$.

2. [Finished?] If $\deg(R) < \deg(B)$ then set $q \leftarrow d^e$, $Q \leftarrow qQ$, $R \leftarrow qR$ and terminate the algorithm.

3. [Find coefficient] Set

$$S \leftarrow \ell(R)X^{\deg(R)-\deg(B)} \ ,$$

then $Q \leftarrow d \cdot Q + S$, $R \leftarrow d \cdot R - S \cdot B$, $e \leftarrow e - 1$ and go to step 2.

Since the algorithm does not use any division, we assume only that K is a ring, for example one can have $K = \mathbb{Z}$. Note also that the final multiplication by $q = d^e$ is needed only to get the exact power of d, and this is necessary for some applications such as the sub-resultant algorithm (see 3.3). If it is only necessary to get some constant multiple of Q and R, one can dispense with e and q entirely.

3.2 Euclid's Algorithms for Polynomials

3.2.1 Polynomials over a Field

Euclid's algorithms given in Section 1.3 can be applied with essentially no modification to polynomials with coefficients in a field K where no coefficient explosion takes place (such as $\mathbb{F}_p$). In fact, these algorithms are even simpler, since it is not necessary to have special versions à la Lehmer for multi-precision numbers. They are thus as follows:

Algorithm 3.2.1 (Polynomial GCD). Given two polynomials A and B over a field K, this algorithm determines their GCD in $K[X]$.

1. [Finished?] If $B = 0$, then output A as the answer and terminate the algorithm.

2. [Euclidean step] Let $A = B \cdot Q + R$ with $\deg(R) < \deg(B)$ be the Euclidean division of A by B. Set $A \leftarrow B$, $B \leftarrow R$ and go to step 1.

The extended version is the following:

Algorithm 3.2.2 (Extended Polynomial GCD). Given two polynomials A and B over a field K, this algorithm determines (U, V, D) such that $AU + BV = D = (A, B)$.

1. [Initialize] Set $U \leftarrow 1$, $D \leftarrow A$, $V_1 \leftarrow 0$, $V_3 \leftarrow B$.

2. [Finished?] If $V_3 = 0$ then let $V \leftarrow (D - AU)/B$ (the division being exact), output (U, V, D) and terminate the algorithm.

3. [Euclidean step] Let $D = QV_3 + R$ be the Euclidean division of D by V_3. Set $T \leftarrow U - V_1 Q$, $U \leftarrow V_1$, $D \leftarrow V_3$, $V_1 \leftarrow T$, $V_3 \leftarrow R$ and go to step 2.

Note that the polynomials U and V given by this algorithm are polynomials of the smallest degree, i.e. they satisfy $\deg(U) < \deg(B/D)$, $\deg(V) < \deg(A/D)$.

If the base field is $\mathbb{R}$ or $\mathbb{C}$, then the condition $B = 0$ of Algorithm 3.2.1 (or $V_3 = 0$ in Algorithm 3.2.2) becomes meaningless since numbers are represented only approximately. In fact, polynomial GCD's over these fields, although mathematically well defined, cannot be used in practice since the coefficients are only approximate. Even if we assume the coefficients to be given by some formula which allows us to compute them as precisely as we desire, the computation cannot usually be done. Consider for example the computation of

$$\gcd(X - \pi, X^2 - 6\zeta(2)) \ ,$$

where $\zeta(s) = \sum_{n \geq 1} n^{-s}$ is the Riemann zeta function. Although we can compute the coefficients to as many decimal places as we desire, algebra alone will not tell us that this GCD is equal to $X - \pi$ since $\zeta(2) = \pi^2/6$. The point of this discussion is that one should keep in mind that it is meaningless in practice to compute polynomial GCD's over $\mathbb{R}$ or $\mathbb{C}$.

On the other hand, if the base field is $\mathbb{Q}$, the above algorithms make perfect sense. Here, as already mentioned for Euclidean division, the practical problem of the coefficient explosion will occur, and since several divisions are performed, it will be much worse.

To be specific, if p is small, the GCD of two polynomials of $\mathbb{F}_p[X]$ of degree 1000 can be computed in a reasonable amount of time, say a few seconds, while the GCD of polynomials in $\mathbb{Q}[X]$ (even with very small integer coefficients) could take incredibly long, years maybe, because of coefficient explosion. Hence in this case it is absolutely necessary to use better algorithms. We will see this in Sections 3.3 and 3.6.1. Before that, we need some important results about polynomials over a Unique Factorization Domain (UFD).

3.2.2 Unique Factorization Domains (UFD's)

Definition 3.2.3. *Let $\mathcal{R}$ be an integral domain (i.e. a commutative ring with unit 1 and no zero divisors). We say that $u \in \mathcal{R}$ is a unit if u has a multiplicative inverse in $\mathcal{R}$. If a and b are elements of $\mathcal{R}$ with $b \neq 0$, we say that b divides a (and write $b \mid a$) if there exists $q \in \mathcal{R}$ such that $a = bq$. Since $\mathcal{R}$ is an integral domain, such a q is unique and denoted by a/b. Finally $p \in \mathcal{R}$ is called an irreducible element or a prime element if q divides p implies that either q or p/q is a unit.*

Definition 3.2.4. *A ring $\mathcal{R}$ is called a unique factorization domain (UFD) if $\mathcal{R}$ is an integral domain, and if every non-unit $x \in \mathcal{R}$ can be written in the form $x = \prod p_i$, where the p_i are (not necessarily distinct) prime elements, and if this form is unique up to permutation and multiplication of the primes by units.*

Important examples of UFD's are given by the following theorem (see [Kap], [Sam]):

Theorem 3.2.5.

(1) *If $\mathcal{R}$ is a principal ideal domain (i.e. $\mathcal{R}$ is an integral domain and every ideal is principal), then $\mathcal{R}$ is a UFD. In particular, Euclidean domains (i.e. those having a Euclidean division) are UFD's.*

(2) *If $\mathcal{R}$ is the ring of algebraic integers of a number field (see Chapter 4), then $\mathcal{R}$ is a UFD if and only if $\mathcal{R}$ is a principal ideal domain.*

(3) *If $\mathcal{R}$ is a UFD, then the polynomial rings $\mathcal{R}[X_1, \dots, X_n]$ are also UFD's.*

Note that the converse of (1) is not true in general: for example the ring $\mathbb{C}[X, Y]$ is a UFD (by (3)), but is not a principal ideal domain (the ideal generated by X and Y is not principal).

We will not prove Theorem 3.2.5 (see Exercise 6 for a proof of (3)), but we will prove some basic lemmas on UFD's before continuing further.

Theorem 3.2.6. *Let $\mathcal{R}$ be a UFD. Then*

(1) *If p is prime, then for all a and b in $\mathcal{R}$, $p \mid ab$ if and only if $p \mid a$ or $p \mid b$.*

(2) *If $a \mid bc$ and a has no common divisor with b other than units, then $a \mid c$.*

(3) *If a and b have no common divisor other than units, then if a and b divide $c \in \mathcal{R}$, then $ab \mid c$.*

(4) *Given a set $S \subset \mathcal{R}$ of elements of $\mathcal{R}$, there exists $d \in \mathcal{R}$ called a greatest common divisor (GCD) of the elements of S, and having the following properties: d divides all the elements of S, and if e is any element of $\mathcal{R}$ dividing all the elements of S, then $e \mid d$. Furthermore, if d and d' are two GCD's of S, then d/d' is a unit.*

Proof. (1) Assume $p \mid ab$. Since $\mathcal{R}$ is a UFD, one can write $a = \prod_{1 \le i \le m} p_i$ and $b = \prod_{m+1 \le i \le m+n} p_i$, the p_i being not necessarily distinct prime elements of $\mathcal{R}$. On the other hand, since $ab/p \in \mathcal{R}$ we can also write $ab = p \prod_j q_j$ with prime elements q_j. By the uniqueness of prime decomposition, since $ab = \prod_{1 \le i \le m+n} p_i$ we deduce that p is equal to a unit times one of the p_i. Hence, if $i \le m$, then $p \mid a$, while if $i > m$, then $p \mid b$, proving (1).

(2) We prove (2) by induction on the number n of prime factors of b, counted with multiplicity. If $n = 0$ then b is a unit and $a \mid c$. Assume the result true for $n - 1$, and let $bc = qa$ with $n \ge 1$. Let p be a prime divisor of b. p divides qa, and by assumption p does not divide a. Hence by (1) p divides q, and we can write $b'c = q'a$ with $b' = b/p$, $q' = q/p$. Since b' has only $n - 1$ prime divisors, (2) follows by induction.

(3) Write $c = qa$ with $q \in \mathcal{R}$. Since $b \mid c$, by (2) we deduce that $b \mid q$, hence $ab \mid c$.

(4) For every element $s \in S$, write

$$s = u \prod_p p^{v_p(s)},$$

where u is a unit, the product is over *all* distinct prime elements of $\mathcal{R}$ up to units, and $v_p(s)$ is the number of times that the prime p occurs in s, hence is 0 for all but finitely many p. Set

$$d = \prod_p p^{\alpha_p} , \qquad \text{where } \alpha_p = \min_{s \in S} v_p(s) .$$

This min is of course equal to 0 for all but a finite number of p, and it is clear that d satisfies the conditions of the theorem.   □

We will say that the elements of S are *coprime* if their GCD is a unit. By definition of a UFD, this is equivalent to saying that no prime element is a common divisor. Note that if $\mathcal{R}$ is not only a UFD but also a principal ideal domain (for example when the UFD $\mathcal{R}$ is the ring of algebraic integers in a number field), then the coprimality condition is equivalent to saying that the ideal generated by the elements is the whole ring $\mathcal{R}$. This is however *not* true in general. For example, in the UFD $\mathbb{C}[X, Y]$, the elements X and Y are coprime, but the ideal which they generate is the set of polynomials P such that $P(0,0) = 0$, and this is not the whole ring.

3.2.3 Polynomials over Unique Factorization Domains

Definition 3.2.7. *Let $\mathcal{R}$ be a UFD, and $A \in R[X]$. We define the content of A and write $\text{cont}(A)$ as a GCD of the coefficients of A. We say that A is primitive if $\text{cont}(A)$ is a unit, i.e. if its coefficients are coprime. Finally, if $A \neq 0$ the polynomial $A/\text{cont}(A)$ is primitive, and is called the primitive part of A, and denoted $\text{pp}(A)$ (in the case $A = 0$ we define $\text{cont}(A) = 0$, $\text{pp}(A) = 0$).*

The fundamental result on these notions, due to Gauss, is as follows:

Theorem 3.2.8. *Let A and B be two polynomials over a UFD $\mathcal{R}$. Then there exists a unit $u \in \mathcal{R}$ such that*

$$\text{cont}(A \cdot B) = u \, \text{cont}(A) \, \text{cont}(B) , \qquad \text{pp}(A \cdot B) = u^{-1} \text{pp}(A) \, \text{pp}(B) .$$

In particular, the product of two primitive polynomials is primitive.

Proof. Since $A = \text{cont}(A) \text{pp}(A)$, it is clear that this theorem is equivalent to the statement that the product of two primitive polynomials A and B is primitive. Assume the contrary. Then there exists a prime $p \in \mathcal{R}$ which divides all the coefficients of AB. Write $A(X) = \sum a_i X^i$ and $B(X) = \sum b_i X^i$. By assumption there exists a j such that a_j is not divisible by p, and similarly a k such that b_k is not divisible by p. Choose j and k as small as possible. The coefficient of X^{j+k} in AB is $a_j b_k + a_{j+1} b_{k-1} + \cdots + a_{j+k} b_0 + a_{j-1} b_{k+1} + \cdots + a_0 b_{k+j}$, and all the terms in this sum are divisible by p except the term $a_j b_k$ (since j and k have been chosen as small as possible), and $a_j b_k$ itself is not divisible by p since p is prime. Hence p does not divide the coefficient of X^{j+k} in AB, contrary to our assumption, and this proves the theorem.   □

Corollary 3.2.9. *Let A and B be two polynomials over a UFD $\mathcal{R}$. Then there exists units u and v in $\mathcal{R}$ such that*

$$\text{cont}(\gcd(A, B)) = u \gcd(\text{cont}(A), \text{cont}(B)) \ ,$$

$$\text{pp}(\gcd(A, B)) = v \gcd(\text{pp}(A), \text{pp}(B)) \ .$$

3.2.4 Euclid's Algorithm for Polynomials over a UFD

We can now give Euclid's algorithm for polynomials defined over a UFD. The important point to notice is that the sequence of operations will be essentially identical to the corresponding algorithm over the quotient field of the UFD, but the algorithm will run much faster. This is because implementing arithmetic in the quotient field (say in $\mathbb{Q}$ if $R = \mathbb{Z}$) will involve taking GCD's in the UFD all the time, many more than are needed to execute Euclid's algorithm. Hence the following algorithm is always to be preferred to Algorithm 3.2.1 when the coefficients of the polynomials are in a UFD. We will however study in the next section a more subtle and efficient method.

Algorithm 3.2.10 (Primitive Polynomial GCD). Given two polynomials A and B with coefficients in a UFD $\mathcal{R}$, this algorithm computes a GCD of A and B, using only operations in $\mathcal{R}$. We assume that we already have at our disposal algorithms for (exact) division and for GCD in $\mathcal{R}$.

1. [Reduce to primitive] If $B = 0$, output A and terminate. Otherwise, set $a \leftarrow \text{cont}(A)$, $b \leftarrow \text{cont}(B)$, $d \leftarrow \gcd(a, b)$, $A \leftarrow A/a$, $B \leftarrow B/b$.

2. [Pseudo division] Compute R such that $\ell(B)^{\deg(A)-\deg(B)+1}A = BQ + R$ using Algorithm 3.1.2. If $R = 0$ go to step 4. If $\deg(R) = 0$, set $B \leftarrow 1$ and go to step 4.

3. [Replace] Set $A \leftarrow B$, $B \leftarrow \text{pp}(R) = R/\text{cont}(R)$ and go to step 2.

4. [Terminate] Output $d \cdot B$ and terminate the algorithm.

In the next section, we will see an algorithm which is in general faster than the above algorithm. There are also other methods which are often even faster, but are based on quite different ideas. Consider the case where $R = \mathbb{Z}$. Instead of trying to control the explosion of coefficients, we simply put ourselves in a field where this does not occur, i.e. in the finite field $\mathbb{F}_p$ for suitable primes p. If one finds that the GCD modulo p has degree 0 (and this will happen often), then if p is suitably chosen it will follow that the initial polynomials are coprime over $\mathbb{Z}$. Even if the GCD is not of degree 0, it is in general quite easy to deduce from it the GCD over $\mathbb{Z}$. We will come back to this question in Section 3.6.1.

3.3 The Sub-Resultant Algorithm

3.3.1 Description of the Algorithm

The main inconvenience of Algorithm 3.2.10 is that we compute the content of R in step 3 each time, and this is a time consuming operation. If we did not reduce R at all, then the coefficient explosion would make the algorithm much slower, and this is also not acceptable. There is a nice algorithm due to Collins, which is a good compromise and which is in general faster than Algorithm 3.2.10, although the coefficients are larger. The idea is that one can give an a priori divisor of the content of R, which is sufficiently large to replace the content itself in the reduction. This algorithm is derived from the algorithm used to compute the resultant of two polynomials (see Section 3.3.2), and is called the sub-resultant algorithm. We could still divide A and B by their content from time to time (say every 10 iterations), but this would be a very bad idea (see Exercise 4).

Algorithm 3.3.1 (Sub-Resultant GCD). Given two polynomials A and B with coefficients in a UFD $\mathcal{R}$, this algorithm computes a GCD of A and B, using only operations in $\mathcal{R}$. We assume that we already have at our disposal algorithms for (exact) division and for GCD in $\mathcal{R}$.

1. [Initializations and reductions] If $\deg(B) > \deg(A)$ exchange A and B. Now if $B = 0$, output A and terminate the algorithm, otherwise, set $a \leftarrow \operatorname{cont}(A)$, $b \leftarrow \operatorname{cont}(B)$, $d \leftarrow \gcd(a,b)$, $A \leftarrow A/a$, $B \leftarrow B/b$, $g \leftarrow 1$ and $h \leftarrow 1$.

2. [Pseudo division] Set $\delta \leftarrow \deg(A) - \deg(B)$. Using Algorithm 3.1.2, compute R such that $\ell(B)^{\delta+1}A = BQ + R$. If $R = 0$ go to step 4. If $\deg(R) = 0$, set $B \leftarrow 1$ and go to step 4.

3. [Reduce remainder] Set $A \leftarrow B$, $B \leftarrow R/(gh^\delta)$, $g \leftarrow \ell(A)$, $h \leftarrow h^{1-\delta}g^\delta$ and go to step 2. (Note that all the divisions which may occur in this step give a result in the ring $\mathcal{R}$.)

4. [Terminate] Output $d \cdot B/\operatorname{cont}(B)$ and terminate the algorithm.

It is not necessary for us to give the proof of the validity of this algorithm, since it is long and is nicely done in [Knu2]. The main points to notice are as follows: first, it is clear that this algorithm gives exactly the same sequence of polynomials as the straightforward algorithm, but multiplied or divided by some constants. Consequently, the only thing to prove is that all the quantities occurring in the algorithm stay in the ring $\mathcal{R}$. This is done by showing that all the coefficients of the intermediate polynomials as well as the quantities h are determinants of matrices whose coefficients are coefficients of A and B, hence are in the ring $\mathcal{R}$.

Another result which one obtains in proving the validity of the algorithm is that in the case $R = \mathbb{Z}$, if $m = \deg(A)$, $n = \deg(B)$, and N is an upper bound for the absolute value of the coefficients of A and B, then the coefficients of the intermediate polynomials are all bounded by the quantity

$$N^{m+n}(m+1)^{n/2}(n+1)^{m/2} \ ,$$

and this is reasonably small. One can then show that the execution time for computing the GCD of two polynomials of degree n over $\mathbb{Z}$ when their coefficients are bounded by N in absolute value is $O(n^4(\ln Nn)^2)$.

I leave as an exercise to the reader the task of writing an extended version of Algorithm 3.3.1 which gives polynomials U and V such that $AU + BV = r(A, B)$, where $r \in \mathcal{R}$. All the operations must of course be done in $\mathcal{R}$ (see Exercise 5). Note that it is not always possible to have $r = 1$. For example, if $A(X) = X$ and $B(X) = 2$, then $(A, B) = 1$ but for any U and V the constant term of $AU + BV$ is even.

3.3.2 Resultants and Discriminants

Let A and B be two polynomials over an integral domain $\mathcal{R}$ with quotient field K, and let $\overline{K}$ be an algebraic closure of K.

Definition 3.3.2. *Let* $A(X) = a(X - \alpha_1) \cdots (X - \alpha_m)$ *and* $B(X) = b(X - \beta_1) \cdots (X - \beta_n)$ *be the decomposition of A and B in $\overline{K}$. Then the* resultant $R(A, B)$ *of A and B is given by one of the equivalent formulas:*

$$R(A, B) = a^n B(\alpha_1) \cdots B(\alpha_m)$$
$$= (-1)^{mn} b^m A(\beta_1) \cdots A(\beta_n)$$
$$= a^n b^m \prod_{1 \le i \le m, 1 \le j \le n} (\alpha_i - \beta_j) \ .$$

Definition 3.3.3. *If* $A \in \mathcal{R}[X]$, *with* $m = \deg(A)$, *the* discriminant $\operatorname{disc}(A)$ *of A is equal to the expression:*

$$(-1)^{m(m-1)/2} R(A, A')/\ell(A) \ ,$$

where A' is the derivative of A.

The main point about these definitions is that resultants and discriminants have coefficients in $\mathcal{R}$. Indeed, by the symmetry in the roots α_i, it is clear that the resultant is a function of the symmetric functions of the roots, hence is in K. It is not difficult to see that the coefficient a^n insures that $R(A, B) \in \mathcal{R}$. Another way to see this is to prove the following lemma.

Lemma 3.3.4. *If* $A(X) = \sum_{0 \le i \le m} a_i X^i$ *and* $B(X) = \sum_{0 \le i \le n} b_i X^i$, *then the resultant $R(A, B)$ is equal to the determinant of the following $(n+m) \times (n+m)$ matrix:*

$$
\begin{pmatrix}
a_m & a_{m-1} & a_{m-2} & \cdots & a_1 & a_0 & 0 & 0 & \cdots & 0 \\
0 & a_m & a_{m-1} & a_{m-2} & \cdots & a_1 & a_0 & 0 & \cdots & 0 \\
0 & 0 & a_m & a_{m-1} & a_{m-2} & \cdots & a_1 & a_0 & \cdots & 0 \\
\vdots & \vdots & \ddots & \ddots & \ddots & \ddots & \ddots & \ddots & \ddots & \vdots \\
0 & 0 & \cdots & 0 & a_m & a_{m-1} & a_{m-2} & \cdots & a_1 & a_0 \\
b_n & b_{n-1} & \cdots & b_2 & b_1 & b_0 & 0 & 0 & \cdots & 0 \\
0 & b_n & b_{n-1} & \cdots & b_2 & b_1 & b_0 & 0 & \cdots & 0 \\
0 & 0 & b_n & b_{n-1} & \cdots & b_2 & b_1 & b_0 & \cdots & 0 \\
\vdots & \vdots & \ddots & \ddots & \ddots & \ddots & \ddots & \ddots & \ddots & \vdots \\
0 & 0 & \cdots & 0 & b_n & b_{n-1} & \cdots & b_2 & b_1 & b_0
\end{pmatrix}
$$

where the coefficients of A are repeated on $n = \deg(B)$ rows, and the coefficients of B are repeated on $m = \deg(A)$ rows.

The above matrix is called Sylvester's matrix. Since the only non-zero coefficients of the first column of this matrix are a_m and b_n, it is clear that $R(A, B)$ is not only in $\mathcal{R}$ but in fact divisible (in $\mathcal{R}$) by $\gcd(\ell(A), \ell(B))$. In particular, if $B = A'$, $R(A, A')$ is divisible by $\ell(A)$, hence $\mathrm{disc}(A)$ is also in $\mathcal{R}$.

Proof. Call M the above matrix. Assume first that the α_i and β_j are all distinct. Consider the $(n + m) \times (n + m)$ Vandermonde matrix $V = (v_{i,j})$ defined by $v_{i,j} = \beta_j^{m+n-i}$ if $j \leq n$, $v_{i,j} = \alpha_{j-n}^{m+n-i}$ if $n + 1 \leq j \leq n + m$. Then the Vandermonde determinant $\det(V)$ is non-zero since we assumed the α_i and β_j distinct, and we have

$$
\det(V) = \prod_{i<j}(\beta_i - \beta_j)\prod_{i<j}(\alpha_i - \alpha_j)\prod_{i,j}(\beta_i - \alpha_j) .
$$

On the other hand, it is clear that

$$
MV = \begin{pmatrix}
\beta_1^{n-1}A(\beta_1) & \cdots & \beta_n^{n-1}A(\beta_n) & 0 & \cdots & 0 \\
\vdots & \vdots & \vdots & \vdots & \vdots & \vdots \\
A(\beta_1) & \cdots & A(\beta_n) & 0 & \cdots & 0 \\
0 & \cdots & 0 & \alpha_1^{m-1}B(\alpha_1) & \cdots & \alpha_m^{m-1}B(\alpha_m) \\
\vdots & \vdots & \vdots & \vdots & \vdots & \vdots \\
0 & \cdots & 0 & B(\alpha_1) & \cdots & B(\alpha_m)
\end{pmatrix},
$$

hence $\det(MV)$ is equal to the product of the two diagonal block determinants, which are again Vandermonde determinants. Hence we obtain:

$$
\det(MV) = A(\beta_1)\cdots A(\beta_n)B(\alpha_1)\cdots B(\alpha_m)\prod_{i<j}(\beta_i - \beta_j)\prod_{i<j}(\alpha_i - \alpha_j) .
$$

Comparing with the formula for $\det(V)$ and using $\det(V) \neq 0$ we obtain

$$\det(M) \prod_{i,j}(\beta_i - \alpha_j) = A(\beta_1)\cdots A(\beta_n)B(\alpha_1)\cdots B(\alpha_m) .$$

Since clearly $A(\beta_1)\cdots A(\beta_n) = a^n \prod_{i,j}(\beta_i - \alpha_j)$, the lemma follows in the case where all the α_j and β_i are distinct, and it follows in general by a continuity argument or by taking the roots as formal variables. $\square$

Note that by definition, the resultant of A and B is equal to 0 if and only if A and B have a common root, hence if and only if $\deg(A, B) > 0$. In particular, the discriminant of a polynomial A is zero if and only if A has a non-trivial square factor, hence if and only if $\deg(A, A') > 0$.

The definition of the discriminant that we have given may seem a little artificial. It is motivated by the following proposition.

Proposition 3.3.5. *Let $A \in R[X]$ with $m = \deg(A)$, and let α_i be the roots of A in $\overline{K}$. Then we have*

$$\mathrm{disc}(A) = \ell(A)^{m-1+\deg(A')} \prod_{1 \le i < j \le m} (\alpha_i - \alpha_j)^2 .$$

Proof. If A has multiple roots, both sides are 0. So we assume that A has only simple roots. Now if $a = \ell(A)$, we have

$$A'(X) = a \sum_i \prod_{j \ne i}(X - \alpha_j)$$

hence

$$A'(\alpha_i) = a \prod_{j \ne i}(\alpha_i - \alpha_j) .$$

Thus we obtain

$$R(A, A') = a^{m+\deg(A')}(-1)^{m(m-1)/2} \prod_{i<j}(\alpha_i - \alpha_j)^2$$

thus proving the proposition. Note that we have $\deg(A') = m-1$, except when the characteristic of R is non-zero and divides m. $\square$

The following corollary follows immediately from the definitions.

Corollary 3.3.6. *We have $R(A_1 A_2, A_3) = R(A_1, A_3)R(A_2, A_3)$ and*

$$\mathrm{disc}(A_1 A_2) = \mathrm{disc}(A_1)\,\mathrm{disc}(A_2)(R(A_1, A_2))^2 .$$

Resultants and discriminants will be fundamental in our handling of algebraic numbers. Now the nice fact is that we have already done essentially all the work necessary to compute them: a slight modification of Algorithm 3.3.1 will give us the resultant of A and B.

Algorithm 3.3.7 (Sub-Resultant). Given two polynomials A and B with co-efficients in a UFD $\mathcal{R}$, this algorithm computes the resultant of A and B.

1. [Initializations and reductions] If $A = 0$ or $B = 0$, output 0 and terminate the algorithm. Otherwise, set $a \leftarrow \text{cont}(A)$, $b \leftarrow \text{cont}(B)$, $A \leftarrow A/a$, $B \leftarrow B/b$, $g \leftarrow 1$, $h \leftarrow 1$, $s \leftarrow 1$ and $t \leftarrow a^{\deg(B)}b^{\deg(A)}$. Finally, if $\deg(A) < \deg(B)$ exchange A and B and if in addition $\deg(A)$ and $\deg(B)$ are odd set $s \leftarrow -1$.

2. [Pseudo division] Set $\delta \leftarrow \deg(A) - \deg(B)$. If $\deg(A)$ and $\deg(B)$ are odd, set $s \leftarrow -s$. Finally, compute R such that $\ell(B)^{\delta+1}A = BQ + R$ using Algorithm 3.1.2.

3. [Reduce remainder] Set $A \leftarrow B$ and $B \leftarrow R/(gh^\delta)$.

4. [Finished?] Set $g \leftarrow \ell(A)$, $h \leftarrow h^{1-\delta}g^\delta$. If $\deg(B) > 0$ go to step 2, otherwise set $h \leftarrow h^{1-\deg(A)}\ell(B)^{\deg(A)}$ output $s \cdot t \cdot h$ and terminate the algorithm.

Proof. Set $A_0 = A$, $A_1 = B$, let A_i be the sequence of polynomials generated by this algorithm, and let R_i be the remainders obtained in step 2. Let t be the index such that $\deg(A_{t+1}) = 0$. Set $d_k = \deg(A_k)$, $\ell_k = \ell(A_k)$, and let g_k and h_k be the quantities g and h in stage k, so that $g_0 = h_0 = 1$. Finally set $\delta_k = d_k - d_{k+1}$. Denoting by β_i the roots of A_k, we clearly have for $k \geq 1$:

$$R(A_{k-1}, A_k) = (-1)^{d_{k-1}d_k}\ell_k^{d_{k-1}} \prod_{1 \leq i \leq d_k} A_{k-1}(\beta_i)$$

$$= (-1)^{d_{k-1}d_k}\ell_k^{d_{k-1}} \prod_{1 \leq i \leq d_k} \frac{R_{k+1}(\beta_i)}{\ell_k^{\delta_{k-1}+1}}$$

$$= (-1)^{d_{k-1}d_k}\ell_k^{d_{k-1}-d_k(\delta_{k-1}+1)} \prod_{1 \leq i \leq d_k} R_{k+1}(\beta_i)$$

$$= (-1)^{d_{k-1}d_k}\ell_k^{d_{k-1}-d_k(\delta_{k-1}+1)-d_{k+1}} R(A_k, g_{k-1}h_{k-1}^{\delta_{k-1}}A_{k+1}) \ .$$

Now using $R(A, cB) = c^{\deg(A)}R(A, B)$ and the identities $g_k = \ell_k$ and $h_k = h_{k-1}^{1-\delta_{k-1}}g_k^{\delta_{k-1}}$ for $k \geq 1$, we see that the expression simplifies to

$$R(A_{k-1}, A_k) = (-1)^{d_{k-1}d_k} \frac{g_{k-1}^{d_k}h_{k-1}^{d_{k-1}-1}}{g_k^{d_{k+1}}h_k^{d_{k-1}}} R(A_k, A_{k+1}) \ .$$

Using $d_{t+1} = 0$, hence $\delta_t = d_t$, we finally obtain

$$R(A, B) = (-1)^{\sum_{1 \leq k \leq t} d_{k-1}d_k} h_t^{1-\delta_t} R(A_t, A_{t+1})$$

$$= (-1)^{\sum_{1 \leq k \leq t} d_{k-1}d_k} h_t^{1-d_t}\ell_{t+1}^{d_t}$$

$$= (-1)^{\sum_{1 \leq k \leq t} d_{k-1}d_k} h_{t+1} \ ,$$

thus proving the validity of the algorithm. $\square$

Note that it is the same kind of argument and simplifications which show that the A_k have coefficients in the same ring $\mathcal{R}$ as the coefficients of A and B, and that the h_k also belong to $\mathcal{R}$. In fact, we have just proved for instance that $h_{t+1} \in \mathcal{R}$.

Finally, to compute discriminants of polynomials, one simply uses Algorithm 3.3.7 and the formula

$$\mathrm{disc}(A) = (-1)^{m(m-1)/2} R(A, A')/\ell(A) \ ,$$

where $m = \deg(A)$.

3.3.3 Resultants over a Non-Exact Domain

Although resultants and GCD's are similar, from the computational point of view, there is one respect in which they completely differ. It does make practical sense to compute (approximate) resultants over $\mathbb{R}$, $\mathbb{C}$ or $\mathbb{Q}_p$, while it does not make sense for GCD's as we have already explained. When dealing with resultants of polynomials with such non-exact coefficients we must however be careful *not* to use the sub-resultant algorithm. For one thing, it is tailored to avoid denominator explosion when the coefficients are, for example, rational numbers or rational functions in other variables. But most importantly, it would simply give wrong results, since the remainders R obtained in the algorithm are only approximate; hence a zero leading coefficient could appear as a very small non-zero number, leading to havoc in the next iteration.

Hence, in this case, the natural solution is to evaluate directly Sylvester's determinant. Now the usual Gaussian elimination method for computing determinants also involves dividing by elements of the ring to which the coefficients belong. In the case of the ring $\mathbb{Z}$, say, this is not a problem since the quotient of two integers will be represented exactly as a rational number. Even for non-exact rings like $\mathbb{R}$, the quotient is another real number given to a slightly worse and computable approximation. On the other hand, in the case where the coefficients are themselves polynomials in another variable over some non-exact ring like $\mathbb{R}$, although one could argue in the same way using rational functions, the final result will not in general simplify to a polynomial as it should, for the same reason as before.

To work around this problem, we must use the Gauss-Bareiss Algorithm 2.2.6 which has exactly the property of keeping all the computations in the initial base ring. Keep in mind, as already mentioned after Algorithm 2.2.6, that if some division of elements of $\mathbb{R}[X]$ (say) is required, then Euclidean division must be used, i.e. we *must* get a polynomial as a result.

Hence to compute resultants we can apply this algorithm to Sylvester's matrix, even when the coefficients are not exact. (In the case of exact coefficients, this algorithm will evidently also work, but will be slower than the sub-resultant algorithm.) Since Sylvester's matrix is an $(n + m) \times (n + m)$ matrix, it is important to note that simple row operations can reduce it to an $n \times n$ matrix to which we can then apply the Gauss-Bareiss algorithm (see Exercise 8).

Remark. The Gauss-Bareiss method and the sub-resultant algorithm are in fact closely linked. It is possible to adapt the sub-resultant algorithm so as to give correct answers in the non-exact cases that we have mentioned (see Exercise 10), but the approach using determinants is probably safer.

3.4 Factorization of Polynomials Modulo p

3.4.1 General Strategy

We now consider the problem of factoring polynomials. In practice, for polynomials in one variable the most important base rings are $\mathbb{Z}$ (or $\mathbb{Q}$), $\mathbb{F}_p$ or $\mathbb{Q}_p$. Factoring over $\mathbb{R}$ or $\mathbb{C}$ is equivalent to root finding, hence belongs to the domain of numerical analysis. We will give a simple but efficient method for this in Section 3.6.3.

Most factorization methods rely on factorization methods over $\mathbb{F}_p$, hence we will consider this first. In Section 1.6, we have given algorithms for finding roots of polynomials modulo p, and explained that no polynomial-time deterministic algorithm is known to do this (if one does not assume the GRH). The more general case of factoring is similar. The algorithms that we will describe are probabilistic, but are quite efficient.

Contrary to the case of polynomials over $\mathbb{Z}$, polynomials over $\mathbb{F}_p$ have a tendency to have several factors. Hence the problem is not only to break up the polynomial into two pieces (at least), but to factor completely the polynomial as a product of powers of irreducible (i.e. prime in $\mathcal{R}[X]$) polynomials. This is done in four steps, in the following way.

Algorithm 3.4.1 (Factor in $\mathbb{F}_p[X]$). Let $A \in \mathbb{F}_p[X]$ be monic (since we are over a field, this does not restrict the generality). This algorithm factors A as a product of powers of irreducible polynomials in $\mathbb{F}_p[X]$.

1. [Squarefree factorization] Find polynomials $A_1, A_2, \ldots, A_k$ in $\mathbb{F}_p[X]$ such that

(1) $A = A_1^1 A_2^2 \cdots A_k^k$,
(2) The A_i are squarefree and coprime.

   (This decomposition of A will be called the squarefree factorization of A).

2. [Distinct degree factorization] For $i = 1, \ldots, k$ find polynomials $A_{i,d} \in \mathbb{F}_p[X]$ such that $A_{i,d}$ is the product of all irreducible factors of A_i of degree d (hence $A_i = \prod_d A_{i,d}$).

3. [Final splittings] For each i and d, factor $A_{i,d}$ into $\deg(A_{i,d})/d$ irreducible factors of degree d.

4. [Cleanup] Group together all the identical factors found, order them by degree, output the complete factorization and terminate the algorithm.

Of course, this is only the skeleton of an algorithm since steps 1, 2 and 3 are algorithms by themselves. We will consider them in turn.

3.4.2 Squarefree Factorization

Let $\overline{\mathbb{F}}_p$ be an algebraic closure of $\mathbb{F}_p$. If $A \in \mathbb{F}_p[X]$ is monic, define $A_i(X) = \prod_j(X - \alpha_j)$ where the α_j are the roots of A in $\overline{\mathbb{F}}_p$ of multiplicity exactly equal to i. Since the Galois group of $\overline{\mathbb{F}}_p/\mathbb{F}_p$ preserves the multiplicity of the roots of A, it permutes the α_j, so all the A_i have in fact coefficients in $\mathbb{F}_p$ (this will also follow from the next algorithm). It is clear that they satisfy the conditions of step 1. It remains to give an algorithm to compute them.

If $A = \prod_i A_i^i$ with A_i squarefree and coprime, then $A' = \sum_i \prod_{j \neq i} A_j^j \cdot i A_i' A_i^{i-1}$. Hence, if $T = \gcd(A, A')$, then for all irreducible P dividing T, the exponent $v_P(T)$ of P in the prime decomposition of T can be obtained as follows: P dividing A must divide an A_m for some m. Hence, for all $i \neq m$ in the sum for A', the v_P of the ith summand is greater than or equal to m and for $i = m$ is equal to $m - 1$ if $p \nmid m$, and otherwise the summand is 0 (note that since A_m is squarefree, A_m' cannot be divisible by P). Hence, we obtain that $v_P(T) = m - 1$ if $p \nmid m$, and $v_P(T) \geq m$, so $v_P(T) = m$ (since T divides A) if $p \mid m$. Finally, we obtain the formula

$$T = (A, A') = \prod_{p \nmid i} A_i^{i-1} \prod_{p \mid i} A_i^i .$$

Note that we could have given a much simpler proof over $\mathbb{Z}$, and in that case the exponent of A_i would be equal to $i - 1$ for all i.

Now we define two sequences of polynomials by induction as follows. Set $T_1 = T$ and $V_1 = A/T = \prod_{p \nmid i} A_i$. For $k \geq 1$, set $V_{k+1} = (T_k, V_k)$ if $p \nmid k$, $V_{k+1} = V_k$ if $p \mid k$, and $T_{k+1} = T_k/V_{k+1}$. It is easy to check by induction that

$$V_k = \prod_{i \geq k,\ p \nmid i} A_i \quad \text{and} \quad T_k = \prod_{i > k,\ p \nmid i} A_i^{i-k} \prod_{p \mid i} A_i^i .$$

From this it follows that $A_k = V_k/V_{k+1}$ for $p \nmid k$. We thus obtain all the A_k for $p \nmid k$, and we continue as long as V_k is a non-constant polynomial. When V_k is constant, we have $T_{k-1} = \prod_{p \mid i} A_i^i$ hence there exists a polynomial U such that $T_{k-1}(X) = U^p(X) = U(X^p)$, and this polynomial can be trivially obtained from T_{k-1}. We then start again recursively the whole algorithm of squarefree decomposition on the polynomial U. Transforming the recursive step into a loop we obtain the following algorithm.

Algorithm 3.4.2 (Squarefree Factorization). Let $A \in \mathbb{F}_p[X]$ be a monic polynomial and let $A = \prod_{i \geq 1} A_i^i$ be its squarefree factorization, where the A_i are squarefree and pairwise coprime. This algorithm computes the polynomials A_i, and outputs the pairs (i, A_i) for the values of i for which A_i is not constant.

1. [Initialize] Set $e \leftarrow 1$ and $T_0 \leftarrow A$.

2. [Initialize e-loop] If T_0 is constant, terminate the algorithm. Otherwise, set $T \leftarrow (T_0, T_0')$, $V \leftarrow T_0/T$ and $k \leftarrow 0$.

3. [Finished e-loop?] If V is constant, T must be of the form $T(X) = \sum_{p \mid j} t_j X^j$, so set $T_0 \leftarrow \sum_{p \mid j} t_j X^{j/p}$, $e \leftarrow pe$ and go to step 2.

4. [Special case] Set $k \leftarrow k + 1$. If $p \mid k$ set $T \leftarrow T/V$ and $k \leftarrow k + 1$.

5. [Compute A_{ek}] Set $W \leftarrow (T, V)$, $A_{ek} \leftarrow V/W$, $V \leftarrow W$ and $T \leftarrow T/V$. If A_{ek} is not constant output (ek, A_{ek}). Go to step 3.

3.4.3 Distinct Degree Factorization

We can now assume that we have a *squarefree* polynomial A and we want to group factors of A of the same degree d. This procedure is known as distinct degree factorization and is quite simple. We first need to recall some results about finite fields. Let $P \in \mathbb{F}_p[X]$ be an irreducible polynomial of degree d. Then the field $K = \mathbb{F}_p[X]/P(X)\mathbb{F}_p[X]$ is a finite field with p^d elements. Hence, every element x of the multiplicative group K^* satisfies the equation $x^{p^d-1} = 1$, therefore every element of K satisfies $x^{p^d} = x$. This shows that P is a divisor of the polynomial $X^{p^d} - X$ in $\mathbb{F}_p[X]$. Conversely, every irreducible factor of $X^{p^d} - X$ which is not a factor of $X^{p^e} - X$ for $e < d$ has degree exactly d. This leads to the following algorithm.

Algorithm 3.4.3 (Distinct Degree Factorization). Given a squarefree polynomial $A \in \mathbb{F}_p[X]$, this algorithm finds for each d the polynomial A_d which is the product of the irreducible factors of A of degree d.

1. [Initialize] Set $V \leftarrow A$, $W \leftarrow X$, $d \leftarrow 0$.

2. [Finished?] Set $e \leftarrow \deg(V)$. If $d + 1 > \frac{1}{2}e$, then if $e > 0$ set $A_e = V$, $A_i = 1$ for all other $i > d$, and terminate the algorithm. If $d + 1 \leq \frac{1}{2}e$, set $d \leftarrow d + 1$, $W \leftarrow W^p \bmod V$.

3. [Output A_d] Output $A_d = (W - X, V)$. If $A_d \neq 1$, set $V \leftarrow V/A_d$, $W \leftarrow W \bmod V$. Go to step 2.

Once the A_d have been found, it remains to factor them. We already know the number of irreducible factors of A_d, which is equal to $\deg(A_d)/d$. In particular, if $\deg(A_d) = d$, then A_d is irreducible.

Note that the distinct degree factorization algorithm above succeeds in factoring A completely quite frequently. With reasonable assumptions, it can be shown that the irreducible factors of A modulo p will have distinct degrees with probability close to $e^{-\gamma} \approx 0.56146$, where γ is Euler's constant, where we assume the degree of A to be large (see [Knu2]).

As a corollary to the above discussion and algorithm, we see that it is easy to determine whether a polynomial is irreducible in $\mathbb{F}_p[X]$. More precisely, we have:

Proposition 3.4.4. *A polynomial $A \in \mathbb{F}_p[X]$ of degree n is irreducible if and only if the following two conditions are satisfied:*

$$X^{p^n} \equiv X \pmod{A(X)},$$

and for each prime q dividing n

$$(X^{p^{n/q}} - X, A(X)) = 1 .$$

Note that to test in practice the second condition of the proposition, one must first compute $B(X) = X^{p^{n/q}} \bmod A(X)$ using the powering algorithm, and then compute $\gcd(B(X) - X, A(X))$. Hence, the time necessary for this irreducibility test, assuming one uses the $O(n^2)$ algorithms for multiplication and division of polynomials of degree n, is essentially $O(n^3 \ln p)$, if the factorization of n is known (since nobody considers polynomials of degree larger, say than 10^9, this is a reasonable assumption).

It is interesting to compare this with the analogous primality test for integers. By Proposition 8.3.1, n is prime if and only if for each prime q dividing $n - 1$ one can find an $a_q \in \mathbb{Z}$ such that $a_q^{n-1} \equiv 1 \pmod{n}$ and $a_q^{(n-1)/q} \not\equiv 1 \pmod{n}$. This takes time $O(\ln^3 n)$, assuming the factorization of $n - 1$ to be known. But this is an unreasonable assumption, since one commonly wants to prove the primality of numbers of 100 decimal digits, and at present it is quite unreasonable to factor a 100 digit number. Hence the above criterion is not useful as a general purpose primality test over the integers.

3.4.4 Final Splitting

Finally we must consider the most important and central part of Algorithm 3.4.1, its step 3, which in fact does most of the work. After the preceding steps we are left with the following problem. Given a polynomial A which is known to be squarefree and equal to a product of irreducible polynomials of degree exactly equal to d, find these factors. Of course, $\deg(A)$ is a multiple of d, and if $\deg(A) = d$ we know that A is itself irreducible and there is nothing to do. A simple and efficient way to do this was found by Cantor and Zassenhaus. Assume first that $p > 2$. Then we have the following lemma:

Proposition 3.4.5. *If A is as above, then for any polynomial $T \in \mathbb{F}_p[X]$ we have the identity:*

$$A = (A, T) \cdot (A, T^{(p^d-1)/2} + 1) \cdot (A, T^{(p^d-1)/2} - 1) \ .$$

Proof. The roots of the polynomial $X^{p^d} - X$, being the elements of $\mathbb{F}_{p^d}$, are all distinct. It follows that for any $T \in \mathbb{F}_p[X]$, the polynomial $T(X)^{p^d} - T(X)$ also has all the elements of $\mathbb{F}_{p^d}$ as roots, hence is divisible by $X^{p^d} - X$. In particular, as we have seen in the preceding section, it is a multiple of every irreducible polynomial of degree d, hence of A, since A is squarefree. The claimed identity follows immediately by noting that

$$T^{p^d} - T = T \cdot (T^{(p^d-1)/2} + 1) \cdot (T^{(p^d-1)/2} - 1)$$

with the three factors pairwise coprime.   □

Now it is not difficult to show that if one takes for T a random monic polynomial of degree less than or equal to $2d - 1$, then $(A, T^{(p^d-1)/2} - 1)$ will be a non-trivial factor of A with probability close to $1/2$. Hence, we can use the following algorithm to factor A:

Algorithm 3.4.6 (Cantor-Zassenhaus Split). Given A as above, this algorithm outputs its irreducible factors (which are all of degree d). This algorithm will be called recursively.

1. [Initialize] Set $k \leftarrow \deg(A)/d$. If $k = 1$, output A and terminate the algorithm.

2. [Try a T] Choose $T \in \mathbb{F}_p[X]$ randomly such that T is monic of degree less than or equal to $2d - 1$. Set $B \leftarrow (A, T^{(p^d-1)/2} - 1)$. If $\deg(B) = 0$ or $\deg(B) = \deg(A)$ go to step 2.

3. [Recurse] Factor B and A/B using the present algorithm recursively, and terminate the algorithm.

Note that, as has already been mentioned after Proposition 3.4.4, to compute B in step 2 one first computes $C \leftarrow T^{(p^d-1)/2} \bmod A$ using the powering algorithm, and then $B \leftarrow (A, C - 1)$.

Finally, we must consider the case where $p = 2$. In that case, the following result is the analog of Proposition 3.4.5:

Proposition 3.4.7. *Set*

$$U(X) = X + X^2 + X^4 + \cdots + X^{2^{d-1}} \ .$$

If $p = 2$ and A is as above, then for any polynomial $T \in \mathbb{F}_2[X]$ we have the identity

$$A = (A, U \circ T) \cdot (A, U \circ T + 1) \ .$$

Proof. Note that $(U \circ T)^2 = T^2 + T^4 + \cdots + T^{2^d}$, hence $(U \circ T) \cdot (U \circ T + 1) = T^{2^d} - T$ (remember that we are in characteristic 2). By the proof of Proposition 3.4.5 we know that this is a multiple of A, and the identity follows. $\square$

Exactly as in the case of $p = 2$, one can show that the probability that $(A, U \circ T)$ is a non-trivial factor of A is close to $1/2$, hence essentially the same algorithm as Algorithm 3.4.6 can be used. Simply replace in step 2 $B \leftarrow (A, T^{(p^d-1)/2} - 1)$ by $B \leftarrow (A, U \circ T)$. Here, however, we can do better than choosing random polynomials T in step 2 as follows.

Algorithm 3.4.8 (Split for $p = 2$). Given $A \in \mathbb{F}_2[X]$ as above, this algorithm outputs its irreducible factors (which are all of degree d). This algorithm will be called recursively.

1. [Initialize] Set $k \leftarrow \deg(A)/d$. If $k = 1$, output A and terminate the algorithm, otherwise set $T \leftarrow X$.

2. [Test T] Set $C \leftarrow T$ and then repeat $d - 1$ times $C \leftarrow T + C^2 \bmod A$. Then set $B \leftarrow (A, C)$. If $\deg(B) = 0$ or $\deg(B) = \deg(A)$ then set $T \leftarrow T \cdot X^2$ and go to step 2.

3. [Recurse] Factor B and A/B using the present algorithm recursively, and terminate the algorithm.

Proof. If this algorithm terminates, it is clear that the output is a factorization of A, hence the algorithm is correct. We must show that it terminates. Notice first that the computation of C done in step 2 is nothing but the computation of $U \circ T$ (note that on page 630 of [Knu2], Knuth gives $C \leftarrow (C + C^2 \bmod A)$, but this should be instead, as above, $C \leftarrow T + C^2 \bmod A$).

Now, since for any $T \in \mathbb{F}_2[X]$, we have by Proposition 3.4.7 $U(T) \cdot (U(T) + 1) \equiv 0 \pmod{A}$, it is clear that $(U(T), A) = 1$ is equivalent to $U(T) \equiv 1 \pmod{A}$. Furthermore, one immediately checks that $U(T^2) \equiv U(T) \pmod{A}$, and that $U(T_1 + T_2) = U(T_1) + U(T_2)$.

Now I claim that the algorithm terminates when $T = X^e$ in step 2 for some odd value of e such that $e \leq 2d - 1$. Indeed, assume the contrary. Then we have for every odd $e \leq 2d - 1$, $(U(X^e), A) = 1$ or A, hence $U(X^e) \equiv 0$ or 1 modulo A. Since $U(T^2) \equiv U(T) \pmod{A}$, this is true also for even values of $e \leq 2d$, and the linearity of U implies that this is true for every polynomial of degree less than or equal to $2d$. Now U is a polynomial of degree 2^{d-1}, and has at most (in fact exactly, see Exercise 15) 2^{d-1} roots in $\mathbb{F}_{2^d}$. Let $\beta \in \mathbb{F}_{2^d}$ not a root of U. The number of irreducible factors of A is at least equal to 2 (otherwise we would have stopped at step 1), and let A_1 and A_2 be two such factors, both of degree d. Let α be a root of A_2 in $\mathbb{F}_{2^d}$ (notice that all the roots of A_2 are in $\mathbb{F}_{2^d}$). Since A_2 is irreducible, α generates $\mathbb{F}_{2^d}$ over $\mathbb{F}_2$. Hence, there exists a polynomial $P \in \mathbb{F}_2[X]$ such that $\beta = P(\alpha)$.

By the Chinese remainder theorem, since A_1 and A_2 are coprime we can choose a polynomial T such that $T \equiv 0 \pmod{A_1}$ and $T \equiv P \pmod{A_2}$, and

T is defined modulo the product $A_1 A_2$. Hence, we can choose T of degree less than $2d$. But

$$U(T) \equiv U(0) \equiv 0 \pmod{A_1}$$

and

$$U(T) \equiv U(P) \not\equiv 0 \pmod{A_2}$$

since

$$U(P(\alpha)) = U(\beta) \neq 0 .$$

This contradicts $U(T) \equiv 0$ or 1 modulo A, thus proving the validity of the algorithm. The same proof applied to $T^{p^d} - T$ instead of $U(T)$ explains why one can limit ourselves to $\deg(T) \leq 2d - 1$ in Algorithm 3.4.6.   □

Proposition 3.4.7 and Algorithm 3.4.8 can be extended to general primes p, but are useful in practice only if p is small (see Exercise 14).

There is another method for doing the final splitting due to Berlekamp which predates that of Cantor-Zassenhaus, and which is better in many cases. This method could be used as soon as the polynomial is squarefree. (In other words, if desirable, we can skip the distinct degree factorization.) It is based on the following proposition.

Proposition 3.4.9. *Let $A \in \mathbb{F}_p[X]$ be a squarefree polynomial, and let*

$$A(X) = \prod_{1 \leq i \leq r} A_i(X)$$

be its decomposition into irreducible factors. The polynomials $T \in \mathbb{F}_p[X]$ with $\deg(T) < \deg(A)$ for which for each i with $1 \leq i \leq r$ there exist $s_i \in \mathbb{F}_p$ such that $T(X) \equiv s_i \pmod{A_i(X)}$, are exactly the p^r polynomials T such that $\deg(T) < \deg(A)$ and $T(X)^p \equiv T(X) \pmod{A(X)}$.

Proof. By the Chinese remainder Theorem 1.3.9 generalized to the Euclidean ring $\mathbb{F}_p[X]$, for each of the p^r possible choices of $s_i \in F_p$ $(1 \leq i \leq r)$, there exists a unique polynomial $T \in \mathbb{F}_p[X]$ such that $\deg(T) < \deg(A)$ and for each i

$$T(X) \equiv s_i \pmod{A_i(X)} .$$

Now if T is a solution of such a system, we have

$$T(X)^p \equiv s_i^p = s_i \equiv T(X) \pmod{A_i(X)}$$

for each i, hence

$$T(X)^p \equiv T(X) \pmod{A(X)} .$$

Conversely, note that we have in $\mathbb{F}_p[X]$ the polynomial identity $X^p - X = \prod_{0 \leq s \leq p-1}(X - s)$, hence

$$T(X)^p - T(X) = \prod_{0 \le s \le p-1} (T(X) - s) .$$

Hence, if $T(X)^p \equiv T(X) \pmod{A(X)}$, we have for all i

$$A_i(X) \mid \prod_{0 \le s \le p-1} (T(X) - s) ,$$

and since the A_i are irreducible this means that $A_i(X) \mid T(X) - s_i$ for some $s_i \in \mathbb{F}_p$ thus proving the proposition. $\square$

The relevance of this proposition to our splitting problem is the following. If T is a solution of such a system of congruences with, say, $s_1 \ne s_2$, then $\gcd(A(X), T(X) - s_1)$ will be divisible by A_1 and not by A_2, hence it will be a non-trivial divisor of A. The above proposition tells us that to look for such nice polynomials T it is not necessary to know the A_i, but simply to solve the congruence $T(X)^p \equiv T(X) \pmod{A(X)}$.

To solve this, write $T(X) = \sum_{0 \le j < n} t_j X^j$, where $n = \deg(A)$, with $t_j \in \mathbb{F}_p$. Then $T(X)^p = \sum_j t_j X^{pj}$, hence if we set

$$X^{pk} \equiv \sum_{0 \le i < n} q_{i,k} X^i \pmod{A(X)}$$

we have

$$T(X)^p \equiv \sum_j t_j \sum_i q_{i,j} X^i \pmod{A(X)}$$

so the congruence $T(X)^p \equiv T(X) \pmod{A(X)}$ is equivalent to

$$\sum_j t_j q_{i,j} = t_i \quad \text{for} \quad 1 \le i < n .$$

If, in matrix terms, we set $Q = (q_{i,j})$, $V = (t_j)$ (column vector), and I the identity matrix, this means that $QV = V$. In other words $(Q - I)V = 0$, so V belongs to the kernel of the matrix $Q - I$.

Algorithm 2.3.1 will allow us to compute this kernel, and it is especially efficient since we work in a finite field where no coefficient explosion or instability occurs.

Thus we will obtain a basis of the kernel of $Q - I$, which will be of dimension r by Proposition 3.4.9. Note that trivially $q_{i,0} = 0$ if $i > 0$ and $q_{0,0} = 1$, hence the column vector $(1, 0, \ldots, 0)^t$ will always be an element of the kernel, corresponding to the trivial choice $T(X) = 1$. Any other basis element of the kernel will be useful. If $T(X)$ is the polynomial corresponding to a V in the kernel of $Q - I$, we compute $(A(X), T(X) - s)$ for $0 \le s \le p - 1$. Since by Proposition 3.4.9 there exists an s such that $T(X) \equiv s \pmod{A_1(X)}$, there will exist an s which will give a non-trivial GCD, hence a splitting of A. We apply this to all values of s and all basis vectors of the kernel until the r irreducible factors of A have been isolated (note that it is better to proceed in this way than to use the algorithm recursively once a split is found as in Algorithm 3.4.6 since it avoids the recomputation of Q and of the kernel of $Q - I$).

This leads to the following algorithm.

Algorithm 3.4.10 (Berlekamp for Small p). Given a squarefree polynomial $A \in \mathbb{F}_p[X]$ of degree n, this algorithm computes the factorization of A into irreducible factors.

1. [Compute Q] Compute inductively for $0 \le k < n$ the elements $q_{i,k} \in \mathbb{F}_p$ such that

$$X^{pk} \equiv \sum_{0 \le i < n} q_{i,k} X^i \pmod{A(X)} .$$

2. [Compute kernel] Using Algorithm 2.3.1, find a basis $V_1, \ldots, V_r$ of the kernel of $Q - I$. Then r will be the number of irreducible factors of A, and $V_1 = (1, 0, \ldots, 0)^t$. Set $E \leftarrow \{A\}$, $k \leftarrow 1$, $j \leftarrow 1$ (E will be a set of polynomials whose product is equal to A, k its cardinality and j is the index of the vector of the kernel which we will use).

3. [Finished?] If $k = r$, output E as the set of irreducible factors of A and terminate the algorithm. Otherwise, set $j \leftarrow j + 1$, and let $T(X)$ be the polynomial corresponding to the vector V_j (i.e. $T(X) \leftarrow \sum_{0 \le i < n} (V_j)_i X^i$).

4. [Split] For each element $B \in E$ such that $\deg(B) > 1$ do the following. For each $s \in \mathbb{F}_p$ compute $(B(X), T(X) - s)$. Let F be the set of such GCD's whose degree is greater or equal to 1. Set $E \leftarrow (E \setminus \{B\}) \cup F$ and $k \leftarrow k - 1 + |F|$. If in the course of this computation we reach $k = r$, output E and terminate the algorithm. Otherwise, go to step 3.

The main drawback of this algorithm is that the running time of step 4 is proportional to p, and this is slower than Algorithm 3.4.6 as soon as p gets above 100 say. On the other hand, if p is small, a careful implementation of this algorithm will be faster than Algorithm 3.4.6. This is important, since in many applications such as factoring polynomials over $\mathbb{Z}$, we will first factor the polynomial over a few fields $\mathbb{F}_p$ for small primes p where Berlekamp's algorithm is superior.

If we use the idea of the Cantor-Zassenhaus split, we can however improve considerably Berlekamp's algorithm when p is large. In steps 3 and 4, instead of considering the polynomials corresponding to the vectors $V_j - sV_1$ for $2 \le j \le r$ and $s \in \mathbb{F}_p$, we instead choose a random linear combination $V = \sum_{i=1}^r a_i V_i$ with $a_i \in \mathbb{F}_p$ and compute $(B(X), T(X)^{(p-1)/2} - 1)$, where T is the polynomial corresponding to V. It is easy to show that this GCD will give a non-trivial factor of $B(X)$ with probability greater than or equal to $4/9$ when $p \ge 3$ and B is reducible (see Exercise 17 and [Knu2] p. 429). This gives the following algorithm.

Algorithm 3.4.11 (Berlekamp). Given a squarefree polynomial $A \in \mathbb{F}_p[X]$ of degree n (with $p \geq 3$), this algorithm computes the factorization of A into irreducible factors.

1. [Compute Q] Compute inductively for $0 \leq k < n$ the elements $q_{i,k} \in \mathbb{F}_p$ such that

$$X^{pk} \equiv \sum_{0 \leq i < n} q_{i,k} X^i .$$

2. [Compute kernel] Using Algorithm 2.3.1, find a basis $V_1, \ldots, V_r$ of the kernel of $Q - I$, and let $T_1, \ldots, T_r$ be the corresponding polynomials. Then r will be the number of irreducible factors of A, and $V_1 = (1, 0, \ldots, 0)^t$ hence $T_1 = 1$. Set $E \leftarrow \{A\}$, $k \leftarrow 1$, (E will be a set of polynomials whose product is equal to A and k its cardinality).

3. [Finished?] If $k = r$, output E as the set of irreducible factors of A and terminate the algorithm. Otherwise, choose r random elements $a_i \in \mathbb{F}_p$, and set $T(X) \leftarrow \sum_{1 \leq i \leq r} a_i T_i(X)$.

4. [Split] For each element $B \in E$ such that $\deg(B) > 1$ do the following. Let $D(X) \leftarrow (B(X), T(X)^{(p-1)/2} - 1)$. If $\deg(D) > 0$ and $\deg(D) < \deg(B)$, set $E \leftarrow (E \setminus \{B\}) \cup \{D, B/D\}$ and $k \leftarrow k+1$. If in the course of this computation we reach $k = r$, output E and terminate the algorithm. Otherwise, go to step 3.

Note that if we precede any of these two Berlekamp algorithms by the distinct degree factorization procedure (Algorithm 3.4.3), we should replace the condition $\deg(B) > 1$ of step 4 by $\deg(B) > d$, since we know that all the irreducible factors of A have degree d.

Using the algorithms of this section, we now have at our disposal several efficient methods for completely factoring polynomials modulo a prime p. We will now consider the more difficult problem of factoring over $\mathbb{Z}$.

3.5 Factorization of Polynomials over $\mathbb{Z}$ or $\mathbb{Q}$

The first thing to note is that factoring over $\mathbb{Q}$ is essentially equivalent to factoring over $\mathbb{Z}$. Indeed if $A = \prod_i A_i$ where the A_i are irreducible over $\mathbb{Q}$, then by multiplying by suitable rational numbers, we have $dA = \prod_i (d_i A_i)$ where the d_i can be chosen so that the $d_i A_i$ have integer coefficients and are primitive. Hence it follows from Gauss's lemma (Theorem 3.2.8) that if $A \in \mathbb{Z}[X]$, then $d = \pm 1$. Conversely, if $A = \prod_i A_i$ with A and the A_i in $\mathbb{Z}[X]$ and the A_i are irreducible over $\mathbb{Z}$, then the A_i are also irreducible over $\mathbb{Q}$, by a similar use of Gauss's lemma.

Therefore in this section, we will consider only the problem of factoring a polynomial A over $\mathbb{Z}$. If $A = BC$ is a splitting of A in $\mathbb{Z}[X]$, then $\bar{A} = \bar{B}\bar{C}$ in $\mathbb{F}_p[X]$, where $\bar{}$ denotes reduction mod p. Hence we can start by reducing mod p for some p, factor mod p, and then see if the factorization over $\mathbb{F}_p$ lifts to one over $\mathbb{Z}$. For this, it is essential to know an upper bound on the absolute value of the coefficients which can occur as a factor of A.

3.5.1 Bounds on Polynomial Factors

The results presented here are mostly due to Mignotte [Mig]. The aim of this section is to prove the following theorem:

Theorem 3.5.1. *For any polynomial* $P = \sum_{0 \le i \le n} p_i X^i \in \mathbb{C}[X]$ *set* $|P| = (\sum_i |p_i|^2)^{1/2}$. *Let* $A = \sum_{0 \le i \le m} a_i X^i$ *and* $B = \sum_{0 \le i \le n} b_i X^i$ *be polynomials with integer coefficients, and assume that* B *divides* A. *Then we have for all* j

$$|b_j| \le \binom{n-1}{j} |A| + \binom{n-1}{j-1} |a_m| .$$

Proof. Let α be any complex number, and let $A = \sum_{0 \le i \le m} a_i X^i$ be any polynomial. Set $G(X) = (X - \alpha)A(X)$ and $H(X) = (\overline{\alpha}X - 1)A(X)$. Then

$$\begin{aligned}
|G|^2 &= \sum |a_{i-1} - \alpha a_i|^2 = \sum (|a_{i-1}|^2 + |\alpha a_i|^2 - 2\operatorname{Re}(\alpha a_i \overline{a_{i-1}})) \\
&= \sum (|\alpha a_{i-1}|^2 + |a_i|^2 - 2\operatorname{Re}(\alpha a_i \overline{a_{i-1}})) \\
&= \sum |\overline{\alpha} a_{i-1} - a_i|^2 = |H|^2 .
\end{aligned}$$

Let now $A(X) = a_m \prod_j (X - \alpha_j)$ be the decomposition of A over $\mathbb{C}$. If we set

$$C(X) = a_m \prod_{|\alpha_j| \ge 1} (X - \alpha_j) \prod_{|\alpha_j| < 1} (\overline{\alpha_j}X - 1) ,$$

it follows that $|C| = |A|$. Hence, taking into account only the coefficient of X^m and the constant term, it follows that

$$|A|^2 = |C|^2 \ge |a_m|^2 (M(A)^2 + m(A)^2) ,$$

where

$$M(A) = \prod_{|\alpha_j| > 1} |\alpha_j| , \qquad m(A) = \prod_{|\alpha_j| < 1} |\alpha_j| .$$

In particular, $M(A) \le |A|/|a_m|$. Now

$$|a_j| = |a_m| \left| \sum \alpha_{i_1} \dots \alpha_{i_{m-j}} \right| \le |a_m| \sum \beta_{i_1} \dots \beta_{i_{m-j}} ,$$

where $\beta_i = \max(1, |\alpha_i|)$. Assume for the moment the following lemma:

Lemma 3.5.2. *If* $x_1 \ge 1, \dots, x_m \ge 1$ *are real numbers constrained by the further condition that their product is equal to* M, *then the elementary symmetric function* $\sigma_{mk} = \sum x_{i_1} \dots x_{i_k}$ *satisfies*

$$\sigma_{mk} \le \binom{m-1}{k-1} M + \binom{m-1}{k} .$$

Since the product of the β_i is by definition $M(A)$, it follows from the lemma that for all j,

$$|a_j| \leq |a_m| \left(\binom{m-1}{m-j-1} M(A) + \binom{m-1}{m-j} \right)$$
$$\leq |a_m| \left(\binom{m-1}{j} M(A) + \binom{m-1}{j-1} \right) .$$

Coming back to our notations and applying the preceding result to the polynomial B, we see that $|b_j| \leq |b_n|(\binom{n-1}{j} M(B) + \binom{n-1}{j-1})$. It follows that $|b_j| \leq |a_m|(\binom{n-1}{j} M(A) + \binom{n-1}{j-1})$ since if B divides A, we must have $M(B) \leq M(A)$ (since the roots of B are roots of A), and $|b_n| \leq |a_m|$ (since in fact b_n divides a_m). The theorem follows from this and the inequality $M(A) \leq |A|/|a_m|$ proved above.

It remains to prove the lemma. Assume without loss of generality that $x_1 \leq x_2 \cdots \leq x_m$. If one changes the pair (x_{m-1}, x_m) into the pair $(1, x_{m-1}x_m)$, all the constraints are still satisfied and it is easy to check that the value of σ_{mk} is increased by

$$\sigma_{(m-2)(k-1)}(x_{m-1} - 1)(x_m - 1) .$$

It follows that if $x_{m-1} > 1$, the point $(x_1, \ldots, x_m)$ cannot be a maximum. Hence a necessary condition for a maximum is that $x_{m-1} = 1$. But this immediately implies that $x_i = 1$ for all $i < m$, and hence that $x_m = M$. It is now a simple matter to check the inequality of the lemma, the term $\binom{m-1}{k-1} M$ corresponding to k-tuples containing x_m, and the term $\binom{m-1}{k}$ to the ones which do not contain x_m. This finishes the proof of Theorem 3.5.1.   □

A number of improvements can be made in the estimates given by this theorem. They do not seriously influence the running time of the algorithms using them however, hence we will be content with this.

3.5.2 A First Approach to Factoring over $\mathbb{Z}$

First note that for polynomials A of degree 2 or 3 with coefficients which are not too large, the factoring problem is easy: if A is not irreducible, it must have a linear factor $qX - p$, and q must divide the leading term of A, and p must divide the constant term. Hence, if the leading term and the constant term can be easily factored, one can check each possible divisor of A. An ad hoc method of this sort could be worked out also in higher degrees, but soon becomes impractical.

A second way of factoring over $\mathbb{Z}$ is to combine information obtained by the mod p factorization methods. For example, if modulo some prime p, $A(X) \bmod p$ is irreducible, then $A(X)$ itself is of course irreducible. A less trivial example is the following: if for some p a polynomial $A(X)$ of degree 4 breaks modulo p into a product of two irreducible polynomials of degree 2, and for another p into a product of a polynomial of degree 1 and an irreducible polynomial of degree 3, then $A(X)$ must be irreducible since these splittings are incompatible. Unfortunately, although this method is useful when combined with other methods, except for polynomials of small degree, when used alone it rarely works. For example, using the quadratic reciprocity law and the identities

$$\begin{aligned}
X^4 + 1 &= (X^2 + \sqrt{-1})(X^2 - \sqrt{-1}) \\
&= (X^2 - X\sqrt{2} + 1)(X^2 + X\sqrt{2} + 1) \\
&= (X^2 + X\sqrt{-2} - 1)(X^2 - X\sqrt{-2} - 1)
\end{aligned}$$

it is easy to check that the polynomial $X^4 + 1$ splits into 4 linear factors if $p = 2$ or $p \equiv 1 \pmod 8$, and into two irreducible quadratic factors otherwise. This is compatible with the possibility that $X^4 + 1$ could split into 2 quadratic factors in $\mathbb{Z}[X]$, and this is clearly not the case.

A third way to derive a factorization algorithm over $\mathbb{Z}$ is to use the bounds given by Theorem 3.5.1 and the mod p factorization methods. Consider for example the polynomial

$$A(X) = X^6 - 6X^4 - 2X^3 - 7X^2 + 6X + 1 .$$

If it is not irreducible, it must have a factor of degree at most 3. The bound of Theorem 3.5.1 shows that for any factor of degree less or equal to 3 and any j, one must have $|b_j| \leq 23$. Take now a prime p greater than twice that bound and for which the polynomial A mod p is squarefree, for example $p = 47$. The mod p factoring algorithms of the preceding section show that modulo 47 we have

$$A(X) = (X - 22)(X - 13)(X - 12)(X + 12)(X^2 - 12X - 4) ,$$

taking as representatives of $\mathbb{Z}/47\mathbb{Z}$ the numbers from -23 to 23. Now the constant term of A being equal to 1, up to sign any factor of A must have that property. This immediately shows that A has no factor of degree 1 over $\mathbb{Z}$ (this could of course have been checked more easily simply by noticing that $A(1)$ and $A(-1)$ are both non-zero), but it also shows that A has no factor of degree 2 since modulo 47 we have $12 \cdot 22 = -18$, $12 \cdot 13 = 15$, $12 \cdot 12 = 3$ and $13 \cdot 22 = 4$. Hence, if A is reducible, the only possibility is that A is a product of two factors of degree 3. One of them must be divisible by $X^2 - 12X - 4$, and hence can be (modulo 47) equal to either $(X^2 - 12X - 4)(X - 12)$ (whose constant term is 1), or to $(X^2 - 12X - 4)(X + 12)$ (whose constant term is -1). Now modulo 47, we have $(X^2 - 12X - 4)(X - 12) = X^3 + 23X^2 - X + 1$ and $(X^2 - 12X - 4)(X + 12) = X^3 - 7X - 1$.

The first case can be excluded a priori because the bound of Theorem 3.5.1 gives $b_2 \leq 12$, hence 23 is too large. In the other case, by the choice made for p, this is the only polynomial in its congruence class modulo 47 satisfying the bounds of Theorem 3.5.1. Hence, if it divides A in $\mathbb{Z}[X]$, we have found the factorization of A, otherwise we can conclude that A is irreducible. Since one checks that $A(X) = (X^3 - 7X - 1)(X^3 + X - 1)$, we have thus obtained the complete factorization of A over $\mathbb{Z}$. Note that the irreducibility of the factors of degree 3 has been proved along the way.

When the degree or the coefficients of A are large however, the bounds of Theorem 3.5.1 imply that we must use a p which is really large, and hence for which the factorization modulo p is too slow. We can overcome this problem by keeping a small p, but factoring modulo p^e for sufficiently large e.

3.5.3 Factorization Modulo p^e: Hensel's Lemma

The trick is that if certain conditions are satisfied, we can "lift" a factorization modulo p to a factorization mod p^e for any desired e, without too much effort. This is based on the following theorem, due to Hensel, and which was one of his motivations for introducing p-adic numbers.

Theorem 3.5.3. *Let p be a prime, and let C, A_e, B_e, U, V be polynomials with integer coefficients and satisfying*

$$C(X) \equiv A_e(X)B_e(X) \pmod{p^e}, \quad U(X)A_e(X)+V(X)B_e(X) \equiv 1 \pmod{p} .$$

Assume that $e \geq 1$, A_e is monic, $\deg(U) < \deg(B_e)$, $\deg(V) < \deg(A_e)$. Then there exist polynomials A_{e+1} and B_{e+1} satisfying the same conditions with e replaced by $e + 1$, and such that

$$A_{e+1}(X) \equiv A_e(X) \pmod{p^e} , \quad B_{e+1}(X) \equiv B_e(X) \pmod{p^e} .$$

Furthermore, these polynomials are unique modulo p^{e+1}.

Proof. Set $D = (C - A_e B_e)/p^e$ which has integral coefficients by assumption. We must have $A_{e+1} = A_e + p^e S$, $B_{e+1} = B_e + p^e T$ with S and T in $\mathbb{Z}[X]$. The main condition needed is $C(X) \equiv A_{e+1}(X)B_{e+1}(X) \pmod{p^{e+1}}$. Since $2e \geq e + 1$, this is equivalent to $A_e T + B_e S \equiv (C - A_e B_e)/p^e = D \pmod{p}$. Since $U A_e + V B_e = 1$ in $\mathbb{F}_p[X]$ and $\mathbb{F}_p$ is a field, the general solution is $S \equiv VD+WA_e \pmod{p}$ and $T \equiv UD - WB_e \pmod{p}$ for some polynomial W. The conditions on the degrees imply that S and T are unique modulo p, hence A_{e+1} and B_{e+1} are unique modulo p^{e+1}. Note that this proof is constructive, and gives a simple algorithm to obtain A_{e+1} and B_{e+1}. $\square$

For reasons of efficiency, it will be useful to have a more general version of Theorem 3.5.3. The proof is essentially identical to the proof of Theorem 3.5.3, and will follow from the corresponding algorithm.

Theorem 3.5.4. *Let p, q be (not necessarily prime) integers, and let $r = (p, q)$. Let C, A, B, U and V be polynomials with integer coefficients satisfying*

$$C \equiv AB \pmod{q}, \qquad UA + VB \equiv 1 \pmod{p},$$

and assume that $(\ell(A), r) = 1$, $\deg(U) < \deg(B)$, $\deg(V) < \deg(A)$ and $\deg(C) = \deg(A) + \deg(B)$. (Note that this last condition is not automatically satisfied since $\mathbb{Z}/q\mathbb{Z}$ may have zero divisors.) Then there exist polynomials A_1 and B_1 such that $A_1 \equiv A \pmod{q}$, $B_1 \equiv B \pmod{q}$, $\ell(A_1) = \ell(A)$, $\deg(A_1) = \deg(A)$, $\deg(B_1) = \deg(B)$ and

$$C \equiv A_1 B_1 \pmod{qr}.$$

In addition, A_1 and B_1 are unique modulo qr if r is prime.

We give the proof as an algorithm.

Algorithm 3.5.5 (Hensel Lift). Given the assumptions and notations of Theorem 3.5.4, this algorithm outputs the polynomials A_1 and B_1. As a matter of notation, we denote by K the ring $\mathbb{Z}/r\mathbb{Z}$.

1. [Euclidean division] Set $f \leftarrow (C - AB)/q \pmod{r} \in K[X]$. Using Algorithm 3.1.1 over the ring K, find $t \in K[X]$ such that $\deg(Vf - At) < \deg(A)$ (this is possible even when K is not a field, since $\ell(A)$ is invertible in K).

2. [Terminate] Let A_0 be a lift of $Vf - At$ to $\mathbb{Z}[X]$ having the same degree, and let B_0 be a lift of $Uf + Bt$ to $\mathbb{Z}[X]$ having the same degree. Output $A_1 \leftarrow A + qA_0$, $B_1 \leftarrow B + qB_0$ and terminate the algorithm.

Proof. It is clear that $BA_0 + AB_0 \equiv f \pmod{r}$. From this, it follows immediately that $C \equiv A_1 B_1 \pmod{qr}$ and that $\deg(B_0) \leq \deg(B)$, thus proving the validity of the algorithm and of Theorem 3.5.4. $\qquad\square$

If $p \mid q$, we can also if desired replace p by $pr = p^2$ in the following way.

Algorithm 3.5.6 (Quadratic Hensel Lift). Assume $p \mid q$, hence $r = p$. After execution of Algorithm 3.5.5, this algorithm finds U_1 and V_1 such that $U_1 \equiv U \pmod{p}$, $V_1 \equiv V \pmod{p}$, $\deg(U_1) < \deg(B_1)$, $\deg(V_1) < \deg(A_1)$ and

$$U_1 A_1 + V_1 B_1 \equiv 1 \pmod{pr}.$$

1. [Euclidean division] Set $g \leftarrow (1 - UA_1 - VB_1)/p \pmod{r}$. Using Algorithm 3.1.1 over the same ring $K = \mathbb{Z}/r\mathbb{Z}$, find $t \in K[X]$ such that $\deg(Vg - A_1 t) < \deg(A_1)$, which is possible since $\ell(A_1) = \ell(A)$ is invertible in K.

2. [Terminate] Let U_0 be a lift of $Ug + B_1 t$ to $\mathbb{Z}[X]$ having the same degree, and let V_0 be a lift of $Vg - A_1 t$ to $\mathbb{Z}[X]$ having the same degree. Output $U_1 \leftarrow U + pU_0$, $V_1 \leftarrow V + pV_0$ and terminate the algorithm.

It is not difficult to see that at the end of this algorithm, (A_1, B_1, U_1, V_1) satisfy the same hypotheses as (A, B, U, V) in the theorem, with (p, q) replaced by (pr, qr).

The condition $p \mid q$ is necessary for Algorithm 3.5.6 (not for Algorithm 3.5.5), and was forgotten by Knuth (page 628). Indeed, if $p \nmid q$, G does not have integral coefficients in general, since after constructing A_1 and B_1, one has only the congruence $UA_1 + VB_1 \equiv 1 \pmod r$ and not $\pmod p$. Of course, this only shows that Algorithm 3.5.6 cannot be used in that case, but it does not show that it is impossible to find U_1 and V_1 by some other method. It is however easy to construct counterexamples. Take $p = 33$, $q = 9$, hence $r = 3$, and $A(X) = X - 3$, $B(X) = X - 4$, $C(X) = X^2 + 2X + 3$, $U(X) = 1$ and $V(X) = -1$. The conditions of the theorem are satisfied, and Algorithm 3.5.5 gives us $A_1(X) = X - 21$ and $B_1(X) = X + 23$. Consider now the congruence that we want, i.e.

$$U_1(X)(X - 21) + V_1(X)(X + 23) \equiv 1 \pmod{99} ,$$

or equivalently

$$U_1(X)(X - 21) + V_1(X)(X + 23) = 1 + 99W(X) ,$$

where all the polynomials involved have integral coefficients. If we set $X = 21$, we obtain $44V_1(21) = 1 + 99W(21)$, hence $0 \equiv 1 \pmod{11}$ which is absurd. This shows that even without any restriction on the degrees, it is not always possible to lift p to pr if $p \nmid q$.

The advantage of using both algorithms instead of one is that we can increase the value of the exponent e much faster. Assume that we start with $p = q$. Then, by using Algorithm 3.5.5 alone, we keep p fixed, and q takes the successive values p^2, p^3, etc .... If instead we use both Algorithms 3.5.5 and 3.5.6, the pair (p, q) takes the successive values (p^2, p^2), (p^4, p^4), etc ... with the exponent doubling each time. In principle this is much more efficient. When the exponent gets large however, the method slows down because of the appearance of multi-precision numbers. Hence, Knuth suggests the following recipe: let E be the smallest power of 2 such that p^E cannot be represented as a single precision number, and e be the largest integer such that p^e is a single precision number. He suggests working successively with the following pairs (p, q):

(p, p), (p^2, p^2), (p^4, p^4), ..., $(p^{E/2}, p^{E/2})$ using both algorithms, then (p^e, p^E) using both algorithms again but a reduced value of the exponent of p (since $e < E$) and finally (p^e, p^{E+e}), (p^e, p^{E+2e}), (p^e, p^{E+3e}), ... using only Algorithm 3.5.5.

Finally, note that by induction, one can extend Algorithms 3.5.5 and 3.5.6 to the case where C is congruent to a product of more than 2 pairwise coprime polynomials mod p.

3.5.4 Factorization of Polynomials over $\mathbb{Z}$

We now have enough ingredients to give a reasonably efficient method for factoring polynomials over the integers as follows.

Algorithm 3.5.7 (Factor in $\mathbb{Z}[X]$). Let $A \in \mathbb{Z}[X]$ be a non-zero polynomial. This algorithm finds the complete factorization of A in $\mathbb{Z}[X]$.

1. [Reduce to squarefree and primitive] Set $c \leftarrow \text{cont}(A)$, $A \leftarrow A/c$, $U \leftarrow A/(A, A')$ where (A, A') is computed using the sub-resultant Algorithm 3.3.1, or the method of Section 3.6.1 below. (Now U will be a squarefree primitive polynomial. In this step, we could also use the squarefree decomposition Algorithm 3.4.2 to reduce still further the degree of U).

2. [Find a squarefree factorization mod p] For each prime p, compute (U, U') over the field $\mathbb{F}_p$, and stop when this GCD is equal to 1. For this p, using the algorithms of Section 3.4, find the complete factorization of U mod p (which will be squarefree). Note that in this squarefree factorization it is not necessary to find each A_j from the U_j: we will have $A_j = U_j$ since $T = (U, U') = 1$.

3. [Find bound] Using Theorem 3.5.1, find a bound B for the coefficients of factors of U of degree less than or equal to $\deg(U)/2$. Choose e to be the smallest exponent such that $p^e > 2\ell(U)B$.

4. [Lift factorization] Using generalizations of Algorithms 3.5.5 and 3.5.6, and the procedure explained in the preceding section, lift the factorization obtained in step 2 to a factorization mod p^e. (One will also have to use Euclid's extended Algorithm 3.2.2.) Let

$$U \equiv \ell(U)U_1 U_2 \ldots U_r \pmod{p^e}$$

be the factorization of U mod p^e, where we can assume the U_i to be monic. Set $d \leftarrow 1$.

5. [Try combination] For every combination of factors $\bar{V} = U_{i_1} \ldots U_{i_d}$, where in addition we take $i_d = 1$ if $d = \frac{1}{2}r$, compute the unique polynomial $V \in \mathbb{Z}[X]$ such that all the coefficients of V are in $[-\frac{1}{2}p^e, \frac{1}{2}p^e[$, and satisfying $V \equiv \ell(U)\bar{V}$ $\pmod{p^e}$ if $\deg(V) \leq \frac{1}{2}\deg(U)$, $V \equiv U/\bar{V} \pmod{p^e}$ if $\deg(V) > \frac{1}{2}\deg(U)$.
   If V divides $\ell(U)U$ in $\mathbb{Z}[X]$, output the factor $F = \text{pp}(V)$, the exponent of F in A, set $U \leftarrow U/F$, and remove the corresponding U_i from the list of factors mod p^e (i.e. remove $U_{i_1} \ldots U_{i_d}$ and set $r \leftarrow r - d$ if $d \leq \frac{1}{2}r$, or leave only these factors and set $r \leftarrow d$ otherwise). If $d > \frac{1}{2}r$ terminate the algorithm by outputting $\text{pp}(U)$ if $\deg(U) > 0$.

6. Set $d \leftarrow d + 1$. If $d \leq \frac{1}{2}r$ go to step 5, otherwise terminate the algorithm by outputting $\text{pp}(U)$ if $\deg(U) > 0$.

Implementation Remarks. To decrease the necessary bound B, it is a good idea to reverse the coefficients of the polynomial U if $|u_0| < |u_n|$ (where of course we have cast out all powers of X so that $u_0 \neq 0$). Then the factors will be the reverse of the factors found.

In step 5, before trying to see whether V divides $\ell(U)U$, one should first test the divisibility of the constant terms, i.e. whether $V(0) \mid (\ell(U)U(0))$, since this will be rarely satisfied in general.

An important improvement can be obtained by using the information gained by factoring modulo a few small primes as mentioned in the second paragraph of Section 3.5.2. More precisely, apply the distinct degree factorization Algorithm 3.4.3 to U modulo a number of primes p_k (Musser and Knuth suggest about 5). If d_j are the degrees of the factors (it is not necessary to obtain the factors themselves) repeated with suitable multiplicity (so that $\sum_j d_j = n = \deg(U)$), build a binary string D_k of length $n+1$ which represents the degrees of all the possible factors mod p_k in the following way: Set $D_k \leftarrow (0\ldots01)$, representing the set with the unique element $\{0\}$. Then, for every d_j set

$$D_k \leftarrow D_k \vee (D_k \upharpoonleft d_j),$$

where $\vee$ is inclusive "or", and $D_k \upharpoonleft d_j$ is D_k shifted left d_j bits. (If desired, one can work with only the rightmost $\lceil (n+1)/2 \rceil$ bits of this string by symmetry of the degrees of the factors.)

Finally compute $D \leftarrow \bigwedge D_k$, i.e. the logical "and" of the bit strings. If the binary string D has only one bit at each end, corresponding to factors of degree 0 and n, this already shows that U is irreducible. Otherwise, choose for p the p_k giving the least number of factors. Then, during the execution of step 5 of Algorithm 3.5.7, keep only those d-uplets $(i_1, \ldots, i_d)$ such that the bit number $\deg(U_{i_1}) + \cdots + \deg(U_{i_d})$ of D is equal to 1.

Note that the prime chosen to make the Hensel lift will usually be small (say less than 20), hence in the modulo p factorization part, it will probably be faster to use Algorithm 3.4.10 than Algorithm 3.4.6 for the final splitting.

3.5.5 Discussion

As one can see, the problem of factoring over $\mathbb{Z}$ (or over $\mathbb{Q}$, which is essentially equivalent) is quite a difficult problem, and leads to an extremely complex algorithm, where there is a lot of room for improvement. Since this algorithm uses factorization mod p as a sub-algorithm, it is probabilistic in nature. Even worse, the time spent in step 5 above can be exponential in the degree. Therefore, a priori, the running time of the above algorithm is exponential in the degree. Luckily, in practice, its average behavior is random polynomial time. One should keep in mind however that in the worst case it *is* exponential time.

An important fact, discovered only relatively recently (1982) by Lenstra, Lenstra and Lovász is that it is possible to factor a polynomial over $\mathbb{Z}[X]$ in polynomial time using a deterministic algorithm. This is surprising in view of the corresponding problem over $\mathbb{Z}/p\mathbb{Z}[X]$ which should be simpler, and for which no such deterministic polynomial time algorithm is known, at least without assuming the Generalized Riemann Hypothesis. Their method uses in a fundamental way the LLL algorithm seen in Section 2.6.

The problem with the LLL factoring method is that, although in theory it is very nice, in practice it seems that it is quite a lot slower than the algorithm presented above. Therefore we will not give it here, but refer the interested reader to [LLL]. Note also that A. K. Lenstra has shown that similar algorithms exist over number fields, and also for multivariate polynomials.

There is however a naïve way to apply LLL which gives reasonably good results. Let A be the polynomial to be factored, and assume as one may, that it is squarefree (but not necessarily primitive). Then compute the roots α_i of A in $\mathbb{C}$ with high accuracy (say 19 decimal digits) (for example using Algorithm 3.6.6 below), then apply Algorithm 2.7.4 to $1, \alpha, \dots, \alpha^{k-1}$ for some $k < n$, where α is one of the α_i. Then if A is not irreducible, and if the constant N of Algorithm 2.7.4 is suitably chosen, α will be a root of a polynomial in $\mathbb{Z}[X]$ of some degree $k < n$, and this polynomial will probably be discovered by Algorithm 2.7.4. Of course, the results of Algorithm 2.7.4 may not correspond to exact relations, so to be sure that one has found a factor, one must algebraically divide A by its tentative divisor.

Although this method does not seem very clean and rigorous, it is certainly the easiest to implement. Hence, it should perhaps be tried before any of the more sophisticated methods above. In fact, in [LLL], it is shown how to make this method into a completely rigorous method. (They use p-adic factors instead of complex roots, but the result is the same.)

3.6 Additional Polynomial Algorithms

3.6.1 Modular Methods for Computing GCD's in $\mathbb{Z}[X]$

Using methods inspired from the factoring methods over $\mathbb{Z}$, one can return to the problem of computing GCD's over the specific UFD $\mathbb{Z}$, and obtain an algorithm which can be faster than the algorithms that we have already seen. The idea is as follows. Let $D = (A, B)$ in $\mathbb{Z}[X]$, and let $Q = (A, B)$ in $\mathbb{F}_p[X]$ where Q is monic. Then $D \bmod p$ is a common divisor of A and B in $\mathbb{F}_p[X]$, hence D divides Q in the ring $\mathbb{F}_p[X]$. (We should put $^-$ to distinguish polynomials in $\mathbb{Z}[X]$ from polynomials in $\mathbb{F}_p[X]$, but the language makes it clear.)

If p does not divide both $\ell(A)$ and $\ell(B)$, then p does not divide $\ell(D)$ and so $\deg(D) \leq \deg(Q)$. If, for example, we find that $Q = 1$ in $\mathbb{F}_p[X]$, it follows that D is constant, hence that $D = (\text{cont}(A), \text{cont}(B))$. This is in general much easier to check than to use any version of the Euclidean algorithm over a UFD (Algorithm 3.3.1 for example). Note also that, contrary to the case of integers, two random polynomials over $\mathbb{Z}$ are in general coprime. (In fact a single random polynomial is in general irreducible.) In general however, we are in a non-random situation so we must work harder. Assume without loss of generality that A and B are primitive.

So as not to be bothered with leading coefficients, instead of D, we will compute an integer multiple $D_1 = c \cdot (A, B)$ such that

$$\ell(D_1) = (\ell(A), \ell(B)) ,$$

(i.e. with $c = \ell(D)/(\ell(A), \ell(B)))$. We can then recover $D = \mathrm{pp}(D_1)$ since we have assumed A and B primitive.

Let M be the smallest of the bounds given by Theorem 3.5.1 for the two polynomials ℓA and ℓB, where $\ell = (\ell(A), \ell(B))$, and where we limit the degree of the factor by $\deg(Q)$. Assume for the moment that we skip the Hensel step, i.e. that we take $p > 2M$ (which in any case is the best choice if this leaves p in single precision). Compute the unique polynomial $Q_1 \in \mathbb{Z}[X]$ such that $Q_1 \equiv \ell Q \pmod{p}$ and having all its coefficients in $[-\frac{1}{2}p, \frac{1}{2}p[$. If $\mathrm{pp}(Q_1)$ is a common divisor of A and B (in $\mathbb{Z}[X]$!), then since D divides Q mod p, it follows that $(A, B) = \mathrm{pp}(Q_1)$. If it is not a common divisor, it is not difficult to see that this will happen only if p divides the leading term of one of the intermediate polynomials computed in the primitive form of Euclid's algorithm over a UFD (Algorithm 3.2.10), hence this will not occur often. If this phenomenon occurs, try again with another prime, and it should quickly work.

If M is really large, then one can use Hensel-type methods to determine D_1 mod p^e for sufficiently large e. The techniques are completely analogous to the ones given in the preceding sections and are left to the reader.

Perhaps the best conclusion for this section is to quote Knuth essentially verbatim:

"The GCD algorithms sketched here are significantly faster than those of Sections 3.2 and 3.3 except when the polynomial remainder sequence is very short. Perhaps the best general procedure would be to start with the computation of (A, B) modulo a fairly small prime p, not a divisor of both $\ell(A)$ and $\ell(B)$. If the result Q is 1, we are done; if it has high degree, we use Algorithm 3.3.1; otherwise we use one of the above methods, first computing a bound for the coefficients of D_1 based on the coefficients of A and B and on the (small) degree of Q. As in the factorization problem, we should apply this procedure to the reverses of A and B and reverse the result, if the trailing coefficients are simpler than the leading ones."

3.6.2 Factorization of Polynomials over a Number Field

This short section belongs naturally in this chapter but uses notions which are introduced only in Chapter 4, so please read Chapter 4 first before reading this section if you are not familiar with number fields.

In several instances, we will need to factor polynomials not only over $\mathbb{Q}$ but also over number fields $K = \mathbb{Q}(\theta)$. Following [Poh-Zas], we give an algorithm for performing this task (see also [Tra]).

Let $A(X) = \sum_{0 \leq i \leq m} a_i X^i \in K[X]$ be a non-zero polynomial. As usual, we can start by computing $A/(A, A')$ so we can transform it into a squarefree polynomial, since $K[X]$ is a Euclidean domain. On the other hand, note that it is not always possible to compute the content of A since the ring of integers $\mathbb{Z}_K$ of K is not always a PID.

Call σ_j the $m = [K : \mathbb{Q}]$ embeddings of K into $\mathbb{C}$. We can extend σ_j naturally to $K[X]$ by acting on the coefficients, and in particular we can define the *norm* of A as follows

$$\mathcal{N}(A) = \prod_{1 \le j \le m} \sigma_j(A) \ ,$$

and it is clear by Galois theory that $\mathcal{N}(A) \in \mathbb{Q}[X]$.

We have the following lemmas. Note that when we talk of factorizations of polynomials, it is always up to multiplication by units of $K[X]$, i.e. by elements of K.

Lemma 3.6.1. *If $A(X) \in K[X]$ is irreducible then $\mathcal{N}(A)(X)$ is equal to the power of an irreducible polynomial of $\mathbb{Q}[X]$.*

Proof. Let $\mathcal{N}(A) = \prod_i N_i^{e_i}$ be a factorization of $\mathcal{N}(A)$ into irreducible factors in $\mathbb{Q}[X]$. Since $A \mid \mathcal{N}(A)$ in $K[X]$ and A is irreducible in $K[X]$, we have $A \mid N_i$ in $K[X]$ for some i. But since $N_i \in \mathbb{Q}[X]$, it follows that $\sigma_j(A) \mid N_i$ for all j, and consequently $\mathcal{N}(A) \mid N_i^m$ in $K[X]$, hence in $\mathbb{Q}[X]$, so $\mathcal{N}(A) = N_i^{m'}$ for some $m' \le m$. $\square$

Lemma 3.6.2. *Let $A \in K[X]$ be a squarefree polynomial, where $K = \mathbb{Q}(\theta)$. Then there exists only a finite number of $k \in \mathbb{Q}$ such that $\mathcal{N}(A(X - k\theta))$ is not squarefree.*

Proof. Denote by $(\beta_{i,j})_{1 \le i \le m}$ the roots of $\sigma_j(A)$. If $k \in \mathbb{Q}$, it is clear that $\mathcal{N}(A(X - k\theta))$ is not squarefree if and only if there exists i_1, i_2, j_1, j_2 such that

$$\beta_{i_1,j_1} + k\sigma_{j_1}(\theta) = \beta_{i_2,j_2} + k\sigma_{j_2}(\theta) \ ,$$

or equivalently $k = (\beta_{i_1,j_1} - \beta_{i_2,j_2})/(\sigma_{j_2}(\theta) - \sigma_{j_1}(\theta))$ and there are only a finite number of such k. $\square$

The following lemma now gives us the desired factorization of A in $K[X]$.

Lemma 3.6.3. *Assume that $A(X) \in K[X]$ and $\mathcal{N}(A)(X) \in \mathbb{Q}[X]$ are both squarefree. Let $\mathcal{N}(A) = \prod_{1 \le i \le g} N_i$ be the factorization of $\mathcal{N}(A)$ into irreducible factors in $\mathbb{Q}[X]$. Then $A = \prod_{1 \le i \le g} \gcd(A, N_i)$ is a factorization of A into irreducible factors in $K[X]$.*

Proof. Let $A = \prod_{1 \le i \le h} A_i$ be the factorization of A into irreducible factors in $K[X]$. Since $\mathcal{N}(A)$ is squarefree, $\mathcal{N}(A_i)$ also hence by Lemma 3.6.1 $\mathcal{N}(A_i) = N_{j(i)}$ for some $j(i)$. Furthermore since for $j \ne i$, $\mathcal{N}(A_i A_j) \mid \mathcal{N}(A)$ hence is squarefree, $\mathcal{N}(A_i)$ is coprime to $\mathcal{N}(A_j)$. So by suitable reordering, we obtain $\mathcal{N}(A_i) = N_i$ and also $g = h$. Finally, since for $j \ne i$, A_j is coprime to N_i it follows that $A_i = \gcd(A, N_i)$ in $K[X]$ (as usual up to multiplicative constants), and the lemma follows. □

With these lemmas, it is now easy to give an algorithm for the factorization of $A \in K[X]$.

Algorithm 3.6.4 (Polynomial Factorization over Number Fields). Let $K = \mathbb{Q}(\theta)$ be a number field, $T \in \mathbb{Q}[X]$ the minimal monic polynomial of θ. Let $A(X)$ be a non-zero polynomial in $K[X]$. This algorithm finds a complete factorization of A in $K[X]$.

1. [Reduce to squarefree] Set $U \leftarrow A/(A, A')$ where (A, A') is computed in $K[X]$ using the sub-resultant Algorithm 3.3.1. (Now U will be a squarefree primitive polynomial. In this step, we could also use the squarefree decomposition Algorithm 3.4.2 to reduce still further the degree of U).

2. [Initialize search] Let $U(X) = \sum_{0 \le i \le m} u_i X^i \in K[X]$ and write $u_i = g_i(\theta)$ for some polynomial $g_i \in \mathbb{Q}[X]$. Set $G(X, Y) \leftarrow \sum_{0 \le i \le m} g_i(Y)X^i \in \mathbb{Q}[X, Y]$ and $k \leftarrow 0$.

3. [Search for squarefree norm] Using the sub-resultant Algorithm 3.3.7 over the UFD $\mathbb{Q}[Y]$, compute $N(X) \leftarrow R_Y(T(Y), G(X - kY, Y))$ where R_Y denotes the resultant with respect to the variable Y. If $N(X)$ is not squarefree (tested using Algorithm 3.3.1), set $k \leftarrow k + 1$ and go to step 3.

4. [Factor norm] (Here $N(X)$ is squarefree) Using Algorithm 3.5.7, let $N \leftarrow \prod_{1 \le i \le g} N_i$ be a factorization of N in $\mathbb{Q}[X]$.

5. [Output factorization] For $i = 1, \ldots, g$ set $A_i(X) \leftarrow \gcd(U(X), N_i(X + k\theta))$ computed in $K[X]$ using Algorithm 3.3.1, output A_i and the exponent of A_i in A (obtained simply by replacing A by A/A_i as long as $A_i \mid A$). Terminate the algorithm.

Proof. The lemmas that we have given above essentially prove the validity of this algorithm, apart from the easily checked fact that the sub-resultant computed in step 3 indeed gives the norm of the polynomial U. □

Remarks.

(1) The norm of U could also be computed using floating point approximations to the roots of T, since (if our polynomials have algebraic integer coefficients) it will have coefficients in $\mathbb{Z}$. This is often faster than sub-resultant computations, but requires careful error bounds.

(2) Looking at the proof of Lemma 3.6.2, it is also clear that floating point computations allow us to give the list of values of k to avoid in step 3, so no trial and error is necessary. However this is not really important since step 3 is in practice executed only once or twice.

(3) The factors that we have found are not necessarily in $\mathbb{Z}_K[X]$, and, as already mentioned, factoring in $\mathbb{Z}_K[X]$ requires a little extra work since $\mathbb{Z}_K$ is not necessarily a PID.

3.6.3 A Root Finding Algorithm over $\mathbb{C}$

In many situations, it is useful to compute explicitly, to some desired approximation, all the complex roots of a polynomial. There exist many methods for doing this. It is a difficult problem of numerical analysis and it is not my intention to give a complete description here, or even to give a description of the "best" method if there is one such. I want to give one reasonably simple algorithm which works most of the time quite well, although it may fail in some situations. In practice, it is quite sufficient, especially if one uses a multi-precision package which allows you to increase the precision in case of failure.

This method is based on the following proposition.

Proposition 3.6.5. *If $P(X) \in \mathbb{C}[X]$ and $x \in \mathbb{C}$, then if $P(x) \neq 0$ and $P'(x) \neq 0$ there exists a positive real number λ such that*

$$\left| P\left(x - \lambda \frac{P(x)}{P'(x)} \right) \right| < |P(x)| \ .$$

Proof. Trivial by Taylor's theorem. In fact, this proposition is valid for any analytic function in the neighborhood of x, and not only for polynomials. $\square$

Note also that as soon as x is sufficiently close to a simple root of P, we can take $\lambda = 1$, and then the formula is nothing but Newton's formula, and as usual the speed of convergence is quadratic.

This leads to the following algorithm, which I call Newton's modified algorithm. Since we will be using this algorithm for irreducible polynomials over $\mathbb{Q}$, we can assume that the polynomial we are dealing with is at least squarefree. The modifications necessary to handle the general case are easy and left to the reader.

Algorithm 3.6.6 (Complex Roots). Given a squarefree polynomial P, this algorithm outputs its complex roots (in a random order). In quite rare cases the algorithm may fail. On the other hand it is absolutely necessary that the polynomial be squarefree (this can be achieved by replacing P by $P/(P, P')$).

1. [Initializations] Set $Q \leftarrow P$, compute P', set $Q' \leftarrow P'$, and set $n \leftarrow \deg(P)$. Finally, set $f \leftarrow 1$ if P has real coefficients, otherwise set $f \leftarrow 0$.

2. [Initialize root finding] Set $x \leftarrow 1.3 + 0.314159i$, $v \leftarrow Q(x)$ and $m \leftarrow |v|^2$.

3. [Initialize recursion] Set $c \leftarrow 0$ and $dx \leftarrow v/Q'(x)$. If $|dx|$ is smaller than the desired absolute accuracy, go to step 5.

4. [Try a λ] Set $y \leftarrow x - dx$, $v_1 \leftarrow Q(y)$ and $m_1 \leftarrow |v_1|^2$. If $m_1 < m$, set $x \leftarrow y$, $v \leftarrow v_1$, $m \leftarrow m_1$ and go to step 3. Otherwise, set $c \leftarrow c+1$, $dx \leftarrow dx/4$. If $c < 20$ go to step 4, otherwise output an error message saying that the algorithm has failed.

5. [Polish root] Set $x \leftarrow x - P(x)/P'(x)$ twice.

6. [Divide] If $f = 0$ or if $f = 1$ and the absolute value of the imaginary part of x is less than the required accuracy, set it equal to 0, output x, set $Q(X) \leftarrow Q(X)/(X - x)$ and $n \leftarrow n - 1$. Otherwise, output x and $\bar{x}$, set $Q(X) \leftarrow Q(X)/(X^2 - 2\operatorname{Re}(x)X + |x|^2)$ and $n \leftarrow n - 2$. Finally, if $n > 0$ then go to step 2, otherwise terminate the algorithm.

Remarks.

(1) The starting value $1.3 + 0.314159i$ given in step 2 is quite arbitrary. It has been chosen so as not to be too close to a trivial algebraic number, and not too far from the real axis, although not exactly on it.

(2) The value 20 taken in step 4, as well as the division by 4, are also arbitrary but correspond to realistic situations. If we find $m_1 \geq m$, this means that we are quite far away from the "attraction zone" of a root. Hence, thanks to Proposition 3.6.5, it is preferable to divide the increment by 4 and not by 2 for example, so as to have a much higher chance of winning next time. Similarly, the limitation of 20 correspond to an increment which is $4^{20} \approx 10^{12}$ times smaller than the Newton increment, and this is in general too small to make any difference. In that case, it will be necessary to increase the working precision.

(3) After each division done in step 6, the quality of the coefficients of Q will deteriorate. Hence, after finding an approximate root, it is essential to polish it, using for example the standard Newton iteration, but with the polynomial P and not Q. It is not necessary to use a factor λ since we are in principle well inside the attraction zone of a root. Two polishing passes will, in principle, be enough.

(4) The divisions in step 6 are simple to perform. If $Q(X) = \sum_{0 \leq i \leq n} q_i X^i$ and $A(X) = \sum_{0 \leq i \leq n-1} a_i X^i = Q(X)/(X - x)$, then set $a_{n-1} \leftarrow q_n$ and for $i = n - 1, \ldots, i = 1$ set $a_{i-1} \leftarrow q_i + xa_i$. Similarly, if $B(X) = \sum_{0 \leq i \leq n-2} b_i X^i = Q(X)/(X^2 - \alpha X + \beta)$, then set $b_{n-2} \leftarrow q_n$, $b_{n-3} \leftarrow q_{n-1} + \alpha b_{n-2}$ and for $i = n - 2, \ldots, i = 2$ set $b_{i-2} \leftarrow q_i + \alpha b_{i-1} - \beta b_i$.

(5) Instead of starting with $\lambda = 1$ as coefficient of $Q(x)/Q'(x)$ in step 3, it may be better to start with

$$\lambda = \min\left(1, \frac{2|Q'(x)|^2}{|Q(x)||Q''(x)|}\right).$$

This value is obtained by looking at the error term in the Taylor expansion proof of Proposition 3.6.5. If this value is too small, then we are probably going to fail, and in fact x is converging to a root of $Q'(X)$ instead of $Q(X)$. If this is detected, the best solution is probably to start again in step 2 with a different starting value. This of course can also be done when $c = 20$ in step 4. We must however beware of doing this too systematically, for failure may indicate that the coefficients of the polynomial P are ill conditioned, and in that case the best remedy is to modify the coefficients of P by a suitable change of variable (typically of the form $X \mapsto aX$). It must be kept in mind that for ill conditioned polynomials, a very small variation of a coefficient can have a drastic effect on the roots.

(6) In step 6, instead going back to step 2 if $n > 0$, we can go back only if $n > 2$, and treat the cases $n = 1$ and $n = 2$ by using the standard formulas. Care must then be taken to polish the roots thus obtained, as is done in step 5.

3.7 Exercises for Chapter 3

1. Write an algorithm for multiplying two polynomials, implicitly based on a recursive use of the splitting formulas explained in Section 3.1.2.

2. Let P be a polynomial. Write an algorithm which computes the coefficients of the polynomial $P(X + 1)$ without using an auxiliary array or polynomial.

3. Let K be a commutative ring which is not necessarily a field. It has been mentioned after Algorithm 3.1.1 that the Euclidean division of A by B is still possible in $K[X]$ if the leading coefficient $\ell(B)$ is invertible in K. Write an algorithm performing this Euclidean division after multiplying A and B by the inverse of $\ell(B)$, and compare the performance of this algorithm with the direct use of Algorithm 3.1.1 in the case $K = \mathbb{Z}/r\mathbb{Z}$.

4. Modify Algorithm 3.3.1 so that A and B are divided by their respective contents every 10 iterations. Experiment and convince yourself that this modification leads to polynomials A and B having much larger coefficients later on in the Algorithm, hence that this is a bad idea.

5. Write an extended version of Algorithm 3.3.1 which computes not only (A, B) but also U and V such that $AU + BV = r \cdot (A, B)$ where r is a non-zero constant (Hint: add a fourth variable in Algorithm 1.3.6 to take care of r). Show that when $(A, B) = 1$ this can always be done with r equal to the resultant of A and B.

6. Show that if A, B and C are irreducible polynomials over a UFD R and if C divides AB but is not a unit multiple of A, then C divides B (Hint: use the preceding exercise). Deduce from this that $R[X]$ is a UFD.

7. Using for example the sub-resultant algorithm, compute explicitly the discriminant of the trinomials $X^3 + aX + b$ and $X^4 + aX + b$. Try to find the general formula for the discriminant of $X^n + aX + b$.

8. Call R_i the i-th row of Sylvester's determinant, for $1 \le i \le n + m$. Show that if we replace for all $1 \le i \le n$ simultaneously R_i by

$$\sum_{k=0}^{i-1}(b_k R_{i-k} - a_k R_{i+m-k})$$

and then suppress the last m rows and columns of the resulting matrix, the $n \times n$ determinant thus obtained is equal to the determinant of Sylvester's matrix.

9. If $Q(X) = (X - a)P(X)$, compute the discriminant of Q in terms of a and of the discriminant of P.

10. Show how to modify the sub-resultant Algorithm 3.3.7 so that it can compute correctly when the coefficients of the polynomials are for example polynomials (in another variable) with real coefficients.

11. Show the following result, due to Eisenstein: if p is prime and $A(X) = \sum_{0 \le i \le n} a_i X^i$ is a polynomial in $\mathbb{Z}[X]$ such that $p \nmid a_n$, $p \mid a_i$ for all $i < n$ and $p^2 \nmid a_0$, then A is irreducible in $\mathbb{Z}[X]$.

12. Using the ideas of Section 3.4, write an algorithm to compute the square root of a mod p, or to determine whether none exist. Implement your algorithm and compare it with Shanks's Algorithm 1.5.1.

13. Using the Möbius inversion formula (see [H-W] Section 16.4) show that the number of monic irreducible polynomials of degree n over $\mathbb{F}_p$ is equal to

$$\frac{1}{n} \sum_{d \mid n} \mu \left(\frac{n}{d} \right) p^d$$

where $\mu(n)$ is the Möbius function (i.e. 0 if n is not squarefree, and equal to $(-1)^k$ if n is a product of k distinct prime factors).

14. Extend Proposition 3.4.7 and Algorithm 3.4.8 to general prime numbers p, using $U_p(X) = X + X^p + \cdots + X^{p^{d-1}}$. Compare in practice the expected speed of the resulting algorithm to that of Algorithm 3.4.6.

15. Show that, as claimed in the proof of Algorithm 3.4.8, the polynomial U has exactly 2^{d-1} roots in $\mathbb{F}_{2^d}$.

16. Generalizing the methods of Section 3.4, write an algorithm to factor polynomials in $\mathbb{F}_q[X]$, where $q = p^d$ and $\mathbb{F}_q$ is given by an irreducible polynomial of degree d in $\mathbb{F}_p[X]$.

17. Let $B(X) \in \mathbb{F}_p[X]$ be a squarefree polynomial with r distinct irreducible factors. Show that if $T(X)$ is a polynomial corresponding to a randomly chosen element of the kernel obtained in step 2 of Algorithm 3.4.10 and if $p \ge 3$, the probability that $(B(X), T(X)^{(p-1)/2} - 1)$ gives a non-trivial factor of B is greater than or equal to $4/9$.

18. Let K be any field, $a \in K$ and p a prime number. Show that the polynomial $X^p - a$ is reducible in $K[X]$ if and only if it has a root in K. Generalize to the polynomials $X^{p^r} - a$.

19. Let p be an odd prime and q a prime divisor of $p-1$. Let $a \in \mathbb{Z}$ be a primitive root modulo p. Using the preceding exercise, show that for any $k \ge 1$ the polynomial

$$X^q + pX^k - a$$

is irreducible in $\mathbb{Q}[X]$.

20. Let p and q be two odd prime numbers. We assume that $q \equiv 2 \pmod 3$ and that p is a primitive root modulo q (i.e. that $p \bmod q$ generates $(\mathbb{Z}/q\mathbb{Z})^*$). Show that the polynomial

$$X^{q+1} - X + p$$

is irreducible in $\mathbb{Q}[X]$. (Hint: reduce mod p and mod 2.)

21. Separating even and odd powers, any polynomial A can be written in the form $A(X) = A_0(X^2) + XA_1(X^2)$. Set $T(A)(X) = A_0(X)^2 - XA_1(X)^2$. With the notations of Theorem 3.5.1, show that for any k

$$|b_j| \le \binom{n-1}{j}|T^k(A)|^{1/2^k} + \binom{n-1}{j-1}|a_m| .$$

What is the behavior of the sequence $|T^k(A)|^{1/2^k}$ as k increases?

22. In Algorithms 3.5.5 and 3.5.6, assume that $p = q$, that A and B are monic, and set $D = AU$, $D_1 = A_1U_1$, $E = BV$, $E_1 = B_1V_1$. Denote by (C, p^2) the ideal of $\mathbb{Z}[X]$ generated by $C(X)$ and p^2. Show that

$$D_1 \equiv 3D^2 - 2D^3 \pmod{(C, p^2)} \quad \text{and} \quad E_1 \equiv 3E^2 - 2E^3 \pmod{(C, p^2)} .$$

Then show that A_1 (resp. B_1) is the monic polynomial of the lowest degree such that $E_1A_1 \equiv 0 \pmod{(C, p^2)}$ (resp. $D_1B_1 \equiv 0 \pmod{(C, p^2)}$).

23. Write a general algorithm for finding all the roots of a polynomial in $\mathbb{Q}_p$ to a given p-adic precision, using Hensel's lemma. Note that multiple roots at the mod p level create special problems which have to be treated in detail.

24. Denote by $(,)_p$ the GCD taken over $\mathbb{F}_p[X]$. Following Weinberger, Knuth asserts that if $A \in \mathbb{Z}[X]$ is a product of exactly k irreducible factors in $\mathbb{Z}[X]$ (not counting multiplicity) then

$$\lim_{x \to \infty} \frac{\sum_{p \le x} \deg(X^p - X, A(X))_p}{\sum_{p \le x} 1} = k .$$

Explore this formula as a heuristic method for determining the irreducibility of a polynomial over $\mathbb{Z}$.

25. Find the complete decomposition into irreducible factors of the polynomial $X^4 + 1$ modulo every prime p using the quadratic reciprocity law and the identities given in Section 3.5.2.

26. Discuss the possibility of computing polynomial GCD's over $\mathbb{Z}$ by computing GCD's of *values* of the polynomials at suitable points. (see [Schön]).

27. Using the ideas of Section 3.4.2, modify the root finding Algorithm 3.6.6 so that it finds the roots of a any polynomial, squarefree or not, with their order of multiplicity. For this question to make practical sense, you can assume that the polynomial has integer coefficients.

28. Let $P(X) = X^3 + aX^2 + bX + c \in \mathbb{R}[X]$ be a monic squarefree polynomial. Let θ_i $(1 \le i \le 3)$ be the roots of P in $\mathbb{C}$ and let

$$\alpha_1 = (\theta_1 + \rho^2\theta_2 + \rho\theta_3)^3 , \quad \alpha_2 = (\theta_1 + \rho\theta_2 + \rho^2\theta_3)^3 .$$

Let $A(X) = (X - \alpha_1)(X - \alpha_2)$.
   a) Compute explicitly the coefficients of $A(X)$.

    b) Show that $-27\,\text{disc}(P) = \text{disc}(A)$, and give an expression for this discriminant.

    c) Show how to compute the roots of P knowing the roots of A.

29. Let $P(X) = X^4 + aX^3 + bX^2 + cX + d \in \mathbb{R}[X]$ be a monic squarefree polynomial. Let θ_i $(1 \le i \le 4)$ be the roots of P in $\mathbb{C}$, and let

$$\alpha_1 = (\theta_1 + \theta_2)(\theta_3 + \theta_4) \quad \alpha_2 = (\theta_1 + \theta_3)(\theta_2 + \theta_4) \quad \alpha_3 = (\theta_1 + \theta_4)(\theta_2 + \theta_3) \ ,$$

and

$$\beta_1 = \theta_1\theta_2 + \theta_3\theta_4 \quad \beta_2 = \theta_1\theta_3 + \theta_2\theta_4 \quad \beta_3 = \theta_1\theta_4 + \theta_2\theta_3 \ .$$

Finally, let $A(X) = (X - \alpha_1)(X - \alpha_2)(X - \alpha_3)$ and $B(X) = (X - \beta_1)(X - \beta_2)(X - \beta_3)$.

    a) Compute explicitly the coefficients of $A(X)$ and $B(X)$ in terms of those of $P(X)$.

    b) Show that $\text{disc}(P) = \text{disc}(A) = \text{disc}(B)$, and give an expression for this discriminant.

    c) Show how to compute the roots of P knowing the roots of A.

30. Recall that the *first case* of Fermat's last "theorem" (FLT) states that if l is an odd prime, then $x^l + y^l + z^l = 0$ implies that $l \mid xyz$. Using elementary arguments (i.e. no algebraic number theory), it is not too difficult to prove the following theorem, essentially due to Sophie Germain.

Theorem 3.7.1. *Let l be an odd prime, and assume that there exists an integer k such that $k \equiv \pm 2 \pmod 6$, $p = lk + 1$ is prime and $p \nmid (k^k - 1)W_k$ where W_k is the resultant of the polynomials $X^k - 1$ and $(X + 1)^k - 1$. Then the first case of FLT is true for the exponent l.*

It is therefore important to compute W_k and in particular its prime factors. Give several algorithms for doing this, and compare their efficiency. Some familiarity with number fields and in particular with cyclotomic fields is needed here.

31. Let $A(X) = a_n X^n + \cdots + a_1 X + a_0$ be a polynomial, with $a_n \ne 0$. Show that for any positive integer k,

$$\text{disc}(A(X^k)) = (-1)^{nk(k+3)/2} k^{nk} (a_n a_0)^{k-1} \text{disc}(A)^k \ .$$

Chapter 4

Algorithms for Algebraic Number Theory I

In this chapter, we give the necessary background on algebraic numbers, number fields, modules, ideals and units, and corresponding algorithms for them. Excellent basic textbooks on these subjects are, for example [Bo-Sh], [Cas-Frö], [Cohn], [Ire-Ros], [Marc], [Sam]. However, they usually have little algorithmic flavor. We will give proofs only when they help to understand an algorithm, and we urge the reader to refer to the above textbooks for the proofs which are not given.

4.1 Algebraic Numbers and Number Fields

4.1.1 Basic Definitions and Properties of Algebraic Numbers

Definition 4.1.1. Let $\alpha \in \mathbb{C}$. Then α is called an algebraic number if there exists $A \in \mathbb{Z}[X]$ such that $A(\alpha) = 0$, and A not identically zero. The number α is called an algebraic integer if, in addition, one can choose A to be monic (i.e. with leading coefficient equal to 1).

Then we have:

Proposition 4.1.2. Let α be an algebraic number, and let A be a polynomial with integer coefficients such that $A(\alpha) = 0$, and assume that A is chosen to have the smallest degree and be primitive with $\ell(A) > 0$. Then such an A is unique, is irreducible in $\mathbb{Q}[X]$, and any $B \in \mathbb{Z}[X]$ such that $B(\alpha) = 0$ is a multiple of A.

Proof. The ring $\mathbb{Q}[X]$ is a principal ideal domain (PID), and the set of $B \in \mathbb{Q}[X]$ such that $B(\alpha) = 0$ is an ideal, hence is the ideal generated by A. If, in addition, B has integral coefficients, Gauss's lemma (Theorem 3.2.8) implies that B is a multiple of A in $\mathbb{Z}[X]$. It is clear that A is irreducible; otherwise A would not be of smallest degree. We will call this A the minimal polynomial of α. □

We will use the notation $\overline{\mathbb{Q}}$ for the set of algebraic numbers, (hence $\overline{\mathbb{Q}} \subset \mathbb{C}$), $\mathbb{Z}_{\overline{\mathbb{Q}}}$ for the set of algebraic integers, and if L is any subset of $\mathbb{C}$ we will set

$$\mathbb{Z}_L = \mathbb{Z}_{\overline{\mathbb{Q}}} \cap L \ ,$$

and call it the set of integers of L. Note that $\overline{\mathbb{Q}}$ is *an* algebraic closure of $\mathbb{Q}$.

For example, we have $\mathbb{Z}_{\mathbb{Q}} = \mathbb{Z}$. Indeed, if $\alpha = p/q \in \mathbb{Q}$ is a root of $A \in \mathbb{Z}[X]$ with A monic, we must have $q \mid \ell(A)$, hence $q = \pm 1$ so α is in $\mathbb{Z}$.

The first important result about algebraic numbers is as follows:

Theorem 4.1.3. *Let $\alpha \in \mathbb{C}$. The following four statements are equivalent.*

(1) α *is an algebraic integer.*
(2) $\mathbb{Z}[\alpha]$ *is a finitely generated additive Abelian group.*
(3) α *belongs to a subring of $\mathbb{C}$ which is finitely generated as an Abelian group.*
(4) *There exists a non-zero finitely generated additive subgroup L of $\mathbb{C}$ such that $\alpha L \subset L$.*

As corollaries we have:

Corollary 4.1.4. *The set of algebraic integers is a ring. In particular, if R is a ring, the set $\mathbb{Z}_R$ of integers of R is a ring.*

Corollary 4.1.5. *If $\alpha \in \mathbb{C}$ is a root of a monic polynomial whose coefficients are algebraic integers (and not simply integers), then α is an algebraic integer.*

Definition 4.1.6. *Let $\alpha \in \mathbb{C}$ be an algebraic number, and A its minimal polynomial. The* conjugates *of α are all the $\deg(A)$ roots of A in $\mathbb{C}$.*

This notion of conjugacy is of course of fundamental importance, but what I would like to stress here is that from an algebraic point of view the conjugates are indistinguishable. For example, any algebraic identity between algebraic numbers is a simultaneous collection of conjugate identities. To give a trivial example, the identity $(1 + \sqrt{2})^2 = 3 + 2\sqrt{2}$ implies the identity $(1 - \sqrt{2})^2 = 3 - 2\sqrt{2}$. This remark is a generalization of the fact that an equality between two complex numbers implies the equality between their conjugates, or equivalently between their real and imaginary parts. The present example is even more striking if one looks at it from a numerical point of view: it says that the identity $(2.41421\ldots)^2 = 5.828427\ldots$ implies the identity $(0.41421\ldots)^2 = 0.171573\ldots$. Of course this is not the correct way to look at it, but the lesson to be remembered is that an algebraic number *always* comes with all of its conjugates.

4.1.2 Number Fields

Definition 4.1.7. *A* number field *is a field containing $\mathbb{Q}$ which, considered as a $\mathbb{Q}$-vector space, is finite dimensional. The number $d = \dim_{\mathbb{Q}} K$ is denoted by $[K : \mathbb{Q}]$ and called the* degree *of the number field K.*

We recall the following fundamental results about number fields:

Theorem 4.1.8. *Let K be a number field of degree n. Then*

(1) *(Primitive element theorem) There exists a $\theta \in K$ such that*

$$K = \mathbb{Q}(\theta).$$

Such a θ is called a primitive element. Its minimal polynomial is an irreducible polynomial of degree n.

(2) *There exist exactly n field embeddings of K in $\mathbb{C}$, given by $\theta \mapsto \theta_i$, where the θ_i are the roots in $\mathbb{C}$ of the minimal polynomial of θ. These embeddings are $\mathbb{Q}$-linear, their images K_i in $\mathbb{C}$ are called the conjugate fields of K, and the K_i are isomorphic to K.*

(3) *For any i, $K_i \subset \overline{\mathbb{Q}}$, in other words all the elements of K_i are algebraic numbers and their degree divides n.*

The assertion made above concerning the indistinguishability of the conjugates can be clearly seen here. The choice of the conjugate field K_i is a priori completely arbitrary. In many cases, this choice is already given. For example, when we speak of "the number field $\mathbb{Q}(2^{1/3})$", this is slightly incorrect, since what we mean by this is that we are considering the number field $K = \mathbb{Q}[X]/(X^3 - 2)\mathbb{Q}[X]$ *together* with the embedding $X \mapsto 2^{1/3}$ of K into $\mathbb{R}$.

Definition 4.1.9. *The signature of a number field is the pair (r_1, r_2) where r_1 is the number of embeddings of K whose image lie in $\mathbb{R}$, and $2r_2$ is the number of non-real complex embeddings, so that $r_1 + 2r_2 = n$ (note that the non-real embeddings always come in pairs since if σ is such an embedding, so is $\bar{\sigma}$, where ‾ denotes complex conjugation). If T is an irreducible polynomial defining the number field K by one of its roots, the signature of K will also be called the signature of T. Here r_1 (resp. $2r_2$) will be the number of real (resp. non-real) roots of T in $\mathbb{C}$. When $r_2 = 0$ (resp. $r_1 = 0$) we will say that K and T are totally real (resp. totally complex).*

It is not difficult to determine the signature of a number field K, but some ways are better than others. If $K = \mathbb{Q}(\theta)$, and if T is the minimal polynomial of θ, we can of course compute the roots of T in $\mathbb{C}$ using, for instance, the root finding Algorithm 3.6.6, and count the number of real roots. This is however quite expensive. A much better way is to use a theorem of Sturm which tells us in essence that the sequence of leading coefficients in the polynomial remainder sequence obtained by applying Euclid's algorithm or its variants to T and T' governs the signature. More precisely, we have the following theorem.

Theorem 4.1.10 (Sturm). *Let T be a squarefree polynomial with real coefficients. Assume that $A_0 = T$, $A_1 = T'$, and that A_i is a polynomial remainder sequence such that for all i with $1 \le i \le k$:*

$$e_i A_{i-1} = Q_i A_i - f_i A_{i+1} ,$$

where the e_i and f_i are real and positive, and A_{k+1} is a constant polynomial (non-zero since T is squarefree). Set $\ell_i = \ell(A_i)$, and $d_i = \deg(A_i)$. Then, if s is the number of sign changes in the sequence $\ell_0, \ell_1, \dots, \ell_{k+1}$, and if t is the number of sign changes in the sequence $(-1)^{d_0}\ell_0, (-1)^{d_1}\ell_1, \dots, (-1)^{d_{k+1}}\ell_{k+1}$, the number of real roots of T is equal to $t - s$.

Proof. For any real a, let $s(a)$ be the number of sign changes, not counting zeros, in the sequence $A_0(a), A_1(a), \dots, A_{k+1}(a)$. We clearly have $\lim_{a \to +\infty} s(a) = s$ and $\lim_{a \to -\infty} s(a) = t$. We are going to prove the following more general assertion: the number of roots of T in the interval $]a, b]$ is equal to $s(a) - s(b)$, which clearly implies the assertion of the theorem.

First, it is clear that a sign sequence at any number a cannot have two consecutive zeros, otherwise these zeros would propagate and we would have $A_{k+1} = 0$. For similar reasons, we cannot have sequences of the form $+$, 0, $+$, or of the form $-$, 0, $-$ since the e_i and f_i are positive. Now the desired formula $s(a) - s(b)$ is certainly valid if $b = a$. We will see that it stays true when b increases. The quantity $s(b)$ can change only when b goes through one of the roots of the A_i, which are finite in number. Let x be a root of such an A_i (maybe of several). If ϵ is sufficiently small, when b goes from $x - \epsilon$ to x, the sign sequence corresponding to indices $i - 1$, i and $i + 1$ goes from $+$, $\pm$, $-$ to $+$, 0, $-$ (or from $-$, $\pm$, $+$ to $-$, 0, $+$) when $i \geq 1$ by what has been said above (no consecutive zeros, and no sequences $+$, 0, $+$ or $-$, 0, $-$). Hence, there is no difference in the number of sign changes not counting zeros if $i \geq 1$. On the other hand, for $i = 0$, the sign sequence corresponding to indices 0 and 1 goes from $+$, $-$ to 0, $-$, or from $-$, $+$ to 0, $+$ since $A_1(b) < 0$ when A_0 is decreasing (recall that A_1 is the derivative of A_0). Hence, the net change in $s(b)$ is equal to -1. This proves our claim and the theorem. $\square$

From this, it is easy to derive an algorithm for computing the signature of a polynomial (hence of a number field). Such an algorithm can of course be written for any polynomial $T \in \mathbb{R}[X]$, but for number-theoretic uses T will have integer coefficients, hence we should use the polynomial remainder sequence given by the sub-resultant Algorithm 3.3.1 to avoid coefficient explosion. This leads to the following algorithm.

Algorithm 4.1.11 (Sturm). Given a polynomial $T \in \mathbb{Z}[X]$, this algorithm determines the signature (r_1, r_2) of T using Sturm's theorem and the sub-resultant Algorithm 3.3.1. If T is not squarefree, it outputs an error message.

1. [Initializations and reductions] If $\deg(T) = 0$, output $(0,0)$ and terminate. Otherwise, set $A \leftarrow \text{pp}(T)$, $B \leftarrow \text{pp}(T')$, $g \leftarrow 1$, $h \leftarrow 1$, $s \leftarrow \text{sign}(\ell(A))$, $n \leftarrow \deg(A)$, $t \leftarrow (-1)^{n-1}s$, $r_1 \leftarrow 1$.

2. [Pseudo division] Set $\delta \leftarrow \deg(A) - \deg(B)$. Using Algorithm 3.1.2, compute R such that $\ell(B)^{\delta+1}A = BQ + R$. If $R = 0$ then T was not squarefree, output

an error message and terminate the algorithm. Otherwise, if $\ell(B) > 0$ or δ is odd, set $R \leftarrow -R$.

3. [Use Sturm] If $\text{sign}(\ell(R)) \neq s$, set $s \leftarrow -s$, $r_1 \leftarrow r_1 - 1$. Then, if $\text{sign}(\ell(R)) \neq (-1)^{\deg(R)} t$, set $t \leftarrow -t$, $r_1 \leftarrow r_1 + 1$.

4. [Finished?] If $\deg(R) = 0$, output $(r_1, (n-r_1)/2)$ and terminate the algorithm. Otherwise, set $A \leftarrow B$, $B \leftarrow R/(gh^\delta)$, $g \leftarrow |\ell(A)|$, $h \leftarrow h^{1-\delta} g^\delta$, and go to step 2.

Another important notion concerning number fields is that of the Galois group of a number field. From now on, we assume that all our number fields are subfields of $\overline{\mathbb{Q}}$.

Definition 4.1.12. *Let K be a number field of degree n. We say that K is Galois (or normal) over $\mathbb{Q}$, or simply Galois, if K is (globally) invariant by the n embeddings of K in $\mathbb{C}$. The set of such embeddings is a group, called the Galois group of K, and denoted $\text{Gal}(K/\mathbb{Q})$.*

Given any number field K, the intersection of all subfields of $\overline{\mathbb{Q}}$ which are Galois and contain K is a finite extension K^s of K called the Galois closure (or normal closure) of K in $\overline{\mathbb{Q}}$. If $K = \mathbb{Q}(\theta)$ where θ is a root of an irreducible polynomial $T \in \mathbb{Z}[X]$, the Galois closure of K can also be obtained as the splitting field of T, i.e. the field obtained by adjoining to $\mathbb{Q}$ all the roots of T. By abuse of language, even when K is not Galois, we will call $\text{Gal}(K^s/\mathbb{Q})$ the Galois group of the number field K (or of the polynomial T).

A special case of the so-called "fundamental theorem of Galois theory" is as follows.

Proposition 4.1.13. *Let K be Galois over $\mathbb{Q}$ and $x \in K$. Assume that for any $\sigma \in \text{Gal}(K/\mathbb{Q})$ we have $\sigma(x) = x$. Then $x \in \mathbb{Q}$. In particular, if in addition x is an algebraic integer then $x \in \mathbb{Z}$.*

The following easy proposition shows that there are only two possibilities for the signature of a Galois extensions. Similarly, we will see (Proposition 4.8.6) that there are only a few possibilities for how primes split in a Galois extension.

Proposition 4.1.14. *Let K be a Galois extension of $\mathbb{Q}$ of degree n. Then, either K is totally real $((r_1, r_2) = (n, 0))$, or K is totally complex $((r_1, r_2) = (0, n/2)$ which can occur only if n is even).*

The computation of the Galois group of a number field (or of its Galois closure) is in general not an easy task. We will study this for polynomials of low degree in Section 6.3.

4.2 Representation and Operations on Algebraic Numbers

It is very important to study the way in which algebraic numbers are represented. There are two completely different problems: that of representing algebraic numbers, and that of representing *sets* of algebraic numbers, e.g. modules or ideals. This will be considered in Section 4.7. Here we consider the problem of representing an individual algebraic number.

Essentially there are four ways to do this, depending on how the number arises. The first way is to represent $\alpha \in \overline{\mathbb{Q}}$ by its minimal polynomial A which exists by Proposition 4.1.2. The three others assume that α is a polynomial with rational coefficients in some fixed algebraic number θ. These other methods are usually preferable, since field operations in $\mathbb{Q}(\theta)$ can be performed quite simply. We will see these methods in more detail in the following sections. However, to start with, we do not always have such a θ available, so we consider the problems which arise from the first method.

4.2.1 Algebraic Numbers as Roots of their Minimal Polynomial

Since A has $n = \deg(A)$ zeros in $\mathbb{C}$, the first question is to determine which of these zeros α is supposed to represent. We have seen that an algebraic number always comes equipped with all of its conjugates, so this is a problem which we must deal with. Since $\mathbb{Q}(\alpha) \simeq \mathbb{Q}[X]/(A(X)\mathbb{Q}[X])$, α may be represented as the class of X in $\mathbb{Q}[X]/(A(X)\mathbb{Q}[X])$, which is a perfectly well defined mathematical quantity. The distinction between α and its conjugates, if really necessary, will then depend not on A but on the specific embedding of $\mathbb{Q}[X]/(A(X)\mathbb{Q}[X])$ in $\mathbb{C}$. In other words, it depends on the numerical value of α as a complex number. This numerical value can be obtained by finding complex roots of polynomials, and we assume throughout that we always take sufficient accuracy to be able to distinguish α from its conjugates. (Recall that since the minimal polynomial of α is irreducible and hence squarefree, the conjugates of α are distinct.)

Hence, we can consider that an algebraic number α is represented by a pair (A, x) where A is the minimal polynomial of α, and x is an approximation to the complex number α (x should be at least closer to α than to any of its conjugates). It is also useful to have numeric approximations to all the conjugates of α. In fact, one can recover the minimal polynomial A of α from this if one knows only its leading term $\ell(A)$, since if one sets $\tilde{A}(X) = \ell(A) \prod_i (X - \tilde{\alpha}_i)$, where the $\tilde{\alpha}_i$ are the approximations to the conjugates of α, then, if they are close enough (and they must be chosen so), A will be the polynomial whose coefficients are the nearest integers to the coefficients of $\tilde{A}$.

With this representation, it is clear that one can now easily work in the subfield $\mathbb{Q}(\alpha)$ generated by α, simply by working modulo A.

More serious problems arise when one wants to do operations between algebraic numbers which are a priori not in this subfield. Assume for instance that $\alpha = (X \bmod A(X))$, and $\beta = (X \bmod B(X))$, where A and B are primitive irreducible polynomials of respective degrees m and n (we omit the $\mathbb{Q}[X]$ for simplicity of notation). How does one compute the sum, difference, product and quotient of α and β? The simplest way to do this is to compute *resultants* of two variable polynomials. Indeed, the resultant of the polynomials $A(X-Y)$ and $B(Y)$ considered as polynomials in Y alone (the coefficient ring being then $\mathbb{Q}[X]$) is up to a scalar factor equal to $P(X) = \prod_{i,j}(X - \alpha_i - \beta_j)$ where the α_i are the conjugates of α, and the β_j are the conjugates of β. Since P is a resultant, it has coefficients in $\mathbb{Q}[X]$, and $\alpha + \beta$ is one of its roots, so $Q = \mathrm{pp}(P)$ is a multiple of the minimal polynomial of $\alpha + \beta$.

If Q is irreducible, then it is the minimal polynomial of $\alpha + \beta$. If it is not irreducible, then the minimal polynomial of $\alpha + \beta$ is one of the irreducible factors of Q which one computes by using the algorithms of Section 3.5. Once again however, it does not make sense to ask which of the irreducible factors $\alpha + \beta$ is a root of, if we do not specify embeddings in $\mathbb{C}$, in other words, numerical approximations to α and β. Given such approximations however, one can readily check in practice which of the irreducible factors of Q is the minimal polynomial that we are looking for.

What holds for addition also holds for subtraction (take the resultant of $A(X + Y)$ and $B(Y)$), multiplication (take the resultant of $Y^m A(X/Y)$ and $B(Y)$), and division (take the resultant of $A(XY)$ with $B(Y)$).

4.2.2 The Standard Representation of an Algebraic Number

Let K be a number field, and let θ_j ($1 \leq j \leq n$) be a $\mathbb{Q}$-basis of K. Let $\alpha \in K$ be any element. It is clear that one can write α in a unique way as

$$\alpha = \frac{\sum_{j=0}^{n-1} a_j \theta_{j+1}}{d} \quad , \text{ with } d > 0, \ a_j \in \mathbb{Z} \text{ and } \gcd(a_0, \dots, a_{n-1}, d) = 1 \ .$$

In the case where $\theta_j = \theta^{j-1}$ for some root θ of a monic irreducible polynomial $T \in \mathbb{Z}[X]$, the $(n+1)$-uplet $(a_0, \dots, a_{n-1}, d) \in \mathbb{Z}^{n+1}$ will be called the *standard representation* of α (with respect to θ). Hence, we can now assume that we know such a primitive element θ. (We will see in Section 4.5 how it can be obtained.)

We must see how to do the usual arithmetic operations on these standard representations. The vector space operations on K are of course trivial. For multiplication, we precompute the standard representation of θ^j for $j \leq 2n-2$ in the following way: if $T(X) = \sum_{i=0}^{n} t_i X^i$ with $t_i \in \mathbb{Z}$ for all i and $t_n = 1$, we have $\theta^n = \sum_{i=0}^{n-1} (-t_i) \theta^i$. If we set $\theta^{n+k} = \sum_{i=0}^{n-1} r_{i,k} \theta^i$, then the standard representation of θ^{n+k} is $(r_{0,k}, r_{1,k}, \dots, r_{n-1,k}, 1)$ and the $r_{i,k}$ are computed by induction thanks to the formulas $r_{i,0} = -t_i$ and

$$r_{k+1,i} = \begin{cases} r_{k,i-1} - t_i r_{k,n-1} & \text{if } i \geq 1, \\ -t_0 r_{k,n-1} & \text{if } i = 0. \end{cases}$$

Now if $(a_0, \ldots, a_{n-1}, d)$ and $(b_0, \ldots, b_{n-1}, e)$ are the standard representations of α and β respectively, then it is clear that

$$\alpha\beta = \frac{\sum_{k=0}^{2n-2} c_k \theta^k}{de} \quad , \qquad \text{where } c_k = \sum_{i+j=k} a_i b_j \ ,$$

hence

$$\alpha\beta = \frac{\sum_{k=0}^{n-1} z_k \theta^k}{de} \quad , \qquad \text{where } z_k = c_k + \sum_{i=0}^{n-2} r_{k,i} c_{n+i} \ .$$

The standard representation of $\alpha\beta$ is then obtained by dividing all the z_k and de by $\gcd(z_0, \ldots, z_{n-1}, de)$.

Note that if we set $A(X) = \sum_{i=0}^{n-1} a_i X^i$ and $B(X) = \sum_{i=0}^{n-1} b_i X^i$, the procedure described above is equivalent to computing the remainder in the Euclidean division of AB by T. Because of the precomputations of the $r_{i,j}$, however, it is slightly more efficient.

The problem of division is more difficult. Here, we need essentially to compute A/B modulo the polynomial T. Hence, we need to invert B modulo T. The simplest efficient way to do this is to use the sub-resultant Algorithm 3.3.1 to obtain U and V (which does not need to be computed explicitly) such that $UB + VT = d$ where d is a constant polynomial. (Note that since T is irreducible and $B \neq 0$, B and T are coprime.) Then the inverse of B modulo T is $\frac{1}{d}U$, and the standard representation of α/β can easily be obtained from this.

4.2.3 The Matrix (or Regular) Representation of an Algebraic Number

A third way to represent algebraic numbers is by the use of integral matrices. If θ_j $(1 \leq j \leq n)$ is a $\mathbb{Q}$-basis of K and if $\alpha \in K$, then multiplication by α is an endomorphism of the $\mathbb{Q}$-vector space K, and we can represent α by the matrix M_α of this endomorphism in the basis θ_j. This will be a matrix with rational entries, hence one can write $M_\alpha = M'/d$ where M' has integral entries, d is a positive integer, and the greatest common divisor of all the entries of M' is coprime to d. This representation is of course unique, and it is clear that the map $\alpha \mapsto M_\alpha$ is an algebra homomorphism from K to the algebra of $n \times n$ matrices over $\mathbb{Q}$. Thus one can compute on algebraic numbers simply by computing with the corresponding matrices. The running time is usually longer however, since more elements are involved. For example, the simple operation of addition takes $O(n^2)$ operations, while it clearly needs only $O(n)$ operations in the standard representation. The matrix representation is clearly more suited for multiplication and division. (Division is performed using the remark following Algorithm 2.2.2.)

4.2.4 The Conjugate Vector Representation of an Algebraic Number

The last method of representing an algebraic number α in a number field $K = \mathbb{Q}(\theta)$ that I want to mention, is to represent α by numerical approximations to its conjugates, repeated with multiplicity. More precisely, let σ_j be the $n = \deg(K)$ distinct embeddings of K in $\mathbb{C}$, ordered in the following standard way: $\sigma_1, \ldots, \sigma_{r_1}$ are the real embeddings, $\sigma_{r_1+r_2+i} = \bar{\sigma}_{r_1+i}$ for $1 \leq i \leq r_2$. If $\alpha = \sum_{i=0}^{n-1} a_i \theta^i$, then

$$\sigma_j(\alpha) = \sum_{i=0}^{n-1} a_i \sigma_j(\theta)^i ,$$

and the $\sigma_j(\alpha)$ are the conjugates of α, but in a specific order (corresponding to the choice of the ordering on the σ_j), and repeated with a constant multiplicity $n/\deg(\alpha)$. We can then represent α as the $(r_1 + r_2)$-uplet of complex numbers

$$(\sigma_1(\alpha), \ldots, \sigma_{r_1+r_2}(\alpha)) ,$$

where the complex numbers $\sigma_j(\alpha)$ are given by a sufficiently good approximation. This will in practice be considered as a $r_1 + 2r_2 = n$-uplet of real numbers. Now operations on this representation are quite trivial since they are done componentwise. In particular, division, which was difficult in the other representations, becomes very simple here. Unfortunately, there is a price to pay: one must be able to go back to one of the exact representations (for example to the standard representation), and hence have good control on the roundoff errors.

For this, we precompute the inverse matrix of the matrix $\Theta = \sigma_i(\theta^{j-1})$. Then, if one knows the conjugate representation of a number α, and an integer d such that $d\alpha \in \mathbb{Z}[\theta]$, one can write $\alpha = (\sum_{j=1}^{n} a_{j-1}\theta^{j-1})/d$ where the a_j are integers, and the column vector $(a_0, \ldots, a_{n-1})^t$ can be obtained as the product $d\Theta^{-1}(\sigma_1(\alpha), \ldots, \sigma_n(\alpha))^t$, and since the a_i are integers, if the roundoff errors have been controlled and are not too large, this gives the a_i exactly (note that in practice one can work with matrices over $\mathbb{R}$ and not over $\mathbb{C}$. The details are left to the reader).

In practice, one can ignore roundoff errors and start with quite precise numerical approximations. Then every operation except division is done using the standard representation, while for division one computes the conjugate representation of the result, converts back, and then check by exact multiplication that the roundoff errors did not accumulate to give us a wrong result. (If they did, this means that one must work with a higher precision.)

4.3 Trace, Norm and Characteristic Polynomial

If α is an algebraic number, the trace (resp. the norm) of α is by definition
the sum (resp. the product) of the conjugates of α. If $A(X) = \sum_{i=0}^{m} a_i X^i$ is
its minimal polynomial, then we clearly have

$$\text{Tr}(\alpha) = -\frac{a_{m-1}}{a_m} \quad \text{and} \quad \mathcal{N}(\alpha) = (-1)^m \frac{a_0}{a_m} \ ,$$

where Tr and $\mathcal{N}$ denote the trace and norm of α respectively. Usually however,
α is considered as an element of a number field K. If $K = \mathbb{Q}(\alpha)$, then the
definitions above are OK, but if $\mathbb{Q}(\alpha) \subsetneq K$, then it is necessary to modify the
definitions so that Tr becomes additive and $\mathcal{N}$ multiplicative. More generally,
we put:

Definition 4.3.1. *Let K be a number field of degree n over $\mathbb{Q}$, and let σ_i be
the n distinct embeddings of K in $\mathbb{C}$.*

(1) *The characteristic polynomial C_α of α in K is*

$$C_\alpha(X) = \prod_{1 \le i \le n} (X - \sigma_i(\alpha)) \ .$$

(2) *If we set*

$$C_\alpha(X) = \sum_{0 \le i \le n} (-1)^{n-i} s_{n-i}(\alpha) X^i \ ,$$

*then $s_k(\alpha)$ is a rational number and will be called the kth symmetric func-
tion of α in K.*

(3) *In particular, $s_1(\alpha)$ is called the trace of α in K and denoted $\text{Tr}_{K/\mathbb{Q}}(\alpha)$,
and similarly $s_n(\alpha)$ is called the norm of α in K and denoted $\mathcal{N}_{K/\mathbb{Q}}(\alpha)$.*

As has already been mentioned, one must be careful to distinguish the
absolute trace of α which we have denoted $\text{Tr}(\alpha)$ from the trace of α in the
field K, denoted $\text{Tr}_{K/\mathbb{Q}}(\alpha)$, and similarly with the norms. More precisely, we
have the following proposition:

Proposition 4.3.2. *Let K be a number field of degree n, σ_i the n distinct
embeddings of K in $\mathbb{C}$.*

(1) *If $\alpha \in K$ has degree m (hence with m dividing n), we have*

$$\text{Tr}_{K/\mathbb{Q}}(\alpha) = \sum_{1 \le i \le n} \sigma_i(\alpha) = \frac{n}{m} \text{Tr}(\alpha) \ ,$$

and

$$\mathcal{N}_{K/\mathbb{Q}}(\alpha) = \prod_{1 \le i \le n} \sigma_i(\alpha) = (\mathcal{N}(\alpha))^{n/m} \ .$$

(2) *For any α and β in K we have*

$$\mathrm{Tr}_{K/\mathbf{Q}}(\alpha + \beta) = \mathrm{Tr}_{K/\mathbf{Q}}(\alpha) + \mathrm{Tr}_{K/\mathbf{Q}}(\beta) \ ,$$

and

$$\mathcal{N}_{K/\mathbf{Q}}(\alpha\beta) = \mathcal{N}_{K/\mathbf{Q}}(\alpha)\,\mathcal{N}_{K/\mathbf{Q}}(\beta) \ .$$

(3) α *is an algebraic integer if and only if $s_k(\alpha) \in \mathbf{Z}$ for all k such that $1 \le k \le n$ (note that $s_0(\alpha) = 1$).*

As usual, we must find algorithms to compute traces, norms and more generally characteristic polynomials of algebraic numbers. Since we have seen four different representations of algebraic numbers (viz. by a minimal polynomial, by the standard representation, by the matrix representation and by the conjugate vector representation), there are at least that many methods to do the job. We will only sketch these methods, except when they involve fundamentally new ideas. We always assume that our number field is given as $K = \mathbf{Q}(\theta)$ where θ is an algebraic integer whose monic minimal polynomial of degree n is denoted $T(X)$. We denote by σ_i the n embeddings of K in $\mathbf{C}$.

In the case where α is represented by its minimal polynomial $A(X)$, then each of the $m = \deg(A)$ embeddings of $\mathbf{Q}(\alpha)$ in $\mathbf{C}$ lifts to exactly n/m embeddings among the σ_i, hence it easily follows that

$$C_\alpha(X) = A(X)^{n/m} \ ,$$

and this immediately implies Proposition 4.3.2 (1), i.e. if we write $A(X) = \sum_{0 \le i \le m} a_i X^i$, then

$$\mathrm{Tr}_{K/\mathbf{Q}}(\alpha) = -\frac{na_{m-1}}{ma_m} \ , \qquad \mathcal{N}_{K/\mathbf{Q}}(\alpha) = (-1)^n \left(\frac{a_0}{a_m}\right)^{n/m} \ .$$

In the case where α is given by its standard representation

$$\alpha = \frac{1}{d}\left(\sum_{0 \le i \le n-1} a_i \theta^i\right) \ ,$$

the only symmetric function which is relatively easy to compute is the trace, since we can precompute the trace of θ^i using Newton's formulas as follows.

Proposition 4.3.3. *Let θ_i be the roots (repeated with multiplicity) of a monic polynomial $T(X) = \sum_{0 \le i \le n} t_i X^i \in \mathbf{C}[X]$ of degree n and set $S_k = \sum_i (\theta_i^k)$. Then*

$$S_k = -kt_{n-k} - \sum_{i=1}^{k-1} t_{n-i} S_{k-i} \qquad \text{(where we set } t_i = 0 \text{ for } i < 0) \ .$$

This result is well known and its proof is left to the reader (Exercise 3).

We can however compute all the symmetric functions, i.e. the characteristic polynomial, by using resultants, as follows.

Proposition 4.3.4. *Let $K = \mathbb{Q}(\theta)$ be a number field where θ is a root of a monic irreducible polynomial $T(X) \in \mathbb{Z}[X]$ of degree n, and let*

$$\alpha = \frac{1}{d}\left(\sum_{0 \leq i \leq n-1} a_i \theta^i\right)$$

be the standard representation of some $\alpha \in K$. Set $A(X) = \sum_{0 \leq i \leq n-1} a_i X^i$. Then the characteristic polynomial $C_\alpha(X)$ of α is given by the formula

$$C_\alpha(X) = d^{-n} R_Y(T(Y), dX - A(Y)) ,$$

where R_Y denotes the resultant taken with respect to the variable Y. In particular, we have

$$N_{K/\mathbb{Q}}(\alpha) = d^{-n} R(T(X), A(X)) .$$

Proof. We have by definition

$$C_\alpha(X) = \prod_i (X - \sigma_i(\alpha)) = \prod_i (X - A(\sigma_i(\theta))/d)$$

$$= d^{-n} \prod_i (dX - A(\theta_i)) = d^{-n} R_Y(T(Y), dX - A(Y)) ,$$

where the θ_i are the conjugates of θ, i.e. the roots of T. The formula for the norm follows immediately on setting $X = 0$. $\qquad\square$

Since the resultant can be computed efficiently by the sub-resultant Algorithm 3.3.7, used here in the UFD's $\mathbb{Z}[X]$ and $\mathbb{Z}$, we see that this proposition gives an efficient way to compute the characteristic polynomial and the norm of an algebraic number given in its standard representation.

In the case where α is given by numerical approximations to its conjugates, as usual we also assume that we know an integer d such that $d\alpha \in \mathbb{Z}[\theta]$. Then we can compute numerically $\prod_i(X - d\sigma_i(\alpha))$, and this must have integer coefficients. Hence, if we have sufficient control on the roundoff errors and sufficient accuracy on the conjugates of α, this enables us to compute $C_{d\alpha}(X)$ exactly, hence $C_\alpha(X) = d^{-n} C_{d\alpha}(dX)$.

Finally, we consider the case where α is given by its matrix representation M_α in the basis $1, \theta, \ldots, \theta^{n-1}$, where dM_α has integral coefficients for some integer d. Then the characteristic polynomial of α is simply equal to the characteristic polynomial of M_α (meaning always $\det(XI_n - M_\alpha)$). In particular, the trace can be read off trivially on the diagonal coefficients, and the norm is, up to sign, equal to the determinant of M_α.

The characteristic polynomial can be computed using one of the algorithms described Section 2.2.4, and the determinant using Algorithm 2.2.6.

In practice, it is not completely clear which representation is preferable. A reasonable choice is probably to use the standard representation and the sub-resultant algorithm. This depends on the context however, and one should always be aware of each of the four possibilities to handle algebraic numbers. Keep in mind that it is usually costly to go from one representation to another, so for a given problem the representation should be fixed.

4.4 Discriminants, Integral Bases and Polynomial Reduction

4.4.1 Discriminants and Integral Bases

We have the following basic result.

Proposition 4.4.1. *Let K be a number field of degree n, σ_i be the n embeddings of K in $\mathbb{C}$, and α_j be a set of n elements of K. Then we have*

$$\det(\sigma_i(\alpha_j))^2 = \det(\mathrm{Tr}_{K/\mathbb{Q}}(\alpha_i\alpha_j)) .$$

This quantity is a rational number and is called the discriminant *of the α_i, and denoted $d(\alpha_1,\ldots,\alpha_n)$. Furthermore, $d(\alpha_1,\ldots,\alpha_n) = 0$ if and only if the α_j are $\mathbb{Q}$-linearly dependent.*

Proof. Consider the $n \times n$ matrix $M = (\sigma_i(\alpha_j))$. Then by definition of matrix multiplication, we have $M^t M = (a_{i,j})$ with

$$a_{i,j} = \sum_k \sigma_k(\alpha_i)\sigma_k(\alpha_j) = \mathrm{Tr}_{K/\mathbb{Q}}(\alpha_i\alpha_j) .$$

Since $\det(M^t) = \det(M)$ the equality of the proposition follows. Since $\mathrm{Tr}_{K/\mathbb{Q}}(\alpha) \in \mathbb{Q}$ the discriminant is a rational number. If the α_j are $\mathbb{Q}$-linearly dependent, it is clear that the columns of the matrix M are also (since $\mathbb{Q}$ is invariant by the σ_i). Therefore the discriminant is equal to 0. Conversely assume that the discriminant is equal to 0. This means that the kernel of the matrix $M^t M$ is non-trivial, and since this matrix has coefficients in $\mathbb{Q}$, there exists $\lambda_i \in \mathbb{Q}$ such that for every j, $\mathrm{Tr}(x\alpha_j) = 0$ where we have set $x = \sum_{1\le i\le n} \lambda_i\alpha_i$. If the α_j were linearly independent over $\mathbb{Q}$, they would generate K as a $\mathbb{Q}$-vector space, and so we would have $\mathrm{Tr}(xy) = 0$ for all $y \in K$ with $x \ne 0$. Taking $y = 1/x$ gives $\mathrm{Tr}(1) = n = 0$, a contradiction, thus showing the proposition. $\qquad\square$

Remark. We have just proved that the quadratic form $\text{Tr}(x^2)$ is non-degenerate on K using that K is of characteristic zero (otherwise $n = 0$ may not be a contradiction). This is the definition of a *separable extension*. It is not difficult to show (see for example Proposition 4.8.11 or Exercise 5) that the signature of this quadratic form (i.e. the number of positive and negative squares after Gaussian reduction) is equal to $(r_1 + r_2, r_2)$ where as usual (r_1, r_2) is the signature of the number field K.

Recall that we denote by $\mathbb{Z}_K$ the ring of (algebraic) integers of K. Then we also have:

Theorem 4.4.2. *The ring $\mathbb{Z}_K$ is a free $\mathbb{Z}$-module of rank $n = \deg(K)$. This is true more generally for any non-zero ideal of $\mathbb{Z}_K$.*

Proof (Sketch). Let α_j be a basis of K as a $\mathbb{Q}$-vector space. Without loss of generality, we can assume that the α_j are algebraic integers. If A is the (free) $\mathbb{Z}$-module generated by the α_j, we clearly have $A \subset \mathbb{Z}_K$, and the formula $M^{-1} = M^{\text{adj}}/\det(M)$ for the inverse of a matrix (see section 2.2.4) shows that $d\mathbb{Z}_K \subset A$, where d is the discriminant of the α_j, whence the result. (Recall that a sub-$\mathbb{Z}$-module of a free module of rank n is a free module of rank less than or equal to n, since $\mathbb{Z}$ is a principal ideal domain, see Theorem 2.4.1.) □

It is important to note that $\mathbb{Z}$ being a PID is crucial in the above proof. Hence, if we consider *relative* extensions, Theorem 4.4.2 will a priori be true only if the base ring is also a PID, and this is not always the case.

Definition 4.4.3. *A $\mathbb{Z}$-basis of the free module $\mathbb{Z}_K$ will be called an* integral basis *of K. The discriminant of an integral basis is independent of the choice of that basis, and is called the* discriminant *of the field K and is denoted by $d(K)$.*

Note that, although the two notions are closely related, the discriminant of K is not in general equal to the discriminant of an irreducible polynomial defining K. More precisely:

Proposition 4.4.4. *Let T be a monic irreducible polynomial of degree n in $\mathbb{Z}[X]$, θ a root of T, and $K = \mathbb{Q}(\theta)$. Denote by $d(T)$ (resp. $d(K)$) the discriminant of the polynomial T (resp. of the number field K).*

(1) We have $d(1, \theta, \ldots, \theta^{n-1}) = d(T)$.
(2) If $f = [\mathbb{Z}_K : \mathbb{Z}[\theta]]$, we have

$$d(T) = d(K)f^2$$

and, in particular, $d(T)$ is a square multiple of $d(K)$.

The proof of this is easy and left to the reader. The number f will be called the *index* of θ in $\mathbb{Z}_K$.

Proposition 4.4.5. *The algebraic numbers $\alpha_1, \ldots, \alpha_n$ form an integral basis if and only if they are algebraic integers and if $d(\alpha_1, \ldots, \alpha_n) = d(K)$, where $d(K)$ is the discriminant of K.*

Proof. If M is the matrix expressing the α_i on some integral basis of K, it is clear that $d(\alpha_1, \ldots, \alpha_n) = d(K) \det(M)^2$ and the proposition follows. $\square$

We also have the following result due to Stickelberger:

Proposition 4.4.6. *Let $\alpha_1, \ldots, \alpha_n$ be algebraic integers. Then*

$$d(\alpha_1, \ldots, \alpha_n) \equiv 0 \ or \ 1 \pmod 4 .$$

Proof. If we expand the determinant $\det(\sigma_i(\alpha_j))$ using the $n!$ terms, we will get terms with a plus sign corresponding to permutations of even signature, and terms with a minus sign. Hence, collecting these terms separately, we can write the determinant as $P - N$ hence

$$d(\alpha_1, \ldots, \alpha_n) = (P - N)^2 = (P + N)^2 - 4PN .$$

Now clearly $P + N$ and PN are symmetric functions of the α_i, hence by Galois theory they are in $\mathbb{Q}$ and in fact in $\mathbb{Z}$ since the α_i are algebraic integers. This proves the proposition, since a square is always congruent to 0 or 1 mod 4. $\square$

The determination of an explicit integral basis and of the discriminant of a number field is not an easy problem, and is one of the main tasks of this course. There is, however one case in which the result is trivial:

Corollary 4.4.7. *Let T be a monic irreducible polynomial in $\mathbb{Z}[X]$, θ a root of T, and $K = \mathbb{Q}(\theta)$. Assume that the discriminant of T is squarefree or is equal to $4d$ where d is squarefree and not congruent to 1 modulo 4. Then the discriminant of K is equal to the discriminant of T, and an integral basis of K is given by $1, \theta, \ldots, \theta^{n-1}$.*

Since a discriminant must be congruent to 0 or 1 mod 4, this immediately follows from the above propositions. $\square$

Unfortunately, this corollary is not of much use, since it is quite rare that the condition on the discriminant of T is satisfied. We will see in Chapter 6 a complete method for finding an integral basis and hence the discriminant of a number field.

Finally, we note without proof the following consequence of the so-called "conductor-discriminant formula".

Proposition 4.4.8. *Let K and L be number fields with $K \subset L$. Then*

$$d(K)^{[L:K]} \mid d(L) .$$

Corollary 4.4.9. *Let $K = \mathbb{Q}(\alpha)$ and $L = \mathbb{Q}(\beta)$ be two number fields, let $m = \deg(K)$, $n = \deg(L)$, $A(X)$ (resp. $B(X)$) the minimal monic polynomial of α (resp. β). Write $d(A)$ and $d(B)$ for the discriminants of the polynomials A and B. Assume that K is conjugate to a subfield of L. Then if p is an odd prime such that $v_p(d(A))$ is odd, we must have $p^{n/m} \mid d(B)$.*

Proof. By Proposition 4.4.4 if $v_p(d(A))$ is odd then $p \mid d(K)$, where $d(K)$ is the discriminant of the field K. By the proposition we therefore have $p^{n/m} \mid d(L) \mid d(B)$, thus proving the corollary.   □

4.4.2 The Polynomial Reduction Algorithm

We will see in Section 4.5 that it is usually not always easy to decide whether two number fields are isomorphic or not. Here we will give a heuristic approach based on the LLL algorithm and ideas of Diaz y Diaz and the author which often gives a useful answer to the following problem: given a number field K, can one find a monic irreducible polynomial defining K which in a certain sense is as simple as possible.

Of course, if this could be done, the isomorphism problem would be completely solved. We will see in Chapters 5 and 6 that it is possible to do this for quadratic fields (in fact it is trivial in that case), and for certain classes of cubic fields, like cyclic cubic fields or pure cubic fields (see Section 6.4). In general, all one can hope for in practice is to find, maybe not *the* simplest, but *a* simple polynomial defining K.

A natural criterion of simplicity would be to take polynomials whose largest coefficients are as small as possible in absolute value (i.e. the L^∞ norm on the coefficients), or such that the sum of the squares of the coefficients is as small as possible (the L^2 norm). Unfortunately, I know of no really efficient way of finding simple polynomials in this sense.

What we will in fact consider is the following "norm" on polynomials.

Definition 4.4.10. *Let $P \in \mathbb{C}[X]$, and let α_i be the complex roots of P repeated with multiplicity. We define the size of P by the formula*

$$\text{size}(P) = \sum |\alpha_i|^2 .$$

This is not a norm in the usual mathematical sense, but it seems reasonable to say that if the size (in this sense) of a polynomial is not large, then the polynomial is simple, and its coefficients should not be too large.

More precisely, we can show (see Exercise 6) that if $P = \sum_{k=0}^{n} a_k X^k$ is a monic polynomial and if $S = \text{size}(P)$, then

$$|a_{n-k}| \leq \binom{n}{k} \left(\frac{S}{n}\right)^{k/2}.$$

Hence, the size of P is related to the size of

$$\max |a_{n-k}|^{2/k}.$$

The reason we take this definition instead of an L^p definition on the coefficients is that we can apply the LLL algorithm to find a polynomial of small size which defines the same number field K as the one defined by a given polynomial P, while I do not know how to achieve this for the norms on the coefficients.

The method is as follows. Let K be defined by a monic irreducible polynomial $P \in \mathbb{Z}[X]$. Using the round 2 Algorithm 6.1.8 which will be explained in Chapter 6, we compute an integral basis $\omega_1, \ldots, \omega_n$ of $\mathbb{Z}_K$. Furthermore, let σ_j denote the n isomorphisms of K into $\mathbb{C}$. If we set

$$x = \sum_{i=1}^{n} x_i \omega_i$$

where the x_i are in $\mathbb{Z}$, then x is an arbitrary algebraic integer in K, hence its characteristic polynomial M_x will be of the form $P_d^{n/d}$ where P_d is the minimal polynomial of x and d the degree of x, and P_d defines a subfield of K. In particular, when $d = n$, this defines an equation for K, and clearly all monic equations for K with integer coefficients (as well as for subfields of K) are obtained in this way.

Now we have by definition

$$M_x(X) = \prod_{k=1}^{n} \left(X - \sum_{i=1}^{n} x_i \sigma_k(\omega_i) \right)$$

hence,

$$\text{size}(M_x) = \sum_{k=1}^{n} \left| \sum_{i=1}^{n} x_i \sigma_k(\omega_i) \right|^2$$

This is clearly a quadratic form in the x_i's, and more precisely

$$\text{size}(M_x) = \sum_{i,j} \left(\sum_{1 \leq k \leq n} \sigma_k(\omega_i) \overline{\sigma_k}(\omega_j) \right) x_i x_j.$$

Note that in the case where K is totally real, that is when all the σ_k are real embeddings, this simplifies to

$$\text{size}(M_x) = \sum_{i,j} \text{Tr}(\omega_i \omega_j) x_i x_j$$

which is now a quadratic form with integer coefficients which can easily be computed from the knowledge of the ω_i.

In any case, whether K is totally real or not, we can apply the LLL algorithm to the lattice $\mathbb{Z}^n$ and the quadratic form $\text{size}(M_x)$. The result will be a set of n vectors x corresponding to reasonably small values of the quadratic form (see Section 2.6 for quantitative statements), hence to polynomials M_x of small size, which is what we want. Note however that the algebraic integers x that we obtain in this way will often have a minimal polynomial of degree less than n, in other words x will define a subfield of K. In particular, $x = 1$ is always obtained as a short vector, and this defines the subfield $\mathbb{Q}$ of K. Practical experiments with this method show however that there will always be at least one element x of degree exactly n, hence defining K, and its minimal polynomial will hopefully be simpler than the polynomial P from which we started.

However the polynomials that we obtain in this way, have sometimes greater coefficients than those of P. This is not too surprising since our definition of "size" of $P(X) = \sum_{0 \leq k \leq n} a_k X^k$ involves the size of the roots of P, hence of the quantities

$$|a_{n-k}|^{1/k}$$

more than the size of the coefficients themselves.

Note that as a by-product of this method, we sometimes also obtain subfields of K. It is absolutely not true however that we obtain all subfields of K in this way. Indeed, the LLL algorithm gives us at most n subfields, while the number of subfields of K may be much larger.

The algorithm, which we name POLRED for polynomial reduction, is as follows (see [Coh-Diaz]).

Algorithm 4.4.11 (POLRED). Let $K = \mathbb{Q}(\theta)$ be a number field defined by a monic irreducible polynomial $P \in \mathbb{Z}[X]$. This algorithm gives a list of polynomials defining certain subfields of K (including $\mathbb{Q}$), which are often simpler than the polynomial P so these can be used to define the field K if they are of degree equal to the degree of K.

1. [Compute the maximal order] Using the round 2 Algorithm 6.1.8 of Chapter 6, compute an integral basis $\omega_1, \ldots, \omega_n$ as polynomials in θ.

2. [Compute matrix] If the field K is totally real (which can be easily checked using Algorithm 4.1.11), set $m_{i,j} \leftarrow \text{Tr}(\omega_i \omega_j)$ for $1 \leq i, j \leq n$, which will be an element of $\mathbb{Z}$.

Otherwise, using Algorithm 3.6.6, compute a reasonably accurate value of θ and its conjugates $\sigma_j(\theta)$ as the roots of P, then the numerical values of $\sigma_j(\omega_k)$, and finally compute a reasonably accurate approximation to

$$m_{i,j} \leftarrow \sum_{1 \leq k \leq n} \sigma_k(\omega_i)\overline{\sigma_k}(\omega_j)$$

(note that this will be a real number).

3. [Apply LLL] Using the LLL Algorithm 2.6.3 applied to the inner product defined by the matrix $M = (m_{i,j})$ and to the standard basis of the lattice $\mathbb{Z}^n$, compute an LLL-reduced basis $b_1, \ldots, b_n$.

4. [Compute characteristic polynomials] For $1 \leq i \leq n$, using the formulas of Section 4.3, compute the characteristic polynomial C_i of the element of $\mathbb{Z}_K$ corresponding to b_i on the basis $\omega_1, \omega_2, \ldots, \omega_n$.

5. [Compute minimal polynomials] For $1 \leq i \leq n$, set $P_i \leftarrow C_i/(C_i, C_i')$ where the GCD is always normalized so as to be monic, and is computed by Euclid's algorithm. Output the polynomials P_i and terminate the algorithm.

From what we have seen in Section 4.3, the characteristic polynomial C_i of an element $x \in \mathbb{Z}_K$ is given by $C_i = P_i^k$, where P_i is the minimal polynomial and k is a positive integer, hence $C_i/(C_i, C_i') = P_i$, thus explaining step 5. In fact, to avoid ambiguities of sign which arise, it is also useful to make the following choice at the end of the algorithm. For each polynomial P_i, set $d_i \leftarrow \deg(P_i)$ and search for the non-zero monomial of largest degree d such that $d \not\equiv d_i \pmod{2}$. If such a monomial exists, make, if necessary, the change $P_i(X) \leftarrow (-1)^{d_i} P_i(-X)$ so that the sign of this monomial is negative.

Let us give an example of the use of the POLRED algorithm. This example is taken from work of M. Olivier. Consider the polynomial

$$T(X) = X^6 + 2X^5 - 7X^4 - 12X^3 + 10X^2 + 17X + 4 .$$

Using the methods of Section 3.5, one easily shows that this polynomial is irreducible over $\mathbb{Q}$, hence defines a number field K of degree 6. Furthermore, using Algorithm 3.6.6, one computes that the complex roots of T are approximately equal to

$$-2.7494482169, -1.7152399972, -0.8531562311, -0.3074682781,$$

$$1.5839340557, 2.0413786677 .$$

Using the methods of the preceding section, it is then easy to check that this field has no proper subfield apart from $\mathbb{Q}$.

From this and the classification of transitive permutation groups of degree 6 which we will see in Section 6.3, we deduce that the Galois group G of the Galois closure of K is isomorphic either to the alternating groups A_5 or A_6,

or to the symmetric groups S_5 or S_6. Now using the sub-resultant Algorithm
3.3.7 or Proposition 3.3.5 one computes that

$$\mathrm{disc}(T) = 11699^2$$

so by Proposition 6.3.1, we have $G \subset A_6$ hence G is isomorphic either to A_5
or to A_6.

Distinguishing between the two is done by using one of the resolvent functions given in Section 6.3, and the resolvent polynomial obtained is

$$R(X) = X^6 + 3694X^5 + 1246830X^4 - 7355817976X^3 - 5140929655107X^2$$
$$+ 3486026298845999X + 2593668315970494361 \ .$$

A computation of the roots of this polynomial shows that it has an integer root
$x = -673$, and the results of Section 6.3 imply that G is isomorphic to A_5. In
addition, $Q(X) = R(X)/(X + 673)$ is an irreducible fifth degree polynomial
which defines a number field with the same discriminant as K. We have

$$Q(X) = X^5 + 3021X^4 - 786303X^3 - 6826636057X^2$$
$$- 546603588746X + 3853890514072057 \ ,$$

and the discriminant of Q (which must be a square by Proposition 6.3.1) has
63 decimal digits. Now if we apply the POLRED algorithm, we obtain five
polynomials, four of which define the same field as Q, and the polynomial
with the smallest discriminant is

$$S(X) = X^5 - 2X^4 - 13X^3 + 37X^2 - 21X - 1 \ ,$$

a polynomial which is much more appealing than Q !

We compute that $\mathrm{disc}(S) = 11699^2$, hence this is the discriminant of the
number field K as well as the number field defined by the polynomial S.

There was a small amount of cheating in the above example: since $\mathrm{disc}(Q)$
is a 63 digit number, the POLRED algorithm, which in particular computes
an integral basis of K hence needs to factor $\mathrm{disc}(Q)$, may need quite a lot
of time to factor this discriminant. We can however in this case "help" the
POLRED algorithm by telling it that $\mathrm{disc}(Q)$ is a square, which we know a
priori, but which is not usually tested for in a factoring algorithm since it is
quite rare an occurrence. This is how the above example was computed in
practice, and the whole computation, including typing the commands, took
only a few minutes on a workstation.

We can slightly modify the POLRED algorithm so as to obtain a defining
polynomial for a number field which is as canonical as possible. One possibility
is as follows.

We first need a notation. If $Q(X) = \sum_{0 \le i \le n} a_i X^i$ is a polynomial of degree
n, we set

$$v(Q) = (|\operatorname{disc}(Q)|, \operatorname{size}(Q), |a_n|, |a_{n-1}|, \ldots, |a_1|, |a_0|) .$$

Algorithm 4.4.12 (Pseudo-Canonical Defining Polynomial). Given a number field K defined by a monic irreducible polynomial $P \in \mathbb{Z}[X]$ of degree n, this algorithm outputs another polynomial defining K which is as canonical as possible.

1. [Apply POLRED] Apply the POLRED algorithm to P, and let P_i (for $i = 1, \ldots, n$) be the n polynomials which are output by the POLRED algorithm. If none of the P_i are of degree n, output a message saying that the algorithm failed, and terminate the algorithm. Otherwise, let $\mathcal{L}$ be the set of i such that P_i is of degree n.

2. [Minimize $v(P_i)$] If $\mathcal{L}$ has a single element, let Q be this element. If not, for each $i \in \mathcal{L}$ compute $v_i \leftarrow v(P_i)$ and let v be the smallest v_i for the lexicographic ordering of the components. Let Q be any P_i such that $v(P_i) = v$.

3. [Possible sign change] Search for the non-zero monomial of largest degree d such that $d \not\equiv n \pmod 2$. If such a monomial exists, make, if necessary, the change $Q(X) \leftarrow (-1)^n Q(-X)$ so that the sign of this monomial is negative.

4. [Terminate] Output Q and terminate the algorithm.

Remarks.

(1) The algorithm may fail, i.e. the POLRED algorithm may give only polynomials of degree less than n. That this is possible in principle has been shown by H. W. Lenstra (private communication), but in practice, on more than 100000 polynomials of various degree, I have never encountered a failure. It seems that failure is very rare.

(2) At the end of step 2 there may be several i such that $v_i = v$. In that case, it may be useful to output all the possibilities (after executing step 3 on each of them) instead of only one. In practice, this is also uncommon.

(3) Although Algorithm 4.4.12 makes an effort towards finding a polynomial defining K with small index $f = [\mathbb{Z}_K : \mathbb{Z}[\theta]]$, it should not be expected that it always finds a polynomial with the smallest possible index. An example is the polynomial $X^3 - X^2 - 20X + 9$ which naturally defines the cyclic cubic field with discriminant 61^2 (see Theorem 6.4.6). Algorithm 4.4.12 finds that this is the pseudo-canonical polynomial defining the cubic field, but it has index equal to 3, while for example the polynomial $X^3 + 12X^2 - 13X + 3$ has index equal to 1. The reason for this behavior is that the notion of "size" of a polynomial is rather indirectly related to the size of the index. See also Exercise 8.

4.5 The Subfield Problem and Applications

Let $K = \mathbb{Q}(\alpha)$ and $L = \mathbb{Q}(\beta)$ be number fields of degree m and n respectively, and let $A(X), B(X) \in \mathbb{Z}[X]$ be the minimal polynomials of α and β respectively. The basic subfield problem is as follows. Determine whether or not K is isomorphic to a subfield of L, or in more down-to-earth terms whether or not some conjugate of α belongs to L. We could of course ask more precisely if α itself belongs to L, and we will see that the answer to this question follows essentially from the answer to the apparently weaker one.

We start by two fast tests. First, if K is conjugate to a subfield of L, then the degree of K clearly must divide the degree of L.

The second test follows from Corollary 4.4.9. We compute $d(A)$ and $d(B)$ and for each odd prime p such that $v_p(d(A))$ is odd, test whether or not $p^{n/m} \mid d(B)$. Note that according to Exercise 15, it is not necessary to assume that A and B are monic, i.e. that α and β are algebraic integers.

We could use the more stringent test $d(K)^{n/m} \mid d(L)$ using Proposition 4.4.8 directly, but this requires the computation of field discriminants, hence essentially of integral bases, and this is often lengthy. So, we do not advise using this more stringent test unless the field discriminants can be obtained cheaply.

We therefore assume that the above tests have been passed successfully. We will give three different methods for solving our problem. The first two require good approximations to the complex roots of the polynomials A and B (computed using for example Algorithm 3.6.6), while the third is purely algebraic, but slower.

4.5.1 The Subfield Problem Using the LLL Algorithm

Let β be an arbitrary, but fixed root of the polynomial B in $\mathbb{C}$. If K is conjugate to a subfield of L, then some root α_i of A is of the form $P(\beta)$ for some $P \in \mathbb{Q}[X]$ of degree less than n. In other words, the complex numbers $1, \beta, \ldots, \beta^{n-1}, \alpha_i$ are $\mathbb{Z}$-linearly dependent. To check this, use the LLL algorithm or one of its variations, as described in Section 2.7.2 on each root of A (or on the root we are specifically interested in as the case may be). Then two things may happen. Either the algorithm gives a linear combination which is not very small in appearance, or it seems to find something reasonable. The reader will notice that in none of these cases have we *proved* anything. If, however, we are in the situation where LLL apparently found a nice relation, this can now be proved: assume the relation gives $\alpha_i = P(\beta)$ for some polynomial P with rational coefficients. (Note that the coefficient of α_i in the linear combination which has been found must be non-zero, otherwise this would mean that the minimal polynomial of β is not irreducible.) To test whether this relation is true, it is now necessary simply to check that

$$A \circ P \equiv 0 \pmod{B} \ ,$$

where A and B are the minimal polynomials of α and β respectively. Indeed, if this is true, this means that $P(\beta)$ is a root of A, i.e. a conjugate of α_i, hence is α_i itself since LLL told us that it was numerically very close to α_i.

To compute $C = A \circ P \pmod{B}$, we use a form of Horner's rule for evaluating polynomials: if $A(X) = \sum_{i=0}^{m} a_i X^i$, then we set $C \leftarrow a_m$, and for $i = m-1, m-2, \ldots, 0$ we compute $C \leftarrow (a_i + P(X)C \bmod B)$.

In the implausible case where one finds that $A \circ P \not\equiv 0 \pmod{B}$, then we must again test for linear dependence with higher precision used for α_i and β.

Remark. There is a better way to test whether each conjugate α_i is or is not a $\mathbb{Q}$-linear combination of $1, \beta, \ldots, \beta^{n-1}$ than to apply LLL to each α_i, each time LLL reducing an $(n+2) \times (n+1)$ matrix (or equivalently a quadratic form in $n+1$ variables). Indeed, keeping with the notations of Remark (2) at the end of Section 2.7.2, the first n columns of that matrix, which correspond to the powers of β, will always be the same. Only the last column depends on α_i. But in LLL reduction, almost all the work is spent LLL reducing the first n columns, the $n+1$-st is done last. Hence, we should first LLL reduce the $(n+2) \times n$ matrix corresponding to the powers of β. Then, for each α_i to be tested, we can now start from the already reduced basis and just add an extra column vector, and since the first n vectors are already LLL reduced, the amount of work which remains to be done to account for the last column will be very small compared to a full LLL reduction. We leave the details to the reader.

If LLL tells us that apparently there is no linear relation, then we suspect that $\alpha \notin \mathbb{Q}(\beta)$. To prove it, the best way is probably to apply one of the two other methods which we are going to explain.

4.5.2 The Subfield Problem Using Linear Algebra over $\mathbb{C}$

A second method is as follows (I thank A.-M. Bergé and M. Olivier for pointing it out to me.) After clearing denominators, we may as well assume that α and β are algebraic integers. We then have the following.

Proposition 4.5.1. *With the above notations, assume that α and β are algebraic integers. Then K is isomorphic to a subfield of L if and only if there exists an n/m to one map ϕ from $[1,n]$ to $[1,m]$ such that for $1 \leq h < n$,*

$$s_h = \sum_{1 \leq i \leq n} \alpha_{\phi(i)} \beta_i^h \in \mathbb{Z} \ ,$$

where the α_j (resp. β_j) denote the roots of $A(X)$ (resp. of $B(X)$) in $\mathbb{C}$.

Proof. Assume first that K is isomorphic to a subfield of L, i.e. that $\alpha_i = P(\beta_1)$ with $P \in \mathbb{Q}[X]$ say. Then, for every i, $P(\beta_i)$ is a root α_j of $A(X) = 0$, and by Galois theory each α_j is obtained exactly n/m times. Therefore the map $i \mapsto j = \phi(i)$ is n/m to one. Furthermore,

$$s_h = \sum_{1 \le i \le n} \alpha_{\phi(i)}\beta_i^h = \sum_{1 \le i \le n} P(\beta_i)\beta_i^h = \mathrm{Tr}_{L/\mathbb{Q}}(P(\beta)\beta^h) \in \mathbb{Q} \;,$$

hence $s_h \in \mathbb{Z}$ since the α_j and β_i are algebraic integers.

Conversely, assume that for some ϕ we have $s_h \in \mathbb{Z}$ for all h such that $1 \le h < n$. Note that $s_0 = (n/m)\,\mathrm{Tr}_{K/\mathbb{Q}}(\alpha) \in \mathbb{Z}$ follows automatically.

Consider the following $n \times n$ linear system:

$$\sum_{0 \le j < n} x_j\,\mathrm{Tr}_{L/\mathbb{Q}}(\beta^j\beta^h) = s_h \;, \qquad 0 \le h < n \;.$$

By Proposition 4.4.4 (1) the determinant of this system is equal to $d(B)$, hence is non-zero. Furthermore, the system has rational coefficients, so the unique solution has coefficients $x_j \in \mathbb{Q}$. If we set $P(X) = \sum_{0 \le j < n} x_j X^j$, we then have $P \in \mathbb{Q}[X]$ and $\sum_{1 \le i \le n} P(\beta_i)\beta_i^h = s_h$. If we set $\gamma = \alpha_{\phi(1)} - P(\beta_1)$, it follows that $\mathrm{Tr}_{L/\mathbb{Q}}(\gamma\beta^h) = 0$ for all h such that $0 \le h < n$ and hence $\mathrm{Tr}_{L/\mathbb{Q}}(\gamma\delta) = 0$ for all $\delta \in L$. Since we have seen that the trace is non-degenerate (see the remark after Proposition 4.4.1), it follows that $\gamma = 0$, thus proving the proposition. $\qquad \square$

Remarks.

(1) The number of maps from $[1, n]$ to $[1, m]$ which are n/m-to-one is equal to $n!/((n/m)!)^m$ hence can be quite large, especially when $m = n$ (which corresponds to the very important *isomorphism problem*). This is to be compared to the number of trials to be done with the LLL method, which is only equal to m. Hence, although LLL is slow, except when n is very small (say $n \le 4$), we suggest starting with the LLL method. If the answer is positive, which will in practice happen quite often, we can stop. If not, use the present method (or the purely algebraic method which is explained below).

(2) To check that $s_h \in \mathbb{Z}$ we must of course compute the roots of $A(X)$ and $B(X)$ sufficiently accurately. Now however the error estimates are trivial (compared to the ones we would need using LLL), and if s_h is sufficiently far away from an integer, it is very easy to prove rigorously that it is so.

(3) We start of course by checking whether $s_1 \in \mathbb{Z}$, since this will eliminate most candidates for ϕ.

The above leads to the following algorithm.

Algorithm 4.5.2 (Subfield Problem Using Linear Algebra). Let $A(X)$ and $B(X)$ be primitive irreducible polynomials in $\mathbb{Z}[X]$ of degree m and n respectively defining number fields K and L. This algorithm determines whether or not K is isomorphic to a subfield of L, and if it is, gives an explicit isomorphism.

1. [Trivial check] If $m \nmid n$, output NO and terminate the algorithm.

2. [Reduce to algebraic integers] Set $a \leftarrow \ell(A)$, $b \leftarrow \ell(B)$ (the leading terms of A and B), and set $A(X) \leftarrow a^{m-1}A(X/a)$ and $B(X) \leftarrow b^{n-1}B(X/b)$.

3. [Check discriminants] For every odd prime p such that $v_p(d(A))$ is odd, check that $p^{n/m} \mid d(B)$ (where $d(A)$ and $d(B)$ are computed using Algorithm 3.3.7). If this is not the case, output NO and terminate the algorithm. If for some reason $d(K)$ and $d(L)$ are known or cheaply computed, replace these checks by the single check $d(K)^{n/m} \mid d(L)$.

4. [Compute roots] Using Algorithm 3.6.6, compute the complex roots α_i and β_i of $A(X)$ and $B(X)$ to a reasonable accuracy (it may be necessary to have more accuracy in the later steps).

5. [Loop on ϕ] For each n/m to one map ϕ from $[1, n]$ to $[1, m]$ execute steps 6 and 7. If all the maps have been examined without termination of the algorithm, output NO and terminate the algorithm.

6. [Check $s_1 \in \mathbb{Z}$] Let $s_1 \leftarrow \sum_{1 \leq i \leq n} \alpha_{\phi(i)}\beta_i$. If s_1 is not close to an integer (this is a rigorous statement, since it depends only on the chosen approximations to the roots), take the next map ϕ in step 5.

   Otherwise, check whether $s_h \leftarrow \sum_{1 \leq i \leq n} \alpha_{\phi(i)}\beta_i^h$ are also close to an integer for $h = 2, \ldots, n-1$. As soon as this is not the case, take the next map ϕ in step 5.

7. [Compute polynomial] (Here the s_h are all close to integers.) Set $s_h \leftarrow \lfloor s_h \rceil$ (the nearest integer to s_h). Compute by induction $t_k \leftarrow \text{Tr}_{L/\mathbb{Q}}(\beta_1^k)$ for $0 \leq k \leq 2n - 2$, and using Algorithm 2.2.1 or a Gauss-Bareiss variant, find the unique solution to the linear system $\sum_{0 \leq j < n} x_j t_{j+h} = s_h$ for $0 \leq h < n$ (note that we know that $d(B)x_j \in \mathbb{Z}$ so we can avoid rational arithmetic), and set $P(X) \leftarrow \sum_{0 \leq j < n} x_j X^j$.

8. [Finished?] Using the variant of Horner's rule explained in Section 4.5.1, check whether $A(P(X)) \equiv 0 \pmod{B(X)}$. If this is the case, then output YES, output also the polynomial $P(bX)/a$ which gives the isomorphism explicitly, and terminate the algorithm. Otherwise, using Algorithm 3.6.6 (or, even more simply, a few Newton iterations to obtain a higher precision) recompute the roots α_i and β_i to a greater accuracy and go to step 6.

4.5.3 The Subfield Problem Using Algebraic Algorithms

The third solution that we give to the subfield problem is usually less efficient but has the advantage that it is guaranteed to work without worrying about complex approximations. The idea is to use Algorithm 3.6.4 which factors polynomials over number fields and the following easy proposition whose proof is left to the reader (Exercise 9).

Proposition 4.5.3. *Let α and β be algebraic numbers with minimal polynomials $A(X)$ and $B(X)$ respectively. Set $K = \mathbb{Q}(\alpha)$, $L = \mathbb{Q}(\beta)$, and let*

$A = \prod_{1 \leq i \leq g} A_i$ be a factorization of A into irreducible factors in $L[X]$. There is a one-to-one correspondence between the A_i of degree equal to one and the conjugates of α belonging to L. In particular, L contains a subfield isomorphic to K if and only if at least one of the A_i is of degree equal to one.

This immediately leads to the following algorithm. Note that we keep the same first three steps of the preceding algorithm.

Algorithm 4.5.4 (Subfield Problem Using Factorization of Polynomials). Let $A(X)$ and $B(X)$ be primitive irreducible polynomials in $\mathbb{Z}[X]$ of degree m and n respectively defining number fields K and L. This algorithm determines whether or not K is isomorphic to a subfield of L, and if it is, gives an explicit isomorphism.

1. [Trivial check] If $m \nmid n$, output NO and terminate the algorithm.

2. [Reduce to algebraic integers] Set $a \leftarrow \ell(A)$, $b \leftarrow \ell(B)$ (the leading terms of A and B), and set $A(X) \leftarrow a^{m-1} A(X/a)$ and $B(X) \leftarrow b^{n-1} B(X/b)$.

3. [Check discriminants] For every odd prime p such that $v_p(d(A))$ is odd, check that $p^{n/m} \mid d(B)$ (where $d(A)$ and $d(B)$ are computed using Algorithm 3.3.7). If this is not the case, output NO and terminate the algorithm. If for some reason $d(K)$ and $d(L)$ are known or cheaply computed, replace these checks by the single check $d(K)^{n/m} \mid d(L)$.

4. [Factor in $L[X]$] Using Algorithm 3.6.4, let $A = \prod_{1 \leq i \leq g} A_i$ be a factorization of A into irreducible factors in $L[X]$, where without loss of generality we may assume the A_i monic.

5. [Conclude] If no A_i is of degree equal to 1, then output NO otherwise output YES, and if we write $A_i = X - g_i(\beta)$ where β is a root of B such that $L = \mathbb{Q}(\beta)$, output also the polynomial $g_i(bX)/a$ which gives explicitly the isomorphism. Terminate the algorithm.

Conclusion. With three different algorithms to solve the subfield problem, it is now necessary to give some practical advice. These remarks are, of course, also valid for the applications of the subfield problem that we will see in the next section, such as the field isomorphism problem.

1) Start by executing steps 1 to 3 of Algorithm 4.5.2. These tests are fast and will eliminate most cases when K is not isomorphic to a subfield of L. If these tests go through, there is now a distinct possibility that the answer to the subfield problem is yes.

2) Apply the LLL method (using the remark made at the end). This is also quite fast, and will give good results if K is indeed isomorphic to a subfield of L. Note that sufficient accuracy should be used in computing the roots of $A(X)$ and $B(X)$ otherwise LLL may miss a dependency. If LLL fails to detect a relation, then especially if the computation has been done to high accuracy it is almost certain that K is *not* isomorphic to a subfield of L.

An alternate method which is numerically more stable is to use Algorithm 4.5.2. However this algorithm is much slower than LLL as soon as n is at all large, hence should be used only for these very small values of n.

3) In the remaining cases, apply Algorithm 4.5.4 which is slow but sure.

4.5.4 Applications of the Solutions to the Subfield Problem

Now that we have seen three methods for solving the subfield problem, we will see that this problem is basic for the solution of a number of other problems. For each of these other problems, we can then choose any method that we like to solve the underlying subfield problem.

The Field Membership Problem.

The first problem that we can now solve is the *field membership problem*. Given two algebraic numbers α and β by their minimal polynomials A and B and suitable complex approximations, determine whether or not $\alpha \in \mathbb{Q}(\beta)$ and if so a polynomial $P \in \mathbb{Q}[X]$ such that $\alpha = P(\beta)$. For this, apply one of the three methods that we have studied for the subfield problem. Note that some steps may be simplified since we have chosen a specific complex root of $A(X)$. For example, if we use LLL, we simply check the linear dependence of α and the powers of β. If we use linear algebra, choosing a numbering of the roots such that $\alpha = \alpha_1$ and $\beta = \beta_1$, we can restrict to maps ϕ such that $\phi(1) = 1$. In the algebraic method on the other hand we must lengthen step 5. For every $A_i = X - g_i(\beta)$ of degree one, we compute $g_i(\beta)$ numerically (it will be a root of $A(X)$) and check whether it is closer to α than to any other root. If this occurs for no i, then the answer is NO, otherwise the answer is YES and we output the correct g_i.

The Field Isomorphism Problem.

The second problem is the *isomorphism problem*. Given two number fields K and L as before, determine whether or not they are isomorphic. This is of course equivalent to K and L having the same degree and K being a subfield of L, so the solution to this problem follows immediately from that of the subfield problem. Since this problem is very important, we give explicitly the two algorithms corresponding to the last two methods (the LLL method can of course also be used). For still another method, see [Poh3].

Algorithm 4.5.5 (Field Isomorphism Using Linear Algebra). Let $A(X)$ and $B(X)$ be primitive irreducible polynomials in $\mathbb{Z}[X]$ of the same degree n defining number fields K and L. This algorithm determines whether or not K is isomorphic to L, and if it is, gives an explicit isomorphism.

1. [Reduce to algebraic integers] Set $a \leftarrow \ell(A)$, $b \leftarrow \ell(B)$ (the leading terms of A and B), and set $A(X) \leftarrow a^{m-1}A(X/a)$ and $B(X) \leftarrow b^{n-1}B(X/b)$.

2. [Check discriminants] Compute $d(A)$ and $d(B)$ using Algorithm 3.3.7), and check whether $d(A)/d(B)$ is a square in $\mathbb{Q}$ using essentially Algorithm 1.7.3.

If this is not the case, output NO and terminate the algorithm. If for some reason $d(K)$ and $d(L)$ are known or cheaply computed, replace this check by $d(K) = d(L)$.

3. [Compute roots] Using Algorithm 3.6.6, compute the complex roots α_i and β_i of $A(X)$ and $B(X)$ to a reasonable accuracy (it may be necessary to have more accuracy in the later steps).

4. [Loop on ϕ] For each permutation ϕ of $[1, n]$ execute steps 5 and 6. If all the permutations have been examined without termination of the algorithm, output NO and terminate the algorithm.

5. [Check $s_1 \in \mathbb{Z}$] Let $s_1 \leftarrow \sum_{1 \le i \le n} \alpha_{\phi(i)} \beta_i$. If s_1 is not close to an integer (this is a rigorous statement, since it depends only on the chosen approximations to the roots), take the next permutation ϕ in step 4.

   Otherwise, check whether $s_h \leftarrow \sum_{1 \le i \le n} \alpha_{\phi(i)} \beta_i^h$ are also close to an integer for $h = 2, \ldots, n-1$. As soon as this is not the case, take the next map ϕ in step 4.

6. [Compute polynomial] (Here the s_h are all close to integers.) Set $s_h \leftarrow \lfloor s_h \rceil$ (the nearest integer to s_h). Compute by induction $t_k \leftarrow \text{Tr}_{L/\mathbb{Q}}(\beta_1^k)$ for $0 \le k \le 2n - 2$, and using Algorithm 2.2.1 or a Gauss-Bareiss variant, find the unique solution to the linear system $\sum_{0 \le j < n} x_j t_{j+h} = s_h$ for $0 \le h < n$. (We know that $d(B)x_j \in \mathbb{Z}$, so we can avoid rational arithmetic.) Now set $P(X) \leftarrow \sum_{0 \le j < n} x_j X^j$.

7. [Finished?] Using the variant of Horner's rule explained in Section 4.5.1, check whether $A(P(X)) \equiv 0 \pmod{B(X)}$. If this is the case, then output YES, and also output the polynomial $P(bX)/a$ which gives the isomorphism explicitly, and terminate the algorithm. Otherwise, using Algorithm 3.6.6 recompute the roots α_i and β_i to a greater accuracy and go to step 5.

Algorithm 4.5.6 (Field Isomorphism Using Polynomial Factorization). Let $A(X)$ and $B(X)$ be primitive irreducible polynomials in $\mathbb{Z}[X]$ of the same degree n defining number fields K and L. This algorithm determines whether or not K is isomorphic to L, and if it is, gives an explicit isomorphism.

1. [Reduce to algebraic integers] Set $a \leftarrow \ell(A)$, $b \leftarrow \ell(B)$ (the leading terms of A and B), and set $A(X) \leftarrow a^{m-1}A(X/a)$ and $B(X) \leftarrow b^{n-1}B(X/b)$.

2. [Check discriminants] Compute $d(A)$ and $d(B)$ using Algorithm 3.3.7), and check whether $d(A)/d(B)$ is a square in $\mathbb{Q}$ using a slightly modified version of Algorithm 1.7.3. If this is not the case, output NO and terminate the algorithm. If for some reason $d(K)$ and $d(L)$ are known or cheaply computed, check instead that $d(K) = d(L)$.

3. [Factor in $L[X]$] Using Algorithm 3.6.4, let $A = \prod_{1 \le i \le g} A_i$ be a factorization of A into irreducible factors in $L[X]$, where without loss of generality we may assume the A_i monic.

4. [Conclude] If no A_i has degree equal to 1, then output NO otherwise output YES, and if we write $A_i = X - g_i(\beta)$ where β is a root of B such that

$L = \mathbb{Q}(\beta)$, also output the polynomial $g_i(bX)/a$ which explicitly gives the isomorphism. Terminate the algorithm.

For the field isomorphism problem, there is a different method which works sufficiently often that it deserves to be mentioned. We have seen that Algorithm 4.4.12 gives a defining polynomial for a number field which is almost canonical. Hence, if we apply this algorithm to two polynomials A and B, then, if the corresponding number fields are isomorphic, there is a good chance that the polynomials output by Algorithm 4.4.12 will be the same. If they are the same, this proves that the fields are isomorphic (and we can easily recover explicitly the isomorphism if desired). If not, it does not prove anything, but we can expect that they are not isomorphic. We must then apply one of the rigorous methods explained above to prove this.

The Primitive Element Problem.

The last application of the subfield problem that we will see is to the *primitive element problem*. This is as follows. Given algebraic numbers $\alpha_1, \ldots, \alpha_m$, set $K = \mathbb{Q}(\alpha_1, \ldots, \alpha_m)$. Then K is a number field, hence it is reasonable (although not always absolutely necessary, see [Duv]) to represent K by a primitive element θ, i.e.

$$K = \mathbb{Q}(\alpha_1, \ldots \alpha_m) = \mathbb{Q}(\theta) \simeq \mathbb{Q}[X]/(T(X)\mathbb{Q}[X]) \ ,$$

where T is the minimal polynomial of θ. Hence, we need an algorithm which finds such a T (which is not unique) given $\alpha_1, \ldots, \alpha_m$. We can do this by induction on m, and the problem boils down to the following: Given α and β by their minimal polynomials A and B (and suitable complex approximations), find a monic irreducible polynomial $T \in \mathbb{Z}[X]$ such that

$$\mathbb{Q}(\alpha, \beta) = \mathbb{Q}(\theta) \ , \quad \text{where } T(\theta) = 0 \ .$$

We can use the solution to the subfield problem to solve this. According to the proof of the primitive element theorem (see [Lang1]), we can take $\theta = k\alpha + \beta$ for a small integer k, and $\mathbb{Q}(\alpha, \beta) = \mathbb{Q}(k\alpha + \beta)$ is equivalent to $\alpha \in \mathbb{Q}(k\alpha + \beta)$ which can be checked using one of the algorithms explained above for the field membership problem.

4.6 Orders and Ideals

4.6.1 Basic Definitions

Definition 4.6.1. *An* order *R in K is a subring of K which as a $\mathbb{Z}$-module is finitely generated and of maximal rank $n = \deg(K)$ (note that we use the*

*"modern" definition of a ring, which includes the existence of the multiplicative
identity 1).*

Proposition 4.1.3 shows that every element of an order R is an algebraic
integer, i.e. that $R \subset \mathbb{Z}_K$. We will see that the ring theory of $\mathbb{Z}_K$ is nicer than
that of an arbitrary order R, but for the moment we let R be an arbitrary order
in a number field K. We emphasize that some of the properties mentioned here
are specific to orders in number fields, and are not usually valid for general
base rings.

Definition 4.6.2. *An ideal I of R is a sub-R-module of R, i.e. a sub-$\mathbb{Z}$-
module of R such that for every $r \in R$ and $i \in I$ we have $ri \in I$.*

Note that the quotient module R/I has a canonical quotient ring structure.
In fact we have:

Proposition 4.6.3. *Let I be a non-zero ideal of R. Then I is a module of
maximal rank. In other words, R/I is a finite ring. Its cardinality is called the
norm of I and denoted $\mathcal{N}(I)$.*

Indeed, if $i \in I$ with $i \neq 0$, then $iR \subset I \subset R$, proving the proposition. $\square$

If I is given by its HNF on a basis of R (or simply by any matrix A), then
Proposition 4.7.4 shows that the norm of I is simply the absolute value of the
determinant of A.

Ideals can be added (as modules), and the sum of two ideals is clearly
again an ideal. Similarly, the intersection of two ideals is an ideal. Ideals can
also be multiplied in the following way: if I and J are ideals, then

$$IJ = \left\{ \sum_i x_i y_i, \text{ where } x_i \in I \text{ and } y_i \in J \right\} .$$

Again, it is clear that this is an ideal. Note that we clearly have the inclusions

$$IJ \subset I \cap J \subset I \subset I + J ,$$

(and similarly with J), and $IR = I$ for all ideals I. It is clearly not always
true that $IJ = I \cap J$ (take $I = J = p\mathbb{Z}$ in $\mathbb{Z}$). We have however the following
easy result.

Proposition 4.6.4. *Let I and J be two ideals in R and assume that $I + J =
R$. (It is then reasonable to say that I and J are coprime.) Then we have the
equality $IJ = I \cap J$.*

Proof. Since $IJ \subset I \cap J$ we need to prove only the reverse inclusion. But since
$I + J = R$, there exists $a \in I$ and $b \in J$ such that $a + b = 1$. If $x \in I \cap J$ it

follows that $x = ax + bx$ and clearly $ax \in IJ$ and $bx \in JI = IJ$ thus proving the proposition. □

Definition 4.6.5. *A fractional ideal I in R is a non-zero submodule of K such that there exists a non-zero integer d with dI ideal of R. An ideal (fractional or not) is said to be a* principal ideal *if there exists $x \in K$ such that $I = xR$. Finally, R is a* principal ideal domain *(PID) if R is an integral domain (this is already satisfied for orders) and if every ideal of R is a principal ideal.*

It is clear that if I is a fractional ideal, then $I \subset R$ if and only if I is an ideal of R, and we will then say that I is an *integral ideal*.

Note that the set-theoretic inclusions seen above remain valid for fractional ideals, except for the one concerning the product. Indeed, if I and J are two fractional ideals, one does not even have $IJ \subset I$ in general: take $I = R$, and J a non-integral ideal.

Definition 4.6.6. *Let I be a fractional ideal of R. We will say that I is* invertible *if there exists a fractional ideal J of R such that $R = IJ$. Such an ideal J will be called an* inverse *of I.*

The following lemma is easy but crucial.

Lemma 4.6.7. *Let I be a fractional ideal, and set*

$$I' = \{x \in K, \ xI \subset R\} \ .$$

Then I is invertible if and only if $II' = R$. Furthermore if this equality is true, then I' is the unique inverse of I and is denoted I^{-1}.

The proof is immediate and left to the reader. □

Remark. It is not true in general that $\mathcal{N}(IJ) = \mathcal{N}(I)\mathcal{N}(J)$. For example, let $\omega = (1 + \sqrt{-7})/2$, take $R = \mathbb{Z} + 3\omega\mathbb{Z}$ and $I = J = 3\mathbb{Z} + 3\omega\mathbb{Z}$. Then one immediately checks that $\mathcal{N}(I) = 3$, but $\mathcal{N}(I^2) = 27$. As the following proposition shows, the equality $\mathcal{N}(IJ) = \mathcal{N}(I)\mathcal{N}(J)$ is however true when either I or J is an invertible ideal in R, and in particular, it is always true when $R = \mathbb{Z}_K$ is the maximal order of K (see Section 4.6.2 for the relevant definitions).

Proposition 4.6.8. *Let R be an order in a number field, and let I and J be two integral ideals of R. If either I or J is invertible, we have $\mathcal{N}(IJ) = \mathcal{N}(I)\mathcal{N}(J)$.*

Proof. (This proof is due to H. W. Lenstra.) Assume for example that I is invertible. We will prove more generally that if $J \subset H$ where J and H are ideals of R, then $[IH : IJ] = [H : J]$. With $H = R$, this gives $[I : IJ] = [R : J]$ hence $\mathcal{N}(IJ) = [R : IJ] = [R : I][I : IJ] = \mathcal{N}(I)\mathcal{N}(J)$ thus proving the proposition.

Let us temporarily say that a pair of ideals (J, H) is a *simple pair* if $[H : J] > 1$ and if there are no ideals containing J and contained in H apart from H and J themselves.

We prove the equality $[IH : IJ] = [H : J]$ by induction on $[H : J]$. For $H = J$ it is trivial, hence assume by induction that $[H : J] > 1$ and that the proposition is true for any pair of ideals such that $[H' : J'] < [H : J]$. Assume that (J, H) is not a simple pair, and let H_1 be an ideal between J and H and distinct from both. By our induction hypothesis we have $[IH : IH_1] = [H : H_1]$ and $[IH_1 : IJ] = [H_1 : J]$ hence $[IH : IJ] = [H : J]$ thus proving the proposition in that case.

Assume now that (J, H) is a simple pair. Then (IJ, IH) is also a simple pair since I is an invertible ideal (in fact multiplication by I gives a one-to-one map from the set of ideals between J and H onto the set of ideals between IJ and IH). Now we have the following lemma.

Lemma 4.6.9. *If (J, H) is a simple pair, then there exists an isomorphism of R-modules from H/J to R/M for some maximal ideal M of R. (Recall that M is a maximal ideal if and only if (M, R) is a simple pair.)*

Indeed, let $x \in H \setminus J$. The ideal $xR + J$ is between J and H but is not equal to J, hence $H = xR + J$. This immediately implies that the map from R to H/J which sends a to the class of ax modulo J is a surjective R-linear map. Call M its kernel, which is an ideal of R. Then by definition R/M is isomorphic to H/J and since (J, H) is a simple pair it follows that (M, R) is a simple pair, in other words that M is a maximal ideal of R, thus proving the lemma. $\qquad\square$

Resuming the proof of the proposition, we see that H/J is isomorphic to R/M and IH/IJ is isomorphic to R/M' for some maximal ideals M and M'. By construction, $MH \subset J$ hence $MIH \subset IJ$, so $M \subset M'$. Since M and M' are maximal ideals (or since I is invertible), it follows that $M = M'$, hence that $[IH : IJ] = \mathcal{N}(M') = \mathcal{N}(M) = [H : J]$ thus showing the proposition. $\quad\square$

Definition 4.6.10. *An ideal $\mathfrak{p}$ of R is called a* prime ideal *if $\mathfrak{p} \neq R$ and if the quotient ring $R/\mathfrak{p}$ is an integral domain (in other words if $xy \in \mathfrak{p}$ implies $x \in \mathfrak{p}$ or $y \in \mathfrak{p}$). The ideal $\mathfrak{p}$ is* maximal *if the quotient ring $R/\mathfrak{p}$ is a field.*

It is easy to see that an ideal $\mathfrak{p}$ is maximal if and only if $\mathfrak{p} \neq R$ and if the only ideals I such that $\mathfrak{p} \subset I \subset R$ are $\mathfrak{p}$ and R, in other words if $(\mathfrak{p}, R)$

form a simple pair in the language used above. Furthermore, it is clear that a maximal ideal is prime. In number fields, the converse is essentially true:

Proposition 4.6.11. *Let $\mathfrak{p}$ be a non-zero prime ideal in R. Then $\mathfrak{p}$ is maximal. (Here it is essential that R be an order in a number field.)*

Indeed, to say that $\mathfrak{p}$ is a prime ideal is equivalent to saying that for every $x \notin \mathfrak{p}$ the maps $y \mapsto xy$ modulo $\mathfrak{p}$ are injections from $A/\mathfrak{p}$ into itself. Since $A/\mathfrak{p}$ is finite, these maps are also bijections, hence $A/\mathfrak{p}$ is a field. $\qquad\square$

Note that $\{0\}$ is indeed a prime ideal, but is not maximal. It will always be excluded, even when this is not explicitly mentioned.

The reason why prime ideals are called "prime" is that the prime ideals of $\mathbb{Z}$ are $\{0\}$, and the ideals $p\mathbb{Z}$ for p a prime number. Prime ideals also satisfy some of the properties of prime numbers. Specifically:

Proposition 4.6.12. *If $\mathfrak{p}$ is a prime ideal and $\mathfrak{p} \supset I_1 \cdots I_k$, where the I_i are ideals, then there exists an i such that $\mathfrak{p} \supset I_i$.*

Proof. By induction on k it suffices to prove the result for $k = 2$. Assume that $\mathfrak{p} \supset IJ$ and $\mathfrak{p} \not\supset I$ and $\mathfrak{p} \not\supset J$. Then there exists $x \in I$ such that $x \notin \mathfrak{p}$, and $y \in J$ such that $y \notin \mathfrak{p}$. Since $\mathfrak{p}$ is a prime ideal, $xy \notin \mathfrak{p}$, but clearly $xy \in IJ$, contradiction. $\qquad\square$

If we interpret $I \supset J$ as meaning $I \mid J$, this says that if $\mathfrak{p}$ divides a product of ideals, it divides one of the factors. Although it is quite tempting to use the notation $I \mid J$, one should be careful with it since it is not true in general that $I \mid J$ implies that there exists an ideal I' such that $J = II'$. As we will see, this will indeed be true if $R = \mathbb{Z}_K$, and in this case it makes perfectly good sense to use that notation.

A variant of the above mentioned phenomenon is that it is not true for general orders R that every ideal is a product of prime ideals. What is always true is that every (non-zero) ideal contains a product of (non-zero) prime ideals. When $R = \mathbb{Z}_K$ however, we will see that everything we want is true at the level of ideals.

Proposition 4.6.13. *If R is an order in a number field (or more generally a Noetherian integral domain), any non-zero integral ideal I in R contains a product of (non-zero) prime ideals.*

This is easily proved by Noetherian induction (see Exercise 11).

An important notion which is weaker than that of PID but almost as useful is that of a *Dedekind domain*. This is by definition a Noetherian integral domain R such that every non-zero prime ideal is maximal, and which is integrally closed. This last condition means that if x is a root of a monic polynomial equation with coefficients in R and if x is in the field of fractions of R, then in fact $x \in R$. This is for example the case of $R = \mathbb{Z}$.

When R is an order in a number field, all the conditions are satisfied except that R must also be integrally closed. Since $R \supset \mathbb{Z}$, it is clear that if R is integrally closed then $R = \mathbb{Z}_K$, and the converse is also true by Proposition 4.1.5. Hence the only order in K which is a Dedekind domain is the ring of integers $\mathbb{Z}_K$. Since we know that every order R is a subring of $\mathbb{Z}_K$, we will also call $\mathbb{Z}_K$ the *maximal order* of K.

We now specialize to the case where $R = \mathbb{Z}_K$.

4.6.2 Ideals of $\mathbb{Z}_K$

In this section, fix $R = \mathbb{Z}_K$. Let $\mathcal{I}(K)$ be the set of fractional ideals of $\mathbb{Z}_K$. We summarize the main properties of $\mathbb{Z}_K$-ideals in the following theorem:

Theorem 4.6.14.

(1) *Every fractional ideal of $\mathbb{Z}_K$ is invertible. In other words, if I is a fractional ideal and if we set $I^{-1} = \{x \in K, xI \subset \mathbb{Z}_K\}$, then $II^{-1} = \mathbb{Z}_K$.*
(2) *The set of fractional ideals of $\mathbb{Z}_K$ is an Abelian group.*
(3) *Every fractional ideal I can be written in a unique way as*

$$I = \prod_{\mathfrak{p}} \mathfrak{p}^{v_{\mathfrak{p}}(I)} \; ,$$

the product being over a finite set of prime ideals, and the exponents $v_{\mathfrak{p}}(I)$ being in $\mathbb{Z}$. In particular, I is an integral ideal (i.e. $I \subset \mathbb{Z}_K$) if and only if all the $v_{\mathfrak{p}}(I)$ are non-negative.
(4) *The maximal order $\mathbb{Z}_K$ is a PID if and only if it is a UFD.*

Hence the ideals of $\mathbb{Z}_K$ behave exactly as the numbers in $\mathbb{Z}$, and can be handled in the same way. Note that (3) is much stronger than Proposition 4.6.13, but is valid only because $\mathbb{Z}_K$ is also integrally closed.

The quantity $v_{\mathfrak{p}}(I)$ is called the $\mathfrak{p}$-adic valuation of I and satisfies the usual properties:

(1) $I \subset \mathbb{Z}_K \iff v_{\mathfrak{p}}(I) \geq 0$ for all prime ideals $\mathfrak{p}$.
(2) $J \subset I \iff v_{\mathfrak{p}}(I) \leq v_{\mathfrak{p}}(J)$ for all prime ideals $\mathfrak{p}$.
(3) $v_{\mathfrak{p}}(I + J) = \min(v_{\mathfrak{p}}(I), v_{\mathfrak{p}}(J))$.
(4) $v_{\mathfrak{p}}(I \cap J) = \max(v_{\mathfrak{p}}(I), v_{\mathfrak{p}}(J))$.
(5) $v_{\mathfrak{p}}(IJ) = v_{\mathfrak{p}}(I) + v_{\mathfrak{p}}(J)$.

Hence the dictionary between fractional ideals and rational numbers is as follows:

Fractional ideals $\longleftrightarrow$ (non-zero) rational numbers.

Integral ideals $\longleftrightarrow$ integers.

Inclusion $\longleftrightarrow$ divisibility (with the reverse order).

Sum $\longleftrightarrow$ greatest common divisor.

Intersection $\longleftrightarrow$ least common multiple.

Product $\longleftrightarrow$ product.

Of course, a few of these notions could be unfamiliar for rational numbers, for example the GCD, but a moment's thought shows that one can give perfectly sensible definitions.

We end this section with the notion of norm of a fractional ideal. We have seen in Proposition 4.6.3 that for an integral ideal I the norm of I is the cardinality of the finite ring R/I. As already mentioned, a corollary of Theorem 4.6.14 is that $\mathcal{N}(IJ) = \mathcal{N}(I)\mathcal{N}(J)$ for ideals I and J of the maximal order $R = \mathbb{Z}_K$ (recall that this is false in general if R is not maximal). This allows us to extend the definition of $\mathcal{N}(I)$ to fractional ideals if desired: any fractional ideal I can be written as a quotient of two integral ideals, say $I = P/Q$ (in fact by definition we can take $Q = dR$ where d is an integer), and we define $\mathcal{N}(I) = \mathcal{N}(P)/\mathcal{N}(Q)$. It is easy to check that this is independent of the choice of P and Q and that it is still multiplicative ($\mathcal{N}(IJ) = \mathcal{N}(I)\mathcal{N}(J)$). Of course, usually it will no longer be an integer.

The notion of norm of an ideal is linked to the notion of norm of an element that we have seen above in the following way:

Proposition 4.6.15. *Let x be a non-zero element of K. Then*

$$\left| \mathcal{N}_{K/\mathbb{Q}}(x) \right| = \mathcal{N}(x\mathbb{Z}_K) \ ,$$

in other words the norm of a principal ideal of $\mathbb{Z}_K$ is equal to the absolute value of the norm (in K) of a generating element.

One should never forget this absolute value. We could in fact have a nicer looking proposition (without absolute values) by using a slight extension of the notion of fractional ideal: because of Theorem 4.6.14 (3), the group of fractional ideals can be identified with the free Abelian group generated by the prime ideals $\mathfrak{p}$. Furthermore, a number field K has *places*, corresponding to equivalence classes of valuations. The finite places, which correspond to non-Archimedean valuations, can be identified with the (non-zero) prime ideals of $\mathbb{Z}_K$. The other (so called infinite places) correspond to Archimedean valuations and can be identified with the embeddings σ_i of K in $\mathbb{C}$, with σ identified with $\bar{\sigma}$ (thus giving $r_1 + r_2$ Archimedean valuations). Hence, we can consider the extended group which is the free Abelian group generated by all valuations, finite or not. One can show that to obtain a sensible definition, the coefficients of the non-real complex embeddings must be considered modulo 1, i.e. can be taken equal to 0, and the coefficients of the real embeddings must be considered modulo 2 (I do not give the justification for these claims). Hence, the group of generalized fractional ideals is

$$\mathbb{Z}[\mathcal{P}(K)] \times \{\pm 1\}^{r_1} \ ,$$

where $\mathcal{P}(K)$ is the set of non-zero prime ideals. The norm of such a generalized ideal is then the norm of its finite part multiplied by the infinite components (i.e. by a sign). Now if $x \in K$, the generalized fractional ideal associated to x is, on the finite part equal to $x\mathbb{Z}_K$, and on the infinite place σ_i (where $1 \le i \le r_1$) equal to the sign of $\sigma_i(x)$. It is then easy to check that these two notions of norm now correspond exactly, including sign.

The discussion above was meant as an aside, but is the beginning of the theory of adeles and ideles (see [Lang2]). In a down to earth way, we can say that most natural questions concerning number fields should treat together the Archimedean and non-Archimedean places (or primes). In addition to the present example, we have already mentioned the parallel between Propositions 4.1.14 and 4.8.6. Similarly, we will see Propositions 4.8.11 and 4.8.10. Maybe the most important consequence is that we will have to compute simultaneously class groups (i.e. the non-Archimedean part) and regulators (the Archimedean part), see Sections 4.9, 5.9 and 6.5.

4.7 Representation of Modules and Ideals

4.7.1 Modules and the Hermite Normal Form

As before, we work in a fixed number field K of degree n, given by $K = \mathbb{Q}(\theta)$, where θ is an algebraic integer whose minimal monic polynomial is denoted $T(X)$.

Definition 4.7.1. *A module in K is a finitely generated sub-$\mathbb{Z}$-module of K of rank exactly equal to n.*

Since $\mathbb{Z}$ is a PID, such a module being torsion free and finitely generated, must be free. Let $\omega_1, \ldots, \omega_n$ be a $\mathbb{Z}$-basis of M. The numbers ω_i are elements of K, hence we can find an integer d such that $d\omega_i \in \mathbb{Z}[\theta]$ for all i. The least such positive d will be called the *denominator* of M with respect to $\mathbb{Z}[\theta]$. More generally, if R is another module (for example $R = \mathbb{Z}_K$), we define the denominator of M with respect to R as the smallest positive d such that $dM \subset R$.

Note that in the context of number fields, the word "module" will always have the above meaning, in other words it will always refer to a submodule of maximal rank n. If as a $\mathbb{Q}$-vector space we identify $K = \mathbb{Q}(\theta)$ with $\mathbb{Q}^n$, and $\mathbb{Z}[\theta]$ with $\mathbb{Z}^n$, the above definition is the same as the one that we have given in Section 2.4.3. In particular, we can use the notions of determinant, HNF and SNF of modules.

We give the following proposition without proof.

Proposition 4.7.2. *Let M be a module in a number field K in the above sense. Then there exists an order R in K and a positive integer d such that dM is an ideal of R. More precisely, there is a maximal such R equal to $R = \{x \in K, xM \subset M\}$, and one can take for d the denominator of M with respect to R.*

Specializing to our case the results of Section 2.4.2, we obtain:

Theorem 4.7.3. *Let $\alpha_1, \ldots, \alpha_n$ be n $\mathbb{Z}$-linearly independent elements of K, and R be the module which they generate. Then for any module M, there exists a unique basis $\omega_1, \ldots, \omega_n$ such that if we write*

$$\omega_j = \frac{1}{d} \left(\sum_{i=1}^{n} w_{i,j} \alpha_i \right) ,$$

where d is the denominator of M with respect to R, then the $n \times n$ matrix $W = (w_{i,j})$ satisfies the following conditions:

(1) *For all i and j the $w_{i,j}$ are integers.*
(2) *W is an upper triangular matrix, i.e. $w_{i,j} = 0$ if $i > j$.*
(3) *For every i, we have $w_{i,i} > 0$.*
(4) *For every $j > i$ we have $0 \le w_{i,j} < w_{i,i}$.*

The corresponding basis $(\omega_i)_{1 \le i \le n}$ will be called the *HNF-basis* of M with respect to R, and the pair (W, d) will be called the *HNF* of M (with respect to R). If $\alpha_i = \theta^{i-1}$, we will call W (or (W, d)) the HNF with respect to θ.

We have already seen in section 2.4.3 how to test equality and inclusion of modules, how to compute the sum of two modules and the product of a module by a constant. In the context of number fields, we can also compute the *product* of two modules. This will be used mainly for ideals.

Recall that

$$MM' = \{\sum_j m_j m'_j, m_j \in M, m'_j \in M'\} .$$

It is clear that MM' is again a module. To obtain its HNF, we proceed as follows: Let $\omega_1, \ldots, \omega_n$ be the basis of M obtained by considering the columns of the HNF of M as the coefficients of ω_i in the standard representation, and similarly for M'. Then the n^2 elements $\omega_i \omega'_j$ form a generating set of MM'. Hence, if we find the HNF of the $n \times n^2$ matrix formed by their coefficients in the standard representation, we will have obtained the HNF of MM'.

Note however that this is quite costly, since n^2 can be pretty large. Another method might be as follows. In the case where M and M' are ideals (of $\mathbb{Z}_K$ say), then M and M' have a $\mathbb{Z}_K$-generating set formed by two elements. In fact, one of these two elements can even be chosen in $\mathbb{Z}$ if desired. Hence it is clear that if $\omega_1, \ldots, \omega_n$ is a $\mathbb{Z}$-basis of M and α, β a $\mathbb{Z}_K$-generating set of M', then $\alpha\omega_1, \ldots, \alpha\omega_n, \beta\omega_1, \ldots, \beta\omega_n$ will be a $\mathbb{Z}$-generating set of MM' (note that M *must* also be an ideal for this to be true). Hence we can obtain the HNF of MM' more simply by finding the HNF of the $n \times 2n$ matrix formed by the coefficients of the above generating set in the standard representation.

We end this section by the following proposition, whose proof is easy and left to the reader (see Exercise 18 of Chapter 2).

Proposition 4.7.4. *Let M be a module with denominator 1 with respect to a given R (i.e. $M \subset R$), and $W = (w_{i,j})$ its HNF with respect to a basis α_1, $\ldots$, α_n of R. Then the product of the $w_{i,i}$ (i.e. the determinant of W) is equal to the index $[R : M]$.*

This will be used, for example, when $R = \mathbb{Z}[\theta]$ or $R = \mathbb{Z}_K$.

4.7.2 Representation of Ideals

The Hermite normal form of an ideal with respect to θ has a special form, as is shown by the following theorem:

Theorem 4.7.5. *Let M be a $\mathbb{Z}[\theta]$-module, let (W, d) be its HNF with respect to the algebraic integer θ, where d is the denominator and $W = (w_{i,j})$ is an integral matrix in upper triangular HNF. Then for every j, $w_{j,j}$ divides all the elements of the $j \times j$ matrix formed by the first j rows and columns. In other words, the HNF basis $\omega_1, \ldots, \omega_n$ of a $\mathbb{Z}[\theta]$-module has the form*

$$\omega_j = \frac{z_j}{d}\left(\theta^{j-1} + \sum_{1 \leq i < j} h_{i,j}\theta^{i-1}\right),$$

where the z_j are positive integers such that $z_j \mid z_i$ for $i < j$, and the $h_{i,j}$ satisfy $0 \leq h_{i,j} < z_i/z_j$ for $i < j$. Furthermore, z_1 is the smallest positive element of $dM \cap \mathbb{Z}$.

Proof. Without loss of generality, we may assume $d = 1$. We prove the theorem by induction on j. It is trivially true for $j = 1$. Assume $j > 1$ and that it is true for $j - 1$. Consider the $(j-1)^{\text{th}}$ basis element ω_{j-1} of M. We have

$$\omega_{j-1} = \sum_{1 \leq i < j} w_{i,j-1}\theta^{i-1}$$

hence $\theta\omega_{j-1} = w_{j-1,j-1}\theta^{j-1} + \sum_{1 \leq i < j-1} w_{i,j-1}\theta^i$. Since M is a $\mathbb{Z}[\theta]$-module, this must be again an element of M, hence it has the form $\theta\omega_{j-1} = \sum_{1 \leq i \leq n} a_i\omega_i$ with integers a_i. Now since we have a triangular basis, identification of coefficients (from θ^{n-1} downwards) shows that $a_i = 0$ for $i > j$

and that $a_j w_{j,j} = w_{j-1,j-1}$. This already shows that $w_{j,j} \mid w_{j-1,j-1}$. But by induction, we know that $w_{j-1,j-1}$ divides $w_{i',j'}$ when i' and j' are less than or equal to $j-1$. It follows that, modulo $w_{j-1,j-1}\mathbb{Z}[\theta]$ we have

$$0 \equiv \theta w_{j-1} \equiv a_j w_j \equiv \frac{w_{j-1,j-1}}{w_{j,j}} \sum_{1 \le i \le j} w_{i,j}\theta^{i-1} \, ,$$

and this means that for every $i \le j$ we have

$$\frac{w_{j-1,j-1}}{w_{j,j}}w_{i,j} \equiv 0 \pmod{w_{j-1,j-1}} \, ,$$

which is equivalent to $w_{j,j} \mid w_{i,j}$ for $i \le j$, thus proving the theorem by induction.   □

Note that the converse of this theorem is false (see Exercise 16).

Theorem 4.7.5 will be mainly used in two cases. First when M is an ideal of $\mathbb{Z}_K$. The second is when M is an order containing θ. In that case one can say slightly more:

Corollary 4.7.6. *Let R be an order in K containing θ (hence containing $\mathbb{Z}[\theta]$). Then the HNF basis $\omega_1, \ldots, \omega_n$ of R with respect to θ has the form*

$$\omega_j = \frac{1}{d_j}\left(\theta^{j-1} + \sum_{1 \le i < j} h_{i,j}\theta^{i-1}\right) \, ,$$

where the d_j are positive integers such that $d_i \mid d_j$ for $i < j$, $d_1 = 1$, and the $h_{i,j}$ satisfy $0 \le h_{i,j} < d_j/d_i$ for $i < j$. In other words, with the notations of Theorem 4.7.5, we have $z_j \mid d$ for all j.

The proof is clear once one notices that the smallest positive integer belonging to an order is 1, hence by Theorem 4.7.5 that $z_1 = d$.   □

If we assume that $R = \mathbb{Z}_K$ is given by an integral basis $\alpha_1, \ldots, \alpha_n$, then the HNF matrix of an ideal I with respect to this basis does *not* usually satisfy the conditions of Theorem 4.7.5. We can always assume that we have chosen $\alpha_1 = 1$, and in that case it is easy to show in a similar manner as above that $w_{1,1}$ is divisible by $w_{i,i}$ for all i, and that if $w_{i,i} = w_{1,1}$, then $w_{j,i} = 0$ for $j \ne i$. This is left as an exercise for the reader (see Exercise 17).

Hence, depending on the context, we will represent an ideal of $\mathbb{Z}_K$ by its Hermite normal form with respect to a fixed integral basis of $\mathbb{Z}_K$, or by its HNF with respect to θ (i.e. corresponding to the standard representations of the basis elements). Please note once again that the special form of the HNF described in Theorem 4.7.5 is valid only in this last case.

Whichever representation is chosen, we have seen in Sections 2.4.3 and 4.7.1 how to compute sums and products of ideals, to test equality and inclusion (i.e. divisibility). Finally, as has already been mentioned several times, the norm is the absolute value of the determinant of the matrix, and in the HNF case this is simply the product of the diagonal elements.

Note that to test whether an element of K is in a given ideal is a special case of the inclusion test, since $x \in I \iff xR \subset I$. Here however it is simpler (although not so much more efficient) to solve a (triangular) system of linear equations: if (W, d) is the HNF of I with respect to θ, then if $x = (\sum_{1 \le i \le n} x_i \theta^{i-1})/e$ is the standard representation of x, we must solve the equation $WA = \frac{d}{e}X$ where X is the column vector of the x_i, and A is the unknown column vector. Since W is triangular, this is especially simple, and $x \in I$ if and only if A has integral coefficients.

To this point, we have considered ideals mainly as $\mathbb{Z}$-modules. There is a completely different way to represent them based on the following proposition.

Proposition 4.7.7. *Let I be an integral ideal of $\mathbb{Z}_K$.*

(1) *For any non-zero element $\alpha \in I$ there exists an element $\beta \in I$ such that $I = \alpha \mathbb{Z}_K + \beta \mathbb{Z}_K$.*
(2) *There exists a non-zero element in $I \cap \mathbb{Z}$. If we denote by $\ell(I)$ the smallest positive element of $I \cap \mathbb{Z}$, then $\ell(I)$ is a divisor of $N(I) = [\mathbb{Z}_K : I]$. In particular, there exists $\beta \in I$ such that $I = \ell(I)\mathbb{Z}_K + \beta \mathbb{Z}_K$.*
(3) *If α and β are in K, then $I = \alpha \mathbb{Z}_K + \beta \mathbb{Z}_K$ if and only if for every prime ideal $\mathfrak{p}$ we have $\min(v_{\mathfrak{p}}(\alpha), v_{\mathfrak{p}}(\beta)) = v_{\mathfrak{p}}(I)$ where $v_{\mathfrak{p}}$ denotes the $\mathfrak{p}$-adic valuation at the prime ideal $\mathfrak{p}$.*

To prove this proposition, we first prove a special case of the so-called approximation theorem valid in any Dedekind domain.

Proposition 4.7.8. *Let S be a finite set of prime ideals of $\mathbb{Z}_K$ and (e_i) a set of non-negative integers indexed by S. There exists a $\beta \in \mathbb{Z}_K$ such that for each $\mathfrak{p}_i \in S$ we have*

$$v_{\mathfrak{p}_i}(\beta) = e_i \ .$$

(Note that there may exist prime ideals $\mathfrak{q}$ not belonging to S such that $v_{\mathfrak{q}}(\beta) > 0$.)

Remark. More generally, S can be taken to be a set of *places* of K, and in particular can contain Archimedean valuations.

Proof. Let $r = |S|$,

$$I = \prod_{i=1}^{r} \mathfrak{p}_i^{e_i+1} \ ,$$

and for each i, set

$$\mathfrak{a}_i = I \cdot \mathfrak{p}_i^{-e_i-1} \, ,$$

which is still an integral ideal. It is clear that $\mathfrak{a}_1 + \mathfrak{a}_2 + \cdots + \mathfrak{a}_r = \mathbb{Z}_K$ (otherwise this sum would be divisible by one of the $\mathfrak{p}_i$, which is clearly impossible). Hence, let $u_i \in \mathfrak{a}_i$ such that $u_1 + u_2 + \cdots + u_r = 1$. Furthermore, for each i choose $\beta_i \in \mathfrak{p}_i^{e_i} \setminus \mathfrak{p}_i^{e_i+1}$ which is possible since $\mathfrak{p}_i$ is invertible. Then I claim that

$$\beta = \sum_{i=1}^{r} \beta_i u_i$$

has the desired property. Indeed, since $\mathfrak{p}_i \mid \mathfrak{a}_j$ for $i \neq j$, it is easy to check from the definition of the $\mathfrak{a}_i$ that

$$v_{\mathfrak{p}_i}(\beta) = v_{\mathfrak{p}_i}(\beta_i u_i) = e_i$$

since $v_{\mathfrak{p}_i}(u_i) = 0$ and $v_{\mathfrak{p}_i}(\beta_i) = e_i$.   □

Proof of Proposition 4.7.7. (1) Let $\alpha \mathbb{Z}_K = \prod_{i=1}^{r} \mathfrak{p}_i^{a_i}$ be the prime ideal decomposition of the principal ideal generated by α. Since $\alpha \in I$, we also have $I = \prod_{i=1}^{r} \mathfrak{p}_i^{e_i}$ for exponents e_i (which may be equal to zero) such that $e_i \leq a_i$. According to Proposition 4.7.8 that we have just proved, there exists a β such that $v_{\mathfrak{p}_i}(\beta) = e_i$ for $i \leq r$. This implies in particular that $I \mid \beta$, i.e. that $\beta \in I$, and furthermore if we set $I' = \alpha \mathbb{Z}_K + \beta \mathbb{Z}_K$ we have for $i \leq r$

$$v_{\mathfrak{p}_i}(I') = \min(v_{\mathfrak{p}_i}(\alpha), v_{\mathfrak{p}_i}(\beta)) = e_i$$

and if $\mathfrak{q}$ is a prime ideal which does not divide α, $v_{\mathfrak{q}}(I') = 0$, from which it follows that $I' = \prod_{i=1}^{r} \mathfrak{p}_i^{e_i} = I$, thus proving (1).

For (2), we note that since $\mathcal{N}(I) = [\mathbb{Z}_K : I]$, any element of the Abelian quotient group $\mathbb{Z}_K/I$ is annihilated by $\mathcal{N}(I)$, in other words we have $\mathcal{N}(I)\mathbb{Z}_K \subset I$. This implies $\mathcal{N}(I) \in I \cap \mathbb{Z}$, and since any subgroup of $\mathbb{Z}$ is of the form $k\mathbb{Z}$, (2) follows.

Finally, for (3) recall that the sum of ideals correspond to taking a GCD, and that the GCD is computed by taking the minimum of the $\mathfrak{p}$-adic valuations.   □

Hence every ideal has a *two element representation* (α, β) where $I = \alpha \mathbb{Z}_K + \beta \mathbb{Z}_K$, and we can take for example $\alpha = \ell(I)$. This two element representation is however difficult to handle: for the sum or product of two ideals, we get four generators over $\mathbb{Z}_K$, and we must get back to two. More generally, it is not very easy to go from the HNF (or more generally any $\mathbb{Z}$-basis n-element representation) to a two element representation.

There are however two cases in which that representation is useful. The first is in the case of quadratic fields ($n = 2$), and we will see this in Chapter 5. The other, which has already been mentioned in Section 4.7.1, is as follows:

we will see in Section 4.9 that prime ideals do not come out of the blue, and that in algorithmic practice most prime ideals $\mathfrak{p}$ are obtained as a two element representation (p, x) where p is a prime number and x is an element of $\mathfrak{p}$. To go from that two element representation to the HNF form is easy, but is not desirable in general. Indeed, what one usually does with a prime ideal is to multiply it with some other ideal I. If $\omega_1, \dots, \omega_n$ is a $\mathbb{Z}$-basis of I (for example the basis obtained from the HNF form of I on the given integral basis of $\mathbb{Z}_K$), then we can build the HNF of the product $\mathfrak{p}I$ by computing the $n \times 2n$ matrix of the generating set $p\omega_1, \dots p\omega_n, x\omega_1, \dots, x\omega_n$ expressed on the integral basis, and then do HNF reduction. As has already been mentioned in Section 4.7.1, this is more efficient than doing a $n \times n^2$ HNF reduction if we used both HNF representations. Note that if one really wants the HNF of $\mathfrak{p}$ itself, it suffices to apply the preceding algorithm to $I = \mathbb{Z}_K$.

Note that if (W, d) (with $W = (w_{i,j})$) is the HNF of I with respect to θ, and if $f = [\mathbb{Z}_K : \mathbb{Z}[\theta]]$, then $\ell(I) = w_{1,1}$ and $d^n N(I) = [\mathbb{Z}_K : dI] = f \prod_{1 \leq i \leq n} w_{i,i}$ so

$$N(I) = d^{-n} f \prod_{1 \leq i \leq n} w_{i,i} \ .$$

Now it often happens that prime ideals are not given by a two element representation but by a larger number of generating elements. If this ideal is going to be used repeatedly, it is worthwhile to find a two element representation for it. As we have already mentioned this is not an easy problem in general, but in the special case of prime ideals we can give a reasonably efficient algorithm. This is based on the following lemma.

Lemma 4.7.9. *Let $\mathfrak{p}$ be a prime ideal above p of norm p^f (f is called the residual degree of $\mathfrak{p}$ as we will see in the next section), and let $\alpha \in \mathfrak{p}$. Then we have $\mathfrak{p} = (p, \alpha) = p\mathbb{Z}_K + \alpha\mathbb{Z}_K$ if and only if $v_p(\mathcal{N}(\alpha)) = f$ or $v_p(\mathcal{N}(\alpha+p)) = f$, where v_p denotes the ordinary p-adic valuation.*

Proof. This proof assumes some results and definitions introduced in the next section. Assume first that $v_p(\mathcal{N}(\alpha)) = f$. Then, since $\alpha \in \mathfrak{p}$ and $\mathcal{N}(\mathfrak{p}) = p^f$, for every prime $\mathfrak{q}$ above p and different from $\mathfrak{p}$ we must have $v_{\mathfrak{q}}(\alpha) = 0$ otherwise $\mathfrak{q}$ would contribute more powers of p to $\mathcal{N}(\alpha)$. In addition and for the same reason we must have $v_{\mathfrak{p}}(\alpha) = 1$. It follows that for any prime ideal $\mathfrak{q}$, $\min(v_{\mathfrak{q}}(p), v_{\mathfrak{q}}(\alpha)) = v_{\mathfrak{q}}(\mathfrak{p})$ and so $\mathfrak{p} = (p, \alpha)$ by Proposition 4.7.7 (3).

If $v_p(\mathcal{N}(\alpha + p)) = f$ we deduce from this that $\mathfrak{p} = p\mathbb{Z}_K + (\alpha + p)\mathbb{Z}_K$, but this is clearly also equal to $p\mathbb{Z}_K + \alpha\mathbb{Z}_K$.

Conversely, let $\mathfrak{p} = p\mathbb{Z}_K + \alpha\mathbb{Z}_K$. Then for every prime ideal $\mathfrak{q}$ above p and different from $\mathfrak{p}$ we have $v_{\mathfrak{q}}(\alpha) = 0$, while for $\mathfrak{p}$ we can only say that $\min(v_{\mathfrak{p}}(p), v_{\mathfrak{p}}(\alpha)) = 1$.

Assume first that $v_{\mathfrak{p}}(\alpha) = 1$. Then clearly $v_p(\mathcal{N}(\alpha)) = v_p(\mathcal{N}(\mathfrak{p})) = f$ as desired. Otherwise we have $v_{\mathfrak{p}}(\alpha) > 1$, and hence $v_{\mathfrak{p}}(p) = 1$. But then we will have $v_{\mathfrak{p}}(\alpha + p) = 1$ (otherwise $v_{\mathfrak{p}}(p) = v_{\mathfrak{p}}((p + \alpha) - \alpha) > 1$), and still

$v_q(\alpha + p) = 0$ for all other primes q above p, and so $v_p(\mathcal{N}(\alpha + p)) = f$ as before, thus proving the lemma. $\qquad\qquad\qquad\qquad\qquad\qquad\qquad\qquad$ □

Note that the condition $v_p(\mathcal{N}(\alpha)) = f$, while sufficient, is not a necessary condition (see Exercise 20).

Note also that if we write $\alpha = \sum_{1 \le i \le k} \lambda_i \gamma_i$ where the γ_i is some generating set of p, we may always assume that $\lceil \lambda_i \rceil \le p/2$ since $p \in \mathfrak{p}$. In addition, if we choose $\gamma_1 = p$, we may assume that $\lambda_1 = 0$.

This suggests the following algorithm, which is simple minded but works quite well.

Algorithm 4.7.10 (Two-Element Representation of a Prime Ideal). Given a prime ideal $\mathfrak{p}$ above p by a system of $\mathbb{Z}_K$-generators γ_i for $(1 \le i \le k)$, this algorithm computes a two-element representation (p, α) for $\mathfrak{p}$.

We assume that one knows the norm p^f of $\mathfrak{p}$ (this is always the case in practice, and in any case it can be obtained by computing the HNF of $\mathfrak{p}$ from the given generators), and that $\gamma_1 = p$ (if this is not the case just add it to the list of generators).

1. [Initialize] Set $R \leftarrow 1$.

2. [Set coefficients] For $2 \le i \le k$ set $\lambda_i \leftarrow R$.

3. [Compute α and check] Let $\alpha \leftarrow \sum_{2 \le i \le m} \lambda_i \gamma_i$, $n \leftarrow \mathcal{N}(\alpha)/p^f$, where the norm is computed, for example, using the sub-resultant algorithm (see Section 4.3). If $p \nmid n$, then output (p, α) and terminate the algorithm. Otherwise, set $n \leftarrow \mathcal{N}(\alpha + p)/p^f$. If $p \nmid n$ then output (p, α) and terminate the algorithm.

4. [Decrease coefficients] Let j be the largest $i \le k$ such that $\lambda_i \ne -R$ (we will always keep $\lambda_2 \ge 0$ so j will exist). Set $\lambda_j \leftarrow \lambda_j - 1$ and for $j + 1 \le i \le m$ set $\lambda_i \leftarrow R$.

5. [Search for first non-zero] Let j be the smallest $i \le k$ such that $\lambda_i \ne 0$. If no such j exists (i.e. if all the λ_i are equal to 0) set $R \leftarrow R + 1$ and go to step 2. Otherwise go to step 3.

Remarks.

(1) Steps 4 and 5 of this algorithm represent a standard backtracking procedure. What we do essentially is to search for $\alpha = \sum_{2 \le i \le k} \lambda_i \gamma_i$, where the λ_i are integers between $-R$ and R. To avoid searching both for α and $-\alpha$, we add the condition that the first non-zero λ should be positive. If the search fails, we start it again with a larger value of R. Of course, some time will be wasted since many old values of α will be recomputed, but in practice this has no real importance, and in fact $R = 1$ or $R = 2$ is usually sufficient. The remark made after Lemma 4.7.9 shows that the algorithm will stop with $R \le p/2$.

(2) It is often the case that one of the γ_i for $2 \leq i \leq k$ will satisfy one of the conditions of step 3. Thus it is useful to test this before starting the backtracking procedure.

We refer to [Poh-Zas] for extensive information on the use of two-element representations.

4.8 Decomposition of Prime Numbers I

For simplicity, we continue to work with a number field K considered as an extension of $\mathbb{Q}$, and not considered as a relative extension. Many of the theorems or algorithms which are explained in that context are still true in the more general case, but some are not. (For example, we have already seen this for the existence of integral bases.) Almost always, these generalizations fail because the ring of integers of the base field is not a PID (or equivalently a UFD).

4.8.1 Definitions and Main Results

The main results concerning the decomposition of primes are as follows. We always implicitly assume that the prime ideals are non-zero.

Proposition 4.8.1.

(1) *If $\mathfrak{p}$ is a prime ideal of K, then $\mathfrak{p} \cap \mathbb{Z} = p\mathbb{Z}$ for some prime number p.*
(2) *If p is a prime number and $\mathfrak{p}$ is a prime ideal of K, the following conditions are equivalent:*

    (i) $\mathfrak{p} \supset p\mathbb{Z}$.
    (ii) $\mathfrak{p} \cap \mathbb{Z} = p\mathbb{Z}$.
    (iii) $\mathfrak{p} \cap \mathbb{Q} = p\mathbb{Z}$.
(3) *For any prime number p we have $p\mathbb{Z}_K \cap \mathbb{Z} = p\mathbb{Z}$.*

More generally, we have $a\mathbb{Z}_K \cap \mathbb{Z} = a\mathbb{Z}$ for any integer a, prime or not.

Definition 4.8.2. *If $\mathfrak{p}$ and p satisfy one of the equivalent conditions of Proposition 4.8.1 (2), we say that $\mathfrak{p}$ is a prime ideal above p, and that p is below $\mathfrak{p}$.*

Theorem 4.8.3. *Let p be a prime number. There exist positive integers e_i such that*

$$p\mathbb{Z}_K = \prod_{i=1}^{g} \mathfrak{p}_i^{e_i} ,$$

where the $\mathfrak{p}_i$ are all the prime ideals above p.

Definition 4.8.4. *The integer e_i is called the* ramification index *of p at $\mathfrak{p}_i$ and is denoted $e(\mathfrak{p}_i/p)$. The degree f_i of the field extension defined by*

$$f_i = [\mathbb{Z}_K/\mathfrak{p}_i : \mathbb{Z}/p\mathbb{Z}]$$

is called the residual degree *(or simply the* degree*) of p and is denoted $f(\mathfrak{p}_i/p)$.*

Note that both $\mathbb{Z}_K/\mathfrak{p}_i$ and $\mathbb{Z}/p\mathbb{Z}$ are finite fields, and f_i is the dimension of the first considered as a vector space over the second.

Theorem 4.8.5. *We have the following formulas:*

$$\mathcal{N}(\mathfrak{p}_i) = p^{f_i} ,$$

and

$$\sum_{i=1}^{g} e_i f_i = n = \deg(K) .$$

In the case when $K/\mathbb{Q}$ is a Galois extension, the result is more specific:

Theorem 4.8.6. *Assume that $K/\mathbb{Q}$ is a Galois extension (i.e. that for all the embeddings σ_i of K in $\mathbb{C}$ we have $\sigma_i(K) = K$). Then, for any p, the ramification indices e_i are equal (say to e), the residual degrees f_i are equal as well (say to f), hence $efg = n$. In addition, the Galois group operates transitively on the prime ideals above p: if $\mathfrak{p}_i$ and $\mathfrak{p}_j$ are two ideals above p, there exists σ in the Galois group such that $\sigma(\mathfrak{p}_i) = \mathfrak{p}_j$.*

Definition 4.8.7. *Let $p\mathbb{Z}_K = \prod_{i=1}^{g} \mathfrak{p}_i^{e_i}$ be the decomposition of a prime p. We will say that p is* inert *if $g = 1$ and $e_1 = 1$, in other words if $p\mathbb{Z}_K = \mathfrak{p}$ (hence $f_1 = n$). We will say that p* splits completely *if $g = n$ (hence for all i, $e_i = f_i = 1$). Finally, we say that p is* ramified *if there is an e_i which is greater than or equal to 2 (in other words if $p\mathbb{Z}_K$ is not squarefree), otherwise we say that p is* unramified*. Those prime ideals $\mathfrak{p}_i$ such that $e_i > 1$ are called the* ramified prime ideals *of $\mathbb{Z}_K$.*

Note that there are intermediate cases which do not deserve a special name. The fundamental theorem about ramification is as follows:

Theorem 4.8.8. *Let p be a prime number. Then p is ramified in K if and only if p divides the discriminant $d(K)$ of K (recall that this is the discriminant of any integral basis of $\mathbb{Z}_K$). In particular, there are only a finite number of ramified primes (exactly $\omega(d(K))$, where $\omega(x)$ is the number of distinct prime divisors of an integer x).*

We can also define the decomposition of the "infinite prime" of $\mathbb{Q}$ in a similar manner, since we are extending valuations. The ordinary primes correspond to the non-Archimedean valuations and the real or complex embeddings correspond to the Archimedean ones. Since we are over $\mathbb{Q}$, there is only the real embedding of $\mathbb{Q}$ to lift, and (as a special case of a general definition), when the signature of K is (r_1, r_2), we will say that the infinite prime of $\mathbb{Q}$ lifts to a product of r_1 real places of K times r_2 non-real places to the power 2. Hence, $g = r_1 + r_2$, $e_i = 1$ for $i \leq r_1$, $e_i = 2$ for $i > r_1$, and $f_i = 1$ for all i.

We also have the following results:

Proposition 4.8.9.

(1) (Hermite). *The set of isomorphism classes of number fields of given discriminant is finite.*

(2) (Minkowski). *If K is a number field different from $\mathbb{Q}$, then $|d(K)| > 1$. In particular, there is at least one ramified prime in K.*

Proposition 4.8.10 (Stickelberger). *If p is an unramified prime in K with $p\mathbb{Z}_K = \prod_{i=1}^{g} \mathfrak{p}_i$, we have*

$$\left(\frac{d(K)}{p} \right) = (-1)^{n-g}$$

including the case $p = 2$ where $\left(\frac{d(K)}{2} \right)$ is to be interpreted as the Jacobi-Kronecker symbol (see Definition 1.4.8).

This shows that the parity of the number of primes above p (i.e. the "Möbius" function of p) can easily be computed.

Note that this proposition is also true for the infinite prime as given above, if we interpret the Legendre symbol as the sign of $d(K)$:

Proposition 4.8.11. *If K is a number field with signature (r_1, r_2), then the sign of the discriminant $d(K)$ is equal to $(-1)^{r_2}$.*

Proof. Since, up to a square, the discriminant $d(K)$ is equal to $\prod_{i<j}(\theta_i - \theta_j)^2$ (with evident notations), then a case by case examination shows that when conjugate terms are paired, all the factors become positive except for

$$\prod_{r_1 < i \leq r_1 + r_2} (\theta_i - \theta_{i+r_2})^2 \ ,$$

whose sign is $(-1)^{r_2}$ since $\theta_i - \theta_{i+r_2}$ is pure imaginary.   □

Corollary 4.8.12. *The decomposition type of a prime number p in a quadratic field K of discriminant D is the following: if $\left(\frac{D}{p} \right) = -1$ then p is inert. If*

$\left(\frac{D}{p}\right) = 0$ then p is ramified (i.e. $p\mathbb{Z}_K = \mathfrak{p}^2$). Finally, if $\left(\frac{D}{p}\right) = +1$, then p splits (completely), i.e. $p\mathbb{Z}_K = \mathfrak{p}_1\mathfrak{p}_2$.

4.8.2 A Simple Algorithm for the Decomposition of Primes

We now consider a more difficult algorithmic problem, that of determining the decomposition of prime numbers in a number field. The basic theorem on the subject, which unfortunately is not completely sufficient, is as follows.

Theorem 4.8.13. Let $K = \mathbb{Q}(\theta)$ be a number field, where θ is an algebraic integer, whose (monic) minimal polynomial is denoted $T(X)$. Let f be the index of θ, i.e. $f = [\mathbb{Z}_K : \mathbb{Z}[\theta]]$. Then for any prime p not dividing f one can obtain the prime decomposition of $p\mathbb{Z}_K$ as follows. Let

$$T(X) \equiv \prod_{i=1}^{g} T_i(X)^{e_i} \pmod{p}$$

be the decomposition of T into irreducible factors in $\mathbb{F}_p[X]$, where the T_i are taken to be monic. Then

$$p\mathbb{Z}_K = \prod_{i=1}^{g} \mathfrak{p}_i^{e_i} ,$$

where

$$\mathfrak{p}_i = (p, T_i(\theta)) = p\mathbb{Z}_K + T_i(\theta)\mathbb{Z}_K .$$

Furthermore, the residual index f_i is equal to the degree of T_i.

Since we have discussed at length in Chapter 3 algorithmic methods for finding the decomposition of polynomials in $\mathbb{F}_p[X]$, we see that this theorem gives us an excellent algorithmic method to find the decomposition of $p\mathbb{Z}_K$ when p does not divide the index f. The hard problems start when $p \mid f$. Of course, one then could try and change θ to get a different index, if possible prime to p, but even this is doomed. There can exist primes, called *inessential discriminantal divisors* which divide any index, no matter which θ is chosen. It can be shown that such exceptional primes are smaller than or equal to $n - 1$, so very few primes if any are exceptional. But the problem still exists: for example it is not difficult to give examples of fields of degree 3 where 2 is exceptional, see Exercise 10 of Chapter 6.

The case when p divides the index is much harder, and will be studied along with an algorithm to find integral bases in Chapter 6.

Proof of Theorem 4.8.13. Set $f_i = \deg(T_i)$ and $\mathfrak{p}_i = p\mathbb{Z}_K + T_i(\theta)\mathbb{Z}_K$. Let us assume that we have proved the following lemma:

Lemma 4.8.14.

(1) *For all i, either $\mathfrak{p}_i = \mathbb{Z}_K$, or $\mathbb{Z}_K/\mathfrak{p}_i$ is a field of cardinality p^{f_i}.*

(2) *If $i \neq j$ then $\mathfrak{p}_i + \mathfrak{p}_j = \mathbb{Z}_K$.*

(3) $p\mathbb{Z}_K \mid \mathfrak{p}_1^{e_1} \cdots \mathfrak{p}_g^{e_g}$.

Then, after reordering the $\mathfrak{p}_i$, we can assume that $\mathfrak{p}_i \neq \mathbb{Z}_K$ for $i \leq s$ and $\mathfrak{p}_i = \mathbb{Z}_K$ for $s < i \leq g$ (we will in fact see that $s = g$). Then by Lemma 4.8.14 (1), the ideals $\mathfrak{p}_i$ are prime for $i \leq s$, and since by definition they contain $p\mathbb{Z}_K$, they are above p (Proposition 4.8.1). (1) also implies that the f_i (for $i \leq s$) are the residual indices of $\mathfrak{p}_i$. By (2) we know that the $\mathfrak{p}_i$ for $i \leq s$ are distinct, and (3) implies that the decomposition of the ideal $p\mathbb{Z}_K$ is

$$p\mathbb{Z}_K = \prod_{i=1}^{s} \mathfrak{p}_i^{d_i} \qquad \text{where } d_i \leq e_i \text{ for all } i \leq s .$$

Hence, by Theorem 4.8.5, we have $n = d_1 f_1 + \cdots + d_s f_s$. Since we also have $\deg(T) = n = e_1 f_1 + \cdots + e_g f_g$ and $d_i \leq e_i$ for all i, this implies that we must have $s = g$ and $e_i = f_i$ for all i, thus proving Theorem 4.8.13. $\square$

Proof of Lemma 4.8.14 (1). Set $K_i = \mathbb{F}_p[X]/(T_i)$. Since T_i is irreducible, K_i is a field. Furthermore, the degree of K_i over $\mathbb{F}_p$ is f_i, and so the cardinality of K_i is p^{f_i}. Thus we need to show that either $\mathfrak{p}_i = \mathbb{Z}_K$ or that $\mathbb{Z}_K/\mathfrak{p}_i \simeq K_i$. Now it is clear that $\mathbb{Z}[X]/(p, T_i) \simeq K_i$, hence (p, T_i) is a maximal ideal of $\mathbb{Z}[X]$. But the kernel of the natural homomorphism ϕ from $\mathbb{Z}[X]$ to $\mathbb{Z}_K/\mathfrak{p}_i$ which sends X to $\theta \bmod \mathfrak{p}_i$ clearly contains this ideal, hence is either $\mathbb{Z}[X]$ or (p, T_i). If we show that ϕ is onto, this will imply that $\mathfrak{p}_i = \mathbb{Z}_K$ or $\mathbb{Z}_K/\mathfrak{p}_i \simeq \mathbb{Z}[X]/(p, T_i) \simeq K_i$, proving (1).

Now to say that ϕ is surjective means that $\mathbb{Z}_K = \mathbb{Z}[\theta] + \mathfrak{p}_i$. By definition, $p\mathbb{Z}_K \subset \mathfrak{p}_i$. Hence

$$[\mathbb{Z}_K : \mathbb{Z}[\theta] + \mathfrak{p}_i] \mid [\mathbb{Z}_K : \mathbb{Z}[\theta] + p\mathbb{Z}_K] = \gcd([\mathbb{Z}_K : \mathbb{Z}[\theta]], [\mathbb{Z}_K : p\mathbb{Z}_K]) .$$

Since we have assumed that p does not divide the index, and since $[\mathbb{Z}_K : p\mathbb{Z}_K] = p^n$, this shows that $[\mathbb{Z}_K : \mathbb{Z}[\theta] + \mathfrak{p}_i] = 1$, hence the surjectivity of ϕ. Note that this is the only part of the whole proof of Theorem 4.8.13 which uses that p does not divide the index of θ.

Proof of Lemma 4.8.14 (2). Since T_i and T_j are coprime in $\mathbb{F}_p[X]$, there exist polynomials U and V such that $UT_i + VT_j - 1 \in p\mathbb{Z}[X]$. It follows that $U(\theta)T_i(\theta) + V(\theta)T_j(\theta) = 1 + pW(\theta)$ for some polynomial $W \in \mathbb{Z}[X]$, and this immediately implies that $1 \in \mathfrak{p}_i + \mathfrak{p}_j$, i.e. that $\mathfrak{p}_i + \mathfrak{p}_j = \mathbb{Z}_K$.

Proof of Lemma 4.8.14 (3). Set $\gamma_i = T_i(\theta)$, so $\mathfrak{p}_i = (p, \gamma_i)$. By distributivity, it is clear that

$$\mathfrak{p}_1^{e_1} \cdots \mathfrak{p}_g^{e_g} \subset (p, \gamma_1^{e_1} \cdots \gamma_g^{e_g}) \; .$$

Now I claim that $(p, \gamma_1^{e_1} \cdots \gamma_g^{e_g}) = p\mathbb{Z}_K$, from which (3) follows. Indeed, $\supset$ is trivial. Conversely we have by definition $T_1^{e_1} \cdots T_g^{e_g} - T \in p\mathbb{Z}[X]$ hence taking $X = \theta$ we obtain

$$\gamma_1^{e_1} \cdots \gamma_g^{e_g} \in p\mathbb{Z}[\theta] \subset p\mathbb{Z}_K \; ,$$

proving our claim and the lemma.   □

Note that in the general case where $p \mid f$ which will be studied in Chapter 6, the prime ideals $\mathfrak{p}_i$ above p are still of the form $p\mathbb{Z}_K + T_i(\theta)\mathbb{Z}_K$, but now $T_i \in \mathbb{Q}[X]$ and does not always correspond to a factor of T modulo p.

4.8.3 Computing Valuations

Once prime ideals are known in a number field K, we will often need to compute the $\mathfrak{p}$-adic valuation v of an ideal I given in its Hermite normal form, where $\mathfrak{p}$ is a prime ideal above p. We may, of course, assume that I is an integral ideal. Then an obvious necessary condition for $v \neq 0$ is that $p \mid \mathcal{N}(I)$. Clearly this condition is not sufficient, since all primes above p must "share" in some way the exponent of p in $\mathcal{N}(I)$.

We assume that our prime ideal is given as $\mathfrak{p} = p\mathbb{Z}_K + \alpha\mathbb{Z}_K$ for a certain $\alpha \in \mathbb{Z}_K$. We will now describe an algorithm to compute $v_\mathfrak{p}(I)$, which was explained to me by H. W. Lenstra, but which was certainly known to Dedekind. It is based on the following proposition.

Proposition 4.8.15. *Let R be an order in K and $\mathfrak{p}$ a prime ideal of R. Then there exists $a \in K \setminus R$ such that $a\mathfrak{p} \subset R$. Furthermore, $\mathfrak{p}$ is invertible in R if and only if $a\mathfrak{p} \not\subset \mathfrak{p}$, and in that case we have $\mathfrak{p}^{-1} = R + aR$.*

Proof. Let $x \in \mathfrak{p}$ be a non-zero element of $\mathfrak{p}$, and consider the non-zero ideal xR. By Proposition 4.6.13, there exist non-zero prime ideals $\mathfrak{q}_i$ such that $xR \supset \prod_{i \in E} \mathfrak{q}_i$ for some finite set E. Assume E is chosen to be minimal in the sense that no proper subset of E can have the same property. Since $\prod \mathfrak{q}_i \subset xR \subset \mathfrak{p}$, by Proposition 4.6.12 we must have $\mathfrak{q}_j \subset \mathfrak{p}$ for some $j \in E$, hence $\mathfrak{q}_j = \mathfrak{p}$ since both are maximal ideals. Set

$$\mathfrak{q} = \prod_{i \in E, i \neq j} \mathfrak{q}_i \; .$$

Then $\mathfrak{p}\mathfrak{q} \subset xR$ and $\mathfrak{q} \not\subset xR$ by the minimality of E. So choose $y \in \mathfrak{q}$ such that $y \notin xR$. Then $y \notin xR$ and $y\mathfrak{p} \subset xR$, hence $a = y/x$ satisfies the conditions of the proposition.

Finally, consider the ideal $\mathfrak{p} + a\mathfrak{p}$. Since it sits between the maximal ideal $\mathfrak{p}$ and R, it must be equal to one of the two. If it is equal to R, we cannot have $a\mathfrak{p} \subset \mathfrak{p}$, and since $(R + aR)\mathfrak{p} = R$, $\mathfrak{p}$ is invertible and $\mathfrak{p}^{-1} = R + aR$. If it is equal to $\mathfrak{p}$, then $a\mathfrak{p} \subset \mathfrak{p}$, and $(R + aR)\mathfrak{p} = R\mathfrak{p}$. This implies that $\mathfrak{p}$ is not invertible since otherwise, by simplifying, we would have $R + aR = R$, hence $a \in R$. This proves the proposition.   □

Knowing this proposition, it is easy to obtain an algorithm for computing a suitable value of a. Note that $a\mathfrak{p} \subset R$ hence $a\mathfrak{p} \in R$, so we write $a = \beta/p$ with $\beta \in R$. The conditions to be satisfied for β are then $\beta \in R \setminus pR$ and $\beta\mathfrak{p} \subset pR$.

Let $\omega_1, \ldots, \omega_n$ be a $\mathbb{Z}$-basis of R, and let $\gamma_1, \ldots, \gamma_m$ be generators of $\mathfrak{p}$ (for example if $\mathfrak{p} = pR + \alpha R$ we take $\gamma_1 = p$ and $\gamma_2 = \alpha$). Then, if we write

$$\beta = \sum_{1 \leq i \leq n} x_i \omega_i \;,$$

we want to find integers x_i which are not all divisible by p such that for all j with $1 \leq j \leq m$ the coordinates of $(\sum x_i \omega_i)\gamma_j$ on the ω_i are all divisible by p. If we set

$$\omega_i \gamma_j = \sum_{1 \leq k \leq n} a_{i,j,k} \omega_k \;,$$

we obtain for all j and k

$$\sum_{1 \leq i \leq n} a_{i,j,k} x_i \equiv 0 \pmod{p}$$

which is a system of mn equations in n unknowns in $\mathbb{Z}/p\mathbb{Z}$ for which we want a non-trivial solution. Since there are many more equations than unknowns (if $m > 1$), there is, a priori, no reason for this system to have a non-trivial solution. The proposition that we have just proved shows that it does, and we can find one by standard Gaussian elimination in $\mathbb{Z}/p\mathbb{Z}$ (for example using Algorithm 2.3.1).

In the frequent special case where $m = 2$, $\gamma_1 = p$ and $\gamma_2 = \alpha$ for some $\alpha \in \mathbb{Z}_K$, the system simplifies considerably. For $j = 1$ the equations are trivial, hence we must simply solve the square linear system

$$\sum_{1 \leq \leq n} a_{i,k} x_i \equiv 0 \pmod{p}$$

where $\omega_i \alpha = \sum_{1 \leq k \leq n} a_{i,k} \omega_k$.

From now on, we assume that $R = \mathbb{Z}_K$ so that all ideals are invertible. Let I be an ideal of $\mathbb{Z}_K$ given by its HNF (M, d) with respect to θ, where M is an $n \times n$ matrix. We want to compute $v_\mathfrak{p}(I)$, where $\mathfrak{p}$ is a prime ideal of $\mathbb{Z}_K$ (hence invertible). By the method explained above, we first compute a such that $a \in K \setminus \mathbb{Z}_K$ and $a\mathfrak{p} \subset \mathbb{Z}_K$, and as above we set $\beta = a\mathfrak{p} \in \mathbb{Z}_K$. We may assume that I is an integral ideal of $\mathbb{Z}_K$. (If $I = I'/d'$ with I' an integral ideal and $d' \in \mathbb{Z}$, then clearly $v_\mathfrak{p}(I) = v_\mathfrak{p}(I') - ev_p(d')$, where e is the ramification index of $\mathfrak{p}$.) Now we have the following lemma which is the *raison d'être* of a.

Lemma 4.8.16. *With the above notations, if I is an integral ideal of $\mathbb{Z}_K$, then $I \subset \mathfrak{p}$ if and only if $aI \subset \mathbb{Z}_K$. In particular, $v_{\mathfrak{p}}(I)$ is the largest integer v such that $a^v I \subset \mathbb{Z}_K$.*

Proof. If $I \subset \mathfrak{p}$, then $aI \subset a\mathfrak{p} \subset \mathbb{Z}_K$. Conversely, assume that $aI \subset \mathbb{Z}_K$, hence $a\mathfrak{p}I \subset \mathfrak{p}$. Since the prime ideal $\mathfrak{p}$ contains the product of the integral ideals $a\mathfrak{p}$ and I, Proposition 4.6.12 shows that $\mathfrak{p}$ contains one of the two. Now since $\mathfrak{p}$ is invertible, $\mathfrak{p}$ cannot contain $a\mathfrak{p}$ by the above proposition, hence $\mathfrak{p} \supset I$. The final claim about the value of $v_{\mathfrak{p}}(I)$ is an immediate consequence of the definitions. $\qquad\square$

If, as above we set $a = \beta/p$ with $\beta \in \mathbb{Z}_K \setminus p\mathbb{Z}_K$, the condition $a^v I \subset \mathbb{Z}_K$ is equivalent to $\beta^v I \subset p^v \mathbb{Z}_K$. Let (N, d) be the HNF of the maximal order $\mathbb{Z}_K$. By Corollary 4.7.6, we may assume that $N_{n,n} = 1$, by choosing $d = d_n$. Now since I is an integral ideal, we have $dI \subset d\mathbb{Z}_K$, and $d\mathbb{Z}_K$ is represented by an integral matrix, hence dI also, so the HNF with respect to θ of any integral ideal can be chosen of the form (M, d) with the same d. Conversely, given (M, d) where M is an integral matrix in Hermite normal form representing a fractional ideal I, we can test whether I is integral by checking $I + \mathbb{Z}_K = \mathbb{Z}_K$, hence by computing the HNF of a $n \times 2n$ matrix as explained in Section 2.4.3. In our situation, a better way is to compute the HNF M' of I with respect to the HNF basis of $\mathbb{Z}_K$ given by the matrix N instead of with respect to θ, where we allow M' to have fractional entries. We clearly have

$$M' = N^{-1} M \ ,$$

except that the non-diagonal entries may have to be reduced, and I is an integral ideal if and only if M' has integral entries.

Hence, let (M_v, d) be the HNF of $\beta^v I$ with respect to θ, $M'_v = N^{-1} M_v$ and set $c_v = (M'_v)_{n,n}$. Then a necessary condition for $\beta^v I$ to be contained in $p^v \mathbb{Z}_K$ is that $p^v \mid c_v$. This condition is in general not sufficient, but very often it is. For example, it is easy to show (see Exercise 21) that the condition is sufficient when p does not divide the index $[\mathbb{Z}_K : \mathbb{Z}[\theta]]$, and in particular if $\mathbb{Z}_K = \mathbb{Z}[\theta]$. In the general case, we have to check the divisibility of all the coefficients of M_v by p^v. This leads to the following algorithm.

Algorithm 4.8.17 (Valuation at a Prime Ideal). Let (N, d) be the HNF of the maximal order $\mathbb{Z}_K$, let $\mathfrak{p}$ be a prime ideal of $\mathbb{Z}_K$ above p given by a generating system $\gamma_1, \dots, \gamma_m$ over $\mathbb{Z}_K$ (for example $\gamma_1 = p$, $\gamma_2 = \alpha$ for some $\alpha \in \mathbb{Z}_K$), and let I be an integral ideal of $\mathbb{Z}_K$ given by its HNF (M, d'). This algorithm computes the p-adic valuation $v_{\mathfrak{p}}(I)$ of the ideal I.

1. [Compute structure constants] Let ω_i be the HNF basis of $\mathbb{Z}_K$ corresponding to (N, d). Compute the integers $a_{i,j,k}$ such that

$$\omega_i \gamma_j = \sum_{1 \leq k \leq n} a_{i,j,k} \omega_k$$

for $1 \leq i \leq n$ and $1 \leq j \leq m$. Note that $\omega_i \gamma_j$ is computed as a polynomial in θ, and since N is an upper triangular matrix it is easy to compute inductively the $a_{i,j,k}$ from $k = n$ down to $k = 1$.

2. [Compute β] Using ordinary Gaussian elimination over $\mathbb{F}_p$ or Algorithm 2.3.1, find a non-trivial solution to the system of congruences

$$\sum_{1 \leq i \leq n} a_{i,j,k} x_i \equiv 0 \pmod{p} .$$

Then set $\beta \leftarrow \sum_i x_i \omega_i$.

3. [Compute $\mathcal{N}(I)$] Set $A \leftarrow d/d' N^{-1} M$ which must be a matrix with integral entries (otherwise I is not an integral ideal). Let P be the product of the diagonal elements of A. If $p \nmid P$, output 0 and terminate the algorithm. Otherwise, set $v \leftarrow 0$.

4. [Multiply] Set $A \leftarrow \beta A$ in the following sense. Each column of A corresponds to an element of K in the basis ω_i, and these elements are multiplied by β and expressed again in the basis ω_i, using the multiplication table for the ω_i.

5. [Simple test] If $p \nmid A_{n,n}$, output v and terminate the algorithm. Otherwise, if p does not divide the index $[\mathbb{Z}_K : \mathbb{Z}[\theta]] = d^n / \det(N)$, set $v \leftarrow v + 1$, $A \leftarrow A/p$ (which will be integral) and go to step 4.

6. [Complete test] Set $A \leftarrow A/p$. If A is not integral, output v and terminate the algorithm. Otherwise, set $v \leftarrow v + 1$ and go to step 4.

Note that steps 1 and 2 depend only on the ideal $\mathfrak{p}$, hence need be done only once if many $\mathfrak{p}$-adic valuations have to be computed for the same prime ideal $\mathfrak{p}$. Hence, a reasonable way to represent a prime ideal $\mathfrak{p}$ is as a quintuplet (p, α, e, f, β). Here p is the prime number over which $\mathfrak{p}$ lies, $\alpha \in \mathbb{Z}_K$ is such that $\mathfrak{p} = p\mathbb{Z}_K + \alpha\mathbb{Z}_K$, e is the ramification index and f the residual index of $\mathfrak{p}$, and β is the element of $\mathbb{Z}_K$ computed by steps 1 and 2 of the above algorithm, given by its coordinates x_i in the basis ω_i. Note also that Proposition 4.8.15 tells us that $\mathfrak{p}\mathfrak{p}^{-1} = p\mathbb{Z}_K + \beta\mathbb{Z}_K$.

4.8.4 Ideal Inversion and the Different

The preceding algorithms will allow us to give a satisfactory answer to a problem which we have not yet studied, that of ideal inversion in $\mathbb{Z}_K$.

Let I be an ideal of $\mathbb{Z}_K$ (which we can assume to be integral without loss of generality) given by a $\mathbb{Z}_K$-generating system $\gamma_1, \ldots, \gamma_m$. We can for example take the HNF basis of I in which case $m = n$, but often I will be given in a simpler way, for example by only 2 elements. We can try to mimic the first two steps of Algorithm 4.8.17 which, as remarked above, amount to computing the inverse of the prime ideal $\mathfrak{p}$.

Hence, let $\omega_1, \ldots, \omega_n$ be an integral basis of $\mathbb{Z}_K$. Then by definition of the inverse, $x \in I^{-1}$ if and only if $x\gamma_j \in \mathbb{Z}_K$ for all $j \leq k$. Fix a positive integer

d belonging to I. Then $dx \in \mathbb{Z}_K$ so we can write $dx = \sum_{1 \leq k \leq n} x_k \omega_k$ with $x_k \in \mathbb{Z}$ and the condition $x \in I^{-1}$ can be written

$$\sum_{1 \leq i \leq n} x_k \gamma_j \omega_k \in d\mathbb{Z}_K \quad \text{for all} \quad j \; .$$

If we define coefficients $u_{i,j,k} \in \mathbb{Z}$ by

$$\gamma_j \omega_k = \sum_{i=1}^{n} u_{i,j,k} \omega_i$$

we are thus led to the $nm \times n$ system of congruences $\sum_{1 \leq k \leq n} x_k u_{i,j,k} \equiv 0$ (mod d) for all i and j.

In the special case where I is a prime ideal as in Algorithm 4.8.17, we can choose $d = p$ a prime number, and hence our system of congruences can be considered as a system of equations in the finite field $\mathbb{F}_p$, and we can apply Algorithm 2.3.1 to find a basis for the set of solutions. Here, I is not a prime ideal in general, and we could try to solve the system of congruences by factoring d and working modulo powers of primes. A better method is probably as follows. Introduce extra integer variables $y_{i,j}$. Then our system is equivalent to the $nm \times (n + nm)$ linear system $\sum_{1 \leq k \leq n} x_k u_{i,j,k} - dy_{i,j} = 0$ for all i and j. We must find a $\mathbb{Z}$-basis of the solutions of this system, and for this we use the integral kernel Algorithm 2.7.2. The kernel will be of dimension n, and a $\mathbb{Z}$-basis of dI^{-1} is then obtained by keeping only the first n rows of the kernel (corresponding to the variables x_k).

In the common case where $m = n$, this algorithm involves $n^2 \times (n^2 + n)$ matrices, and this becomes large rather rapidly. Thus the algorithm is very slow as soon as n is at all large, and hence we must find a better method. For this, we introduce an important notion in algebraic number theory, the different, referring to the introductory books mentioned at the beginning of this chapter for more details.

Definition 4.8.18. *Let K be a number field. The different $\mathfrak{d}(K)$ of K is defined as the inverse of the ideal (called the* codifferent*)*

$$\{x \in K, \quad \mathrm{Tr}_{K/\mathbb{Q}}(x\mathbb{Z}_K) \subset \mathbb{Z}\} \; .$$

It is clear that the different $\mathfrak{d}(K)$ is an integral ideal. What makes the different interesting in our context is the following proposition.

Proposition 4.8.19. *Let $(\omega_i)_{1 \leq i \leq n}$ be an integral basis and let I be an ideal of $\mathbb{Z}_K$ given by an $n \times n$ matrix M whose columns give the coordinates of a $\mathbb{Z}$-basis $(\gamma_i)_{1 \leq i \leq n}$ of I on the chosen integral basis. Let $T = (t_{i,j})$ be the $n \times n$ matrix such that $t_{i,j} = \mathrm{Tr}_{K/\mathbb{Q}}(\omega_i \omega_j)$. Then the columns of the matrix $(M^t T)^{-1}$*

(again considered as coordinates on our integral basis) form a $\mathbb{Z}$-basis of the ideal $I^{-1}\mathfrak{d}(K)^{-1}$.

Proof. First, note that by definition of M, the coefficient of row i and column j in $M^t T$ is equal to $\text{Tr}_{K/\mathbb{Q}}(\gamma_i \omega_j)$. Furthermore, if $V = (v_i)$ is a column vector, then V belongs to the lattice spanned by the columns of $(M^t T)^{-1}$ if and only if $M^t T V$ has integer coefficients. This implies that for all i $\text{Tr}_{K/\mathbb{Q}}(\gamma_i(\sum_j v_j \omega_j)) \in \mathbb{Z}$, in other words that $\text{Tr}_{K/\mathbb{Q}}(xI) \subset \mathbb{Z}$, where we have set $x = \sum_j v_j \omega_j$. Since $xI = xI\mathbb{Z}_K$, the proposition follows. $\square$

In particular, when $I = \mathbb{Z}_K$ and $\gamma_i = \omega_i$ is an integral basis, this proposition shows that a $\mathbb{Z}$-basis of $\mathfrak{d}(K)^{-1}$ is obtained by computing the inverse of the matrix $\text{Tr}_{K/\mathbb{Q}}(\omega_i \omega_j)$. Since the determinant of this matrix is by definition equal to $d(K)$, this also shows that $\mathcal{N}(\mathfrak{d}(K)) = |d(K)|$.

The following theorem is a refinement of Theorem 4.8.8 (see [Mar]).

Theorem 4.8.20. *The prime ideals dividing the different are exactly the ramified prime ideals, i.e. the prime ideals whose ramification index is greater than 1.*

To compute the inverse of an ideal I given by a $\mathbb{Z}$-basis γ_j represented by an $n \times n$ matrix M on the integral basis as above, we thus proceed as follows. Computing T^{-1} we first obtain a basis of the codifferent $\mathfrak{d}(K)^{-1}$. We then compute the *ideal* product $I\mathfrak{d}(K)^{-1}$ by Hermite reduction of an $n \times n^2$ matrix as explained in Section 4.7. If N is the HNF matrix of this ideal product, then by Proposition 4.8.19, the columns $(N^t T)^{-1}$ will form a $\mathbb{Z}$-basis of the ideal $(I\mathfrak{d}(K)^{-1})^{-1}\mathfrak{d}(K)^{-1} = I^{-1}$, thus giving the inverse of I after another HNF. In paractice, it is better to work only with integral ideals, and since we know that $\det(T) = d(K)$, this means that we will replace $\mathfrak{d}(K)^{-1}$ by $d(K)\mathfrak{d}(K)^{-1}$ which is an integral ideal.

This leads to the following algorithm.

Algorithm 4.8.21 (Ideal Inversion). Given an integral basis $(\omega_i)_{1 \le i \le n}$ of the ring of integers of a number field K and an integral ideal I given by an $n \times n$ matrix M whose columns give the coordinates of a $\mathbb{Z}$-basis γ_j of I on the ω_i, this algorithm computes the HNF of the inverse ideal I^{-1}.

1. [Compute $d(K)\mathfrak{d}(K)^{-1}$] Compute the $n \times n$ matrix $T = (t_{i,j})$ such that $t_{i,j} = \text{Tr}_{K/\mathbb{Q}}(\omega_i \omega_j)$. Set $d \leftarrow \det(T)$ (this is the determinant $d(K)$ of K hence is usually available with the ω_i already). Finally, call δ_j the elements of $\mathbb{Z}_K$ whose coordinates on the ω_i are the columns of dT^{-1} (thus the δ_j will be a $\mathbb{Z}$-basis of the integral ideal $d(K)\mathfrak{d}(K)^{-1}$).

2. [Compute $d(K)I\mathfrak{d}(K)^{-1}$] Let N be the HNF of the $n \times n^2$ matrix whose columns are the coordinates on the integral basis of the n^2 products $\gamma_i \delta_j$ (the columns of N will form a $\mathbb{Z}$-basis of $d(K)I\mathfrak{d}(K)^{-1}$).

3. [Compute I^{-1}] Set $P \leftarrow d(K)(N^t T)^{-1}$, and let e be a common denominator for the entries of the matrix P. Let W be the HNF of eP. Output (W, e) as the HNF of I^{-1} and terminate the algorithm.

The proof of the validity of the algorithm is easy and left to the reader. □

Remarks.

(1) If many ideal inversions are to be done in the same number field, step 1 should of course be done only once. In addition, it may be useful to find a two-element representation for the integral ideal $d(K)\mathfrak{d}(K)^{-1}$ since this will considerably speed up the ideal multiplication of step 2. Algorithm 4.7.10 cannot directly be used for that purpose since it is valid only for prime ideals, but similar algorithms exist for general ideals (see Exercise 30). In addition, if $\mathbb{Z}_K = \mathbb{Z}[\theta]$ and if $P[X]$ is the minimal monic polynomial of θ, then one can prove (see Exercise 33) that $\mathfrak{d}(K)$ is the principal ideal generated by $P'(\theta)$, so the ideal multiplication of step 2 is even simpler.
(2) If we want to compute the HNF of the different $\mathfrak{d}(K)$ itself, we apply the above algorithm to the integral ideal $d(K)\mathfrak{d}(K)^{-1}$ (with $M = d(K)T^{-1}$) and multiply the resulting inverse by $d(K)$ to get $\mathfrak{d}(K)$.

Now that we know how to compute the inverse of an ideal, we can give an algorithm to compute intersections. This is based on the following formula, which is valid if I and J are integral ideals of $\mathbb{Z}_K$:

$$I \cap J = I \cdot J \cdot (I + J)^{-1} .$$

This corresponds to the usual formula $\mathrm{lcm}(a, b) = a \cdot b \cdot (\gcd(a, b))^{-1}$. We have seen above how to compute the HNF of sums and products of modules, and in particular of ideals, knowing the HNF of each operand. Since we have just seen an algorithm to compute the inverse of an ideal, this gives an algorithm for the intersection of two ideals.

However, a more direct (and usually better) way to compute the intersection of two ideals is described in Exercise 18.

4.9 Units and Ideal Classes

4.9.1 The Class Group

Definition 4.9.1. *Let K be a number field and $\mathbb{Z}_K$ be the ring of integers of K. We say that two (fractional) ideals I and J of K are equivalent if there exists $\alpha \in K^*$ such that $J = \alpha I$. The set of equivalence classes is called the class group of $\mathbb{Z}_K$ (or of K) and is denoted $Cl(K)$.*

Since fractional ideals of $\mathbb{Z}_K$ form a group it follows that $Cl(K)$ is also a group. The main theorem concerning $Cl(K)$ is that it is finite:

Theorem 4.9.2. *For any number field K, the class group $Cl(K)$ is a finite Abelian group, whose cardinality, called the* class number, *is denoted $h(K)$.*

Denote by $\mathcal{I}(K)$ the set of fractional ideals of K, and $\mathcal{P}(K)$ the set of principal ideals. We clearly have the exact sequence

$$1 \longrightarrow \mathcal{P}(K) \longrightarrow \mathcal{I}(K) \longrightarrow Cl(K) \longrightarrow 1 \ .$$

The determination of the structure of $Cl(K)$ and in particular of the class number $h(K)$ is one of the main problems in algorithmic algebraic number theory. We will study this problem in the case of quadratic fields in Chapter 5 and for general number fields in Chapter 6.

Note that $h(K) = 1$ if and only if $\mathbb{Z}_K$ is a PID which in turn is if and only if $\mathbb{Z}_K$ is a UFD. Hence the class group is the obstruction to $\mathbb{Z}_K$ being a UFD.

We can also define the class group for an order in K which is not the maximal order. In this case however, since not every ideal is invertible, we must slightly modify the definition.

Definition 4.9.3. *Let R be an order in K which is not necessarily maximal. We define the class group of R and denote by $Cl(R)$ the set of equivalence classes of* invertible *ideals of R (the equivalence relation being the same as before).*

Since all fractional ideals of $\mathbb{Z}_K$ are invertible, this does generalize the preceding definition. The class group is still a finite Abelian group whose cardinality is called the class number of R and denoted $h(R)$. Furthermore, it follows immediately from the definitions that the map $I \mapsto I\mathbb{Z}_K$ from R-ideals to $\mathbb{Z}_K$-ideals induces a homomorphism from $Cl(R)$ to $Cl(K)$ and that this homomorphism is *surjective*. In particular, $h(R)$ is a multiple of $h(K)$.

Since the discovery of the class group in 1798 by Gauss, many results have been obtained on class groups. Our ignorance however is still enormous. For example, although widely believed to be true, it is not even known if there exist an infinite number of isomorphism classes of number fields having class number 1 (i.e. with trivial class group, or again such that $\mathbb{Z}_K$ is a PID). Numerical and heuristic evidence suggests that already for real quadratic fields $\mathbb{Q}(\sqrt{p})$ with p prime and $p \equiv 1 \pmod 4$, not only should there be an infinite number of PID's, but their proportion should be around 75.446% (see [Coh-Len1], [Coh-Mar] and Section 5.10).

Class numbers and class groups arise very often in number theory. We give two examples. In the work on Fermat's last "theorem" (FLT), it was soon discovered that the obstruction to a proof was the failure of unique factorization in the cyclotomic fields $\mathbb{Q}(\zeta_p)$ where ζ_p is a primitive p^{th} root of unity (a number field of degree $p - 1$, generated by the polynomial $X^{p-1} + \cdots + X + 1$), where p is an odd prime. It was Kummer who essentially introduced the notion of ideals, and who showed how to replace unique factorization of elements by unique factorization of ideals, which as we have seen, is always satisfied in a Dedekind domain. It is however necessary to come back to the elements themselves in order to finish the argument—that is to obtain a principal ideal. What is obtained is that $\mathfrak{a}^p$ is principal for some ideal $\mathfrak{a}$. Now, by definition of the class group, we also know that $\mathfrak{a}^h$ is principal, where h is the class number of our cyclotomic field. Hence, we can deduce that $\mathfrak{a}$ itself is principal if p does not divide h. This fortunately seems to happen quite often (for example, for 22 out of the 25 primes less than 100); this proves FLT in many cases (the so-called regular primes). One can also prove FLT in other cases by more sophisticated methods.

The second use of class groups, which we will see in more detail in Chapters 8 and 10, is for factoring large numbers. In that case one uses class groups of quadratic fields. For example, the knowledge of the class group (in fact only of the 2-Sylow subgroup) of $\mathbb{Q}(\sqrt{-N})$ is essentially equivalent to knowing the factors of N, hence if we can find an efficient method to compute this class group or its 2-Sylow subgroup, we obtain a method for factoring N. This is the basis of work initiated by Shanks ([Sha1]) and followed by many other people (see for example [Sey1], [Schn-Len] and [Bue1]).

4.9.2 Units and the Regulator

Recall that a unit x in K is an algebraic integer such that $1/x$ is also an algebraic integer, or equivalently is an algebraic integer of norm ± 1.

Definition 4.9.4. *The set of units in K form a multiplicative group which we will denote by $U(K)$. The torsion subgroup of $U(K)$, i.e. the group of roots of unity in K, will be denoted by $\mu(K)$.*

(Note that some people write $E(K)$ because of the German word "Einheiten" for units, but we will keep the letter E for elliptic curves.)

It is clear that we have the exact sequence

$$1 \longrightarrow U(K) \longrightarrow K^* \longrightarrow \mathcal{P}(K) \longrightarrow 1 \ ,$$

where as before $\mathcal{P}(K)$ denotes the set of principal ideals in K. If we combine this exact sequence with the preceding one, we can complete a commutative diagram in the context of ideles, by introducing a generalization of the class group, called the idele class group $C(K)$. We will not consider these subjects in this course, but without explaining the notations (see [Lang2]) I give the diagram:

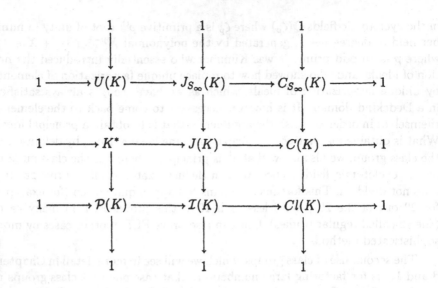

The main result concerning units is the following theorem

Theorem 4.9.5 (Dirichlet). *Let (r_1, r_2) be the signature of K. Then the group $U(K)$ is a finitely generated Abelian group of rank $r_1 + r_2 - 1$. In other words, we have a group isomorphism*

$$U(K) \simeq \mu(K) \times \mathbb{Z}^{r_1 + r_2 - 1} ,$$

and $\mu(K)$ is a finite cyclic group.

If we set $r = r_1 + r_2 - 1$, we see that there exist units $u_1, \ldots, u_r$ such that every element x of $U(K)$ can be written in a unique way as

$$x = \zeta u_1^{n_1} \cdots u_r^{n_r} ,$$

where $n_i \in \mathbb{Z}$ and ζ is a root of unity in K. Such a family (u_i) will be called a *system of fundamental units* of K. It is not unique, but since changing a $\mathbb{Z}$-basis of $\mathbb{Z}^r$ into another involves multiplication by a matrix of determinant ± 1, the absolute value of the determinant of the u_i in some appropriate sense, is independent of the choice of the u_i, and this is what we will call the regulator of K. The difficulty in defining the determinant comes because the units form a multiplicative group. To use determinants, one must linearize the problem, i.e. take logarithms.

Let $\sigma_1, \ldots, \sigma_{r_1}, \sigma_{r_1+1}, \ldots, \sigma_{r_1+r_2}$ be the first $r_1 + r_2$ embeddings of K in $\mathbb{C}$, where the σ_i for $i \leq r_1$ are the real embeddings, and the other embeddings are the σ_i and $\overline{\sigma_i} = \sigma_{r_2+i}$ for $i > r_1$.

Definition 4.9.6. *The* **logarithmic embedding** *of K^* in $\mathbb{R}^{r_1+r_2}$ is the map L which sends x to*

$$L(x) = (\ln|\sigma_1(x)|, \ldots, \ln|\sigma_{r_1}(x)|, 2\ln|\sigma_{r_1+1}(x)|, \ldots, 2\ln|\sigma_{r_1+r_2}(x)|) \ .$$

It is clear that L is an Abelian group homomorphism. Furthermore, we clearly have $\ln|\mathcal{N}_{K/\mathbb{Q}}(x)| = \sum_{1 \le i \le r_1+r_2} L_i(x)$ where $L_i(x)$ denotes the ith component of $L(x)$. It follows that the image of the subgroup of K^* of elements of norm equal to ± 1 is contained in the hyperplane $\sum_{1 \le i \le r_1+r_2} x_i = 0$ of $\mathbb{R}^{r_1+r_2}$.

The first part of the following theorem is essentially a restatement of Theorem 4.9.5, and the second part is due to Kronecker (see Exercise 25).

Theorem 4.9.7.

(1) *The image of the group of units $U(K)$ under the logarithmic embedding is a lattice (of rank $r_1 + r_2 - 1$) in the hyperplane $\sum_{1 \le i \le r_1+r_2} x_i = 0$ of $\mathbb{R}^{r_1+r_2}$.*

(2) *The kernel of the logarithmic embedding is exactly equal to the group $\mu(K)$ of the roots of unity in K.*

Definition 4.9.8. *The volume of this lattice, i.e. the absolute value of the determinant of any $\mathbb{Z}$-basis of the above defined lattice is called the* regulator *of K and denoted $R(K)$. If $u_1, \ldots, u_r$ is a system of fundamental units of K (where $r = r_1 + r_2 - 1$), $R(K)$ can also be defined as the absolute value of the determinant of any of the $r \times r$ matrices extracted from the $r \times (r+1)$ matrix*

$$\ln\|\sigma_j(u_i)\|_{1 \le i \le r, 1 \le j \le r+1} \ ,$$

where $\|x\|$ denotes the absolute value of x if x is real and the square of the modulus of x if x is complex (note that $L(x) = (\ln\|\sigma_i(x)\|)_{1 \le i \le r+1})$.

The problem of computing regulators (or fundamental units) is closely linked to the problem of computing class numbers, and is one of the other main tasks of computational algebraic number theory.

On the other hand, the problem of computing the subgroup of roots of unity $\mu(K)$ is not difficult. Note, for example, that if $r_1 > 0$ then $\mu(K) = \{\pm 1\}$ since all other roots of unity are non-real. Hence, we can assume $r_1 = 0$, and by the above theorem we must find integers x_i such that for every j, $|\sigma_j(\sum_{1 \le i \le n} x_i\omega_i)|^2 = 1$ where ω_i is an integral basis of $\mathbb{Z}_K$. If we set $x = (x_1, \ldots, x_n)$, this implies that

$$Q(x) = \sum_j |\sigma_j(\sum_{1 \le i \le n} x_i\omega_i)|^2 = n \ .$$

Conversely, the inequality between arithmetic and geometric mean shows that if $\rho \in \mathbb{Z}_K \setminus \{0\}$, then

$$\sum_j |\sigma_j(\rho)|^2 \geq n(\prod_j |\sigma_j(\rho)|)^{2/n} \geq n$$

with equality if and only if all $|\sigma_j(\rho)|^2$ are equal. It follows that n is the minimum non-zero value of the quadratic form Q on $\mathbb{Z}^n$, and that this minimum is attained when $|\sigma_j(\rho)| = 1$ for all j, where $\rho = \sum_i x_i \omega_i$. Finally, Theorem 4.9.7 (2) tells us that such a ρ is a root of unity (see Exercise 25). Hence, the computation of the minimal vectors of the lattice $(\mathbb{Z}^n, Q)$ using, for example, the Fincke-Pohst Algorithm 2.7.7, will give us quite rapidly the set of roots of unity in K. Thus we have the following algorithm.

Algorithm 4.9.9 (Roots of Unity Using Fincke-Pohst). Let $K = \mathbb{Q}(\theta)$ be a number field of degree n and T the minimal monic polynomial of θ over $\mathbb{Q}$. This algorithm computes the order $w(K)$ of the group of roots of unity $\mu(K)$ of K (hence $\mu(K)$ will be equal to the set of powers of a primitive $w(K)$-th root of unity).

1. [Initialize] Using Algorithm 4.1.11 compute the signature (r_1, r_2) of K. If $r_1 > 0$, output $w(K) = 2$ and terminate the algorithm. Otherwise, using Algorithm 6.1.8 of Chapter 6, compute an integral basis $\omega_1, \ldots, \omega_n$ of K as polynomials in θ.

2. [Compute matrix] Using Algorithm 3.6.6, compute a reasonably accurate value of θ and its conjugates $\sigma_j(\theta)$ as the roots of T, then the numerical values of $\sigma_j(\omega_k)$. Finally, compute a reasonably accurate approximation to

$$a_{i,j} \leftarrow \sum_{1 \leq k \leq n} \sigma_k(\omega_i)\overline{\sigma_k}(\omega_j)$$

(note that this will be a real number), and let A be the symmetric matrix $A = (a_{i,j})_{1 \leq i,j \leq n}$.

3. [Apply Fincke-Pohst] Apply Algorithm 2.7.7 to the matrix A and the constant $C = n + 0.1$.

4. [Final check] Set $s \leftarrow 0$. For each pair $(\mathbf{x}, -\mathbf{x})$ with $(x_1, \ldots, x_n)$ which is output by Algorithm 2.7.7, set $\rho \leftarrow \sum_{1 \leq i \leq n} x_i \omega_i$, and set $s \leftarrow s + 1$ if ρ is a root of unity (this can be checked exactly in several easy ways, see Exercise 26).

5. Output $w(K) \leftarrow 2s$ and terminate the algorithm.

Remark. The quadratic form Q considered here is, not surprisingly, the same as the one that we used for the polynomial reduction Algorithm 4.4.11. Note however that in POLRED we only wanted small vectors in the lattice, corresponding to algebraic numbers of degree exactly equal to n, while here we want the smallest vectors, and they correspond in general to algebraic numbers of degree less than n. Note also that in practice all the vectors output in step 4 correspond to roots of unity.

We can also give an algorithm based on those of Section 4.5.3 as follows.

Algorithm 4.9.10 (Roots of Unity Using the Subfield Problem). Let $K = \mathbb{Q}(\theta)$ be a number field of degree n and T the minimal monic polynomial of θ over $\mathbb{Q}$. This algorithm computes the order $w(K)$ of the group of roots of unity $\mu(K)$ of K (hence $\mu(K)$ will be equal to the set of powers of a primitive $w(K)$-th root of unity).

1. [Initialize] Using Algorithm 4.1.11 compute the signature (r_1, r_2) of K. If $r_1 > 0$, output $w(K) = 2$ and terminate the algorithm. Otherwise, using Algorithm 6.1.8 of Chapter 6, compute the discriminant $d(K)$ of K, and set $w \leftarrow 1$.

2. [Compute primes] Let $\mathcal{L}$ be the list of primes p such that $(p - 1) \mid n$ (since n is very small, $\mathcal{L}$ can be simply obtained by trial division). Let c be the number of elements of $\mathcal{L}$, and set $i \leftarrow 0$.

3. [Get next prime and exponent] Set $i \leftarrow i + 1$. If $i > c$ output w and terminate the algorithm. Otherwise, let p be the i-th element in the list $\mathcal{L}$, set

$$k \leftarrow \left\lfloor \frac{v_p(d(K))}{n} + \frac{1}{p-1} \right\rfloor \;,$$

and set $j \leftarrow 0$.

4. [Test cyclotomic polynomials] Set $j \leftarrow j + 1$. If $j > k$, go to step 3. Otherwise, applying Algorithm 4.5.4 to $A(X) = \Phi_{p^j}(X)$ and $B(X) = T(X)$ (where $\Phi_{p^j}(X) = \sum_{i=0}^{p-1} X^{ip^{j-1}}$ is the p^j-th cyclotomic polynomial) determine whether K has a subfield isomorphic to $\mathbb{Q}(\zeta_{p^j})$ (where ζ_{p^j} is some root of $\Phi_{p^j}(X)$, i.e. a primitive p^j-th root of unity). If it does, go to step 4, and if not, set $w \leftarrow wp^{j-1}$ and go to step 3.

Remarks.

(1) The validity of the check in step 3 follows from Exercise 24 and Proposition 4.4.8. We can avoid the computation of the discriminant of K, and skip this step, at the expense of spending more time in step 4.

(2) We refer to [Was] or [Ire-Ros] for cyclotomic fields (which we will meet again in Chapter 9) and cyclotomic polynomials. The (general) cyclotomic polynomials can be computed either by induction or by the explicit formula

$$\Phi_m(X) = \prod_{d \mid m} (X^d - 1)^{\mu(m/d)}$$

where $\mu(n)$ is the Möbius function, but in our case this simplifies to the formula

$$\Phi_{p^j}(X) = \sum_{i=0}^{p-1} X^{ip^{j-1}}$$

used in the algorithm.

(3) Although Algorithm 4.9.10 is more pleasing to the mind, Algorithm 4.9.9 is considerably faster and should therefore be preferred in practice. Care should be taken however to be sufficiently precise in the computation of the numerical values of the coefficients of Q. We have given in detail Algorithm 4.9.10 to show that an exact algorithm also exists.

All the quantities that we have defined above are tied together if we view them analytically.

Definition 4.9.11. *Let K be a number field. We define for $\mathrm{Re}(s) > 1$ the Dedekind zeta function $\zeta_K(s)$ of K by the formulas*

$$\zeta_K(s) = \sum_{\mathfrak{a}} \frac{1}{\mathcal{N}(\mathfrak{a})^s} = \prod_{\mathfrak{p}} \frac{1}{1 - \dfrac{1}{\mathcal{N}(\mathfrak{p})^s}}$$

where the sum is over all non-zero integral ideals of $\mathbb{Z}_K$ and the product is over all non-zero prime ideals of $\mathbb{Z}_K$.

The equality between the two definitions follows from unique factorization into prime ideals (Theorem 4.6.14), and the convergence for $\mathrm{Re}(s) > 1$ is proved in Exercise 22.

The basic theorem concerning this function is the following.

Theorem 4.9.12 (Dedekind). *Let K be a number field of degree n having r_1 real places and r_2 complex ones (so $r_1 + 2r_2 = n$). Denote by $d(K)$, $h(K)$, $R(K)$ and $w(K)$ the discriminant, class number, regulator and number of roots of unity of K respectively.*

(1) *The function $\zeta_K(s)$ can be analytically continued to the whole complex plane into a meromorphic function having a single pole at $s = 1$ which is simple.*

(2) *If we set*

$$\Lambda(s) = |d(K)|^{s/2} \left(\pi^{-\frac{s}{2}} \Gamma\left(\frac{s}{2}\right)\right)^{r_1 + r_2} \left(\pi^{-\frac{s+1}{2}} \Gamma\left(\frac{s+1}{2}\right)\right)^{r_2} \zeta_K(s)$$

we have the functional equation

$$\Lambda(1 - s) = \Lambda(s) \ .$$

(3) *If we set $r = r_1 + r_2 - 1$ (which is the rank of the unit group), $\zeta_K(s)$ has a zero of order r at $s = 0$ and we have*

$$\lim_{s \to 0} s^{-r} \zeta_K(s) = -h(K)R(K)w(K)^{-1} \ .$$

(4) *Equivalently by the functional equation, the residue of $\zeta_K(s)$ at $s = 1$ is given by*

$$\lim_{s \to 1}(s - 1)\zeta_K(s) = 2^{r_1}(2\pi)^{r_2}\frac{h(K)R(K)}{w(K)\sqrt{|d(K)|}} \ .$$

This theorem shows one among numerous instances where $h(K)$ and $R(K)$ are inextricably linked.

Remarks.

(1) From this theorem it is easily shown (see Exercise 23) that if $N_K(x)$ denotes the number of integral ideals of norm less than or equal to x, then

$$\lim_{x \to \infty}\frac{N_K(x)}{x} = 2^{r_1}(2\pi)^{r_2}\frac{h(K)R(K)}{w(K)\sqrt{|d(K)|}} \ .$$

(2) It is also possible to prove the following generalization of the prime number theorem (see [Lang2]).

Theorem 4.9.13. *Let $\pi_K(x)$ (resp. $\pi_K^{(1)}(x)$) be the number of prime ideals (resp. prime ideals of degree 1) whose norm is less than equal to x. Then*

$$\lim_{x \to \infty}\frac{\pi_K(x)}{x/\ln(x)} = \lim_{x \to \infty}\frac{\pi_K^{(1)}(x)}{x/\ln(x)} = 1 \ .$$

Dedekind's Theorem 4.9.12 shows that the behavior of $\zeta_K(s)$ at $s = 0$ and $s = 1$ is linked to fundamental arithmetic invariants of the number field K. Siegel proved that the values at negative integers are rational numbers, hence they also have some arithmetic significance. From the functional equation it is immediately clear that $\zeta_K(s)$ vanishes for all negative integers s if K is not totally real, and for even negative integers if K is totally real. Hence, the only interesting values are the $\zeta_K(1 - 2m)$ for totally real fields K ($r_2 = 0$) and positive integral m. There are special methods, essentially due to Siegel, for computing these values using the theory of Hilbert modular forms. As an example, we give the following result, which also shows the arithmetic significance of these values (see [Coh], [Zag1]).

Theorem 4.9.14. *Let $K = \mathbb{Q}(\sqrt{D})$ be a real quadratic field of discriminant D. Define $\sigma(n)$ to be equal to the sum of the positive divisors of n if n is positive, and equal to 0 otherwise. Then*

(1)
$$\zeta_K(-1) = \frac{1}{6} \sum_{s \equiv D \ (\mathrm{mod}\ 2)} \sigma\left(\frac{D - s^2}{4}\right)$$

(this is a finite sum).
(2) *The number $r_5(D)$ of representations of D as a sum of 5 squares of elements of $\mathbb{Z}$ (counting representations with a different ordering as distinct) is given by*

$$r_5(D) = 48\left(5 - 2\left(\frac{D}{2}\right)\right)\zeta_K(-1)$$

(this formula must be slightly modified if D is not the discriminant of a real quadratic field, see [Coh2]).

I have already mentioned how little we know about class numbers. The same can be said about regulators. For example, we can define the regulator of a number field in a p-adic context, essentially by replacing the real logarithms by p-adic ones. In that case, even an analogue of Dirichlet's theorem that the regulator does not vanish is not known. This is a famous unsolved problem known as Leopoldt's conjecture. It is known to be true for some classes of fields, for example Abelian extensions of $\mathbb{Q}$ (see [Was] Section 5.5).

We do have a theorem which gives a quantitative estimate for the product of the class number and the regulator (see [Sie], [Brau] and [Lang2]):

Theorem 4.9.15 (Brauer-Siegel). *Let K vary in a family of number fields such that $|d(K)|^{1/\deg(K)}$ tends to infinity, where $d(K)$ is the discriminant of K. Assume, in addition, that these fields are Galois over $\mathbb{Q}$. Then, we have the following asymptotic relation:*

$$\ln(h(K)R(K)) \sim \ln(|d(K)|^{1/2}) \ .$$

This shows that the product $h(K)R(K)$ behaves roughly as the square root of the discriminant. The main problem with this theorem is that it is *non-effective*, meaning that nobody knows how to give explicit constants to make the $\sim$ sign disappear. For example, for imaginary quadratic fields, $r = 0$ hence $R(K) = 1$, and although the Brauer-Siegel theorem tells us that $h(K)$ tends to infinity with $|d(K)|$, and even much more, the problem of finding an explicit function $f(d)$ tending to infinity with d and such that $h(K) \geq f(|d(K)|)$ is extremely difficult and was only solved recently using sophisticated methods involving elliptic curves and modular forms, by Goldfeld, Gross and Zagier ([Gol], [Gro-Zag2]).

Note that one conjectures that the theorem is still true without the hypothesis that the fields are Galois extensions. This would follow from Artin's conjecture on non-Abelian L-functions and on certain Generalized

Riemann Hypotheses. On the other hand, one can prove that the hypothesis on $|d(K)|^{1/\deg(K)}$ is necessary. The following is a simple corollary of the Brauer-Siegel Theorem 4.9.15:

Corollary 4.9.16. *Let K vary over a family of number fields of fixed degree over $\mathbb{Q}$. Then, as $|d(K)| \to \infty$, we have*

$$\ln(h(K)R(K)) \sim \ln(|d(K)|^{1/2}) \ .$$

4.9.3 Conclusion: the Main Computational Tasks of Algebraic Number Theory

From the preceding definitions and results, it can be seen that the main computational problems for a number field $K = \mathbb{Q}(\theta)$ are the following:

(1) Compute an integral basis of $\mathbb{Z}_K$, determine the decomposition of prime numbers in $\mathbb{Z}_K$ and p-adic valuations for given ideals or elements.
(2) Compute the Galois group of the Galois closure of K.
(3) Compute a system of fundamental units of K and/or the regulator $R(K)$. Note that these two problems are not completely equivalent, since for many applications, only the approximate value of the real number $R(K)$ is desired. In most cases, by the Brauer-Siegel theorem, the fundamental units are too large even to write down, at least in a naïve manner (see Section 5.8.3 for a representation which avoids this problem).
(4) Compute the class number and the structure of the class group $Cl(K)$. It is essentially impossible to do this without also computing the regulator.
(5) Given an ideal of $\mathbb{Z}_K$, determine whether or not it is principal, and if it is, compute $\alpha \in K$ such that $I = \alpha\mathbb{Z}_K$.

In the rest of this book, we will give algorithms for these tasks, placing special emphasis on the case of quadratic fields.

Although they are all rather complex, some sophisticated versions are quite efficient. With fast computers and careful implementations, it is possible to tackle the above tasks for quadratic number fields whose discriminant has 50 or 60 decimal digits (less for general number fields). Work on this subject is currently in progress in several places.

4.10 Exercises for Chapter 4

1. (J. Martinet) Let $P(X) = X^4 + aX^3 + bX^2 + cX + d \in \mathbb{R}[X]$ be a squarefree polynomial. Set $D \leftarrow \operatorname{disc}(P)$, $A \leftarrow 8b - 3a^2$, $B \leftarrow b^2 - a^2b + (3/16)a^4 + ac - 4d$. Show that the signature of P is given by the following formulas. $(r_1, r_2) = (2, 1)$ iff $D < 0$, $(r_1, r_2) = (4, 0)$ iff $D > 0$, $A > 0$ and $B > 0$, and $(r_1, r_2) = (0, 4)$ iff $D > 0$ and $AB < 0$. (Hint: use Exercise 29 of Chapter 3.)

2. If α and θ are two algebraic numbers of degree n generating the same number field K over $\mathbb{Q}$, write an algorithm to find the standard representation of θ in terms of α knowing the standard representation of α in terms of θ.

3. Prove Newton's formulas (i.e. Proposition 4.3.3).

4. Compute the minimal polynomial of $\alpha = 2^{1/4} + 2^{1/2}$ using several methods, and compare their efficiency.

5. Let K be a number field of signature (r_1, r_2). Using the canonical isomorphism

    $$K \otimes \mathbb{R} \simeq \mathbb{R}^{r_1} \times \mathbb{C}^{r_2}$$

    show that the quadratic form $\mathrm{Tr}_{K/\mathbb{Q}}(x^2)$ has signature $(r_1 + r_2, r_2)$.

6. Prove that if $P = \sum_{k=0}^{n} a_k X^k$ is a monic polynomial and if $S = \mathrm{size}(P)$ in the sense of Section 4.4.2, then

    $$|a_{n-k}| \le \binom{n}{k} \left(\frac{S}{n} \right)^{k/2} \;,$$

    and that the constant is best possible if P is assumed to be with complex (as opposed to integral) coefficients (hint: use a variational principle).

7. (D. Shanks.) Using for example Algorithm 4.4.11, show the following "incredible identity" $A = B$, where

    $$A = \sqrt{5} + \sqrt{22 + 2\sqrt{5}}$$

    and

    $$B = \sqrt{11 + 2\sqrt{29}} + \sqrt{16 - 2\sqrt{29} + 2\sqrt{55 - 10\sqrt{29}}} \;.$$

    See [Sha4] for an explanation of this phenomenon and other examples. See also [BFHT] and [Zip] for the general problem of radical simplification.

8. Consider modifying the POLRED algorithm as follows. Instead of the quadratic form $\mathrm{size}(P)$, we take instead

    $$f(P) = \sum_{i<j} |\alpha_i - \alpha_j|^2 \;,$$

    which is still a quadratic form in the n variables x_i when we write $\alpha = \sum_{i=1}^{n} x_i \omega_i$. Experiment on this to compare it with POLRED, and in particular see whether it gives a larger number of proper subfields of K or a smaller index.

9. Prove Proposition 4.5.3.

10. Write an algorithm which outputs all quadratic subfields of a given number field.

11. Let R be a Noetherian integral domain. Show that any non-zero ideal of R contains a product of non-zero prime ideals.

12. Let d_1 and d_2 be coprime integers such that $d = d_1 d_2 \in I$, where I is an integral ideal in a number field K. Show that $I = I_1 I_2$ where $I_i = I + d_i \mathbb{Z}_K$, and show that this is false in general if d_1 and d_2 are not assumed to be coprime.

13. Let R be an order in a number field, and let I and J be two ideals in R. Assume that I is a maximal (i.e. non-zero prime) ideal. Show that $\mathcal{N}(I)\mathcal{N}(J) \mid \mathcal{N}(IJ)$

and that $\mathcal{N}(I^2) = \mathcal{N}(I)^2$ if and only if I is invertible. (Note that these two results are not true anymore if I is not assumed maximal.)

14. Let R be an order in a number field. For any non-zero integral ideal of R, set $f(I) = [R : II']$ where as in Lemma 4.6.7 we set $I' = \{x \in K, \, xI \subset R\}$. This function can be considered as a measure of the non-invertibility of the ideal I.

    a) If I is a maximal ideal, show that either I is invertible (in which case $f(I) = 1$) or else $f(I) = \mathcal{N}(I)$.

    b) Generalizing Proposition 4.6.8, show that if I and J are two ideals such that $f(I)$ and $f(J)$ are coprime, we still have $\mathcal{N}(IJ) = \mathcal{N}(I)\mathcal{N}(J)$.

15. (H. W. Lenstra) Let α be an algebraic number which is not necessarily an algebraic integer, and let $a_n X^n + a_{n-1} X^{n-1} + \cdots + a_0$ be its minimal polynomial. Set

$$\mathbb{Z}[\alpha] = \mathbb{Z} + (a_n\alpha)\mathbb{Z} + (a_n\alpha^2 + a_{n-1}\alpha)\mathbb{Z} + \cdots$$

    a) Show that $\mathbb{Z}[\alpha]$ is an order of K, and that its definition coincides with the usual one when α is an algebraic integer.

    b) Show that Proposition 4.4.4 (2) remains valid if $T \in \mathbb{Z}[X]$ is not assumed to be monic, if we use this generalized definition for $\mathbb{Z}[\theta]$. How should Proposition 4.4.4 (1) be modified?

16. Show that the converse of Theorem 4.7.5 is not always true, in other words if (W, d) is a HNF representation of a $\mathbb{Z}$-module M satisfying the properties given in the theorem, show that M is not always a $\mathbb{Z}[\theta]$-module.

17. Assume that W is a HNF of an ideal I of R with respect to a basis $\alpha_1 = 1$, $\alpha_2, \ldots, \alpha_n$ of R. Show that it is still true that $w_{i,i} \mid w_{1,1}$ for all i, and that if $w_{i,i} = w_{1,1}$ then $w_{j,i} = 0$ for $j \neq i$.

18. Show that by using Algorithms 2.4.10 or 2.7.2 instead of Algorithm 2.3.1, Algorithm 2.3.9 can be used to compute the intersection of two $\mathbb{Z}$-modules, and in particular of two ideals. Compare the efficiency of this method with that given in the text.

19. Let $\mathfrak{p}$ be a (non-zero) prime ideal in $\mathbb{Z}_K$ for some number field K, and assume that $\mathfrak{p}$ is not above 2. If $x \in \mathbb{Z}_K$, show that there exists a unique $\varepsilon \in \{-1, 0, +1\}$ such that

$$x^{(\mathcal{N}(\mathfrak{p})-1)/2} \equiv \varepsilon \pmod{\mathfrak{p}},$$

where we write $x \equiv y \pmod{\mathfrak{p}}$ if $x - y \in \mathfrak{p}$. This ε is called a "generalized Legendre symbol" and denoted $\left(\frac{x}{\mathfrak{p}}\right)$. Study the generalization to this symbol of the properties of the ordinary Legendre symbol seen in Chapter 1.

20. Show that the condition $v_\mathfrak{p}(\mathcal{N}(\alpha)) = f$ of Lemma 4.7.9 is not a necessary condition for $\mathfrak{p}$ to be equal to (p, α) (hint: decompose α and $p\mathbb{Z}_K$ as a product of prime ideals).

21. Using the notation of Algorithm 4.8.17, show that if the prime p does not divide the index $[R : \mathbb{Z}[\theta]]$, then $p^v \mid A_{n,n}$ is equivalent to p^v divides all the coefficients of the matrix A.

22. Let s be a *real* number such that $s > 1$. Show that if K is a number field of degree n we have $\zeta_K(s) \leq \zeta^n(s)$ where $\zeta(s) = \zeta_{\mathbb{Q}}(s)$ is the usual Riemann zeta function, and hence that the product and series defining $\zeta_K(s)$ converge absolutely for $\mathrm{Re}(s) > 1$.

23. If K is a number field, let $N_K(x)$ be the number of integral ideals of $\mathbb{Z}_K$ of norm less than or equal to x. Using Theorem 4.9.12, and a suitable Tauberian theorem, find the limit as x tends to infinity of $N_K(x)/x$.

24. Let $K = \mathbb{Q}(\zeta_{p^k})$ where p is a prime and ζ_m denotes a primitive m-th root of unity. One can show that $\mathbb{Z}_K = \mathbb{Z}[\zeta_{p^k}]$. Using this, compute the discriminant of the field K, and hence show the validity of the formula in Step 3 of Algorithm 4.9.10.

25. Let α be an algebraic integer of degree d all of whose conjugates have absolute value 1.

    a) Show that for every positive integer k, the monic minimal polynomial of α^k in $\mathbb{Z}[X]$ has all its coefficients bounded in absolute value by 2^d.

    b) Deduce from this that there exists only a finite number of distinct powers of α, hence that α is a root of unity. (This result is due to Kronecker.)

26. Let $\rho \in \mathbb{Z}_K$ be an algebraic integer given as a polynomial in θ, where $K = \mathbb{Q}(\theta)$ and T is the minimal monic polynomial of θ in $\mathbb{Z}[X]$. Give algorithms to check exactly whether or not ρ is a root of unity, and compare their efficiency.

27. Let $K = \mathbb{Q}[\theta]$ where θ is a root of the polynomial $X^4 + 1$. Show that the subgroup of roots of unity of K is the group of 8-th roots of unity. Show that $1 + \sqrt{2}$ is a generator of the torsion-free part of the group of units of K. What is the regulator of K? (Warning: it is not equal to $\ln(1 + \sqrt{2})$).

28. Let $\mathfrak{p}$ be a (non-zero) prime ideal in $\mathbb{Z}_K$ for some number field K, let $e = e(\mathfrak{p}/p)$ be its ramification index, let $\mathfrak{p} = p\mathbb{Z}_K + \alpha\mathbb{Z}_K$ be a two-element representation of $\mathfrak{p}$, and finally let $v = v_{\mathfrak{p}}(\alpha)$. Let $a \geq 1$ and $b \geq 1$ be integers. By computing q-adic valuations for each prime ideal $\mathfrak{q}$, show that

$$p^a\mathbb{Z}_K + \alpha^b\mathbb{Z}_K = \mathfrak{p}^{\min(ae,bv)} .$$

Deduce from this formulas for computing explicitly $\mathfrak{p}^k$ for any $k \geq 1$.

29. Let I be an integral ideal in a number field K and let $\ell(I)$ be the positive generator of $I \cap \mathbb{Z}$.

    a) Show that

$$\ell(I) = \prod_{p|\mathcal{N}(I)} p^{\max_{\mathfrak{p}|p} \lceil v_{\mathfrak{p}}(I)/e(\mathfrak{p}/p) \rceil} .$$

    b) Let $\alpha \in I$ be such that $(\mathcal{N}(I), \mathcal{N}(\alpha)/\mathcal{N}(I)) = 1$. Show that

$$I = \ell(I)\mathbb{Z}_K + \alpha\mathbb{Z}_K = \mathcal{N}(I)\mathbb{Z}_K + \alpha\mathbb{Z}_K$$

(this is a partial generalization of Lemma 4.7.9).

    c) Deduce from this an algorithm for finding a two-element representation of I analogous to Algorithm 4.7.10.

30. Let $K = \mathbb{Q}[\theta]$ be a number field, where θ is an algebraic integer whose minimal monic polynomial is $P(X) \in \mathbb{Z}[X]$. Assume that $\mathbb{Z}_K = \mathbb{Z}[\theta]$. Show that the different $\mathfrak{d}(K)$ is the principal ideal generated by $P'(\theta)$.

31. Let I and J be two integral ideals in a number field K given by their HNF matrices M_I and M_J. Assume that I and J are coprime, i.e. that $I + J = \mathbb{Z}_K$. Give an algorithm which finds $i \in I$ and $j \in J$ such that $i + j = 1$.

32.   a) Using the preceding exercise, give an algorithm wich finds explicitly the element $\beta \in \mathbb{Z}_K$ whose existence is proven in Proposition 4.7.7.

b) Deduce from this an algorithm which finds a two-element representation $I = \alpha \mathbb{Z}_K + \beta \mathbb{Z}_K$ of an integral ideal I given a non-zero element $\alpha \in I$.

c) In the case where $\alpha = \ell(I)$, compare the theoretical and practical performance of this algorithm with the one given in Exercise 29.

33. Let α and β be non-zero elements of K^*. Show that there exist u and v in $\mathbb{Z}_K$ such that $\alpha\beta = u\alpha^2 + v\beta^2$, and give an algorithm for computing u and v.

34. Modify Proposition 4.3.4 so that it is still valid when $T(X) \in \mathbb{Q}[X]$ and not necessarily monic.

Chapter 5

Algorithms for Quadratic Fields

5.1 Discriminant, Integral Basis and Decomposition of Primes

In this chapter, we consider the simplest of all number fields that are different from $\mathbb{Q}$, i.e. quadratic fields. Since $n = 2 = r_1 + 2r_2$, the signature (r_1, r_2) of a quadratic field K is either $(2, 0)$, in which case we will speak of *real* quadratic fields, or $(0, 1)$, in which case we will speak of *imaginary* (or complex) quadratic fields. By Proposition 4.8.11 we know that imaginary quadratic fields are those of negative discriminant, and that real quadratic fields are those with positive discriminant.

Furthermore, by Dirichlet's unit theorem, the rank of the group of units is $r_1 + r_2 - 1$, hence it can be equal to zero only in two cases: either $r_1 = 1$, $r_2 = 0$, hence $n = 1$ so $K = \mathbb{Q}$, a rather uninteresting case (see below however). Or, $r_1 = 0$ and $r_2 = 1$, hence $n = 2$, and this corresponds to imaginary quadratic fields. One reason imaginary quadratic fields are simple is that they are the only number fields (apart from $\mathbb{Q}$) with a finite number of units (almost always only 2). We consider them first in what follows. However, a number of definitions and simple results can be given uniformly.

Since a quadratic field K is of degree 2 over $\mathbb{Q}$, it can be given by $K = \mathbb{Q}(\theta)$ where θ is a root of a monic irreducible polynomial of $\mathbb{Z}[X]$, say $T(X) = X^2 + aX + b$. If we set $\theta' = 2\theta + a$, then θ' is a root of $X^2 = a^2 - b = d$. Hence, $K = \mathbb{Q}(\sqrt{d})$ where d is an integer, and the irreducibility of T means that d is not a square. Furthermore, it is clear that $\mathbb{Q}(\sqrt{df^2}) = \mathbb{Q}(\sqrt{d})$, hence we may assume d squarefree. The discriminant and integral basis problem is easy.

Proposition 5.1.1. *Let $K = \mathbb{Q}(\sqrt{d})$ be a quadratic field with d squarefree and not a square (i.e. different from 1). Let $1, \omega$ be an integral basis and $d(K)$ the discriminant of K. Then, if $d \equiv 1 \pmod 4$, we can take $\omega = (1 + \sqrt{d})/2$, and we have $d(K) = d$, while if $d \equiv 2$ or $3 \pmod 4$, we can take $\omega = \sqrt{d}$ and we have $d(K) = 4d$.*

This is well known and left as an exercise. Note that we can, for example, appeal to Corollary 4.4.7, which is much more general.

For several reasons, in particular to avoid making unnecessary case distinctions, it is better to consider quadratic fields as follows.

Definition 5.1.2. *An integer D is called a* fundamental discriminant *if D is the discriminant of a quadratic field K. In other words, $D \neq 1$ and either $D \equiv 1 \pmod 4$ and is squarefree, or $D \equiv 0 \pmod 4$, $D/4$ is squarefree and $D/4 \equiv 2$ or $3 \pmod 4$.*

If K is a quadratic field of discriminant D, we will use the following as standard notations: $K = \mathbb{Q}(\sqrt{D})$, where D is a fundamental discriminant. Hence $D = d(K)$, and an integral basis of K is given by $(1, \omega)$, where

$$\omega = \frac{D + \sqrt{D}}{2} ,$$

and therefore $\mathbb{Z}_K = \mathbb{Z}[\omega]$.

Proposition 5.1.3. *If K is a quadratic field of discriminant D, then every order R of K has discriminant Df^2 where f is a positive integer called the* conductor *of the order. Conversely, if A is any non-square integer such that $A \equiv 0$ or $1 \pmod 4$, then A is uniquely of the form $A = Df^2$ where D is a fundamental discriminant, and there exists a unique order R of discriminant A (and R is an order of the quadratic field $\mathbb{Q}(\sqrt{D})$).*

Again this is very easy and left to the reader.

A consequence of this is that it is quite natural to consider quadratic fields together with their orders, since their discriminants form a sequence which is almost a union of two arithmetic progressions. It is however necessary to separate the positive from the negative discriminants, and for positive discriminants we should add the squares to make everything uniform. This corresponds to considering the sub-orders of the étale algebra $\mathbb{Q} \times \mathbb{Q}$ (which is not a field) as well. We will see applications of these ideas later in this chapter.

To end this section, note that Theorem 4.8.13 immediately shows how prime numbers decompose in a quadratic field:

Proposition 5.1.4. *Let $K = \mathbb{Q}(\sqrt{D})$ where as usual $D = d(K)$, $\mathbb{Z}_K = \mathbb{Z}[\omega]$ where $\omega = (D + \sqrt{D})/2$ its ring of integers, and let p be a prime number. Then*

(1) *If $p \mid D$, i.e. if $\left(\frac{D}{p}\right) = 0$, then p is ramified, and we have $p\mathbb{Z}_K = \mathfrak{p}^2$, where*

$$\mathfrak{p} = p\mathbb{Z}_K + \omega\mathbb{Z}_K$$

except when $p = 2$ and $D \equiv 12 \pmod{16}$. In this case, $\mathfrak{p} = p\mathbb{Z}_K + (1 + \omega)\mathbb{Z}_K$.

(2) *If $\left(\frac{D}{p}\right) = -1$, then p is inert, hence $\mathfrak{p} = p\mathbb{Z}_K$ is a prime ideal.*

(3) *If $\left(\frac{D}{p}\right) = 1$, then p is split, and we have $p\mathbb{Z}_K = \mathfrak{p}_1 \mathfrak{p}_2$, where*

$$\mathfrak{p}_1 = p\mathbb{Z}_K + (\omega - \frac{D+b}{2})\mathbb{Z}_K \text{ and } \mathfrak{p}_2 = p\mathbb{Z}_K + (\omega - \frac{D-b}{2})\mathbb{Z}_K ,$$

and b is any solution to the congruence $b^2 \equiv D \pmod{4p}$.

Recall that in Section 1.5 we gave an efficient algorithm to compute square roots modulo p. To obtain the number b occurring in (3) above, it is only necessary, when p is an odd prime and the square root obtained is not of the same parity as D, to add p to it. When $p = 2$, one can always take $b = 1$ since $D \equiv 1 \pmod 8$.

5.2 Ideals and Quadratic Forms

Let D be a non-square integer congruent to 0 or 1 modulo 4, R the unique quadratic order of discriminant D, $(1, \omega)$ the standard basis of R (i.e. with $\omega = (D + \sqrt{D})/2$) and K be the unique quadratic field containing R (i.e. the quotient field of R). We denote by σ real or complex conjugation in K, i.e. the $\mathbb{Q}$-linear map sending $\sqrt{D}$ to $-\sqrt{D}$. From the general theory, we have:

Proposition 5.2.1. *Any integral ideal $\mathfrak{a}$ of R has a unique Hermite normal form with denominator equal to 1, and with matrix*

$$\begin{pmatrix} a & b \\ 0 & c \end{pmatrix}$$

with respect to ω, where c divides a and b and $0 \le b < a$. In other words, $\mathfrak{a} = a\mathbb{Z} + (b + c\omega)\mathbb{Z}$. Furthermore, $a = \ell(\mathfrak{a})$ is the smallest positive integer in $\mathfrak{a}$ and $\mathcal{N}(\mathfrak{a}) = ac$.

Definition 5.2.2. *We will say that an integral ideal $\mathfrak{a}$ of R is* primitive *if $c = 1$, in other words if $\mathfrak{a}/n$ is not an integral ideal of R for any integer $n > 1$.*

We also need some definitions about binary quadratic forms.

Definition 5.2.3. *A binary quadratic form f is a function $f(x, y) = ax^2 + bxy + cy^2$ where a, b and c are integers, which is denoted more briefly by (a, b, c). We say that f is* primitive *if $\gcd(a, b, c) = 1$. If f and g are two quadratic forms, we say that f and g are* equivalent *if there exists a matrix $\begin{pmatrix} \alpha & \beta \\ \gamma & \delta \end{pmatrix} \in \mathrm{SL}_2(\mathbb{Z})$ (i.e. an integral matrix of determinant equal to 1), such that $g(x, y) = f(\alpha x + \beta y, \gamma x + \delta y)$.*

It is clear that equivalence preserves the discriminant $D = b^2 - 4ac$ of the quadratic form (in fact it would also be preserved by matrices of determinant equal to -1 but as will be seen, the use of these matrices would lead to the wrong notion of equivalence). One can also easily check that equivalence preserves primitivity. It is also clear that if D is a fundamental discriminant, then any quadratic form of discriminant $D = b^2 - 4ac$ is primitive.

Note that the action of $A \in SL_2(\mathbb{Z})$ is the same as the action of $-A$, hence the natural group which acts on quadratic forms (as well as on complex numbers by linear fractional transformations) is the group $PSL_2(\mathbb{Z})$ where we identify γ and $-\gamma$. By abuse of notation, we will consider an element of $PSL_2(\mathbb{Z})$ as a matrix instead of an equivalence class of matrices.

We will now explain why computing on ideals and on binary quadratic forms is essentially the same. Since certain algorithms are more efficient in the context of quadratic forms, it is important to study this in detail.

As above let D be a non-square integer congruent to 0 or 1 modulo 4 and R be the unique order of discriminant D. We consider the following quotient sets.

$$F = \{(a, b, c), \ b^2 - 4ac = D\}/\Gamma_\infty$$

where $\Gamma_\infty = \left\{ \begin{pmatrix} 1 & m \\ 0 & 1 \end{pmatrix}, \ m \in \mathbb{Z} \right\}$ is a multiplicative group (isomorphic to the additive group of $\mathbb{Z}$) which acts on binary quadratic forms by the formula

$$\begin{pmatrix} 1 & m \\ 0 & 1 \end{pmatrix} \cdot (a, b, c) = (a, b + 2am, c + bm + am^2)$$

which is induced by the action of $SL_2(\mathbb{Z})$.

The second set is

$$I = \{\mathfrak{a} \text{ fractional ideal of } R\}/\mathbb{Q}^*$$

where $\mathbb{Q}^*$ is understood to act multiplicatively on fractional ideals.

The third set is

$$Q = \left\{ \tau = \frac{-b + \sqrt{D}}{2a}, \ a > 0 \text{ and } 4a \mid (D - b^2) \right\}/\mathbb{Z} \ ,$$

where $\mathbb{Z}$ is understood to act additively on quadratic numbers τ. We also define maps as follows. If (a, b, c) is a quadratic form, we set

$$\phi_{FI}(a, b, c) = \left(a\mathbb{Z} + \frac{-b + \sqrt{D}}{2}\mathbb{Z}, \text{sign}(a) \right) \ .$$

If $\mathfrak{a}$ is a fractional ideal and $s = \pm 1$, choose a $\mathbb{Z}$-basis (ω_1, ω_2) of $\mathfrak{a}$ with $\omega_1 \in \mathbb{Q}$ and $(\omega_2 \sigma(\omega_1) - \omega_1 \sigma(\omega_2))/\sqrt{D} > 0$ (this is possible by Proposition 5.2.1), and set

$$\phi_{IF}(\mathfrak{a}, s) = s\frac{\mathcal{N}(x\omega_1 - sy\omega_2)}{\mathcal{N}(\mathfrak{a})} \ .$$

If $\mathfrak{a}$ is a fractional ideal, choose a $\mathbb{Z}$ basis (ω_1, ω_2) as above, and set

$$\phi_{IQ}(\mathfrak{a}) = \frac{\omega_2}{\omega_1} \ .$$

Finally, if $\tau = (-b + \sqrt{D})/2a$ is a quadratic number, set

$$\phi_{QI}(\tau) = a(\mathbb{Z} + \tau\mathbb{Z}) \ .$$

The following theorem, while completely elementary, is fundamental to understanding the relationships between quadratic forms, ideals and quadratic numbers. We always identify the group $\mathbb{Z}/2\mathbb{Z}$ with ± 1.

Theorem 5.2.4. *With the above notations, the maps that we have given can be defined at the level of the equivalence classes defining F, I and Q, and are then set isomorphisms (which we denote in the same way). In other words, we have the following isomorphisms:*

$$F \simeq I \times \mathbb{Z}/2\mathbb{Z} \ , \quad I \simeq Q \ , \quad F \simeq Q \times \mathbb{Z}/2\mathbb{Z} \ .$$

Proof. The proof is a simple but tedious verification that everything works. We comment only on the parts which are not entirely trivial.

(1) ϕ_{FI} sends a quadratic form to an ideal. Indeed, if a and b are integers with $b \equiv D \pmod 2$, the $\mathbb{Z}$-module $a\mathbb{Z} + ((-b + \sqrt{D})/2)\mathbb{Z}$ is an ideal if and only if $4a \mid (b^2 - D)$.

(2) ϕ_{FI} depends only on the equivalence class modulo Γ_∞ hence induces a map from F to I.

(3) ϕ_{IF} sends a pair (a, s) to an integral quadratic form. Indeed, by homogeneity, if we multiply a by a suitable element of $\mathbb{Q}$, we may assume that a is a primitive integral ideal. If $\omega_1 < 0$, we can also change (ω_1, ω_2) into $(-\omega_1, -\omega_2)$. In that case, by Proposition 5.2.1 (or directly), we have $\mathcal{N}(a) = \omega_1$ and $\omega_2 - \sigma(\omega_2) = \sqrt{D}$. Finally, since a is an integral ideal, $\omega_1 \mid \omega_2\sigma(\omega_2)$, and a simple calculation shows that we obtain an integral binary quadratic form of discriminant D.

(4) ϕ_{IF} does not depend on the equivalence class of a, nor on the choice of ω_1 and ω_2. Indeed, if ω_1 is given, then ω_2 is defined modulo ω_1, and this corresponds precisely to the action of Γ_∞ on quadratic forms.

(5) ϕ_{IF} and ϕ_{FI} are inverse maps. This is left to the reader, and is the only place where we must really use the $\mathrm{sign}(a)$ component.

(6) I also leave to the reader the easy proof that ϕ_{IQ} and ϕ_{QI} are well defined and are inverse maps.

$\square$

We now need to identify precisely the invertible ideals in R so as to be able to work in the class group.

Proposition 5.2.5. *Let $a = a\mathbb{Z} + ((-b + \sqrt{D})/2)\mathbb{Z}$ be an ideal of R, and let (a, b, c) be the corresponding quadratic form. Then a is invertible in R if and only if (a, b, c) is primitive. In that case, we have $a^{-1} = \mathbb{Z} + ((b + \sqrt{D})/(2a))\mathbb{Z}$.*

Proof. From Lemma 4.6.7 we know that $\mathfrak{a}$ is invertible if and only if $\mathfrak{a}\mathfrak{b} = R$ where $\mathfrak{b} = \{z \in K,\ z\mathfrak{a} \subset R\}$. Writing $\mathfrak{a} = a\mathbb{Z} + ((-b+\sqrt{D})/2)\mathbb{Z}$, from $a \in \mathfrak{a}$ we see that such a z must be the form $z = (x+y\sqrt{D})/(2a)$ with x and y in $\mathbb{Z}$ such that $x \equiv yD \pmod 2$. From $(-b + \sqrt{D})/2 \in \mathfrak{a}$, we obtain the congruences $bx \equiv Dy \pmod{2a}$, $x \equiv by \pmod{2a}$ and $(Dy - bx)/(2a) \equiv D(x - by)/(2a)$ $\pmod 2$. An immediate calculation gives us $\mathfrak{b} = \mathbb{Z} + ((b + \sqrt{D})/(2a))\mathbb{Z}$ as claimed.

Now the $\mathbb{Z}$-module $\mathfrak{a}\mathfrak{b}$ is generated by the four products of the generators, i.e. by a, $(b + \sqrt{D})/2$, $(-b + \sqrt{D})/2$ and $-c$. We obtain immediately

$$\mathfrak{a}\mathfrak{b} = \gcd(a,b,c)\mathbb{Z} + \frac{-b+\sqrt{D}}{2}\mathbb{Z}.$$

hence this is equal to $R = \mathbb{Z} + ((-b+\sqrt{D})/2)\mathbb{Z}$ if and only if $\gcd(a,b,c) = 1$, thus proving the proposition. $\qquad\square$

Corollary 5.2.6. *Denote by F_0 the subset of classes of primitive forms in F, I_0 the subset of classes of invertible ideals in I and Q_0 the subset of classes of primitive quadratic numbers in Q (where $\tau \in Q$ is said to be primitive if $(a, b, c) = 1$ where a, b and c are as in the definition of Q). Then the maps ϕ_{FI} and ϕ_{IQ} also give isomorphisms:*

$$F_0 \simeq I_0 \times \mathbb{Z}/2\mathbb{Z}\ ,\quad I_0 \simeq Q_0\ ,\quad F_0 \simeq Q_0 \times \mathbb{Z}/2\mathbb{Z}\ .$$

Theorem 5.2.4 gives set isomorphisms between ideals and quadratic forms at the level of equivalence classes of quadratic forms modulo Γ_∞. As we shall see, this will be useful in the real quadratic case. When considering the class group however, we need the corresponding theorem at the level of equivalence classes of quadratic forms modulo the action of the whole group $\mathrm{PSL}_2(\mathbb{Z})$. Since we must restrict to invertible ideals in order to define the class group, the above proposition shows that we will have to consider only primitive quadratic forms.

Here, it is slightly simpler to separate the case $D < 0$ from the case $D > 0$. We begin by defining the sets with which we will work.

Definition 5.2.7. *Let D be a non-square integer congruent to 0 or 1 modulo 4, and R the unique quadratic order of discriminant D.*

(1) *We will denote by $\mathcal{F}(D)$ the set of equivalence classes of primitive quadratic forms of discriminant D modulo the action of $\mathrm{PSL}_2(\mathbb{Z})$, and in the case $D < 0$, $\mathcal{F}^+(D)$ will denote those elements of $\mathcal{F}(D)$ represented by a positive definite quadratic form (i.e. a form (a, b, c) with $a > 0$).*

(2) *We will denote by $Cl(D)$ the class group of R, and in the case $D > 0$, $Cl^+(D)$ will denote the narrow class group of R, i.e. the group of equivalence classes of R-ideals modulo the group $\mathcal{P}^+$ of principal ideals generated by an element of positive norm.*

(3) *Finally, we will set $h(D) = |Cl(D)|$ and $h^+(D) = |Cl^+(D)|$.*

We then have the following theorems.

Theorem 5.2.8. *Let D be a negative integer congruent to 0 or 1 modulo 4. The maps*

$$\psi_{FI}(a,b,c) = a\mathbb{Z} + \frac{-b+\sqrt{D}}{2}\mathbb{Z} \ ,$$

and

$$\psi_{IF}(\mathfrak{a}) = \frac{\mathcal{N}(x\omega_1 - y\omega_2)}{\mathcal{N}(\mathfrak{a})}$$

where $\mathfrak{a} = \omega_1\mathbb{Z} + \omega_2\mathbb{Z}$ with

$$\frac{\omega_2\sigma(\omega_1) - \omega_1\sigma(\omega_2)}{\sqrt{D}} > 0$$

induce inverse bijections from $\mathcal{F}^+(D)$ to $Cl(D)$.

Theorem 5.2.9. *Let D be a non-square positive integer congruent to 0 or 1 modulo 4. The maps*

$$\psi_{FI}(a,b,c) = \left(a\mathbb{Z} + \frac{-b+\sqrt{D}}{2}\mathbb{Z}\right)\alpha \ ,$$

where α is any element of K^* such that $\operatorname{sign}(\mathcal{N}(\alpha)) = \operatorname{sign}(a)$, and

$$\psi_{IF}(\mathfrak{a}) = \frac{\mathcal{N}(x\omega_1 - y\omega_2)}{\mathcal{N}(\mathfrak{a})}$$

where $\mathfrak{a} = \omega_1\mathbb{Z} + \omega_2\mathbb{Z}$ with

$$\frac{\omega_2\sigma(\omega_1) - \omega_1\sigma(\omega_2)}{\sqrt{D}} > 0$$

induce inverse bijections from $\mathcal{F}(D)$ to $Cl^+(D)$.

Proof. As for Theorem 5.2.4, the proofs consist of a series of simple verifications.

(1) The map ψ_{FI} is well defined on classes modulo $PSL_2(\mathbb{Z})$. If $\begin{pmatrix} A & B \\ U & V \end{pmatrix} \in$ $PSL_2(\mathbb{Z})$ acts on (a,b,c), then the quantity $\tau = (-b+\sqrt{D})/(2a)$ becomes $\tau' = (V\tau - B)/(-U\tau + A)$, and a becomes $a\mathcal{N}(-U\tau + A)$, hence since $\mathbb{Z} + \tau'\mathbb{Z} = (\mathbb{Z} + \tau\mathbb{Z})/(-U\tau + A)$, it follows immediately that ψ_{FI} is well defined.

(2) Similarly, ψ_{IF} is well defined, and we can check that it gives an integral quadratic form of discriminant D as for the map ϕ_{IF} of Theorem 5.2.4. This form is primitive since we restrict to invertible ideals.

(3) Finally, the same verification as in the preceding theorem shows that ψ_{IF} and ψ_{FI} are inverse maps.

$\square$

Remarks.

(1) Although we have given the bijections between classes of forms and ideals, we could, as in Theorem 5.2.4, give bijections with classes of quadratic numbers modulo the action of $PSL_2(\mathbb{Z})$. This is left to the reader (Exercise 3).

(2) In the case $D < 0$, a quadratic form is either positive definite or negative definite, hence the set F breaks up naturally into two disjoint pieces. The map ψ_{FI} is induced by the restriction of ϕ_{FI} to the positive piece, and ψ_{IF} is induced by ϕ_{IF} and forgetting the factor $\mathbb{Z}/2\mathbb{Z}$.

(3) In the case $D > 0$, there is no such natural breaking up of F. In this case, the maps ϕ_{FI} and ϕ_{IF} induce inverse isomorphisms between $\mathcal{F}(D)$ and

$$\mathcal{I}(D) = (\mathcal{I} \times \mathbb{Z}/2\mathbb{Z})/\tilde{\mathcal{P}} \ ,$$

where $\tilde{\mathcal{P}}$ is the quotient of K^* by the subgroup of units of positive norm, and $\beta \in \tilde{\mathcal{P}}$ acts by sending $(\mathfrak{a}, s)$ to $(\beta\mathfrak{a}, s \cdot \text{sign}(\mathcal{N}(\beta)))$. (Note also the exact sequence

$$1 \longrightarrow \mathcal{P}^+ \longrightarrow \tilde{\mathcal{P}} \longrightarrow \mathbb{Z}/2\mathbb{Z} \longrightarrow 1 \ ,$$

where the map to $\mathbb{Z}/2\mathbb{Z}$ is induced by the sign of the norm map.) The maps ψ_{FI} and ψ_{IF} are obtained by composition of the above isomorphisms with the isomorphisms between $\mathcal{I}(D)$ and $Cl^+(D)$ given as follows. The class of $(\mathfrak{a}, s)$ representing an element of $\mathcal{I}(D)$ is sent to the class of $\beta\mathfrak{a}$ in $Cl^+(D)$, where $\beta \in K^*$ is any element such that $\text{sign}(\mathcal{N}(\beta)) = s$. Conversely, the class of $\mathfrak{a} \in Cl^+(D)$ is sent to the class of $(\mathfrak{a}, 1)$ in $\mathcal{I}(D)$.

Although F, I and Q are defined as quotient sets, it is often useful to use precise representatives of classes in these sets. We have already implicitly done so when we defined all the maps ϕ_{IF} etc ... above, but we make our choice explicit.

An element of F will be represented by the unique element (a, b, c) in its class chosen as follows. If $D < 0$, then $-|a| < b \le |a|$. If $D > 0$, then $-|a| < b \le |a|$ if $a > \sqrt{D}$, $\sqrt{D} - 2|a| < b < \sqrt{D}$ if $a < \sqrt{D}$.

An element of I will be represented by the unique primitive integral ideal in its class.

An element of Q will be represented by the unique element τ in its class such that $-1 < \tau + \sigma(\tau) \le 1$, where σ denotes (complex or real) conjugation in K.

The tasks that remain before us are that of computing the class group or class number, and in the real case, that of computing the fundamental unit. It is now time to separate the two cases, and in the next sections we shall examine in detail the case of imaginary quadratic fields.

5.3 Class Numbers of Imaginary Quadratic Fields

Until further notice, all fields which we consider will be imaginary quadratic fields. First, let us solve the problem of units. From the general theory, we know that the units of an imaginary quadratic field are the (finitely many) roots of unity inside the field. An easy exercise is to show the following:

Proposition 5.3.1. *Let $D < 0$ congruent to 0 or 1 modulo 4. Then the group $\mu(R)$ of units of the unique quadratic order of discriminant D is equal to the group of $w(D)^{th}$ roots of unity, where*

$$w(D) = \begin{cases} 2, & \text{if } D < -4 \\ 4, & \text{if } D = -4 \\ 6, & \text{if } D = -3 \end{cases}.$$

Let us now consider the problem of computing the class group. For this, the correspondences that we have established above between classes of quadratic forms and ideal class groups will be very useful. Usually, the ideals will be used for conceptual (as opposed to computational) proofs, and quadratic forms will be used for practical computation.

Thanks to Theorem 5.2.8, we will use interchangeably the language of ideal classes or of classes of quadratic forms. One of the advantages is that the algorithms are simpler. For example, we now consider a simple but still reasonable method for computing the class *number* of an imaginary quadratic field.

5.3.1 Computing Class Numbers Using Reduced Forms

Definition 5.3.2. *A positive definite quadratic form (a, b, c) of discriminant D is said to be* reduced *if $|b| \le a \le c$ and if, in addition, when one of the two inequalities is an equality (i.e. either $|b| = a$ or $a = c$), then $b \ge 0$.*

This definition is equivalent to saying that the number $\tau = (-b + \sqrt{D})/(2a)$ corresponding to (a, b, c) as above is in the standard fundamental domain $\mathcal{D}$ of $\mathcal{H}/\operatorname{PSL}_2(\mathbb{Z})$ (where $\mathcal{H} = \{\tau \in \mathbb{C}, \operatorname{Im}(\tau) > 0\}$), defined by

$$\mathcal{D} = \left\{ \tau \in \mathcal{H}, \operatorname{Re}(\tau) \in [-\tfrac{1}{2}, \tfrac{1}{2}[, |\tau| > 1 \text{ or } |\tau| = 1 \text{ and } \operatorname{Re}(\tau) \le 0 \right\}.$$

The nice thing about this notion is the following:

Proposition 5.3.3. *In every class of positive definite quadratic forms of discriminant $D < 0$ there exists exactly one reduced form. In particular $h(D)$ is equal to the number of primitive reduced forms of discriminant D.*

An equivalent form of this proposition is that the set $\mathcal{D}$ defined above is a fundamental domain for $\mathcal{H}/\operatorname{PSL}_2(\mathbb{Z})$.

Proof. Among all forms (a, b, c) in a given class, consider one for which a is minimal. Note that for any such form we have $c \geq a$ since (a, b, c) is equivalent to $(c, -b, a)$ (change (x, y) into $(-y, x)$). Changing (x, y) into $(x + ky, y)$ for a suitable integer k (precisely for $k = \lfloor (a - b)/(2a) \rfloor$) will not change a and put b in the interval $]-a, a]$. Since a is minimal, we will still have $a \leq c$, hence the form that we have obtained is essentially reduced. If $c = a$, changing (a, b, c) again in $(c, -b, a)$ sets $b \geq 0$ as required. This shows that in every class there exists a reduced form.

Let us show the converse. If (a, b, c) is reduced, I claim that a is minimal among all the forms equivalent to (a, b, c). Indeed, every other a' has the form $a' = am^2 + bmn + cn^2$ with m and n coprime integers, and the identities

$$am^2 + bmn + cn^2 = am^2\left(1 + \frac{b}{a}\frac{n}{m}\right) + cn^2 = am^2 + cn^2\left(1 + \frac{b}{c}\frac{m}{n}\right)$$

immediately imply our claim, since $|b| \leq a \leq c$. Now in fact these same identities show that the only forms equivalent to (a, b, c) with $a' = a$ are obtained by changing (x, y) into $(x + ky, y)$ (corresponding to $m = 1$ and $n = 0$), and this finishes the proof of the proposition.   □

We also have the following lemma.

Lemma 5.3.4. *Let $f = (a, b, c)$ be a positive definite binary quadratic form of discriminant $D = b^2 - 4ac < 0$.*

(1) *If f is reduced, we have the inequality*

$$a \leq \sqrt{|D|/3} \ .$$

(2) *Conversely, if*

$$a < \sqrt{|D|/4} \quad and \quad -a < b \leq a$$

then f is reduced.

Proof. For (1) we note that if f is reduced then $|D| = 4ac - b^2 \geq 4a^2 - a^2$ hence $a \leq \sqrt{|D|/3}$. For (2), we have $c = (b^2 + |D|)/(4a) \geq |D|/(4a) > a^2/a = a$, therefore f is reduced.   □

As a consequence, we deduce that when $D < 0$ the class number $h(D)$ of $\mathbb{Q}(\sqrt{D})$ can be obtained simply by counting reduced forms of discriminant D (since in that case all forms of discriminant D are primitive), using the inequalities $|b| \leq a \leq \sqrt{|D|/3}$. This leads to the following algorithm.

Algorithm 5.3.5 ($h(D)$ Counting Reduced Forms). Given a negative discriminant D, this algorithm outputs the class number of quadratic forms of discriminant D, i.e. $h(D)$ when D is a fundamental discriminant.

1. [Initialize b] Set $h \leftarrow 1$, $b \leftarrow D \bmod 2$ (i.e. 0 if $D \equiv 0 \pmod 4$, 1 if $D \equiv 1$ $\pmod 4$)), $B \leftarrow \left\lfloor \sqrt{|D|/3} \right\rfloor$.

2. [Initialize a] Set $q \leftarrow (b^2 - D)/4$, $a \leftarrow b$, and if $a \leq 1$ set $a \leftarrow 1$ and go to step 4.

3. [Test] If $a \mid q$ then if $a = b$ or $a^2 = q$ or $b = 0$ set $h \leftarrow h + 1$, otherwise (still in the case $a \mid q$) set $h \leftarrow h + 2$.

4. [Loop on a] Set $a \leftarrow a + 1$. If $a^2 \leq q$ go to step 3.

5. [Loop on b] Set $b \leftarrow b + 2$. If $b \leq B$ go to step 2, otherwise output h and terminate the algorithm.

It can easily be shown that this algorithm indeed counts reduced forms. One must be careful in the formulation of this algorithm since the extra boundary conditions which occur if $|b| = a$ or $a = c$ complicate things. It is also easy to give some cosmetic improvements to the above algorithm, but these have little effect on its efficiency.

The running time of this algorithm is clearly $O(|D|)$, but the O constant is very small since very few computations are involved. Hence it is quite a reasonable algorithm to use for discriminants up to a few million in absolute value. The typical running time for a discriminant of the order of 10^6 is at most a few seconds on modern microcomputers.

Remark. If we want to compute $h(D)$ for a non-fundamental discriminant D, we must only count primitive forms. Therefore the above algorithm must be modified by replacing the condition "if $a \mid q$" of Step 3 by "if $a \mid q$ and $\gcd(a, b, q/a) = 1$".

A better method is as follows. Write $D = D_0 f^2$ where D_0 is a fundamental discriminant. The general theory seen in Chapter 4 tells us that $h(D)$ is a multiple of $h(D_0)$, but in fact Proposition 5.3.12 implies the following precise formula:

$$\frac{h(D)}{w(D)} = \frac{h(D_0)}{w(D_0)} f \prod_{p \mid f} \left(1 - \frac{\left(\frac{D_0}{p}\right)}{p} \right).$$

Hence, we compute $h(D_0)$ using the above algorithm, and deduce $h(D)$ from this formula.

Reduced forms are also very useful for making *tables* of class numbers of quadratic fields or forms up to a certain discriminant bound. Although each individual computation takes time $O(|D|)$, hence for $|D| \leq M$ the time would be $O(M^2)$, it is easy to see that a simultaneous computation (needing of course $O(M)$ memory locations to hold the class numbers) takes only $O(M^{3/2})$, hence an average of $O(|D|^{1/2})$ per class number.

Since class numbers of imaginary quadratic fields occur so frequently, it is useful to have a small table available. Such a table can be found in Appendix B. Some selected values are:

- Class number 1 occurs only for $D = -3, -4, -7, -8, -11, -19, -43,$ -67 and -163.
- Class number 2 occurs only for $D = -15, -20, -24, -35, -40, -51,$ $-52, -88, -91, -115, -123, -148, -187, -232, -235, -267, -403, -427$.
- Class number 3 occurs only for $D = -23, -31, -59, -83, -107, -139,$ $-211, -283, -307, -331, -379, -499, -547, -643, -883, -907$.
- Class number 4 occurs for $D = -39, -55, -56, -68, \ldots, -1555$.
- Class number 5 occurs for $D = -47, -79, -103, -127, \ldots, -2683$.
- Class number 6 occurs for $D = -87, -104, -116, -152, \ldots, -3763$.
- Class number 7 occurs for $D = -71, -151, -223, -251, \ldots, -5923$.

etc ...

Note that the first two statements concerning class numbers 1 and 2 are very difficult theorems proved in 1952 by Heegner and in 1968-1970 by Stark and Baker (see [Cox]). The general problem of determining all imaginary quadratic fields with a given class number has been solved in principle by Goldfeld-Gross-Zagier ([Gol], [Gro-Zag2]), but the explicit computations have been carried to the end only for class numbers up to 7 and all odd numbers up to 23 (see [ARW], [Wag]).

The method using reduced forms is a very simple method to implement and is eminently suitable for computing tables of class numbers or for computing class numbers of reasonable discriminant, say less than a few million in absolute value. Since it is only a simple counting process, it does not give the structure of the class group. Also, it becomes too slow for larger discriminants, therefore we must find better methods.

5.3.2 Computing Class Numbers Using Modular Forms

I do not intend to explain why the theory of modular forms (specifically of weight 3/2 and weight 2) is closely related to class numbers of imaginary quadratic fields, but I would like to mention formulas which enable us to compute tables of class numbers essentially as fast as the method using reduced forms. First we need a definition.

Definition 5.3.6. *Let N be a non-negative integer. The* Hurwitz *class number $H(N)$ is defined as follows.*

(1) *If $N \equiv 1$ or $2 \pmod 4$ then $H(N) = 0$.*
(2) *If $N = 0$ then $H(0) = -1/12$.*
(3) *Otherwise (i.e. if $N \equiv 0$ or $3 \pmod 4$ and $N > 0$) we define $H(N)$ as the class number of not necessarily primitive (positive definite) quadratic forms of discriminant $-N$, except that forms equivalent to $a(x^2 + y^2)$ should be counted with coefficient $1/2$, and those equivalent to $a(x^2 + xy + y^2)$ with coefficient $1/3$.*

Let us denote by $h(D)$ the class number of *primitive* positive definite quadratic forms of discriminant D. (This agrees with the preceding definition when D is a fundamental discriminant since in that case every form is primitive.) Next, we define $h(D) = 0$ when D is not congruent to 0 or 1 modulo 4. Then we have the following lemma.

Lemma 5.3.7. *Let $w(D)$ be the number of roots of unity in the quadratic order of discriminant D, hence $w(-3) = 6$, $w(-4) = 4$ and $w(D) = 2$ for $D < -4$, and set $h'(D) = h(D)/(w(D)/2)$ (hence $h'(D) = h(D)$ for $D < -4$). Then for $N > 0$ we have*

(1)
$$H(N) = \sum_{d^2 | N} h'(-N/d^2)$$

and in particular if $-N$ is a fundamental discriminant, we have $H(N) = h(-N)$ except in the special cases $N = 3$ ($H(3) = 1/3$ and $h(-3) = 1$) and $N = 4$ ($H(4) = 1/2$ and $h(-4) = 1$).
(2) *Conversely, we have*

$$h'(-N) = \sum_{d^2 | N} \mu(d) H(N/d^2)$$

where $\mu(d)$ is the Möbius function defined by $\mu(d) = (-1)^k$ if d is equal to a product of k distinct primes (including $k = 0$), and $\mu(d) = 0$ otherwise.

Proof. The first formula follows immediately from the definition of $H(N)$. The second formula is a direct consequence of the Möbius inversion formula (see [H-W]). □

From this lemma, it follows that the computation of a table of the function $H(N)$ is essentially equivalent to the computation of a table of the function $h(D)$.

For $D = -N$, Algorithm 5.3.5 computes a quantity similar to $H(N)$ but without the denominator $w(-N/d^2)/2$ in the formula given above. Hence, it can be readily adapted to compute $H(N)$ itself by replacing step 3 with the following:

3'. [Test] If $a \nmid q$ go to step 4. Now if $a = b$ then if $ab = q$ set $h \leftarrow h + 1/3$ otherwise set $h \leftarrow h + 1$ and go to step 4. If $a^2 = q$, then if $b = 0$ set $h \leftarrow h + 1/2$, otherwise set $h \leftarrow h + 1$. In all other cases (i.e. if $a \neq b$ and $a^2 \neq q$) set $h \leftarrow h + 2$.

The theory of modular forms of weight 3/2 tells us that the Fourier series

$$\sum_{N=0}^{\infty} H(N) e^{2i\pi N\tau}$$

has a special behavior when one changes τ by a linear fractional transformation $\tau \mapsto \frac{a\tau+b}{c\tau+d}$ in $PSL_2(\mathbb{Z})$. Combined with other results, this gives many nice recursion formulas for $H(N)$ which are very useful for practical computation.

Let $\sigma(n) = \sum_{d|n} d$ be the sum of divisors function, and define

$$\lambda(n) = \frac{1}{2} \sum_{d|n} \min(d, n/d) = {\sum_{d|n, d \leq \sqrt{n}}}' d \ ,$$

where $\sum'$ means that if the term $d = \sqrt{n}$ is present it should have coefficient $1/2$. In addition we define $\sigma(n) = \lambda(n) = 0$ if n is not integral. Then (see [Eic2], [Zag1]):

Theorem 5.3.8 (Hurwitz, Eichler). *We have the following relations, where it is understood that the summation variable s takes positive, zero or negative values:*

$$\sum_{s^2 \leq 4N} H(4N - s^2) = 2\sigma(N) - 2\lambda(N) \ ,$$

and if N is odd,

$$\sum_{s^2 \leq N, s \equiv (N+1)/2 \ (\text{mod} \ 2)} H(N - s^2) = \frac{\sigma(N)}{3} - \lambda(N) \ .$$

From a computational point of view, the second formula is better. It is used in the following way:

Corollary 5.3.9. *If $N \equiv 3$ (mod 4), then*

$$H(N) = \frac{\sigma(N)}{3} - \lambda(N) - 2 \sum_{1 \leq s < \sqrt{N/4}} H(N - 4s^2) \ ,$$

and if $N \equiv 0$ (mod 4), then

$$H(N) = \frac{\sigma(N+1)}{6} - \frac{\lambda(N+1)}{2} - \sum_{1 \leq s \leq (\sqrt{N+1}-1)/2} H(N - 4s(s+1)) \ .$$

This corollary allows us to compute a table of class numbers up to any given bound M in time $O(M^{3/2})$, hence is comparable to the method using reduced forms. It is slightly simpler to implement, but has the disadvantage that individual class numbers cannot be computed without knowing the preceding ones. It has an advantage, however, in that the computation of a block of class numbers can be done simply using the table of the lower ones, while this cannot be done with the reduced forms technique, at least without wasting a lot of time.

Remark. The above theorem is similar to Theorem 4.9.14 and can be proved similarly. While $\zeta_K(-1)$ is closely linked to $r_5(D)$ when $D > 0$, $\zeta_K(0)$ (or essentially $h(D)$) is closely linked to $r_3(-D)$ when $D < 0$. More precisely we have (see [Coh2]):

Proposition 5.3.10. *Let $D < -4$ be the discriminant of an imaginary quadratic field K. Then the number $r_3(|D|)$ of representations of $|D|$ as a sum of 3 squares of elements of $\mathbb{Z}$ (counting representations with a different ordering as distinct) is given by*

$$r_3(|D|) = -24\left(1 - \left(\frac{D}{2}\right)\right)\zeta_K(0) = 12\left(1 - \left(\frac{D}{2}\right)\right)h(D) .$$

(This formula must be slightly modified if D is not the discriminant of an imaginary quadratic field, see [Coh2].)

5.3.3 Computing Class Numbers Using Analytic Formulas

It would carry us too far afield to enter into the details of the analytic theory of L-functions, hence we just recall a few definitions and results.

Proposition 5.3.11 (Dirichlet). *Let D be a negative fundamental discriminant, and define*

$$L_D(s) = \sum_{n \geq 1}\left(\frac{D}{n}\right)n^{-s} .$$

This series converges for $\mathrm{Re}(s) > 1$, and defines an analytic function which can be analytically continued to the whole complex plane to an entire function satisfying

$$\Lambda_D(1 - s) = \Lambda_D(s) ,$$

where we have set

$$\Lambda_D(s) = |D/\pi|^{(s+1)/2}\Gamma((s + 1)/2)L_D(s) .$$

The link with class numbers is the following result also due to Dirichlet:

Proposition 5.3.12. *If D is a negative discriminant (not necessarily fundamental), then*

$$L_D(1) = \frac{2\pi h(D)}{w(D)\sqrt{|D|}}$$

and in particular $L_D(1) = \pi h(D)/\sqrt{|D|}$ if $D < -4$.

Note that these results are special cases of Theorem 4.9.12 since it immediately follows from Proposition 5.1.4 that if $K = \mathbb{Q}(\sqrt{D})$, then

$$\zeta_K(s) = \zeta(s)L_D(s) \ .$$

The series $L_D(1)$ is only conditionally convergent, hence it is not very reasonable to compute $L_D(1)$ directly using Dirichlet's theorem. A suitable transformation of the series however gives the following:

Corollary 5.3.13. *If $D < -4$ is a fundamental discriminant, then*

$$h(D) = \frac{1}{D} \sum_{1 \le r < |D|} r\left(\frac{D}{r}\right) = \frac{1}{2 - \left(\frac{D}{2}\right)} \sum_{1 \le r < |D|/2} \left(\frac{D}{r}\right) \ .$$

This formula is aesthetically very pleasing, and it can be transformed into even simpler expressions. It is unfortunately totally useless from a computational point of view since one must compute D terms each involving the computation (admittedly rather short) of a Kronecker symbol. Hence, the execution time would be $O(|D|^{1+\epsilon})$, worse than the preceding methods.

A considerable improvement can be obtained if we also use the functional equation. This leads to a formula which is less pleasing, but which is much more efficient:

Proposition 5.3.14. *Let $D < -4$ be a fundamental discriminant. Then*

$$h(D) = \sum_{n \ge 1} \left(\frac{D}{n}\right) \left(\operatorname{erfc}\left(n\sqrt{\frac{\pi}{|D|}}\right) + \frac{\sqrt{|D|}}{\pi n} e^{-\pi n^2/|D|}\right) \ ,$$

where

$$\operatorname{erfc}(x) = \frac{2}{\sqrt{\pi}} \int_x^\infty e^{-t^2}\, dt$$

is the complementary error function.

Note that the function $\operatorname{erfc}(x)$ can be computed efficiently using the following formulas.

Proposition 5.3.15.

(1) *We have for all x*

$$\operatorname{erfc}(x) = 1 - \frac{2}{\sqrt{\pi}} \sum_{k \ge 0} (-1)^k \frac{x^{2k+1}}{k!(2k+1)} \ ,$$

and this should be used when x is small, say $x \le 2$.

(2) *We have for all $x > 0$*

$$\text{erfc}(x) = \frac{e^{-x^2}}{x\sqrt{\pi}} \left(1 - \cfrac{1/2}{2 + X - \cfrac{1 \cdot 3/2}{4 + X - \cfrac{2 \cdot 5/2}{6 + X - \cdots}}} \right),$$

where $X = x^2 - 1/2$, and this should be used for x large, say $x \geq 2$.

Implementation Remark. When implementing these formulas it is easy to make a mistake in the computation of $\text{erfc}(x)$, and tables of this function are not always at hand. One good check is of course that the value found for $h(D)$ must be close to an integer, and for small D equal to the values found by the slower methods. Another check is that, although we have given the most rapidly convergent series for $h(D)$ which can be obtained from the functional equation, we can get a one parameter family of formulas:

$$h(D) = \sum_{n \geq 1} \left(\frac{D}{n} \right) \left(\text{erfc}\left(n\sqrt{\frac{\pi}{A|D|}} \right) + \frac{\sqrt{|D|}}{\pi n} e^{-\pi n^2 A/|D|} \right).$$

The sum of the series must be independent of $A > 0$.

The above results show that the series given in Proposition 5.3.14 for $h(D)$ converges exponentially, and since $h(D)$ is an integer it is clear that the computation time of $h(D)$ by this method is $O(|D|^{1/2+\epsilon})$ for any $\epsilon > 0$, however with a large O constant. In fact it is not difficult to show the following precise result:

Corollary 5.3.16. *With the same notations as in Proposition 5.3.14, $h(D)$ is the closest integer to the n-th partial sum of the series of Proposition 5.3.14 for $h(D)$, where $n = \left\lfloor \sqrt{|D| \ln |D|/(2\pi)} \right\rfloor$.*

Hence, we see that this method is considerably faster than the two preceding methods, at least for sufficiently large discriminants. In addition, it is possible to avoid completely the computation of the higher transcendental function erfc, and this makes the method even more attractive (See Exercise 28).

It is reasonable to compute class numbers of discriminants having 12 to 15 digits by this method, but not much more. We must therefore find still better methods. In addition, we still have not given any method for computing the class *group*.

5.4 Class Groups of Imaginary Quadratic Fields

It was noticed by Shanks in 1968 that if one tries to obtain the class group structure and not only the class number, this leads to an algorithm which is much faster than the preceding algorithms, in time $O(|D|^{1/4+\epsilon})$ or even $O(|D|^{1/5+\epsilon})$ if the Generalized Riemann Hypothesis is true, for any $\epsilon > 0$. Hence not only does one get much more information, i.e. the whole group structure, but even if one is interested only in the class number, this is a much better method.

Before entering into the details of the algorithm, we will describe a method introduced (for this purpose) by Shanks and which is very useful in many group-theoretic and similar contexts.

5.4.1 Shanks's Baby Step Giant Step Method

We first explain the general idea. Let G be a finite Abelian group and g an element of G. We want to compute the order of g in G, i.e. the smallest positive integer n such that $g^n = 1$, where we denote by 1 the identity element of G. One way of doing this is simply to compute g, g^2, g^3, $\ldots$, until one gets 1. This clearly takes $O(n)$ group operations. In certain cases, it is impossible to do much better. In most cases however, one knows an upper bound, say B on the number n, and in that case one can do much better, using Shanks's baby-step giant-step strategy. One proceeds as follows. Let $q = \left\lceil \sqrt{B} \right\rceil$. Compute $1, g, \ldots, g^{q-1}$, and set $g_1 = g^{-q}$. Then if the order n of g is written in the form $n = aq + r$ with $0 \le r < q$, by the choice of q we must also have $a \le q$. Hence, for $a = 1, \ldots, q$ we compute g_1^a and check whether or not it is in our list of g^r for $r < q$. If it is, we have $g^{aq+r} = 1$, hence n is a divisor of $aq + r$, and the exact order can easily be obtained by factoring $aq + r$, at least if $aq + r$ is of factorable size (see Chapter 10). This method clearly requires only $O\left(B^{1/2}\right)$ group operations, and this number is much smaller than $O(n)$ if B is a reasonable upper bound.

There is however one pitfall to avoid in this algorithm: we need to search (at most q times) if an element belongs to a list having q elements. If this is done naïvely, this will take $O(q^2) = O(B)$ comparisons, and even if group operations are much slower than comparisons, this will ultimately dominate the running time and render useless the method. To avoid this, we can first *sort* the list of q elements, using a $O(q \ln q)$ sorting method such as heapsort (see [Knu3]). A search in a sorted list will then take only $O(\ln q)$ comparisons, bringing the total time down to $O(q \ln q)$. We can also use hashing techniques (see [Knu3] again).

This simple instance of Shanks's method involves at most q "giant steps" (i.e. multiplication by g_1), each of size q. Extra information on n can be used to improve the efficiency of the algorithm. We give two basic examples. Assume that in addition to an upper bound B, we also know a lower bound C, say, so that $C \le n \le B$. Then, by starting our list with g^C instead of $g^0 = 1$, we can reduce both the maximum number of giant steps and the size of the giant steps (and of the list) to $\lceil \sqrt{B-C} \rceil$.

As a second example, assume that we know that n satisfies some congruence condition $n \equiv n_0 \pmod{b}$. Then it is easily seen that one can reduce the size and number of giant steps to $\lceil \sqrt{B/b} \rceil$.

Shanks's method is usually used not only to find the order of an element of the group G, but the order of the group itself. If g is a generator of G, the preceding algorithm does the trick. In general however this will not be the case, and in addition G may be non-cyclic (although cyclic groups occur much more often than one expects, see Section 5.10). In this case we must use the whole group structure, and not only one cyclic part. To do this, we can use the following algorithm.

Algorithm 5.4.1 (Shanks's Baby-Step Giant-Step Method). Given that one can compute in G, and the inequalities $B/2 < C \leq h \leq B$ on the order h of G, this algorithm finds h. We denote by 1 the identity element of G and by $\cdot$ the product operation in G. The variables S and L will represent subsets of G.

1. [Initialize] Set $h \leftarrow 1$, $C_1 \leftarrow C$, $B_1 \leftarrow B$, $S \leftarrow \{1\}$, $L \leftarrow \{1\}$.

2. [Take a new g] (Here we know that the order of G is a multiple of h). Choose a new random $g \in G$, $q \leftarrow \lceil \sqrt{B_1 - C_1} \rceil$.

3. [Compute small steps] Set $x_0 \leftarrow 1$, $x_1 \leftarrow g^h$ and if $x_1 = 1$ set $n \leftarrow 1$ and go to step 6. Otherwise, for $r = 2$ to $r = q - 1$ set $x_r \leftarrow x_1 \cdot x_{r-1}$. For each r with $0 \leq r < q$ set $S_{1,r} \leftarrow x_r \cdot S$, $S_1 \leftarrow \bigcup_{0 \leq r < q} S_{1,r}$, and sort S_1 so that a search in S_1 is easy. If during this computation one finds $1 \in S_{1,r}$ for $r > 0$, set $n \leftarrow r$ (where r is the smallest) and go to step 6. Otherwise, set $y \leftarrow x_1 \cdot x_{q-1}$, $z \leftarrow x_1^{C_1}$, $n \leftarrow C_1$.

4. [Compute giant steps] For each $w \in L$, set $z_1 \leftarrow z \cdot w$ and search for z_1 in the sorted list S_1. If z_1 is found and $z_1 \in S_{1,r}$, set $n \leftarrow n - r$ and go to step 6.

5. [Continue] Set $z \leftarrow y \cdot z$, $n \leftarrow n + q$. If $n \leq B_1$ go to step 4. Otherwise output an error message stating that the order of G is larger than B and terminate the algorithm.

6. [Initialize order] Set $n \leftarrow hn$.

7. [Compute the order of g mod $L \cdot S$] (Here we know that $g^n \in L \cdot S$). For each prime p dividing n, do the following: set $S_1 \leftarrow g^{n/p} \cdot S$ and sort S_1. If there exists a $z \in L$ such that $z \in S_1$, set $n \leftarrow n/p$ and go to step 7.

8. [Finished?]. Set $h \leftarrow hn$. If $h \geq C$ then output h and terminate the algorithm. Otherwise, set $B_1 \leftarrow \lfloor B_1/n \rfloor$, $C_1 \leftarrow \lceil C_1/n \rceil$, $q \leftarrow \lceil \sqrt{n} \rceil$, $S \leftarrow \bigcup_{0 \leq r < q} g^r \cdot S$, $y \leftarrow g^q$, $L \leftarrow \bigcup_{0 \leq a \leq q} y^a \cdot L$ and go to step 2.

This is of course a probabilistic algorithm. The correctness of the result depends in an essential way on the correctness of the bounds C and B. Since during the algorithm the order of G is always a multiple of h, and since

$C > B/2$, the stopping criterion $h \geq C$ in step 8 is correct (any multiple of h larger than h would be larger than B). In practice however we may not be so lucky as to have a lower bound C such that $C > B/2$. In that case, one cannot easily give any stopping criteria, and my advice is to stop as soon as h has not changed after 10 passes through step 8. Note however that this is no longer an algorithm, since nothing guarantees the correctness of the result.

Note that if g_i are elements of G of respective orders e_i, then the exponent of G is a multiple of the least common multiple (LCM) of the e_i. Hence, if one expects the exponent of the group to be not too much lower than the order h, one can use a much simpler method in which one simply computes the LCM of sufficiently many random elements of G, and then taking the multiple of this LCM which is between the given bounds C and B. For this to succeed, the bounds have to be close enough. In practice, it is advised to first use this method to get a tentative order, then to use the rigorous algorithm given above to prove it, since a knowledge of the exponent of G can clearly be used to improve the efficiency of Algorithm 5.4.1.

Let us explain why Algorithm 5.4.1 works. Let H be the true order of G. Consider the first g. We have $g^H = 1$, and if we write $H - C = aq - r$ with $0 \leq r < q$ and $q = \lceil \sqrt{B - C} \rceil$, then also $a - 1 < (H - C)/q \leq (B - C)/q \leq q$ hence $a \leq q$. This implies that we have an equality of the form

$$g^C \cdot (g^q)^a = g^r$$

with $0 \leq r < q$ and $1 \leq a \leq q$. This is detected in step 4 of the algorithm, where we have $x_r = g^r$, $y = g^q$ and $z_1 = g^C \cdot (g^q)^a$. When we arrive in step 6 we know that $g^n = 1$ with $n = C + aq - r$, hence the order of g is a divisor of n, and step 7 is the standard method for computing the order of an element in a group.

After that, h is set to the order of g, and by a similar baby step giant step construction, S and L are constructed so that $S \cdot L = \langle g \rangle$, the subgroup generated by g. We also know that the order H of G is a multiple of h. Hence, for a new g_1, instead of writing $g_1^H = 1$ and $H - C = aq - r$ we will write $(g_1^h)^{H_1} \in \langle g \rangle$ and $H_1 - C_1 = aq_1 - r_1$, where $H_1 = H/h$ is known to be between $C_1 = \lceil C/h \rceil$ and $B_1 = \lfloor B/h \rfloor$, whence the modifications given in the algorithm when we start with a new g. $\square$

Note that as we have already mentioned, it is essential to do some kind of ordering on the x_r in step 3, otherwise the search time in step 4 would dominate the total time. In practical implementations, the best method is probably not to sort completely, but to use hashing techniques (see [Knu3]).

The expected running time of this algorithm is $O((B - C)^{1/2})$ group operations, and this is usually $O(B^{1/2+\epsilon})$ for all $\epsilon > 0$. For obvious reasons, the method above is called Shanks's baby-step giant-step method, and it can be profitably used in many contexts. For example, it can be used to compute class numbers and class groups (see Algorithm 5.4.10), regulators (see Algorithm 5.8.5), or the number of points of an elliptic curve over a finite field (see Algorithm 7.4.12).

We must now explain how to obtain the whole group structure. Call g_1, ..., g_k the elements of G which are chosen in step 2. Then when a match is found in step 3 or 4, we must record not only the exponent of g which occurs, but the specific exponents of the preceding g_i. In other words, one must keep track of the multi-index exponents in the lists L and S. If at step i we have a relation of the form $g_1^{k_{1,i}} \cdots g_{i-1}^{k_{i-1,i}} g_i^{k_{i,i}} = 1$, with $g = g_i$ and $k_{i,i} = n$ after step 7 in the notation of the algorithm, we then consider the matrix $K = (k_{i,j})$ where we set $k_{i,j} = 0$ if $i > j$. Then we compute the Smith normal form of this matrix using Algorithm 2.4.14, and if d_i are the diagonal elements of the Smith normal form, we have

$$G \simeq \bigoplus_{1 \le i \le k} (\mathbb{Z}/d_i\mathbb{Z}) \, ,$$

i.e. the group structure of G.

5.4.2 Reduction and Composition of Quadratic Forms

Before being able to apply the above algorithm (or any other algorithm using the group structure) to the class group, it is absolutely essential to be able to compute in the class group. As already mentioned, we could do this by using HNF computations on ideals. Although theoretically equivalent, it is more practical however to work on classes of quadratic forms. In Theorem 5.2.8 we have seen that the set of classes of quadratic forms is in a natural bijection with the class group. Hence, we can easily transport this group structure so as to give a group structure to classes of quadratic forms. This operation, introduced by Gauss in 1798 is called *composition* of quadratic forms. Also, since we will want to work with a class of forms, we will have a *reduction procedure* which, given any quadratic form, will give us the unique reduced form in its class. I refer the reader to [Len1] and [Bue] for more details on this subject.

The reduction algorithm is a variant of Euclid's algorithm:

Algorithm 5.4.2 (Reduction of Positive Definite Forms). Given a positive definite quadratic form $f = (a, b, c)$ of discriminant $D = b^2 - 4ac < 0$, this algorithm outputs the unique reduced form equivalent to f.

1. [Initialize] If $-a < b \le a$ go to step 3.

2. [Euclidean step] Let $b = 2aq + r$ with $0 \le r < 2a$ be the Euclidean division of b by $2a$. If $r > a$, set $r \leftarrow r - 2a$ and $q \leftarrow q + 1$. (In other words, we want $b = 2aq + r$ with $-a < r \le a$.) Then set $c \leftarrow c - \frac{1}{2}(b + r)q$, $b \leftarrow r$.

3. [Finished?] If $a > c$ set $b \leftarrow -b$, exchange a and c and go to step 2. Otherwise, if $a = c$ and $b < 0$, set $b \leftarrow -b$. Output (a, b, c) and terminate the algorithm.

The proof of the validity of this algorithm follows from the proof of Proposition 5.3.3. Note that in step 2 we could have written $c \leftarrow c - bq + aq^2$, but writing it the way we have done avoids one multiplication per loop.

This algorithm has exactly the same behavior as Euclid's algorithm which we have analyzed in Chapter 1, hence is quite fast. In fact, we have the following.

Proposition 5.4.3. *The number of Euclidean steps in Algorithm 5.4.2 is at most equal to*

$$2 + \left\lceil \lg \left(\frac{a}{\sqrt{|D|}} \right) \right\rceil .$$

Proof. Consider the form (a, b, c) at the beginning of step 3. Note first that if $a > \sqrt{|D|}$, then

$$c = \frac{b^2 + |D|}{4a} \leq \frac{a^2 + a^2}{4a} = \frac{a}{2} ,$$

hence, since in step 3 a and c are exchanged, a decreases by a factor at least equal to 2. Hence, after at most $\lceil \lg(a/\sqrt{|D|}) \rceil$ steps, we obtain at the beginning of step 3 a form with $a < \sqrt{|D|}$. Now we have the following lemma.

Lemma 5.4.4. *Let (a, b, c) is a positive definite quadratic form of discriminant $D = b^2 - 4ac < 0$ such that $-a < b \leq a$ and $a < \sqrt{|D|}$. Then either (a, b, c) is already reduced, or the form (c, r, s) where $-b = 2cq + r$ with $-c < r \leq c$ obtained by one reduction step of Algorithm 5.4.2 will be reduced.*

Proof. If (a, b, c) is already reduced, there is nothing to prove. Assume it is not. Since $-a < b \leq a$, this means that $a > c$ or $a = c$ and $b < 0$. This last case is trivial since at the next step we obtain the reduced form $(a, -b, a)$. Hence, assume $a > c$. If $-c < -b \leq c$, then $q = 0$ and so $(c, r, s) = (c, -b, a)$ is reduced. If $a \geq 2c$, then $c < \sqrt{|D|/4}$, and hence (c, r, s) is reduced by Lemma 5.3.4. So we may assume $c < a < 2c$ and $-b \leq -c$ or $-b > c$. Since $|b| \leq a$, it follows that in the Euclidean division of $-b$ by $2c$ we must have $q = \pm 1$, the sign being the sign of $-b$. Now we have $s = a - bq + cq^2$, hence when $q = \pm 1$, $s = a \mp b + c \geq c$ since $|b| \leq a$. This proves that (c, r, s) is reduced, except perhaps when $s = c$. In that case however we must have $a = \pm b$, hence $a = b$ so $b > 0$, $q = -1$ and $r = 2c - b \geq 0$. Therefore (c, r, s) is also reduced in this case. This proves the lemma, and hence Proposition 5.4.3.   □

We will now consider composition of forms. Although the group structure on ideal classes carries over only to classes of quadratic forms via the maps ϕ_{FI} and ϕ_{IF} defined in Section 5.2, we can define an operation between forms, which we call composition, which becomes a group law only at the level of classes modulo $PSL_2(\mathbb{Z})$. Hence we will usually work on the level of forms.

Let (a_1, b_1, c_1) and (a_2, b_2, c_2) be two quadratic forms with the *same* discriminant D, and consider the corresponding ideals

$$I_k = a_k \mathbb{Z} + \frac{-b_k + \sqrt{D}}{2} \mathbb{Z} \qquad (k = 1, 2)$$

given by the map ϕ_{FI} of Theorem 5.2.4. We have the following lemma

Lemma 5.4.5. *Let I_1 and I_2 be two ideals as above, set $s = (b_1 + b_2)/2$, $d = \gcd(a_1, a_2, s)$, and let u, v, w be integers such that $ua_1 + va_2 + ws = d$. Then we have*

$$I_1 \cdot I_2 = d \left(A\mathbb{Z} + \frac{-B + \sqrt{D}}{2} \mathbb{Z} \right),$$

where

$$A = d_0 \frac{a_1 a_2}{d^2}, \quad B = b_2 + \frac{2a_2}{d}(v(s - b_2) - wc_2)$$

and $d_0 = 1$ if at least one of the forms (a_1, b_1, c_1) or (a_2, b_2, c_2) is primitive and in general $d_0 = \gcd(a_1, a_2, s, c_1, c_2, n)$ where $n = (b_1 - b_2)/2$.

Proof. The ideal $I_3 = I_1 \cdot I_2$ is generated as a $\mathbb{Z}$-module by the four products of the generators of I_1 and I_2, i.e. by $g_1 = a_1 a_2$, $g_2 = (-a_1 b_2 + a_1\sqrt{D})/2$, $g_3 = (-a_2 b_1 + a_2\sqrt{D})/2$ and $g_4 = ((b_1 b_2 + D)/2 - s\sqrt{D})/2$. Now by Proposition 5.2.1 we know that we can write

$$I_3 = C \left(A\mathbb{Z} + \frac{-B + \sqrt{D}}{2} \mathbb{Z} \right)$$

for some integers A, B and C. It is clear that C is the smallest positive coefficient of $\sqrt{D}/2$ in I_3, hence is equal to the GCD of a_1, a_2 and s, so $C = d$ as stated. If one of the forms is primitive, or equivalently by Proposition 5.2.5 if one of the ideals is invertible, then by Proposition 4.6.8, we have $\mathcal{N}(I_3) = \mathcal{N}(I_1)\mathcal{N}(I_2) = a_1 a_2$ and since $\mathcal{N}(I_3) = AC^2$ we have $A = a_1 a_2/d^2$. (By Exercise 14 of Chapter 4, this will in fact still be true if $\gcd(a_1, b_1, c_1, a_2, b_2, c_2) = 1$, which is a slightly stronger condition than $d_0 = 1$.) This will also follow from the more general result where we make no assumptions of primitivity.

Let us directly determine the value of AC, i.e. the least positive integer belonging to I_3. Any element of I_3 being of the form $u_1 g_1 + u_2 g_2 + u_3 g_3 + u_4 g_4$ for integers u_i, the set $I_3 \cap \mathbb{Z}$ is the set of such elements with $u_2 a_1 + u_3 a_2 - u_4 s = 0$. Using Exercise 11, the general solution to this is given by $u_2 = a_2/(a_1, a_2)\nu - s/(a_1, s)\mu$, $u_3 = s/(a_2, s)\lambda - a_1/(a_1, a_2)\nu$, $u_4 = a_2/(a_2, s)\lambda - a_1/(a_1, s)\mu$ for integers λ, μ, ν. After a short calculation, we see that $I_3 \cap \mathbb{Z} = e\mathbb{Z}$ where

$$e = \gcd\left(\frac{a_1 a_2 c_1}{(a_2, s)}, \frac{a_1 a_2 c_2}{(a_1, s)}, \frac{a_1 a_2 n}{(a_1, a_2)}, a_1 a_2 \right).$$

Another computation (see Exercise 8) shows that

$$e = \frac{a_1 a_2}{(a_1, a_2, s)} \gcd(a_1, a_2, s, c_1, c_2, n).$$

thus giving the claimed value for $A = e/C = e/d$. Since $b_1 = s + n$ and $b_2 = s - n$, it is clear that if one of the forms is primitive then $d_0 = \gcd(a_1, a_2, s, c_1, c_2, n) = 1$ thus proving the statement made above.

Finally, if $d = ua_1 + va_2 + ws$, one possible value of B is clearly

$$B = \frac{ua_1 b_2 + va_2 b_1 + w(b_1 b_2 + D)/2}{d} = \frac{db_2 + va_2(b_1 - b_2) - 2a_2 c_2 w}{d} ,$$

thus proving the lemma.   □

Note that if one writes $I_i = a_i \mathbb{Z} + \tau_i \mathbb{Z}$, then we can reformulate the above lemma by saying that (with the same definitions of d, u, v and w) we have $a_3 = a_1 a_2 d_0/d$ and $\tau_3 = ua_1 \tau_2 + va_2 \tau_1 + w\tau_1 \tau_2$.

This leads to the following basic definition of the composite of two forms.

Definition 5.4.6. *Let $f_1 = (a_1, b_1, c_1)$ and $f_2 = (a_2, b_2, c_2)$ be two quadratic forms of the same discriminant D. Set $s = (b_1 + b_2)/2$, $n = (b_1 - b_2)/2$ and let u, v, w and d be such that*

$$ua_1 + va_2 + ws = d = \gcd(a_1, a_2, s)$$

(obtained by two applications of Euclid's extended algorithm), and let $d_0 = \gcd(d, c_1, c_2, n)$. We define the composite of the two forms f_1 and f_2 as the form

$$(a_3, b_3, c_3) = \left(d_0 \frac{a_1 a_2}{d^2}, b_2 + \frac{2a_2}{d}(v(s - b_2) - wc_2), \frac{b_3^2 - D}{4a_3} \right) .$$

modulo the action of Γ_∞, i.e. viewed as a form in the set F introduced in Section 5.2.

Since composition comes from the product of ideals, using the isomorphism given in Section 5.2, it is clear that the class in F of (a_3, b_3, c_3) does not depend on the particular choices of u, v and w. This can of course also be checked directly (see Exercise 12). Note that if we do not take the class modulo Γ_∞, the result is not at all canonical. Therefore when we speak of composition of quadratic forms we will always implicitly assume that we are working modulo the action of Γ_∞, i.e. in the set F, and not on quadratic forms themselves.

To obtain the reduced composite of two forms, it is usually necessary to reduce the form obtained by composition. By abuse of language, in the case of negative discriminants we will also call this reduced form the composite of the two forms. (In the case of positive discriminants, there is in general more than one reduced form equivalent to a given form, hence this abuse of language is not permitted.)

Although the raw formulas given in the definition can be used directly, they can be improved by careful rearrangements. This leads to the following algorithm, due to Shanks [Sha1]. Since imprimitive forms are almost never used, for the sake of efficiency we will restrict to the case of primitive forms. Note also that the composite of two primitive forms is still primitive (Exercise 9).

Algorithm 5.4.7 (Composition of Positive Definite Forms). Given two *primitive* positive definite quadratic forms $f_1 = (a_1, b_1, c_1)$ and $f_2 = (a_2, b_2, c_2)$ with the same discriminant, this algorithm computes the composite $f_3 = (a_3, b_3, c_3)$ of f_1 and f_2.

1. [Initialize] If $a_1 > a_2$ exchange f_1 and f_2. Then set $s \leftarrow \frac{1}{2}(b_1 + b_2)$, $n \leftarrow b_2 - s$.

2. [First Euclidean step] If $a_1 \mid a_2$, set $y_1 \leftarrow 0$ and $d \leftarrow a_1$. Otherwise, using Euclid's extended algorithm compute (u, v, d) such that $ua_2 + va_1 = d = \gcd(a_2, a_1)$, and set $y_1 \leftarrow u$.

3. [Second Euclidean step] If $d \mid s$, set $y_2 \leftarrow -1$, $x_2 \leftarrow 0$ and $d_1 \leftarrow d$. Otherwise, using Euclid's extended algorithm compute (u, v, d_1) such that $us + vd = d_1 = \gcd(s, d)$, and set $x_2 \leftarrow u$, $y_2 \leftarrow -v$.

4. [Compose] Set $v_1 \leftarrow a_1/d_1$, $v_2 \leftarrow a_2/d_1$, $r \leftarrow (y_1 y_2 n - x_2 c_2 \bmod v_1)$, $b_3 \leftarrow b_2 + 2v_2 r$, $a_3 \leftarrow v_1 v_2$, $c_3 \leftarrow (c_2 d_1 + r(b_2 + v_2 r))/v_1$ (or $c_3 \leftarrow (b_3^2 - D)/(4a_3)$), then reduce the form $f = (a_3, b_3, c_3)$ using Algorithm 5.4.2, output the result and terminate the algorithm.

Note that this algorithm should be implemented as written: in step 2 we first consider the special case $a_1 \mid a_2$ because it occurs very often (at least each time one squares a form, and this is the most frequent operation when one raises a form to a power.) Therefore, it should be considered separately for efficiency's sake, although the general Euclidean step would give the same result. Similarly, in step 3 it often happens that $d \mid s$ because $d = 1$ also occurs quite often. Finally, note that the computation of c_3 in step 4 can be done using any of the two formulas given.

The generalization of this algorithm to imprimitive forms is immediate (see Exercise 10).

Since we have $|b_3| \leq a_3 \leq \sqrt{|D|/3}$ and since c_3 can be computed from a_3 and b_3, it seems plausible that one can make most of the computations in Algorithm 5.4.7 using numbers only of size $O(\sqrt{|D|})$ and not $O(D)$ or worse. That this is the case was noticed comparatively recently by Shanks and published only in 1989 [Sha2]. The improvement is considerable since in multi-precision situations it may gain up to a factor of 4, while in the case where $\sqrt{|D|}$ is single precision while D is not, the gain is even larger.

This modified algorithm (called NUCOMP by Shanks) was modified again by Atkin [Atk1]. As mentioned above, squaring of a form is important and simpler, so Atkin gives two algorithms, one for duplication and one for composition.

Algorithm 5.4.8 (NUDUPL). Given a primitive positive definite quadratic form $f = (a, b, c)$ of discriminant D, this algorithm computes the square $f^2 = f_2 = (a_2, b_2, c_2)$ of f. We assume that the constant $L = \lfloor |D/4|^{1/4} \rfloor$ has been precomputed.

1. [Euclidean step] Using Euclid's extended algorithm, compute (u, v, d_1) such that $ub + va = d_1 = \gcd(b, a)$. Then set $A \leftarrow a/d_1$, $B \leftarrow b/d_1$, $C \leftarrow (-cu \bmod A)$, $C_1 \leftarrow A - C$ and if $C_1 < C$, set $C \leftarrow -C_1$.

2. [Partial reduction] Execute Sub-algorithm PARTEUCL(A, C) below (this is an extended partial Euclidean algorithm).

3. [Special case] If $z = 0$, set $g \leftarrow (Bv_3 + c)/d$, $a_2 \leftarrow d^2$, $c_2 \leftarrow v_3^2$, $b_2 \leftarrow b + (d + v_3)^2 - a_2 - c_2$, $c_2 = c_2 + gd_1$, reduce the form $f_2 = (a_2, b_2, c_2)$, output the result and terminate the algorithm.

4. [Final computations] Set $e \leftarrow (cv + Bd)/A$, $g \leftarrow (ev_2 - B)/v$ (these divisions are both exact and $v = 0$ has been dealt with in step 3), then $b_2 \leftarrow ev_2 + vg$. Then, if $d_1 > 1$, set $b_2 \leftarrow d_1 b_2$, $v \leftarrow d_1 v$, $v_2 \leftarrow d_1 v_2$. Finally, in order, set $a_2 \leftarrow d^2$, $c_2 \leftarrow v_3^2$, $b_2 \leftarrow b_2 + (d + v_3)^2 - a_2 - c_2$, $a_2 \leftarrow a_2 + ev$, $c_2 \leftarrow c_2 + gv_2$, reduce the form $f_2 = (a_2, b_2, c_2)$, output the result and terminate the algorithm.

Sub-algorithm PARTEUCL(a, b). This algorithm does an extended partial Euclidean algorithm on a and b, but uses the variables v and v_2 instead of u and v_1 in Algorithm 1.3.6.

1. [Initialize] Set $v \leftarrow 0$, $d \leftarrow a$, $v_2 \leftarrow 1$, $v_3 \leftarrow b$, $z \leftarrow 0$.

2. [Finished?] If $|v_3| > L$ go to step 3. Otherwise, if z is odd, set $v_2 \leftarrow -v_2$ and $v_3 \leftarrow -v_3$. Terminate the sub-algorithm.

3. [Euclidean step] Let $d = qv_3 + t_3$ be the Euclidean division of d by v_3 with $0 \leq t_3 < |v_3|$. Set $t_2 \leftarrow v - qv_2$, $v \leftarrow v_2$, $d \leftarrow v_3$, $v_2 \leftarrow t_2$, $v_3 \leftarrow t_3$, $z \leftarrow z + 1$ and go to step 2.

I have given the gory details in steps 3 and 4 of Algorithm 5.4.8 just to show how a careful implementation can save time: the formula for b_2 in step 4 could have simply been written $b_2 \leftarrow b_2 + 2dv_3$. This would involve one multiplication and 2 additions. Since we need the quantities d^2 and v_3^2 for a_2 and c_2 anyway, the way we have written the formula involves 3 additions and one squaring. By a suitable implementation of a method analogous to the splitting method for polynomials explained in Chapter 3, this will be faster than 2 additions and one multiplication. Of course the gain is slight and the lazy reader may implement this in the more straightforward way, but it should be remembered that we are programming a basic operation in a group which will be used a large number of times, so any gain, even small, is worth taking.

Note also that the final reduction of f_2 will be very short, usually one or two Euclidean steps at most.

The proof of the validity of the algorithm is not difficult (see [Sha2]) and is left to the reader. It can also be checked that all the iterations (Euclid and reductions) are done on numbers less than $O(\sqrt{|D|})$, and that only a small and fixed number of operations are done on larger numbers.

Let us now look at the general algorithm for composition.

Algorithm 5.4.9 (NUCOMP). Given two primitive positive definite quadratic forms with the same discriminant $f_1 = (a_1, b_1, c_1)$ and $f_2 = (a_2, b_2, c_2)$, this algorithm computes the composite $f_3 = (a_3, b_3, c_3)$ of f_1 and f_2. As in NUDUPL (Algorithm 5.4.8) we assume already precomputed the constant $L = \lfloor |D/4|^{1/4} \rfloor$. Note that the values of a_1 and a_2 may get changed, so they should be preserved if needed.

1. [Initialize] If $a_1 < a_2$ exchange f_1 and f_2. Then set $s \leftarrow \frac{1}{2}(b_1 + b_2)$, $n \leftarrow b_2 - s$.

2. [First Euclidean step] Using Euclid's extended algorithm, compute (u, v, d) such that $ua_2 + va_1 = d = \gcd(a_1, a_2)$. If $d = 1$, set $A \leftarrow -un$, $d_1 \leftarrow d$ and go to step 5. If $d \mid s$ but $d \neq 1$, set $A \leftarrow -un$, $d_1 \leftarrow d$, $a_1 \leftarrow a_1/d_1$, $a_2 \leftarrow a_2/d_1$, $s \leftarrow s/d_1$ and go to step 5.

3. [Second Euclidean step] (here $d \nmid s$) Using Euclid's extended algorithm again, compute (u_1, v_1, d_1) such that $u_1 s + v_1 d = d_1 = \gcd(s, d)$. Then, if $d_1 > 1$, set $a_1 \leftarrow a_1/d_1$, $a_2 \leftarrow a_2/d_1$, $s \leftarrow s/d_1$ and $d \leftarrow d/d_1$.

4. [Initialization of reduction] Compute $l \leftarrow -u_1(uc_1 + vc_2) \bmod d$ by first reducing c_1 and c_2 (which are large) modulo d (which is small), doing the operation, and reducing again. then set $A \leftarrow -u(n/d) + l(a_1/d)$.

5. [Partial reduction] Set $A \leftarrow (A \bmod a_1)$, $A_1 \leftarrow a_1 - A$ and if $A_1 < A$ set $A \leftarrow -A_1$, then execute Sub-algorithm PARTEUCL(a_1, A) above.

6. [Special case] If $z = 0$, set $Q_1 \leftarrow a_2 v_3$, $Q_2 \leftarrow Q_1 + n$, $f \leftarrow Q_2/d$, $g \leftarrow (v_3 s + c_2)/d$, $a_3 \leftarrow da_2$, $c_3 \leftarrow v_3 d + g d_1$, $b_3 \leftarrow 2Q_1 + b_2$, reduce the form $f_3 = (a_3, b_3, c_3)$, output the result and terminate the algorithm.

7. [Final computations] Set $b \leftarrow (a_2 d + nv)/a_1$, $Q_1 \leftarrow bv_3$, $Q_2 \leftarrow Q_1 + n$, $f \leftarrow Q_2/d$, $e \leftarrow (sd + c_2 v)/a_1$, $Q_3 \leftarrow ev_2$, $Q_4 \leftarrow Q_3 - s$, $g \leftarrow Q_4/v$ (the case $v = 0$ has been dealt with in step 6), and if $d_1 > 1$ set $v_2 \leftarrow d_1 v_2$, $v \leftarrow d_1 v$. Finally, set $a_3 \leftarrow db + ev$, $c_3 \leftarrow v_3 f + gv_2$, $b_3 \leftarrow Q_1 + Q_2 + d_1(Q_3 + Q_4)$, reduce the form $f_3 = (a_3, b_3, c_3)$, output the result and terminate the algorithm.

Note that all the divisions which are performed in this algorithm are exact, and that the final reduction step, as in NUDUPL, will be very short, usually one or two Euclidean steps at most. As for NUDUPL, we leave to the reader the proof of the validity of this algorithm.

Implementation Remark. We have used the basic Algorithm 1.3.6 as a template for Sub-algorithm PARTEUCL. In practice, when dealing with multi-precision numbers, it is preferable to use one of its variants such as Algorithm 1.3.7 or 1.3.8.

5.4.3 Class Groups Using Shanks's Method

From the Brauer-Siegel theorem, we know that the class number $h(D)$ of an imaginary quadratic field grows roughly like $|D|^{1/2}$. This means that the baby-step giant-step algorithm given above allows us to compute $h(D)$ in time $O(|D|^{1/4+\epsilon})$, which is much better than the preceding methods. In fact, suitably implemented, one can reasonably expect to compute class numbers and class groups of discriminants having up to 20 or 25 decimal digits. For taking powers of the quadratic forms one should use the powering algorithm of Section 1.2, using if possible NUDUPL for the squarings and NUCOMP for general composition, or else using Shanks less optimized but simpler Algorithm 5.4.7. To be able to use the baby-step giant-step Algorithm 5.4.1 however, we need bounds for the class number $h(D)$. Now rigorous and explicit bounds are difficult to obtain, even assuming the GRH. Hence, we will push our luck and give only *tentative* bounds. Of course, this completely invalidates the rigor of the algorithm. To be sure that the result is correct, one should start with proven bounds like $C = 0$ and $B = \frac{1}{\pi}\sqrt{|D|}\ln|D|$ (see Exercise 27), however the performance is much worse.

Now the series giving $L_D(1)$ is only conditionally convergent, as is the corresponding Euler product

$$L_D(s) = \prod_p \left(1 - \frac{\left(\frac{D}{p}\right)}{p^s}\right)^{-1} .$$

However this Euler product is faster to compute to a given accuracy, since only the primes are needed. Hence, to start Shanks's algorithm, we take a large prime number bound P (say $P = 2^{18}$), and guess that $h(D)$ will be close to

$$\tilde{h} = \left\lfloor \frac{\sqrt{|D|}}{\pi} \prod_{p \le P} \left(1 - \frac{\left(\frac{D}{p}\right)}{p}\right)^{-1} \right\rceil .$$

Assuming GRH, one can show that

$$h(D) - \tilde{h} = O(\tilde{h}P^{-1/2}\ln(P|D|)) ,$$

and one can give explicit values for the O constant. In practice, Shanks noticed experimentally that the relative error is around $1/1000$ when $P = 2^{17}$. Hence, if we use these numerical bounds combined with the baby-step giant-step method, we will correctly compute $h(D)$ unless the exponent of the group is very small compared to the order.

A very important speedup in computing $h(D)$ by Shanks's method is obtained by noticing that the inverse for composition of the form (a, b, c) is the form $(a, -b, c)$, hence requires no calculation. Hence, one can double the size of the giant steps (by setting $y \leftarrow x_1^{2q}$ instead of $y \leftarrow x_1^q$ in step 3 of Algorithm 5.4.1). Therefore the optimal value for q is no longer $\sqrt{B - C}$ but rather $\sqrt{(B - C)/2}$.

Finally, note that during the computation of the Euler product leading to $\bar{h}$, we will also have found the primes p for which $\left(\frac{D}{p}\right) = 1$. For the first few such p, we compute the square root b_p of D mod $4p$ by a simple modification of Algorithm 1.5.1, and we store the forms (p, b_p, c_p) where $c_p = (b_p^2 - D)/(4p)$. These will be used as our "random" x in step 2 of the algorithm.

Putting all these ideas together leads to the following method:

Heuristic Algorithm 5.4.10 ($h(D)$ Using Baby-Step Giant-Step). If $D < -4$ is a discriminant, this algorithm tries to compute $h(D)$ using a simpleminded version of Shanks's baby-step giant-step method. We denote by $\cdot$ the operation of composition of quadratic forms, and by 1 the unit element in the class group. We choose a small bound b (for example $b = 10$).

1. [Compute Euler product] For $P = \max(2^{18}, |D|^{1/4})$, compute the product

$$Q \leftarrow \left\lfloor \frac{\sqrt{|D|}}{\pi} \prod_{p \leq P} \left(1 - \frac{\left(\frac{D}{p}\right)}{p}\right)^{-1} \right\rfloor .$$

Then set $B \leftarrow \lfloor Q(1 + 1/(2\sqrt{P})) \rfloor$, $C \leftarrow \lceil Q(1 - 1/(2\sqrt{P})) \rceil$. For the first b values of p such that $\left(\frac{D}{p}\right) = 1$, compute b_p such that $b_p^2 \equiv D \pmod{4p}$ using Algorithm 1.5.1 (and modifying the result to get the correct parity). Set $f_p \leftarrow (p, b_p, (b_p^2 - D)/(4p))$.

2. [Initialize] Set $e \leftarrow 1$, $c \leftarrow 0$, $B_1 \leftarrow B$, $C_1 \leftarrow C$, $Q_1 \leftarrow Q$.

3. [Take a new g] (Here we know that the exponent of $Cl(D)$ is a multiple of e). Set $g \leftarrow f_p$ for the first new f_p, and set $c \leftarrow c+1$, $q \leftarrow \lceil \sqrt{(B_1 - C_1)/2} \rceil$.

4. [Compute small steps] Set $x_0 \leftarrow 1$, $x_1 \leftarrow g^e$ then for $r = 2$ to $r = q - 1$ set $x_r \leftarrow x_1 \cdot x_{r-1}$. If, during this computation one finds $x_r = 1$, then set $n \leftarrow r$ and go to step 7. Otherwise, sort the x_r so that searching among them is easy, and set $y \leftarrow x_1 \cdot x_{q-1}$, $y \leftarrow y^2$, $z \leftarrow x_1^{Q_1}$, $n \leftarrow Q_1$.

5. [Compute giant steps] Search for z or z^{-1} in the sorted list of x_r for $0 \leq r < q$ (recall that if $z = (a, b, c)$, $z^{-1} = (a, -b, c)$). If a match $z = x_r$ is found, set $n \leftarrow n - r$ and go to step 7. If a match $z^{-1} = x_r$ is found, set $n \leftarrow n + r$ and go to step 7.

6. [Continue] Set $z \leftarrow y \cdot z$, $n \leftarrow n + 2q$. If $n \leq B_1$ go to step 5. Otherwise output an error message stating that the order of G is larger than B and terminate the algorithm.

7. [Compute the order of g] (Here we know that $g^{en} = x_1^n = 1$). For each prime p dividing n, do the following: if $x_1^{n/p} = 1$, then set $n \leftarrow n/p$ and go to step 7.

8. [Finished?] (Here n is the exact order of x_1). Set $e \leftarrow en$. If $e > B - C$, then set $h \leftarrow e\lfloor B/e \rfloor$, output h and terminate the algorithm. If $c \geq b$ output a message saying that the algorithm fails to find an answer and terminate the algorithm. Otherwise set $B_1 \leftarrow \lfloor B_1/n \rfloor$, $C_1 \leftarrow \lceil C_1/n \rceil$ and go to step 3.

This is *not* an algorithm, in the sense that the output may be false. One should compute the whole group structure using Algorithm 5.4.1 to be sure that the result is valid. It almost always gives the right answer however, and thus should be considered as a first step.

5.5 McCurley's Sub-exponential Algorithm

We now come to an algorithm discovered in 1988 by McCurley [McCur, McCur-Haf] and which is much faster than the preceding algorithms for large discriminants. Several implementations of this algorithm have been done, for example by Düllmann, ([Buc-Dül]) and it is now reasonable to compute the class group for a discriminant of 50 decimal digits. Such examples have been computed by Düllmann and Atkin.

Incidentally, unlike almost all other algorithms in this book, little has been done to optimize the algorithm that we give, and there is plenty of room for (serious) improvements. This is, in fact, a subject of active research.

5.5.1 Outline of the Algorithm

Before giving the details of the algorithm, let us give an outline of the main ideas. First, instead of trying to obtain the class number and class group "from below", by finding relations $x^e = 1$, and hence *divisors* of the class number, we will find it "from above", i.e. by finding *multiples* of the class number.

Let $\mathcal{P}$ be a finite set of primes p such that $\left(\frac{D}{p}\right) = 1$ for all $p \in \mathcal{P}$. Then, as in Shanks's method, we can find reduced forms $f_p = (p, b_p, c_p)$, which we will call *prime forms*, for each $p \in \mathcal{P}$. Now, assuming GRH, one can prove that there exists a constant c which can be computed effectively such that if $\mathcal{P}$ contains all the primes p such that $\left(\frac{D}{p}\right) = 1$ and $p \leq c \ln^2 |D|$, then the classes of the forms f_p for $p \in \mathcal{P}$ generate the class group. This means that if we set $n = |\mathcal{P}|$, the map

$$\phi : \mathbb{Z}^n \to Cl(D)$$
$$(x_p)_{p \in \mathcal{P}} \mapsto \prod_{p \in \mathcal{P}} f_p^{x_p}$$

is a surjective group homomorphism. Hence, the kernel Λ of ϕ is a sublattice of $\mathbb{Z}^n$, and we have

$$\mathbb{Z}^n / \Lambda \simeq Cl(D) \qquad \text{and} \qquad |\det(\Lambda)| = h(D) \ ,$$

denoting by $\det(\Lambda)$ the determinant of any $\mathbb{Z}$-basis of Λ. The lattice Λ is the lattice of relations among the f_p. If one finds any system of n independent elements in this lattice, it is clear that the determinant of this system will be a multiple of the determinant of Λ, hence of $h(D)$. This is how we obtain multiples of the class number.

Now there remains the question of obtaining (many) relations between the f_p. To do this, one uses the following lemma:

Lemma 5.5.1. *Let (a, b, c) be a primitive positive definite quadratic form of discriminant $D < 0$, and $a = \prod_p p^{v_p}$ be the prime decomposition of a. Then we have up to equivalence:*

$$(a, b, c) = \prod_p f_p^{\epsilon_p v_p} ,$$

where $f_p = (p, b_p, c_p)$ is the prime form corresponding to p, and $\epsilon_p = \pm 1$ is defined by the congruence

$$b \equiv \epsilon_p b_p \pmod{2p} .$$

In fact, all the possible choices for the ϵ_p correspond exactly to the possible square roots b of D mod $4a$, with b defined modulo $2a$.

Proof. This lemma follows immediately from the raw formulas for composition that we have given in Section 5.4.2. In terms of ideals, using the correspondence given by Theorem 5.2.8, if $I = \psi_{FI}(\bar{f})$, the factorization of $a = \mathcal{N}(I)$ corresponds to a factorization $I = \prod \mathfrak{p}^{v_p}$ where $\mathfrak{p}$ is an ideal above $p\mathbb{Z}_K$, and ϵ_p must be chosen as stated so that $\mathfrak{p} \supset I$. $\qquad\square$

This leads immediately to the following idea for generating relations in Λ: choose random integer exponents e_p, and compute the reduced form (a, b, c) equivalent to $\prod_{p \in \mathcal{P}} f_p^{e_p}$. If all the factors of a are in $\mathcal{P}$, we keep the form (a, b, c), otherwise we take other random exponents. If the form is kept, we will have the relation

$$\prod_{p \in \mathcal{P}} f_p^{e_p - \epsilon_p v_p} = 1 ,$$

giving the element

$$(e_p - \epsilon_p v_p)_{p \in \mathcal{P}} \in \Lambda \subset \mathbb{Z}^n .$$

Continuing in this way, one may reasonably hope to generate Λ if $\mathcal{P}$ has been chosen large enough, and this is indeed what one proves, under suitable hypotheses.

The crucial point is the choice of $\mathcal{P}$. We will take

$$\mathcal{P} = \left\{ p \leq P, \left(\frac{D}{p} \right) \neq -1 \right\}$$

for a suitable P, but one must see how large this P must be to optimize the algorithm. If P is chosen too small, numbers a produced as above will almost never factor into primes less than P. If P is too large, then the factoring time of a becomes prohibitive, as does the memory required to keep all the relations and the f_p. To find the right compromise, one must give the algorithm in much greater detail and analyze its behavior. This is done in [Haf-McCur1], where it is shown that P should be taken of the order of $L(|D|)^\alpha$, where $L(x)$ is a very important function defined by

$$L(x) = e^{\sqrt{\ln x \ln \ln x}} ,$$

and α depends on the particular implementation, one possible value being $1/\sqrt{8}$. We will meet this very important function $L(x)$ again in Chapter 10 in connection with modern factoring methods.

In addition we must have $P \geq c \ln^2 |D|$ so that (assuming GRH) the classes of prime forms f_p with $p \in \mathcal{P}$ generate the class group. Unfortunately, at present, the best known bound for the constant c, due to Bach, is 6, although practical experience shows that this is much too pessimistic. (In fact it is believed that $O(\ln^{1+\epsilon} |D|)$ generators should suffice for any $\epsilon > 0$). Hence, we will choose

$$P = \max \left(6 \ln^2 |D|, L(|D|)^{1/\sqrt{8}} \right) .$$

Note that, although the $\ln^2$ function grows asymptotically much more slowly than the $L(|D|)$ function, in practice the constants 6 and $1/\sqrt{8}$ will make the $\ln^2$ term dominate. More precisely, the $L(|D|)$ term will start to dominate only for discriminants having at least 103 digits, well outside the range of practical applicability of this method. Even if one could reduce the constant 6 to 1, the $\ln^2$ term would still dominate for numbers having up to 70 digits.

Let n be the number of $p \in \mathcal{P}$. To give a specific numerical example, for D of the order of -10^{40}, with the above formula P will be around 50900, and n around 2600, while if D is of the order of -10^{50}, P will be around 79500 and n around 3900. Since we will be handling determinants of $n \times n$ matrices, many problems become serious, in particular the storage problems, though they are perhaps still manageable. In any case, the computational load becomes very great. In particular, for matrices of this size it is essential to use special techniques adapted to the type of matrices which we have, i.e. sparse matrices. Since we are over $\mathbb{Z}$ and not over a field, the use of methods such as Wiedemann's *coordinate recurrence method* (see [Wie]) is possible only through the use of the Chinese remainder theorem, and is quite painful. An easier approach is to use "intelligent Hermite reduction", analogous to the intelligent Gaussian elimination technique used by LaMacchia and Odlyzko (see [LaM-Odl]). This method has been implemented by Düllmann ([Buc-Dül]) and by Cohen, Diaz y Diaz and Olivier ([CohDiOl]), and is described below.

5.5.2 Detailed Description of the Algorithm

We first make a few remarks.

The first important remark is that although one should generate random relations using Lemma 5.5.1, one may hope to obtain a non-trivial relation as soon as $\prod_p p^{e_p} > \sqrt{|D|/3}$ since the resulting form obtained by multiplication without reduction will not be reduced. Hence, instead of taking the whole of $\mathcal{P}$ to compute the products, we take a much smaller subset $\mathcal{P}_0$ not containing any prime dividing D and such that

$$\prod_{p \in \mathcal{P}_0} p > \sqrt{|D|/3} .$$

Then $\mathcal{P}_0$ will be *very* small, typically of cardinality 10 or 20, even for discriminants in the 40 to 50 digit range. In fact, by the prime number theorem, the cardinality of $\mathcal{P}_0$ should be of the order of $\ln|D| / \ln\ln|D|$. For similar reasons, although the exponents e_p should be chosen randomly up to $|D|$ as McCurley's analysis shows, in practice it suffices to take very small random exponents, say $1 \le e_p \le 20$.

A second remark is that, even if we use intelligent Hermite reduction as will be described, the size of the matrix involved will be very large. Hence, we must try to make it smaller even before we start the reduction. One way to do this is to decide to take a lower value of P, say one corresponding to the constant $c = 1$ (i.e. the split primes of norm less than $\ln^2|D|$ instead of $6\ln^2|D|$). This would probably work, but even under the GRH the result may be false since we may not have enough generators. There is however one way out of this. For every prime q such that $\ln^2|D| < q < 6\ln^2|D|$, let g_q be a reduced form equivalent to $f_q \prod_{p \in \mathcal{P}_0} f_p^{e_p}$ with small random exponents e_p as before. If $g_q = (a, b, c)$, then, if a factors over our factor base $\mathcal{P}$, since q is quite large, with a little luck after a few trials we will find an a which not only factors, but whose prime factors are all less than q. This means that f_q belongs to the subgroup generated by the other f_p's, hence can be discarded as a generator of the class group. Doing this for all the $q > \ln^2|D|$ is fast and does not involve any matrix handling, and in effect reduces the problem to taking the constant 1 instead of 6 in the definition of P, giving much smaller matrices. Note that the constant 1 which we have chosen is completely arbitrary, but it must not be chosen too small, otherwise it will become very difficult to eliminate the big primes q. In practice, values between 0.5 and 2 seem reasonable.

These kind of ideas can be pushed further. Instead of taking products using only powers of forms f_p with $p \in \mathcal{P}_0$, we can systematically multiply such a relation by a prime q larger than the ones in $\mathcal{P}_0$, with the hope that this extra prime will still occur non-trivially in the resulting relation.

A third remark is that ambiguous forms (i.e. whose square is principal) have to be treated specially in the factor base, since only the parity of the exponents will count. (This is why we have excluded primes dividing D in

$\mathcal{P}_0$.) In fact, it would be better to add the free relations $f_p^2 = 1$ for all $p \in \mathcal{P}$ dividing D. On the other hand, when D is not a fundamental discriminant, one must exclude from $\mathcal{P}$ the primes p dividing D to a power higher than the first (except for $p = 2$ which one keeps if $D/4$ is congruent to 2 or 3 modulo 4). For our present exposition, such primes will be called *bad*, the others *good*.

Algorithm 5.5.2 (Sub-Exponential Imaginary Class Group). If $D < 0$ is a discriminant, this algorithm computes the class number $h(D)$ and the class group $Cl(D)$. As before, in practice we work with binary quadratic forms. We also choose a positive real constant b.

1. [Compute primes and Euler product] Set $m \leftarrow b \ln^2 |D|$, $M \leftarrow L(|D|)^{1/\sqrt{8}}$, $P \leftarrow \lfloor \max(m, M) \rfloor$

$$\mathcal{P} \leftarrow \left\{ p \leq P, \left(\frac{D}{p} \right) \neq -1 \text{ and } p \text{ good} \right\}$$

and compute the product

$$B \leftarrow \left\lfloor \frac{\sqrt{|D|}}{\pi} \prod_{p \leq P} \left(1 - \frac{\left(\frac{D}{p} \right)}{p} \right)^{-1} \right\rfloor .$$

2. [Compute prime forms] Let $\mathcal{P}_0$ be the set made up of the smallest primes of $\mathcal{P}$ not dividing D such that $\prod_{p \in \mathcal{P}_0} p > \sqrt{|D|/3}$. For the primes $p \in \mathcal{P}$ do the following. Compute b_p such that $b_p^2 \equiv D \pmod{4p}$ using Algorithm 1.5.1 (and modifying the result to get the correct parity). If $b_p > p$, set $b_p \leftarrow 2p - b_p$. Set $f_p \leftarrow (p, b_p, (b_p^2 - D)/(4p))$. Finally, let n be the number of primes $p \in \mathcal{P}$.

3. [Compute powers] For each $p \in \mathcal{P}_0$ and each integer e such that $1 \leq e \leq 20$ compute and store the unique reduced form equivalent to f_p^e. Set $k \leftarrow 0$.

4. [Generate random relations] Let f_q be the primeform number $k + 1 \bmod n$ in the factor base. Choose random e_p between 1 and 20, and compute the unique reduced form (a, b, c) equivalent to

$$f_q \prod_{p \in \mathcal{P}_0} f_p^{e_p}$$

until $v_q(a) \neq 1$ (note that the $f_p^{e_p}$ have already been computed in step 3). Set $e_p \leftarrow 0$ if $p \notin \mathcal{P}_0$ then $e_q \leftarrow e_q + 1$.

5. [Factor a] Factor a using trial division. If a prime factor of a is larger than P, do not continue the factorization and go to step 4. Otherwise, if $a = \prod_{p \leq P} p^{v_p}$, set $k \leftarrow k + 1$, and for $i \leq n$

$$a_{i,k} \leftarrow e_{p_i} - \epsilon_{p_i} v_{p_i}$$

where $\epsilon_{p_i} = +1$ if $(b \bmod 2p_i) \leq p_i$, $\epsilon_{p_i} = -1$ otherwise.

6. [Enough relations?] If $k < n + 10$ go to step 4.

7. [Be honest] For each prime q such that $P < q \leq 6\ln^2|D|$ do the following. Choose random e_p between 1 and 20 (say) and compute the primeform f_q corresponding to q and the unique reduced form (a, b, c) equivalent to $f_q \prod_{p \in P_0} f_p^{e_p}$. If a does not factor into primes less than q, choose other exponents e_p and continue until a factors into such primes. Then go on to the next prime q until the list is exhausted.

8. [Simple HNF] Perform a preliminary simple Hermite reduction on the $n \times k$ matrix $A = (a_{i,j})$ as described below, thus obtaining a much smaller matrix A_1.

9. [Compute determinant] Using standard Gaussian elimination techniques, compute the determinant of the lattice generated by the columns of the matrix A_1 modulo small primes p. Then compute the determinant d exactly using the Chinese remainder theorem and Hadamard's inequality (see also Exercise 13). If the matrix is not of rank equal to its number of rows, get 5 more relations (in steps 4 and 5) and go to step 8.

10. [HNF reduction] Using Algorithm 2.4.8 compute the Hermite normal form $H = (h_{i,j})$ of the matrix A_1 using modulo d techniques. Then, for every i such that $h_{i,i} = 1$, suppress row and column i. Let W be the resulting matrix.

11. [Finished?] Let $h \leftarrow \det(W)$ (i.e. the product of the diagonal elements). If $h \geq B\sqrt{2}$, get 5 more relations (in steps 4 and 5) and go to step 8. (It will not be necessary to recompute the whole HNF, but only to take into account the last 5 columns.) Otherwise, output h as the class number.

12. [Class group] Compute the Smith normal form of W using Algorithm 2.4.14. Output those diagonal elements d_i which are greater than 1 as the invariants of the class group (i.e. $Cl(D) = \bigoplus \mathbb{Z}/d_i\mathbb{Z}$) and terminate the algorithm.

Implementation Remarks.

(1) The constant b used in step 1 is important mainly to control the size of the final matrix A on which we are going to work. As mentioned above however, b must not be chosen too small, otherwise we will have a lot of trouble in the factoring stages. Practice shows that values between 0.5 and 2.0 are quite reasonable.

   With such a choice of b, we could of course avoid step 7 entirely since it seems highly implausible that the class group is not generated by the first $0.5\ln^2|D|$ primeforms. Including step 7, however, makes the correctness of the result depend only on the GRH and nothing else. Note also that strictly speaking the above algorithm could run indefinitely, either because it does not find enough relations, or because the condition of step 7 is never satisfied for some prime q. In practice this never occurs.

(2) The simple Hermite reduction which is needed in step 8 is the following. We first scan all the rows of the $n \times k$ matrix A to detect if some have a

single ± 1, the other coefficients being equal to zero. If this is the case and we find that $a_{i,j} = \pm 1$ is the only non-zero element of its row, we exchange rows i and n and columns j and k, and scan the matrix formed by the first $n - 1$ rows and $k - 1$ columns. We continue in this way until no such rows are found. We are now reduced to the study of a $(n - s) \times (k - s)$ matrix A', where s is the number of rows found.

In the second stage, we scan A' for rows having only 0 and ± 1. In this case, simple arithmetic is needed to eliminate the ± 1 as one does in ordinary HNF reduction, and, in particular, one may hope to work entirely with ordinary (as opposed to multi-precision) integers. The second stage ends when either all rows have been scanned, or if a coefficient exceeds half the maximal possible value for ordinary integers.

In a third and last stage before starting the modulo d HNF reduction of step 10, we can proceed as follows (see [Buc-Dül]). We apply the ordinary HNF reduction Algorithm 2.4.5 keeping track of the size of the coefficients which are encountered. In this manner, we Hermite-reduce a few rows (corresponding to the index j in Algorithm 2.4.5) until some coefficient becomes in absolute value larger than a given bound (for example as soon as a coefficient does not fit inside a single-precision number). If the first non-Hermite-reduced row has index j, we use the MLLL Algorithm 2.6.8 or an all-integer version on the matrix formed by the first j rows. The effect of this will be to decrease the size of the coefficients, and since as in Hermite reduction only column operations are involved, the LLL reduction is allowed. We now start again Hermite-reducing a few rows using Algorithm 2.4.5, and we continue until either the matrix is completely reduced, or until the LLL reduction no longer improves matters (i.e. the partial Hermite reduction reduced no row at all).

After these reductions are performed, practical experience shows that the size of the matrix will have been considerably reduced, and this is essential since otherwise the HNF reduction would have to be performed on matrices having up to several thousand rows and columns, and this is almost impossible in practice.

(3) If Hermite reduction is performed carefully as described above, by far the most costly part of the algorithm is the search for relations. This part can be considerably improved by using the *large prime variation* idea common to many modern factoring methods (see Remark (2) in Section 10.1) as follows. In step 5, all a with a prime factor greater than P will be rejected. But assume that all prime factors of a are less than or equal to P, except one prime factor p_a which is larger. The corresponding quadratic form cannot be used directly without increasing the value of P. But assume that for two values of a, i.e. for two quadratic forms $f = (a, b, c)$ and $g = (a', b', c')$, the large prime p_a is the same. Then either the form fg^{-1} or the form fg (depending on whether $b' \equiv b \pmod{p_a}$ or not) will give us a relation in which no primes larger than P will occur, hence a useful relation. The coincidence of two values of p_a will not be a rare phenomenon, and for

large discriminants the improvement will be considerable. See Exercise 14 for some hints on how to implement the large prime variation.

(4) Note that the '10' and '5' which occur in the algorithm are quite arbitrary, but are usually sufficient in practice. Note also that the correctness of the result is guaranteed only if one assumes GRH. Hence, this is a conditional algorithm, but in a much more precise sense than Algorithm 5.4.10.

(5) In step 5, we need to factor a using trial division. Now a can be as large as $\sqrt{|D|/3}$, hence a may have more than 20 digits in the region we are aiming for, and factoring by trial division may seem too costly. We have seen however that M is a few thousand at most in this region, so using trial divisors up to M is reasonable. We can improve on this by using the early abort strategy which will be explained in Chapter 10.

(6) Step 9 requires computing a determinant using the Chinese remainder theorem (although as seen in Exercise 13 we can also compute it directly). This means that we first compute it modulo sufficiently many small primes. Then, by using the Chinese remainder Algorithm 1.3.12, we can obtain it modulo the product of these primes. Finally, Hadamard's inequality (Proposition 2.2.4) gives us an upper bound on the result. Hence, if the product of our primes is greater than twice this upper bound, we find the value of the determinant exactly. We have already mentioned this method in Section 4.3 for computing norms of algebraic integers.

The Hadamard bound may, however, be extremely large, and in that case it is preferable to proceed as follows. We take many more extra relations than needed (say 100 instead of 10) and we must assume that we will obtain the class number itself and not a multiple of it. Then the quantity $B\sqrt{2}$ is an upper bound for the determinant and can be used instead of the Hadamard bound. Once the class group is obtained, we must then check that it is correct, and this can be done without too much difficulty (or we can stop and assume that the result is correct).

(7) Finally, the main point of this method is, of course, its speed since under reasonable hypotheses one can prove that the expected asymptotic average running time is

$$O(L(|D|)^\alpha)$$

with $\alpha = \sqrt{2}$, and perhaps even $\alpha = \sqrt{9/8}$. This is much faster than any of the preceding methods. Furthermore, it can be hoped that one can bring down the constant α to 1. This seems to be the limit of what one can expect to achieve on the subject for the following reason. Many fast factoring methods are known, using very different methods. To mention just a few, there is one using the 2-Sylow subgroup of the class group, one using elliptic curves (ECM), and a sieve type method (MPQS). All these methods have a common expected running time of the order of $O(L(N))$. In 1989, the discovery of the number field sieve lowered this running time to $O(e^{\ln(N)^{1/3+\epsilon}})$ (see Chapter 10), but this becomes better than the preceding methods for special numbers having more than 100 digits, and for general numbers having more than (perhaps) 130 digits,

hence does not concern us here. Since computing the class group is at
least as difficult as factoring, one cannot expect to find a significantly faster
method than McCurley's algorithm without fundamentally new ideas. It is
plausible, however, that using ideas from the number field sieve would give
an $O(e^{\ln(N)^{1/3+\epsilon}})$ algorithm, but nobody knows how to do this at the time
of this writing. In practice, using Section 6.5, we may speedup Algorithm
5.5.2 by finding some of the relations using the basic number field sieve
idea (see remark (3) after Algorithm 6.5.9).

5.5.3 Atkin's Variant

A variant of the above algorithm has been proposed by Atkin. It has the
advantage of being faster, but the disadvantage of not always giving the class
group. Atkin's idea is as follows.

Instead of taking P_0, which is already a small subset of the factor base of
prime forms, to generate the relations, we choose a *single form* f. Of course,
there is now no reason for f to generate the class group, but at least when
the discriminant is prime this often happens, as tables and the heuristics of
[Coh-Len1] show (see Section 5.10).

We then determine the order of f in the class group, using a method
which is more efficient than the baby-step giant-step Algorithm 5.4.1 for large
discriminants, since it is also a sub-exponential algorithm. The improvement
comes, as in McCurley's algorithm, from the use of a factor base. (The phi-
losophy being that any number-theoretic algorithm which can be made to
efficiently use factor bases automatically becomes sub-exponential thanks to
the theorem of Canfield-Erdős-Pomerance 10.2.1 that we will see in Chapter
10.)

To compute the order of f, we start with the same two steps as Algorithm
5.5.2. In particular, we set n equal to the number of primeforms in our factor
base.

We now compute the reduced forms equivalent to f, f^2, f^3, ... For each
such form (a, b, c) we execute step 5 of Algorithm 5.5.2, i.e. we check whether
the form factors on our factor base, and if it does, we keep the corresponding
relation.

We continue in this way until exactly $n + 1$ relations have been obtained,
i.e. one more than the cardinality of the factor base. Let $e_1, e_2, \ldots, e_{n+1}$ be
the exponents of f for which we have obtained a relation. Since we have now
an $n \times (n + 1)$ matrix with integral entries, there exists a non-trivial linear
relation between the columns with integral coefficients, and this relation can
be obtained by simple linear algebra, *not* by using number-theoretic methods
such as Hermite normal form computations which are much slower. We can
for example use a special case of Algorithm 2.3.1.

Now, if C_i is column number i of our matrix, for $1 \leq i \leq n + 1$, and if x_i
are the coefficients of our relation, so that $\sum_{1 \leq i \leq n+1} x_i C_i = 0$, then clearly

$$f^N = 1 \ , \quad \text{where} \quad N = \sum_{1 \le i \le n+1} x_i e_i \ .$$

This is exactly the kind of relation that one obtains by using the baby-step giant-step method, but the running time can be shown to be sub-exponential as in McCurley's algorithm.

The relation may of course be trivial, i.e. we may have $N = 0$. This happens rarely however. Furthermore, if it does happen, we may have at our disposal more independent relations between the columns of our $n \times (n + 1)$ matrix, which are also given by Algorithm 2.3.1. If not, we take higher powers of f until we obtain a non-trivial relation.

As soon as we have a non-zero N such that $f^N = 1$, we can compute the exact order of f in the class group as in Algorithm 5.4.1, after having factored N. Of course, this factorization may not be easy, but N is probably of similar size as the class number, hence about $\sqrt{|D|}$, so even if D has 60 digits, we probably will have to factor a number having around 30 digits, which is not too difficult.

If e is the exact order of f, we know that e divides the class number. If e already satisfies the lower bound inequalities given by the Euler product, that is if

$$e > \frac{1}{\sqrt{2}} \frac{\sqrt{|D|}}{\pi} \prod_{p \le P} \left(1 - \frac{\left(\frac{D}{p} \right)}{p} \right)^{-1} ,$$

then assuming GRH, we must have $e = h(D)$, and the class group is cyclic and generated by f. When it applies, this gives a faster method to compute the class number and class group than McCurley's algorithm. If the inequality is not satisfied, we can proceed with another form, as in Algorithm 5.4.1. The details are left to the reader.

Note that according to tables and the heuristic conjectures of [Coh-Len1] (see Section 5.10), the odd part of the class group should very often be cyclic (probability greater than 97%). Hence, if the discriminant D is prime, so that the class number is odd, there is a very good chance that $Cl(D)$ is cyclic. Furthermore, the number of generators of a cyclic group with h elements is $\phi(h)$, and this is also quite large, so there is a good chance that our randomly chosen f will generate the class group.

The implementation details of Atkin's algorithm are left to the reader (see Exercise 15).

5.6 Class Groups of Real Quadratic Fields

We now consider the problem of computing the class group and the regulator of a real quadratic field $K = \mathbb{Q}(\sqrt{D})$, and more generally of the unique real quadratic order of discriminant D. We will consider the problem of computing the regulator in Section 5.7, so we assume that we already have computed the regulator which we will denote by $R(D)$.

5.6.1 Computing Class Numbers Using Reduced Forms

Thanks to Theorem 5.2.9, we still have a correspondence between the narrow ideal class group and equivalence classes of quadratic forms of the same discriminant D. It is not difficult to have a correspondence with the ideal class group itself.

Proposition 5.6.1. *If D is a non-square positive integer congruent to 0 or 1 modulo 4, the maps ψ_{FI} and ψ_{IF} of Theorem 5.2.9 induce inverse isomorphisms between $Cl(D)$ and the quotient set of $\mathcal{F}(D)$ obtained by identifying the class of (a, b, c) with the class of $(-a, b, -c)$.*

The proof is easy and left to the reader (Exercise 18).

The big difference between forms of negative and positive discriminant however is that, although one can define the notion of a reduced form (differently from the negative case), there will in general not exist only one reduced form per equivalence class, but several, which are naturally organized in a cycle structure.

Definition 5.6.2. *Let $f = (a, b, c)$ be a quadratic form with positive discriminant D. We say that f is* reduced *if we have*

$$|\sqrt{D} - 2|a|| < b < \sqrt{D} .$$

The justification for this definition, as well as for the definition in the case of negative discriminants, is given in Exercise 16.

Note immediately the following proposition.

Proposition 5.6.3. *Let (a, b, c) be a reduced form with positive discriminant D. Then*

(1) *$|a|$, b and $|c|$ are less than $\sqrt{D}$ and a and c are of opposite signs.*
(2) *More precisely we have $|a| + |c| < \sqrt{D}$.*
(3) *Finally, (a, b, c) is reduced if and only if $|\sqrt{D} - 2|c|| < b < \sqrt{D}$.*

Proof. The result for b is trivial, and since $ac = (b^2 - D)/4 < 0$ it is clear that a and c are of opposite signs. Now we have

$$|a| + |c| - \sqrt{D} = \frac{D - 4|a|\sqrt{D} + 4a^2 - b^2}{4|a|} = \frac{(\sqrt{D} - 2|a|)^2 - b^2}{4|a|} \, ,$$

hence by definition of reduced we have $|a| + |c| - \sqrt{D} < 0$, which implies (2) and hence (1).

To prove (3), we note that we have the identity

$$2|c| - \sqrt{D} = \frac{(\sqrt{D} - |a|)^2 - a^2 - b^2}{2|a|} \, ,$$

hence if $\epsilon = \pm 1$, we have

$$b - \epsilon(2|c| - \sqrt{D}) = \frac{(\sqrt{D} + \epsilon b)(b + \epsilon(2|a| - \sqrt{D}))}{2|a|}$$

which is positive by definition. Since a and c play symmetrical roles, this proves (3) and hence the proposition. $\square$

If $\tau = (-b + \sqrt{D})/(2|a|)$ is the quadratic number associated to the form (a, b, c) as in Section 5.2, it is not difficult to show that (a, b, c) is reduced if and only if $0 < \tau < 1$ and $-\sigma(\tau) > 1$.

We now need a reduction algorithm on quadratic forms of positive discriminant. It is useful to give a preliminary definition:

Definition 5.6.4. *Let $D > 0$ be a discriminant. If $a \neq 0$ and b are integers, we define $r(b, a)$ to be the unique integer r such that $r \equiv b \pmod{2a}$ and $-|a| < r \leq |a|$ if $|a| > \sqrt{D}$, $\sqrt{D} - 2|a| < r < \sqrt{D}$ if $|a| < \sqrt{D}$. In addition, we define the reduction operator ρ on quadratic forms (a, b, c) of discriminant $D > 0$ by*

$$\rho(a, b, c) = \left(c, r(-b, c), \frac{r(-b, c)^2 - D}{4c} \right) .$$

The reduction algorithm is then simply as follows.

Algorithm 5.6.5 (Reduction of Indefinite Quadratic Forms). Given a quadratic form $f = (a, b, c)$ with positive discriminant D, this algorithm finds a reduced form equivalent to f.

1. [Iterate] If (a, b, c) is reduced, output (a, b, c) and terminate the algorithm. Otherwise, set $(a, b, c) \leftarrow \rho(a, b, c)$ and go to step 1.

We must show that this algorithm indeed produces a reduced form after a finite number of iterations. In fact, we have the following stronger result:

Proposition 5.6.6.

(1) *The number of iterations of ρ which are necessary to reduce a form (a, b, c) is at most $2 + \lceil \lg(|c|/\sqrt{D}) \rceil$.*
(2) *If $f = (a, b, c)$ is a reduced form, then $\rho(a, b, c)$ is again a reduced form.*
(3) *The reduced forms equivalent to f are exactly the forms $\rho^n(f)$, for n sufficiently large (i.e. n greater than or equal to the least n_0 such that $\rho^{n_0}(f)$ is reduced) and are finite in number.*

Proof. The proof of (1) is similar in nature to that of Proposition 5.4.3. Set $\rho(f) = (a', b', c')$. I first claim that if $|c| > \sqrt{D}$ then $|c'| \leq |c|/2$. Indeed, in that case $|r(-b, c)| \leq |c|$, hence

$$|c'| = \frac{|r(-b, c)^2 - D|}{4|c|} \leq \frac{2c^2}{4|c|} \leq \frac{|c|}{2}$$

since $D < c^2$. So, after at most $\lceil \lg(|c|/\sqrt{D}) \rceil$ iterations, we will end up with a form where $|c| < \sqrt{D}$. As in the imaginary case one can then check that the form is almost reduced, in the sense that after another iteration of ρ we will have $|a|$, $|b|$ and $|c|$ less than $\sqrt{D}$, and then either the form is reduced, or it will be after one extra iteration. The details are left as an exercise for the reader.

For (2), note that if (a, b, c) is reduced, then

$$r(-b, c) = -b + 2|c| \left\lfloor \frac{b + \sqrt{D}}{2|c|} \right\rfloor ,$$

since this is clearly in the interval $[\sqrt{D} - 2|c|, \sqrt{D}]$. If $|c| < \sqrt{D}/2$, this implies that $\rho(a, b, c)$ is reduced by definition. If $|c| > \sqrt{D}/2$, it is clear that

$$r(-b, c) = -b + 2|c| > 2|c| - \sqrt{D} = |\sqrt{D} - 2|c|| ,$$

proving again that $\rho(a, b, c)$ is reduced.

Finally, to prove (3), set $\sigma(a, b, c) = (c, b, a)$. Using again Proposition 5.6.3 (3), it is clear that σ is an involution on reduced forms. Furthermore, one checks immediately that $\rho\sigma$ and $\sigma\rho$ are both involutions on the set of reduced forms, thus proving that ρ is a permutation of this set, the inverse of ρ being $\rho^{-1} = \sigma\rho\sigma$.

Another way to see this is to check directly that the inverse of ρ on reduced forms is given explicitly by

$$\rho^{-1}(a, b, c) = \left(\frac{r(-b, a)^2 - D}{4a}, r(-b, a), a \right) ,$$

and ρ^{-1} can be used instead of ρ to reduce a form, although one must take care that for non-reduced forms, it will *not* be the inverse of ρ since ρ is not one-to-one. $\square$

We can summarize Proposition 5.6.6 by saying that if we start with any form f, the sequence $\rho^n(f)$ is ultimately periodic, and we arrive inside the period exactly when the form is reduced.

Finally, note that it follows from Proposition 5.6.3 that the set of reduced forms of discriminant D has cardinality at most D (the possible number of pairs (a, b)), but a closer analysis shows that its cardinality is $O(D^{1/2} \ln D)$.

It follows from the above discussion and results that in every equivalence class of quadratic forms of discriminant $D > 0$, there is not only one reduced form, but a cycle of reduced forms (cycling under the operation ρ), and so the class number is the number of such *cycles*.

It is not necessary to formally write an algorithm analogous to Algorithm 5.3.5 for computing the class number using reduced forms. We make a list of all the reduced forms of discriminant D by testing among all pairs (a, b) such that $|a| < \sqrt{D}$, $|\sqrt{D} - 2|a|| < b < \sqrt{D}$ and $b \equiv D \pmod 2$, those for which $b^2 - D$ is divisible by $4a$. Then we count the number of orbits under the permutation ρ, and the result is the narrow class number $h^+(D)$. If, in addition, we identify the forms (a, b, c) and $(-a, b, -c)$, then, according to Proposition 5.6.1 we obtain the class number $h(D)$ itself.

As for Algorithm 5.3.5, this is an algorithm with $O(D)$ execution time, so is feasible only for discriminants up to 10^6, say. Hence, as in the imaginary case, it is necessary to find better methods.

For future reference, let us determine the exact correspondence between the action of ρ and the continued fraction expansion of a quadratic irrationality.

In Section 5.2 we have defined maps ϕ_{FI} and ϕ_{IQ}, and by composition, Theorem 5.2.4 tells us that the map ϕ_{FQ} from I to $Q \times \mathbb{Z}/2\mathbb{Z}$ defined by

$$\phi_{FQ}(a, b, c) = \left(\frac{-b + \sqrt{D}}{2|a|}, \operatorname{sign}(a) \right)$$

is an isomorphism. (Note the absolute value of a, coming from the necessity of choosing an oriented basis for our ideals.)

From this, one checks immediately that if $f = (a, b, c)$ is reduced, and if $\phi_{FQ}(f) = (\tau, s)$, then

$$\phi_{FQ}(\rho(f)) = \left(\frac{1}{\tau} - \left\lfloor \frac{1}{\tau} \right\rfloor, -s \right) ,$$

where by abuse of notation we still use the notation ϕ_{FQ} for the map at the level of forms and not at the level of classes of forms modulo Γ_∞.

For ρ^{-1} we define

$$\psi_{FQ}(a, b, c) = \left(\frac{b + \sqrt{D}}{2|a|}, \operatorname{sign}(a) \right) .$$

Then, if $f = (a, b, c)$ is reduced and $\psi_{FQ}(f) = (\tau', s)$, we have

$$\psi_{FQ}(\rho^{-1}(f)) = \left(\frac{1}{\tau' - \lfloor \tau' \rfloor}, -s\right) .$$

Thus the action of ρ and ρ^{-1} on reduced forms correspond exactly to the continued fraction expansion of τ and $\tau' = -\sigma(\tau)$ respectively, with in addition a ± 1 variable which gives the parity of the number of reduction steps.

In addition, since ρ and ρ^{-1} are inverse maps on reduced forms, we obtain as a corollary of Proposition 5.6.6 the following.

Corollary 5.6.7. *Let $\tau = (-b + \sqrt{D})/(2|a|)$ corresponding to a reduced quadratic form (a, b, c). Then the continued fraction expansion of τ is purely periodic, and the period of the continued fraction expansion of $-\sigma(\tau) = (b + \sqrt{D})/(2|a|)$ is the reverse of that of τ.*

5.6.2 Computing Class Numbers Using Analytic Formulas

We will follow closely Section 5.3.3. The definition of $L_D(s)$ is the same, but the functional equation is slightly different:

Proposition 5.6.8. *Let D be a positive fundamental discriminant, and define*

$$L_D(s) = \sum_{n \geq 1} \left(\frac{D}{n}\right) n^{-s} .$$

This series converges for $\mathrm{Re}(s) > 1$, and defines an analytic function which can be analytically continued to the whole complex plane to an entire function satisfying

$$\Lambda_D(1 - s) = \Lambda_D(s) ,$$

where we have set

$$\Lambda_D(s) = (D/\pi)^{s/2} \Gamma(s/2) L_D(s) .$$

Note that the special case $D = 1$ of this proposition (which is excluded since it is not the discriminant of a quadratic field) is still true if one adds the fact that the function has a simple pole at $s = 1$. In that case, we simply recover the usual functional equation of the Riemann zeta function. The link with the class number and the regulator is as follows. (Recall that the regulator $R(D)$ is in our case the logarithm of the unique generator greater than 1 of the torsion free part of the unit group.)

Proposition 5.6.9. *If D is a positive fundamental discriminant, then*

$$L_D(1) = \frac{2h(D)R(D)}{\sqrt{D}} \ .$$

Note that as in the imaginary case, these results are special cases of Theorem 4.9.12 using the identity $\zeta_K(s) = \zeta(s)L_D(s)$ for $K = \mathbb{Q}(\sqrt{D})$.

Also, as in the imaginary case, it is not very reasonable to compute $L_D(1)$ directly from this formula since its defining series converges so slowly. However, a suitable reordering of the series gives the following:

Corollary 5.6.10. *If D is a positive fundamental discriminant, then*

$$h(D)R(D) = - \sum_{r=1}^{\lfloor (D-1)/2 \rfloor} \left(\frac{D}{r}\right) \ln \sin \left(\frac{r\pi}{D}\right) \ .$$

As usual, this kind of formula, although a finite sum, is useless from a computational point of view, and is worse than the method of reduced forms, although maybe slightly simpler to program. If we also use the functional equation we obtain a considerable improvement, leading to a complicated but much more efficient formula:

Proposition 5.6.11. *If D is a positive fundamental discriminant, then*

$$2h(D)R(D) = \sum_{n\geq 1} \left(\frac{D}{n}\right) \left(\frac{\sqrt{D}}{n} \operatorname{erfc}\left(n\sqrt{\frac{\pi}{D}}\right) + E_1\left(\frac{\pi n^2}{D}\right)\right) \ ,$$

where $\operatorname{erfc}(x)$ *is the complementary error function (see Propositions 5.3.14 and 5.3.15), and $E_1(x)$ is the exponential integral function defined by*

$$E_1(x) = \int_x^{\infty} \frac{e^{-t}}{t} \, dt \ .$$

Note that the function $E_1(x)$ can be computed efficiently using the following formulas.

Proposition 5.6.12.

(1) *We have for all x*

$$E_1(x) = -\gamma - \ln(x) + \sum_{k\geq 1}(-1)^{k-1}\frac{x^k}{k!k} \ ,$$

where $\gamma = 0.57721566490153286\ldots$ is Euler's constant, and this should be used when x is small, say $x \leq 4$.

(2) *We have for all $x > 0$*

$$E_1(x) = \frac{e^{-x}}{x}\left(1 - \cfrac{1}{2 + x - \cfrac{1 \cdot 2}{4 + x - \cfrac{2 \cdot 3}{6 + x - \cdots}}}\right),$$

and this should be used for x large, say $x \geq 4$.

Implementation Remark. The remark made after Proposition 5.3.15 is also valid here, the general formula being here

$$2h(D)R(D) = \sum_{n \geq 1} \left(\frac{D}{n}\right)\left(\frac{AD}{n} \operatorname{erfc}\left(n\sqrt{\frac{\pi}{AD}}\right) + E_1\left(\frac{\pi n^2 A}{D}\right)\right).$$

These results show that the series given in Proposition 5.6.11 converges exponentially, and since $h(D)$ is an integer and $R(D)$ has been computed beforehand, it is clear that the computation time of $h(D)$ by this method is $O(D^{1/2+\epsilon})$ for any $\epsilon > 0$. As in the case $D < 0$ it would be easy to give an upper bound for the number of terms that one must take in the series. This is left as an exercise for the reader. See also Exercise 28 for a way to avoid computing the transcendental functions erfc and E_1.

5.6.3 A Heuristic Method of Shanks

An examination of the heuristic conjectures of [Coh-Len1] (see Section 5.10) shows that one must expect that, on average, the class number $h(D)$ will be quite small for positive discriminants, in contrast to the case of negative discriminants. Hence, one can use the following method, which is of course not an algorithm, but has a very good chance of giving the correct result quite quickly.

Heuristic Algorithm 5.6.13 (Class Number for $D > 0$). Given a positive fundamental discriminant D, this algorithm computes a value which has a pretty good chance of being equal to the class number $h(D)$. As always, we assume that the regulator $R(D)$ has already been computed. We denote by p_i the ith prime number.

1. [Regulator small?] If $R(D) < D^{1/4}$, then output a message saying that the algorithm will probably not work, and terminate the algorithm.
2. [Initialize] Set $h_1 \leftarrow \sqrt{D}/(2R(D))$, $h \leftarrow 0$, $c \leftarrow 0$, $k \leftarrow 0$.
3. [Compute block] Set

$$h_1 \leftarrow h_1 \prod_{500k < i \leq 500(k+1)} \left(1 - \frac{\left(\frac{D}{p_i}\right)}{p_i}\right)^{-1},$$

$m \leftarrow \lfloor h_1 \rfloor$, $k \leftarrow k + 1$.

4. [Seems integral?] If $|m - h_1| > 0.1$ set $c \leftarrow 0$ and go to step 3.

5. [Seems constant?] If $m \neq h$, set $h \leftarrow m$ and $c \leftarrow 1$ and go to step 3. Otherwise, set $c \leftarrow c + 1$. If $c \leq 5$ go to step 3, otherwise output h as the tentative class number and terminate the algorithm.

The reason for the frequent success of this algorithm is clear. Although we use the slowly convergent Euler product for $L_D(1)$, if the regulator is not too small, the integer m computed in step 3 has a reasonable chance of being equal to the class number. The heuristic criterion that we use, due to Shanks, is that if the Euler product is less than 0.1 away from the same integer h for 6 consecutive blocks of 500 prime numbers, we assume that h is the class number. In fact, assuming GRH, this heuristic method can be made completely rigorous. I refer to [Mol-Wil] for details. In practice it works quite well, except of course for the quite rare cases in which the regulator is too small.

We still have not given any method for computing the structure of the class group. Before considering this point, we now consider the question of computing the regulator of a real quadratic field.

5.7 Computation of the Fundamental Unit and of the Regulator

As we have seen, reduced forms are grouped into $h(D)$ cycles under the permutation ρ. We will see that one can define a distance between forms which, in particular, has the property that the length of each cycle is the same, and equal to the regulator. Note that this is absolutely *not* true for the naïve length defined as the number of forms.

5.7.1 Description of the Algorithms

The action of ρ and ρ^{-1} corresponding to the continued fraction expansion of the quadratic irrationals τ and $-\sigma(\tau)$ respectively, it is clear that we must be able to compute the fundamental unit and the regulator from these expansions. From Corollary 5.6.7, we know that one of these expansions will be reverse of the other, so we can choose as we like between the two.

It is slightly simpler to use the expansion of $-\sigma(\tau)$, and this leads to the following algorithm whose validity will be proved in the next section. Note that in this algorithm we assume $a > 0$, but it is easy to modify it so that it stays valid in general (Exercise 20).

Algorithm 5.7.1 (Fundamental Unit Using Continued Fractions). Given a quadratic irrational $\tau = (-b + \sqrt{D})/(2a)$ where $4a \mid (D - b^2)$ and $a > 0$, corresponding to a *reduced* form $(a, b, (b^2 - D)/(4a))$, this algorithm computes the fundamental unit ε of $\mathbb{Q}(\sqrt{D})$ using the ordinary continued fraction expansion of $-\sigma(\tau)$.

1. [Initialize] Set $u_1 \leftarrow -b$, $u_2 \leftarrow 2a$, $v_1 \leftarrow 1$, $v_2 \leftarrow 0$, $p \leftarrow b$ and $q \leftarrow 2a$. Precompute $d \leftarrow \lfloor \sqrt{D} \rfloor$.

2. [Euclidean step] Set $A \leftarrow \lfloor (p + d)/q \rfloor$, then in that order, set $p \leftarrow Aq - p$ and $q \leftarrow (D - p^2)/q$. Finally, set $t \leftarrow Au_2 + u_1$, $u_1 \leftarrow u_2$, $u_2 \leftarrow t$, $t \leftarrow Av_2 + v_1$, $v_1 \leftarrow v_2$, and $v_2 \leftarrow t$.

3. [End of period?] If $q = 2a$ and $p \equiv b \pmod{2a}$, set $u \leftarrow |u_2/a|$, $v \leftarrow |v_2/a|$ (both divisions being exact), output $\varepsilon \leftarrow (u + v\sqrt{D})/2$, and terminate the algorithm. Otherwise, go to step 2.

As will be proved in the next section, the result of this algorithm is the fundamental unit, independently of the initial reduced form. Hence, the simplest solution is to start with the unit reduced form, i.e. with $\tau = (-b + \sqrt{D})/2$ and $b = d$ if $d \equiv D \pmod 2$, $b = d - 1$ otherwise, where as in the algorithm $d = \lfloor \sqrt{D} \rfloor$.

Also, note that the form corresponding to $(p + \sqrt{D})/q$ at step i is

$$((-1)^i q/2, p, (-1)^i (p^2 - D)/(2q)) .$$

If we had wanted the exact action of ρ^{-1}, we would have to put $q \leftarrow (p^2 - D)/q$ instead of $q \leftarrow (D - p^2)/q$ in step 2 of the algorithm, and then q would alternate in sign instead of always being positive.

Now the continued fraction expansion of the quadratic irrational corresponding to the unit reduced form is not only periodic, but in fact symmetric. This is true more generally for forms belonging to *ambiguous* cycles, i.e. forms whose square lie in the principal cycle (see Exercise 22). Hence, it is possible to divide by two the number of iterations in Algorithm 5.7.1. This leads to the following algorithm, whose proof is left to the reader.

Algorithm 5.7.2 (Fundamental Unit). Given a fundamental discriminant $D > 0$, this algorithm computes the fundamental unit of $\mathbb{Q}(\sqrt{D})$.

1. [Initialize] Set $d \leftarrow \lfloor \sqrt{D} \rfloor$. If $d \equiv D \pmod 2$, set $b \leftarrow d$ otherwise set $b \leftarrow d - 1$. Then set $u_1 \leftarrow -b$, $u_2 \leftarrow 2$, $v_1 \leftarrow 1$, $v_2 \leftarrow 0$, $p \leftarrow b$ and $q \leftarrow 2$.

2. [Euclidean step] Set $A \leftarrow \lfloor (p + d)/q \rfloor$, $t \leftarrow p$ and $p \leftarrow Aq - p$. If $t = p$ and $v_2 \neq 0$, then go to step 4, otherwise set $t \leftarrow Au_2 + u_1$, $u_1 \leftarrow u_2$, $u_2 \leftarrow t$, $t \leftarrow Av_2 + v_1$, $v_1 \leftarrow v_2$, and $v_2 \leftarrow t$, $t \leftarrow q$, $q \leftarrow (D - p^2)/q$.

3. [Odd period?] If $q = t$ and $v_2 \neq 0$, set $u \leftarrow |(u_1 u_2 + D v_1 v_2)/q|$, $v \leftarrow |(u_1 v_2 + u_2 v_1)/q|$ (both divisions being exact), output $\varepsilon \leftarrow (u + v\sqrt{D})/2$ and terminate the algorithm. Otherwise, go to step 2.

4. [Even period] Set $u \leftarrow |(u_2^2 + v_2^2 D)/q|$, $v \leftarrow |2u_2 v_2/q|$ (both divisions being exact), output $\varepsilon \leftarrow (u + v\sqrt{D})/2$ and terminate the algorithm.

The performance of both these algorithms is quite reasonable for discriminants up to 10^6. It can be proved that the number of steps is $O(D^{1/2+\epsilon})$ for all $\epsilon > 0$. Furthermore, all the computations on p and q are done with numbers less than $2\sqrt{D}$, hence of reasonable size. The main problem is that the fundamental unit itself has coefficients u and v which are of unreasonable size. One can show that $\ln u$ and $\ln v$ can be as large as $\sqrt{D}$. Hence, although the number of steps is $O(D^{1/2+\epsilon})$, this does not correctly reflect the practical execution time, since multi-precision operations become predominant. In fact, it is easy to see that the only bound one can give for the execution time itself is $O(D^{1+\epsilon})$.

The problem is therefore not so much in computing the numbers u and v, which do not make much sense when they are so large, but in computing the regulator itself to some reasonable accuracy, since after all, this is all we need in the class number formula. It would seem that it is not possible to compute $R(D)$ without computing ε exactly, but luckily this is not the case, and there is a variant of Algorithm 5.7.2 (or 5.7.1) which gives the regulator instead of the fundamental unit. This variant uses floating point numbers, which must be computed to sufficient accuracy (but not unreasonably so: double precision, i.e. 15 decimals, is plenty). The advantage is that no numbers will become large.

5.7.2 Analysis of the Continued Fraction Algorithm

To do this, we must analyze the behavior of the continued fraction algorithm, and along the way we will prove the validity of Algorithm 5.7.1. We assume for the sake of simplicity that $a > 0$ (hence $c < 0$), although the same analysis holds in general.

Call p_i, q_i, A_i, $u_{1,i}$, $u_{2,i}$, $v_{1,i}$, $v_{2,i}$ the quantities occurring in step i of the algorithm, where the initializations correspond to step 0, and set for $i \geq -1$, $a_i = u_{1,i+1}$, $b_i = v_{1,i+1}$. Then we can summarize the recursion implicit in the algorithm by the following formulas:

For all $i \geq 0$, $u_{1,i} = a_{i-1}$, $u_{2,i} = a_i$, $v_{1,i} = b_{i-1}$, $v_{2,i} = b_i$. Furthermore:

$p_0 = b$, $q_0 = 2a$, $a_{-1} = -b$, $a_0 = 2a$, $b_{-1} = 1$, $b_0 = 0$ (recall that $a = 1$ in Algorithm 5.7.2), and for $i \geq 0$:

$A_i = \lfloor (p_i + d)/q_i \rfloor$, $p_{i+1} = A_i q_i - p_i$, $q_{i+1} = (D - p_{i+1}^2)/q_i$, $a_{i+1} = A_i a_i + a_{i-1}$, $b_{i+1} = A_i b_i + b_{i-1}$.

By the choice of b, we know that $q_0 \mid D - p_0^2$, and if by induction we assume that all the above quantities are integers and that $q_i \mid D - p_i^2$, one sees that $D - p_{i+1}^2 \equiv D - p_i^2 \equiv 0 \pmod{q_i}$, hence q_{i+1} is an integer. In addition, we clearly have $q_{i+1} \mid D - p_{i+1}^2$ since the quotient is simply q_i, thus proving our claim by induction. We also have $q_{i+1} - q_{i-1} = (D - p_{i+1}^2)/q_i - (D - p_i^2)/q_i = (p_i - p_{i+1})(p_i + p_{i+1})/q_i$, hence we obtain the formula

$$q_{i+1} = q_{i-1} - A_i(p_{i+1} - p_i) \ ,$$

which is in general computationally simpler than the formula used in the algorithms.

That the algorithms above correspond to the continued fraction expansion of $(b+\sqrt{D})/(2a)$ (where in Algorithm 5.7.2 it is understood that we take $a = 1$) is quite clear. Set $\zeta_i = (p_i + \sqrt{D})/q_i$. Then we have $\zeta_0 = (b + \sqrt{D})/(2a)$, $A_i = \lfloor \zeta_i \rfloor$, and hence

$$\frac{1}{\zeta_i - \lfloor \zeta_i \rfloor} = \frac{q_i}{p_i - A_i q_i + \sqrt{D}} = \frac{A_i q_i - p_i + \sqrt{D}}{(D - (A_i q_i - p_i)^2)/q_i} = \zeta_{i+1} \ ,$$

thus giving the above formulas.

This is of course nothing other than the translation of the formula giving $\psi_{FQ}(\rho^{-1}(f))$ in terms of $\psi_{FQ}(f)$.

Note that in practice the computations on the pair (p, q) should be done in the following way: use three extra variables r and p_1, q_1. Replace steps 1 and 2 of Algorithm 5.7.2 by

1'. [Initialize] Set $d \leftarrow \lfloor \sqrt{D} \rfloor$. If $d \equiv D \pmod 2$, set $b \leftarrow d$ otherwise set $b \leftarrow d - 1$. Then set $u_1 \leftarrow -b$, $u_2 \leftarrow 2$, $v_1 \leftarrow 1$, $v_2 \leftarrow 0$, $p \leftarrow b$ and $q \leftarrow 2$, $q_1 \leftarrow (D - p^2)/q$.

2'. [Euclidean step] Let $p + d = qA + r$ with $0 \leq r < q$ be the Euclidean division of $p + d$ by q, and set $p_1 \leftarrow p$, $p \leftarrow d - r$. If $p_1 = p$ and $v_2 \neq 0$, then go to step 4, otherwise set $t \leftarrow Au_2 + u_1$, $u_1 \leftarrow u_2$, $u_2 \leftarrow t$, $t \leftarrow Av_2 + v_1$, $v_1 \leftarrow v_2$, and $v_2 \leftarrow t$, $t \leftarrow q$, $q \leftarrow q_1 - A(p - p_1)$, $q_1 \leftarrow t$.

This has the same effect as steps 1 and 2 of Algorithm 5.7.2, but avoids one division in each loop. Note that this method can also be used in general.

Now that we have seen that we are computing the continued fraction expansion of $(b + \sqrt{D})/(2a)$, we must study the behavior of the sequences a_i and b_i. This is summarized in the following proposition.

Proposition 5.7.3. *With the above notations, we have*

(1)

$$\frac{a_{i+1} + b_{i+1}\sqrt{D}}{a_i + b_i\sqrt{D}} = \frac{p_{i+1} + \sqrt{D}}{q_i} \ ,$$

(2)

$$a_i b_{i-1} - a_{i-1} b_i = (-1)^i 2a \ ,$$

(3)

$$a_i^2 - b_i^2 D = (-1)^i 2a q_i \ ,$$

(4)

$$a_i a_{i-1} - b_i b_{i-1} D = (-1)^{i-1} 2a p_i \ ,$$

(5)
$$\sqrt{D} = \frac{a_i \zeta_i + a_{i-1}}{b_i \zeta_i + b_{i-1}} \ ,$$

where as before $\zeta_i = (p_i + \sqrt{D})/q_i$.

Proof. Denote real conjugation $\sqrt{D} \mapsto -\sqrt{D}$ in the field $\mathbb{Q}(\sqrt{D})$ by σ, and set $\rho_i = (p_i + \sqrt{D})/q_{i-1}$. Then $\rho_{i+1} = A_i - \sigma(\zeta_i)$ and since $\zeta_{i+1} = 1/(\zeta_i - A_i)$ we have by applying σ,

$$\sigma(\zeta_{i+1}) = 1/(\sigma(\zeta_i) - A_i) = -1/\rho_{i+1} \ .$$

Therefore $\rho_{i+1} = A_i - \sigma(\zeta_i) = A_i + 1/\rho_i$. On the other hand, to be compatible with the recursions, we must define $q_{-1} = (D - b^2)/(2a)$. Thus we see that $\rho_0 = 2a/(\sqrt{D} - b)$ (which comes also from the formula $\rho_i = -1/\sigma(\zeta_i)$). If we set $\alpha_i = a_i + b_i\sqrt{D}$, the recursions show that $\alpha_{i+1} = A_i\alpha_i + \alpha_{i-1}$. Therefore if we set $\beta_i = \alpha_i/\alpha_{i-1}$, we have $\beta_{i+1} = A_i + 1/\beta_i$, and this is the same recursion satisfied by ρ_i. Since we have $\beta_0 = 2a/(\sqrt{D} - b) = \rho_0$, this shows that $\beta_i = \rho_i$ for all i, thus showing (1).

Formula (2) is a standard formula in continued fraction expansions: we have the matrix recursion

$$\begin{pmatrix} a_{i+1} & b_{i+1} \\ a_i & b_i \end{pmatrix} = \begin{pmatrix} A_i & 1 \\ 1 & 0 \end{pmatrix} \begin{pmatrix} a_i & b_i \\ a_{i-1} & b_{i-1} \end{pmatrix} \ ,$$

hence formula (2) follows trivially on taking determinants and noticing that $a_0 b_{-1} - a_{-1}b_0 = 2a$.

To prove (3), we take the norm (with respect to $\mathbb{Q}(\sqrt{D})/\mathbb{Q}$) of formula (1). We obtain:

$$\frac{a_{i+1}^2 - b_{i+1}^2 D}{a_i^2 - b_i^2 D} = \frac{p_{i+1}^2 - D}{q_i^2} = -\frac{q_{i+1}}{q_i} \ ,$$

hence by multiplying out we obtain

$$a_i^2 - b_i^2 D = (-1)^i q_i \frac{a_0^2 - b_0^2 D}{q_0} = (-1)^i 2aq_i \ ,$$

showing (3).

Finally, to prove (4) we take the trace (with respect to $\mathbb{Q}(\sqrt{D})/\mathbb{Q}$) of formula (1). We obtain:

$$\frac{a_{i+1} + b_{i+1}\sqrt{D}}{a_i + b_i\sqrt{D}} + \frac{a_{i+1} - b_{i+1}\sqrt{D}}{a_i - b_i\sqrt{D}} = \frac{2p_{i+1}}{q_i} \ ,$$

hence grouping and using (3) we get

$$\frac{2a_{i+1}a_i - 2b_{i+1}b_i D}{(-1)^i 2aq_i} = \frac{2p_{i+1}}{q_i} \ ,$$

and this proves (4).

Formula (5) follows easily from (1) and its proof is left to the reader. □

Corollary 5.7.4. Set $c = (b^2 - D)/(4a)$, so $D = b^2 - 4ac$. Define two sequences c_i and d_i by $c_{-1} = 0$, $c_0 = 1$, $c_{i+1} = A_i c_i + c_{i-1}$, and $d_{-1} = -2c$, $d_0 = b$ and $d_{i+1} = A_i d_i + d_{i-1}$. Then the five formulas of Proposition 5.7.3 hold with (a, a_i, b_i) replaced by (c, d_i, c_i).

The proof is easy and left to the reader.

Now for simplicity let us consider the case of Algorithm 5.7.1. Let $i = k$ be the stage at which we stop, i.e. for which $q_k = 2a$ and $p_k \equiv b \pmod{2a}$. Then we output $\varepsilon = (|a_k| + |b_k|\sqrt{D})/|2a|$. We are going to show that this is indeed the correct result. First, I claim that ε is a unit. Indeed, notice that using (3), the norm of ε is equal to $(-1)^k$. Hence, to show that ε is a unit, it is only necessary to show that it is an algebraic integer. Moreover, since its norm is equal to ± 1, hence integral, we must only show that the trace of ε is integral, i.e. that $a_k \equiv 0 \pmod{a}$.

For this, we use the sequence c_i defined in Corollary 5.7.4. It is clear that we have $a_i = 2ac_i - bb_i$. From Proposition 5.7.3 (3) an easy computation gives

$$b_k(cb_k - bc_k) = a((-1)^k \frac{q_k}{2a} - c_k^2) \equiv 0 \pmod{a} ,$$

since $q_k = 2a$. Similarly, since $p_k \equiv b \pmod{2a}$, from (4) a similar computation gives

$$b_k(cb_{k-1} - bc_{k-1}) = a \left((-1)^{k-1} \frac{p_k - b}{2a} - c_k c_{k-1} \right) \equiv 0 \pmod{a} .$$

If we set $\delta_i = cb_i - bc_i$, it is clear by induction that

$$\gcd(\delta_k, \delta_{k-1}) = \gcd(\delta_{k-1}, \delta_{k-2}) = \cdots = \gcd(\delta_0, \delta_{-1}) = \gcd(b, c) .$$

From the two congruences proved above and the existence of u and v such that $u\delta_k + v\delta_{k-1} = \gcd(b, c)$, it follows that

$$b_k \gcd(b, c) \equiv 0 \pmod{a} .$$

But since D is a fundamental discriminant, the quadratic form (a, b, c) is primitive, hence $\gcd(a, b, c) = 1 = \gcd(\gcd(b, c), a)$, so we obtain $b_k \equiv 0 \pmod{a}$, hence also $a_k = 2ac_k - bb_k \equiv 0 \pmod{a}$ as was to be shown.

Now that we know that ε is a unit, we will show it is the fundamental unit. Since clearly $\varepsilon > 1$, this will follow from the following more general result. We say that an algebraic integer α is *primitive* if for any integer n, α/n is an algebraic integer only for $n = \pm 1$. Then we have:

Proposition 5.7.5. Let us keep all the above notations. Let $N \geq 1$ be a squarefree integer such that $\gcd(a, N) = 1$. Assume that $2|a|N < \sqrt{D}$.

Then the solutions (A, B) of the Diophantine equation

$$A^2 - B^2 D = \pm 4N, \quad \text{with } A > 0, \ B > 0 \text{ and } \frac{A + B\sqrt{D}}{2} \text{ primitive}$$

are given by $(A, B) = (|a_n/a|, |b_n/a|)$, for every n such that $q_n = 2|a|N$ and $p_n \equiv b \pmod{2a}$.

Proof. We have proven above that ε was an algebraic *integer* using only $q_k \equiv 0 \pmod{2a}$ and not precisely the value $q_k = 2a$. This shows that if the conditions of Proposition 5.7.5 are satisfied, we will have $a \mid a_n$ and $a \mid b_n$, and since by Proposition 5.7.3 (3) we have $a_n^2 - b_n^2 D = \pm 2a q_n = \pm 4a^2 N$, the pair $(A, B) = (|a_n/a|, |b_n/a|)$ is indeed a solution to our Diophantine equation with $A > 0$, $B > 0$, and since $\mathcal{N}((a_n + b_n\sqrt{D})/(2a)) = \pm N$ and that N is squarefree, $(A + B\sqrt{D})/2$ is primitive.

We must now show the converse. Assume that $A^2 - B^2 D = 4sN$ with $s = \pm 1$. Let $\tau' = -\sigma(\tau) = (b + \sqrt{D})/(2a)$ as in Algorithm 5.7.1. Then an easy calculation gives

$$\left| \tau' - \frac{A + bB}{2aB} \right| = \left| \frac{4N}{2aB^2 |\sqrt{D} + A/B|} \right| .$$

Now, $A/B = \sqrt{D \pm 4N/B^2} \geq \sqrt{D - 4N/B^2}$, hence

$$\left| \tau' - \frac{(A + bB)/2}{aB} \right| \leq \frac{4N}{2|a|B^2(\sqrt{D} + \sqrt{D - 4N/B^2})} .$$

We also have the following lemma whose proof is left to the reader. (See [H-W] for a slightly weaker version, but the proof is the same, see Exercise 21.)

Lemma 5.7.6. *If p and q are integers such that*

$$\left| \tau' - \frac{p}{q} \right| \leq \frac{1}{q(\max(2q - 1, 2))}$$

then p/q is a convergent in the continued fraction expansion of τ'.

Consider first the case $|a| > 1$. One easily checks that $4a^2 N^2 - 4N/B^2 > (2|a|N - 2N/B)^2$ is equivalent to $2|a|BN > N + 1$ which is clearly true. Hence, since $\sqrt{D} > 2|a|N$, we have $\sqrt{D} + \sqrt{D - 4N/B^2} > 4|a|N - 2N/B$ and therefore

$$\left| \tau' - \frac{(A + bB)/2}{aB} \right| < \frac{1}{|a|B(2|a|B - 1)} .$$

Since $b \equiv D \pmod 2$ and $A \equiv BD \pmod 2$, $(A + bB)/2$ is an integer, and so we can apply the lemma. This shows that $\frac{(A+bB)/2}{aB}$ is a convergent to τ'. A similar proof applies to the case $|a| = 1$, except when $B = 1$. But in the case $|a| = B = 1$, we have $D - 2\sqrt{D} < A^2 < D + 2\sqrt{D}$ hence either $\sqrt{D} - 1 < A < \sqrt{D} + 1$, and hence $|\tau' - (A+b)/2| < 1/2$ and we can conclude as before that $(A+b)/2$ is a convergent, or else $D - 2\sqrt{D} < A^2 < D - 2\sqrt{D} + 1$ which implies that $\tau' - 1 < (b + A)/2 < \tau'$ hence $(b + A)/2 = \lfloor \tau' \rfloor$ is also a convergent to τ'.

By definition, the convergents to τ' are c_n/b_n, and the equation $(A + bB)/(2aB) = c_n/b_n$ is equivalent to $A/B = a_n/b_n$.

Now we have the following lemma:

Lemma 5.7.7. *We have for all i,*

$$\left(\frac{p_i - b}{2}, \frac{q_i}{2}, a \right) = (b_i, a) \ ,$$

and

$$\frac{a_i + b_i \sqrt{D}}{2(b_i, a)}$$

is a primitive algebraic integer.

Proof. We know that

$$b_i(cb_i - bc_i) = (-1)^i \frac{q_i}{2} - ac_i^2 \quad \text{and} \quad b_i(cb_{i-1} - bc_{i-1}) = (-1)^{i-1} \frac{p_i - b}{2} - ac_i c_{i-1}$$

hence as above $b_i(b, c) \equiv 0 \pmod{((p_i - b)/2, q_i/2, a)}$, and since $(a, b, c) = 1$, we obtain $((p_i - b)/2, q_i/2, a) \mid (b_i, a)$. Conversely, the same relations show immediately that $(b_i, a) \mid ((p_i - b)/2, q_i/2, a)$, thus giving the first formula of the lemma. For the second, we note that $a_i = 2ac_i - bb_i$, hence $(b_i, a) \mid a_i$, and since by Proposition 5.7.3 (3) $a_i^2 - b_i^2 D = (-1)^i 2aq_i$, we see that $4(b_i, a)^2 \mid a_i^2 - b_i^2 D$ since we have proved that $(b_i, a) \mid q_i/2$, and these two divisibility conditions show that $\alpha = (a_i + b_i \sqrt{D})/(2(b_i, a))$ is an algebraic integer.

Let us show that it is primitive. Note first that since $a_i = 2ac_i - bb_i$ and $(c_i, b_i) = 1$, we have $(a_i, b_i) = (b_i, 2a)$. This shows that if we write $\alpha = (A' + B'\sqrt{D})/2$, we have $(A', B') = (b_i, 2a)/(b_i, a)$ and therefore $(A', B') \mid 2$. If $D \equiv 1 \pmod 4$, it can easily be seen that this is the only required condition for primitivity. If $D \equiv 0 \pmod 4$, we must show that A' is even and that $(A'/2, B') = 1$. In this case however, $b \equiv D \equiv 0 \pmod 2$, hence $a_i/2 = ac_i - (b/2)b_i$ showing that $A' = a_i/(b_i, a)$ is even, and $(a_i/2, b_i) = (a, b_i)$ so $(A'/2, B') = 1$ as was to be shown.

Now that we have this lemma, we can finish the proof of Proposition 5.7.5. We have shown that $A/B = a_n/b_n$, and since $(A + B\sqrt{D})/2$ was assumed primitive, we obtain from the lemma the equalities $A = |a_n|/(b_n, a)$, $B = |b_n|/(b_n, a)$. Plugging this in the Diophantine equation gives, using Proposition 5.7.3 (3), $\pm 4N = 2aq_n/(b_n, a)^2$ or in other words since it is clear by induction that $aq_i > 0$ for all i:

$$N = \frac{a}{(b_n, a)} \frac{q_n/2}{(b_n, a)} .$$

Since we have assumed $(N, a) = 1$, it follows that $a/(b_n, a) = \pm 1$, so that $a \mid b_n$, hence also $a \mid a_n$, and hence $q_n = 2aN$, thus finishing the proof of Proposition 5.7.5. □

Although we have proved a lot, we are still not finished. We need to show that we do indeed obtain the fundamental unit and not a power of it for every reduced (a, b, c), and not simply for $2|a| < \sqrt{D}$. To do this, it would be necessary to relax the condition $2|a|N < \sqrt{D}$ to $|a|N < \sqrt{D}$ for instance, but this is false as can easily be seen (take for example $D = 136$, $(a, b, c) = (5, 6, -5)$ and $N = 2$. This is only a random example). In the special case $N = 1$ however, which is the case we are most interested in, we can prove our claim by using the symmetry between a and c, i.e. by also using Corollary 5.7.4. First, we note the proposition which is symmetric to Proposition 5.7.5.

Proposition 5.7.8. *Let us keep all the above notations, and, in particular, those of Corollary 5.7.4. Let $N \geq 1$ be a squarefree integer such that $\gcd(c, N) = 1$ and $2|c|N < \sqrt{D}$.*
Then the solutions (A, B) of the Diophantine equation

$$A^2 - B^2 D = \pm 4N, \quad \text{with } A > 0, \ B > 0 \text{ and } \frac{A + B\sqrt{D}}{2} \text{ primitive}$$

are given by $(A, B) = (|d_n/c|, |c_n/c|)$, for every n such that $q_n = 2|c|N$ and $p_n \equiv -b \pmod{2c}$.

The proof is identical to that of Proposition 5.7.5, but uses the formulas of Corollary 5.7.4 instead of those of Proposition 5.7.3. □

Now we can prove:

Proposition 5.7.9. *The conclusion of Proposition 5.7.5 is valid for $N = 1$, with the only needed condition being that (a, b, c) is a reduced quadratic form.*

Proof. If $|a| < \sqrt{D}/2$, then the result follows from Proposition 5.7.5. Assume now $|a| > \sqrt{D}/2$. By Proposition 5.6.3 (2), we have $|c| < \sqrt{D}/2$, hence we can apply Proposition 5.7.8. We obtain $(A, B) = (|d_n/c|, |c_n/c|)$ for an n such that $p_n \equiv -b \pmod{2c}$ and $q_n = 2|c|$. This implies that $p_{n+1} = A_n q_n - p_n \equiv b \pmod{2c}$ and furthermore, by definition of A_n, that $\sqrt{D} - 2|c| < p_{n+1} < \sqrt{D}$. Hence, since $|c| < \sqrt{D}/2$ and (a, b, c) is reduced, we have $p_{n+1} = b$, so $q_{n+1} = 2|a|$. Now from Proposition 5.7.3 and Corollary 5.7.4, we obtain immediately that

$$\frac{d_{n+1} + c_{n+1}\sqrt{D}}{a_{n+1} + b_{n+1}\sqrt{D}} = \frac{d_n + c_n\sqrt{D}}{a_n + b_n\sqrt{D}},$$

hence by induction

$$\frac{d_n + c_n\sqrt{D}}{a_n + b_n\sqrt{D}} = \frac{b + \sqrt{D}}{2a} \quad,$$

and from this, Proposition 5.7.3 (1), Lemma 5.7.7 and its analog for c instead of a, we obtain the identities $|a_{n+1}/a| = |d_n/c|$ and $|b_{n+1}/a| = |c_n/c|$, proving the proposition. □

5.7.3 Computation of the Regulator

We have already mentioned that the fundamental unit ε itself can involve huge coefficients, and that what one usually needs is only the regulator to a reasonable degree of accuracy. Note first that for all $i \geq 1$, we have $a_i/b_i > 0$. This is an amusing exercise left to the reader (hint: consider separately the four cases $a > 0$ and $a < 0$, and $2|a| < \sqrt{D}$, $2|a| > \sqrt{D}$). Hence we have

$$R(D) = \ln\varepsilon = \ln\left(\frac{|a_k + b_k\sqrt{D}|}{|2a|}\right) = \sum_{i=0}^{k-1}\ln\left(\frac{|a_{i+1} + b_{i+1}\sqrt{D}|}{|a_i + b_i\sqrt{D}|}\right) \quad,$$

so by Proposition 5.7.3,

$$R(D) = \sum_{i=0}^{k-1}\ln\left(\frac{p_{i+1} + \sqrt{D}}{|q_i|}\right) = \sum_{i=1}^{k}\ln\left(\frac{p_i + \sqrt{D}}{|q_i|}\right) \quad,$$

since $q_k = q_0 = 2a$, and since the p_i and $|q_i|$ are always small (less than $2\sqrt{D}$), this enables us to compute the regulator to any given accuracy without handling huge numbers. The computation of a logarithm is a time consuming operation however, and hence it is preferable to write

$$R(D) = \ln\left(\prod_{i=1}^{k}\frac{p_i + \sqrt{D}}{|q_i|}\right) \quad,$$

the product being computed to a given numerical accuracy. In most cases, this method will again not work, because the *exponents* of the floating point numbers become too large. The trick is to keep the exponent in a separate variable which is updated either at each multiplication, or as soon as there is the risk of having an exponent overflow in the multiplication. Note that we have the trivial inequality $(p_i + \sqrt{D})/|q_i| < \sqrt{D}$, hence exponent overflow can easily be checked. This leads to the following algorithm, analogous to Algorithm 5.7.1.

Algorithm 5.7.10 (Regulator). Given a quadratic irrational $\tau = \dfrac{-b + \sqrt{D}}{2a}$ where $4a \mid (D - b^2)$ and $a > 0$, corresponding to a reduced form $(a, b, (b^2 -$

$D)/(4a))$, this algorithm computes the regulator $R(D)$ of $\mathbb{Q}(\sqrt{D})$ using the ordinary continued fraction expansion of $-\sigma(\tau)$.

1. [Initialize] Precompute $f \leftarrow \sqrt{D}$ to the desired accuracy, and set $d \leftarrow \lfloor f \rfloor$, $e \leftarrow 0$, $R \leftarrow 1$, $p \leftarrow b$, $q \leftarrow 2a$, and $q_1 \leftarrow (D - p^2)/q$. Finally, let 2^L be the highest power of 2 such that $2^L f$ does not give an exponent overflow.

2. [Euclidean step] Let $p + d = qA + r$ with $0 \leq r < |q|$ be the Euclidean division of $p + d$ by q, and set $p_1 \leftarrow p$, $p \leftarrow d - r$, $t \leftarrow q$, $q \leftarrow q_1 - A(p - p_1)$, $q_1 \leftarrow t$ and $R \leftarrow R(p + f)/q$. If $R \geq 2^L$, set $R \leftarrow R/2^L$, $e \leftarrow e + 1$.

3. [End of period?] If $q = 2a$ and $p \equiv b \pmod{2a}$, output $R(D) \leftarrow \ln R + eL \ln 2$ and terminate the algorithm. Otherwise, go to step 2.

In the case where we start with the unit form, we can use the symmetry of the period to obtain an algorithm similar to Algorithm 5.7.2. We leave this as an exercise for the reader (Exercise 23). We can also modify the algorithm so that it works for reduced forms with $a < 0$.

The running time of this algorithm is $O(D^{1/2+\epsilon})$ for all $\epsilon > 0$, but here this corresponds to the actual behavior since no multi-precision variables are being used. Although this is reasonable, we will now see that we can adapt Shanks's baby-step giant-step method to obtain a $O(D^{1/4+\epsilon})$ algorithm, bringing down the computation time to one similar to the case of imaginary quadratic fields.

Remark. If the regulator is computed to sufficient accuracy and is not too large, we can recover the fundamental unit by exponentiating. It is clear that it is impossible to find a sub-exponential algorithm for the fundamental unit in general, since, except when the regulator is very small, it already takes exponential time just to print it in the form $\epsilon = a + b\sqrt{D}$. It is possible however to write down explicitly the fundamental unit itself if we use a different representation, which H. Williams calls a *compact representation*. We will see in Section 5.8.3 how this is achieved.

5.8 The Infrastructure Method of Shanks

5.8.1 The Distance Function

The fundamental new idea introduced by Shanks in the theory of real quadratic fields is that one can introduce a distance function between quadratic forms or between ideals, and that this function will enable us to consider the principal cycle pretty much like a cyclic group. The initial theory is explained in [Sha3], and the refined theory which we will now explain can be found in [Len1].

Definition 5.8.1. *Let $\mathcal{O}$ be the quadratic order of discriminant D, and denote as usual by σ real conjugation in $\mathcal{O}$. If $\mathfrak{a}$ and $\mathfrak{b}$ are fractional ideals of $\mathcal{O}$, we* *define the* distance *of* $\mathfrak{a}$ *to* $\mathfrak{b}$ *as follows. If* $\mathfrak{a}$ *and* $\mathfrak{b}$ *are not equivalent (modulo principal ideals), the distance is not defined. Otherwise, write*

$$\mathfrak{b} = \gamma \mathfrak{a}$$

for some $\gamma \in K$. *We define the distance* $\delta(\mathfrak{a}, \mathfrak{b})$ *by the formula*

$$\delta(\mathfrak{a}, \mathfrak{b}) = \frac{1}{2} \ln \left| \frac{\gamma}{\sigma(\gamma)} \right|$$

where δ *is considered to be defined only modulo the regulator* R *(i.e.* $\delta \in \mathbb{R}/R\mathbb{Z}$).

Note that this distance is well defined (modulo R) since if we take another γ' such that $\mathfrak{b} = \gamma'\mathfrak{a}$, then $\gamma' = \epsilon\gamma$ where ϵ is a unit, hence the distance does not change modulo R. Note also that if $\mathfrak{a}$ is multiplied by a rational number, its distance to any other ideal does not change, hence in fact this distance carries over to the set I of ideal classes defined in Section 5.2. This remark will be important later on.

In a similar manner, we can define the distance between two quadratic forms of positive discriminant D as follows.

Definition 5.8.2. *Let* f *and* g *be two quadratic forms of discriminant* D, *and set* $(\mathfrak{a}, s) = \phi_{FI}(f)$, $(\mathfrak{b}, t) = \phi_{FI}(g)$ *as in Section 5.2, where* $s, t = \pm 1$. *If* f *and* g *are not equivalent modulo* $\mathrm{PSL}_2(\mathbb{Z})$, *the distance is not defined. If* f *and* g *are equivalent, then by Theorem 5.2.9 there exists* $\gamma \in K$ *such that*

$$\mathfrak{b} = \gamma \mathfrak{a} \quad \text{and} \quad t = s \cdot \mathrm{sign}(\mathcal{N}(\gamma)) \ .$$

We then define as above

$$\delta(f, g) = \frac{1}{2} \ln \left| \frac{\gamma}{\sigma(\gamma)} \right|$$

where δ *is now considered to be defined modulo the regulator in the narrow sense* R^+, *i.e. the logarithm of the smallest unit greater than 1 which is of positive norm.*

Note once again that this distance is well defined, but this time modulo R^+, since if we take another γ' we must have $\gamma' = \epsilon\gamma$ with ϵ a unit of positive norm.

Ideals are usually given by a $\mathbb{Z}$-basis, hence it is not easy to show that they are equivalent or not. Even if one knows for some reason that they are, it is still not easy to find a $\gamma \in K$ sending one into the other. In other words, it is not easy to compute the distance of two ideals (or of two quadratic forms) directly from the definition.

Luckily, we can bypass this problem in practice for the following reason. The quadratic forms which we will consider will almost always be obtained either by reduction of other quadratic forms (using the reduction step ρ a number of times), or by composition of quadratic forms. Hence, it suffices to give transformation formulas for the distance δ under these two operations.

Composition is especially simple if one remembers that it corresponds to ideal multiplication. If, for $k = 1, 2$, we have $\mathfrak{b}_k = \gamma_k \mathfrak{a}_k$, then $\mathfrak{b}_1 \mathfrak{b}_2 = \gamma_1 \gamma_2 \mathfrak{a}_1 \mathfrak{a}_2$. This means that (*before any reduction step*), the distance function δ is exactly additive

$$\delta(\mathfrak{b}_1 \mathfrak{b}_2, \mathfrak{a}_1 \mathfrak{a}_2) = \delta(\mathfrak{b}_1, \mathfrak{a}_1) + \delta(\mathfrak{b}_2, \mathfrak{a}_2)$$

when all distances are defined. This is true for the distance function on ideals as well as for the distance function between quadratic forms *since δ does not change when one multiplies an ideal with a rational number.*

In the case of reduction, it is easier to work with quadratic forms. Let $f = (a, b, c)$ be a quadratic form of discriminant D. Then

$$\phi_{FI}(f) = \left(a\mathbb{Z} + \frac{-b + \sqrt{D}}{2}\mathbb{Z}, \operatorname{sign}(a) \right) \ .$$

Furthermore, $\rho(f) = (c, b', a')$ where $b' \equiv -b \pmod{2c}$, hence

$$\phi_{FI}(\rho(f)) = \left(c\mathbb{Z} + \frac{b + \sqrt{D}}{2}\mathbb{Z}, \operatorname{sign}(c) \right) \ ,$$

since changing b' modulo $2c$ does not change the ideal. Now clearly

$$c\mathbb{Z} + \frac{b + \sqrt{D}}{2}\mathbb{Z} = \gamma \left(a\mathbb{Z} + \frac{-b + \sqrt{D}}{2}\mathbb{Z} \right)$$

where

$$\gamma = \frac{b + \sqrt{D}}{2a} \ .$$

Hence we obtain

Proposition 5.8.3. *If $f = (a, b, c)$ is a quadratic form of discriminant D, then*

$$\delta(f, \rho(f)) = \frac{1}{2} \ln \left| \frac{b + \sqrt{D}}{b - \sqrt{D}} \right| \ .$$

Of course, the map ϕ_{IF} of Section 5.2 enables us also to compute distances between ideals.

If we have two quadratic forms f and g such that $g = \rho^n(f)$ for n not too large, then by using the formula

$$\delta(f,g) = \sum_{i=1}^{n} \delta(\rho^{i-1}(f), \rho^{i}(f))$$

and this proposition, we can compute the distance of f and g. When n is large however, this formula, which takes time at least $O(n)$, becomes impractical. This is where we need to use composition.

For simplicity, we now assume that our forms are in the principal cycle, i.e. are equivalent to the unit form which we denote by $\mathbf{1}$. We then have the following proposition

Proposition 5.8.4. *Let f_1 and f_2 be two reduced forms in the principal cycle, and let $\mathbf{1}$ be the unit form. Then if we define $g = f_1 \cdot f_2$ by the composition algorithm given in Section 5.4.2, g may not be reduced, but let f_3 be a (non-unique) form obtained from g by the reduction algorithm, i.e. by successive applications of ρ. Then we have*

$$\delta(\mathbf{1}, f_3) = \delta(\mathbf{1}, f_1) + \delta(\mathbf{1}, f_2) + \delta(g, f_3) ,$$

and furthermore

$$|\delta(g, f_3)| < 2\ln(D) .$$

This proposition follows at once from the property that δ is exactly additive under composition (before any reductions are made).   □

If we assume that we know $\delta(\mathbf{1}, f_1)$ and $\delta(\mathbf{1}, f_2)$, then it is easy to compute $\delta(\mathbf{1}, f_3)$ since the number of reduction steps needed to go from g to f_3 is very small.

Important Remark. In the preceding section we have computed the regulator by adding $\ln((p_i + \sqrt{D})/|q_i|)$ over a cycle (or a half cycle). This corresponds to choosing a modified distance such that $\delta'(f, \rho(f)) = \ln((b + \sqrt{D})/(2|a|))$, and this clearly corresponds to defining

$$\delta'(a, b) = \ln|\gamma|$$

instead of $\delta(a, b) = \frac{1}{2}\ln|\gamma/\sigma(\gamma)|$ if $b = \gamma a$. This distance, which was the initial one suggested by Shanks, can also be used for regulator computations since it is also additive. Note however that it is no longer defined on the set I of ideals modulo the multiplicative action of $\mathbb{Q}^*$, but on the ideals themselves. In particular, with reference to Lemma 5.4.5, we must subtract $\ln(d)$ to the sum of the distances of I_1 and I_2 before starting the reduction of our composed quadratic form (A, B, C). It also introduces extra factors when one computes the inverse of a form. For example, this would introduce many unnecessary complications in Buchmann's sub-exponential algorithm that we will study below (Section 5.9).

On the other hand, although Shanks's distance is less natural, it is computationally slightly better since it is simpler to multiply by $(b + \sqrt{D})/(2|a|)$ than by $|(b + \sqrt{D})/(b - \sqrt{D})|$. Note also that Proposition 5.8.4 is valid with δ replaced by δ', if we take care to subtract the $\ln(d)$ value after composition as we have just explained.

Hence, for simplicity, we will use the distance δ instead of Shanks's δ', except in the baby-step giant-step Algorithm 5.8.5 where the use of δ' gives a slightly more efficient algorithm.

5.8.2 Description of the Algorithm

We consider the set S of pairs (f, z), where f is a reduced form of discriminant D in the principal cycle, and $z = \delta(1, f)$. We can transfer the action of ρ to S by setting $\rho(f, z) = (\rho(f), z + \ln|(b + \sqrt{D})/(b - \sqrt{D})|/2)$ if $f = (a, b, c)$, using the above notations. Furthermore, we can transfer the composition operation by setting

$$(f_1, z_1) \cdot (f_2, z_2) = (f_3, z_1 + z_2 + \delta(g, f_3)) \ ,$$

using the notations of Proposition 5.8.4. Similar formulas are valid with δ replaced by δ'. Recall that f_3 is not uniquely defined, but this does not matter for our purposes as long as we choose f_3 not too far away from the first reduced form that one meets after applying ρ to $f_1 \cdot f_2$.

Using these notations, we can apply Shanks's baby-step giant-step method to compute $R(D)$. Indeed, although the principal cycle is not a group, because of the set S we can follow the value of δ through composition and reduction. This means that Shanks's method allows us to find the regulator in $O(D^{1/4+\epsilon})$ steps instead of the usual $O(D^{1/2+\epsilon})$. If, as for negative discriminants, we also use that the inverse of a form (a, b, c) is a form equivalent to $(a, -b, c)$, i.e. $(a, r(-b, a), (r(-b, a)^2 - D)/4a)$, we obtain the following algorithm, due in essence to Shanks, and modified by Williams. Note that we give the algorithm using Shanks's distance δ' instead of δ since it is slightly more efficient, and also we use the language of continued fractions as in Algorithm 5.7.10, in other words, instead of (a, b, c) we use $(p, q) = (b, 2|a|)$.

Algorithm 5.8.5 (Regulator Using Infrastructure). Given a positive fundamental discriminant D, this algorithm computes $R(D)$. We assume that all the real numbers involved are computed with a finite and reasonably small accuracy. We make use of an auxiliary table T of quadruplets (q, p, e, R) where p, q, e are integers and R is a real number.

1. [Initialize] Precompute $f \leftarrow \sqrt{D}$, and set $d \leftarrow \lfloor\sqrt{D}\rfloor$, $e \leftarrow 0$, $R \leftarrow 1$, $s \leftarrow \lceil 1.5\sqrt{d}\rceil$, $T \leftarrow s + \lceil \ln(4d)/(2\ln((1 + \sqrt{5})/2))\rceil$ and $q \leftarrow 2$. If $d \equiv D$ (mod 2), set $p \leftarrow d$, otherwise set $p \leftarrow d - 1$. Set $q_1 = (D - p^2)/q$, $i \leftarrow 0$, and store the (q, p, e, R) in T. Finally, let 2^L be the highest power of 2 such that $2^L f$ does not give an exponent overflow.

2. [Small steps] Set $i \leftarrow i + 1$, and let $p + d = Aq + r$ with $0 \leq r < q$ be the Euclidean division of $p + d$ by q. Set $p_1 \leftarrow p$, $p \leftarrow d - r$, $t \leftarrow q$,

$q \leftarrow q_1 - A(p - p_1)$, $q_1 \leftarrow t$, $R \leftarrow R(p + f)/q_1$. If $R \geq 2^L$, set $R \leftarrow R/2^L$, $e \leftarrow e + 1$. If $q \leq d$, store (q, p, e, R) in $\mathcal{T}$.

3. [Finished already?] If $p_1 = p$ and $i > 1$, then output

$$R(D) = 2(\ln(R) + eL \ln(2)) - \ln(q_1/2)$$

and terminate the algorithm. If $q_1 = q$ and $i > 1$, then output

$$R(D) = 2(\ln(R) + eL \ln(2)) - \ln((p + f)/2)$$

and terminate the algorithm. If $i = s$, then if $q \leq d$ set $(Q, P, E, R_1) \leftarrow (q, p, e, R)$ otherwise (still if $i = s$) set $s \leftarrow s + 1$ and $T \leftarrow T + 1$. Finally, if $i < T$ go to step 2.

4. [Initialize for giant steps] Sort table $\mathcal{T}$ lexicographically (or in any other way). Then using the composition Algorithm 5.8.6 given below, compute

$$(Q, P, E, R_1) \leftarrow (Q, P, E, R_1) \cdot (Q, P, E, R_1) \ ,$$

and set $R \leftarrow 1$, $e \leftarrow 0$, $j \leftarrow 1$, and $q \leftarrow Q$, $p \leftarrow P$.

5. [Match found?] If $(q, p) = (q_1, p_1)$ for some $(q_1, p_1, e_1, r_1) \in \mathcal{T}$, output

$$R(D) = j(\ln(R_1) + EL \ln(2)) + \ln(R) + eL \ln(2) - \ln(r_1) - e_1 L \ln(2)$$

and terminate the algorithm.
If $(q, r(-p, q)) = (q_1, p_1)$ for some $(q_1, p_1, e_1, r_1) \in \mathcal{T}$, output

$$R(D) = j(\ln(R_1) + EL \ln(2)) + \ln(R) + eL \ln(2) + \ln(r_1) + e_1 L \ln(2) - \ln(q_1/2)$$

and terminate the algorithm.

6. [Giant steps] Using the composition Algorithm 5.8.6 below, compute

$$(q, p, e, R) \leftarrow (q, p, e, R) \cdot (Q, P, E, R_1) \ ,$$

set $j \leftarrow j + 1$ and go to step 5.

We need to compose two quadratic forms of positive discriminant D, expressed as quadruplets (q, p, e, R), where the pair (e, R) keeps track of the distance from 1 (more precisely $\delta'(1, f) = eL \ln 2 + \ln R$), and the form itself is $(q, p, (p^2 - D)/q)$ or $(-q, p, (D - p^2)/q)$. The algorithm is identical to the positive definite case (Algorithm 5.4.7), except that the reduction in step 4 must be done using Algorithm 5.6.5 (i.e. powers of ρ) instead of Algorithm 5.4.2. We must also keep track of the distance function, and, since we use δ' instead of δ, we must subtract a $\ln(d_1)$ (i.e. divide by d_1) where d_1 is the computed GCD.

This leads to the following algorithm.

Algorithm 5.8.6 (Composition of Indefinite Forms with Distance Function).
Given two quadruplets (q_1, p_1, e_1, R_1) and (q_2, p_2, e_2, R_2) as above (in particular
with q_i even and positive), this algorithm computes the composition

$$(q_3, p_3, e_3, R_3) = (q_1, p_1, e_1, R_1) \cdot (q_2, p_2, e_2, R_2) .$$

We assume $f \leftarrow \sqrt{D}$ already computed to sufficient accuracy.

1. [Initialize] If $q_1 > q_2$, exchange the quadruplets. Then set $s \leftarrow \frac{1}{2}(p_1 + p_2)$,
   $n \leftarrow p_2 - s$.

2. [First Euclidean step] If $q_1 \mid q_2$, set $y_1 \leftarrow 0$ and $d \leftarrow q_1/2$. Otherwise, using
   Euclid's extended algorithm, compute (u, v, d) such that $uq_2/2 + vq_1/2 = d = \gcd(q_2/2, q_1/2)$ and set $y_1 \leftarrow u$.

3. [Second Euclidean step] If $d \mid s$, set $y_2 \leftarrow -1$, $x_2 \leftarrow 0$ and $d_1 \leftarrow d$. Otherwise,
   using Euclid's extended algorithm, compute (u, v, d_1) such that $us + vd = d_1 = \gcd(s, d)$, and set $x_2 \leftarrow u$, $y_2 \leftarrow -v$.

4. [Compose] Set $v_1 \leftarrow q_1/(2d_1)$, $v_2 \leftarrow q_2/(2d_1)$, $r \leftarrow ((y_1 y_2 n - x_2(p_2^2 - D))/(2q_2) \bmod v_1$, $p_3 \leftarrow p_2 + 2v_2 r$, $q_3 \leftarrow 2v_1 v_2$.

5. [initialize reduction] Set $e_3 \leftarrow e_1 + e_2$ and $R_3 \leftarrow R_1 R_2/d_1$. If $R_3 \geq 2^L$, set
   $R_3 \leftarrow R_3/2^L$ and $e_3 \leftarrow e_3 + 1$.

6. [Reduced?] If $|f - q_3| < p_3$, then output (q_3, p_3, e_3, R_3) and terminate the
   algorithm. Otherwise, set $p_3 \leftarrow r(-p_3, q_3/2)$, $R_3 \leftarrow R_3(p_3 + f)/q_3$, $q_3 \leftarrow (D - p_3^2)/q_3$, and if $R_3 \geq 2^L$ set $R_3 \leftarrow R_3/2^L$ and $e_3 \leftarrow e_3 + 1$. Finally, go
   to step 6.

Note that $r(-p_3, q_3/2)$ is easily computed by a suitable Euclidean division.

This algorithm performs very well, and one can compute regulators of
real quadratic fields with discriminants with up to 20 digits in reasonable
time. To go beyond this requires new ideas which are essentially the same as
the ones used in McCurley's sub-exponential algorithm and will in fact give
us simultaneously the regulator and the class group. We will study this in
Section 5.9.

5.8.3 Compact Representation of the Fundamental Unit

The algorithms that we have seen above allow us to compute the regulator of
a real quadratic field to any desired accuracy. If this accuracy is high, however,
and in particular if we want infinite accuracy (i.e. the fundamental unit itself
and not its logarithm), we must not apply the algorithms exactly as they are
written. The reason for this is that by using the infrastructure ideas of Shanks
(essentially the distance function), the knowledge of a crude approximation
to the regulator $R(D)$ (say only its integer part) allows us to compute it very
fast to any desired accuracy. Let us see how this is done.

Let f be the form $\rho(1)$. It is the first form encountered in the principal cycle
when we start at the unit form, and in particular has the smallest distance to

1. Assume that after applying one of the regulator algorithms we know that $R_1 < R(D) < R_2$ (this can be a very crude estimate, for example we could ask that $R_2 - R_1 < 1$). By using the same idea as in Exercise 4 of Chapter 1, it is easy to find in time $O(\ln(D))$ composition operations, an integer n such that $\delta(1, f^n) \leq R_1$ and $\delta(1, f^{n+1}) > R_1$. This implies that f^n is before the unit form in the principal cycle (counting in terms of increasing distances), but not much before since $R_2 - R_1$ is small. Hence, there exists a small $k \geq 0$ which one finds by simply trying $k = 0, 1, \ldots$ such that $1 = \rho^k(f^n)$. Note that this is checked on the exact components of the forms, not on the distance. Hence, we now assume that k and n have been found.

If we want the regulator very precisely, we recompute $f = \rho(1)$ to the desired accuracy, and then the distance component of $\rho^k(f^n)$ will give us the regulator to the accuracy that we want.

If we want the fundamental unit itself, note that by Proposition 5.8.4 the composition of two forms implies the addition of three distances, or equivalently the multiplication of three quadratic numbers. For the ρ operator, only one such multiplication is required. Finally, note that k will be $O(\ln(D))$ and n will be $O(\sqrt{D})$ hence only $O(\ln(D))$ composition or reduction steps are required to compute $\rho^k(f^n)$. This implies that we can express the fundamental unit as a product of at most $O(\ln(D))$ terms of the form $(b + \sqrt{D})/(2|a|)$ (or $|(b + \sqrt{D})/(b - \sqrt{D})|$ if we use the distance δ instead of δ') and this is a compact way of keeping the fundamental unit even when D is very large.

Let us give a numerical example. Take $D = 10209$. A rough computation using one of the regulator algorithms shows that $R(D) \approx 67.7$. Furthermore, one computes that $f = \rho(1) = (-2, 99, 51)$. The binary algorithm gives $f^{14} = (1, 101, -2) = 1$ with $\delta'(1, f^{14}) \approx 67.7$. Note that this exponent 14 is not at all canonical and depends on the number of reduction steps performed at each composition, and on the order in which the compositions steps are made. Here, we assume that we stop applying ρ as soon as the form is reduced, and that f^n is computed using the right-left binary powering Algorithm 1.2.1.

We now start again recomputing f and f^{14}, keeping the quantities $(b + \sqrt{D})/(2|a|)|$ that are multiplied, along with their exponents. If ϵ is the fundamental unit, we obtain

$$\epsilon = \left(\frac{101 + \sqrt{D}}{2}\right)^{14} \left(\frac{111 + \sqrt{D}}{32}\right)^3 \frac{1}{3} \frac{219 + \sqrt{D}}{242}$$
$$\frac{351 + \sqrt{D}}{264} \frac{77 + \sqrt{D}}{428} \frac{93 + \sqrt{D}}{780} .$$

The lonely $1/3$ in the middle is due to the use of the imperfect distance function δ' which as we have already mentioned introduces extra quantities $- \ln d$ in the compositions.

If we instead use the distance δ, we obtain $\epsilon^2 = \tau/\overline{\tau}$ with

$$\tau = (101 + \sqrt{D})^{14}(111 + \sqrt{D})^3(219 + \sqrt{D})(197 + \sqrt{D})(103 + \sqrt{D}) .$$

Hence, to represent ϵ, we could simply keep the pairs $(101, 14)$, $(111, 3)$, $(219, 1)$, $(197, 1)$ and $(103, 1)$. It is a matter of taste which of the two representations above is preferable. Note that in fact

$$\epsilon = 1309694962454302631594431778775 + 129621951366321815797594195 6\sqrt{D}$$

which does not really take more space, but for larger discriminants this kind of explicit representation becomes impossible, while the compact one survives without any problem since there are only $O(\ln(D))$ terms of size $O(\ln(D))$ to be kept.

In [Buc-Thi-Wil], the authors have given a slightly more elegant compact representation of the fundamental unit, but the basic principle is the same. This idea can be generalized to the representation of algebraic numbers (and not only units), and to any number field.

5.8.4 Other Application and Generalization of the Distance Function

An important aspect of the distance function should be stressed at this point. Not only does it give us a fundamental hold on the fine structure of units, but it also allows us to solve the *principal ideal* problem which is the following. Assume that $\mathfrak{a}$ is an integral ideal of $\mathbb{Z}_K$ which is known to be a principal ideal (for example because $\mathfrak{a} = \mathfrak{b}^h$ for some ideal $\mathfrak{b}$, where h is the class number of K). Assume that we know the distance function $\delta(1, \mathfrak{a})$. Then it is easy to find an element γ such that $\mathfrak{a} = \gamma \mathbb{Z}_K$ using the formulas

$$\gamma = \pm\sqrt{\mathcal{N}(\mathfrak{a})}e^{\delta(1,\mathfrak{a})} \ , \sigma(\gamma) = \pm\sqrt{\mathcal{N}(\mathfrak{a})}e^{-\delta(1,\mathfrak{a})} \ .$$

This leaves only 2 possibilities for $\pm\gamma$, and usually only one will belong to K. Note that since δ is defined only in $\mathbb{R}/R\mathbb{Z}$, γ will be defined up to multiplication by a unit.

Similarly, if the distance function $\delta'(1, \mathfrak{a})$ is known, we use the formulas

$$\gamma = \pm e^{\delta'(1,\mathfrak{a})} \ , \sigma(\gamma) = \pm\mathcal{N}(\mathfrak{a})e^{-\delta'(1,\mathfrak{a})} \ .$$

The distance function δ can be naturally generalized to arbitrary number fields K as follows. Let

$$L(x) = (\ln|\sigma_1(x)|, \ldots, \ln|\sigma_{r_1}(x)|, 2\ln|\sigma_{r_1+1}(x)|, \ldots, 2\ln|\sigma_{r_1+r_2}(x)|)$$

be the logarithmic embedding of K^* into $\mathbb{R}^{r_1+r_2}$ seen in Definition 4.9.6, where (r_1, r_2) is the signature of K. If $n = r_1 + 2r_2$ is the degree of K, we will set

$$\Delta(\mathfrak{a}, \gamma\mathfrak{a}) = L(\gamma/|\mathcal{N}_{K/\mathbb{Q}}(\gamma)|^{1/n}) \ ,$$

where it is understood that the σ_i act trivially on the n-th roots of the norms.

Then Δ belongs to the hyperplane $\sum_{1 \leq i \leq r_1 + r_2} x_i = 0$ of $\mathbb{R}^{r_1 + r_2}$ and is defined modulo the lattice which is the image of the group of units $U(K)$ under the embedding $L(x)$.

In the case where K is a real quadratic field, then clearly $\Delta = (\delta, -\delta)$, so this is a reasonable generalization of δ. If K is an imaginary quadratic field, we have $\Delta = 0$.

The principal ideal problem can, of course, be asked in general number fields and it is clear that Δ cannot help us to solve it in general since it cannot do so even for imaginary quadratic fields. For this specific application, the logarithmic embedding L should be replaced by the ordinary embedding

$$(\sigma_1(x), \ldots, \sigma_{r_1}(x), \sigma_{r_1+1}(x), \ldots, \sigma_{r_1+r_2}(x))$$

of K into $\mathbb{R}^{r_1} \times \mathbb{C}^{r_2}$.

The components of this embedding are in general too large to be represented exactly, hence we will preferably choose the complex logarithmic embedding

$$L_C(x) = (\ln \sigma_1(x), \ldots, \ln \sigma_{r_1}(x), 2 \ln \sigma_{r_1+1}(x), \ldots, 2 \ln \sigma_{r_1+r_2}(x)) \ ,$$

where the logarithms are defined up to addition of an integer multiple of $2i\pi$. Note that this requires only twice as much storage space as the embedding L, and also that the first r_1 components have an imaginary part which is a multiple of π. Let $V = (n_i)_{1 \leq i \leq r_1 + r_2}$ be the vector such that $n_i = 1$ for $i \leq r_1$ and $n_i = 2$ otherwise. We can then define

$$\Delta_C(\mathfrak{a}, \gamma \mathfrak{a}) = L_C(\gamma) - \frac{\ln(\mathcal{N}(\gamma))}{n} V \ ,$$

and it is clear that the sum of the $r_1 + r_2$ components of Δ_C is an integral multiple of $2i\pi$. We will see the use of this function in Section 6.5.

5.9 Buchmann's Sub-exponential Algorithm

We will now describe a fast algorithm for computing the class group and the regulator of a real quadratic field, which uses essentially the same ideas as Algorithm 5.5.2.

Although the main ideas are in McCurley and Shanks, I have seen this algorithm explained only in manuscripts of J. Buchmann whom I heartily thank for the many conversations which we have had together. The first implementation of this algorithm is due to Cohen, Diaz y Diaz and Olivier (see [CohDiOl]).

5.9.1 Outline of the Algorithm

We will follow very closely Algorithm 5.5.2, and use the distance function δ and *not* Shanks's distance δ' which we used in Algorithm 5.8.5.

As we have already explained, in the quadratic case it is simpler to work with forms instead of directly with ideals. Note however that because of Theorem 5.2.9, we will be computing the *narrow* ideal class group and the regulator in the narrow sense, since this is the natural correspondence with quadratic forms. If, on the other hand, we want the ideal class group and the regulator in the ordinary sense, then, according to Proposition 5.6.1, we will have to identify the form (a, b, c) with the form $(-a, b, -c)$. (This is implicitly what we did in Algorithm 5.8.5.) Although it is very easy to combine both procedures into a single algorithm, note that the computations are independent. More precisely, to the best of my knowledge it does not seem to be easy, given the ideal class group and regulator in one sense (narrow or ordinary) to deduce the ideal class group and regulator in the other sense, although of course only a factor of 2 is involved. We will describe the algorithm for the class group and regulator in the ordinary sense, leaving to the reader the simple modifications that must be made to obtain the class group and regulator in the narrow sense (see Exercise 26).

We now describe the outline of the algorithm. As in Algorithm 5.8.5, we keep track of the distance function as a pair (e, R), but this time we will keep all three coefficients of the quadratic form. Also, we are going to use the distance δ instead of δ', and since there is a factor $1/2$ in the definition of δ, we will use the correspondence $\delta(1, f) = (eL \ln 2 + \ln R)/2$.

In other words, in this section a quadratic form of positive discriminant will be a quintuplet $f = (a, b, c, e, R)$ where a, b, c and e are integers and R is a real number such that $1 \le R < 2^L$.

We can compose two such forms by using the following algorithm, which is a trivial modification of Algorithm 5.8.6.

Algorithm 5.9.1 (Composition of Indefinite Forms with Distance Function). Given two primitive quadratic forms $(a_1, b_1, c_1, e_1, R_1)$ and $(a_2, b_2, c_2, e_2, R_2)$ as above, this algorithm computes the composition

$$(a_3, b_3, c_3, e_3, R_3) = (a_1, b_1, c_1, e_1, R_1) \cdot (a_2, b_2, c_2, e_2, R_2) \ .$$

We assume $f \leftarrow \sqrt{D}$ already computed to sufficient accuracy.

1. [Initialize] If $|a_1| > |a_2|$ exchange the quintuplets. Then set $s \leftarrow \frac{1}{2}(b_1 + b_2)$, $n \leftarrow b_2 - s$.

2. [First Euclidean step] If $a_1 \mid a_2$, set $y_1 \leftarrow 0$ and $d \leftarrow |a_1|$. Otherwise, using Euclid's extended algorithm, compute u, v and d such that $ua_2 + va_1 = d = \gcd(a_2, a_1)$ and set $y_1 \leftarrow u$.

3. [Second Euclidean step] If $d \mid s$, set $y_2 \leftarrow -1$, $x_2 \leftarrow 0$ and $d_1 \leftarrow d$. Otherwise, again using Euclid's extended algorithm, compute (u, v, d_1) such that $us + vd = d_1 = \gcd(s, d)$, and set $x_2 \leftarrow u$ and $y_2 \leftarrow -v$.

4. [Compose] Set $v_1 \leftarrow a_1/d_1$, $v_2 \leftarrow a_2/d_1$, $r \leftarrow (y_1 y_2 n - x_2 c_2 \bmod v_1)$, $b_3 \leftarrow b_2 + 2 v_2 r$ and $a_3 \leftarrow v_1 v_2$.

5. [Initialize reduction] Set $e_3 \leftarrow e_1 + e_2$, $R_3 \leftarrow R_1 R_2$. If $R_3 \geq 2^L$, set $R_3 \leftarrow R_3/2^L$ and $e_3 \leftarrow e_3 + 1$.

6. [Reduced?] If $|f - 2|a_3|| < b_3 < f$, then output $(a_3, b_3, c_3, e_3, R_3)$ and terminate the algorithm.

7. [Apply ρ] Set $R_3 \leftarrow R_3|(b_3 + f)/(b_3 - f)|$ and if $R_3 \geq 2^L$, set $R_3 \leftarrow R_3/2^L$ and $e_3 \leftarrow e_3 + 1$. Then set $a_3 \leftarrow c_3$, $b_3 \leftarrow r(-b_3, c_3)$, $c_3 \leftarrow (b_3^2 - D)/a_3$ and go to step 6.

Note that, apart from some absolute value signs, steps 1 to 4 are identical to the corresponding steps in Algorithm 5.4.7, but the reduction operation is quite different since it involves iterating the function ρ in step 7 of the algorithm and the bookkeeping necessary for the distance function.

Returning to Buchmann's algorithm, what we will do is essentially, instead of keeping track only of $f_p = (p, b_p, (b_p^2 - D)/(4p))$, we also keep track of the distance function. Hence, in step 3 of Algorithm 5.5.2, we compute the product $\prod_{p \leq P} f_p^{e_p}$, doing the reduction at each product (of course the reduction being non-unique), and keeping track of the distance function thanks to Theorem 5.8.4. In this way we obtain a reduced form $f = (a, b, c)$ equivalent to the above product, and also the value of $\delta(1, f)$. Since we identify (a, b, c) with $(-a, b, -c)$, we will replace (a, b, c) by $(|a|, b, -|c|)$.

If a does not factor easily, in step 5 we have the option of doing more reduction steps instead of going back to step 4 in the hope of getting an easily factorable a. Since this is much faster than recomputing a new product, we will use this method as much as possible. Note that, although we have extra computations to make because of the distance function, the basic computational steps will be *faster* than in the imaginary quadratic case, hence this algorithm will be faster than the corresponding one for imaginary quadratic fields.

This behavior is to be expected since on heuristic and experimental grounds class numbers of real quadratic fields are much smaller than those of imaginary quadratic fields.

Finally, if a factors easily, in step 5 we compute not only $a_{i,k}$ for $1 \leq i \leq n$, but also $a_{n+1,k} \leftarrow \delta(1, fg^{-1})$ where $g = \prod_{p \leq P} f_p^{v_p}$ and $\delta(1, fg^{-1})$ is computed as usual at the same time as the product is done, using Theorem 5.8.4.

We thus obtain a matrix $A = (a_{i,j})$ with $n + 1$ rows and k columns, whose entries in the first n rows are integers and the entries in the last row are real numbers. Note that by definition, for every $j \leq k$ we have

$$\delta \left(1, \prod_{1 \leq i \leq n} f_{p_i}^{a_{i,j}} \right) \equiv a_{n+1,j} \pmod{R(D)} .$$

Since the distance function that we have chosen is exactly additive, it follows that when performing column operations on the complete matrix A, this relation between the $n + 1$-st component and the others is preserved.

Hence we apply Hermite reduction to the matrix formed by the first n rows, but performing the corresponding column operations also the entries of the last row. The first $k - n$ columns of the resulting matrix will thus have only zero entries, except perhaps for the entry in the $n + 1$-st row. By the remark made above, for $1 \leq j \leq k - n$ we will thus have

$$a_{n+1,j} = \delta(1,1) \equiv 0 \pmod{R(D)} ,$$

in other words $a_{n+1,j}$ is equal to a multiple of the regulator $R(D)$ for $1 \leq j \leq k - n$.

If k is large enough, it follows that in a certain sense the GCD of the $a_{n+1,j}$ for $1 \leq j \leq k - n$ should be exactly equal to $R(D)$. We must be careful in the computation of this "GCD" since we are dealing with inexact real numbers. For this purpose, we can either use the LLL algorithm which will give us a small linear combination of the $a_{n+1,j}$ for $1 \leq j \leq k - n$ with integral coefficients, which should be the regulator $R(D)$, or use the "real GCD" Algorithm 5.9.3 as described below.

The rest of the algorithm will compute the class group structure in essentially the same way, except of course that in step 1 one must use the analytic class number formula for positive discriminants (Proposition 5.6.9).

5.9.2 Detailed Description of Buchmann's Sub-exponential Algorithm

A practical implementation of this algorithm should take into account at least two remarks. First, note that most of the time is spent in looking for relations. Hence, it is a waste of time to compute with the distance function during the search for relations: we do the search only with the components (a, b, c) of the quadratic forms, and only in the rare cases where a relation is obtained do we recompute the relation with the distance function. The slight loss of time due to the recomputation of each relation is more than compensated by the gain obtained by not computing the distance function during the search for relations.

The second remark is that, as in McCurley's sub-exponential algorithm, the Hermite reduction of the first n rows must be performed modulo a multiple of the determinant, which can be computed before starting the reduction. In other words, we will use Algorithm 2.4.8. The reduction of the last row is however another problem, and in the implementation due to the author, Diaz y Diaz and Olivier, the best method found was to compute the integer kernel of the integer matrix formed by the first n rows using Algorithm 2.7.2, and multiply the $n + 1$-st row of distances by this kernel, thus obtaining a vector whose components are (approximately) small multiples of the regulator, and we find the regulator itself using one of the methods explained above, for example the LLL algorithm.

These remarks lead to the following algorithm.

Algorithm 5.9.2 (Sub-Exponential Real Class Group and Regulator). If $D > 0$ is a non-square discriminant, this algorithm computes the class number $h(D)$, the class group $Cl(D)$ and the regulator $R(D)$. As before, in practice we work with binary quadratic forms. We also choose at will a positive real constant b.

1. [Compute primes and Euler product] Set $m \leftarrow b \ln^2 D$, $M \leftarrow L(D)^{1/\sqrt{8}}$, $P \leftarrow \lfloor \max(m, M) \rfloor$

$$\mathcal{P} \leftarrow \left\{ p \leq P, \left(\frac{D}{p}\right) \neq -1 \text{ and } p \text{ good} \right\}$$

and compute the product

$$B \leftarrow \frac{\sqrt{D}}{2} \prod_{p \leq P} \left(1 - \frac{\left(\frac{D}{p}\right)}{p}\right)^{-1}.$$

2. [Compute prime forms] Let $\mathcal{P}_0$ be the set made up of the smallest primes of $\mathcal{P}$ not dividing D such that $\prod_{p \in \mathcal{P}_0} p > \sqrt{D}$. For the primes $p \in \mathcal{P}$ do the following. Compute b_p such that $b_p^2 \equiv D \pmod{4p}$ using Algorithm 1.5.1 (and modifying the result to get the correct parity). If $b_p > p$, set $b_p \leftarrow 2p - b_p$. Set $f_p \leftarrow (p, b_p, (b_p^2 - D)/(4p))$ and $g_p \leftarrow (p, b_p, (b_p^2 - D)/(4p), 0, 1.0)$ Finally, let n be the number of primes $p \in \mathcal{P}$.

3. [Compute powers] For each $p \in \mathcal{P}_0$ and each integer e such that $1 \leq e \leq 20$ compute and store a reduced form equivalent to f_p^e. Set $k \leftarrow 0$.

4. [Generate random relations] Let f_q be the primeform number $k + 1 \bmod n$ in the factor base. Choose random e_p between 1 and 20, and compute a reduced form (a, b, c) equivalent to

$$f_q \prod_{p \in \mathcal{P}_0} f_p^{e_p}$$

by using the composition algorithm for positive binary quadratic forms, replacing the final reduction step by a sufficient number of applications of the ρ operator (note that $f_p^{e_p}$ has already been computed in step 3). Set $e_p \leftarrow 0$ if $p \notin \mathcal{P}_0$ then $e_q \leftarrow e_q + 1$. Set $(a_0, b_0, c_0) \leftarrow (a, b, c)$, $r \leftarrow 0$ and go to step 6.

5. [Apply ρ] Set $(a, b, c) \leftarrow \rho(a, b, c)$ and $r \leftarrow r + 1$. If $|a| = |a_0|$ and r is odd, or if $b = b_0$ and r is even, go to step 4.

6. [Factor $|a|$] Factor $|a|$ using trial division. If a prime factor of $|a|$ is larger than P, do not continue the factorization and go to step 5. Otherwise, if $|a| = \prod_{p \leq P} p^{v_p}$, set $k \leftarrow k + 1$, and for $i \leq n$ set

$$a_{i,k} \leftarrow e_{p_i} - \epsilon_{p_i} v_{p_i}$$

where $\epsilon_{p_i} = +1$ if $(b \bmod 2p_i) \leq p_i$, $\epsilon_{p_i} = -1$ otherwise.

7. [Recompute relation with distance] Compute

$$(a_0, b_0, c_0, e_0, R_0) \leftarrow g_q \prod_{p \in P_0} g_p^{e_p}$$

by mimicking the order of squarings, compositions and reductions done to compute (a_0, b_0, c_0), but this time using Algorithm 5.9.1 for composition. Then compute $(a, b, c, e, R) \leftarrow \rho^r(a_0, b_0, c_0, e_0, R_0)$ by applying the formulas of step 7 of Algorithm 5.9.1 to our forms. Finally, set $a_{n+1,k} \leftarrow (eL \ln 2 + \ln R)/2$.

8. [Enough relations?] If $k < n + 10$ go to step 4.

9. [Be honest] For each prime q such that $P < q \leq 6 \ln^2 D$ do the following. Choose random e_p between 1 and 20 (say), compute the primeform f_q corresponding to q and some reduced form (a, b, c) equivalent to $f_q \prod_{p \in P_0} f_p^{e_p}$. If a does not factor into primes less than q, choose other exponents e_p and continue until a factors into such primes (or apply the ρ operator as in step 5). Then go on to the next prime q until the list is exhausted.

10. [Simple HNF] Perform a preliminary simple Hermite reduction on the $(n + 1) \times k$ matrix $A = (a_{i,j})$ as described in the remarks following Algorithm 5.5.2. In this reduction, only the first n rows should be examined, but column operations should of course be done also with the $n+1$-st row. Let A_1 be the matrix thus obtained without its last row, and let V be the last row (whose components are linear combinations of distances).

11. [Compute regulator] Using Algorithm 2.7.2, compute the LLL-reduced integral kernel M of A_1 as a rectangular matrix, and set $V \leftarrow VM$. Let s be the number of elements of V. Set $R \leftarrow |V_1|$, and for $i = 2, \ldots, s$ set $R \leftarrow RGCD(R, |V_i|)$ where RGCD is the real GCD algorithm described below. (Now R is probably the regulator.)

12. [Compute determinant] Using standard Gaussian elimination techniques, compute the determinant of the lattice generated by the columns of the matrix A_1 modulo small primes p. Then compute the determinant d exactly using the Chinese remainder theorem and Hadamard's inequality (see also Exercise 13).

13. [HNF reduction] Using Algorithm 2.4.8 compute the Hermite normal form $H = (h_{i,j})$ of the matrix A_1 using modulo d techniques. Then for every i such that $h_{i,i} = 1$, suppress row and column i. Let W be the resulting matrix.

14. [Finished?] Let $h \leftarrow \det(W)$ (i.e. the product of the diagonal elements). If $hR \geq B\sqrt{2}$, get 5 more relations (in steps 4, 5 and 6) and go to step 10. (It will not be necessary to recompute the whole HNF, only that which takes into account the last 5 columns.) Otherwise, output h as the class number and R as the regulator.

15. [Class group] Compute the Smith normal form of W using Algorithm 2.4.14. Output those among the diagonal elements d_i which are greater than 1 as the invariants of the class group (i.e. $Cl(D) = \bigoplus \mathbb{Z}/d_i\mathbb{Z}$) and terminate the algorithm.

The real GCD algorithm is copied on the ordinary Euclidean algorithm, as follows. We use in an essential way that the regulator is bounded from below (by 1 for real quadratic fields of discriminant greater than 8) so as to have a reasonable stopping criterion. Since we will also use it for general number fields, we use 0.2 as a lower bound of the regulators of all number fields (see [Zim1], [Fri]).

Algorithm 5.9.3 (Real GCD). Given two non-negative real numbers a and b which are known to be approximate integer multiples of some positive real number $R > 0.2$, this algorithm outputs the real GCD (RGCD) of a and b, i.e. a non-negative real number d which is an approximate integer multiple of R and divisor of a and b, and is the largest with this property. The algorithm also outputs an estimate on the absolute error for d.

1. [Finished?] If $b < 0.2$, then output a as the RGCD, and b as the absolute error and terminate the algorithm.
2. [Euclidean step] Let $r \leftarrow a - b\lfloor a/b \rfloor$, $a \leftarrow b$, $b \leftarrow r$ and go to step 1.

Remarks.

(1) It should be noted that not only does Algorithm 5.9.2 compute the class number and class group in sub-exponential time, but it is the only algorithm which is able to compute the regulator in sub-exponential time, even if we are not interested in the class number. In fact, in all the preceding algorithms, we first had to compute the regulator (for example using the infrastructure Algorithm 5.8.5), and combining this with the analytic class number formula giving the product $h(D)R(D)$, we could then embark on the computation of $h(D)$ and $Cl(D)$. The present algorithm goes the other way: we can in fact compute a small multiple of the class number alone, without using distances at all, and then compute the distances and the regulator, and at that point use the analytic class number formula to check that we have the correct regulator and class number, and not multiples.

(2) In an actual implementation of this algorithm, one should keep track of the absolute error of each real number. First, in the distance computation in step 7, the precision with which the computations are done gives a bound on the absolute error. Then, during steps 10 and 11, $\mathbb{Z}$-linear combinations of distances will be computed, and the errors updated accordingly (with suitable absolute value signs everywhere). Finally, in the last part of step 11 where real GCD's are computed, one should use the errors output by Algorithm 5.9.3.

(3) Essentially all the implementation details given for Algorithm 5.5.2 apply also here.

5.10 The Cohen-Lenstra Heuristics

The purpose of this section is to explain a number of observations which have been made on tables of class groups and regulators of quadratic fields. As already mentioned very few theorems exist (in fact essentially only the theorem of Brauer-Siegel and the theorem of Goldfeld-Gross-Zagier) so most of the explanations will be conjectural. These conjectures are however based on solid heuristic grounds so they may well turn out to be correct. As usual, we first start with imaginary quadratic fields.

5.10.1 Results and Heuristics for Imaginary Quadratic Fields

In this subsection K will denote the unique imaginary quadratic field of discriminant $D < 0$. As we have seen, the only problem here is the behavior of the class group $Cl(D)$ and hence of the class number $h(D)$, all other basic problems being trivial to solve.

Here the Brauer-Siegel theorem says that $\ln(h(D)) \sim \ln(\sqrt{|D|})$ as $D \to -\infty$, which shows that $h(D)$ tends to infinity at least as fast as $|D|^{1/2-\epsilon}$ and at most as fast as $|D|^{1/2+\epsilon}$ for every $\epsilon > 0$. The main problem is that this is not effective in a very strong sense, and this is why one has had to wait for the Gross-Zagier result to get any kind of effective result, and a very weak one at that since using their methods one can show only that

$$h(D) > \frac{1}{K} \ln(|D|) \prod_{p|D}^{*} \left(1 - \frac{2\sqrt{p}}{p+1} \right) ,$$

where $K = 55$ if $(D, 5077) = 1$ and $K = 7000$ otherwise, and the star indicates that the product is taken over all prime divisors p of D with the exception of the largest prime divisor (see [Oes]). This is of course much weaker than the Brauer-Siegel theorem.

Results in the other direction are much easier. For example, one can show that for all $D < -4$, we have

$$h(D) < \frac{1}{\pi} \sqrt{|D|} \ln(|D|)$$

(see Exercise 27). Similarly, it is very easy to obtain *average* results, which were known since Gauss. The result is as follows (see [Ayo]).

$$\sum_{|D| \le x} h(D) \sim \frac{x^{3/2}}{3\pi} C$$

where the sum runs over fundamental discriminants and

$$C = \prod_p \left(1 - \frac{1}{p^2(p+1)} \right) \approx 0.881538397 .$$

Since by Exercise 1 the number of fundamental discriminants up to x is asymptotic to $(3/\pi^2)x$, this shows that on average, $h(D)$ behaves as $C\pi/6\sqrt{|D|} \approx$ $0.461559\sqrt{|D|}$, and shows that the upper bound given for $h(D)$ is at most off by a factor $O(\ln(D))$.

All the above results deal with the size of $h(D)$. If we consider problems concerning its arithmetic properties (for example divisibility by small primes) or properties of the class group $Cl(D)$ itself, very little is known. If we make however the heuristic assumption that class groups behave as random groups except that they must be weighted by the inverse of the number of their automorphisms (this is a very common weighting factor in mathematics), then it is possible to make precise quantitative predictions about class numbers and class groups. This was done by H. W. Lenstra and the author in [Coh-Len1]. We summarize here some of the conjectures which are obtained in this way and which are well supported by numerical evidence.

It is quite clear that the prime 2 behaves in a special way, so we exclude it from the class group. More precisely, we will denote by $Cl_o(D)$ the odd part of the class group, i.e. the subgroup of elements of odd order. We then have the following conjectures.

Conjecture 5.10.1 (Cohen-Lenstra). *For any odd prime p and any integer r including $r = \infty$ set $(p)_r = \prod_{1 \le k \le r}(1 - p^{-k})$, and let $A = \prod_{k \ge 2} \zeta(k) \approx$ 2.29486, where $\zeta(s)$ is the ordinary Riemann zeta function.*

(1) *The probability that $Cl_o(D)$ is cyclic is equal to*

$$\zeta(2)\zeta(3)/(3(2)_\infty A\zeta(6)) \approx 0.977575 .$$

(2) *If p is an odd prime, the probability that $p \mid h(D)$ is equal to*

$$f(p) = 1 - (p)_\infty = \frac{1}{p} + \frac{1}{p^2} - \frac{1}{p^5} - \cdots$$

For example, $f(3) \approx 0.43987$, $f(5) \approx 0.23967$, $f(7) \approx 0.16320$.

(3) *If p is an odd prime, the probability that the p-Sylow subgroup of $Cl(D)$ is isomorphic to a given finite Abelian p-group G is equal to $(p)_\infty/|\operatorname{Aut}(G)|$, where $\operatorname{Aut}(G)$ denotes the group of automorphisms of G.*

(4) *If p is an odd prime, the probability that the p-Sylow subgroup of $Cl(D)$ has rank r (i.e. is isomorphic to a product of r cyclic groups) is equal to $p^{-r^2}(p)_\infty/((p)_r)^2$.*

These conjectures explain the following qualitative observations which were made by studying the tables.

(1) The odd part of the class group is quite rarely non-cyclic. In fact, it was only in the sixties that the first examples of class groups with 3-rank greater or equal to 3 were discovered.

(2) Higher ranks are even more difficult to find, and the present record for $p = 3$, due to Quer (see [Llo-Quer] and [Quer]) is 3-rank equal to 6. Note that there is a very interesting connection with elliptic curves of high rank over $\mathbb{Q}$ (see Chapter 7), and Quer's construction indeed gives curves of rank 12.

(3) If p is a small odd prime, the probability that $p \mid h(D)$ is substantially higher than the expected naïve value $1/p$. Indeed, it should be very close to $1/p + 1/p^2$.

5.10.2 Results and Heuristics for Real Quadratic Fields

Because of the presence of non-trivial units, the situation in this case is completely different and even less understood than the imaginary quadratic case. Here the Brauer-Siegel theorem tells us that $\ln(R(D)h(D)) \sim \ln(\sqrt{D})$ as $D \to \infty$, where $R(D)$ is the regulator. Unfortunately, we have little control on $R(D)$, and this is the main source of our ignorance about real quadratic fields. It is conjectured that $R(D)$ is "usually" of the order of $\sqrt{D}$, hence that $h(D)$ is usually very small, and this is what the tables show. For example, there should exist an infinite number of D such that $h(D) = 1$, but this is not known to be true and is a famous conjecture. In fact, it is not even known whether there exists an infinite number of non-isomorphic number fields K (all degrees taken together) with class number equal to one.

As in the imaginary case however, we can give an upper bound $h(D) < \sqrt{D}$ when $D > 0$, and the following average for $R(D)h(D)$:

$$\sum_{D \le x} R(D)h(D) \sim \frac{x^{3/2}}{6} C$$

where the sum runs over fundamental discriminants and the constant C is as before.

It is possible to generalize the heuristic method used in the imaginary case. In fact, we could reinterpret Shanks's infrastructure idea as saying that the class group of a real quadratic field is equal to the quotient of the "group" of reduced forms by the "cyclic subgroup" formed by the principal cycle. This of course does not make any direct sense since the reduced forms form a group only in an approximate sense, and similarly for the principal cycle. It suggests however that we could consider the (odd part) of the class group of a real quadratic field as the quotient of a random finite Abelian group of odd order (weighted as before) by a random cyclic subgroup. This indeed works out very well and leads to the following conjectures.

Conjecture 5.10.2 (Cohen-Lenstra). *Let D be a positive fundamental discriminant.*

(1) *If p is an odd prime, the probability that $p \mid h(D)$ is equal to*

$$1 - \frac{(p)_\infty}{1 - 1/p} = \frac{1}{p^2} + \frac{1}{p^3} + \frac{1}{p^4} - \cdots$$

(2) *The probability that $Cl_o(D)$ is isomorphic to a given finite Abelian group G of odd order g is equal to $m(G) = 1/(2g(2)_\infty A|\operatorname{Aut}(G)|)$. For example $m(\{0\}) \approx 0.75446$, $m(\mathbb{Z}/3\mathbb{Z}) \approx 0.12574$, $m(\mathbb{Z}/5\mathbb{Z}) \approx 0.03772$.*

(3) *If p is an odd prime, the probability that the p-Sylow subgroup of $Cl(D)$ has rank r is equal to $p^{-r(r+1)}(p)_\infty/((p)_r(p)_{r+1})$.*

(4) *We have*

$$\sum_{p \leq x} h(p) \sim \frac{x}{8} \ ,$$

where the sum runs over primes congruent to 1 modulo 4.

These conjectures explain in particular the experimental observation that most quadratic fields of prime discriminant p (in fact more than three fourths) have class number one.

These heuristic conjectures have been generalized to arbitrary number fields by J. Martinet and the author (see [Coh-Mar1], [Coh-Mar2]). Note that contrary to what was claimed in these papers, apparently all the primes dividing the degree of the Galois closure should be considered as non-random (see [Coh-Mar3]), hence the numerical values given in [Coh-Mar1] should be corrected accordingly (e.g. by removing the 2-part for non-cyclic cubic fields or the 3-part for quartic fields of type A_4 or S_4).

5.11 Exercises for Chapter 5

1. Show that the number of imaginary quadratic fields with discriminant D such that $|D| \leq x$ is asymptotic to $3x/\pi^2$, and similarly for real quadratic fields.

2. Compute the probability that the discriminant of a quadratic field is divisible by a given prime number p (beware: the result is not what you may expect).

3. Complete Theorem 5.2.9 by giving explicitly the correspondences between ideal classes, classes of quadratic forms and classes of quadratic numbers, at the level of $\mathrm{PSL}_2(\mathbb{Z})$.

4. Let K be a quadratic field and p a prime. Generalizing Theorem 1.4.1, find the structure of the multiplicative group $(\mathbb{Z}_K/p\mathbb{Z}_K)^*$, and in particular compute its cardinality.

5. (H.W. Lenstra and D. Knuth) Let D denote the discriminant of an imaginary quadratic field. If $x \geq 0$, let $f(x, D)$ be the probability that a quadratic form (a, b, c) with $-a < b \leq a$ and $a < x\sqrt{|D|}$ is reduced. From Lemma 5.3.4, we know that $f(x, D) = 1$ if $x \leq 1/2$ and $f(x, D) = 0$ if $x \geq 1/\sqrt{3}$. Show that $f(x, D)$ has a limit $f(x)$ as $|D| \to \infty$, and give a closed formula for $f(x)$, assuming that a quadratic number behaves like a random irrational number. Note that this exercise is difficult, and the complete result without the randomness assumption has only recently been proved by Duke (see [Duk]).

6. If D_0 is a fundamental negative discriminant and $D = D_0 f^2$, show directly from the formula given in the text that $h(D_0) \mid h(D)$.

7. Let p be a prime number such that $p \equiv 3 \pmod 4$. Using Dirichlet's class number formula (Corollary 5.3.13) express $h(-p)$ as a function of

    $$\sum_{1 \leq n \leq (p-1)/2} \left\lfloor \frac{n^2}{p} \right\rfloor .$$

    Is this algorithmically better than Dirichlet's formula?

8. Carry out in detail the GCD computations of the proof of Lemma 5.4.5.

9. Show that the composite of two primitive forms is primitive, and also that primitivity is preserved under reduction (both for complex quadratic fields and real ones). Prove these results first using the interpretation in terms of ideals, then directly on the formulas.

10. Show that, in order to generalize Algorithm 5.4.7 to imprimitive forms, we can replace the assignment $v_1 \leftarrow a_1/d_1$ of Step 4 by $v_1 \leftarrow \gcd(d_1, c_1, c_2, n)a_1/d_1$.

11. Let A, B and C be integers, and assume that at most one of them is equal to zero. Show that the general integral solution to the equation

    $$uA + vB + wC = 0$$

    is given by

    $$u = \frac{B}{(A,B)}\nu - \frac{C}{(A,C)}\mu, \quad v = \frac{C}{(B,C)}\lambda - \frac{A}{(A,B)}\nu, \quad w = \frac{A}{(A,C)}\mu - \frac{B}{(B,C)}\lambda$$

    where λ, μ and ν are arbitrary integers.

12. Using the preceding exercise, show that as claimed after Definition 5.4.6 the class of (a_3, b_3, c_3) modulo Γ_∞ is well defined.

13. In step 9 of Algorithm 5.5.2, it is suggested to compute the determinant of the lattice generated by the columns of a rectangular matrix A_1 of full rank by computing this determinant modulo p and using the Chinese remainder theorem together with Hadamard's inequality. Show that it is possible to modify the Gauss-Bareiss Algorithm 2.2.6 so as to compute this determinant directly, and compare the efficiency of the two methods, in theory as well as in practice (in the author's experience, the direct method is usually superior). Hint: use flags c_k and/or d_k as in Algorithm 2.3.1.

14. Implement the large prime variation explained after Algorithm 5.5.2 in the following manner. Choose some integer k (say $k = 500$) and use k lists of quadratic forms as follows. Each time that some p_a is encountered, we store p_a and the corresponding quadratic form in the n-th list, where $n = p_a \bmod k$. If p_a is already in the list, we have a relation, otherwise we do nothing else. Study the efficiency of this method and the choice of k. (Note: this method is a special case of a well known method used in computer science called *hashing*, see [Knu3].)

15. Implement Atkin's variant of McCurley's algorithm assuming that the discriminant D is a prime number and that the order of f is larger than the bound given by the Euler product.

16. Let $\mathfrak{a}$ be an integral ideal in a number field K, $\ell(\mathfrak{a})$ the smallest positive rational integer belonging to $\mathfrak{a}$, and σ_i the embeddings of K into $\mathbb{C}$. We will say that $\mathfrak{a}$ is *reduced* if $\mathfrak{a}$ is primitive and if the conditions $\alpha \in \mathfrak{a}$ and for all i, $|\sigma_i(\alpha)| < \ell(\mathfrak{a})$ imply that $\alpha = 0$.

    a) If $(\mathfrak{a}, s) = \phi_{FI}(a, b, c)$, show that $\mathfrak{a}$ is reduced if and only if there exists a (unique) quadratic form in the Γ_∞-class of (a, b, c) which is reduced. (Since the cases K real and imaginary must be treated separately, this is in fact two exercises in one.)

    b) In the case where $K = \mathbb{Q}(\sqrt{D})$ is a real quadratic field, show that $\mathfrak{a}$ is reduced if and only if there exists integers a_1 and a_2 such that $a_1 \equiv a_2 \equiv b$ (mod $2a$), $0 < a_1 < \sqrt{D}$ and $-\sqrt{D} < a_2 < 0$.

    c) Let $\mathfrak{a}$ be an ideal in the number field K. Show that there exists an $\alpha \in \mathfrak{a}$ such that $|\sigma_i(\beta)| < |\sigma_i(\alpha)|$ for all i implies that $\beta = 0$. By considering the ideal $(d/\alpha)\mathfrak{a}$ for a suitable integer d, deduce from this that, as in the quadratic case, every ideal is equivalent to a (not necessarily unique) reduced ideal.

17. Show that in any cycle of reduced quadratic forms of discriminant $D > 0$, there exists a form (a, b, c) with $|a| \le \sqrt{D/5}$. In other words, show that in any ideal class there exists an ideal $\mathfrak{a}$ such that $\mathcal{N}(\mathfrak{a}) \le \sqrt{D/5}$. (Hint: use Theorem 454 in [H-W].)

18. Prove Proposition 5.6.1.

19. Using Definition 4.9.11 and Proposition 5.1.4, show that if K is a (real or imaginary) quadratic field of discriminant D we have $\zeta_K(s) = \zeta(s)L_D(s)$, and hence that Propositions 5.3.12 and 5.6.9 are special cases of Dedekind's Theorem 4.9.12.

20. Modify Algorithm 5.7.1 so that it is still valid for $a < 0$.

21. Prove the following precise form of Lemma 5.7.6. If p and q are coprime integers, denote by p' the inverse of p modulo q such that $1 \le p' \le q$. Let α be a real number. Then p/q is a convergent in the continued fraction expansion of α if and only if

$$-\frac{1}{q(q+p')} < \alpha - \frac{p}{q} < \frac{1}{q(2q-p')} \ .$$

22. Show that the period of the continued fraction expansion of the quadratic irrational corresponding to the inverse of a reduced quadratic form f of positive discriminant is the reverse of the period of the quadratic number corresponding to f. Conclude that for ambiguous forms, the period is symmetric.

23. Write an algorithm corresponding to Algorithm 5.7.2 as Algorithm 5.7.10 corresponds to Algorithm 5.7.1 for computing the regulator of a real quadratic field using the symmetry of the period when we start with the unit form instead of any reduced form.

24. Assume that one has computed the regulator of a real quadratic field using the method explained in Section 5.9 to a given precision which need not be very high. Show that one can then compute the regulator to any desired accuracy in a small extra amount of time (hint: using the distance function, we now know where to look in the cycle).

25. Similarly to the preceding exercise, show that one can also compute the p-adic regulator to any desired accuracy in a small extra amount of time.

26. Let D be a fundamental discriminant.

   a) Show that $h^+(D)R^+(D) = 2h(D)R(D)$ and that $R^+(D) = 2R(D)$ if and only if the fundamental unit is of norm equal to -1.

   b) What modifications can be made to Algorithm 5.9.2 so that it computes the regulator and the class number in the narrow sense?

27. Let $D < -4$ be a fundamental discriminant, and set $f = |D|$.

   a) Set $s(x) = \sum_{1 \le n \le x} \left(\frac{D}{n}\right)$. Show that $|s(x)| \le f/2$ and by Abel summation that $|\sum_{n>f} \left(\frac{D}{n}\right)/n| < 1/2$.

   b) Show that $h(D) < \frac{1}{\pi}\sqrt{f}\ln f$.

   c) Using the Polya-Vinogradov inequality (see Exercise 8 of Chapter 9), give a better explicit upper bound for $h(D)$, asymptotic to $\frac{1}{2\pi}\sqrt{f}\ln f$.

28. (S. Louboutin) Using again the function $s(x)$ defined in Exercise 27 and Abel summation, show that we can avoid the computation of the function $\mathrm{erfc}(x)$ in Proposition 5.3.14 using the fact that $h(D)$ is an integer whose parity can be computed in advance ($h(D)$ is odd if and only if $D = -4$, $D = -8$ or $D = -p$ where p is a prime congruent to 3 modulo 4). Apply a similar method in Proposition 5.6.11.

Chapter 6

Algorithms for Algebraic Number Theory II

We now leave the realm of quadratic fields where the main computational tasks of algebraic number theory mentioned at the end of Chapter 4 were relatively simple (although as we have seen many conjectures remain), and move on to general number fields.

We first discuss practical algorithms for computing an integral basis and for the decomposition of primes in a number field K, essentially following a paper of Buchmann and Lenstra [Buc-Len], except that we avoid the explicit use of Artinian rings. We then discuss algorithms for computing Galois groups (up to degree 7, but see also Exercise 15). As examples of number fields of higher degree we then treat cyclic and pure cubic fields. Finally, in the last section of this chapter, we give a complete algorithm for class group and regulator computation which is sufficient for dealing with fields having discriminants of reasonable size. This algorithm also gives a system of fundamental units if desired.

6.1 Computing the Maximal Order

Let $K = \mathbb{Q}[\theta]$ be a number field, where θ is a root of a monic polynomial $T(X) \in \mathbb{Z}[X]$. Recall that $\mathbb{Z}_K$ has been defined as the set of algebraic integers belonging to K, and that it is called the maximal order since it is an order in K containing every order of K. We will build it up by starting from a known order (in fact from $\mathbb{Z}[\theta]$) and by successively enlarging it.

6.1.1 The Pohst-Zassenhaus Theorem

The main tool that we will use for enlarging an order is the Pohst-Zassenhaus Theorem 6.1.3 below. We first need a few basic results and definitions.

Definition 6.1.1. *Let $\mathcal{O}$ be an order in a number field K and let p be a prime number.*

(1) We will say that $\mathcal{O}$ is p-maximal if $[\mathbb{Z}_K : \mathcal{O}]$ is not divisible by p.
(2) We define the p-radical I_p as follows.

$$I_p = \{x \in \mathcal{O} \mid \exists m \geq 1 \text{ such that } x^m \in p\mathcal{O}\}$$

Proposition 6.1.2. *Let $\mathcal{O}$ be an order in a number field K and let p be a prime number.*

(1) *The p-radical I_p is an ideal of $\mathcal{O}$.*
(2) *We have*

$$I_p = \prod_{1 \le i \le g} \mathfrak{p}_i$$

the product being over all distinct prime ideals $\mathfrak{p}_i$ of $\mathcal{O}$ which lie above p.
(3) *There exists an integer m such that $I_p^m \subset p\mathcal{O}$.*

Proof. For (1), the only thing which is not completely trivial is that I_p is stable under addition. If $x^m \in p\mathcal{O}$ and $y^n \in p\mathcal{O}$, then clearly $(x+y)^{n+m} \in p\mathcal{O}$ as we see by using the binomial theorem.

For (2) note that since $\mathfrak{p}_i$ lies above p then $p\mathcal{O} \subset \mathfrak{p}_i$. So, if $x \in I_p$ there exists an m such that $x^m \in p\mathcal{O} \subset \mathfrak{p}_i$, and hence $x \in \mathfrak{p}_i$ by definition of a prime ideal. By Proposition 4.6.4 this shows that $x \in \bigcap_{1 \le i \le g} \mathfrak{p}_i = \prod_{1 \le i \le g} \mathfrak{p}_i$ since the distinct maximal ideals $\mathfrak{p}_i$ are pairwise coprime.

Conversely, assume that $x \in \prod_{1 \le i \le g} \mathfrak{p}_i$. By definition, the set of ideals of $\mathcal{O}$ containing $p\mathcal{O}$ is in canonical one-to-one correspondence with the ideals of the finite quotient ring $R = \mathcal{O}/p\mathcal{O}$. We will use this at length later. For now, note that it implies that this set is finite, and in particular the ideals $\alpha^n R$ are finite in number, where α is the class of x in R. In particular, there exists an n such that $\alpha^n R = \alpha^{n+1} R$, i.e. $\alpha^n(1 - \alpha\beta) = 0$ for some $\beta \in R$. By assumption, α belongs to all the maximal ideals $\overline{\mathfrak{p}}_i$ of R hence $(1 - \alpha\beta)$ cannot belong to any of them, otherwise 1 would also, which is impossible. It follows that the ideal $(1 - \alpha\beta)R$, not being contained in any maximal ideal, must be equal to R, i.e. $1 - \alpha\beta$ is invertible R. The equality $\alpha^n(1 - \alpha\beta) = 0$ thus implies that $\alpha^n = 0$ in R, i.e. that $x^n \in p\mathcal{O}$ or again that $x \in I_p$ as was to be proved.

Finally, for (3) note that since I_p is an ideal of an order in a number field it has a finite $\mathbb{Z}$-basis x_i for $1 \le i \le n$. For each x_i there exists an m_i such that $x_i^{m_i} \in p\mathcal{O}$, and if we set $m = \sum_{1 \le i \le n} m_i$ it is clear that $I_p^m \subset p\mathcal{O}$, again by the binomial theorem. $\qquad\square$

The procedure that we will use to obtain the maximal order is to start with $\mathcal{O} = \mathbb{Z}[\theta]$ and enlarge it for successive primes so as to get an order which is p-maximal for every p, hence which will be the maximal order. The enlarging procedure which we will use, due to Pohst and Zassenhaus, is based on the following theorem.

Theorem 6.1.3. *Let $\mathcal{O}$ be an order in a number field K and let p be a prime number. Set*

$$O' = \{x \in K | xI_p \subset I_p\} .$$

Then either $O' = \mathcal{O}$, in which case $\mathcal{O}$ is p-maximal, or $O' \supsetneq \mathcal{O}$ and $p \mid [O' : \mathcal{O}] \mid p^n$.

Proof. Since I_p is an ideal, it is clear that $\mathcal{O}'$ is a ring containing $\mathcal{O}$. Furthermore, since $p \in I_p$, $x \in \mathcal{O}'$ implies that $xp \in I_p \subset \mathcal{O}$ and hence $\mathcal{O} \subset \mathcal{O}' \subset \frac{1}{p}\mathcal{O}$. This shows that $\mathcal{O}'$ has maximal rank, i.e. is an order in K, and it also shows that $[\mathcal{O}' : \mathcal{O}] | p^n$.

We now assume that $\mathcal{O}' = \mathcal{O}$. Define

$$\mathcal{O}_p = \{x \in \mathbb{Z}_K | \exists j \geq 1, p^j x \in \mathcal{O}\} \ .$$

It is clear that $\mathcal{O} \subset \mathcal{O}_p$ and that $\mathcal{O}_p$ is an order. Furthermore, $\mathcal{O}_p$ is p-maximal. Indeed, if p divides the index $[\mathbb{Z}_K : \mathcal{O}_p]$, then there exists $x \in \mathbb{Z}_K$ such that $x \notin \mathcal{O}_p$ but $px \in \mathcal{O}_p$. The definition of $\mathcal{O}_p$ shows that this is impossible.

We are now going to show that $\mathcal{O}_p = \mathcal{O}$. Since $\mathcal{O}_p$ is an order, it is finitely generated over $\mathbb{Z}$. Hence there exists an $r \geq 1$ such that $p^r \mathcal{O}_p \subset \mathcal{O}$ (take r to be the maximum of the j such that $p^j x_i \in \mathcal{O}$ for a finite generating set (x_i) of $\mathcal{O}_p$). Since $I_p^m \subset p\mathcal{O}$ it follows that $\mathcal{O}_p I_p^{mr} \subset \mathcal{O}$. Assume by contradiction that $\mathcal{O}_p \neq \mathcal{O}$, hence $\mathcal{O}_p \not\subset \mathcal{O}$. Let n be the largest index such that $\mathcal{O}_p I_p^n \not\subset \mathcal{O}$ (hence n exists and $0 \leq n < mr$). We thus also have $\mathcal{O}_p I_p^{n+1} \subset \mathcal{O}$. Choose any $x \in \mathcal{O}_p I_p^n \setminus \mathcal{O}$. Then $x I_p \subset \mathcal{O}$. Since $\mathcal{O}_p I_p^{n+m+1} \subset I_p^m \subset p\mathcal{O}$ it follows that if $y \in I_p$, then $(xy)^{n+m+1} \in p\mathcal{O}$ hence that $xy \in I_p$, so $x I_p \subset I_p$ thus showing that $x \in \mathcal{O}'$. This is a contradiction since $x \notin \mathcal{O}$ and we have assumed that $\mathcal{O}' = \mathcal{O}$. This finishes the proof of Theorem 6.1.3. □

(I thank D. Bernardi for the final part of the proof.)

6.1.2 The Dedekind Criterion

From the Pohst-Zassenhaus theorem, starting from a number field $K = \mathbb{Q}(\theta)$ defined by a monic polynomial $T \in \mathbb{Z}[X]$, we will enlarge the order $\mathbb{Z}[\theta]$ for every prime p such that p^2 divides the discriminant of T until we obtain an order which is p-maximal for every p, i.e. the maximal order. In practice however, even when the discriminant has square factors, $\mathbb{Z}[\theta]$ is quite often p-maximal for a number of primes p, and it is time consuming to have to compute $\mathcal{O}'$ as in Theorem 6.1.3 just to notice that $\mathcal{O}' = \mathbb{Z}[\theta]$, i.e. that $\mathbb{Z}[\theta]$ is p-maximal. Fortunately, there is a simple and important criterion due to Dedekind which allows us to decide, without the more complicated computations explained in the next section, whether $\mathbb{Z}[\theta]$ is p-maximal or not for prime numbers p, and if it is not, it will give us a larger order, which of course may still not be p-maximal.

It must be emphasized that this will work *only* for $\mathbb{Z}[\theta]$, or for any order $\mathcal{O}$ containing $\mathbb{Z}[\theta]$ with $[\mathcal{O} : \mathbb{Z}[\theta]]$ prime to p, but not for an order which has already been enlarged for the prime p itself.

This being said the basic theorem that we will prove, of which Dedekind's criterion is a special case, is as follows.

Theorem 6.1.4 (Dedekind). *Let $K = \mathbb{Q}(\theta)$ be a number field, $T \in \mathbb{Z}[X]$ the monic minimal polynomial of θ and let p be a prime number. Denote by $\bar{\ }$ reduction modulo p (in $\mathbb{Z}$, $\mathbb{Z}[X]$ or $\mathbb{Z}[\theta]$). Let*

$$\overline{T}(X) = \prod_{i=1}^{k} \overline{t_i}(X)^{e_i}$$

be the factorization of $T(X)$ modulo p in $\mathbb{F}_p[X]$, and set

$$g(X) = \prod_{i=1}^{k} t_i(X)$$

where the $t_i \in \mathbb{Z}[X]$ are arbitrary monic lifts of $\overline{t_i}$. Then

(1) The p-radical I_p of $\mathbb{Z}[\theta]$ at p is given by

$$I_p = p\mathbb{Z}[\theta] + g(\theta)\mathbb{Z}[\theta] \ .$$

In other words, $x = A(\theta) \in I_p$ if and only if $\overline{g} \mid \overline{A}$.
(2) Let $h(X) \in \mathbb{Z}[X]$ be a monic lift of $\overline{T}(X)/\overline{g}(X)$ and set

$$f(X) = (g(X)h(X) - T(X))/p \in \mathbb{Z}[X] \ .$$

Then $\mathbb{Z}[\theta]$ is p-maximal if and only if

$$(\overline{f}, \overline{g}, \overline{h}) = 1 \quad in \ \mathbb{F}_p[X] \ .$$

(3) More generally, let $\mathcal{O}'$ be the order given by Theorem 6.1.3 when we start with $\mathcal{O} = \mathbb{Z}[\theta]$. Then, if U is a monic lift of $\overline{T}/(\overline{f}, \overline{g}, \overline{h})$ to $\mathbb{Z}[X]$ we have

$$\mathcal{O}' = \mathbb{Z}[\theta] + \frac{1}{p} U(\theta)\mathbb{Z}[\theta]$$

and if $m = \deg(\overline{f}, \overline{g}, \overline{h})$, then $[\mathcal{O}' : \mathbb{Z}[\theta]] = p^m$, hence $\mathrm{disc}(\mathcal{O}') = \mathrm{disc}(T)/p^{2m}$.

Proof of (1). $p \in I_p$ trivially, and since the exponents e_i are at most equal to $n = [K : \mathbb{Q}] = \deg(T)$, we have $\overline{T} \mid \overline{g}^n$ hence $g^n(\theta) \equiv 0 \pmod{p\mathbb{Z}[\theta]}$ so $g(\theta) \in I_p$, thus proving that $I_p \supset p\mathbb{Z}[\theta] + g(\theta)\mathbb{Z}[\theta]$.

Now the minimal polynomial over $\mathbb{F}_p$ of θ in $\mathbb{Z}[\theta]/p\mathbb{Z}[\theta]$ (which is not a field in general) is clearly the polynomial $\overline{T}$. Indeed, it clearly divides $\overline{T}$, but it is of degree at least n since $1, \theta, \ldots, \theta^{n-1}$ are $\mathbb{F}_p$-linearly independent.

Conversely let $x \in I_p$. Then $x = A(\theta)$ for $A \in \mathbb{Z}[X]$, and so there exists an integer m such that $x^m \equiv 0 \pmod{p\mathbb{Z}[\theta]}$, in other words $\overline{A}^m(\theta) = 0$ in $\mathbb{Z}[\theta]/p\mathbb{Z}[\theta]$. Hence $\overline{T} \mid \overline{A}^m$. Since $e_i \geq 1$ for all i, this implies that $\overline{t_i} \mid \overline{A}^m$ hence $\overline{t_i} \mid \overline{A}$ since t_i is irreducible in $\mathbb{F}_p[X]$, and since the $\overline{t_i}$ are pairwise coprime, we get $\overline{g} \mid \overline{A}$ which means that $x \in p\mathbb{Z}[\theta] + g(\theta)\mathbb{Z}[\theta]$ thus proving (1).

Since $\overline{T}$ is the minimal polynomial of θ in $\mathbb{Z}[\theta]/p\mathbb{Z}[\theta]$, it is clear that (2) follows from (3).

Let us now prove (3). Recall that $\mathcal{O}' = \{x \in K | x I_p \subset I_p\}$. From (1) we have that $x \in \mathcal{O}'$ if and only if $xp \in I_p$ and $xg(\theta) \in I_p$. Since $I_p \subset \mathbf{Z}[\theta]$, $xp \in I_p$ implies that

$$x = A_1(\theta)/p$$

where $A_1 \in \mathbf{Z}[X]$. Part (3) of the theorem will immediately follow from the following lemma.

Lemma 6.1.5. *Let* $x = A_1(\theta)/p$ *with* $A_1 \in \mathbf{Z}[X]$. *Then*

(1) $xp \in I_p$ *if and only if*

$$\overline{g} \mid \overline{A_1} \ .$$

(2) *Let* $\overline{k} = \overline{g}/(\overline{f}, \overline{g})$, *where (here as elsewhere in this section)* k *is implicitly considered to be a monic lift of* $\overline{k}$ *to* $\mathbf{Z}[X]$. *Then* $xg(\theta) \in I_p$ *if and only if*

$$\overline{hk} \mid \overline{A_1} \ .$$

Proof of the Lemma. Part (1) of the lemma is an immediate consequence of part (1) of the theorem. Let us prove part (2).

From part (1) of the theorem, $xg(\theta) \in I_p$ if and only if there exist polynomials A_2 and A_3 in $\mathbf{Z}[X]$ such that

$$A_1(\theta)g(\theta) = p(pA_2(\theta) + g(\theta)A_3(\theta)) \ ,$$

and since T is the minimal polynomial of θ, this is true if and only if there exists $A_4 \in \mathbf{Z}[X]$ such that

$$A_1(X)g(X) = p^2 A_2(X) + pg(X)A_3(X) + A_4(X)T(X) \ .$$

For the rest of this proof, we will work only with polynomials (in $\mathbf{Z}[X]$ or $\mathbb{F}_p[X]$), and not any more in K.

Reducing modulo p, the above equation implies that $\overline{A_1} = \overline{A_4 h}$. Hence write

$$A_1 = hA_4 + pA_5$$

with $A_5 \in \mathbf{Z}[X]$. We have that $xg(\theta) \in I_p$ if and only if there exist polynomials $A_i \in \mathbf{Z}[X]$ such that

$$(gh - T)A_4 = p^2 A_2 + pg(A_3 - A_5) \ ,$$

hence if and only if there exist A_i such that

$$fA_4 = pA_2 + gA_6 \ .$$

This last condition is equivalent to $\overline{g} \mid \overline{fA_4}$ so to $\overline{k} \mid \overline{A_4}$ where $\overline{k} = \overline{g}/(\overline{f}, \overline{g})$, and this is equivalent to the existence of A_7 and A_8 in $\mathbf{Z}[X]$ such that $A_4 = kA_7 + pA_8$.

To sum up, we see that if $x = A_1(\theta)/p$, then $xg(\theta) \in I_p$ if and only if there exist polynomials A_5, A_7 and A_8 in $\mathbb{Z}[X]$ such that

$$A_1 = hkA_7 + p(hA_8 + A_5) \ ,$$

and this is true if and only if there exist $A_9 \in \mathbb{Z}[X]$ such that $A_1 = hkA_7 + pA_9$ or equivalently $\overline{hk} \mid \overline{A_1}$, thus proving the lemma.   □

We can now prove part (3) of the theorem. From the lemma, we have that $x = A_1(\theta)/p \in \mathcal{O}'$ if and only if both $\overline{g}$ and $\overline{hk}$ divide $\overline{A_1}$ in the PID $\mathbb{F}_p[X]$, hence if and only if the least common multiple of $\overline{g}$ and $\overline{hk}$ divides A_1. Since in any PID, $\mathrm{lcm}(x, y) = xy/(x, y)$ and $\mathrm{lcm}(zx, zy) = z\,\mathrm{lcm}(x, y)$, we have

$$\mathrm{lcm}(\overline{g}, \overline{hk}) = \overline{k}\,\mathrm{lcm}(\gcd(\overline{f}, \overline{g}), \overline{h}) = \frac{\overline{g}}{(\overline{f}, \overline{g})}\frac{\overline{h}(\overline{f}, \overline{g})}{(\overline{f}, \overline{g}, \overline{h})} = \frac{\overline{T}}{(\overline{f}, \overline{g}, \overline{h})} = \overline{U}$$

thus proving that $\mathcal{O}' = \mathbb{Z}[\theta] + (U(\theta)/p)\mathbb{Z}[\theta]$. Now it is clear that a system of representatives of $\mathcal{O}'$ modulo $\mathbb{Z}[\theta]$ is given by $A(\theta)U(\theta)/p$ where A runs over uniquely chosen representatives in $\mathbb{Z}[X]$ of polynomials in $\mathbb{F}_p[X]$ such that $\deg(A) < \deg(T) - \deg(U) = m$, thus finishing the proof of the theorem.   □

An important remark is that the proof of this theorem is *local* at p, in other words we can copy it essentially verbatim if we everywhere replace $\mathbb{Z}[\theta]$ by any overorder $\mathcal{O}$ of $\mathbb{Z}[\theta]$ such that $[\mathcal{O} : \mathbb{Z}[\theta]]$ is coprime to p. The final result is then that the new order enlarged at p is

$$\mathcal{O} + \frac{U(\theta)}{p}\mathcal{O} \ ,$$

and $[\mathcal{O}' : \mathcal{O}] = p^m$.

6.1.3 Outline of the Round 2 Algorithm

From the Pohst-Zassenhaus theorem it is easy to obtain an algorithm for computing the maximal order. We will of course use the Dedekind criterion to simplify the first steps for every prime p.

Let $K = \mathbb{Q}(\theta)$ be a number field, where θ is an algebraic integer. Let T be the minimal polynomial of θ. We can write $\mathrm{disc}(T) = df^2$, where d is either 1 or a fundamental discriminant. If $\mathbb{Z}_K$ is the maximal order which we are looking for, then the index $[\mathbb{Z}_K : \mathbb{Z}[\theta]]$ has only primes dividing f as prime divisors because of Proposition 4.4.4. We are going to compute $\mathbb{Z}_K$ by successive enlargements from $\mathcal{O} = \mathbb{Z}[\theta]$, one prime dividing f at a time. For every p dividing f we proceed as follows. By using Dedekind's criterion, we check whether $\mathcal{O}$ is p-maximal and if it is not we enlarge it once using Theorem 6.1.4 (3) applied to $\mathcal{O}$. If the new discriminant is not divisible by p^2, then we are done, otherwise we compute $\mathcal{O}'$ as described in Theorem 6.1.3. If $\mathcal{O}' = \mathcal{O}$, then $\mathcal{O}$ is p-maximal and we are finished with the prime p, so we move on to the next prime, if any. (Here again we can start using Dedekind's criterion.) Otherwise, replace $\mathcal{O}$ by $\mathcal{O}'$, and use the method of Theorem 6.1.3 again. It is clear that this algorithm is valid and will lead quite rapidly to the maximal order. This algorithm was the second one invented by Zassenhaus for maximal order computations, and so it has become known as the round 2 algorithm (the latest and most efficient is round 4).

What remains is to explain how to carry out explicitly the different steps of the algorithm, when we apply Theorem 6.1.3.

First, θ is fixed, and all ideals and orders will be represented by their upper triangular HNF as explained in Section 4.7.2. We must explain how to compute the HNF of I_p and of $\mathcal{O}'$ in terms of the HNF of $\mathcal{O}$. It is simpler to compute in $R = \mathcal{O}/p\mathcal{O}$. To compute the radical of R, we note the following lemma:

Lemma 6.1.6. *If $n = [K : \mathbb{Q}]$ and if $j \geq 1$ is such that $p^j \geq n$, then the radical of R is equal to the kernel of the map $x \mapsto x^{p^j}$, which is the jth power of the Frobenius homomorphism.*

Proof. It is clear that the map in question is the jth power of the Frobenius homomorphism, hence talking about its kernel makes sense. By definition of the radical, it is clear that this kernel is contained in the radical. Conversely, let x be in the radical. Then x induces a nilpotent map defined by multiplication by x from R to R, and considering R as an $\mathbb{F}_p$-vector space, this means that the eigenvalues of this map in $\overline{\mathbb{F}_p}$ are all equal to 0. Hence, its characteristic polynomial must be X^n (since $n = \dim_{\mathbb{F}_p} R$), and by the Cayley-Hamilton theorem this shows that $x^n = 0$, and hence that $x^{p^j} = 0$, proving the lemma. $\square$

Let $\omega_1, \ldots, \omega_n$ be the HNF basis of $\mathcal{O}$. Then it is clear that $\overline{\omega}_1, \ldots, \overline{\omega}_n$ is an $\mathbb{F}_p$-basis of R. For $k = 1, \ldots, n$, we compute $\overline{a}_{i,k}$ such that

$$\overline{\omega}_k^{p^j} = \sum_{i=1}^n \overline{a}_{i,k} \overline{\omega}_i \ ,$$

the left hand side being computed as a polynomial in θ by the standard representation algorithms, and the coefficients $\overline{a}_{i,k}$ being easily found inductively since an HNF matrix is triangular. Hence, if $\overline{A}$ is the matrix of the $\overline{a}_{i,k}$, the radical is simply the kernel of this matrix.

Hence, if we apply Algorithm 2.3.1, we will obtain a basis of $\overline{I}_p$, the radical of R, in terms of the standard representation. Since I_p is generated by pullbacks of a basis of $\overline{I}_p$ and $p\omega_1, \ldots, p\omega_n$, to obtain the HNF of I_p we apply the HNF reduction algorithm to the matrix whose columns are the standard representations of these elements.

Now that we have I_p, we must compute $\mathcal{O}'$. For this, we use the following lemma:

Lemma 6.1.7. *With the notations of Theorem 6.1.3, if U is the kernel of the map*

$$\alpha \longmapsto (\bar{\beta} \mapsto \overline{\alpha\beta})$$

from $\mathcal{O}$ to $\mathrm{End}(I_p/pI_p)$, then $\mathcal{O}' = \frac{1}{p}U$.

Proof. Trivial and left to the reader. Note that $\mathrm{End}(I_p/pI_p)$ is considered as a $\mathbb{Z}$-module. $\qquad\square$

Hence, we first need to find a basis of I_p/pI_p. There are two methods to do this. From the HNF reduction above, we know a basis of I_p, and it is clear that the image of this basis in I_p/pI_p is a basis of I_p/pI_p. The other method is as follows. We use only the $\mathbb{F}_p$-basis $\bar{\beta}_1, \ldots, \bar{\beta}_l$ of $\bar{I}_p$ found above. Using Algorithm 2.3.6, we can supplement this basis into a basis $\bar{\beta}_1, \ldots, \bar{\beta}_l, \bar{\beta}_{l+1}, \ldots, \bar{\beta}_n$ of $\mathcal{O}/p\mathcal{O}$, and then $\tilde{\beta}_1, \ldots, \tilde{\beta}_l, p\tilde{\beta}_{l+1}, \ldots, p\tilde{\beta}_n$ will be an $\mathbb{F}_p$-basis of I_p/pI_p, where $\tilde{\ }$ denotes reduction modulo pI_p, and β_i denotes any pull-back of $\bar{\beta}_i$ in $\mathcal{O}$. (Note that the basis which one obtains depends on the pull-backs used.)

This method for finding a basis of I_p/pI_p has the advantage of staying at the mod p level, hence avoids the time consuming Hermite reduction, so it is preferable.

Now that we have a basis of I_p/pI_p, the elementary matrices give us a basis of $\mathrm{End}(I_p/pI_p)$. Hence, we obtain explicitly the matrix of the map whose kernel is U, and it is a $n^2 \times n$ matrix. Algorithm 2.3.1 makes sense only over a field, so we must first compute the kernel $\overline{U}$ of the map from $\mathcal{O}/p\mathcal{O}$ into $\mathrm{End}(I_p/pI_p)$ which can be done using Algorithm 2.3.1. If $\bar{v}_1, \ldots, \bar{v}_k$ is the basis of this kernel, to obtain U, we apply Hermite reduction to the matrix whose column vectors are $v_1, \ldots, v_k, pw_1, \ldots, pw_n$. In fact, we can apply Hermite reduction modulo the prime p, i.e. take $D = p$ in Algorithm 2.4.8.

Finally, note that to obtain the $n^2 \times n$ matrix above, if the $\overline{\gamma}_i$ form a basis of I_p/pI_p one computes

$$\omega_k \overline{\gamma}_i = \sum_{1 \leq j \leq n} a_{k,i,j} \overline{\gamma}_j \ ,$$

and k is the column number, while (i, j) is the row index. Unfortunately, in the round 2 algorithm, it seems unavoidable to use such large matrices. Note that to obtain the $a_{k,i,j}$, the work is much simpler if the matrix of the $\overline{\gamma}_j$ is triangular, and this is not the case in general if we complete the basis as explained above. On the other hand, this would be the case if we used the first method consisting of applying Hermite reduction to get the HNF of I_p itself. Tests must be made to see which method is preferable in practice.

6.1.4 Detailed Description of the Round 2 Algorithm

Using what we have explained, we can now give in complete detail the round 2 algorithm.

Algorithm 6.1.8 (Zassenhaus's Round 2). Let $K = \mathbb{Q}(\theta)$ be a number field given by an algebraic integer θ as root of its minimal monic polynomial T of degree n. This algorithm computes an integral basis $\omega_1 = 1, \omega_2, \ldots, \omega_n$ of the maximal order $\mathbb{Z}_K$ (as polynomials in θ) and the discriminant of the field. All the computations in K are implicitly assumed to be done using the standard representation of numbers as polynomials in θ.

1. [Factor discriminant of polynomial] Using Algorithm 3.3.7, compute $D \leftarrow \operatorname{disc}(T)$. Then using a factoring algorithm (see Chapters 8 to 10) factor D in the form $D = D_0 F^2$ where D_0 is either equal to 1 or to a fundamental discriminant.

2. [Initialize] For $i = 1, \ldots, n$ set $\omega_i \leftarrow \theta^{i-1}$.

3. [Loop on factors of F] If $F = 1$, output the integral basis ω_i (which will be in HNF with respect to θ), compute the product G of the diagonal elements of the matrix of the ω_i (which will be the inverse of an integer by Corollary 4.7.6), set $d \leftarrow D \cdot G^2$, output the field discriminant d and terminate the algorithm. Otherwise, let p be the smallest prime factor of F.

4. [Factor modulo p] Using the mod p factoring algorithms of Section 3.4, factor T modulo p as $\overline{T} = \prod \overline{t_i}^{e_i}$ where the $\overline{t_i}$ are distinct irreducible polynomials in $\mathbb{F}_p[X]$ and $e_i > 0$ for all i. Set $\overline{g} \leftarrow \prod \overline{t_i}, \overline{h} \leftarrow \overline{T}/\overline{g}, f \leftarrow (gh - T)/p$, $\overline{Z} \leftarrow (\overline{f}, \overline{g}, \overline{h}), \overline{U} \leftarrow \overline{T}/\overline{Z}$ and $m \leftarrow \deg(\overline{Z})$.

5. [Apply Dedekind] If $m = 0$, then $\mathcal{O}$ is p-maximal so while $p \mid F$ set $F \leftarrow F/p$, then go to step 3. Otherwise, for $1 \leq i \leq m$, let v_i be the column vector of the components of $\omega_i U(\theta)$ on the standard basis $1, \theta, \ldots, \theta^{n-1}$ and set $v_{m+j} = p\omega_j$ for $1 \leq j \leq n$.
   Apply the Hermite reduction Algorithm 2.4.8 to the $n \times (n + m)$ matrix whose column vectors are the v_i. (Note that the determinant of the final matrix is known to divide D.) If H is the $n \times n$ HNF reduced matrix which we obtain, set for $1 \leq i \leq n$, $\omega_i \leftarrow H_i/p$ where H_i is the i-th column of H.

6. [Is the new order p-maximal?] If $p^{m+1} \nmid F$, then the new order is p-maximal so while $p \mid F$ set $F \leftarrow F/p$, then go to step 3.

7. [Compute radical] Set $q \leftarrow p$, and while $q < n$ set $q \leftarrow qp$. Then compute the $n \times n$ matrix $A = (a_{i,j})$ over $\mathbb{F}_p$ such that $\omega_j^q \equiv \sum_{1 \leq i \leq n} a_{i,j}\omega_i$. Note that the matrix of the ω_i will stay triangular, so the $a_{i,j}$ are easy to compute.
   Finally, using Algorithm 2.3.1, compute a basis $\overline{\beta_1}, \ldots, \overline{\beta_l}$ of the kernel of the matrix A over $\mathbb{F}_p$ (this will be a basis of $I_p/p\mathcal{O}$).

8. [Compute new basis mod p] Using the known basis $\overline{\omega_1}, \ldots, \overline{\omega_n}$ of $\mathcal{O}/p\mathcal{O}$, supplement the linearly independent vectors $\overline{\beta_1}, \ldots, \overline{\beta_l}$ to a basis $\overline{\beta_1}, \ldots, \overline{\beta_n}$ of $\mathcal{O}/p\mathcal{O}$ using Algorithm 2.3.6.

9. [Compute big matrix] Set $\alpha_i \leftarrow \beta_i$ for $1 \leq i \leq l$, $\alpha_i \leftarrow p\beta_i$ for $l < i \leq n$, where β_i is a lift to $\mathcal{O}$ of $\overline{\beta_i}$. Compute coefficients $c_{i,j,k} \in \mathbb{F}_p$ such that $\omega_k \alpha_j \equiv \sum_{1 \leq i \leq n} c_{i,j,k} \alpha_i \pmod{p}$. Let C be the $n^2 \times n$ matrix over $\mathbb{F}_p$ such that $C_{(i,j),k} = c_{i,j,k}$.

10. [Compute new order] Using Algorithm 2.3.1, compute a basis $\gamma_1, \ldots \gamma_m$ for the kernel of C (these are vectors in $\mathbb{F}_p^n$, and m can be as large as n^2). For $1 \leq i \leq m$ let v_i be a lift of γ_i to $\mathbb{Z}^n$, and set $v_{m+j} = p\omega_j$ for $1 \leq j \leq n$. Apply the Hermite reduction Algorithm 2.4.8 to the $n \times (n+m)$ matrix whose column vectors are the v_i. (Note again that the determinant of the final matrix is known to divide D.) If H is the $n \times n$ HNF reduced matrix which we obtain, set for $1 \leq i \leq n$, $\omega_i' \leftarrow H_i/p$ where H_i is the i-th column of H.

11. [Finished with p?] If there exists an i such that $\omega_i' \neq \omega_i$, then for every i such that $1 \leq i \leq n$ set $\omega_i \leftarrow \omega_i'$ and go to step 7. Otherwise, $\mathcal{O}$ is p-maximal, so while $p \mid F$ set $F \leftarrow F/p$, and go to step 3.

This finishes our description of the round 2 algorithm. This algorithm seems complicated at first. Although it has been superseded by the round 4 algorithm, it is much simpler to implement and it performs very well. The major bottleneck is perhaps not where the reader expects it to be, i.e. in the handling of large matrices. It is, in fact, in the very first step which consists in factoring disc(T) in the form $D_0 F^2$. Indeed, as we will see in Chapter 10, factoring an 80 digit number takes a considerable amount of time, and factoring a 50 digit one is already not that easy. One can refine the methods given above to the case where one does not suppose p to be necessarily prime (see [Buc-Len] and [Buc-Len2]), but unfortunately this does *not* avoid finding the largest square dividing disc(T), which is apparently almost as difficult as factoring it completely.

6.2 Decomposition of Prime Numbers II

As we shall see, the general problem of decomposing prime numbers in an algebraic number field is closely related to the problem of computing the maximal order. Consequently, we have already given most of the theory and auxiliary algorithms that we will need. As we have already seen, the problem is as follows. Given a prime p and a p-maximal order $\mathcal{O}$, for example the maximal order $\mathbb{Z}_K$ itself, determine the maximal ideals $\mathfrak{p}_i$ and the exponents e_i such that

$$p\mathcal{O} = \prod_{i=1}^{g} \mathfrak{p}_i^{e_i} .$$

As usual $\mathcal{O}$ will be given by its HNF on a power basis $1, \theta, \ldots, \theta^{n-1}$, and we want the HNF basis of the $\mathfrak{p}_i$. The determinant of the corresponding matrix is equal to $\mathcal{N}(\mathfrak{p}_i) = p^{f_i}$ in the traditional notation. For practical applications, it will also be useful to have a two-element representation of the ideals $\mathfrak{p}_i$ (see Proposition 4.7.7).

In Theorem 4.8.13 we saw how to obtain this decomposition when p does not divide the index $[\mathcal{O} : \mathbb{Z}[\theta]]$. Hence we will concentrate on the case where p divides the index.

6.2.1 Newton Polygons

Historically the first method to deal with this problem is the so-called *Newton polygon method*. When it applies, it is very easy to use, but it must be stressed that it is not a general method. We will give a completely general method in the next section.

I am grateful to F. Diaz y Diaz and M. Olivier for the presentation of Newton polygons given here, which follows [Ore] and [Mon-Nar]. Essentially no proofs are given.

We may assume without loss of generality that the minimal polynomial $T(X)$ of θ is in $\mathbb{Z}[X]$ and is monic.

The first result tells us what survives of Theorem 4.8.13 in the case where p divides the index.

Proposition 6.2.1. *Let*

$$T(X) \equiv \prod_{i=1}^{g} \overline{T_i(X)}^{e_i} \pmod{p}$$

be the decomposition of T into irreducible factors in $\mathbb{F}_p[X]$, where the T_i are taken to be arbitrary monic lifts of $\overline{T_i(X)}$ in $\mathbb{Z}[X]$. Then

$$p\mathbb{Z}_K = \prod_{i=1}^{g} \mathfrak{a}_i \; ,$$

where

$$\mathfrak{a}_i = (p, T_i^{e_i}(\theta)) = p\mathbb{Z}_K + T_i^{e_i}(\theta)\mathbb{Z}_K$$

and the $\mathfrak{a}_i$ are pairwise coprime (i.e. $\mathfrak{a}_i + \mathfrak{a}_j = \mathbb{Z}_K$ for $i \neq j$). Furthermore, if n_i is the degree of T_i we have $\mathcal{N}(\mathfrak{a}_i) = p^{e_i n_i}$, and all prime ideals dividing $\mathfrak{a}_i$ are of residual degree divisible by n_i.

Proof. The proof follows essentially the same lines as that of Theorem 4.8.13. It is useful to also prove that the inverse of $\mathfrak{a}_i$ is given explicitly as

$$\mathfrak{a}_i^{-1} = (1, \prod_{j \neq i} T_j^{e_j}(\theta)/p)$$

(see Exercise 5). $\qquad \square$

The problem is that the ideals $\mathfrak{a}_i$ are not necessarily of the form $\mathfrak{p}_i^{e_i}$ as in Theorem 4.8.13 (the reader can also check via examples that it would not do any good to set $\mathfrak{p}_i = (p, T_i(\theta))$). We must therefore try to split the ideals $\mathfrak{a}_i$ some more. For this we can proceed as follows. By successive Euclidean divisions of T by T_i, we can write T in a unique way in the form

$$T(X) = \sum_{j=0}^{\lfloor n/n_i \rfloor} Q_{i,j} T_i^j$$

with $\deg(Q_{i,j}) < n_i$. We will call this the T_i-expansion of T. We will write $d_i = \lfloor n/n_i \rfloor$.

If $Q = \sum_{0 \le k \le m} a_k X^k \in \mathbb{Z}[X]$, we will set

$$v_p(Q) = \min_k(v_p(a_k)) \ ,$$

where we set $v_p(0) = +\infty$ (or in other words we ignore coefficients equal to zero). The basic definition is as follows.

Definition 6.2.2. *With the above notations, for a fixed i, the convex hull of the set of points $(j, v_p(Q_{i,d_i-j}))$ for each j such that $Q_{i,d_i-j} \ne 0$, is called the Newton polygon of T relative to T_i and the prime number p (since p is always fixed, we will in fact simply say "relative to T_i").*

Note that $Q_{i,j} = 0$ for $j < 0$ or $j > d_i$, hence the Newton polygon is bounded laterally by two infinite vertical half lines. Furthermore, since T and the T_i are monic, so is Q_{i,d_i} hence $v_p(Q_{i,d_i}) = 0$. It follows that the first vertex of the Newton polygon is the origin $(0,0)$. Let a be the largest real number (which is of course an integer) such that $(a,0)$ is still on the Newton polygon (we may have $a = 0$ or $a = d_i$). The part of the Newton polygon from the origin to $(a,0)$ is either empty (if $a = 0$) or is a horizontal segment. The rest of the Newton polygon, i.e. the points whose abscissa is greater than or equal to a, is called the *principal part* of the Newton polygon, and $(a,0)$ is its first vertex.

We assume now that i is fixed.

Let V_j for $0 \le j \le r$ be the vertices of the principal part of the Newton polygon of T relative to T_i (in the strict sense: if a point on the convex hull lies on the segment joining two other points, it is not a vertex), and set $V_j = (x_j, y_j)$. The *sides* of the polygon are the segments joining two consecutive vertices (not counting the infinite vertical lines), and the *slopes* are the slopes of these sides, i.e. the positive rational numbers $(y_j - y_{j-1})/(x_j - x_{j-1})$ for $1 \le j \le r$ (note that they cannot be equal to zero since we are in the principal part).

The second result gives us a more precise decomposition of $p\mathbb{Z}_K$ than the one given by Proposition 6.2.1 above, whose notations we keep. We refer to [Ore] for a proof.

Proposition 6.2.3. *Let i be fixed.*

(1) *To each side $[V_{j-1}, V_j]$ of the principal part of the Newton polygon of T relative to T_i we can associate an ideal $\mathfrak{q}_{i,j}$ such that the $\mathfrak{q}_{i,j}$ are pairwise coprime and*

$$\mathfrak{a}_i = \prod_{j=1}^{r} \mathfrak{q}_{i,j} \;.$$

(2) *Set $h_j = y_j - y_{j-1}$ and $k_j = x_j - x_{j-1}$. If h_j and k_j are coprime for some j, then the corresponding ideal $\mathfrak{q}_{i,j}$ is of the form $\mathfrak{q}_{i,j} = \mathfrak{p}^{k_j}$ where $\mathfrak{p}$ is a prime ideal of degree n_i.*

(3) *In the special case when the principal part of the Newton polygon has a single side and $h_1 = y_1 - y_0 = y_1$ is equal to 1, then $\mathfrak{a}_i = \mathfrak{p}^{e_i}$ where $\mathfrak{p} = (p, T_i(\theta))$ is a prime ideal of degree n_i.*

Corollary 6.2.4. *Let $T \in \mathbb{Z}[X]$ be an Eisenstein polynomial with respect to a prime number p, i.e. a monic polynomial $T(X) = \sum_{i=0}^{n} a_i X^i$ with $p \mid a_i$ for all $i < n$ and $p^2 \nmid a_0$ (see Exercise 11 of Chapter 3). In the number field $K = \mathbb{Q}[\theta]$ defined by T the prime p is totally ramified, and more precisely $p\mathbb{Z}_K = \mathfrak{p}^n$ with $\mathfrak{p} = (p, \theta)$.*

Proof. In this case we have $T \equiv X^n \pmod{p}$, hence $T_1(X) = X$, $Q_{i,j} = a_j$, and since $p \mid a_i$ for all $i < n$, the principal part of the Newton polygon is the whole polygon, and since $p^2 \nmid a_0$ we are in the special case (3) of the proposition, so the corollary follows. $\qquad\square$

Although Proposition 6.2.3 gives results in a number of cases, and can be generalized further (see [Ore] and [Mon-Nar]), it is far from being satisfactory from an algorithmic point of view.

6.2.2 Theoretical Description of the Buchmann-Lenstra Method

The second method for decomposing primes in number fields, which is completely general, is due to Buchmann and Lenstra ([Buc-Len]). We proceed as follows. (The reader should compare this to the method used for factoring polynomials modulo p given in Chapter 3.) Write I_p for the p-radical of $\mathcal{O}$. We know that $I_p = \prod_{i=1}^{g} \mathfrak{p}_i$. Set for any $j \geq 0$:

$$K_j = I_p^j + p\mathcal{O} \;.$$

It is clear that the valuation at $\mathfrak{p}_i$ of K_j is equal to $\min(e_i, j)$, hence

$$K_j = \prod_{i=1}^{g} \mathfrak{p}_i^{\min(e_i, j)} \;.$$

It is also clear that $K_j \subset K_{j-1}$. Hence, if we set

$$J_j = K_j(K_{j-1})^{-1} \, ,$$

then J_j is an integral ideal, and in fact $J_j = \prod_{e_i \geq j} \mathfrak{p}_i$ so in particular $J_j \subset J_{j+1}$. Finally, if we define

$$H_j = J_j(J_{j+1})^{-1} \, ,$$

we have

$$H_j = \prod_{e_i = j} \mathfrak{p}_i \, .$$

This exactly corresponds to the squarefree decomposition procedure of Section 3.4.2, the H_i playing the role of the A_i, and without the inseparability problems. In other words, if we set $e = \max_i(e_i)$, we have

$$p\mathcal{O} = \prod_{j=1}^{e} H_j^j \, ,$$

and the H_j are pairwise coprime and are products of distinct maximal ideals. To find the splitting of $p\mathcal{O}$, it is of course sufficient to find the splitting of each H_j.

Now, since H_j is a product of distinct maximal ideals, i.e. is squarefree, the $\mathbb{F}_p$-algebra $\mathcal{O}/H_j$ is separable. Therefore, by the primitive element theorem there exists $\overline{\alpha}_j \in \mathcal{O}/H_j$ such that $\mathcal{O}/H_j = \mathbb{F}_p[\overline{\alpha}_j]$. Let $\overline{h}_j$ be the characteristic polynomial of $\overline{\alpha}_j$ over $\mathbb{F}_p$, and h_j be any pullback in $\mathbb{Z}[X]$. Then exactly the same proof as in Section 4.8.2 shows that, if

$$h_j(X) \equiv \prod_{i=1}^{g_j} q_{i,j}(X) \pmod{p}$$

is the decomposition modulo p of the polynomial h_j, then the ideals

$$\mathfrak{q}_{i,j} = H_j + q_{i,j}(\alpha_j)\mathcal{O}$$

are maximal and that

$$H_j = \prod_{i=1}^{g_j} \mathfrak{q}_{i,j}$$

is the desired decomposition of H_j into a product of prime ideals.

We must now give algorithms for all the steps described above. Essentially, the two new things that we need are operations on ideals in our special case, and splitting of a separable algebra over $\mathbb{F}_p$.

6.2.3 Multiplying and Dividing Ideals Modulo p

Although the most delicate step in the decomposition of $p\mathbb{Z}_K$ is the final splitting of the ideals H_j, experiment (and complexity analysis) shows that this is paradoxically the fastest part. The conceptually easier steps of multiplying and dividing ideals take, in fact, most of the time and so must be speeded up as much as possible.

Looking at what is needed, it is clear that we use only the reductions modulo $p\mathcal{O}$ of the ideals involved. Hence, although for ease of presentation we have implicitly assumed that the ideals are represented by their HNF, we will in fact consider only ideals $I/p\mathcal{O}$ of $\mathcal{O}/p\mathcal{O}$ which will be represented by an $\mathbb{F}_p$-basis. All the difficulties of HNF (Euclidean algorithm, coefficient explosion) disappear and are replaced by simple linear algebra algorithms. Moreover, we are working with coefficients in a field which is usually of small cardinality. (Recall that p divides the index, otherwise the much simpler algorithm of Section 4.8.2 can be used.)

If I is given by its HNF with respect to θ (this will not happen in our case since we start working directly modulo p), then, since $I \supset p\mathcal{O} \supset p\mathbb{Z}[\theta]$, the diagonal elements of the HNF will be equal to 1 or p. Therefore, to find a basis of $\overline{I}$, we simply take the basis elements corresponding to the columns whose diagonal element is equal to 1.

The algorithm for multiplication is straightforward.

Algorithm 6.2.5 (Ideal Multiplication Modulo $p\mathcal{O}$). Given two ideals $I/p\mathcal{O}$ and $J/p\mathcal{O}$ by $\mathbb{F}_p$-bases $(\alpha_i)_{1 \leq i \leq r}$ and $(\beta_j)_{1 \leq j \leq m}$ respectively, where the α_i and β_j are expressed as $\mathbb{F}_p$-linear combinations of a fixed integral basis $\omega_1, \ldots, \omega_n$ of $\mathcal{O}$, this algorithm computes an $\mathbb{F}_p$-basis of the ideal $IJ/p\mathcal{O}$.

1. [Compute matrix] Using the multiplication table of the ω_i, let M be the $n \times rm$ matrix M with coefficients in $\mathbb{F}_p$ whose columns express the products $\alpha_i\beta_j$ on the integral basis.

2. [Compute image] Using Algorithm 2.3.2 compute a matrix M_1 whose columns form an $\mathbb{F}_p$-basis of the image of M. Output the columns of M_1 and terminate the algorithm.

Ideal division modulo $p\mathcal{O}$ is slightly more difficult. We first need a lemma.

Lemma 6.2.6. *Denote by $^-$ reduction mod p. Let I and J two integral ideals of $\mathcal{O}$ containing $p\mathcal{O}$ and assume that $I \subset J$. Then, as a $\mathbb{Z}/p\mathbb{Z}$-vector space, $\overline{IJ^{-1}}$ is equal to the kernel of the map ϕ from $\mathcal{O}/p\mathcal{O}$ to $\mathrm{End}(J/I)$ given by*

$$\phi(\overline{\beta}) = (\overline{\alpha} \longmapsto \overline{\alpha\beta}) \ .$$

Indeed, $\phi(\overline{\beta})$ is equal to 0 if and only if $\alpha\beta \in I$ for every $\alpha \in J$, i.e. if $\beta J \subset I$, or in other words if $\beta \in IJ^{-1}$, proving the lemma.   □

This leads to the following algorithm.

Algorithm 6.2.7 (Ideal Division Modulo $p\mathcal{O}$). Given two ideals $I/p\mathcal{O}$ and $J/p\mathcal{O}$ by $\mathbb{F}_p$ bases $(\alpha_i)_{1\le i\le r}$ and $(\beta_j)_{1\le j\le m}$ respectively, where the α_i and β_j are expressed as $\mathbb{F}_p$-linear combinations of a fixed integral basis $\omega_1, \ldots, \omega_n$ of $\mathcal{O}$, this algorithm computes an $\mathbb{F}_p$-basis of the ideal $IJ^{-1}/p\mathcal{O}$ assuming that $I \subset J$.

1. [Find basis of J/I] Apply Algorithm 2.3.7 to the subspaces $I/p\mathcal{O}$ and $J/p\mathcal{O}$ of $\mathbb{F}_p^n$, thus obtaining a basis $(\gamma_j)_{1\le j\le m-r}$ of a supplement of $I/p\mathcal{O}$ in $J/p\mathcal{O}$.

2. [Setup ideal division] By using the multiplication table of the ω_i and Algorithm 2.3.5, compute elements $a_{i,j,k}$ and $b_{i,j,k}$ in $\mathbb{F}_p$ such that

$$\overline{\omega}_k\gamma_i = \sum_j a_{i,j,k}\gamma_j + \sum_j b_{i,j,k}\alpha_j \; ,$$

   and let M be the $(m-r)^2 \times n$ matrix formed by the $a_{i,j,k}$ for $1 \le i,j \le m-r$ and $1 \le k \le n$ (we can forget the $b_{i,j,k}$).

3. [Compute $IJ^{-1}/p\mathcal{O}$] Using Algorithm 2.3.1, compute a matrix M_1 whose columns form an $\mathbb{F}_p$-basis of the kernel of M, output M_1 and terminate the Algorithm.

Indeed, M is clearly equal to the matrix of ϕ in the standard basis of $\text{End}(J/I)$. $\qquad\qquad\square$

6.2.4 Splitting of Separable Algebras over $\mathbb{F}_p$

To avoid unnecessary indices, we set simply $H = H_j$. Using the above algorithms, it is straightforward to compute an $\mathbb{F}_p$-basis $\overline{\beta}_1, \ldots, \overline{\beta}_m$ of $\overline{H} = H/p\mathcal{O}$. Using Algorithm 2.3.6, we can supplement this basis to a basis $\overline{\beta}_1, \ldots, \overline{\beta}_n$ of $\mathcal{O}/p\mathcal{O}$ It is then clear that the images of $\beta_{m+1}, \ldots, \beta_n$ in $\mathcal{O}/H$ form an $\mathbb{F}_p$-basis of $\mathcal{O}/H$.

In order to finish the decomposition, there remains the problem of splitting the separable algebra $A = \mathcal{O}/H$ given by this $\mathbb{F}_p$-basis. As explained above, one method is to start by finding a primitive element $\overline{\alpha}$. Finding a primitive element is not, however, a completely trivial task. Perhaps the best way is to choose at random an element $x \in A \backslash \mathbb{F}_p$ (note that $\mathbb{F}_p$ can be considered naturally embedded in A), compute its *minimal* polynomial $P(X)$ over $\mathbb{F}_p$ (which need not be irreducible), and check whether $\deg(P) = \dim(A)$. Although practical, this method has the disadvantage of being completely non-deterministic, although it is easy to give estimates for the number of trials that one has to perform before succeeding in finding a suitable x, see Exercise 6.

We give another method which does not have this disadvantage. It is based on the following proposition.

Proposition 6.2.8. *Let A be a finite separable algebra over $\mathbb{F}_p$. There exists an efficient probabilistic algorithm which either shows that A is a field, or finds a non-trivial idempotent in A, i.e. an element $\varepsilon \in A$ such that $\varepsilon^2 = \varepsilon$ with $\varepsilon \neq 0$ and $\varepsilon \neq 1$.*

Proof. Since A is a finite separable algebra, A is isomorphic to a finite product of fields, say $A \simeq A_1 \times \cdots \times A_k$. Write any element α of A as $(\alpha_1, \ldots, \alpha_k)$ where $\alpha_i \in A_i$. Consider the map ϕ from A to A defined by $\phi(x) = x^p - x$. It is clear that $\mathbb{F}_p$, considered as embedded in A, is in the kernel V of ϕ. By Algorithm 2.3.1, we can easily compute a basis for V, and, in particular, its dimension. Note that $\alpha = (\alpha_1, \ldots, \alpha_k) \in V$ if and only if for all i such that $1 \le i \le k$, $\alpha_i \in \mathbb{F}_p$ where $\mathbb{F}_p$ is considered embedded in A_i. It follows that $\dim(V) = k$, and hence $\dim(V) = 1$ if and only if A is a field.

Therefore assume that $\dim(V) > 1$, and let $\alpha \in V \setminus \mathbb{F}_p$. By computing successive powers of α, we can find the minimal polynomial $m_\alpha(X)$ of α in A. If $\alpha = (\alpha_1, \ldots, \alpha_k)$, it is clear that $m_\alpha(X)$ is the least common multiple of the $m_{\alpha_i}(X)$, and since $\alpha \in V$, the polynomials $m_{\alpha_i}(X)$ are polynomials of degree 1. It follows that $m_\alpha(X)$ is a squarefree polynomial equal to a product of at least two linear factors (since $\alpha \notin \mathbb{F}_p$). Write $m_\alpha(X) = m_1(X)m_2(X)$ where m_1 and m_2 are non-constant polynomials in $\mathbb{F}_p[X]$. Since m_α is squarefree, m_1 and m_2 are coprime, so we can find polynomials $U(X)$ and $V(X)$ in $\mathbb{F}_p[X]$ such that $U(X)m_1(X) + V(X)m_2(X) = 1$. We now choose $\varepsilon = Um_1(\alpha)$. Since $m_1m_2(\alpha) = 0$, ε is an idempotent. In addition, it is clear that $(U, m_2) = (V, m_1) = 1$ and m_1, m_2 non-constant imply that $\varepsilon \neq 0$ and $\varepsilon \neq 1$. $\square$

Remark. Note that it is not necessary to compute the complete basis of the kernel of ϕ in order to obtain the result. We need only, either show that the kernel V is of dimension 1 (proving that A is a field), or give an element of V which is not in the one-dimensional subspace $\mathbb{F}_p$. Hence, we can stop algorithm 2.3.1 as soon as such an element is found.

Using this proposition, it is easy to finish the splitting of our ideals $H = H_j$. Set $A = \mathcal{O}/H$ as before. Using the above proposition, either we have shown that A is a field (hence H is a prime ideal, so we have shown that the splitting is trivial), or we have found a non-trivial idempotent ε. Set $H_1 = H + e\mathcal{O}$, $H_2 = H + (1 - e)\mathcal{O}$ where e is any lift to $\mathcal{O}$ of ε. I claim that $H = H_1 \cdot H_2$. Indeed, since $e(1 - e) \in H$, it is clear that $H_1 \cdot H_2 \subset H$. Conversely, if $x \in H$ we can write $x = ex + (1 - e)x$, and $ex \in e\mathcal{O} \cdot H$, $(1 - e)x \in (1 - e)\mathcal{O} \cdot H$ so $x \in H_1 \cdot H_2$ as claimed.

Hence, we have split H non-trivially (since e is a non-trivial idempotent) and we can continue working on H_1 and H_2 separately. This process terminates in at most k steps, where k is the number of prime factors of H.

A more efficient method would be to use the complete splitting of $m_\alpha(X)$ (in the notation of the proof of Proposition 6.2.8) which gives a corresponding splitting of H as a product of more than two ideals. This will be done in the algorithm given below.

Remark. For some applications, such as computing the values of zeta and L-functions, it is not necessary to obtain the explicit decomposition of $p\mathcal{O}$, but only the ramification indices and residual degrees e_i and f_i. Once the H_j above have been computed, this can be done without much further work, as explained in Exercise 8 (this remark is due to H. W. Lenstra).

Once H has been shown to be a maximal ideal by successive splittings, what remains is the problem of representing H. Since we will have computed an $\mathbb{F}_p$-basis $(\alpha_i)_{1 \le i \le m}$ of $H/p\mathcal{O}$, to obtain the HNF of H we arbitrarily lift the α_i to $a_i \in \mathcal{O}$, and then do an HNF reduction of the matrix whose first m columns are the components of the a_i on the ω_j, and whose last n columns form p times the $n \times n$ identity matrix. It is obviously possible to do this HNF reduction modulo p (Algorithm 2.4.8), so no coefficient explosion can take place.

Even after finding the HNF of H we should still not be satisfied, because in practice, it is much more efficient to represent prime ideals by a two-element representation. To obtain this, we apply Algorithm 4.7.10. Note that we know the degree of H (the number f in the notation of Algorithm 4.7.10), which is simply equal to $n - m$ (since $p^n = [\mathcal{O} : p\mathcal{O}] = [\mathcal{O} : H][H : p\mathcal{O}] = p^f p^m$). Also we do not need to compute the HNF of H at all to apply Algorithm 4.7.10 since (together with p) the a_i clearly form a $\mathbb{Z}_K$-generating set.

6.2.5 Detailed Description of the Algorithm for Prime Decomposition

We can summarize the preceding discussions in the following algorithm

Algorithm 6.2.9 (Prime Decomposition).   Let $K = \mathbb{Q}(\theta)$ be a number field given by an algebraic integer θ as root of its minimal monic polynomial T of degree n. We assume that we have already computed an integral basis $\omega_1 = 1, \ldots, \omega_n$ and the discriminant $d(K)$ of K, for example, by using the round 2 Algorithm 6.1.8.

Given a prime number p, this algorithm outputs the decomposition $p\mathbb{Z}_K = \prod_{1 \le i \le g} \mathfrak{p}_i^{e_i}$ by giving for each i the values of e_i, $f_i = \deg(\mathfrak{p}_i)$ and a two-element representation $\mathfrak{p}_i = (p, \alpha_i)$. All the ideals I which we will use (except for the final $\mathfrak{p}_i$) will be represented by $\mathbb{F}_p$ bases of $I/p\mathcal{O}$.

1. [Check if easy] If $p \nmid \operatorname{disc}(T)/d(K)$, then by applying the algorithms of Section 3.4 factor the polynomial $T(X)$ modulo p, output the decomposition of $p\mathbb{Z}_k$ given by Theorem 4.8.13 and terminate the algorithm.

2. [Compute radical] Set $q \leftarrow p$, and while $q < n$ set $q \leftarrow qp$. Now compute the $n \times n$ matrix $A = (a_{i,j})$ over $\mathbb{F}_p$ such that $\omega_j^q \equiv \sum_{1 \le i \le n} a_{i,j}\omega_i$. Note that the matrix of the ω_i will stay triangular, so the $a_{i,j}$ are easy to compute.
    Finally, using Algorithm 2.3.1, compute a basis $\overline{\beta_1}, \ldots, \overline{\beta_l}$ of the kernel of the matrix A over $\mathbb{F}_p$ (this will be a basis of $I_p/p\mathcal{O}$). (Note that this step

has already been performed as step 7 of the round 2 algorithm, so if the result has been kept it is not necessary to recompute this again.)

3. [Compute $\overline{K_i}$] Set $\overline{K_1} \leftarrow I_p/p\mathcal{O}$ (computed in step 2), $i \leftarrow 1$ and while $\overline{K_i} \neq \{0\}$ set $i \leftarrow i+1$ and $\overline{K_i} \leftarrow \overline{K_1 K_{i-1}}$ computed using Algorithm 6.2.5.

4. [Compute $\overline{J_j}$] Set $\overline{J_1} \leftarrow \overline{K_1}$ and for $j = 2, \ldots, i$ set $\overline{J_j} \leftarrow \overline{K_j K_{j-1}^{-1}}$ using Algorithm 6.2.7.

5. [Compute $\overline{H_j}$] For $j = 1, \ldots, i-1$ set $\overline{H_j} \leftarrow \overline{J_j J_{j+1}^{-1}}$ using Algorithm 6.2.7, and set $\overline{H_i} \leftarrow \overline{J_i}$.

6. [Initialize loop] Set $j \leftarrow 0$, $c \leftarrow 0$.

7. [Finished?] If $c = 0$ do the following: if $j = i$ terminate the algorithm, otherwise set $j \leftarrow j+1$ and if $\dim_{\mathbb{F}_p}(\overline{H_j}) < n$ set $\mathcal{L} \leftarrow \{\overline{H_j}\}$ and $c \leftarrow 1$, else go to step 7 ($\mathcal{L}$ will be a list of c ideals of $\mathcal{O}/p\mathcal{O}$).

8. [Compute separable algebra A] Let $\overline{H}$ be an element of $\mathcal{L}$. Compute an $\mathbb{F}_p$-basis of $A = \mathcal{O}/H = (\mathcal{O}/p\mathcal{O})/(H/p\mathcal{O})$ in the following way. If $\beta_1, \ldots, \beta_r$ is the given $\mathbb{F}_p$-basis of $\overline{H}$, set $\beta_{r+1} \leftarrow (1, 0, \ldots, 0)^t$ (which will be linearly independent of the β_i for $i \leq r$ since $1 \notin H$), supplement this family of vectors using Algorithm 2.3.6 to a basis $\beta_1, \ldots, \beta_n$ of $\mathcal{O}/p\mathcal{O}$. Then, as an $\mathbb{F}_p$-basis of A, take $\beta_{r+1}, \ldots, \beta_n$. (This insures that the first vector of our basis of A is always $(1, 0, \ldots, 0)^t$, which would not be the case if we applied Algorithm 2.3.6 directly.)

9. [Compute multiplication table] Denote by $\gamma_1, \ldots, \gamma_s$ the $\mathbb{F}_p$-basis of A just obtained (hence $\gamma_i = \beta_{r+i}$ and $s = n - r$). By using the multiplication table of the ω_i and Algorithm 2.3.5, compute elements $a_{i,j,k}$ and $b_{i,j,k}$ in $\mathbb{F}_p$ such that

$$\gamma_i \gamma_j = \sum_{1 \leq j \leq s} a_{i,j,k} \gamma_j + \sum_{1 \leq j \leq r} b_{i,j,k} \beta_j \ .$$

The multiplication table of the γ_i (which will be used implicitly from now on) is given by the $a_{i,j,k}$ (we can forget the $b_{i,j,k}$).

10. [Compute $V = \ker(\phi)$] Let M be the matrix of the map $\alpha \mapsto \alpha^p - \alpha$ from A to A on the $\mathbb{F}_p$ basis that we have found. Compute a basis M_1 of the kernel of M using Algorithm 2.3.1. Note that if some other algorithm is used to find the kernel, we should nonetheless insure that the first column of M_1 is equal to $(1, 0, \ldots, 0)^t$.

11. [Do we have a field?] If M_1 has at least two columns (i.e. if the kernel of M is not one-dimensional), go to step 12. Otherwise, set $f \leftarrow \dim_{\mathbb{F}_p}(A)$, let (p, α) be the two-element representation of H obtained by applying Algorithm 4.7.10 to $\overline{H}$. Output j as ramification index, f as residual degree of H, and the prime ideal (p, α). Then remove H from the list $\mathcal{L}$, set $c \leftarrow c - 1$ and go to step 7.

12. [Find $m(X)$] Let $\alpha \in A$ correspond to a column of M_1 which is not proportional to $(1, 0, \ldots, 0)^t$. By computing the successive powers of α in A, let $m(X) \in \mathbb{F}_p[X]$ be the minimal monic polynomial of α in A.

13. [Factor $m(X)$] (We know that $m(X)$ is a squarefree product of linear polynomials.) By using one of the final splitting methods described in Section 3.4, or simply by trial and error if p is small, factor $m(X)$ into linear factors as $m(X) = m_1(X) \cdots m_k(X)$.

14. [Split H] Let $d = \dim_{\mathbb{F}_p}(\overline{H})$. For $r = 1, \ldots, k$ do as follows. Set $\beta_r \leftarrow m_r(\alpha)$, let M_r be the $n \times (d+n)$ matrix over $\mathbb{F}_p$ whose first d columns give the basis of $\overline{H}$ and the last n express $\omega_i \beta_r$ on the integral basis. Finally, let $\overline{H_r}$ be the image of M_r computed using Algorithm 2.3.2.

15. [Update list] Remove $\overline{H}$ and add $\overline{H_1}, \ldots, \overline{H_k}$ to the list $\mathcal{L}$, set $c \leftarrow c + k - 1$ and go to step 8.

The dimension condition in step 7 was added so as to avoid considering values of j such that there are no prime ideals over p whose ramification index is equal to j.

The validity of steps 14 and 15 of the algorithm is left as an exercise for the reader (Exercise 27).

Remark. If we want to avoid writing routines for ideal multiplication and division, we can also proceed as follows. After step 2 of the above algorithm set $\mathcal{L} \leftarrow \{\overline{I_p}\}$ and go directly to step 8 to compute the decomposition of the separable algebra $A = \mathcal{O}/I_p$. In step 11, we must compute the ramification index j of each prime ideal found, and this is easily done by using Algorithm 4.8.17. We leave the details of these modifications to the reader (Exercise 11). This method is in practice much faster than the method using ideal multiplication and division.

6.3 Computing Galois Groups

6.3.1 The Resolvent Method

I am indebted to Y. Eichenlaub for help in writing this section.

Let $K = \mathbb{Q}(\theta)$ be a number field of degree n, where θ is an algebraic integer whose minimal monic polynomial is denoted $T(X)$. An important algebraic question is to compute the *Galois group* $\mathrm{Gal}(T)$ of the polynomial T, in other words the Galois group of the splitting field of T, or equivalently of the Galois closure of K in $\overline{\mathbb{Q}}$. Since by definition elements of $\mathrm{Gal}(T)$ act as permutations on the roots of T, once an ordering of the roots is given, $\mathrm{Gal}(T)$ can naturally be considered as a subgroup of S_n, the symmetric group on n letters. Changing the ordering of the roots clearly transforms $\mathrm{Gal}(T)$ into a conjugate group, and since the ordering is not canonical, the natural objects to consider are subgroups of S_n up to conjugacy. It will be important in what follows to remember that we have chosen a specific, but arbitrary ordering, since it will sometimes be necessary to change it.

Furthermore, since the polynomial T is irreducible, the group $\mathrm{Gal}(T)$ is a *transitive* subgroup of S_n, i.e. there is a single orbit for the action of $\mathrm{Gal}(T)$ on the roots θ_i of T (each orbit corresponding to an irreducible factor of T). Hence, the first task is to classify transitive subgroups of S_n up to conjugacy. This is a non-trivial (but purely) group-theoretical question. It has been solved up to $n = 32$ (see [But-McKay] and [Hül]), but the number of groups becomes unwieldy for higher degrees. We will give the classification for $n \leq 7$.

Note that since the cardinality of an orbit divides the order of $\mathrm{Gal}(T)$, the cardinality of a transitive subgroup of S_n is divisible by n.

Once the transitive groups are classified, we must still determine which corresponds to our Galois group $\mathrm{Gal}(T)$. We first note the following simple, but important proposition.

Proposition 6.3.1. *Let A_n be the alternating group on n letters corresponding to the even permutations. Then $\mathrm{Gal}(T) \subset A_n$ if and only if $\mathrm{disc}(T)$ is a square.*

Proof. Let θ_i be the roots of T. By Proposition 3.3.5, we know that

$$\mathrm{disc}(T) = f^2 , \quad \text{where} \quad f = \prod_{1 \leq i < j \leq n} (\theta_j - \theta_i) .$$

Clearly f is an algebraic integer, and for any $\sigma \in \mathrm{Gal}(T)$ we have

$$\sigma(f) = \epsilon(\sigma)f ,$$

where $\epsilon(\sigma)$ denotes the signature of σ. Hence, if $\mathrm{Gal}(T) \subset A_n$, all permutations of $\mathrm{Gal}(T)$ are even, so f is invariant under $\mathrm{Gal}(T)$. Thus by Galois theory, $f \in \mathbb{Z}$. Conversely, if $f \in \mathbb{Z}$, we have $f \neq 0$ since the roots of T are distinct. Therefore $\epsilon(\sigma) = 1$ for all $\sigma \in \mathrm{Gal}(T)$, so $\mathrm{Gal}(T) \subset A_n$. Note that since A_n is a normal subgroup, that a group is a subgroup of A_n depends only on its conjugacy class, and not on the precise conjugate. $\square$

We now need to introduce a definition which will be basic to our work.

Definition 6.3.2. *Let G be a subgroup of S_n containing $\mathrm{Gal}(T)$ (not up to conjugacy, but for the given numbering of the roots), and let $F(X_1, X_2, \ldots, X_n)$ be a polynomial in n variables with coefficients in $\mathbb{Z}$. If H is the stabilizer of F in G, i.e.*

$$H = \{\sigma \in G, F\left(X_{\sigma(1)}, X_{\sigma(2)}, \ldots, X_{\sigma(n)}\right) = F(X_1, X_2, \ldots, X_n)\} ,$$

we define the resolvent polynomial *$R_G(F, T)$ with respect to G, F and the polynomial T by*

$$R_G(F, T)(X) = \prod_{\sigma \in G/H} \left(X - F\left(\theta_{\sigma(1)}, \theta_{\sigma(2)}, \ldots, \theta_{\sigma(n)}\right)\right) ,$$

where G/H denotes any set of left coset representatives of G modulo H.

When $G = S_n$, we will omit the subscript in the notation.

It is clear from elementary Galois theory that $R_G(F,T) \in \mathbb{Z}[X]$. The main theorem which we will use concerning resolvent polynomials is as follows.

Theorem 6.3.3. *With the notation of the preceding definition, set $m = [G : H] = \deg(R_G(F,T))$. Then, if $R_G(F,T)$ is squarefree, its Galois group (as a subgroup of S_m) is equal to $\phi(\mathrm{Gal}(T))$, where ϕ is the natural group homomorphism from G to S_m given by the natural left action of G on G/H. In particular, the list of degrees of the irreducible factors of $R_G(F,T)$ in $\mathbb{Z}[X]$ is the same as the list of the length of the orbits of the action of $\phi(\mathrm{Gal}(T))$ on $[1, \ldots, m]$. For example, $R_G(F,T)$ has a root in $\mathbb{Z}$ if and only if $\mathrm{Gal}(T)$ is conjugate under G to a subgroup of H.*

For the proof, see [Soi].

Note that it is important to specify that $\mathrm{Gal}(T)$ is conjugate under G, since this is a stronger condition than being conjugate under S_n.

Now it will often happen that $R_G(F,T)$ is not squarefree. In that case, to be able to apply the theorem, we use the following algorithm.

Algorithm 6.3.4 (Tschirnhausen Transformation). Given a monic irreducible polynomial T defining a number field $K = \mathbb{Q}(\theta)$, we find another such polynomial U defining the same number field.

1. [Choose random polynomial] Let $n \leftarrow \deg(T)$. Choose at random a polynomial $A \in \mathbb{Z}[X]$ of degree less than or equal to $n - 1$.

2. [Compute characteristic polynomial] Using the method explained in Section 4.3, compute the characteristic polynomial U of $\alpha = A(\theta)$. In other words, using the sub-resultant Algorithm 3.3.7, set $U \leftarrow R_Y(T(Y), X - A(Y))$.

3. [Check degree] Using Euclid's algorithm, compute $V \leftarrow \gcd(U, U')$. If V is constant, then output U and terminate the algorithm, otherwise go to step 1.

The validity of this algorithm is clear.

Modifying T if necessary by using such a Tschirnhausen transformation, it is always easy to reduce to the case where $R_G(F,T)$ is squarefree.

Finally, we need some notation. The elements of the set G/H will be given as products of disjoint cycles, with I denoting the identity permutation. Usually, apart from I, G/H will contain only transpositions.

We denote by C_n the cyclic group $\mathbb{Z}/n\mathbb{Z}$, and by D_n the dihedral group of order $2n$, isomorphic to the isometries of a regular n-gon. As before, A_n and S_n denote the alternating group and symmetric group on n letters respectively. Finally, $A \rtimes B$ denotes the semi-direct product of the groups A and B, where the action of B on A is understood.

When we compute a group, we will output not only the isomorphism class of the group, but also a sign expressing whether the group is contained in A_n (+ sign) or not (– sign). This will help resolve a number of ambiguities since isomorphic groups are not always conjugate in S_n.

Let us now examine in turn each degree up to degree 7. The particular choices of resolvents that we give are in no way canonical, although we have tried to give the ones which are the most efficient. The reader can find many other choices in the literature ([Stau], [Gir], [Soi] and [Soi-McKay], [Eic1]). The validity of the algorithms given can be checked using Theorem 6.3.3.

In degrees 1 and 2 there is of course nothing to say since the only possible group is S_n in these cases, so we always output $(S_n, -)$.

6.3.2 Degree 3

In degree 3, it is obvious that the only transitive subgroups of S_3 are $C_3 \simeq A_3$ and $S_3 \simeq D_3$ which may be separated by the discriminant. In other words:

Proposition 6.3.5. *If $n = 3$, we have either* $\mathrm{Gal}(T) \simeq C_3$ *or* $\mathrm{Gal}(T) \simeq S_3$ *depending on whether* $\mathrm{disc}(T)$ *is a square or not.*

Thus we output $(C_3, +)$ or $(S_3, -)$ depending on $\mathrm{disc}(T)$.

6.3.3 Degree 4

In degree 4, there are (up to conjugacy) five transitive subgroups of S_4. These are C_4 (the cyclic group), $V_4 = C_2^2$ (the Klein 4-group), D_4 (the dihedral group of order 8, group of isometries of the square), A_4 and S_4.

Some inclusions are $V_4 \subset D_4 \cap A_4$, and $C_4 \subset D_4$.

Important remark: note that although we consider the groups only up to conjugacy, the notion of inclusion for two groups G_1 and G_2 can reasonably be defined by saying that $G_1 \subset G_2$ only when G_1 is a subgroup of some conjugate of G_2. On the other hand, when we consider *abstract* groups such as V_4, D_4, etc ... , the notion of inclusion is much more delicate since some subgroups of S_n can be isomorphic as abstract groups but not conjugate in S_n. In this case, we write $G_1 \subset G_2$ only if this is valid for all conjugacy classes isomorphic to G_1 and G_2 respectively.

A simple algorithm is as follows.

Algorithm 6.3.6 (Galois Group for Degree 4). Given an irreducible monic polynomial $T \in \mathbb{Z}[X]$ of degree 4, this algorithm computes its Galois group.

1. [Compute resolvent] Using Algorithm 3.6.6, compute the roots θ_i of T in $\mathbb{C}$. Let

$$F \leftarrow X_1 X_2^2 + X_2 X_3^2 + X_3 X_4^2 + X_4 X_1^2$$

and let $R \leftarrow R(F, T)$, where a system of representatives of G/H is given by

$$G/H = \{I, (12), (13), (14), (23), (34)\} \ .$$

Then round the coefficients of R to the nearest integer (note that the roots θ_i must be computed to a sufficient accuracy for this rounding to be correct, and the needed accuracy is easily determined, see Exercise 13).

2. [Squarefree?] Compute $V \leftarrow (R, R')$ using the Euclidean algorithm. If V is non-constant, replace T by the polynomial obtained by applying a Tschirnhausen transformation using Algorithm 6.3.4 and go to step 1.

3. [Factor resolvent] Using Algorithm 3.5.7, factor R over $\mathbb{Z}$. Let L be the list of the degrees of the irreducible factors sorted in increasing order.

4. [Conclude] If R is irreducible, i.e. if $L = (6)$, then output $(A_4, +)$ or $(S_4, -)$ depending on whether $\operatorname{disc}(T)$ is a perfect square or not. Otherwise, output $(C_4, -)$, $(V_4, +)$ or $(D_4, -)$ depending on whether $L = (1, 1, 4)$, $L = (2, 2, 2)$ or $L = (2, 4)$ respectively. Terminate the algorithm.

Note that with this choice of resolvent, we have $H = C_4 = <(1234)>$, the group of cyclic permutations, but this fact is needed in checking the correctness of the algorithm, not in the algorithm itself, where only G/H is used.

Another algorithm which is computationally slightly simpler is as follows. We give it also to illustrate the importance of the root ordering.

Algorithm 6.3.7 (Galois Group for Degree 4). Given an irreducible monic polynomial $T \in \mathbb{Z}[X]$ of degree 4, this algorithm computes its Galois group.

1. [Compute resolvent] Using Algorithm 3.6.6, compute the roots θ_i of T in $\mathbb{C}$. Let

$$F \leftarrow X_1 X_3 + X_2 X_4$$

and let $R \leftarrow R(F, T)$, where a system of representatives of G/H is given by

$$G/H = \{I, (12), (14)\} \ .$$

Round the coefficients of R to the nearest integer.

2. [Squarefree?] Compute $V \leftarrow (R, R')$ using the Euclidean algorithm. If V is non-constant, replace T by the polynomial obtained by applying a Tschirnhausen transformation using Algorithm 6.3.4 and go to step 1.

3. [Integral root?] Check whether R has an integral root by explicitly computing them in terms of the θ_i. (This is usually much faster than using the general factoring procedure 3.5.7.)

4. [Can one conclude?] If R does not have an integral root (so R is irreducible), then output $(A_4, +)$ or $(S_4, -)$ depending on whether $\operatorname{disc}(T)$ is a perfect square or not and terminate the algorithm. Otherwise, if $\operatorname{disc}(T)$ is a square, output $(V_4, +)$ and terminate the algorithm.

5. [Renumber] (Here R has an integral root and $\mathrm{disc}(T)$ is not a square. The Galois group must be isomorphic either to C_4 or to D_4.) Let σ be the element of S_4 corresponding to the integral root of R, and set $(t_i) \leftarrow (t_{\sigma(i)})$ (i.e. we renumber the roots of T according to σ).

6. [Use new resolvent] Set

$$d \leftarrow ((\theta_1 - \theta_3)(\theta_2 - \theta_4)(\theta_1 + \theta_3 - \theta_2 - \theta_4))^2$$

rounded to the nearest integer (with the same remarks as before about the accuracy needed for the θ_i). If $d \neq 0$, output $(C_4, -)$ or $(D_4, -)$ depending on whether d is a perfect square or not and terminate the algorithm.

7. [Replace] (Here $d = 0$.) Replace T by the polynomial obtained by applying a Tschirnhausen transformation A using Algorithm 6.3.4. Set $\theta_i \leftarrow A(\theta_i)$ (which will be the roots of the new T). Reorder the θ_i so that $\theta_1\theta_3 + \theta_2\theta_4 \in \mathbb{Z}$, (only the 3 elements of G/H given in step 1 need to be tried), then go to step 6.

In principle, this algorithm involves factoring polynomials of degree 3, hence is computationally simpler than the preceding algorithm, although its structure is more complicated due to the implicit use of two different resolvents. The first resolvent corresponds to $G = S_4$ and $H = D_4 =< (1234), (13) >$. The second resolvent corresponds to $F = X_1X_2^2 + X_2X_3^2 + X_3X_4^2 + X_4X_1^2$, $G = D_4$, $H = C_4$ and $G/H = \{I, (13)\}$, hence the polynomial of degree 2 need not be explicitly computed in order to find its arithmetic structure.

Remark. (This remark is valid in any degree.) As can be seen from the preceding algorithm, it is not really necessary to compute the resolvent polynomial R explicitly, but only a sufficiently close approximation to its roots (which are known explicitly by definition). To check whether R is squarefree or not can also be done by simply checking that R does not have any multiple root (to sufficient accuracy). In fact, we have the following slight strengthening of Theorem 6.3.3 which can be proved in the same way.

Proposition 6.3.8. *We keep the notations of Theorem 6.3.3, but we do not necessarily assume that $R_G(F, T)$ is squarefree. If $R_G(F, T)$ has a simple root in $\mathbb{Z}$, then $\mathrm{Gal}(T)$ is conjugate under G to a subgroup of H.*

This proposition shows that it is not necessary to assume $R_G(F, T)$ squarefree in order to apply the above algorithms, as well as any other which depend only on the existence of an integral root and not more generally on the degrees of the irreducible factors of $R_G(F, T)$. (This is the case for the algorithms that we give in degree 4 and 5.) This remark should of course be used when implementing these algorithms.

6.3.4 Degree 5

In degree 5 there are also (up to conjugacy) five transitive subgroups of S_5. These are C_5 (the cyclic group), D_5 (the dihedral group of order 10), M_{20} (the metacyclic group of degree 5), A_5 and S_5.

Some inclusions are

$$C_5 \subset D_5 \subset A_5 \cap M_{20} .$$

The algorithm that we suggest is as follows.

Algorithm 6.3.9 (Galois Group for Degree 5). Given an irreducible monic polynomial $T \in \mathbb{Z}[X]$ of degree 5, this algorithm computes its Galois group.

1. [Compute resolvent] Using Algorithm 3.6.6, compute the roots θ_i of T in $\mathbb{C}$. Let

$$F \leftarrow X_1^2(X_2X_5 + X_3X_4) + X_2^2(X_1X_3 + X_4X_5) + X_3^2(X_1X_5 + X_2X_4)$$
$$+ X_4^2(X_1X_2 + X_3X_5) + X_5^2(X_1X_4 + X_2X_3)$$

and let $R \leftarrow R(F, T)$, where a system of representatives of G/H is given by

$$G/H = \{I, (12), (13), (14), (15), (25)\} .$$

Round the coefficients of R to the nearest integer.

2. [Squarefree?] Compute $V \leftarrow (R, R')$ using the Euclidean algorithm. If V is non-constant, replace T by the polynomial obtained by applying a Tschirnhausen transformation using Algorithm 6.3.4 and go to step 1.

3. [Factor resolvent] Factor R using Algorithm 3.5.7. (Note that one can show that either R is irreducible or R has an integral root. So, as in the algorithm for degree 4, it may be better to compute the roots of R which are known explicitly.)

4. [Can one conclude?] If R is irreducible, then output $(A_5, +)$ or $(S_5, -)$ depending on whether $\mathrm{disc}(T)$ is a perfect square or not, and terminate the algorithm. Otherwise, if $\mathrm{disc}(T)$ is not a perfect square, output $(M_{20}, -)$ and terminate the algorithm.

5. [Renumber] (Here R has an integral root and $\mathrm{disc}(T)$ is a square. The Galois group must be isomorphic either to C_5 or to D_5.) Let σ be the element of S_5 corresponding to the integral root of R, and set $(t_i) \leftarrow (t_{\sigma(i)})$ (i.e. we renumber the roots of T according to σ).

6. [Compute discriminant of new resolvent] Set

$$d \leftarrow (\theta_1\theta_2(\theta_2 - \theta_1) + \theta_2\theta_3(\theta_3 - \theta_2) + \theta_3\theta_4(\theta_4 - \theta_3)$$
$$+ \theta_4\theta_5(\theta_5 - \theta_4) + \theta_5\theta_1(\theta_1 - \theta_5))^2$$

rounded to the nearest integer (with the same remarks as before about the accuracy needed for the θ_i). If $d \neq 0$, output $(C_5, +)$ or $(D_5, +)$ depending on whether d is a perfect square or not, and terminate the algorithm.

7. [Replace] (Here $d = 0$.) Replace T by the polynomial obtained by applying a Tschirnhausen transformation A using Algorithm 6.3.4. Set $\theta_i \leftarrow A(\theta_i)$ (which will be the roots of the new T). Reorder the θ_i so that $F(\theta_1, \theta_1, \theta_3, \theta_4, \theta_5) \in \mathbb{Z}$ where F is as in step 1, (only the 6 elements of G/H given in step 1 need to be tried), then go to step 6.

The first resolvent corresponds to $G = S_5$ and

$$H = M_{20} = <(12345), (2354)> .$$

Step 6 corresponds implicitly to the use of the second degree resolvent obtained with $F = X_1 X_2^2 + X_2 X_3^2 + X_3 X_4^2 + X_4 X_5^2 + X_5 X_1^2$, $G = D_5$, $H = C_5$ and $G/H = \{I, (12)(35)\}$.

6.3.5 Degree 6

In degree 6 there are up to conjugation, 16 transitive subgroups of S_6. The inclusion diagram is complicated, and the number of resolvent polynomials is high. The best way to study this degree is to work using *relative extensions*, that is study the number field K as a quadratic or cubic extension of a cubic or quadratic subfield respectively, if they exist. This is done in [Oli2] and [BeMaOl].

In this book we have not considered relative extensions. Furthermore, when a sextic field is given by a sixth degree polynomial over $\mathbb{Q}$, it is not immediately obvious, even if it is theoretically possible, how to express it as a relative extension, although the POLRED Algorithm 4.4.11 often gives such information. Hence, we again turn to the heavier machinery of resolvent polynomials.

It is traditional to use the notation G_k to denote a group of cardinality k. Also, special care must be taken when considering abstract groups. For example, the group S_4 occurs as two different conjugacy classes of S_6, one which is in A_6, the other which is not (the traditional notation would then be S_4^+ and S_4^- respectively).

We will describe the groups as we go along the algorithm. There are many possible resolvents which can be used. The algorithm that we suggest has the advantage of needing a single resolvent, except in one case, similarly to degrees 4 and 5.

Algorithm 6.3.10 (Galois Group for Degree 6). Given an irreducible monic polynomial $T \in \mathbb{Z}[X]$ of degree 6, this algorithm computes its Galois group.

1. [Compute resolvent] Using Algorithm 3.6.6, compute the roots θ_i of T in $\mathbb{C}$. Let

$$F \leftarrow X_1^2 X_5^2 (X_2 X_4 + X_3 X_6) + X_2^2 X_4^2 (X_1 X_5 + X_3 X_6) + X_3^2 X_6^2 (X_1 X_5 + X_2 X_4)$$
$$+ X_1^2 X_6^2 (X_2 X_5 + X_3 X_4) + X_2^2 X_5^2 (X_1 X_6 + X_3 X_4) + X_3^2 X_4^2 (X_1 X_6 + X_2 X_5)$$
$$+ X_1^2 X_3^2 (X_2 X_6 + X_4 X_5) + X_2^2 X_6^2 (X_1 X_3 + X_4 X_5) + X_4^2 X_5^2 (X_1 X_3 + X_2 X_6)$$
$$+ X_1^2 X_4^2 (X_2 X_3 + X_5 X_6) + X_2^2 X_3^2 (X_1 X_4 + X_5 X_6) + X_5^2 X_6^2 (X_1 X_4 + X_2 X_3)$$
$$+ X_1^2 X_2^2 (X_3 X_5 + X_4 X_6) + X_3^2 X_5^2 (X_1 X_2 + X_4 X_6) + X_4^2 X_6^2 (X_1 X_2 + X_3 X_5)$$

and let $R \leftarrow R(F, T)$, where a system of representatives of G/H is given by

$$G/H = \{I, (12), (13), (14), (15), (16)\} \; .$$

Round the coefficients of R to the nearest integer.

2. [Squarefree?] Compute $V \leftarrow (R, R')$ using the Euclidean algorithm. If V is non-constant, replace T by the polynomial obtained by applying a Tschirnhausen transformation using Algorithm 6.3.4 and go to step 1.

3. [Factor resolvent] Factor R using Algorithm 3.5.7. If R is irreducible, then go to step 5, otherwise let L be the list of the degrees of the irreducible factors sorted in increasing order.

4. [Conclude]
     a) If $L = (1, 2, 3)$, let f_1 be the irreducible factor of R of degree equal to 3. Output $(C_6, -)$ or $(D_6, -)$ depending on whether $\mathrm{disc}(f_1)$ is a square or not.
     b) If $L = (3, 3)$, let f_1 and f_2 be the irreducible factors of R. If both $\mathrm{disc}(f_1)$ and $\mathrm{disc}(f_2)$ are not squares output $(G_{36}, -)$, otherwise output $(G_{18}, -)$. Note that $G_{36}^- = C_3^2 \rtimes C_2^2 \simeq D_3 \times D_3$, and $G_{18} = C_3^2 \rtimes C_2 \simeq C_3 \times D_3$.
     c) If $L = (2, 4)$ and $\mathrm{disc}(T)$ is a square, output $(S_4, +)$. Otherwise, if $L = (2, 4)$ and $\mathrm{disc}(T)$ is not a square, let f_1 be the irreducible factor of degree 4 of R. Then output $(A_4 \times C_2, -)$ or $(S_4 \times C_2, -)$ depending on whether $\mathrm{disc}(f_1)$ is a square or not.
     d) If $L = (1, 1, 4)$ then output $(A_4, +)$ or $(S_4, -)$ depending on whether $\mathrm{disc}(T)$ is a square or not.
     e) If $L = (1, 5)$, then output $(\mathrm{PSL}_2(\mathbb{F}_5), +)$ or $(\mathrm{PGL}_2(\mathbb{F}_5), -)$ depending on whether $\mathrm{disc}(T)$ is a square or not. Note that $\mathrm{PSL}_2(\mathbb{F}_5) \simeq A_5$ and that $\mathrm{PGL}_2(\mathbb{F}_5) \simeq S_5$.
     f) Finally, if $L = (1, 1, 1, 3)$, output $(S_3, -)$.
     Then terminate the algorithm.

5. [Compute new resolvent] (Here our preceding resolvent was irreducible. Note that we do *not* have to reorder the roots.) Let

$$F \leftarrow X_1 X_2 X_3 + X_4 X_5 X_6$$

and let $R \leftarrow R(F, T)$, where a system of representatives of G/H is now given by

$$G/H = \{I, (14), (15), (16), (24), (25), (26), (34), (35), (36)\} \; .$$

Round the coefficients of R to the nearest integer.

6. [Squarefree?] Compute $V \leftarrow (R, R')$ using the Euclidean algorithm. If V is non-constant, replace T by the polynomial obtained by applying a Tschirnhausen transformation using Algorithm 6.3.4 and go to step 5.

7. [Factor resolvent] Factor R using Algorithm 3.5.7 (Note that in this case either R is irreducible, or it has an integral root, so again it is probably better to compute these 10 roots directly from the roots of T and check whether they are integral.)

8. [Conclude] If R is irreducible (or has no integral root), then output $(A_6, +)$ or $(S_6, -)$ depending on whether $\mathrm{disc}(T)$ is a square or not. Otherwise, output $(G_{36}, +)$ or $(G_{72}, -)$ depending on whether $\mathrm{disc}(T)$ is a square or not. Then terminate the algorithm. Note that $G_{36}^+ = C_3^2 \rtimes C_4$ and $G_{72} = C_3^2 \rtimes D_4$.

The first resolvent corresponds to $G = S_6$ and

$$H = \mathrm{PGL}_2(\mathbb{F}_5) = < (12345), (16)(23)(45) > \ .$$

The second resolvent, used in step 5, corresponds to $G = S_6$ and

$$H = G_{72} = < (123), (14)(25)(36), (1524)(36) > \ .$$

Remark. It can be shown that a sextic field has a quadratic subfield if and only if its Galois group is isomorphic to a (transitive) subgroup of G_{72}. This corresponds to the groups $(C_6, -)$, $(S_3, -)$, $(D_6, -)$, $(G_{18}, -)$, $(G_{36}, -)$, $(G_{36}, +)$ and $(G_{72}, -)$.

Similarly, it has a cubic subfield if and only if its Galois group is isomorphic to a (transitive) subgroup of $S_4 \times C_2$. This corresponds to the groups $(C_6, -)$, $(S_3, -)$, $(D_6, -)$, $(A_4, +)$, $(S_4, +)$, $(S_4, -)$, $(A_4 \times C_2, -)$ and $(S_4 \times C_2, -)$.

Hence, it has both a quadratic and a cubic subfield if and only if its Galois group is isomorphic to $(C_6, -)$, $(S_3, -)$ or $(D_6, -)$.

If the field is primitive, i.e. does not have quadratic or cubic subfields, this implies that its Galois group can only be $\mathrm{PSL}_2(\mathbb{F}_5) \simeq A_5$, $\mathrm{PGL}_2(\mathbb{F}_5) \simeq S_5$, A_6 or S_6.

6.3.6 Degree 7

In degree 7, there are seven transitive subgroups of S_7 which are C_7, D_7, M_{21}, M_{42}, $\mathrm{PSL}_2(\mathbb{F}_7) \simeq \mathrm{PSL}_3(\mathbb{F}_2)$, A_7 and S_7.

Some inclusions are

$$C_7 \subset D_7 \subset M_{42} \ , \quad C_7 \subset M_{21} \subset \mathrm{PSL}_2(\mathbb{F}_7) \subset A_7 \quad \text{and} \quad M_{21} \subset M_{42} \ .$$

In this case there exists a remarkably simple algorithm.

Algorithm 6.3.11 (Galois Group for Degree 7). Given an irreducible monic polynomial $T \in \mathbb{Z}[X]$ of degree 7, this algorithm computes its Galois group.

1. [Compute resolvent] Using Algorithm 3.6.6, compute the roots θ_i of T in $\mathbb{C}$. Let

$$R \leftarrow \prod_{1 \leq i < j < k \leq 7} (X - (\theta_i + \theta_j + \theta_k))$$

   which is a polynomial of degree 35, and round the coefficients of R to the nearest integer.

2. [Squarefree?] Compute $V \leftarrow (R, R')$ using the Euclidean algorithm. If V is nonconstant, replace T by the polynomial obtained by applying a Tschirnhausen transformation using Algorithm 6.3.4 and go to step 1.

3. [Factor resolvent and conclude] Factor R using Algorithm 3.5.7. If R is irreducible, then output $(A_7, +)$ or $(S_7, -)$ depending on whether $\mathrm{disc}(T)$ is a square or not. Otherwise, let L be the list of the degrees of the irreducible factors sorted in increasing order. Output $(\mathrm{PSL}_2(\mathbb{F}_7), +)$, $(M_{42}, -)$, $(M_{21}, +)$, $(D_7, -)$ or $(C_7, +)$ depending on whether $L = (7, 28)$, $L = (14, 21)$, $L = (7, 7, 21)$, $L = (7, 7, 7, 14)$ or $L = (7, 7, 7, 7, 7)$ respectively. Then terminate the algorithm.

Note that this algorithm does not exactly correspond to the framework based on Theorem 6.3.3 but it has the advantage of being very simple, and computationally not too inefficient. It does involves factoring a polynomial of degree 35 over $\mathbb{Z}$ however, and this can be quite slow. (To give some idea of the speed: on a modern workstation the algorithms take a few seconds for degrees less than or equal to 6, while for degree 7, a few minutes may be required using this algorithm.)

Several methods can be used to improve this basic algorithm in practice. First of all, one expects that the overwhelming majority of polynomials will have S_7 as their Galois group, and hence that our resolvent will be irreducible. We can test for irreducibility, without actually factoring the polynomial, by testing this modulo p for small primes p. If it is already irreducible modulo p for some p, then there is no need to go any further. Of course, this is done automatically if we use Algorithm 3.5.7, but that algorithm will start by doing the distinct degree factorization 3.4.3, when it is simpler here to use Proposition 3.4.4.

Even if one expects that the resolvent will factor, we can use the divisibility by 7 of the degrees of its irreducible factors in almost every stage of the factoring Algorithm 3.5.7.

Another idea is to use the resolvent method as explained at the beginning of this chapter. Instead of factoring polynomials having large degrees, we simply find the list of all cosets σ of G modulo H such that

$$F\left(\theta_{\sigma(1)}, \theta_{\sigma(2)}, \ldots, \theta_{\sigma(n)}\right) \in \mathbb{Z} .$$

If there is more than one coset, this means that the resolvent is not squarefree, hence we must apply a Tschirnhausen transformation. If there is exactly one, then the Galois group is isomorphic to a subgroup of H, and the coset gives the permutation of the roots which must be applied to go further down the tree of subgroups. If there are none, the Galois group is not isomorphic to a subgroup of H. Of course, all this applies to any degree, not only to degree 7.

As the reader can see, I do not give explicitly the resolvents and cosets for degree 7. The resolvents themselves are as simple as the ones that we have given in lower degrees. On the other hand, the list of cosets is long. For example for the pair (S_7, M_{42}) we need 120 elements. This is cumbersome to write down. It should be noted however that the resulting algorithm is much more efficient than the preceding one (again at most a few seconds on a modern workstation). These cosets and resolvents in degree 7, 8, 9, 10 and 11 may be obtained in electronic form upon request from M. Olivier (same address as the author).

6.3.7 A List of Test Polynomials

As a first check of the correctness of an implementation of the above algorithms, we give a polynomial for each of the possible Galois groups occurring in degree less than or equal to 7. This list is taken from [Soi-McKay]. Note that for many of the given polynomials, it will be necessary to apply a Tschirnhausen transformation. We list first the group as it is output by the algorithm, then a polynomial having this as Galois group.

$(S_1, -)$: X

$(S_2, -)$: $X^2 + X + 1$

$(C_3, +)$: $X^3 + X^2 - 2X - 1$

$(S_3, -)$: $X^3 + 2$

$(C_4, -)$: $X^4 + X^3 + X^2 + X + 1$

$(V_4, +)$: $X^4 + 1$

$(D_4, -)$: $X^4 - 2$

$(A_4, +)$: $X^4 + 8X + 12$

$(S_4, -)$: $X^4 + X + 1$

$(C_5, +)$: $X^5 + X^4 - 4X^3 - 3X^2 + 3X + 1$

$(D_5, +)$: $X^5 - 5X + 12$

$(M_{20}, -)$: $X^5 + 2$

$(A_5, +)$: $X^5 + 20X + 16$

$(S_5, -)$: $X^5 - X + 1$

$(C_6, -)$: $X^6 + X^5 + X^4 + X^3 + X^2 + X + 1$

$(S_3, -)$: $X^6 + 108$

$(D_6, -)$: $X^6 + 2$

$(A_4, +)$: $X^6 - 3X^2 - 1$

$(G_{18}, -)$: $X^6 + 3X^3 + 3$

$(A_4 \times C_2, -)$: $X^6 - 3X^2 + 1$

$(S_4, +)$: $X^6 - 4X^2 - 1$

$(S_4, -)$: $X^6 - 3X^5 + 6X^4 - 7X^3 + 2X^2 + X - 4$
$(G_{36}, -)$: $X^6 + 2X^3 - 2$
$(G_{36}, +)$: $X^6 + 6X^4 + 2X^3 + 9X^2 + 6X - 4$
$(S_4 \times C_2, -)$: $X^6 + 2X^2 + 2$
$(\mathrm{PSL}_2(\mathbb{F}_5), +) \simeq (A_5, +)$: $X^6 - 2X^5 - 5X^2 - 2X - 1$
$(G_{72}, -)$: $X^6 + 2X^4 + 2X^3 + X^2 + 2X + 2$
$(\mathrm{PGL}_2(\mathbb{F}_5), -) \simeq (S_5, -)$: $X^6 - X^5 - 10X^4 + 30X^3 - 31X^2 + 7X + 9$
$(A_6, +)$: $X^6 + 24X - 20$
$(S_6, -)$: $X^6 + X + 1$
$(C_7, +)$: $X^7 + X^6 - 12X^5 - 7X^4 + 28X^3 + 14X^2 - 9X + 1$
$(D_7, -)$: $X^7 + 7X^3 + 7X^2 + 7X - 1$
$(M_{21}, +)$: $X^7 - 14X^5 + 56X^3 - 56X + 22$
$(M_{42}, -)$: $X^7 + 2$
$(\mathrm{PSL}_2(\mathbb{F}_7), +) \simeq (\mathrm{PSL}_3(\mathbb{F}_2), +)$: $X^7 - 7X^3 + 14X^2 - 7X + 1$
$(A_7, +)$: $X^7 + 7X^4 + 14X + 3$
$(S_7, -)$: $X^7 + X + 1$

6.4 Examples of Families of Number Fields

6.4.1 Making Tables of Number Fields

It is important to try to describe the family of all number fields (say of a given degree, Galois group of the Galois closure and signature) up to isomorphism. Unfortunately, this is a hopeless task except for some special classes of fields such as quadratic fields, cyclic cubic fields, cyclotomic fields, etc. We could, however, ask for a list of such fields whose discriminant is in absolute value bounded by a given constant, i.e. ask for *tables* of number fields. We first explain briefly how this can be done, referring to [Mart] and [Poh1] for complete details.

We need two theorems. The first is an easy result of the geometry of numbers (which we already used in Section 2.6 to show that the LLL algorithm terminates) which we formulate as follows.

Proposition 6.4.1. *There exists a positive constant γ_n having the following property. In any lattice (L, q) of $\mathbb{R}^n$, there exists a non-zero vector x such that $q(x) \le \gamma_n D^{2/n}$ where $D = \det(L) = \det(Q)^{1/2}$ is the determinant of the lattice (here Q is the matrix of q in some $\mathbb{Z}$-basis of L, see Section 2.5).*

See for example [Knu2] (Section 3.3.4, Exercise 9) for a proof.

The best possible constant γ_n is called Hermite's constant, and is known only for $n \le 8$:

$$\gamma_1 = 1, \ \gamma_2^2 = \frac{4}{3}, \ \gamma_3^3 = 2, \ \gamma_4^4 = 4, \ \gamma_5^5 = 8, \ \gamma_6^6 = \frac{64}{3}, \ \gamma_7^7 = 64, \ \gamma_8^8 = 256 \ .$$

For larger values of n, the recursive upper bound

$$\gamma_n^n \leq \gamma_{n-1}^{(n-1)n/(n-2)}$$

gives useful results. The best known bounds are given for $n \leq 24$ in [Con-Slo], Table 1.2 and Formula (47).

The basic theorem, due to Hunter (see [Hun] and Exercise 26), is as follows.

Theorem 6.4.2 (Hunter). *Let K be a number field of degree n over $\mathbb{Q}$. There exists $\theta \in \mathbb{Z}_K \setminus \mathbb{Z}$ having the following property. Call θ_i the conjugates of θ in K. Then*

$$\sum_{i=1}^{n} |\theta_i|^2 \leq \frac{1}{n} \operatorname{Tr}(\theta)^2 + \gamma_{n-1} \left(\frac{|d(K)|}{n} \right)^{1/(n-1)} ,$$

where $d(K)$ is the discriminant of K and $\operatorname{Tr}(\theta) = \sum_{i=1}^{n} \theta_i$ is the trace of θ over $\mathbb{Q}$. In addition, we may assume that $0 \leq \operatorname{Tr}(\theta) \leq n/2$.

This theorem is used as follows. Assume that we want to make a table of number fields of degree n and having a given signature, with discriminant $d(K)$ satisfying $|d(K)| \leq M$ for a given bound M. Then replacing $d(K)$ by M in Hunter's theorem gives an upper bound for the $|\theta_i|$ and hence for the coefficients of the characteristic polynomial of θ in K.

If K is primitive, i.e. if the only subfields of K are $\mathbb{Q}$ and K itself, then since $\theta \notin \mathbb{Z}$ we know that $K = \mathbb{Q}(\theta)$, and thus we obtain a finite (although usually large) collection of polynomials to consider. Most of these polynomials can be discarded because their roots will not satisfy Hunter's inequality. Others can be discarded because they are reducible, or because they do not have the correct signature. Note that a given signature will give several inequalities between the coefficients of acceptable polynomials, and these should be checked before using Sturm's Algorithm 4.1.11 which is somewhat longer. (We are talking of millions if not billions of candidate polynomials here, depending on the degree and, of course, the size of M.)

Finally, using Algorithm 6.1.8 compute the discriminant of the number fields corresponding to each of the remaining polynomials. This is the most time-consuming part. After discarding the polynomials which give a field discriminant which is larger than M in absolute value, we have a list of polynomials which define all the number fields that we are interested in. Many polynomials may give the same number field, so this is the next thing to check. Since we have computed an integral basis for each polynomial during the computation of the discriminant of the corresponding number field, we can use the POLRED algorithm (or more precisely Algorithm 4.4.12) to give a pseudo-canonical polynomial for each number field. This will eliminate practically all the coincidences.

When two distinct polynomials give the same field discriminant, we must now check whether or not the corresponding number fields are isomorphic, and this is done by using one of the algorithms given in Section 4.5.4. Note that this will now occur very rarely (since most cases have been dealt with using Algorithm 4.4.12).

If the field K is not primitive, we must use a relative version of Hunter's theorem due to Martinet (see [Mart]), and make a separate table of imprimitive fields.

In the rest of this chapter we will give some examples of families of number fields.

The simplest of all number fields (apart from $\mathbb{Q}$ itself) are quadratic fields. This case has been studied in detail in Chapter 5, and we have also seen that there exist methods for computing regulators and class groups which do not immediately generalize to higher degree fields. Note also that higher degree fields are not necessarily Galois.

The next simplest case is probably that of cyclic cubic fields, which we now consider.

6.4.2 Cyclic Cubic Fields

Let K be a number field of degree 3 over $\mathbb{Q}$, i.e. a cubic field. If K is Galois over $\mathbb{Q}$, its Galois group must be isomorphic to the cyclic group $\mathbb{Z}/3\mathbb{Z}$, hence we say that K is a cyclic cubic field. The Galois group has, apart from its identity element, two other elements which are inverses. We denote them by σ and $\sigma^{-1} = \sigma^2$. The first proposition to note is as follows.

Proposition 6.4.3. *Let $K = \mathbb{Q}(\theta)$ be a cubic field, where θ is an algebraic integer whose minimal monic polynomial will be denoted $P(X)$. Then K is a cyclic cubic field if and only if the discriminant of P is a square.*

Proof. This is a restatement of Proposition 6.3.5. □

This proposition clearly gives a trivial algorithm to check whether a cubic field is Galois or not.

In the rest of this (sub)section, we assume that K is a cyclic cubic field. Our first task is to determine a general equation for such fields. Let θ be an algebraic integer such that $K = \mathbb{Q}(\theta)$, and let $P(X) = X^3 - SX^2 + TX - N$ be the minimal monic polynomial of θ, with integer coefficients S, T and N.

Note first that since any cubic field has at least one real embedding (as does any odd degree field) and since K is Galois, all the roots of P must be in K hence they must all be real, so a cyclic cubic field must be totally real (i.e. $r_1 = 3$ real embeddings, and $r_2 = 0$ complex ones). Of course, this also follows because the discriminant of P is a square.

In what follows, we set $\zeta = e^{2i\pi/3}$, i.e. a primitive cube root of unity. Since K is totally real, $\zeta \notin K$, hence the extension field $K(\zeta)$ is a sixth degree field over $\mathbb{Q}$. It is easily checked that it is still Galois, with Galois group generated by commuting elements σ and τ, where σ acts on K as above and trivially on ζ, and τ denotes complex conjugation.

The first result that we need is as follows.

Lemma 6.4.4. *Set* $\gamma = \theta + \zeta^2\sigma(\theta) + \zeta\sigma^2(\theta) \in K(\zeta)$, *and* $\beta = \gamma^2/\tau(\gamma)$. *Then* $\beta \in \mathbb{Q}(\zeta)$ *and we have*

$$P(X) = X^3 - SX^2 + \frac{S^2 - e}{3}X - \frac{S^3 - 3Se + eu}{27} \ ,$$

where we have set $e = \beta\tau(\beta)$ *and* $u = \beta + \tau(\beta)$ *(i.e.* e *and* u *are the norm and trace of* β *considered as an element of* $\mathbb{Q}(\zeta)$*).*

Proof. We have $\tau(\gamma) = \theta + \zeta\sigma(\theta) + \zeta^2\sigma^2(\theta)$. One sees immediately that $\sigma(\gamma) = \zeta\gamma$ and $\sigma(\tau(\gamma)) = \zeta^2\tau(\gamma)$ hence β is invariant under the action of σ, so by Galois theory β must belong to the quadratic subfield $\mathbb{Q}(\zeta)$ of $K(\zeta)$. In particular, e and u as defined above are in $\mathbb{Q}$. Now we have the matrix equation

$$\begin{pmatrix} S \\ \gamma \\ \tau(\gamma) \end{pmatrix} = \begin{pmatrix} 1 & 1 & 1 \\ 1 & \zeta^2 & \zeta \\ 1 & \zeta & \zeta^2 \end{pmatrix} \begin{pmatrix} \theta \\ \sigma(\theta) \\ \sigma^2(\theta) \end{pmatrix} \ ,$$

so it follows by inverting the matrix that

$$\begin{pmatrix} \theta \\ \sigma(\theta) \\ \sigma^2(\theta) \end{pmatrix} = \frac{1}{3} \begin{pmatrix} 1 & 1 & 1 \\ 1 & \zeta & \zeta^2 \\ 1 & \zeta^2 & \zeta \end{pmatrix} \begin{pmatrix} S \\ \gamma \\ \tau(\gamma) \end{pmatrix} \ .$$

From the formulas $T = \theta\sigma(\theta) + \theta\sigma^2(\theta) + \sigma(\theta)\sigma^2(\theta)$ and $N = \theta\sigma(\theta)\sigma^2(\theta)$, a little computation gives the result of the lemma. $\qquad\square$

We will now modify θ (hence its minimal polynomial $P(X)$) so as to obtain a unique equation for each cyclic cubic field. First note that replacing γ by $(b + c\zeta)\gamma$ is equivalent to changing θ into $b\theta + c\sigma(\theta)$, and β is changed into

$$\beta\frac{(b + c\zeta)^2}{b + c\zeta^2} \ .$$

Let p_k be the primes which split in $\mathbb{Q}(\zeta)$ (as $p_k = \pi_k\overline{\pi_k}$), i.e. such that $p_k \equiv 1 \pmod 3$, let q_k be the inert primes, i.e. such that $q_k \equiv 2 \pmod 3$, and let $\rho = 1 + 2\zeta = \sqrt{-3}$ be a ramified element (i.e. a prime element above the prime 3). We can write

$$b + c\zeta = (-\zeta)^g \rho^f \prod \pi_k^{e_k} \prod \overline{\pi_k}^{f_k} \prod q_k^{g_k} \ .$$

Hence, since $b + c\zeta^2 = \overline{b + c\zeta}$, we have

$$\frac{(b + c\zeta)^2}{b + c\zeta^2} = (-1)^{g+f} \rho^f \prod \pi_k^{2e_k - f_k} \prod \overline{\pi_k}^{2f_k - e_k} \prod q_k^{g_k} .$$

If the decomposition of β (which is in $\mathbb{Q}(\zeta)$ but perhaps not in $\mathbb{Z}[\zeta]$) is

$$\beta = (-\zeta)^n \rho^m \prod \pi_k^{l_k} \prod \overline{\pi_k}^{m_k} \prod q_k^{n_k}$$

then we can choose $g_k = -n_k$ and $f = -m$. Furthermore, for each k consider the quantity $m_k + 2l_k$. If it is congruent to 0 or 1 modulo 3, we will choose $e_k = \lfloor (-m_k - 2l_k + 1)/3 \rfloor$ and $f_k = l_k + 2e_k$. If it is congruent to 2 modulo 3, then $l_k + 2m_k \equiv 1 \pmod 3$ and we choose $f_k = \lfloor (-l_k - 2m_k + 1)/3 \rfloor$ and $e_k = m_k + 2f_k$.

It is easy to check that, with this choice of exponents, the new value of β is an element of $\mathbb{Z}[\zeta]$ (and not only of $\mathbb{Q}(\zeta)$), is not divisible by any inert or ramified prime, and is divisible by split primes only to the first power. Also, at most one of π_k or $\overline{\pi_k}$ divides β. In other words, if $e = \beta\tau(\beta)$ is the new value of the norm of β, then e is equal to a product of distinct primes congruent to 1 modulo 3.

Finally, since $1 + \zeta + \zeta^2 = 0$, if we change θ into $a + \theta$ with $a \in \mathbb{Q}$, then γ does not change and so neither do β or e. Taking $a = S/3$, we obtain a new value of θ whose trace is equal to 0. Putting all this together we have almost proved the following lemma.

Lemma 6.4.5. *For any cyclic cubic field K, there exists a unique pair of integers e and u such that e is equal to a product of distinct primes congruent to 1 modulo 3, $u \equiv 2 \pmod 3$ and such that $K = \mathbb{Q}(\theta')$ where θ' is a root of the polynomial*

$$Q(X) = X^3 - \frac{e}{3}X - \frac{eu}{27} ,$$

or equivalently $K = \mathbb{Q}(\theta)$ where θ is a root of

$$P(X) = 27Q(X/3) = X^3 - 3eX - eu .$$

Proof. Since $\beta = (u + v\sqrt{-3})/2$, u cannot be divisible by 3 since β is not divisible by the ramified prime. Hence, by suitably choosing the exponent g above (which amounts to changing β into $-\beta$ if necessary), we may assume $u \equiv 2 \pmod 3$.

For the uniqueness statement, note that all the possible choices of generators of K are of the form $a + b\theta + c\sigma(\theta)$, and since we want a trace equal to 0, this gives us the value of a as a function of b and c, where these last values are determined because we want e to be equal to a product of primes congruent to 1 modulo 3, hence β is unique. The last statement is trivial. $\square$

We can now state the main theorem of this section.

Theorem 6.4.6. *All cyclic cubic fields K are given exactly once (up to isomorphism) in the following way.*

(1) *If the prime 3 is ramified in K, then $K = \mathbb{Q}(\theta)$ where θ is a root of the equation with coefficients in $\mathbb{Z}$*

$$P(X) = X^3 - \frac{e}{3}X - \frac{eu}{27}, \quad \text{where}$$

$$e = \frac{u^2 + 27v^2}{4}, \quad u \equiv 6 \pmod 9, \ 3 \nmid v, \ u \equiv v \pmod 2, \ v > 0$$

and $e/9$ is equal to the product of distinct primes congruent to 1 modulo 3.

(2) *If the prime 3 is unramified in K, then $K = \mathbb{Q}(\theta)$ where θ is a root of the equation with coefficients in $\mathbb{Z}$*

$$P(X) = X^3 - X^2 + \frac{1-e}{3}X - \frac{1 - 3e + eu}{27}, \quad \text{where}$$

$$e = \frac{u^2 + 27v^2}{4}, \quad u \equiv 2 \pmod 3, \ u \equiv v \pmod 2, \ v > 0$$

and e is equal to the product of distinct primes congruent to 1 modulo 3.

In both cases, the discriminant of P is equal to e^2v^2 and the discriminant of the number field K is equal to e^2.

(3) *Conversely, if e is equal to 9 times the product of $t - 1$ distinct primes congruent to 1 modulo 3, (resp. is equal to the product of t distinct primes congruent to 1 modulo 3), then there exists up to isomorphism exactly 2^{t-1} cyclic cubic fields of discriminant e^2 defined by the polynomials $P(X)$ given in (1) (resp. (2)).*

To prove this theorem, we will need in particular to compute explicitly integral bases and discriminants of cyclic cubic fields. Although there are other (essentially equivalent) methods, we will apply the round 2 algorithm to do this.

So, let K be a cyclic cubic field. By Lemma 6.4.5, we have $K = \mathbb{Q}(\theta)$ where θ is a root of the equation

$$P(X) = X^3 - 3eX - eu, \quad \text{where} \quad e = \frac{u^2 + 3v^2}{4}, \quad u \equiv 2 \pmod 3$$

and e is equal to a product of distinct primes congruent to 1 modulo 3.

We first prove a few lemmas.

Lemma 6.4.7. *Let $p \mid e$. Then the order $\mathbb{Z}[\theta]$ is p-maximal.*

Proof. We apply Dedekind's criterion. Since $p \mid e$, $\overline{P}(X) = X^3$, therefore with the notations of Theorem 6.1.4, $t_1(X) = X$, $g(X) = X$, $h(X) = X^2$

and $f(X) = (3e/p)X + eu/p$. Since $p \mid e$ we cannot have $p \mid u$, otherwise $p \mid v$, hence $p^2 \mid e$ which was assumed not to be true. Therefore, $p \nmid eu/p$ so $(\bar{f}, \overline{gh}) = 1$, showing that $\mathbb{Z}[\theta]$ is p-maximal. $\square$

Corollary 6.4.8. *The discriminant of $P(X)$ is equal to $81e^2v^2$. The discriminant of the number field K is divisible by e^2.*

Proof. The discriminant of $X^3 + aX + b$ is equal to $-(4a^3 + 27b^2)$ (see Exercise 7 of Chapter 3), hence the discriminant of P is equal to

$$-(4(-3e)^3 + 27e^2u^2) = -27e^2(u^2 - 4e) = 81e^2v^2$$

thus proving the first formula. For the second, we know that the discriminant of the field K is a square divisor of $81e^2v^2$. By the preceding lemma, $\mathbb{Z}[\theta]$ is p-maximal for all primes dividing e, and since e is coprime to $81v^2$, the primes for which $\mathbb{Z}[\theta]$ may not be p-maximal are divisors of $81v^2$, hence the discriminant of K is divisible by e^2. $\square$

Since, as we will see, the prime divisors of v other than 3 are irrelevant, what remains is to look at the behavior of the prime 3.

Lemma 6.4.9. *Assume that $3 \nmid v$. Then $\mathbb{Z}[\theta]$ is 3-maximal.*

Proof. Again we use Dedekind's criterion. Since $eu \equiv 2 \pmod 3$, we have $\bar{P} = (X + 1)^3$ in $\mathbb{F}_3[X]$ hence $t_1(X) = X + 1$, $g(X) = X + 1$, $h(X) = (X + 1)^2$ and $f(X) = X^2 + (e + 1)X + (1 + eu)/3 = (X + 1)(X + e) + (eu + 1 - 3e)/3$ hence

$$(\bar{f}, \bar{g}, \bar{h}) = (X + 1, \bar{f}) = (X + 1, \overline{(eu + 1 - 3e)/3}) .$$

Now we check that

$$r = \frac{eu + 1 - 3e}{3} = \frac{(u^2 + 3v^2)(u - 3) + 4}{12} = \frac{(u - 2)^2(u + 1) + 3v^2(u - 3)}{12}$$

hence, since $u \equiv 2 \pmod 3$, $4r \equiv v^2(u - 3) \pmod 9$ and, in particular, since $3 \nmid v$, $r \equiv 1 \pmod 3$ so $(\bar{f}, \bar{g}, \bar{h}) = 1$, which proves the lemma. $\square$

Lemma 6.4.10. *With the above notation, let θ be a root of $P(X) = X^3 - 3eX - eu$, where $e = (u^2 + 3v^2)/4$ and $u \equiv 2 \pmod 3$. The conjugates of θ are given by the formulas*

$$\sigma(\theta) = \frac{-2e}{v} - \frac{u + v}{2v}\theta + \frac{1}{v}\theta^2 ,$$

$$\sigma^2(\theta) = \frac{2e}{v} + \frac{u - v}{2v}\theta - \frac{1}{v}\theta^2 .$$

Proof. From the proof of Proposition 6.4.3, we have $f = (\theta - \theta_2)(\theta_2 - \theta_3)(\theta_3 - \theta) = \pm 9ev$ (since the discriminant is equal to $81e^2v^2$). If necessary, by exchanging θ_2 and θ_3, we may assume that $\theta_2 - \theta_3 = 9ev/(\theta - \theta_2)(\theta - \theta_3) = 9ev/P'(\theta) = 9ev/(3\theta^2 - 3e)$. Using the extended Euclidean algorithm with $A(X) = X^3 - 3eX - eu$ and $B(X) = X^2 - e$, one finds immediately that the inverse of B modulo A is equal to $(2X^2 - uX - 4e)/(3v^2e)$ hence

$$\theta_2 - \theta_3 = \frac{1}{v}(2\theta^2 - u\theta - 4e) \ .$$

On the other hand, since the trace of θ is equal to 0, we have $\theta_2 + \theta_3 = -\theta$, and the formulas for $\theta_2 = \sigma(\theta)$ and $\theta_3 = \sigma^2(\theta)$ follow immediately.

It would of course have been simple, but less natural, to check directly with the given formulas that $(X - \theta)(X - \sigma(\theta))(X - \sigma^2(\theta)) = X^3 - 3eX - eu$. □

We can now prove a theorem which immediately implies the first two statements of Theorem 6.4.6.

Theorem 6.4.11. *Let $K = \mathbb{Q}(\theta)$ be a cyclic cubic field where θ is a root of $X^3 - 3eX - eu = 0$ and where, as above, $e = (u^2 + 3v^2)/4$ is equal to a product of distinct primes congruent to 1 modulo 3.*

(1) *Assume that $3 \nmid v$. Then $(1, \theta, \sigma(\theta))$ (where $\sigma(\theta)$ is given by the above formula) is an integral basis of K and the discriminant of K is equal to $(9e)^2$.*
(2) *Assume now that $3 \mid v$. Then, if $\theta' = (\theta + 1)/3$, $(1, \theta', \sigma(\theta'))$ is an integral basis of K and the discriminant of K is equal to e^2.*

Proof. 1) Since $\theta^2 = v\sigma(\theta) + ((u+v)/2)\theta + 2e$, the $\mathbb{Z}$-module $\mathcal{O}$ generated by $(1, \theta, \sigma(\theta))$ contains $\mathbb{Z}[\theta]$. One computes immediately (in fact simply from the formula that we have just given for θ^2) that $\mathbb{Z}[\theta]$ is of index v in $\mathcal{O}$. Hence, the discriminant of $\mathcal{O}$ is equal to $81e^2$. Since we know that $\mathbb{Z}[\theta]$, and a fortiori that $\mathcal{O}$ is 3-maximal and p-maximal for every prime dividing e, it follows that $\mathcal{O}$ is the maximal order, thus proving the first part of the theorem.

2) We now consider the case where $3 \mid v$. The field K can then be defined by the polynomial

$$Q(X) = P(3X - 1)/27 = X^3 - X^2 + \frac{1 - e}{3}X - \frac{1 - 3e + eu}{27} \ .$$

Since $e \equiv 1 \pmod 3$, $u \equiv 2 \pmod 3$ and $3 \mid v$, a simple calculation shows that $Q \in \mathbb{Z}[X]$. Furthermore, from Proposition 3.3.5 the discriminant of Q is equal to the discriminant of P divided by 3^6, i.e. to $e^2(v/3)^2$. Set $\theta' = (\theta + 1)/3$, which is a root of Q, and let $\mathcal{O}$ be the $\mathbb{Z}$-module generated by $(1, \theta', \sigma(\theta'))$. We compute that

$$\sigma(\theta') = \frac{2 + u + 3v - 4e}{6v} - \frac{4 + u + v}{2v}\theta' + \frac{3}{v}\theta'^2$$

and so, as in the proof of the first part, one checks that $\mathcal{O} \supset \mathbb{Z}[\theta']$ and $[\mathcal{O} : \mathbb{Z}[\theta']] = v/3$. Therefore the discriminant of $\mathcal{O}$ is equal to e^2. By Corollary 6.4.8 the discriminant of K must also be divisible by e^2, and so the theorem follows. $\square$

Proof of Theorem 6.4.6. First, we note that the polynomials given in Theorem 6.4.6 are irreducible in $\mathbb{Q}[X]$ (see Exercise 17).

From Theorem 6.4.11, one sees immediately that 3 is ramified in K (i.e. 3 divides the discriminant of K) if and only if $3 \nmid v$. Hence, Lemma 6.4.5 tells us that K is given by an equation $P(X) = X^3 - 3eX - eu$ (with several conditions on e and u). If we set $u_1 = 3u$, $v_1 = v$ and $e_1 = 9e$, we have $e_1 = (u_1^2 + 27v_1^2)/4$, $u_1 \equiv 6 \pmod 9$, $3 \nmid v_1$, and $P(X) = X^3 - (e_1/3)X - (e_1 u_1)/27$ as claimed in Theorem 6.4.6 (1).

Assume now that 3 is not ramified, i.e. that $3 \mid v$. From the proof of the second part of Theorem 6.4.11, we know that K can be defined by the polynomial $X^3 - X^2 + ((1 - e)/3)X - (1 - 3e + eu)/27 \in \mathbb{Z}[X]$ and this time setting $e_1 = e$, $v_1 = v/3$ and $u_1 = u$, it is clear that the second statement of Theorem 6.4.6 follows.

We still need to prove that any two fields defined by different polynomials $P(X)$ given in (1) or (2) are not isomorphic, i.e. that the pair (e, u) determines the isomorphism class. This follows immediately from the uniqueness statement of Lemma 6.4.5. (Note that the e and u in Lemma 6.4.5 are either equal to the e and u of the theorem (in case (2)), or to $e/9$ and $u/3$ (in case (1)).)

Let us prove (3). Assume that e is equal to a product of t distinct primes congruent to 1 modulo 3 (the case where e is equal to 9 times the product of $t - 1$ distinct primes congruent to 1 modulo 3 is dealt with similarly, see Exercise 18). Let $A = \mathbb{Z}[(1 + \sqrt{-3})/2]$ be the ring of algebraic integers of $\mathbb{Q}(\sqrt{-3})$. It is trivial to check (and in fact we have already implicitly used this in the proof of (2)) that if $\alpha \in A$ with $3 \nmid \mathcal{N}(\alpha)$, there exists a unique α' associate to α (i.e. generating the same principal ideal) such that

$$\alpha' = (u + 3v\sqrt{-3})/2 , \qquad u \equiv 2 \pmod 3 .$$

Furthermore, since A is a Euclidean domain and in particular a PID, Proposition 5.1.4 shows that if p_i is a prime congruent to 1 modulo 3, then $p_i = \alpha_i \overline{\alpha_i}$ for a unique $\alpha_i = (u_i + 3v_i\sqrt{-3})/2$ with $u_i \equiv 2 \pmod 3$ and $v_i > 0$.

Hence, if $e = \prod_{1 \leq i \leq t} p_i$, then $e = (u^2 + 27v^2)/4 = \mathcal{N}(u + 3v\sqrt{-3})/2$ if and only if

$$(u + 3v\sqrt{-3})/2 = \prod_{1 \leq i \leq t} \beta_i$$

where $\beta_i = \alpha_i$ or $\beta_i = \overline{\alpha_i}$, and this gives 2^t solutions to the equation $e = (u^2 + 27v^2)/4$. (Note that using associates of β_i do not give any new solutions.)

But, we have seen above that the isomorphism class of a cyclic cubic field is determined uniquely by the pair (e, u) satisfying appropriate conditions. Since $e = (u^2 + 27(-v)^2)/4$ gives the same field as $e = (u^2 + 27v^2)/4$, this shows, as claimed, that there exist exactly 2^{t-1} distinct values of u, hence 2^{t-1} non-isomorphic fields of discriminant e^2. This finishes the proof of Theorem 6.4.6.

$\square$

Corollary 6.4.12. *With the notation of Theorem 6.4.6 (i.e. not those of Theorem 6.4.11), the conjugates of θ are given by the formula*

$$\sigma^{\pm 1}(\theta) = \mp\frac{2e}{9v} + \frac{-3v \pm u}{6v}\theta \pm \frac{1}{v}\theta^2$$

when 3 is ramified in K (i.e. in case (1)), and by the formula

$$\sigma^{\pm 1}(\theta) = \frac{9v \pm (u + 2 - 4e)}{18v} + \frac{-3v \mp (u + 4)}{6v}\theta \pm \frac{1}{v}\theta^2$$

when 3 is not ramified in K (i.e. in case (2)).

In addition, in all cases the discriminant of the polynomial P is equal to e^2v^2, the discriminant of the field K is equal to e^2 and $(1, \theta, \sigma(\theta))$ is an integral basis of K.

The proof of this corollary follows immediately from Lemma 6.4.10 and the proof of Theorems 6.4.11 and 6.4.6.

$\square$

For another way to describe cyclic cubic fields parametrically see Exercise 21.

6.4.3 Pure Cubic Fields

Another class of fields which is easy to describe is the class of pure cubic fields, i.e. fields $K = \mathbb{Q}(\sqrt[3]{m})$ where m is an integer which we may assume not to be divisible by a cube other than ± 1.

The defining polynomial is $P(X) = X^3 - m$ whose discriminant is equal to $-27m^2$. Let θ be the root of this polynomial which is in K.

As in the case of cyclic cubic fields, we must compute the maximal order of K. This is very easy to do using Dedekind's criterion (see Exercise 2). I would like to show however how the Pohst-Zassenhaus Theorem 6.1.3 is really used in the round 2 algorithm, so I will deliberately skip the steps of Algorithm 6.1.8 which use the Dedekind criterion. This will of course make the computations longer, but will illustrate the full use of the round 2 algorithm.

Let p be a prime dividing m and not equal to 3. Then p^2 divides the discriminant of P. Let r be 1 if $p \equiv 1 \pmod 3$, $r = 2$ if not. Then, clearly $\theta^p = m^{(p-r)/3}\theta^r$. Hence, in the basis 1, θ, θ^2 the matrix of the Frobenius at p (or of its square if $p = 2$) is clearly equal to

$$\begin{pmatrix} 1 & 0 & 0 \\ 0 & 0 & 0 \\ 0 & 0 & 0 \end{pmatrix} .$$

This implies that a basis of the p-radical is given by (θ, θ^2). Hence, in step 9 we take $\alpha_1 = \theta$, $\alpha_2 = \theta^2$ and $\alpha_3 = p$.

The 9 by 3 matrix C is obtained by stacking the following three matrices:

$$\begin{pmatrix} 1 & 0 & 0 \\ 0 & 1 & 0 \\ 0 & 0 & m/p \end{pmatrix} , \begin{pmatrix} 0 & 0 & 0 \\ 1 & 0 & 0 \\ 0 & m/p & 0 \end{pmatrix} , \begin{pmatrix} 0 & 0 & 0 \\ 0 & 0 & 0 \\ 1 & 0 & 0 \end{pmatrix} .$$

It follows from the first three equations that, if $p^2 \nmid m$, the kernel of C is trivial, hence that $\mathbb{Z}[\theta]$ is p-maximal. Therefore, we will write

$$m = ab^2 , \quad a \text{ and } b \text{ squarefree} , (a, b) = 1 .$$

Indeed, a is chosen squarefree, but since m is cubefree the other conditions follow.

With these notations, we have just shown that if $p \mid a$ then $\mathbb{Z}[\theta]$ is p-maximal. Take now $p \mid b$ (still with $p \neq 3$). The kernel of the matrix C is now clearly generated over $\mathbb{F}_p$ by the column vector $(0, 0, 1)$ corresponding to θ^2, hence in step 10 we will compute the Hermite normal form of the matrix

$$\begin{pmatrix} 0 & p & 0 & 0 \\ 0 & 0 & p & 0 \\ 1 & 0 & 0 & p \end{pmatrix} .$$

This is clearly equal to the matrix

$$\begin{pmatrix} p & 0 & 0 \\ 0 & p & 0 \\ 0 & 0 & 1 \end{pmatrix} ,$$

thus enlarging the order $\mathbb{Z}[\theta]$ to the order whose $\mathbb{Z}$-basis is $(1, \theta, \theta^2/p)$. If we apply the round 2 algorithm again to this new order, one checks immediately that the new matrix C will be the same as the one above with m/p replaced by m/p^2. Since m is cubefree, this is not divisible by p which shows that the kernel is trivial and so the new order is p-maximal.

Putting together all the local pieces, we can enlarge our order to $(1, \theta, \theta^2/b')$ where $b' = b$ if $3 \nmid b$, $b' = b/3$ if $3 \mid b$. This order will then be p-maximal for every prime p except perhaps the prime 3, which we now consider.

We start from the order $(1, \theta, \theta^2/b')$ and consider separately the cases where $3 \mid m$ and $3 \nmid m$.

Assume first that $3 \mid m$. The matrix of the Frobenius with respect to the basis $(1, \theta, \theta^2/b')$ is equal to

$$\begin{pmatrix} 1 & m & a^2b^4/b'^3 \\ 0 & 0 & 0 \\ 0 & 0 & 0 \end{pmatrix},$$

and modulo 3 both m and a^2b^4/b'^3 are equal to 0. Hence, as in the case $p \neq 3$, the kernel of the Frobenius is generated by $(\theta, \theta^2/b')$. Therefore, in step 9 we take $\alpha_1 = \theta$, $\alpha_2 = \theta^2/b'$ and $\alpha_3 = 3$. The matrix C is then obtained by stacking the following three matrices:

$$\begin{pmatrix} 1 & 0 & 0 \\ 0 & b' & 0 \\ 0 & 0 & m/(3b') \end{pmatrix}, \begin{pmatrix} 0 & 0 & m/b'^2 \\ 1 & 0 & 0 \\ 0 & m/(3b') & 0 \end{pmatrix}, \begin{pmatrix} 0 & 0 & 0 \\ 0 & 0 & 0 \\ 1 & 0 & 0 \end{pmatrix}.$$

Since $3 \nmid b'$ but $3 \mid m$, we have $m/b'^2 \equiv 0 \pmod 3$. On the other hand, $m/(3b')$ is equal to 0 modulo 3 if and only if $3^2 \mid m$, i.e. $3 \mid b$. Hence, we consider two sub-cases.

If $3 \nmid b$, the first three relations show that the kernel of C is equal to 0 and so our order is 3-maximal. Thus, in that case $b' = b$ so an integral basis is $(1, \theta, \theta^2/b)$ and the discriminant of the field K is equal to $-27a^2b^2$.

If $3 \mid b$, the kernel of C is generated by $(0, 0, 1)$ corresponding to θ^2/b'. The Hermite normal form obtained in step 10 is, as for $p \neq 3$, equal to the matrix

$$\begin{pmatrix} 3 & 0 & 0 \\ 0 & 3 & 0 \\ 0 & 0 & 1 \end{pmatrix},$$

giving the larger order $(1, \theta, \theta^2/b'/3) = (1, \theta, \theta^2/b)$.

Since the discriminant of this order is still divisible by 9, we must start again. A similar computation shows that the matrix C is obtained by stacking the following 3 matrices:

$$\begin{pmatrix} 1 & 0 & 0 \\ 0 & 0 & 0 \\ 0 & 0 & ab/3 \end{pmatrix}, \begin{pmatrix} 0 & 0 & a \\ 1 & 0 & 0 \\ 0 & ab/3 & 0 \end{pmatrix}, \begin{pmatrix} 0 & 0 & 0 \\ 0 & 0 & 0 \\ 1 & 0 & 0 \end{pmatrix},$$

and since $3 \nmid ab/3$, the first, third and sixth relation show that the kernel of C is trivial, hence that our order is now 3-maximal. So if $3 \mid b$, an integral basis is $(1, \theta, \theta^2/b)$ and the discriminant of K is equal to $-27a^2b^2$, giving exactly the same result as when $3 \nmid b$.

We now assume that $3 \nmid m$, and so in particular we have $b' = b$. The matrix of the Frobenius is equal to

$$\begin{pmatrix} 1 & ab^2 & a^2b \\ 0 & 0 & 0 \\ 0 & 0 & 0 \end{pmatrix}.$$

Since $a^2 \equiv b^2 \equiv 1 \pmod 3$, this shows that the kernel of the Frobenius is equal to the set of elements $x + y\theta + z\theta^2/b$ such that $x + ay + bz \equiv 0 \pmod 3$. Hence modulo 3 it is, for example, generated by $(\theta - a, \theta^2/b - b)$. This means that in step 9 we can take $\alpha_1 = \theta - a$, $\alpha_2 = \theta^2/b - b$ and $\alpha_3 = 3$. The matrix C is obtained by stacking the following three matrices:

$$\begin{pmatrix} 1 & -a & 0 \\ 0 & b & -a \\ 0 & (b^2 - a^2)/3 & 0 \end{pmatrix}, \begin{pmatrix} 0 & -b & a \\ 1 & 0 & -b \\ 0 & 0 & (a^2 - b^2)/3 \end{pmatrix}, \begin{pmatrix} 0 & 0 & 0 \\ 0 & 0 & 0 \\ 1 & a & b \end{pmatrix}.$$

We consider two subcases. First assume that $a^2 \not\equiv b^2 \pmod 9$. Then from the first, third and sixth relation we see that the kernel of C is trivial, hence that our order is 3-maximal. This means, as in the case $3 \mid m$, that $(1, \theta, \theta^2/b)$ is an integral basis and the discriminant of K is equal to $-27a^2b^2$.

Assume now that $a^2 \equiv b^2 \pmod 9$. In this case, one sees easily that the kernel of C is generated by $(b, ab, 1)$ corresponding to $\theta^2/b + ab\theta + b$, and the computation of the Hermite normal form of the matrix

$$\begin{pmatrix} b & 3 & 0 & 0 \\ ab & 0 & 3 & 0 \\ 1 & 0 & 0 & 3 \end{pmatrix}$$

leads to the matrix

$$\begin{pmatrix} 3 & 0 & b \\ 0 & 3 & ab \\ 0 & 0 & 1 \end{pmatrix},$$

thus giving a larger order generated by $(1, \theta, (\theta^2 + ab^2\theta + b^2)/(3b))$, and the discriminant of this order being equal to $-3a^2b^2$, hence not divisible by 3^2, this enlarged order is 3-maximal.

We summarize what we have proved in the following theorem.

Theorem 6.4.13. *Let $K = \mathbb{Q}(\sqrt[3]{m})$ be a pure cubic field, where m is cubefree and not equal to ± 1. Write $m = ab^2$ with a and b squarefree and coprime. Let θ be the cube root of m belonging to K. Then*

(1) *If $a^2 \not\equiv b^2 \pmod 9$ then*

$$\left(1, \theta, \frac{\theta^2}{b}\right)$$

is an integral basis of K and the discriminant of K is equal to $-27a^2b^2$.

(2) *If $a^2 \equiv b^2 \pmod 9$ then*

$$\left(1, \theta, \frac{\theta^2 + ab^2\theta + b^2}{3b}\right)$$

is an integral basis of K and the discriminant of K is equal to $-3a^2b^2$.

Proof. Simply note that since a and b are coprime, when $3 \mid m$ we cannot have $a^2 \equiv b^2 \pmod 9$. $\square$

Remark. The condition $a^2 \equiv b^2 \pmod 9$ is clearly equivalent to the condition $m \equiv \pm 1 \pmod 9$.

6.4.4 Decomposition of Primes in Pure Cubic Fields

As examples of applications of Algorithm 6.2.9, we will give explicitly the decomposition of primes in pure cubic fields. We could also treat the case of cyclic cubic fields, but the results would be a little more complicated.

Let θ be the real root of the polynomial $X^3 - m$, and let $K = \mathbb{Q}(\theta)$. First consider the case of "good" prime numbers p, i.e. such that p does not divide the index $[\mathbb{Z}_K : \mathbb{Z}[\theta]]$ (which, by Theorem 6.4.13 is equal to $3b$ or b depending on whether $a^2 \equiv b^2 \pmod 9$ or not). In this case we can directly apply Theorem 4.8.13. In other words the decomposition of $p\mathbb{Z}_K$ mimics that of the polynomial $T(X) = X^3 - m$ modulo p.

Now this decomposition is obtained as follows (compare with Section 1.4.2 where the Legendre symbol is defined).

Proposition 6.4.14. *Let p be a prime number not dividing m. The decomposition of $X^3 - m$ modulo p is of the following type.*

(1) *If $p \equiv 2 \pmod 3$, then $X^3 - m \equiv (X - u)(X^2 - vX + w) \pmod p$ (where it is of course implicitly understood that the polynomial $X^2 - vX + w$ is irreducible in $\mathbb{F}_p[X]$).*

(2) *If $p \equiv 1 \pmod 3$ and $m^{(p-1)/3} \equiv 1 \pmod p$ then $X^3 - m \equiv (X - u_1)(X - u_2)(X - u_3) \pmod p$, where u_1, u_2 and u_3 are distinct elements of $\mathbb{F}_p$.*

(3) *If $p \equiv 1 \pmod 3$ and $m^{(p-1)/3} \not\equiv 1 \pmod p$, then $X^3 - m$ is irreducible in $\mathbb{F}_p[X]$.*

(4) *If $p = 3$, then $X^3 - m \equiv (X - a)^3 \pmod p$.*

Proof. Consider the group homomorphism ϕ such that $\phi(x) = x^3$ from $\mathbb{F}_p^*$ into itself. It is clear that if $\phi(x) = 1$, then $(x - 1)(x^2 + x + 1) = 0$ (in $\mathbb{F}_p$) hence

$$(x - 1)((2x + 1)^2 + 3) = 0 \ .$$

If $p \equiv 2 \pmod 3$ the quadratic reciprocity law 1.4.7 shows that $\left(\frac{-3}{p}\right) = -1$, hence -3 is not equal to a square in $\mathbb{F}_p$. This shows that $(2x + 1)^2 + 3 = 0$ is impossible, hence that the function ϕ is injective, hence bijective. In particular, there exists a unique $u \in \mathbb{F}_p^*$ such that $\phi(u) = m$, hence a unique root of $X^3 - m$ in $\mathbb{F}_p$, proving (1).

For (2) and (3), by quadratic reciprocity we have $\left(\frac{-3}{p}\right) = 1$, hence there exists $z \in \mathbb{F}_p^*$ such that $z^2 = -3$. This immediately implies that the kernel of ϕ has exactly 3 elements, and hence that the image of ϕ has $(p-1)/3$ elements.

Furthermore, if g is a primitive root modulo p, then clearly the image of ϕ is the set of elements x of the form g^{3k} for $0 \le k < (p-1)/3$, and these are exactly those elements such that $x^{(p-1)/3} = 1$ in $\mathbb{F}_p$, proving (2) and (3).

Finally, (4) is trivial. $\qquad\qquad\qquad\qquad\qquad\qquad\qquad\qquad\qquad\qquad$ $\square$

When $p \mid m$ we trivially have $X^3 - m \equiv X^3 \pmod{p}$, so we immediately obtain the following corollary in the "easy" cases where p does not divide the index.

Corollary 6.4.15. *As above let $K = \mathbb{Q}(\sqrt[3]{m})$ and recall that we have set $m = ab^2$. Assume that $p \nmid b$ and that if $a^2 \equiv b^2 \pmod{9}$, then also $p \ne 3$. Then the decomposition of $p\mathbb{Z}_K$ is given as follows.*

(1) *If $p \mid a$, then $p\mathbb{Z}_K = \mathfrak{p}^3$ where $\mathfrak{p} = p\mathbb{Z}_K + \theta\mathbb{Z}_K$.*

(2) *If $p \nmid a$ and $p \equiv 2 \pmod{3}$, then $p\mathbb{Z}_K = \mathfrak{p}_1\mathfrak{p}_2$ where $\mathfrak{p}_1 = p\mathbb{Z}_K + (\theta - u)\mathbb{Z}_K$ is an ideal of degree 1 and $\mathfrak{p}_2 = p\mathbb{Z}_K + (\theta^2 - v\theta + w)\mathbb{Z}_K$ is an ideal of degree 2.*

(3) *If $p \nmid a$, $p \equiv 1 \pmod{3}$ and $m^{(p-1)/3} \equiv 1 \pmod{p}$, then $p\mathbb{Z}_K = \mathfrak{p}_1\mathfrak{p}_2\mathfrak{p}_3$ where $\mathfrak{p}_i = p\mathbb{Z}_K + (\theta - u_i)\mathbb{Z}_K$ are three distinct ideals of degree 1.*

(4) *If $p \nmid a$, $p \equiv 1 \pmod{3}$ and $m^{(p-1)/3} \not\equiv 1 \pmod{p}$, then the ideal $p\mathbb{Z}_K$ is inert.*

(5) *If $p = 3$ and $p \nmid a$, then $p\mathbb{Z}_K = \mathfrak{p}^3$, where $\mathfrak{p} = p\mathbb{Z}_K + (\theta - a)\mathbb{Z}_K$ is an ideal of degree 1.*

We must now consider the more difficult cases where p divides the index. Here we will follow the Algorithm 6.2.9 more closely, and we will skip the detailed computations of products and quotients of ideals, which are easy but tedious.

Assume first that $a^2 \not\equiv b^2 \pmod{9}$. Then Theorem 6.4.13 tells us that 1, θ, θ^2/b is an integral basis, and according to the algorithm described in Section 6.2 we start by computing the p-radical of $\mathbb{Z}_K$, assuming that $p \mid b$. It is easily seen that the matrix of the Frobenius at p (or its square for $p = 2$) is always equal to the matrix

$$\begin{pmatrix} 1 & 0 & 0 \\ 0 & 0 & 0 \\ 0 & 0 & 0 \end{pmatrix}$$

in $\mathbb{F}_p$. Therefore $(\theta, \theta^2/b)$ is an $\mathbb{F}_p$-basis of $\overline{I_p}$. From this, using Algorithm 6.2.5 we obtain the following $\mathbb{F}_p$ bases.

$\overline{K_1} = (\theta, \theta^2/b)$, $\overline{K_2} = (\theta)$ and $\overline{K_j} = \{0\}$ for $j \ge 3$.

As a consequence, using Algorithm 6.2.7 we obtain

$\overline{J_1} = \overline{J_2} = \overline{J_3} = (\theta, \theta^2/b)$, and $\overline{J_j} = (1, \theta, \theta^2/b)$ for $j \ge 4$.

From this, it is clear that we have $H_1 = H_2 = \mathbb{Z}_K$, $H_3 = K_1$ and $H_j = \mathbb{Z}_K$ for $j \ge 4$, from which it follows that

$$p\mathbb{Z}_K = K_1^3 \ .$$

Since K is a field of degree equal to 3, this implies that K_1 is a prime ideal (which of course can be checked directly since it is of norm p). This shows that p is totally ramified, and the unique prime ideal $\mathfrak{p}$ above p is generated over $\mathbb{Z}$ by $(p, \theta, \theta^2/b)$.

Note that most of these computations can be avoided. Indeed, once we know a $\mathbb{Z}$-basis of I_p, a trivial determinant computation shows that I_p is of norm p, hence is a prime ideal of degree 1. Using the notations of Section 6.2, it follows that $g = 1$ and that $p\mathbb{Z}_K = I_p^{e_1}$ and since we are in a field of degree 3, the relation $\sum e_i f_i = 3$ tells us that $e_1 = 3$, thus showing that p is totally ramified.

We have kept the computations however, so that the reader can check his implementation of ideal multiplication and division.

Assume now that $a^2 \equiv b^2 \pmod 9$. Recall that in this case we have $3 \nmid b$. Then Theorem 6.4.13 tells us that 1, θ, $(\theta^2 + ab^2\theta + b^2)/(3b)$ is an integral basis, and we must first compute the p-radical of $\mathbb{Z}_K$, assuming that $p \mid 3b$.

Consider first the case where $p \neq 3$, i.e. $p \mid b$. It is easily seen that the matrix of the Frobenius at p (or its square for $p = 2$) is still equal to the matrix

$$\begin{pmatrix} 1 & 0 & 0 \\ 0 & 0 & 0 \\ 0 & 0 & 0 \end{pmatrix}$$

in $\mathbb{F}_p$ hence we obtain that a $\mathbb{F}_p$-basis of $\overline{I_p}$ is $(\theta, (\theta^2 + ab^2\theta + b^2)/(3b))$. As in the preceding case, one checks trivially that I_p has norm equal to p so is a prime ideal of degree 1, so as before p is totally ramified and $p\mathbb{Z}_K = I_p^3$. For the sake of completeness (or again as exercises), we give the computations as they would have been carried out without noticing this.

By Algorithm 6.2.5 we obtain the following $\mathbb{F}_p$-bases.
$\overline{K_1} = (\theta, (\theta^2 + ab^2\theta + b^2)/(3b))$, $\overline{K_2} = (\theta)$ and $\overline{K_j} = \{0\}$ for $j \geq 3$.

As a consequence, using Algorithm 6.2.7, we obtain
$\overline{J_1} = \overline{J_2} = \overline{J_3} = (\theta, (\theta^2 + ab^2\theta + b^2)/(3b))$, and $\overline{J_j} = (1, \theta, (\theta^2 + ab^2\theta + b^2)/(3b))$ for $j \geq 4$.

From this, as before, we have $H_1 = H_2 = \mathbb{Z}_K$, $H_3 = K_1$ and $H_j = \mathbb{Z}_K$ for $j \geq 4$, from which it follows that

$$p\mathbb{Z}_K = K_1^3 \ .$$

Therefore p is totally ramified, and the unique prime ideal $\mathfrak{p}$ above p is generated over $\mathbb{Z}$ by $(p, \theta, (\theta^2 + ab^2\theta + b^2)/(3b))$.

Finally, still assuming $a^2 \equiv b^2 \pmod 9$, consider the case $p = 3$. The matrix of the Frobenius at 3 is now equal to the matrix

$$\begin{pmatrix} 1 & ab^2 & \frac{b(a^2-2b^2+3a^2b^2-3a^2b^4+a^4b^4)}{27} \\ 0 & 0 & ab\frac{1-a^2b^4}{9} \\ 0 & 0 & b^2\frac{1+a^2+a^2b^2}{3} \end{pmatrix}$$

with coefficients in $\mathbb{F}_3$. Since $a^2 \equiv b^2 \pmod 9$ and $3 \nmid ab$, we have

$$1 + a^2 + a^2b^2 \equiv 1 + a^2 + a^4 \equiv (1 - a^2)(1 + 2a^2) + 3a^4 \equiv 3a^4 \pmod 9 ,$$

hence $3 \nmid (1 + a^2 + a^2b^2)/3$. This shows that $(3, \theta - a, (\theta^2 + ab^2\theta + b^2)/b)$ is a $\mathbb{Z}$-basis of I_p, and hence $(\theta - a)$ is an $\mathbb{F}_p$-basis of $\overline{I_p}$. Here the norm of I_p is equal to 9, so we cannot obtain the decomposition of $3\mathbb{Z}_K$ directly, and it is really necessary to do the computations of Algorithm 6.2.9.

By Algorithm 6.2.5, we obtain the following $\mathbb{F}_3$-bases.
$\overline{K_1} = (\theta - a)$ and $\overline{K_j} = \{0\}$ for $j \geq 2$.

As a consequence, using Algorithm 6.2.7, we obtain
$\overline{J_1} = (\theta - a)$, $\overline{J_2} = (\theta - a, (\theta^2 + ab^2\theta + a^2b^4)/(3b))$ and $\overline{J_j} = (1, \theta, (\theta^2 + ab^2\theta + b^2)/(3b))$ for $j \geq 3$.

From this we obtain (after lifting to $\mathcal{O}$) that $H_1 = (3, \theta - a, (\theta^2 + ab^2\theta - b^2(1 + a^2))/(3b))$, $H_2 = J_2 = (3, \theta - a, (\theta^2 + ab^2\theta + a^2b^4)/(3b))$ and $H_j = \mathbb{Z}_K$ for $j \geq 3$. It is immediately checked (for example using the determinant of the matrix of H_j) that H_1 and H_2 are of norm equal to 3, hence are prime ideals. Thus, we obtain that the prime ideal decomposition of $3\mathbb{Z}_K$ is given by

$$3\mathbb{Z}_K = H_1 H_2^2$$

where H_1 and H_2 are distinct prime ideals with $\mathbb{Z}$-basis given above. Hence, 3 is ramified (as it must be since the discriminant of the field is divisible by 3), but not totally ramified as in the case $a^2 \not\equiv b^2 \pmod 9$.

We summarize the above in the following theorem.

Theorem 6.4.16. *Let $(1, \theta, \omega)$ be the integral basis of $\mathbb{Z}_K$ given by Theorem 6.4.13 (hence $\omega = \theta^2/b$ if $a^2 \not\equiv b^2 \pmod 9$, $\omega = (\theta^2 + ab^2\theta + b^2)/(3b)$ if $a^2 \equiv b^2 \pmod 9$). Then*

(1) *If $p \mid b$, then p is totally ramified, and we have $p\mathbb{Z}_K = \mathfrak{p}^3$, where $\mathfrak{p}$ is a prime ideal of degree 1 given by*

$$\mathfrak{p} = p\mathbb{Z} + \theta\mathbb{Z} + \omega\mathbb{Z} = p\mathbb{Z}_K + \omega\mathbb{Z}_K .$$

(2) *If $p = 3$ and $a^2 \equiv b^2 \pmod 9$, then 3 is partially ramified and we have $3\mathbb{Z}_K = \mathfrak{p}_1\mathfrak{p}_2^2$ where $\mathfrak{p}_1$ and $\mathfrak{p}_2$ are prime ideals of degree 1 given by*

$$\mathfrak{p}_1 = 3\mathbb{Z} + (\theta - a)\mathbb{Z} + (\omega - b(2 + a^2)/3)\mathbb{Z} = 3\mathbb{Z}_K + (\omega - b(a^2 + 2)/3)\mathbb{Z}_K$$

and

$$\mathfrak{p}_2 = 3\mathbb{Z} + (\theta - a)\mathbb{Z} + (\omega - b(a^2 - 1)/3)\mathbb{Z} = 3\mathbb{Z}_K + \alpha\mathbb{Z}_K$$

where

$$\alpha = \omega - b(a^2 - 1)/3 \quad \text{if} \quad a^2b^4 \not\equiv 1 \pmod{27} ,$$

$$\alpha = \omega + \theta - a - b(a^2 - 1)/3 \quad \text{if} \quad a^2b^4 \equiv 1 \pmod{27} .$$

Proof. We have shown everything except the generating systems over $\mathbb{Z}_K$. If $p \mid b$, a simple HNF computation shows that one has $p\mathbb{Z}_K + \omega\mathbb{Z}_K = (p, \theta, \omega)$.

If $p = 3$ and $a^2 \equiv b^2$ (mod 9), we could also check the result via a HNF computation. Another method is to notice that $3\mathbb{Z}_K = \mathfrak{p}_1\mathfrak{p}_2^2$ and that if we set $\alpha_1 = \omega - b(a^2 + 2)/3$, then $\alpha_1 \in \mathfrak{p}_1$, but $\alpha_1 \notin \mathfrak{p}_2$ otherwise $\mathfrak{p}_1 \subset \mathfrak{p}_2$ which is absurd, so that $\alpha_1 = \mathfrak{p}_1^e q$ with q prime to 3, so $3\mathbb{Z}_K + \alpha_1\mathbb{Z}_K = \mathfrak{p}_1$.

For $\mathfrak{p}_2$, if we set $\alpha_2 = \omega - b(a^2 - 1)/3$, then again $\alpha_2 \in \mathfrak{p}_2$ and $\alpha_2 \notin \mathfrak{p}_1$. Hence $\alpha_2 = \mathfrak{p}_2^e q$ with q prime to 3. This implies that $3\mathbb{Z}_K + \alpha_2\mathbb{Z}_K = \mathfrak{p}_2^{\min(e,2)}$ hence this can be equal to $\mathfrak{p}_2$ or to its square. To distinguish the two cases, we must compute the norm of α_2, whose 3-adic valuation will be equal to e. As it happens, it is simpler to work with the norm of $\alpha_2' = \alpha_2 + b(a^2b^2 + a^2 - 2)/3$ (note that $a^2b^2 + a^2 - 2 \equiv (a^2 - 1)(a^2 + 2)$ (mod 9) hence $3\mathbb{Z}_K + \alpha_2\mathbb{Z}_K = 3\mathbb{Z}_K + \alpha_2'\mathbb{Z}_K$).

One computes that $n = \mathcal{N}(\alpha_2') = a^2b(1 - a^2b^4)^2/27$. Hence, if $a^2b^4 \not\equiv 1$ (mod 27), the 3-adic valuation of n is equal to 1, therefore $3\mathbb{Z}_K + \alpha_2\mathbb{Z}_K = \mathfrak{p}_2$.

If $a^2b^4 \equiv 1$ (mod 2)7, a similar computation shows that the 3-adic valuation of $\mathcal{N}(\alpha_2' + \theta - a)$ is equal to 1, thus proving the theorem. $\qquad\square$

6.4.5 General Cubic Fields

In this section, we give without proof a few results concerning the decomposition of primes in general cubic extensions of $\mathbb{Q}$.

Let K be a cubic field. The discriminant $d(K)$ of the number field K can (as any discriminant) be written in a unique way in the form $d(K) = df^2$ where d is either a fundamental discriminant or is equal to 1. The field $k = \mathbb{Q}(\sqrt{d})$ is either $\mathbb{Q}$ if $d = 1$, or is a quadratic field, and is the unique subfield of index 3 of the Galois closure of K.

In particular, cyclic cubic fields correspond to $d = 1$, i.e. $k = \mathbb{Q}$, and pure cubic fields correspond to $d = -3$, i.e. $k = \mathbb{Q}(\sqrt{-3})$ the cyclotomic field of third roots of unity.

Let p be a prime number. If $p \nmid d(K)$, then p is unramified. Therefore by Proposition 4.8.10 we have the following cases.

(1) If $\left(\frac{d(K)}{p}\right) = -1$, then $g = 2$. Hence, we have a decomposition of p in the form $p\mathbb{Z}_K = \mathfrak{p}_1\mathfrak{p}_2$ where $\mathfrak{p}_1$ is a prime ideal of degree 1 and $\mathfrak{p}_2$ is a prime ideal of degree 2.

(2) If $\left(\frac{d(K)}{p}\right) = 1$, then g is odd. Hence, either p is inert or $p\mathbb{Z}_K$ is equal to the product of three prime ideals of degree 1.

If p does not divide the index $[\mathbb{Z}_K : \mathbb{Z}[\theta]]$ where $K = \mathbb{Q}(\theta)$, then the two cases are distinguished by the splitting modulo p of the minimal polynomial $T(X)$ of θ.

If p divides the index, then T has at least a double root modulo p. If T has a double root, but not a triple root, then T also has a simple root which corresponds to a prime ideal of degree 1. In this case $p\mathbb{Z}_K$ is the product of three ideals of degree 1. Finally, if T has a triple root modulo p, we must apply other techniques such as the ones in Section 6.2.

Assume now that $p \mid d(K) = df^2$, hence that p is ramified. Then the result is as follows.

(1) If $p \mid f$, then p is totally ramified. In other words, $p\mathbb{Z}_K = \mathfrak{p}^3$ where $\mathfrak{p}$ is a prime ideal of degree 1.

(2) If $p \mid d$ and $p \nmid f$, then p is partially ramified. In other words, $p\mathbb{Z}_K = \mathfrak{p}_1\mathfrak{p}_2^2$, where $\mathfrak{p}_1$ and $\mathfrak{p}_2$ are distinct prime ideals of degree 1.

(3) Furthermore, if there exists a p such that $p \mid (d, f)$, then we must have $p = 3$ (and we are in case (1), since $p \mid f$).

See for example [Has] for proofs of these results.

6.5 Computing the Class Group, Regulator and Fundamental Units

In this section, we shall give a practical generalization of Buchmann's sub-exponential Algorithm 5.9.2 to an arbitrary number field. This algorithm computes the class group, the regulator and also if desired a system of fundamental units, for a number field whose discriminant is not too large. Although based on essentially the same principles as Algorithm 5.9.2, we do not claim that its running time is sub-exponential, even assuming some reasonable conjectures. On the other hand it performs very well in practice. The algorithm originates in an unpublished paper of J. Buchmann, but the present formulation is due to F. Diaz y Diaz, M. Olivier and myself. As almost all other algorithms in this book, this algorithm has been fully implemented in the author's PARI package (see Appendix A). It is still in an experimental state, hence many refinements need to be made to achieve optimum performance.

We assume that our number field K is given as usual as $K = \mathbb{Q}[\theta]$ where θ is an algebraic integer. Let $T(X)$ be the minimal monic polynomial of θ. Let $n = [K : \mathbb{Q}] = r_1 + 2r_2$, denote by σ_i the complex embeddings of K ordered as usual, and finally let $\omega_1, \ldots, \omega_n$ be an integral basis of $\mathbb{Z}_K$, found using for example the round 2 Algorithm 6.1.8.

6.5.1 Ideal Reduction

The only notion that we have not yet introduced and that we will need in an essential way in our algorithm is that of ideal reduction.

Definition 6.5.1. *Let I be a fractional ideal and α a non-zero element of I. We will say that α is a minimum in I if, for all $\beta \in I$, we have*

$$(\forall i \quad |\sigma_i(\beta)| < |\sigma_i(\alpha)|) \implies \beta = 0 .$$

We will say that the ideal I is reduced if $\ell(I)$ is a minimum in I, where $I \cap \mathbb{Q} = \ell(I)\mathbb{Z}$.

The reader can check that this definition of reduction coincides with the definitions given for the imaginary and real quadratic case (see Exercise 16 of Chapter 5).

Definition 6.5.2. *Let $v = (v_i)_{1 \leq i \leq n}$ be a vector of real numbers such that $v_{r_2+i} = v_i$ for $r_1 < i \leq r_1+r_2$. We define the v-norm $\|\alpha\|_v$ of α by the formula*

$$\|\alpha\|_v^2 = \sum_{i=1}^n e^{v_i} |\sigma_i(\alpha)|^2 .$$

If $\alpha_1, \dots, \alpha_n$ is a $\mathbb{Z}$-basis for the ideal I, then $\| \sum_j x_j \alpha_j \|_v^2$ defines a positive definite quadratic form on I.

Definition 6.5.3. *We say that a $\mathbb{Z}$-basis $\alpha_1, \dots, \alpha_n$ of an ideal I is LLL-reduced along the vector v if it is LLL-reduced for the quadratic form defined by $\|\alpha\|_v^2$.*

Thanks to the LLL algorithms seen in Section 2.6 we can efficiently LLL-reduce along v any given basis.

The main point of these definitions is the following.

Proposition 6.5.4. *If $\alpha \in I$ is a (non-zero) minimum for the quadratic form $\|\alpha\|_v^2$, then α is a minimum of I in the sense of Definition 6.5.1 above, and I/α is a reduced ideal.*

Proof. If $\beta \in I$ is such that for all i, $|\sigma_i(\beta)| < |\sigma_i(\alpha)|$, then clearly $\|\beta\|_v^2 < \|\alpha\|_v^2$. Hence, since α is a minimum non-zero value of the quadratic form, we must have $\beta = 0$ so α is a minimum in I. Let us show that I/α is a reduced ideal. First, I claim that $I/\alpha \cap \mathbb{Q} = \mathbb{Z}$. Indeed, if $r \in \mathbb{Q}^*$, $r \in I/\alpha$ is equivalent to $r\alpha \in I$ and since α is a minimum and r is invariant under the σ_i, this implies that $|r| \geq 1$. Since $1 \in I/\alpha$, this proves my claim, hence $\ell(I/\alpha) = 1$. The proposition now follows since α minimum in I is clearly equivalent to 1 minimum in I/α. $\square$

The LLL-algorithm allows us to find a small vector for our quadratic form, corresponding to an $\alpha \in I$. This α may not be a true minimum, but the inequalities proved in Chapter 2 show that it will in any case be a small vector. If we choose this α instead of a minimum, the ideal I/α will not be necessarily reduced, but it will be sufficient for our needs. For lack of a better term, we will say that I/α is *LLL-reduced in the direction v*.

To summarize, this gives the following algorithm for reduction.

Algorithm 6.5.5 (LLL-Reduction of an Ideal Along a Direction v).   Given a vector v as above and an ideal I by a $\mathbb{Z}$-basis $\alpha_1, \dots, \alpha_n$, this algorithm computes $\alpha \in I$ and a new ideal $J = I/\alpha$ such that the v-norm of α is small.

1. [Set up quadratic form] Let

$$q_{i,j} = \sum_{k=1}^n e^{v_k} \overline{\sigma_k(\alpha_i)} \sigma_k(\alpha_j)$$

   (note that these are all real numbers), and let Q be the quadratic form on $\mathbb{R}^n$ whose matrix is $(q_{i,j})$.

2. [Apply LLL] Using the LLL Algorithm 2.6.3, compute an LLL-reduced basis $\beta_1, \dots, \beta_n$ of I corresponding to this quadratic form, and let $\alpha \leftarrow \beta_1$.

3. [Compute J] Output α and the $\mathbb{Z}$-basis β_i/α of the ideal $J = I/\alpha$ and terminate the algorithm.

Remarks.

(1) The ideal J is a fractional ideal. If desired, we can multiply it by a suitable rational number to make it integral and primitive.

(2) In practice the basis elements α_i are given in terms of a fixed basis $\mathcal{B}$ of K (for example either a power basis or an integral basis of $\mathbb{Z}_K$). If we compute once and for all the quadratic form $Q_{\mathcal{B}}$ attached to $\mathcal{B}$, it is then easier to compute the quadratic form attached to the ideal I. Note however that this argument is only valid for a fixed choice of the vector v.

6.5.2 Computing the Relation Matrix

As in the quadratic case we choose a suitable integer L such that non-inert prime ideals of norm less than or equal to L generate the class group. The GRH implies that we can take $L = 12 \ln^2 |D|$ where D is the discriminant of K (see [Bach]). This is only twice the special value used for quadratic fields. However, if we allow ourselves to be not completely rigorous, we could choose a lower value.

To obtain relations, we will compute random products I of powers of prime ideals. Let $J = I/\alpha$ be an LLL-reduced ideal along a certain direction v, obtained using Algorithm 6.5.5. If J factors on a given factor base, as in the quadratic case we will obtain a relation of the type $\prod_i \mathfrak{p}_i^{x_i} = \alpha \mathbb{Z}_K$. This relation will be stored in two parts. The non-Archimedean information (x_i) will be stored as a column of an integral relation matrix M. The Archimedean information α will be stored as an $r_1 + r_2$-component column vector, by using the complex logarithmic embedding $L_C(\alpha) - \frac{\ln(\mathcal{N}(\alpha))}{n} V$ defined in Section 5.8.4.

Note that, by definition, the sum of the $r_1 + r_2$ components of this vector is an integral multiple of $2i\pi$.

We now give the algorithm which computes the factor bases and the relation matrix.

Algorithm 6.5.6 (Computation of the Relation Matrix). Given a number field K as above, this algorithm computes integers k and k_2 with $k_2 > k$, a $k \times k_2$ integral relation matrix M, an $(r_1 + r_2) \times k_2$ complex logarithm matrix M_C and an Euler product z. These objects will be needed in the class group and unit Algorithm 6.5.9 below. We set $r_u \leftarrow r_1 + r_2$ (this is equal to the unit rank plus one). We choose at will a positive real number B_1 and we set $B_2 \leftarrow 12$.

1. [Compute integral basis and limits] Using Algorithm 6.1.8 compute the field discriminant $D = D(K)$ and an integral basis $\omega_1 = 1, \ldots, \omega_n$. Set $L_1 \leftarrow B_1 \ln^2 |D|$, $L_2 \leftarrow B_2 \ln^2 |D|$ and $L_s \leftarrow (4/\pi)^{r_2} n! / n^n \sqrt{|D|}$.

2. [Compute small factor base] Set $u \leftarrow 1$, $S \leftarrow \emptyset$ and for each prime p such that $p \nmid D$ (i.e. p unramified) do the following until $u > L_s$. Let $p\mathbb{Z}_K = \prod_{1 \leq i \leq g} \mathfrak{p}_i$ be the prime ideal decomposition of $p\mathbb{Z}_K$ obtained using Algorithm 6.2.9. For each $i \leq g - 1$ such that $\mathcal{N}(\mathfrak{p}_i) \leq L_2$, set $S \leftarrow S \cup \{\mathfrak{p}_i\}$ and $u \leftarrow u\mathcal{N}(\mathfrak{p}_i)$. Then S will be a set of prime ideals which we call the small factor base. Let s be its cardinality.

3. [Compute and store powers] For each $\mathfrak{p} \in S$ and each integer e such that $0 \leq e \leq 20$, compute and store an LLL-reduced ideal equivalent to $\mathfrak{p}^e$, where the reduction is done using Algorithm 6.5.5 with v equal to the zero vector. Note that the Archimedean information must also be stored, using the function L_C as explained above.

4. [Compute factor bases and Euler product] For all primes $p \leq L_2$ compute the prime ideal decomposition of $p\mathbb{Z}_K$ using Algorithm 6.2.9, and let the large factor base LFB be the list of all non-inert prime ideals of norm less than or equal to L_2 (where if necessary we also add the elements of S), and let the factor base FB be the subset of LFB containing only those primes of norm less than or equal to L_1 as well as the elements of S. Set k equal to the cardinality of FB, and set $k_2 \leftarrow k + r_u + 10$. Finally, using the prime ideal decompositions, compute the Euler product

$$ z \leftarrow \prod_{p \leq L_2} \frac{1 - 1/p}{\prod_{\mathfrak{p}|p}(1 - 1/\mathcal{N}(\mathfrak{p}))} . $$

5. [Store trivial relations] Set $m \leftarrow 0$. For each $p \leq L_1$ such that all the prime ideals above $\mathfrak{p}$ are in FB, set $m \leftarrow m + 1$ and store the relation $p\mathbb{Z}_K = \prod_{1 \leq i \leq g} \mathfrak{p}_i^{e_i}$ found in step 4 as the m-th column of the matrices M and M_C as explained above.

6. [Generate random power products] Call S_i the elements of the small factor base S. Let q be the ideal number $m + 1 \bmod k$ in FB. Choose random nonnegative integers $v_i \leq 20$ for $i \leq s + r_u$, set $v_{i+r_2} \leftarrow v_i$ for $s < i \leq s + r_u$,

compute the ideal $I \leftarrow \mathfrak{q} \prod_{1 \leq i \leq s} S_i^{v_i}$ and let $J = I/\alpha$ be the ideal obtained by LLL-reducing I along the direction determined by the v_i for $s < i \leq s+n$ using Algorithm 6.5.5. Note that the $S_i^{v_i}$ have been precomputed in step 4.

7. [Relation found?] Using Algorithm 4.8.17, try to factor α (or equivalently the ideal J) on the factor base FB. If it factors, set $m \leftarrow m+1$ and store the relation $IJ^{-1} = \alpha \mathbb{Z}_K$ as the m-th column of the matrices M and M_C as explained above.

8. [Enough relations?] If $m \leq k_2$ go to step 6.

9. [Be honest] For all prime ideals $\mathfrak{q}$ in the large factor base LFB and not belonging to FB, do as follows. Choose randomly integers v_i as in step 6, compute $I \leftarrow \mathfrak{q} \prod_{1 \leq i \leq s} S_i^{v_i}$ and let $J = I/\alpha$ be the ideal obtained by LLL-reducing I along the direction determined by the v_i for $s \leq s+n$. If all the prime ideals dividing J belong to FB or have been already checked in this test, then $\mathfrak{q}$ is OK, otherwise choose other random integers v_i until $\mathfrak{q}$ passes this test.

10. [Eliminate spurious factors] For each ramified prime ideal $\mathfrak{q}$ which belongs to the factor base FB, check whether the GCD of the coefficients occurring in the matrix M in the row corresponding to $\mathfrak{q}$ is equal to 1 (this is always true if $\mathfrak{q}$ is unramified). If not, as in step 9, choose random v_i, compute $I \leftarrow \mathfrak{q} \prod_{1 \leq i \leq s} S_i^{v_i}$, LLL-reduce along the v_i for $i > s$ and see if the resulting ideal factors on FB. If it does, add the relation to the matrices M and M_C, set $k_2 \leftarrow k_2 + 1$, and continue doing this until the GCD of the coefficients occurring in the row corresponding to $\mathfrak{q}$ is equal to 1.

Remarks.

(1) The constant B_1 is usually chosen between 0.1 and 0.8, and controls the execution speed of the general algorithm, as in the quadratic case. On the other hand, the constant B_2 must be taken equal to 12 according to Bach's result. It can be taken equal to B_1 for maximum speed, but in this case, the result may not be correct even under the GRH. This is useful for long searches.

(2) As in the quadratic case, the constants 10 and 20 used in this algorithm are quite arbitrary but usually work.

(3) Step 10 of this algorithm was added only after the implementation was finished since it was noticed that for number fields of small discriminant, the class number was usually a multiple of the correct value due to the presence of ramified primes.

(4) The Euler product that is computed is closely linked to $h(K)R(K)$ since

$$\frac{h(K)R(K)}{w(K)} = 2^{-r_1}(2\pi)^{-r_2}\sqrt{|d(K)|}\prod_p \frac{(1-1/p)}{\prod_{\mathfrak{p}|p}(1-1/\mathcal{N}(\mathfrak{p}))} ,$$

where the outer product runs over all primes p and the innermost product runs over the prime ideals above p (see Exercise 23).

6.5.3 Computing the Regulator and a System of Fundamental Units

Before giving the complete algorithm, we need to explain how to extract from the Archimedean information that we have computed, both the regulator and a system of fundamental units of K.

After suitable column operations on the matrices M and M_C as explained below in Algorithm 6.5.9, we will obtain a complex matrix C whose columns correspond to the Archimedean information associated to zero exponents, i.e. to a relation of the form $\mathbb{Z}_K = \alpha\mathbb{Z}_K$. In other words, the columns are complex logarithmic embeddings of units. As in the real quadratic case, we can obtain the regulator of the subgroup spanned by these units (which hopefully is equal to the field regulator) by computing a real GCD of $(r_u - 1) \times (r_u - 1)$ sub-determinants as follows.

Algorithm 6.5.7 (Computation of the Regulator and Fundamental Unit Matrix). Given a $r_u \times r$ complex matrix C whose columns are the complex logarithmic embeddings of units, this algorithm computes the regulator R of the subgroup spanned by these units as well as an $r_u \times (r_u - 1)$ complex matrix F whose columns give a basis of the lattice spanned by the columns of C. As usual we denote by C_j the columns of the matrix C and we assume that the real part of C is of rank equal to $r_u - 1$.

1. [Initialize] Let $R \leftarrow 0$ and $j \leftarrow r_u - 2$.

2. [Loop] Set $j \leftarrow j + 1$. If $j > r$, let F be the matrix formed by the last $r_u - 1$ columns of C, output R and F and terminate the algorithm.

3. [Compute determinant] Let A be the $(r_u - 1) \times (r_u - 1)$ matrix obtained by extracting from C any $r_u - 1$ rows, columns $j - r_u + 2$ to j, and taking the real part. Let $R_1 \leftarrow \det(A)$. Using the real GCD Algorithm 5.9.3, compute the RGCD d of R and R_1 as well as integers u and v such that $uR + vR_1 = d$ (note that Algorithm 5.9.3 does not give u and v, but it can be easily extended to do so, as in Algorithm 1.3.6).

4. [Replace] Set $R \leftarrow d$, $C_j \leftarrow vC_j + (-1)^{r_u}uC_{j-r_u+1}$ (where C_0 is to be understood as the zero column) and go to step 2.

The proof of the validity of this algorithm is immediate once we notice that the GCD and replacement operations in steps 3 and 4 correspond to computing the sum of two sub-lattices of the unit lattice given by two $\mathbb{Z}$-bases differing by a single element. The sign $(-1)^{r_u}$ is the signature of the cyclic permutation that is performed. Note also that the real GCD Algorithm 5.9.3 may be applied since by [Zim1] and [Fri] we know that regulators of number fields are uniformly bounded from below by 0.2. $\square$

To compute the regulator, we have only used the real part of the matrix C. We now explain how the use of the imaginary part, and more precisely of the matrix F output by this algorithm, allows us in principle to compute a system of fundamental units. Note that, by construction, the columns of F are the complex logarithmic embeddings of a system of fundamental units of $\mathbb{Z}_K$. However this may be a very badly skewed basis of units, hence the first thing is to compute a nice basis using the LLL algorithm. This leads to the following algorithm.

Algorithm 6.5.8 (Computation of a System of Fundamental Units). Given the regulator R and the $r_u \times (r_u - 1)$ matrix F output by Algorithm 6.5.7, this algorithm computes a system of fundamental units, expressing them on an integral basis ω_i. We let $f_{i,j}$ be the coefficients of F.

1. [Build matrix] Set $r \leftarrow r_u - 1$. For $j = 1, \ldots, j = r$ set $b_{i,j} \leftarrow f_{i,j}$ if $i \leq r_1$, $b_{i,j} \leftarrow f_{i,j}/2$ if $r_1 < i \leq r_u$ and $b_{i,j} \leftarrow \overline{f_{i-r_2,j}}/2$ if $r_u < i \leq n$. Let B be the $n \times r$ matrix with coefficients $b_{i,j}$.

2. [LLL reduce] Using the LLL Algorithm 2.6.3 on the real part of the matrix B, compute a $r \times r$ unimodular matrix U such that the real part of BU is LLL-reduced. Let $E = (e_{i,j})$ be the $n \times r$ matrix such that $e_{i,j} = \exp(b'_{i,j})$, where $BU = (b'_{i,j})$. (Note that the exponential taken here may overflow the possibilities of the implementation, in which case the algorithm must be aborted.)

3. [Solve linear system] Let $\Omega = (w_{i,j})$ be the $n \times n$ matrix such that $w_{i,j} = \sigma_j(\omega_i)$ (where, as before, (ω_i) is an integral basis of $\mathbb{Z}_K$). Set $F_u \leftarrow \Omega^{-1}E$.

4. [Round] The coefficients of F_u should be close to rational integers. If this is not the case, then either the precision used to make the computations was insufficient or the units are too large, and the algorithm fails. Otherwise, round all the coefficients of F_u to the nearest integer.

5. [Check] Check that the columns of F_u correspond to units and that the usual regulator determinant constructed using the columns of F_u is equal to R. If this is the case, output the matrix F_u and terminate the algorithm (the columns of this matrix gives the coefficients of a system of fundamental units expressed on the integral basis ω_i). Otherwise, output an error message saying that the accuracy is insufficient to compute the fundamental units.

6.5.4 The General Class Group and Unit Algorithm

We are now ready to give a general algorithm for class group, regulator and fundamental unit computation.

Algorithm 6.5.9 (Class Group, Regulator and Units for General Number Fields). Let $K = \mathbb{Q}[\theta]$ be a number field of degree n given by a primitive algebraic number θ, let T be the minimal monic polynomial of θ. We assume that we have already computed the signature (r_1, r_2) of K using Algorithm 4.1.11. This algorithm computes the class number $h(K)$, the class group $Cl(K)$, the order of the subgroup of roots of unity $w(K)$, the regulator $R(K)$ and a system of fundamental units of $\mathbb{Z}_K$.

1. [Compute relation matrices and Euler product] Using Algorithm 6.5.6, compute the discriminant $D(K)$, a $k \times k_2$ integral relation matrix M, a $r_u \times k_2$ complex logarithm matrix M_C and an Euler product z.

2. [Compute roots of unity] Using Algorithm 4.9.9 compute the order $w(K)$ of the group of roots of unity in K. Output $w(K)$ and set

$$z \leftarrow 2^{-r_1}(2\pi)^{-r_2} w(K) \sqrt{|D(K)|} \cdot z$$

(now z should be close to $h(K)R(K)$).

3. [Simple HNF] Perform a preliminary simple Hermite reduction on the matrix M as described in the remarks after Algorithm 5.5.2. All column operations done on the matrix M should also be done on the corresponding columns of the matrix M_C. Denote by M' and M_C' the matrices obtained in this way.

4. [Compute probable regulator and units] Using Algorithm 2.7.2, compute the LLL-reduced integral kernel A of M' as a rectangular matrix, and set $C \leftarrow M_C'A$. By applying Algorithm 6.5.7 and if desired also Algorithm 6.5.8, compute a probable value for the regulator R and the corresponding system of units which will be fundamental if R is correct.

5. [HNF reduction] Using Algorithm 2.4.8, compute the Hermite normal form $H = (h_{i,j})$ of the matrix M' using modulo d techniques, where d can be computed using standard Gaussian elimination (or simply use Algorithm 2.4.5). If the matrix is not of maximal rank, get 10 more relations as in steps 6 and 7 of Algorithm 6.5.6 and go to step 3. (It will not be necessary to recompute the whole HNF.)

6. [Simplify H] For every i such that $h_{i,i} = 1$, suppress row and column i, and let W be the resulting matrix.

7. [Finished?] Let $h \leftarrow \det(W)$ (i.e. the product of the diagonal elements). If $hR \geq z\sqrt{2}$, get 10 more relations in steps 6 and 7 of Algorithm 6.5.6 and go to step 3 (same remark as above). Otherwise, output h as the class number, R as the regulator, and the system of fundamental units if it has been computed.

8. [Class group] Compute the Smith normal form of W using Algorithm 2.4.14. Output those among the diagonal elements d_i which are greater than 1 as the invariants of the class group (i.e. $Cl(K) = \bigoplus \mathbb{Z}/d_i\mathbb{Z}$) and terminate the algorithm.

Remarks.

(1) Most implementation remarks given after Algorithm 5.5.2 also apply here. In particular the correctness of the results given by this algorithm depends on the validity of GRH and the constant $B_2 = 12$ chosen in Algorithm 6.5.6. To speed up this algorithm, one can take B_2 to be a much lower value, and practice shows that this works well, but the results are not

anymore guaranteed to be correct even under GRH until someone improves Bach's bounds.

(2) The randomization of the direction of ideal reduction performed in step 6 of Algorithm 6.5.6 is absolutely essential for the correct performance of the algorithm. Intuitively the first s values of v_i correspond to randomization of the non-Archimedean components, while the last r_u values randomize the Archimedean components. If the reduction was always done using the zero vector for instance, we would almost never obtain a relation matrix giving us the correct class number and regulator.

(3) An important speedup can be obtained by generating some relations in a completely different way. Assume that we can generate many elements $\alpha \in \mathbb{Z}_K$ of reasonably small norm. Then it is reasonable to expect that $\alpha \mathbb{Z}_K$ will factor on the factor base FB, thus giving us a relation. To obtain elements of small norm we can use the Fincke-Pohst Algorithm 2.7.7 on the quadratic form $\|\alpha\|_0^2$ defined on the lattice $\mathbb{Z}_K$, where 0 denotes the zero vector. If $\|\alpha\|_0^2 \leq n B^{2/n}$ then the inequality between arithmetic and geometric mean easily shows that $|\mathcal{N}(\alpha)| \leq B$, hence this indeed allows us to find elements of small norm. The reader is warned however that the relations that may be obtained in this way will in general not be random and may generate sub-lattices of the correct lattice.

(4) It is often useful, not only to compute the class group as an abstract group $Cl(K) = \bigoplus \mathbb{Z}/d_i\mathbb{Z}$, but to compute explicitly a generating set of ideal classes $\overline{g_i}$ such that $\overline{g_i}$ is of order d_i. This can easily be done by keeping track of the Smith reduction matrices in the above algorithm.

6.5.5 The Principal Ideal Problem

As in the real quadratic case, we can now solve the principal ideal problem for general number fields. In other words, given an ideal I of $\mathbb{Z}_K$, determine whether I is a principal ideal, and if this is the case, find an $\alpha \in K$ such that $I = \alpha \mathbb{Z}_K$.

To do this, we need to keep some information that was discarded in Algorithm 6.5.9. More precisely, we must keep better track of the Hermite reduction which is performed, including the simple Hermite reduction stage. If we do so, we will have kept a matrix M'' of relations which will be of the form

$$M'' = \begin{pmatrix} 0 & W & B \\ 0 & 0 & I \end{pmatrix} ,$$

where 0 denotes the zero matrix, I is some identity matrix and W is the square matrix in Hermite normal form computed in Step 6 of Algorithm 6.5.9. Together with this matrix, we must also compute the corresponding complex matrix M_C'', so that each column of M'' and M_C'' still corresponds to a relation. Finally, in Step 8 of Algorithm 6.5.9, we also keep the unimodular matrix U such that $D = UWV$ is in Smith normal form (it is not necessary to keep the unimodular matrix V).

Now given an ideal I we can first compute the norm of I. If it is small, then I will factor on the factor base FB chosen in Algorithm 6.5.6. Otherwise, as in Algorithm 6.5.6, we choose random exponents v_i and compute $I \prod_{1 \leq i \leq s} S_i^{v_i}$ and reduce this ideal (along the direction 0 for instance, here it does not matter). Since this reduced ideal has a reasonably small norm, we may hope to factor it on our factor base, thus expressing I in the form $I = \alpha \prod_{1 \leq i \leq k} \mathfrak{p}_i^{x_i}$, where we denote by $\mathfrak{p}_i$ the elements of FB.

Once such an equality is obtained, we proceed as follows. Since the columns of M'' generate the lattice of relations among the $\mathfrak{p}_i$ in the class group, it is clear that I is a principal ideal if and only if the column vector of the x_i is in the image of M''. Let r (resp. c) be the number of rows (resp. columns) of the matrix B occuring in M'' as described above, and let c_1 be the number of initial columns of zeros in M''. Then if X (resp. Y) is the column vector of the x_i for $1 \leq i \leq r$ (resp. $r < i \leq k$), then I is a principal ideal if and only if there exists an integral column vector Z such that $WZ + BY = X$. This is equivalent to $U^{-1}DV^{-1}Z = X - BY$, and since V is unimodular this is equivalent to the existence of an integral column vector Z_1 such that

$$DZ_1 = U(X - BY) .$$

Since D is a diagonal matrix, this means that the j-th element of $U(X - BY)$ must be divisible by the j-th diagonal element of D.

If I is found in this way to be a principal ideal, the use of the complex matrix M_C'' allows us to find α such that $I = \alpha \mathbb{Z}_K$.

This gives the following algorithm.

Algorithm 6.5.10 (Principal Ideal Testing). Given an ideal I of $\mathbb{Z}_K$, this algorithm tests whether I is a principal ideal, and if it is, computes an $\alpha \in K$ such that $I = \alpha \mathbb{Z}_K$. We assume computed the matrices M'' and M_C'' (and hence the matrices W and B), as well as the unimodular matrices U and V and the diagonal matrix D such that $UWV = D$ is in Smith normal form, as explained above. We keep the notations of Algorithm 6.5.6.

1. [Reduce to primitive] If I is not a primitive integral ideal, compute a rational number a such that I/a is primitive integral, and set $I \leftarrow I/a$.

2. [Small norm?] If $\mathcal{N}(I)$ is divisible only by prime numbers below the prime ideals in the factor base FB (i.e. less than or equal to L_1) set $v_i \leftarrow 0$ for $i \leq s$, $\beta \leftarrow a$ and go to step 4.

3. [Generate random relations] Choose random nonnegative integers $v_i \leq 20$ for $i \leq s$, compute the ideal $I_1 \leftarrow I \prod_{1 \leq i \leq s} S_i^{v_i}$, and let $J = I_1/\gamma$ be the ideal obtained by LLL-reducing I_1 along the direction of the zero vector. If $\mathcal{N}(J)$ is divisible only by the prime numbers less than equal to L_1, set $I \leftarrow J$, $\beta \leftarrow a\gamma$ and go to step 4. Otherwise, go to step 3.

4. [Factor I] Using Algorithm 4.8.17, factor I on the factor base FB. Let $I = \prod_{1 \leq i \leq k} \mathfrak{p}_i^{x_i}$. Let X (resp. Y) be the column vector of the $x_i - v_i$ for $i \leq r$

(resp. $i > r$), where r is the number of rows of the matrix B, as above, and where we set $v_i = 0$ for $i > s$.

5. [Check if principal] Let $Z \leftarrow D^{-1}U(X - BY)$ (since D is a diagonal matrix, no matrix inverse must be computed here). If some entry of Z is not integral, output a message saying that the ideal I is not a principal ideal and terminate the algorithm.

6. [Use Archimedean information] Let A be the $(c_1 + k)$-column vector whose first c_1 elements are zero, whose next r elements are the elements of Z, and whose last $k - r$ elements are the elements of Y. Let $A_C = (a_i)_{1 \leq i \leq r_u} \leftarrow M_C'' A$.

7. [Restore correct information] Set $s \leftarrow (\ln \mathcal{N}(I))/n$, and let $A' = (a_i')_{1 \leq i \leq n}$ be defined by $a_i' \leftarrow \exp(s + a_i)$ if $i \leq r_1$, $a_i' \leftarrow \exp(s + (a_i/2))$ if $r_1 < i \leq r_u$ and $a_i' \leftarrow \exp(s + \overline{(a_{i-r_2}/2)})$ if $r_u < i \leq n$. (As in Algorithm 6.5.8, the exponential which is computed here may overflow the possibilities of the implementation, in which case the algorithm must be aborted.)

8. [Round] Set $A'' \leftarrow \Omega^{-1} A'$, where $\Omega = \sigma_j(\omega_i)$ as in Algorithm 6.5.8. The coefficients of A'' must be close to rational integers. If this is not the case, then either the precision used to make the computation was insufficient or the desired α is too large. Otherwise, round the coefficients of A'' to the nearest integer.

9. [Terminate] Let α' be the element of $\mathbb{Z}_K$ whose coordinates in the integral basis are given by the vector A''. Set $\alpha \leftarrow \beta \alpha'$ (product computed in K). If $I \neq \alpha \mathbb{Z}_K$, output an error message stating that the accuracy is not sufficient to compute α. Otherwise, output α and terminate the algorithm.

Note that, since we chose the complex logarithmic embedding $L_C(\alpha) - \frac{\ln(\mathcal{N}(\alpha))}{n} V$ as defined in Section 5.8.4, we must adjust the components by $s = (\ln \mathcal{N}(I))/n$ before computing the exponential in Step 7.

Remark. It is often useful in step 5 to give more information than just the negative information that I is not a principal ideal. Indeed, if as suggested in Remark (4) after Algorithm 6.5.9, the explicit generators $\overline{g_i}$ of order d_i of the class group $Cl(K)$ have been computed, we can easily compute α and k_i such that $I = \alpha \prod_i g_i^{k_i}$ and $0 \leq k_i < d_i$. The necessary modifications to the above algorithm are easy and left to the reader.

6.6 Exercises for Chapter 6

1. By Theorem 6.1.4, $\mathbb{Z}[\theta] + (U(\theta)/p)\mathbb{Z}[\theta]$ is an order, hence a ring. Clearly the only non-trivial fact to check about this is that $(U(\theta)/p)^2$ is still in this order. Using the notations of Theorem 6.1.4, show how to compute polynomials A and B in $\mathbb{Z}[X]$ such that

$$\frac{U(\theta)^2}{p^2} = A(\theta) + \frac{U(\theta)}{p} B(\theta) .$$

2. Compute the maximal order of pure cubic fields using only Dedekind's criterion (Theorem 6.1.4) instead of the Pohst-Zassenhaus theorem.

3. (F. Diaz y Diaz.) With the notations of Theorem 6.1.4, show that a restatement of the Dedekind criterion is the following. Let $r_i(X)$ be the remainder of the Euclidean division of $T(X)$ by $t_i(X)$. We have evidently $r_i \in p\mathbb{Z}[X]$. Set $d_i = 1$ if $e_i \geq 2$ and $r_i \in p^2\mathbb{Z}[X]$, $d_i = 0$ otherwise. Then in (3) we can take $U(X) = \prod_{1 \leq i \leq k} t_i^{e_i - d_i}$. In particular, $\mathbb{Z}[\theta]$ is p-maximal if and only if $r_i \notin p^2\mathbb{Z}[X]$ for every i such that $e_i \geq 2$.

4. Let $\mathcal{O}$ be an order in a number field K and let p be a prime number. Show that $\mathcal{O}$ is p-maximal if and only if every ideal $\mathfrak{p}_i$ of $\mathcal{O}$ which lies above p is invertible in $\mathcal{O}$.

5. Prove Proposition 6.2.1 by first proving the formula for $\mathfrak{a}_i^{-1}$ given in the text.

6. Given a finite separable algebra A over F_p isomorphic to a product of k fields A_i, compute the probability that a random element x of A is a generator of A in terms of the dimensions d_i of the A_i (hint: use Exercise 13 of Chapter 3).

7. Let m and n be distinct squarefree (positive or negative) integers different from 1. Compute an integral basis for the quartic field $K = \mathbb{Q}(\sqrt{n}, \sqrt{m})$. Find also the explicit decomposition of prime numbers in K.

8. (H. W. Lenstra)

   a) Let A be a separable algebra of degree n over $\mathbb{F}_p$ (for example $A = \mathcal{O}/H_j$ in the notation of Section 6.2). Then A is isomorphic to a product of fields K, and let χ_m be the number of such fields which are of degree m over $\mathbb{F}_p$ (if $A = \mathcal{O}/H_j$, then χ_m is the number of prime ideals of $\mathcal{O}$ of degree m dividing H_j). Show that for all d such that $1 \leq d \leq n$ one has

$$\sum_{1 \leq m \leq n} \gcd(d, m)\chi_m = \dim_{\mathbb{F}_p}(\ker(\sigma^d - 1)) ,$$

   where σ denotes the Frobenius homomorphism $x \mapsto x^p$ from A to A.

   b) Compute explicitly the inverse of the matrix $M_n = (\gcd(i,j))_{1 \leq i,j \leq n}$ and give an algorithm which computes the local Euler factor

$$L_p = \prod_{\mathfrak{p} | p} (1 - \mathcal{N}(\mathfrak{p})^{-s})^{-1}$$

   without splitting explicitly the H_j of Section 6.2.

9. Using the ideas used in decomposing prime numbers into a product of prime ideals, write a general algorithm for factoring polynomials over $\mathbb{Q}_p$. You may assume that the coefficients are known to any necessary accuracy (for example that they are in $\mathbb{Q}$), and that the required p-adic precision for the result is sufficiently high. (Hint: If $K = \mathbb{Q}[\theta]$ with $T(\theta) = 0$ and if $p\mathbb{Z}_K = \prod_i \mathfrak{p}_i^{e_i}$, consider the characteristic polynomial of the map multiplication by θ in the $\mathbb{Z}/p^k\mathbb{Z}$-module $\mathbb{Z}_K/\mathfrak{p}_i^{k e_i}$.)

10. (Dedekind) Let $K = \mathbb{Q}(\theta)$ be the cubic field defined by the polynomial $P(X) = X^3 + X^2 - 2X + 8$.

    a) Compute the discriminant of $P(X)$.

    b) Show that $(1, \theta, (\theta + \theta^2)/2)$ is an integral basis of $\mathbb{Z}_K$ and that the discriminant of K is equal to -503.

    c) Using Algorithm 6.2.9 show that the prime 2 is totally split in K.

    d) Conclude from Theorem 4.8.13 that 2 is an inessential discriminantal divisor, i.e. that it divides the index $[\mathbb{Z}_K : \mathbb{Z}[\alpha]]$ for any $\alpha \in \mathbb{Z}_K$.

11. So as to avoid ideal multiplication and division, implement the idea given in the remark after Algorithm 6.2.9, and compare the efficiency of this modified algorithm with Algorithm 6.2.9.

12. Compute the Galois group of the fields generated by the polynomials $X^3 - 2$, $X^3 - X^2 - 2X + 1$ and $X^4 - 10X^2 + 1$.

13. Compute the accuracy needed for the roots of T so that the rounding procedures used in computing the resolvents in all the Galois group finding algorithms given in the text be correct.

14. Implement the Galois group algorithms and check your implementation with the list of 37 polynomials given at the end of Section 6.3.

15.   a) Using Proposition 4.5.3, give an algorithm which determines whether or not a number field K is Galois over $\mathbb{Q}$ (without explicitly computing its Galois group).

    b) Using the methods of Section 4.5 write an algorithm which finds explicitly the conjugates of an element of a number field K belonging to K. The correctness of the results given by your algorithm should *not* depend on approximations, that is once a tentative formula has been found it must be checked exactly. Note that this algorithm may allow to compute the Galois group of K if K is Galois over $\mathbb{Q}$, even when the degree of K is larger than 7.

16. Determine the decomposition of prime numbers dividing the index in cyclic cubic fields by using the method of Algorithm 6.2.9. (Note: if the reader wants to find also the explicit decomposition of prime numbers not dividing the index, which is given by Theorem 4.8.13, he will first need to solve Exercise 28 of Chapter 1.)

17. Show that the polynomials $P(X)$ given in Theorem 6.4.6 (1) and (2) are irreducible in $\mathbb{Q}[X]$.

18. Complete the proof of Theorem 6.4.6 (3) in the case where e is equal to 9 times a product of $t - 1$ primes congruent to 1 modulo 3.

19. Check that the fields defined in Theorem 6.4.6 (2) are not isomorphic for distinct pairs (e, u) (the proof was given explicitly in the text only for case (1)).

20. Generalize the formulas and results of Section 6.4.2 to cyclic quartic fields, replacing $\mathbb{Q}(\zeta)$ by $\mathbb{Q}(i)$. (Hint: start by showing that such a field has a unique quadratic subfield, which is real.)

21. Using the notations of Theorem 6.4.6, find the minimal equation of $\alpha = (\sigma(\theta) - \theta)/3$, and deduce from this another complete parametrization of cyclic cubic fields.

22. Let K be a cubic field.

    a) Show that there exists a $\theta \in \mathbb{Z}_K$ and a, b and c in $\mathbb{Z}$ such $(1, \theta, (\theta^2 + a\theta + b)/c)$ is an integral basis, and give an algorithm for finding θ, a, b and c.

    b) Such a θ being found, show that there exists a $k \in \mathbb{Z}$ such that if we set $\omega = \theta + k$, then $(1, \omega, (\omega^2 + a_2\omega)/a_3)$ is an integral basis of $\mathbb{Z}_K$ for some integers a_2 and a_3.

   c) Deduce from this that for any cubic field K there exists $\alpha \in K$ which is not necessarily an algebraic integer such that $\mathbb{Z}_K = \mathbb{Z}[\alpha]$ in the sense of Exercise 15 of Chapter 4.

   d) Generalize this result to the case of an arbitrary order in a cubic field K by allowing the polynomial used in Exercise 15 of Chapter 4 to have a content larger than 1.

23. Prove that, as claimed in the text, Theorem 4.9.12 (4) implies the formula

$$\frac{h(K)R(K)}{w(K)} = 2^{-r_1}(2\pi)^{-r_2}\sqrt{|d(K)|}\prod_{p}\frac{(1-1/p)}{\prod_{\mathfrak{p}|p}(1-1/\mathcal{N}(\mathfrak{p}))} \ .$$

24. Using Algorithm 6.5.9 compute the class group, the regulator and a system of fundamental units for the number fields defined by the polynomials $T(X) = X^4 + 6$, $T(X) = X^4 - 3X + 5$ and $T(X) = X^4 - 3X - 5$.

25. Compute the different of pure cubic fields and of cyclic cubic fields using Proposition 4.8.19 and Algorithm 4.8.21.

26. Let $(\omega_i)_{1 \le i \le n}$ be an integral basis for a number field K of degree n such that $\omega_n = 1$, and set $t_i = \operatorname{Tr}_{K/\mathbb{Q}}(\omega_i)/n$. Consider the lattice $\mathbb{Z}^{n-1}$ together with the quadratic form

$$q(\mathbf{x}) = \sum_{k=1}^{n}\left|\sigma_k\left(\sum_{1 \le i \le n-1} x_i(\omega_i - t_i)\right)\right|^2 \ .$$

   a) Show that the determinant of this lattice is equal to $\sqrt{|d(K)|/n}$.

   b) Setting $\theta = \sum_{i=1}^{n-1} x_i\omega_i - \lfloor\sum_{i=1}^{n-1} x_i t_i\rceil$ prove Hunter's Theorem 6.4.2.

27. Let $m(X) = m_1(X)\cdots m_k(X)$ be the decomposition of $m(X)$ obtained in step 13 of Algorithm 6.2.9. For $1 \le r \le k$, let e_r be a lift to $\mathcal{O}$ of $m_r(\alpha)$, and set $H_r = H + e_r\mathcal{O}$. Show that $H = H_1 \cdots H_r$, and hence that steps 14 and 15 of Algorithm 6.2.9 are valid. (Note: the e_r are *not* orthogonal idempotents.)

Chapter 7

Introduction to Elliptic Curves

7.1 Basic Definitions

7.1.1 Introduction

The aim of this chapter is to give a brief survey of results, essentially without proofs, about elliptic curves, complex multiplication and their relations to class groups of imaginary quadratic fields. A few algorithms will be given (in Section 7.4, so as not to interrupt the flow of the presentation), but, unlike other chapters, the main emphasis will be on the theory (some of which will be needed in the next chapters). We also describe the superb landscape that is emerging in this theory, although much remains conjectural. It is worth noting that many of the recent advances on the subject (in particular the Birch and Swinnerton-Dyer conjecture) were direct consequences of number-theoretical experiments. This lends further support to the claim that number theory, even in its sophisticated areas, is an experimental as well as a theoretical science.

As elsewhere this book, we have tried to keep the exposition as self-contained as possible. However, for mastering this information, it would be useful if the reader had some knowledge of complex variables and basic algebraic geometry. Nonetheless, the material needed for the applications in the later chapters is fully described here.

As suggestions for further reading, I heartily recommend Silverman's books [Sil] and [Sil3], as well as [Cas], [Hus], [Ire-Ros], [Lang3] and [Shi]. Finally, the algorithms and tables contained in [LN476] (commonly called Antwerp IV) and [Cre] are invaluable.

7.1.2 Elliptic Integrals and Elliptic Functions

Historically, the word elliptic (in the modern sense) came from the theory of elliptic integrals, which occur in many problems, for example in the computation of the length of an arc of an ellipse (whence the name), or in physical problems such as the movement of a pendulum. Such integrals are of the form

$$\int R(x,y)\, dx \ ,$$

where $R(x, y)$ is a rational function in x and y, and y^2 is a polynomial in x of degree 3 or 4 having no multiple root. It is not our purpose here to explain the theory of these integrals (for this see e.g. [W-W], Ch. XXII). They have served as a motivation for the theory of elliptic *functions*, developed in particular by Abel, Jacobi and Weierstraß.

Elliptic functions can be defined as inverse functions of elliptic integrals, but the main property that interests us here is that these functions $f(x)$ are doubly periodic. More precisely we have:

Definition 7.1.1. *An* elliptic function *is a meromorphic function $f(x)$ on the whole complex plane, which is doubly periodic, i.e. such that there exist complex numbers ω_1 and ω_2 such that $\omega_2/\omega_1 \notin \mathbb{R}$ and for all x which is not a pole, $f(x + \omega_1) = f(x + \omega_2) = f(x)$.*

If

$$L = \{m\omega_1 + n\omega_2 | m, n \in \mathbb{Z}\}$$

is the lattice generated by ω_1 and ω_2, it is clear that f is elliptic if and only if $f(x + \omega) = f(x)$ for all $x \in \mathbb{C}$ and all $\omega \in L$. The lattice L is called the period lattice of f. It is clear that every element of $\mathbb{C}$ is equivalent modulo a translation by an element of L to a unique element of the set $F = \{x\omega_1 + y\omega_2, \ 0 \le x, y < 1\}$. Such a set will be called a *fundamental domain* for $\mathbb{C}/L$.

Standard residue calculations immediately show the following properties:

Theorem 7.1.2. *Let $f(x)$ be an elliptic function with period lattice L, let $\{z_i\}$ be the set of zeros and poles of f in a fundamental domain for $\mathbb{C}/L$, and n_i be the order of f at z_i ($n_i > 0$ when z_i is a zero, $n_i < 0$ if z_i is a pole). Then*

(1) *The sum of the residues of f in a fundamental domain is equal to 0.*
(2) $\sum_i n_i = 0$, *in other words f has as many zeros as poles (counted with multiplicity).*
(3) *If f is non-constant, counting multiplicity, f must have at least 2 poles (and hence 2 zeros) in a fundamental domain.*
(4) $\sum_i n_i z_i \in L$. *Note that this makes sense since z_i is defined modulo L.*

Note that the existence of non-constant elliptic functions is not a priori obvious from Definition 7.1.1. In fact, we have the following general theorem, due to Abel and Jacobi:

Theorem 7.1.3. *Assume that z_i and n_i satisfy the above properties. Then there exists an elliptic function f with zeros and poles at z_i of order n_i.*

The simplest construction of non-constant elliptic functions is due to Weierstraß. One defines

$$\wp(z) = \frac{1}{z^2} + \sum_{\omega \in L \setminus \{0\}} \left(\frac{1}{(z+\omega)^2} - \frac{1}{\omega^2} \right) ,$$

and one easily checks that this is an absolutely convergent series which defines an elliptic function with a double pole at 0. Since non-constant elliptic functions must have poles, it is then a simple matter to check that if we define

$$g_2 = 60 \sum_{\omega \in L \setminus \{0\}} \frac{1}{\omega^4} \quad \text{and} \quad g_3 = 140 \sum_{\omega \in L \setminus \{0\}} \frac{1}{\omega^6} ,$$

then $\wp(z)$ satisfies the following differential equation:

$$\wp'^2 = 4\wp^3 - g_2\wp - g_3 .$$

In more geometric terms, one can say that the map

$$z \mapsto \begin{cases} (\wp(z) : \wp'(z) : 1) & \text{for } z \notin L \\ (0 : 1 : 0) & \text{for } z \in L \end{cases}$$

from $\mathbb{C}$ to the projective complex plane gives an isomorphism between the torus $\mathbb{C}/L$ and the projective algebraic curve $y^2t = 4x^3 - g_2xt^2 - g_3t^3$. This is in fact a special case of a general theorem of Riemann which states that all compact Riemann surfaces are algebraic. Note that it is easy to prove that the field of elliptic functions is generated by $\wp$ and $\wp'$ subject to the above algebraic relation.

Since $\mathbb{C}/L$ is non-singular, the corresponding algebraic curve must also be non-singular, and this is equivalent to saying that the discriminant

$$\Delta = 16(g_2^3 - 27g_3^2)$$

of the cubic polynomial is non-zero. This leads directly to the definition of elliptic curves.

7.1.3 Elliptic Curves over a Field

From the preceding section, we see that there are at least two ways to generalize the above concepts to an arbitrary field: we could define an elliptic curve as a curve of genus 1 or as a non-singular plane cubic curve. Luckily, the Riemann-Roch theorem shows that these two definitions are equivalent, hence we set:

Definition 7.1.4. *Let K be a field. An elliptic curve over K is a non-singular projective plane cubic curve E together with a point with coordinates in K. The (non-empty) set of projective points which are on the curve and with coordinates in K will be called the set of K-rational points of E and denoted $E(K)$.*

Up to a suitable birational transformation, it is a simple matter to check that such a curve can always be given by an equation of the following (affine) type:

$$y^2 + a_1 xy + a_3 y = x^3 + a_2 x^2 + a_4 x + a_6 \ ,$$

the point defined over K being the (unique) point at infinity, and hence this can be taken as an alternative definition of an elliptic curve (see Algorithm 7.4.10 for the explicit formulas for the transformation). This will be called a (generalized) Weierstraß equation for the curve.

Note that this equation is not unique. Over certain number fields K such as $\mathbb{Q}$, it can be shown however that there exists an equation which is minimal, in a well defined sense. We will call it *the* minimal Weierstraß equation of the curve. Note that such a minimal equation does not necessarily exist for any number field K. For example, it can be shown (see [Sil], page 226) that the elliptic curve $y^2 = x^3 + 125$ has no minimal Weierstraß equation over the field $\mathbb{Q}(\sqrt{-10})$.

Theorem 7.1.5. *An elliptic curve over* $\mathbb{C}$ *has the form* $\mathbb{C}/L$ *where* L *is a lattice. In other words, if* g_2 *and* g_3 *are any complex numbers such that* $g_2^3 - 27 g_3^2 \neq 0$, *then there exist* ω_1 *and* ω_2 *with* $\mathrm{Im}(\omega_2/\omega_1) > 0$ *and* $g_2 = 60 \sum_{(m,n)\neq(0,0)} (m\omega_1 + n\omega_2)^{-4}$, $g_3 = 140 \sum_{(m,n)\neq(0,0)} (m\omega_1 + n\omega_2)^{-6}$.

A fundamental property of elliptic curves is that they are commutative algebraic groups. This is true over any base field. Over $\mathbb{C}$ this follows immediately from Theorem 7.1.5. The group law is then simply the quotient group law of $\mathbb{C}$ by L. On the other hand, it is not difficult to prove the addition theorem for the Weierstraß $\wp$ function, given by:

$$\wp(z_1 + z_2) = \begin{cases} -\wp(z_1) - \wp(z_2) + \dfrac{1}{4}\left(\dfrac{\wp'(z_1) - \wp'(z_2)}{\wp(z_1) - \wp(z_2)}\right)^2 \ , & \text{if } z_1 \neq z_2; \\[3mm] -2\wp(z_1) + \dfrac{1}{4}\left(\dfrac{\wp''(z_1)}{\wp'(z_1)}\right)^2 \ , & \text{if } z_1 = z_2 \ . \end{cases}$$

From this and the isomorphism given by the map $z \mapsto (\wp(z), \wp'(z))$, one obtains immediately:

Proposition 7.1.6. *Let* $y^2 = 4x^3 - g_2 x - g_3$ *be the equation of an elliptic curve. The neutral element for the group law is the point at infinity* $(0:1:0)$. *The inverse of a point* (x_1, y_1) *is the point* $(x_1, -y_1)$ *i.e. the symmetric point with respect to the x-axis. Finally, if* $P_1 = (x_1, y_1)$ *and* $P_2 = (x_2, y_2)$ *are two non-opposite points on the curve, their sum* $P_3 = (x_3, y_3)$ *is given by the following formulas. Set*

$$m = \begin{cases} \dfrac{y_1 - y_2}{x_1 - x_2} \ , & \text{if } P_1 \neq P_2; \\[3mm] \dfrac{12 x_1^2 - g_2}{2 y_1} \ , & \text{if } P_1 = P_2 \ . \end{cases}$$

Then

$$x_3 = -x_1 - x_2 + m^2/4 \ , \qquad y_3 = -y_1 - m(x_3 - x_1) \ .$$

It is easy to see that this theorem enables us to define an addition law on an elliptic curve over any base field of characteristic zero, and in fact in any characteristic different from 2 and 3. Furthermore, it can be checked that this indeed defines a group law.

More generally one can define such a law over any field, in the following way.

Proposition 7.1.7. *Let*

$$y^2 + a_1 xy + a_3 y = x^3 + a_2 x^2 + a_4 x + a_6$$

be the equation of an elliptic curve defined over an arbitrary base field. Define the neutral element as the point at infinity $(0 : 1 : 0)$, *the opposite of a point* (x_1, y_1) *as the point* $(x_1, -y_1 - a_1 x_1 - a_3)$. *Finally, if* $P_1 = (x_1, y_1)$ *and* $P_2 = (x_2, y_2)$ *are two non-opposite points on the curve, define their sum* $P_3 = (x_3, y_3)$ *by the following. Set*

$$m = \begin{cases} \dfrac{y_1 - y_2}{x_1 - x_2} \, , & \text{if } P_1 \neq P_2; \\[2ex] \dfrac{3x_1^2 + 2a_2 x_1 + a_4 - a_1 y_1}{2y_1 + a_1 x_1 + a_3} \, , & \text{if } P_1 = P_2 \, , \end{cases}$$

and put

$$x_3 = -x_1 - x_2 - a_2 + m(m + a_1) \, , \quad y_3 = -y_1 - a_3 - a_1 x_3 + m(x_1 - x_3) \, .$$

Then these formulas define an (algebraic) Abelian group law on the curve.

The only non-trivial thing to check in this theorem is the associativity of the law. This can most easily be seen by interpreting the group law in terms of divisors, but we will not do this here.

The geometric interpretation of the formulas above is the following. Let P_1 and P_2 be points on the (projective) curve. The line D from P_1 to P_2 (the tangent to the curve if $P_1 = P_2$) intersects the curve at a third point R, say. Then, if O is the point at infinity on the curve, the sum of P_1 and P_2 is the third point of intersection with the curve of the line from O to R. One checks easily that this leads to the above formulas.

For future reference, given a general equation as above, we define the following quantities:

$$b_2 = a_1^2 + 4a_2, \quad b_4 = a_1 a_3 + 2a_4$$
$$b_6 = a_3^2 + 4a_6, \quad b_8 = a_1^2 a_6 + 4a_2 a_6 - a_1 a_3 a_4 + a_2 a_3^2 - a_4^2$$
$$c_4 = b_2^2 - 24 b_4, \quad c_6 = -b_2^3 + 36 b_2 b_4 - 216 b_6 \qquad (7.1)$$
$$\Delta = -b_2^2 b_8 - 8 b_4^3 - 27 b_6^2 + 9 b_2 b_4 b_6, \quad j = c_4^3/\Delta$$
$$\omega = dx/(2y + a_1 x + a_3) = dy/(3x^2 + 2a_2 x + a_4 - a_1 y) \, .$$

Then it is easy to see that if we set $Y = 2y + a_1 x + a_3$, on a field of characteristic different from 2, the equation becomes

$$Y^2 = 4x^3 + b_2 x^2 + 2b_4 x + b_6 .$$

Setting $X = x + b_2/12$, if the characteristic of the field is different from 2 and 3 the equation becomes

$$Y^2 = 4X^3 - (c_4/12)X - (c_6/216) .$$

7.1.4 Points on Elliptic Curves

Consider an abstract equation $y^2 + a_1 xy + a_3 y = x^3 + a_2 x^2 + a_4 x + a_6$, where the coefficients a_i are in $\mathbb{Z}$. Since for any field K there exists a natural homomorphism from $\mathbb{Z}$ to K, this equation can be considered as defining a curve over any field K. Note that even if the initial curve was non-singular, in positive characteristic the curve can become singular.

We shall consider successively the case where $K = \mathbb{R}$, $K = \mathbb{F}_q$, where q is a power of a prime p, and $K = \mathbb{Q}$.

Elliptic Curves over $\mathbb{R}$. In the case where the characteristic is different from 2 and 3, the general equation can be reduced to the following Weierstraß form:

$$y^2 = x^3 + a_4 x + a_6 .$$

(We could put a 4 in front of the x^3 as in the equation for the $\wp$ function, but this introduces unnecessary constant factors in the formulas). The discriminant of the cubic *polynomial* is $-(4a_4^3 + 27a_6^2)$, however the y^2 term must be taken into account, and general considerations show that one must take

$$-16(4a_4^3 + 27a_6^2)$$

as the definition of the discriminant of the elliptic curve.

Several cases can occur. Let $Q(x) = x^3 + a_4 x + a_6$ and $\Delta = -16(4a_4^3 + 27a_6^2)$.

(1) $\Delta < 0$. Then the equation $Q(x) = 0$ has only one real root, and the graph of the curve has only one connected component.

(2) $\Delta > 0$. Then the equation $Q(x) = 0$ has three distinct real roots, and the graph of the curve has two connected components: a non-compact one, which is the component of the zero element of the curve (i.e. the point at infinity), and a compact one, oval shaped.

From the geometric construction of the group law, one sees that the roots of $Q(x) = 0$ are exactly the points of order 2 on the curve (the points of order 3 correspond to the inflection points).

(3) $\Delta = 0$. The curve is no longer an elliptic curve, since it now has a singular point. This case splits into three sub-cases. Since the polynomial $Q(x)$ has at least a double root, write

$$Q(x) = (x - a)^2(x - b) .$$

Note that $2a + b = 0$.

(3a) $a > b$. Then the curve has a unique connected component, which has a double point at $x = a$. The tangents at the double point have distinct real slopes.

(3b) $a < b$. Then the curve has two connected components: a non-compact one, and the single point of coordinates $(a, 0)$. In fact this point is again a double point, but with distinct *purely imaginary* tangents.

(3c) $a = b$. (In this case $a = b = 0$ since $2a + b = 0$). Then the curve has a cusp at $x = 0$, i.e. the tangents at the singular point are the same.

See Fig. 7.1 for the different possible cases. Note that case (1) is subdivided into the case where the curve does not have any horizontal tangent ($a_4 > 0$), and the case where it does ($a_4 \leq 0$).

In case 3, one says that the curve is a degenerate elliptic curve. One easily checks that the group law still exists, but on the curve minus the singular point. This leads to the following terminology: in cases 3a, the group is naturally isomorphic to $\mathbb{R}^*$, and this is called the case of split multiplicative degeneracy. In case 3b, the group is isomorphic to the group S^1 of complex numbers of modulus equal to 1, and this is called non-split multiplicative degeneracy. Finally, in case 3c, the group is isomorphic to the additive group $\mathbb{R}$, and this case is called additive degeneracy.

These notions can be used, not only for $\mathbb{R}$, but for any base field K. In that case, the condition $a > b$ is replaced by $a - b$ is a (non-zero) square in K.

Elliptic Curves over a Finite Field. To study curves (or more general algebraic objects) over $\mathbb{Q}$, it is very useful to study first the reduction of the curve modulo primes. This leads naturally to elliptic curves over $\mathbb{F}_p$, and more generally over an arbitrary finite field $\mathbb{F}_q$, where q is a power of p. Note that when one reduces an elliptic curve mod p, the resulting curve over $\mathbb{F}_p$ may be singular, hence no longer an elliptic curve. Such p are called primes of bad reduction, and are finite in number since they must divide the discriminant of the curve. According to the terminology introduced in the case of $\mathbb{R}$, we will say that the reduction mod p is (split or non-split) multiplicative or additive, according to the type of degeneracy of the curve over $\mathbb{F}_p$. The main theorem concerning elliptic curves over finite fields, due to Hasse, is as follows:

Theorem 7.1.8 (Hasse). *Let p be a prime, and E an elliptic curve over $\mathbb{F}_p$. Then there exists an imaginary quadratic integer α_p such that*

(1) *If $q = p^n$ then*

$$|E(\mathbb{F}_q)| = q + 1 - \alpha_p{}^n - \overline{\alpha_p}{}^n$$

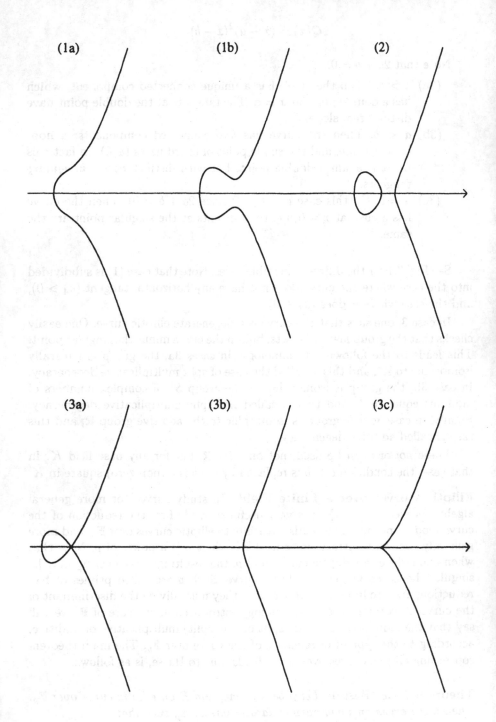

Figure 7.1. Non-Degenerate and Degenerate Elliptic Curves over **R**.

(2)
$$\alpha_p \overline{\alpha_p} = p, \text{ or equivalently } |\alpha_p| = \sqrt{p} \; .$$

(3) *In particular, we have*

$$|E(\mathbb{F}_p)| = p + 1 - a_p \quad \text{with } |a_p| < 2\sqrt{p} \; ,$$

and α_p is a root of the equation

$$\alpha_p{}^2 - a_p \alpha_p + p = 0 \; .$$

The numbers a_p are very important and are (conjecturally) coefficients of a modular form of weight 2. We will come back to this subject in Section 7.3.

The second important result gives some information on the group structure of $E(\mathbb{F}_q)$, and is as follows.

Proposition 7.1.9. *If E is an elliptic curve over a finite field $\mathbb{F}_q$, then $E(\mathbb{F}_q)$ is either cyclic or isomorphic to a product of two cyclic groups. Furthermore, in the case where it is not cyclic, if we write $E(\mathbb{F}_q) \simeq \mathbb{Z}/d_1\mathbb{Z} \times \mathbb{Z}/d_2\mathbb{Z}$ with $d_1 \mid d_2$, then $d_1 \mid q - 1$.*

Elliptic Curves over $\mathbb{Q}$. From a number theorist's point of view, this is of course the most interesting base field. The situation in this case and in the case of more general number fields is much more difficult. The first basic theorem, due to Mordell and later generalized by Weil to the case of number fields and of Abelian varieties, is as follows:

Theorem 7.1.10 (Mordell). *Let E be an elliptic curve over $\mathbb{Q}$. The group of points of E with coordinates in $\mathbb{Q}$ (denoted naturally $E(\mathbb{Q})$) is a finitely generated Abelian group. In other words,*

$$E(\mathbb{Q}) \simeq E(\mathbb{Q})_{\text{tors}} \oplus \mathbb{Z}^r \; ,$$

where r is a non-negative integer called the rank of the curve, and $E(\mathbb{Q})_{\text{tors}}$ is the torsion subgroup of $E(\mathbb{Q})$, which is a finite Abelian group.

The torsion subgroup of a given elliptic curve is easy to compute. On the other hand the study of possible torsion subgroups for elliptic curves over $\mathbb{Q}$ is a difficult problem, solved only in 1977 by Mazur ([Maz]). His theorem is as follows:

Theorem 7.1.11 (Mazur). *Let E be an elliptic curve over $\mathbb{Q}$. The torsion subgroup $E(\mathbb{Q})_{\text{tors}}$ of E can be isomorphic only to one of the 15 following groups:*

$$\mathbb{Z}/m\mathbb{Z} \quad \text{for } 1 \leq m \leq 10 \text{ or } m = 12 \; ,$$

$$\mathbb{Z}/2\mathbb{Z} \times \mathbb{Z}/2m\mathbb{Z} \quad \text{for } 1 \leq m \leq 4 \ .$$

In particular, its cardinality is at most 16.

Note that all of the 15 groups above do occur for an infinite number of non-isomorphic elliptic curves. The corresponding theorem for all quadratic fields (even allowing the discriminant to vary) was proved in 1990 by Kamienny ([Kam]) (with more groups of course), and finally for all number fields in 1994 by Merel ([Mer]).

The other quantity which occurs in Mordell's theorem is the rank r, and is a much more difficult number to compute, even for an individual curve. There is no known mathematically proven algorithm to compute r in general. Even the apparently simpler question of deciding whether r is zero or not (or equivalently whether the curve has a finite or an infinite number of rational points) is still not solved. This is the subject of active research, and we will come back in more detail to this question in Section 7.4.

Let us give an example of a down to earth application of Mordell's theorem. Consider the curve

$$y^2 = x^3 - 36x \ .$$

It is easy to show (see Exercise 3) that the only torsion points are the points of order 1 or 2, i.e. the point at infinity and the three points $(0,0), (6,0), (-6,0)$. But the point $(-2,8)$ is also on the curve. Therefore we must have $r > 0$, hence an infinite number of points, a fact which is not a priori evident. What Mordell's theorem tells us is that r is *finite*, and in fact one can show in this case that $r = 1$, and that the *only* rational points on the curve are integral multiples of the point $(-2,8)$ added to one of the four torsion points.

This curve is in fact closely related to the so-called congruent number problem, and the statement that we have just made means, in this context, that there exists an infinite number of non-equivalent right angled triangles with all three sides rational and area equal to 6, the simplest one (corresponding to the point $(-2,8)$) being the well known $(3,4,5)$ Pythagorean triangle.

As an exercise, the reader can check that twice the point $(-2,8)$ is the point $\left(\frac{25}{4}, \frac{35}{8}\right)$, and that this corresponds to the right-angled triangle of area 6 with sides $\left(\frac{120}{7}, \frac{7}{10}, \frac{1201}{70}\right)$. See [Kob] for the (almost) complete story on the congruent number problem.

7.2 Complex Multiplication and Class Numbers

In this section, we will study maps between elliptic curves. We begin by the case of curves over $\mathbb{C}$.

7.2.1 Maps Between Complex Elliptic Curves

Recall that a complex elliptic curve E has the form $\mathbb{C}/L$ where L is a lattice. Let $E = \mathbb{C}/L$ and $E' = \mathbb{C}/L'$ be two elliptic curves. A map ϕ from E to E' is by definition a holomorphic $\mathbb{Z}$-linear map from E to E'. Since $\mathbb{C}$ is the universal covering of E', ϕ lifts to a holomorphic $\mathbb{Z}$-linear map f from $\mathbb{C}$ to $\mathbb{C}$, and such a map has the form $f(z) = \alpha z$ for some complex number α, which induces a map from E to E' iff $\alpha L \subset L'$. Thus we have:

Proposition 7.2.1. Let $E = \mathbb{C}/L$ and $E' = \mathbb{C}/L'$ be two elliptic curves over $\mathbb{C}$. Then

(1) E is isomorphic to E' if and only if $L' = \alpha L$ for a certain non-zero complex number α.
(2) The set of maps from E to E' can be identified with the set of complex numbers α such that $\alpha L \subset L'$. In particular, the set $\mathrm{End}(E)$ of endomorphisms of E is a commutative ring isomorphic to the set of α such that $\alpha L \subset L$.

In terms of the Weierstraß equation of the curves, this theorem gives the following. Recall that the equation of E (resp E') is $y^2 = 4x^3 - g_2 x - g_3$ (resp. $y^2 = 4x^3 - g_2' x - g_3'$) where

$$g_2 = 60 \sum_{\omega \in L \setminus \{0\}} \omega^{-4}, \quad g_3 = 140 \sum_{\omega \in L \setminus \{0\}} \omega^{-6},$$

and similarly for g_2' and g_3'. Hence, the first part of the theorem says that if $E \simeq E'$, there exists α such that

$$g_2' = \alpha^{-4} g_2, \quad g_3' = \alpha^{-6} g_3.$$

The converse is also clear from the Weierstraß equation. Now, since E is a non-singular curve, the discriminant $g_2^3 - 27g_3^2$ is non-zero, so we can define

$$j(E) = 1728 g_2^3 / (g_2^3 - 27g_3^2),$$

and we obtain:

Proposition 7.2.2. The function $j(E)$ characterizes the isomorphism class of E over $\mathbb{C}$. More precisely, $E \simeq E'$ if and only if $j(E) = j(E')$.

The quantity $j(E)$ is called the modular invariant of the elliptic curve E. The number $1728 = 12^3$ will be explained later. Although we have been working over $\mathbb{C}$, Proposition 7.2.2 is still valid over any algebraically closed field of characteristic different from 2 and 3 (it is also valid in characteristic 2 or 3, for a slightly generalized definition of $j(E)$). On the other hand, it is false if the field is not algebraically closed (consider for example $y^2 = 4x^3 - 4x$ and $y^2 = 4x^3 + 4x$ over $\mathbb{R}$).

Remark. It is easy to construct an elliptic curve with a given modular invariant j. We give the formulas when the characteristic is different from 2 and 3 since we have not given the definition otherwise.

(1) If $j = 0$, one can take $y^2 = x^3 - 1$.
(2) If $j = 1728$, one can take $y^2 = x^3 - x$.
(3) Otherwise, one sets $c = j/(j - 1728)$, and then one can take $y^2 = x^3 - 3cx + 2c$. (If one wants equations with a coefficient of 4 in front of x^3, multiply by 4 and replace y by $y/2$.)

Now let $E = \mathbb{C}/L$ be an elliptic curve over $\mathbb{C}$. Then, as a $\mathbb{Z}$-module, L can be generated by two $\mathbb{R}$-linearly independent complex numbers ω_1 and ω_2, and by suitably ordering them, we may assume that $\operatorname{Im} \tau > 0$, where $\tau = \omega_2/\omega_1$. Since multiplying a lattice by a non-zero complex number does not change the isomorphism class of E, we have $j(E) = j(E_\tau)$, where $E_\tau = \mathbb{C}/L_\tau$ and L_τ is the lattice generated by 1 and τ. By abuse of notation, we will write $j(\tau) = j(E_\tau)$. This defines a complex function j on the upper half-plane $\mathcal{H} = \{\tau \in \mathbb{C}, \operatorname{Im} \tau > 0\}$. If a, b, c and d are integers such that $ad - bc = 1$ (i.e. if $\begin{pmatrix} a & b \\ c & d \end{pmatrix} \in \mathrm{SL}_2(\mathbb{Z})$), then the lattice generated by $a\tau + b$ and $c\tau + d$ is equal to L_τ. This implies the *modular invariance* of $j(\tau)$:

Theorem 7.2.3. *For any* $\begin{pmatrix} a & b \\ c & d \end{pmatrix} \in \mathrm{SL}_2(\mathbb{Z})$, *we have*

$$j\left(\frac{a\tau + b}{c\tau + d}\right) = j(\tau) \ .$$

In particular, $j(\tau)$ is periodic of period 1. Hence it has a Fourier expansion, and one can prove the following theorem:

Theorem 7.2.4. *There exist positive integers* c_n *such that, if we set* $q = e^{2i\pi\tau}$, *we have for all complex* τ *with* $\operatorname{Im} \tau > 0$:

$$j(\tau) = \frac{1}{q} + 744 + \sum_{n \geq 1} c_n q^n \ .$$

The factor 1728 used in the definition of j is there to avoid denominators in the Fourier expansion of $j(\tau)$, and more precisely to have a residue equal to 1 at infinity (the local variable at infinity being taken to be q). These theorems show that j is a meromorphic function on the compactification (obtained by adding a point at infinity) of the quotient $\mathcal{H}/\mathrm{SL}_2(\mathbb{Z})$.

Proposition 7.2.5. *The function j is a one-to-one mapping from the compactification of $\mathcal{H}/\operatorname{SL}_2(\mathbb{Z})$ onto the projective complex plane $\mathbb{P}_1(\mathbb{C})$ (which is naturally isomorphic to the Riemann sphere S^2). In other words, $j(\tau)$ takes once and only once every possible value (including infinity) on $\mathcal{H}/\operatorname{SL}_2(\mathbb{Z})$.*

Note that this proposition is obtained essentially by combining the remark made after Proposition 7.2.2 (surjectivity) with Proposition 7.2.1 (injectivity).

Since the field of meromorphic functions on the sphere is the field of rational functions, we deduce that the field of *modular functions*, i.e. meromorphic functions which are meromorphic at infinity and invariant under $\operatorname{SL}_2(\mathbb{Z})$, is the field of rational functions in j. In particular, modular functions which are holomorphic outside the point at infinity of the Riemann sphere are simply polynomials in j. Finally, if we want to have such a function which is one to one as in Theorem 7.2.5, the only possibilities are linear polynomials $aj + b$. As mentioned above, the constant 1728 has been chosen so that the residue at infinity is equal to one. If we want to keep this property, we must have $a = 1$. This leaves only the possibility $j + b$ for a function having essentially the same properties as j. In other words, the only freedom that we really have in the choice of the modular function j is the constant term 744 in its Fourier expansion.

Although it is a minor point, I would like to say that the normalization of j with constant term 744 is not the correct one for several reasons. The "correct" constant should be 24, so the "correct" j function should in fact be $j - 720$. Maybe the most natural reason is as follows: there exists a rapidly convergent series due to Rademacher for the Fourier coefficients c_n of j. For $n = 0$, this series gives 24, not 744. Other good reasons are due to Atkin and Zagier (unpublished).

7.2.2 Isogenies

We now come back to the case of elliptic curves over an arbitrary field.

Definition 7.2.6. *Let E and E' be two elliptic curves defined over a field K. An isogeny from E to E' is a map of algebraic curves from E to E' sending the zero element of E to the zero element of E'. The curves are said to be isogenous if there exists a non-constant isogeny from E to E'.*

The following theorem summarizes the main properties of non-constant isogenies:

Theorem 7.2.7. *Let ϕ be a non-constant isogeny from E to E'. Then:*

(1) *If K is an algebraically closed field, ϕ is a surjective map.*
(2) *ϕ is a finite map, in other words the fiber over any point of E' is constant and finite.*

(3) ϕ preserves the group laws of the elliptic curves (note that this was not required in the definition), i.e. it is a map of algebraic groups.

From these properties, one can see that ϕ induces an injective map from the corresponding function field of E' to that of E (over some algebraic closure of the base field). The degree of the corresponding field extensions is finite and called the degree of ϕ.

Note that if the above extension of fields is separable, for example if the base field has characteristic zero, then the degree of ϕ is also equal to the cardinality of a fiber, i.e. to the cardinality of its kernel $\phi^{-1}(O)$, but this is not true in general.

Theorem 7.2.8. *Let E be an elliptic curve over a field K, and let m be a positive integer. Then the map $[m]$ (multiplication by m) is an endomorphism of E with the following properties:*

(1) $\deg[m] = m^2$.
(2) *Let $E[m]$ denote the kernel of $[m]$ in some algebraic closure of K, i.e. the group of points of order dividing m. If the characteristic of K is prime to m (or if it is equal to 0), we have*

$$E[m] \simeq (\mathbb{Z}/m\mathbb{Z}) \times (\mathbb{Z}/m\mathbb{Z}) \ .$$

Another important point concerning isogenies is the following:

Theorem 7.2.9. *Let ϕ be an isogeny from E to E'. There exists a unique isogeny $\hat{\phi}$ from E' to E called the dual isogeny, such that*

$$\hat{\phi} \circ \phi = [m] \ ,$$

where m is the degree of ϕ. In addition, we also have

$$\phi \circ \hat{\phi} = [m]' \ ,$$

where $[m]'$ denotes multiplication by m on E'.

Note also the following:

Theorem 7.2.10. *Let E be an elliptic curve and Φ a finite subgroup of E. Then there exists an elliptic curve E' and an isogeny ϕ from E to E' whose kernel is equal to Φ. The elliptic curve E' is well defined up to isomorphism and is denoted E/Φ.*

We end this section by giving a slightly less trivial example of an isogeny: Let E and E' be two elliptic curves over a field of characteristic different from 2, given by the equations

$$y^2 = x^3 + ax^2 + bx \quad \text{and} \quad y^2 = x^3 - 2ax^2 + (a^2 - 4b)x \ ,$$

where we assume that b and $a^2 - 4b$ are both non-zero. Then the map ϕ from E to E' given by

$$\phi(x, y) = \left(\frac{y^2}{x^2}, \frac{y(x^2 - b)}{x^2} \right)$$

is an isogeny of degree 2 with kernel $\{O, (0, 0)\}$.

7.2.3 Complex Multiplication

Let E be an elliptic curve. To make life simpler, we will assume that the base field has characteristic zero. We have seen that the maps $[m]$ are elements of $\mathrm{End}(E)$. Usually, they are the only ones, and since they are distinct, $\mathrm{End}(E) \simeq \mathbb{Z}$. It may however happen that $\mathrm{End}(E)$ is larger than $\mathbb{Z}$.

Definition 7.2.11. *We say that E has* complex multiplication *if $\mathrm{End}(E)$ contains elements other than $[m]$, i.e. if as a ring it is strictly larger than $\mathbb{Z}$.*

The theory of complex multiplication is vast, and we can just give a glimpse of its contents. The first result is as follows:

Proposition 7.2.12. *Let E be an elliptic curve defined over a field of characteristic zero, and assume that E has complex multiplication. Then the ring $\mathrm{End}(E)$ is an order in an imaginary quadratic field, i.e. has the form $\mathbb{Z} + \mathbb{Z}\tau$ where τ is a complex number with positive imaginary part and which is an algebraic integer of degree 2 (that is, satisfies an equation of the form*

$$\tau^2 - s\tau + n = 0 \ ,$$

with s and n in $\mathbb{Z}$ and $s^2 - 4n < 0$).

Proof. We shall give the proof in the case where the base field is $\mathbb{C}$. Then $E \simeq \mathbb{C}/L$ for a certain lattice L, and we know that $\mathrm{End}(E)$ is canonically isomorphic to the set of α such that $\alpha L \subset L$. After division by one of the generators of L, we can assume that L is generated by 1 and τ for a certain $\tau \in \mathcal{H}$, where we recall that $\mathcal{H}$ is the upper half-plane. Then if α stabilizes L, there must exist integers a, b, c and d such that $\alpha = a + b\tau$, $\alpha\tau = c + d\tau$. In other words, α is an eigenvalue of the matrix $\begin{pmatrix} a & b \\ c & d \end{pmatrix}$, hence is an algebraic integer of degree 2 (with $s = a + d$, $n = ad - bc$). Since $\alpha = a + b\tau$, this shows that $\mathbb{Q}(\tau) = \mathbb{Q}(\alpha)$ is a fixed imaginary quadratic extension k of $\mathbb{Q}$, and hence $\mathrm{End}(E)$ is (canonically isomorphic to) a subring of $\mathbb{Z}_k$, the ring of integers of k, and hence is an order in k if it is larger than $\mathbb{Z}$. $\qquad\square$

Example. The curves $y^2 = x^3 - ax$ all have complex multiplication by $\mathbb{Z}[i]$ (map (x, y) to $(-x, iy)$). Similarly, the curves $y^2 = x^3 + b$ all have complex multiplication by $\mathbb{Z}[\rho]$, where ρ is a primitive cube root of unity (map (x, y) to $(\rho x, y)$). For a less trivial example, one can check that the curve

$$y^2 = x^3 - (3/4)x^2 - 2x - 1$$

has complex multiplication by $\mathbb{Z}[\omega]$, where $\omega = \frac{1+\sqrt{-7}}{2}$, multiplication by ω sending (x, y) to (u, v), where

$$u = \omega^{-2} \frac{x^2 - \omega}{x - a}$$

$$v = \omega^{-3} y \frac{x^2 - 2ax + \omega}{(x - a)^2},$$

where we have set $a = (\omega - 3)/4$ (I thank D. Bernardi for these calculations). For a simple algorithm which makes these computations easy to perform see [Star].

Remark. Note that if the base field is a finite field, $\text{End}(E)$ is either isomorphic to an order in an imaginary quadratic field or to the maximal order in a definite quaternion algebra of dimension 4 over $\mathbb{Z}$. In this last case, which is the only case where $\text{End}(E)$ is non-commutative, we say that the elliptic curve E is *supersingular*.

The next theorem concerning complex multiplication is as follows:

Theorem 7.2.13. *Let τ be a quadratic algebraic number with positive imaginary part. Then the elliptic curve $E_\tau = \mathbb{C}/(\mathbb{Z} + \mathbb{Z}\tau)$ has complex multiplication by an order in the quadratic field $\mathbb{Q}(\tau)$, and the j-invariant $j(E_\tau) = j(\tau)$ is an algebraic integer.*

Note that although the context (and the proof) of this theorem involves elliptic curves, its statement is simply that a certain explicit function $j(\tau)$ on $\mathcal{H}$ takes algebraic integer values at quadratic imaginary points.

Examples. Here are a few selected values of j.

$$j((1 + i\sqrt{3})/2) = 0 = 1728 - 3(24)^2$$
$$j(i) = 1728 = 12^3 = 1728 - 4(0)^2$$
$$j((1 + i\sqrt{7})/2) = -3375 = (-15)^3 = 1728 - 7(27)^2$$
$$j(i\sqrt{2}) = 8000 = 20^3 = 1728 + 8(28)^2$$
$$j((1 + i\sqrt{11})/2) = -32768 = (-32)^3 = 1728 - 11(56)^2$$
$$j((1 + i\sqrt{19})/2) = -884736 = (-96)^3 = 1728 - 19(216)^2$$
$$j((1 + i\sqrt{43})/2) = -884736000 = (-960)^3 = 1728 - 43(4536)^2$$
$$j((1 + i\sqrt{67})/2) = -147197952000 = (-5280)^3 = 1728 - 67(46872)^2$$
$$j((1 + i\sqrt{163})/2) = -262537412640768000 = (-640320)^3$$
$$= 1728 - 163(40133016)^2$$
$$j(i\sqrt{3}) = 54000 = 2(30)^3 = 1728 + 12(66)^2$$
$$j(2i) = 287496 = (66)^3 = 1728 + 8(189)^2$$
$$j((1 + 3i\sqrt{3})/2) = -12288000 = -3(160)^3 = 1728 - 3(2024)^2$$
$$j(i\sqrt{7}) = 16581375 = (255)^3 = 1728 + 7(1539)^2$$
$$j((1 + i\sqrt{15})/2) = \frac{-191025 - 85995\sqrt{5}}{2}$$
$$= \frac{1 - \sqrt{5}}{2}\left(\frac{75 + 27\sqrt{5}}{2}\right)^3 = 1728 - 3\left(\frac{273 + 105\sqrt{5}}{2}\right)^2$$
$$j((1 + i\sqrt{23})/2) = -(820750\theta^2 + 1084125\theta + 616750)$$
$$= -(25\theta^2 + 55\theta + 35)^3$$
$$= 1728 - (3\theta^2 - 4)(406\theta^2 + 511\theta + 273)^2 ,$$

where θ is the real root of the cubic equation $X^3 - X - 1 = 0$.

The reason for the special values chosen will become clear later.

An amusing consequence of the above results is the following. We know that if $q = e^{2i\pi\tau}$ then $j(\tau) = 1/q + 744 + O(|q|)$. Hence when $|q|$ is very small (i.e. when the imaginary part of τ is large), it can be expected that $j(\tau)$ is well approximated by $1/q + 744$. Taking the most striking example, this implies that $e^{\pi\sqrt{163}}$ should be close to an integer, and that $(e^{\pi\sqrt{163}} - 744)^{1/3}$ should be even closer. This is indeed what one finds:

$$e^{\pi\sqrt{163}} = 262537412640768743.99999999999925007259\ldots$$

$$(e^{\pi\sqrt{163}} - 744)^{1/3} = 640319.99999999999999999999999939031735\ldots$$

Note that by well known transcendence results, although these quantities are very close to integers, they cannot be integers and they are in fact transcendental numbers.

7.2.4 Complex Multiplication and Hilbert Class Fields

The following theorem gives more precise information on the nature of the algebraic integer $j(\tau)$ and will be one of our basic tools in our study of Atkin's primality test (see Section 9.2). We define the *discriminant* of a quadratic number τ as the discriminant of the unique primitive positive definite quadratic form (a, b, c) such that τ is a root of the equation $ax^2 + bx + c = 0$.

Theorem 7.2.14. *Let $\tau \in \mathcal{H}$ be a quadratic imaginary number, and let D be its discriminant as just defined. Then $j(\tau)$ is an algebraic integer of degree exactly equal to $h(D)$, where $h(D)$ is the class number of the imaginary quadratic order of discriminant D. More precisely, the minimal polynomial of $j(\tau)$ over $\mathbb{Z}$ is the equation $\prod(X - j(\alpha)) = 0$, where α runs over the quadratic numbers associated to the reduced forms of discriminant D.*

Note that $j(\tau)$ is indeed a root of this polynomial, since any quadratic form of discriminant D is equivalent to a reduced form, and since the j function is $SL_2(\mathbb{Z})$-invariant. The difficult part of this theorem is that the polynomial has integral coefficients.

I can now explain the reason for the selection of j-values given in the preceding section. From Theorem 7.2.14, we see that $j(\tau)$ is rational (in fact integral) if and only if $h(D) = 1$ (we assume of course that τ is a quadratic number). Hence, by the Heegner-Stark-Baker theorem (see Section 5.3.1), this corresponds to only 9 quadratic fields. There are 4 more corresponding to non-maximal orders: -12 and -27 (in the field $\mathbb{Q}(\sqrt{-3})$), -16 (in the field $\mathbb{Q}(\sqrt{-4})$), and -28 (in the field $\mathbb{Q}(\sqrt{-7})$).

The first 13 values of our little table above correspond to these 13 quadratic orders, and the last two are for $D = -15$ and $D = -23$, which are the first values for which the class number is 2 and 3 respectively.

Now if τ corresponds to a maximal order in an imaginary quadratic field K, Theorem 7.2.14 tells us that the field $H = K(j(\tau))$ obtained by adjoining $j(\tau)$ to K is an algebraic extension of degree $h(D)$ (this is not strictly true: it tells us this for $K = \mathbb{Q}$, but the statement holds nonetheless). Now in fact much more is true: it is a Galois extension, with Abelian Galois group isomorphic to the class group of the imaginary quadratic field K. Furthermore, it is unramified, and it is the maximal Abelian unramified extension of K. By definition, such a field H is called the Hilbert class field of K. One sees that in the case of imaginary quadratic fields, the Hilbert class field can be obtained by adjoining a value of the j-function. This kind of construction is lacking for other types of fields (except of course for $\mathbb{Q}$). See [Shi] for the relevant definitions and theorems about class fields.

A cursory glance at the table of j-values which we have given reveals many other interesting aspects. For example, in most cases, it seems that $j(\tau)$ is a cube. Furthermore, it can be checked that no large prime factors occur in the values of $j(\tau)$ (or of its norm when it is not in $\mathbb{Q}$). These properties are indeed quite general, with some restrictions. For example, if D is not divisible by 3, then up to multiplication by a unit, $j(\tau)$ is a cube in H. One can also check that (still up to units) $j(\tau) - 1728$ is a square in K if $D \equiv 1$ (mod 4). Finally, not only the values of $j(\tau)$, but more generally the differences $j(\tau_1) - j(\tau_2)$ have only small prime factors (the case of $j(\tau_1)$ is recovered by taking $\tau_2 = \rho = (-1 + \sqrt{-3})/2$). All these properties have been proved by Gross-Zagier [Gro-Zag1].

The other property of an elliptic curve with complex multiplication, which will also be basic to Atkin's primality test, is that it is easy to compute the number of its points in a finite field, i.e. its L-function (see Section 7.3 for the definition). We state only the special cases which we will need (see [Deu]).

Theorem 7.2.15. *Let E be an elliptic curve with complex multiplication by an imaginary quadratic order of discriminant D, and let p be a prime number. Then we have*

$$|E(\mathbb{F}_p)| = p + 1 - a_p ,$$

where a_p is given as follows.

(1) *If p is inert (i.e. if $\left(\frac{D}{p}\right) = -1$), then $a_p = 0$.*

(2) *If p splits into a product of prime elements, say $p = \pi\bar{\pi}$, then $a_p = \pi + \bar{\pi}$ for a suitable choice of π.*

Remarks.

(1) If $D < -4$, there exist only two (opposite) choices for π since the order has only 2 units. These choices give two opposite values of a_p, one of these values giving the correct a_p for E, the other one giving the a_p for the curve E "twisted" by a quadratic non-residue (see Section 7.4.3). On the other hand if $D = -4$ or $D = -3$, there exist 4 (resp. 6) choices for π, also corresponding to twisted curves.

(2) If p is ramified or splits into a product of prime ideals which are not principal, then one can still give the value of a_p, but the recipe is more involved. In terms of L-functions, the general result says that there exists a Hecke character ψ on the field $\mathbb{Q}(\sqrt{D})$ such that

$$L(E, s) = L(\psi, s)L(\bar{\psi}, s) .$$

7.2.5 Modular Equations

Another remarkable property of the j-function, which is not directly linked to complex multiplication, but rather to the role that j plays as a modular invariant, is that the functions $j(N\tau)$ for N integral (or more generally rational) are algebraic functions of $j(\tau)$. The minimal equation of the form

$\Phi_N(j(\tau), j(N\tau)) = 0$ satisfied by $j(N\tau)$ is called the modular equation of level N. This result is not difficult to prove. We will prove it explicitly in the special case $N = 2$. Set

$$P(X) = (X - j(2\tau))(X - j(\frac{\tau}{2}))(X - j(\frac{\tau+1}{2})) = X^3 - s(\tau)X^2 + t(\tau)X - n(\tau) .$$

I claim that the functions s, t and n are polynomials in j. Since they are clearly meromorphic, and in fact holomorphic outside infinity, from Section 7.2.1 it is enough to prove that they are modular functions (i.e. invariant under $\mathrm{SL}_2(\mathbb{Z})$). Since the action of $\mathrm{SL}_2(\mathbb{Z})$ on $\mathcal{H}$ is generated by $\tau \mapsto \tau + 1$, and $\tau \mapsto -1/\tau$, it suffices to show the invariance of s, t and n under these transformations, and this is easily done using the modular invariance of j itself. This shows the existence of a cubic equation satisfied by $j(2\tau)$ over the field $\mathbb{C}(j(\tau))$. If one wants the equation explicitly, one must compute the first few coefficients of the Fourier expansion of $s(\tau)$, $t(\tau)$, and $n(\tau)$, using the Fourier expansion of $j(\tau)$:

$$j(\tau) = \frac{1}{q} + 744 + 196884q + 21493760q^2 + 864299970q^3 + \cdots$$

The result is as follows:

$$s = j^2 - 2^4 3 \cdot 31 j - 2^4 3^4 5^3 ,$$

$$t = 2^4 3 \cdot 31 j^2 + 3^4 5^3 4027 j + 2^8 3^7 5^6 ,$$

$$n = -j^3 + 2^4 3^4 5^3 j^2 - 2^8 3^7 5^6 j + 2^{12} 3^9 5^9 .$$

This gives as modular polynomial of level 2 the polynomial

$$\Phi_2(X, Y) = X^3 + Y^3 - X^2 Y^2 + 2^4 3 \cdot 31(X^2 Y + XY^2) - 2^4 3^4 5^3(X^2 + Y^2)$$
$$+ 3^4 5^3 4027 XY + 2^8 3^7 5^6(X + Y) - 2^{12} 3^9 5^9.$$

As we can see from this example, the modular polynomials are symmetric in X and Y. They have many other remarkable properties that tie them closely to complex multiplication and class numbers, but we will not pursue this subject any further here. See for example [Her], [Mah] and [Coh3] for results and more references on the polynomials Φ_N.

7.3 Rank and L-functions

We have seen in Theorem 7.1.10 that if E is an elliptic curve defined over $\mathbb{Q}$, then

$$E(\mathbb{Q}) \simeq E(\mathbb{Q})_{\mathrm{tors}} \oplus \mathbb{Z}^r ,$$

where $E(\mathbb{Q})_{\mathrm{tors}}$ is a finite group which is easy to compute for a given curve, and r is an integer called the rank. As has already been mentioned, r is very difficult to compute, even for a specific curve. Most questions here have conjectural answers, but very few are proved. In this section, we try to give some indications on the status of the subject at the time of this writing.

7.3.1 The Zeta Function of a Variety

I heartily recommend reading [Ire-Ros] for detailed and concrete examples on this subject.

After clearing the denominators of the coefficients, we may assume that our curve has coefficients in $\mathbb{Z}$. Now it is a classical technique to look at the equation modulo primes p, and to gather this information to obtain results on the equation over $\mathbb{Q}$ or over $\mathbb{Z}$. This can be done more generally for any smooth projective algebraic variety (and more general objects if needed), and not only for elliptic curves. Although it carries us a little away, I believe it worthwhile to do it in this more general context first.

Let V be a (smooth projective) variety of dimension d, defined by equations with coefficients in $\mathbb{Z}$. For any prime p, we can consider the variety V_p obtained by reducing the coefficients modulo p (it may, of course, not be smooth any more). For any $n \geq 1$, let $N_n(p)$ be the number of points of V_p defined over the finite field $\mathbb{F}_{p^n}$ and consider the following formal power series in the variable T:

$$Z_p(T) = \exp\left(\sum_{n \geq 1} \frac{N_n(p)}{n} T^n\right) .$$

Then we have the following very deep theorem, first conjectured by Weil (and proved by him for curves and Abelian varieties, see [Weil]), and proved completely by Deligne in 1974 [Del]:

Theorem 7.3.1. *Let V_p be a smooth projective variety of dimension d over $\mathbb{F}_p$. Then:*

(1) *The series $Z_p(T)$ is a rational function of T, i.e. $Z_p(T) \in \mathbb{Q}(T)$.*
(2) *There exists an integer e (called the Euler characteristic of V_p), such that*

$$Z_p(1/(p^d T)) = \pm p^{de/2} T^e Z_p(T) .$$

(3) *The rational function $Z_p(T)$ factors as follows:*

$$Z_p(T) = \frac{P_1(T) \cdots P_{2d-1}(T)}{P_0(T) P_2(T) \cdots P_{2d}(T)} ,$$

where for all i, $P_i(T) \in \mathbb{Z}[T]$, $P_0(T) = 1 - T$, $P_{2d}(T) = 1 - p^d T$, and for all other i,

$$P_i(T) = \prod_j (1 - \alpha_{ij} T) \quad \text{with } |\alpha_{ij}| = p^{i/2} .$$

The first assertion was actually proved by Dwork a few years before Deligne using relatively elementary methods, but by far the hardest part of this theorem is the last assertion, that $|\alpha_{ij}| = p^{i/2}$. This is called the Riemann hypothesis for varieties over finite fields.

Now given all the local $Z_p(T)$, we can form a global zeta function by setting for s complex with $\operatorname{Re} s$ sufficiently large:

$$\zeta(V, s) = \prod_p Z_p(p^{-s}) \ .$$

This should be taken with a grain of salt, since there are some p (finite in number) such that V_p is not smooth. In fact, given the underlying cohomological interpretation of the P_i, it is more reasonable to consider the global L-functions defined by

$$L_i(V, s) = \prod_p P_i(p^{-s})^{-1} \quad \text{for } 0 \le i \le 2d \ ,$$

and recover the zeta function as

$$\zeta(V, s) = \prod_{0 \le i \le 2d} L_i(V, s)^{(-1)^i} \ .$$

Very little is known about these general zeta function and L-functions. It is believed (can one say conjectured when so few cases have been closely examined?) that these functions can be analytically continued to meromorphic functions on the whole complex plane. When the local factors at the bad primes p are correctly chosen, they should have a functional equation and the L-functions should satisfy the Riemann hypothesis, i.e. apart from "trivial" zeros, all the other complex zeros of $L_i(V, s)$ should lie on the vertical line $\operatorname{Re} s = (i + 1)/2$.

One recovers the ordinary Riemann zeta function by taking for V the single point 0. More generally, one can recover the Dedekind zeta function of a number field by taking for V the 0-dimensional variety defined in the projective line by $P(X) = 0$, where P is a monic polynomial with integer coefficients defining the field over $\mathbb{Q}$.

7.3.2 L-functions of Elliptic Curves

Let us now consider the special case where V is an elliptic curve E. In that case, Hasse's Theorem 7.1.8 gives us all the information we need about the number of points of E over a finite field. This leads to the following corollary:

Corollary 7.3.2. *Let E be an elliptic curve over $\mathbb{Q}$, and let p be a prime of good reduction (i.e. such that E_p is still smooth). Then*

$$Z_p(E) = \frac{1 - a_p T + p T^2}{(1 - T)(1 - pT)} \ ,$$

where a_p is as in Theorem 7.1.8.

In fact, Hasse's theorem is simply the special case of the Weil conjectures for elliptic curves (and can be proved quite simply, see e.g. [Sil] pp 134–136).

Ignoring for the moment the question of bad primes, the general definition of zeta and L-functions gives us

$$\zeta(E, s) = \frac{\zeta(s)\zeta(s-1)}{L(E, s)} \ ,$$

where

$$L(E, s) = L_1(E, s) = \prod_p (1 - a_p p^{-s} + p^{1-2s})^{-1} \ .$$

The function $L(E, s)$ will be called the Hasse-Weil L-function of the elliptic curve E. To give a precise definition, we also need to define the local factors at the bad primes p. This can be done, and finally leads to the following definition.

Definition 7.3.3. Let E be an elliptic curve over $\mathbb{Q}$, and let $y^2 + a_1 xy + a_3 y = x^3 + a_2 x^2 + a_4 x + a_6$ be a minimal Weierstraß equation for E (see 7.1.3). When E has good reduction at p, define $a_p = p + 1 - N_p$ where N_p is the number of (projective) points of E over $\mathbb{F}_p$. If E has bad reduction, define

$$\epsilon(p) = \begin{cases} 1 & \text{if } E \text{ has split multiplicative reduction at } p; \\ -1 & \text{if } E \text{ has non-split multiplicative reduction at } p; \\ 0 & \text{if } E \text{ has additive reduction at } p. \end{cases}$$

Then we define the L-function of E as follows, for $\operatorname{Re} s > 3/2$:

$$L(E, s) = \prod_{\text{bad } p} \frac{1}{1 - \epsilon(p) p^{-s}} \prod_{\text{good } p} \frac{1}{1 - a_p p^{-s} + p^{1-2s}} \ .$$

Note that in this definition it is crucial to take a minimal Weierstraß equation for E: taking another equation could increase the number of primes of bad reduction, and hence change a finite number of local factors. On the other hand, one can prove that $L(E, s)$ depends only on the isogeny class of E.

By expanding the product, it is clear that $L(E, s)$ is a Dirichlet series, i.e. of the form $\sum_{n \geq 1} a_n n^{-s}$ (this of course is the case for all zeta functions of varieties). We will set

$$f_E(\tau) = \sum_{n \geq 1} a_n q^n \ , \quad \text{where as usual } q = e^{2i\pi\tau} \ .$$

We can now state the first conjecture on L-functions of elliptic curves:

Conjecture 7.3.4. *The function $L(E, s)$ can be analytically continued to the whole complex plane to an entire function. Furthermore, there exists a positive integer N, such that if we set*

$$\Lambda(E, s) = N^{s/2} (2\pi)^{-s} \Gamma(s) L(E, s) \ ,$$

then we have the following functional equation:

$$\Lambda(E, 2 - s) = \pm \Lambda(E, s) \ .$$

In this case, the Riemann hypothesis states that apart from the trivial zeros at non-positive integers, the zeros of $L(E, s)$ all lie on the critical line $\mathrm{Re}\, s = 1$.

The number N occurring in Conjecture 7.3.4 is a very important invariant of the curve. It is called the (analytic) conductor of E. From work of Carayol [Car], it follows that it must be equal to the (geometric) conductor of E which can be defined without reference to any conjectures. It suffices to say that it has the form $\prod_p p^{e_p}$, where the product is over primes of bad reduction, and for $p > 3$, $e_p = 1$ if E has multiplicative reduction at p, $e_p = 2$ if E has additive reduction. For $p \leq 3$, the recipe is more complicated and is given in Section 7.5.

One can also give a recipe for the $\pm$ sign occurring in the functional equation.

7.3.3 The Taniyama-Weil Conjecture

Now if the reader has a little acquaintance with modular forms, he will notice that the conjectured form of the functional equation of $L(E, s)$ is the same as the functional equation for the Mellin transform of a modular form of weight 2 over the group

$$\Gamma_0(N) = \left\{ \begin{pmatrix} a & b \\ c & d \end{pmatrix} \in \mathrm{SL}_2(\mathbb{Z}), c \equiv 0 \pmod{N} \right\}$$

(see [Lang4], [Ogg] or [Zag] for all relevant definitions about modular forms). Indeed, one can prove the following

Theorem 7.3.5. *Let f be a modular cusp form of weight 2 on the group $\Gamma_0(N)$ (equivalently $f \frac{dq}{q}$ is a differential of the first kind on $X_0(N) = \overline{\mathcal{H}/\Gamma_0(N)}$). Assume that f is a normalized newform (hence, in particular,* an eigenform for the Hecke operators) and that f has rational Fourier coefficients. Then there exists an elliptic curve E defined over $\mathbb{Q}$ such that $f = f_E$, i.e. such that the Mellin transform of $f(it/\sqrt{N})$ is equal to $\Lambda(E, s)$.

Such a curve E is called a modular elliptic curve, and is a natural quotient of the Jacobian of the curve $X_0(N)$. Since analytic continuation and functional equations are trivial consequences of the modular invariance of modular forms we obtain:

Corollary 7.3.6. *Let E be a modular elliptic curve, and let $f = \sum_{n\geq 1} a_n q^n$ be the corresponding cusp form. Then Conjecture 7.3.4 is true for the curve E. In addition, it is known from Atkin-Lehner theory that one must have $f(-1/(N\tau)) = -\varepsilon N\tau^2 f(\tau)$ with $\varepsilon = \pm 1$. Then the functional equation is*

$$\Lambda(E, 2 - s) = \varepsilon\Lambda(E, s) .$$

(Please note the minus sign in the formula for $f(-1/(N\tau))$ which causes confusion and many mistakes in tables.) The number ε is called the sign of the functional equation.

With Theorem 7.3.5 in mind, it is natural to ask if the converse is true, i.e. whether every elliptic curve over $\mathbb{Q}$ is modular. This conjecture was first set forth by Taniyama. Its full importance and plausibility was understood only after Weil proved the following theorem, which we state only in an imprecise form (the precise statement can be found e.g. in [Ogg]):

Theorem 7.3.7 (Weil). *Let $f(\tau) = \sum_{n\geq 1} a_n q^n$, and for all primitive Dirichlet characters χ of conductor m set*

$$L(f, \chi, s) = \sum_{n\geq 1} \frac{a_n\chi(n)}{n^s} ,$$

$$\Lambda(f, \chi, s) = |Nm^2|^{s/2}(2\pi)^{-s}\Gamma(s)L(f, \chi, s) .$$

Assume that these functions satisfy functional equations of the following form:

$$\Lambda(f, \chi, 2 - s) = w(\chi)\Lambda(f, \overline{\chi}, s) ,$$

where $w(\chi)$ has modulus one, and assume that as χ varies, $w(\chi)$ satisfies certain compatibility conditions (being precise here would carry us a little too far). Then f is a modular form of weight 2 over $\Gamma_0(N)$.

Because of this theorem, the above conjecture becomes much more plausible. The Taniyama-Weil conjecture is then as follows:

Conjecture 7.3.8 (Taniyama-Weil). *Let E be an elliptic curve over $\mathbb{Q}$, let $L(E, s) = \sum_{n\geq 1} a_n n^{-s}$ be its L-series, and let $f_E(\tau) = \sum_{n\geq 1} a_n q^n$, so that* the Mellin transform of $f_E(it/\sqrt{N})$ is equal to $\Lambda(E,s)$. Then f is a cusp form of weight 2 on $\Gamma_0(N)$ which is an eigenfunction of the Hecke operators. Furthermore, there exists a morphism ϕ of curves from $X_0(N)$ to E, defined over $\mathbb{Q}$, such that the inverse image by ϕ of the differential $dx/(2y + a_1 x + a_3)$ is the differential $c(2i\pi)f(\tau)d\tau = cf(\tau)dq/q$, where c is some constant.

Note that the constant c, called Manin's constant, is conjectured to be always equal to ± 1 when ϕ is a "strong Weil parametrization" of E (see [Sil]).

A curve satisfying the Taniyama-Weil conjecture was called above a modular elliptic curve. Since this may lead to some confusion with modular curves (the curves $X_0(N)$) which are in general not elliptic, they are called Weil curves (which incidentally seems a little unfair to Taniyama).

The main theorem concerning this conjecture is Wiles's celebrated theorem, which states than when N is squarefree, the conjecture is true (see [Wiles], [Tay-Wil]). This result has been generalized by Diamond to the case where N is only assumed not to be divisible by 9 and 25. In addition, using Weil's Theorem 7.3.7, it was proved long ago by Shimura (see [Shi1] and [Shi2]) that it is true for elliptic curves with complex multiplication.

There is also a recent conjecture of Serre (see [Ser1]), which roughly states that any odd 2-dimensional representation of the Galois group $\mathrm{Gal}(\overline{\mathbb{Q}}/\mathbb{Q})$ over a finite field must come from a modular form. It can be shown that Serre's conjecture implies the Taniyama-Weil conjecture.

The Taniyama-Weil conjecture, and hence the Taylor-Wiles proof, is mainly important for its own sake. However, it has attracted a lot of attention because of a deep result due to Ribet [Rib], saying that the Taniyama-Weil conjecture for squarefree N implies the full strength of Fermat's last "theorem" (FLT): if $x^n + y^n = z^n$ with x, y, z non-zero integers, then one must have $n \geq 2$. Thanks to Wiles, this is now really a theorem. Although it is not so interesting in itself, FLT has had amazing consequences on the development of number theory, since it is in large part responsible for the remarkable achievements of algebraic number theorists in the nineteenth century, and also as a further motivation for the study of elliptic curves, thanks to Ribet's result.

7.3.4 The Birch and Swinnerton-Dyer Conjecture

The other conjecture on elliptic curves which is of fundamental importance was stated by Birch and Swinnerton-Dyer after doing quite a lot of computer calculations on elliptic curves (see [Bir-SwD1], [Bir-SwD2]). For the remaining of this paragraph, we assume that we are dealing with a curve E defined over $\mathbb{Q}$ and satisfying Conjecture 7.3.4, for example a curve with complex multiplication, or more generally a Weil curve. (The initial computations of Birch and Swinnerton-Dyer were done on curves with complex multiplication).

Recall that we defined in a purely algebraic way the rank of an elliptic curve. A weak version of the Birch and Swinnerton-Dyer Conjecture (BSD) is that the rank is positive (i.e. $E(\mathbb{Q})$ is infinite) if and only if $L(E,1) = 0$. This is quite remarkable, and illustrates the fact that the function $L(E, s)$ which is obtained by putting together local data for every prime p, conjecturally gives information on global data, i.e. on the rational points.

The precise statement of the Birch and Swinnerton-Dyer conjecture is as follows:

Conjecture 7.3.9 (Birch and Swinnerton-Dyer). *Let E be an elliptic curve over $\mathbb{Q}$, and assume that Conjecture 7.3.4 (analytic continuation essentially) is true for E. Then if r is the rank of E, the function $L(E, s)$ has a zero of order exactly r at $s = 1$, and in addition*

$$\lim_{s \to 1} (s - 1)^{-r} L(E, s) = \Omega |\text{III}(E/\mathbb{Q})| R(E/\mathbb{Q}) |E(\mathbb{Q})_{\text{tors}}|^{-2} \prod_p c_p ,$$

where Ω is a real period of E, $R(E/\mathbb{Q})$ is the so-called regulator of E, which is an $r \times r$ determinant formed by pairing in a suitable way a basis of the non-torsion points, the product is over the primes of bad reduction, c_p are small integers, and $\text{III}(E/\mathbb{Q})$ is the so-called Tate-Shafarevitch group of E.

It would carry us too far to explain in detail these quantities. Note only that the only quantity which is difficult to compute (in addition to the rank r) is the Tate-Shafarevitch group. In Sections 7.4 and 7.5 we will give algorithms to compute all the quantities which enter into this conjecture, except for $|\text{III}(E/\mathbb{Q})|$ which is then obtained by division (the result must be an integer, and in fact even a square, and this gives a check on the computations). More precisely, Section 7.5.3 gives algorithms for computing $\lim_{s \to 1}(s-1)^{-r} L(E, s)$, the quantities Ω and $|E(\mathbb{Q})_{\text{tors}}|$ are computed using Algorithms 7.4.7 and 7.5.5, the regulator $R(E/\mathbb{Q})$ is obtained by computing a determinant of height pairings of a basis of the torsion-free part of $E(\mathbb{Q})$, these heights being computed using Algorithms 7.5.6 and 7.5.7. Finally, the c_p are obtained by using Algorithm 7.5.1 if $p \geq 5$ and Algorithm 7.5.2 if $p = 2$ or 3.

Note that the above computational descriptions assume that we know a basis of the torsion-free part of $E(\mathbb{Q})$ and hence, in particular, the rank r, and that this is in general quite difficult.

The reader should compare Conjecture 7.3.9 with the corresponding result for the 0-dimensional case, i.e. Theorem 4.9.12. Dedekind's formula at $s = 0$ is very similar to the BSD formula, with the regulator and torsion points playing the same role, and with the class group replaced by the Tate-Shafarevitch group, the units of K being of course analogous to the rational points.

Apart from numerous numerical verifications of BSD, few results have been obtained on BSD, and all are very deep. For example, only in 1987 was it proved by Rubin and Kolyvagin (see [Kol1], [Kol2], [Rub]) that III is finite for certain elliptic curves. The first result on BSD was obtained in 1977 by Coates and Wiles [Coa-Wil] who showed that if E has complex multiplication and if $E(\mathbb{Q})$ is infinite, then $L(E, 1) = 0$. Further results have been obtained, in particular by Gross-Zagier, Rubin and Kolyvagin (see [Gro-Zag2], [GKZ], [Kol1], [Kol2]). For example, the following is now known:

Theorem 7.3.10. *Let E be a Weil curve. Then*

(1) *If $L(E,1) \neq 0$ then $r = 0$.*
(2) *If $L(E,1) = 0$ and $L'(E,1) \neq 0$ then $r = 1$.*

Furthermore, in both these cases $|\text{III}|$ is finite, and up to some simple factors divides the conjectural $|\text{III}|$ involved in BSD.

The present status of BSD is essentially that very little is known when the rank is greater than or equal to 2.

Another conjecture about the rank is that it is unbounded. This seems quite plausible. Using a construction of J.-F. Mestre (see [Mes3] and Exercise 9), Nagao has obtained an infinite family of curves of rank greater or equal to 13 (see [Nag]), and Mestre himself has just obtained an infinite family of curves of rank greater or equal to 14 (see [Mes5]). Furthermore, using Mestre's construction, several authors have obtained individual curves of much higher rank, the current record being rank 22 by Fermigier (see [Mes4], [Fer1], [Nag-Kou] and [Fer2]).

7.4 Algorithms for Elliptic Curves

The previous sections finish up our survey of results and conjectures about elliptic curves. Although the only results which we will need in what follows are the results giving the group law, and Theorems 7.2.14 and 7.2.15 giving basic properties of curves with complex multiplication, elliptic curves are a fascinating field of study *per se*, so we want to describe a number of algorithms to work on them. Most of the algorithms will be given without proof since this would carry us too far. Note that these algorithms are for the most part scattered in the literature, but others are part of the folklore or are new. I am particularly indebted to J.-F. Mestre and D. Bernardi for many of the algorithms of this section. The most detailed collection of algorithms on elliptic curves can be found in the recent book of Cremona [Cre].

7.4.1 Algorithms for Elliptic Curves over $\mathbb{C}$

The problems that we want to solve here are the following.

(1) Given ω_1 and ω_2, compute the coefficients g_2 and g_3 of the Weierstraß equation of the corresponding curve.
(2) Given ω_1 and ω_2 and a complex number z, compute $\wp(z)$ and $\wp'(z)$.
(3) Conversely given g_2 and g_3 such that $g_2^3 - 27g_3^2 \neq 0$, compute ω_1 and ω_2 (which are unique only up to an element of $\text{SL}_2(\mathbb{Z})$).

(4) Similarly, given g_2, g_3 and a point (x, y) on the corresponding Weierstraß curve, compute the complex number z (unique up to addition of an element of the period lattice generated by ω_1 and ω_2) such that $x = \wp(z)$ and $y = \wp'(z)$.

If necessary, after exchanging ω_1 and ω_2, we may assume that $\mathrm{Im}(\omega_2/\omega_1) > 0$, i.e. if we set $\tau = \omega_2/\omega_1$ then $\tau \in \mathcal{H}$. As usual, we always set $q = e^{2i\pi\tau}$, and we have $|q| < 1$ when $\tau \in \mathcal{H}$. Then we have the following proposition:

Proposition 7.4.1. *We have*

$$g_2 = \frac{1}{12}\left(\frac{2\pi}{\omega_1}\right)^4\left(1 + 240\sum_{n\geq 1}\frac{n^3 q^n}{1 - q^n}\right)$$

and also

$$g_3 = \frac{1}{216}\left(\frac{2\pi}{\omega_1}\right)^6\left(1 - 504\sum_{n\geq 1}\frac{n^5 q^n}{1 - q^n}\right).$$

This could already be used to compute g_2 and g_3 reasonably efficiently, but it would be slow when τ is close to the real line. In this case, one should first find the complex number τ' belonging to the fundamental domain $\mathcal{F}$ which is equivalent to τ, compute g_2 and g_3 for τ', and then come back to τ using the (trivial) transformation laws of g_2 and g_3, i.e. $g_k(a\omega_1 + b\omega_2, c\omega_1 + d\omega_2) = g_k(\omega_1, \omega_2)$ when $\begin{pmatrix} a & b \\ c & d \end{pmatrix} \in \mathrm{SL}_2(\mathbb{Z})$. This leads to the following algorithms.

Algorithm 7.4.2 (Reduction). Given $\tau \in \mathcal{H}$, this algorithm outputs the unique τ' equivalent to τ under the action of $\mathrm{SL}_2(\mathbb{Z})$ and which belongs to the standard fundamental domain $\mathcal{F}$, as well as the matrix $A \in \mathrm{SL}_2(\mathbb{Z})$ such that $\tau' = A\tau$.

1. [Initialize] Set $A \leftarrow \begin{pmatrix} 1 & 0 \\ 0 & 1 \end{pmatrix}$.

2. [Reduce real part] Let $n \leftarrow \lfloor \mathrm{Re}(\tau) \rceil$, $\tau \leftarrow \tau - n$, $A \leftarrow \begin{pmatrix} 1 & -n \\ 0 & 1 \end{pmatrix} \cdot A$.

3. [Finished] Set $m \leftarrow \tau\bar{\tau}$. If $m \geq 1$, output τ and A and terminate the algorithm. Otherwise set $\tau \leftarrow -\bar{\tau}/m$, $A \leftarrow \begin{pmatrix} 0 & -1 \\ 1 & 0 \end{pmatrix} \cdot A$ and go to step 2.

This is of course closely related to the reduction algorithm for positive definite quadratic forms (Algorithm 5.4.2), as well as to Gauss's lattice reduction algorithm in dimension 2 (Algorithm 1.3.14).

We can now give the algorithm for computing g_2 and g_3.

Algorithm 7.4.3 (g_2 and g_3). Given ω_1 and ω_2 generating a lattice L, this algorithm computes the coefficients g_2 and g_3 of the Weierstraß equation of the elliptic curve $\mathbb{C}/L$.

1. [Initialize] If $\mathrm{Im}(\omega_2/\omega_1) < 0$, exchange ω_1 and ω_2. Then set $\tau \leftarrow \omega_2/\omega_1$.

2. [Reduce] Using Algorithm 7.4.2, find a matrix $A = \begin{pmatrix} a & b \\ c & d \end{pmatrix} \in \mathrm{SL}_2(\mathbb{Z})$ such that $\tau' = A\tau$ is in the fundamental domain $\mathcal{F}$. Set $q' = e^{2i\pi\tau'}$.

3. [Compute] Compute g_2 and g_3 using the formulas given in Proposition 7.4.1, replacing q by q' and ω_1 by $c\omega_2 + d\omega_1$, and terminate the algorithm.

Since $\tau' \in \mathcal{F}$, we have $\mathrm{Im}\,\tau' \geq \sqrt{3}/2$ hence $|q| \leq e^{-\pi\sqrt{3}} \approx 4.33 \cdot 10^{-3}$, so the convergence of the series, although linear, will be very fast.

We can also use the power series expansions to compute $\wp(z)$ and $\wp'(z)$:

Proposition 7.4.4. *Set* $\tau = \omega_2/\omega_1 \in \mathcal{H}$, $q = e^{2i\pi\tau}$ *and* $u = e^{2i\pi z/\omega_1}$. *Then*

$$\wp(z) = \left(\frac{2i\pi}{\omega_1}\right)^2 \left(\frac{1}{12} + \frac{u}{(1-u)^2}\right.$$
$$\left. + \sum_{n=1}^{\infty} q^n \left(u\left(\frac{1}{(1-q^n u)^2} + \frac{1}{(q^n - u)^2}\right) - \frac{2}{(1-q^n)^2}\right)\right)$$

and

$$\wp'(z) = \left(\frac{2i\pi}{\omega_1}\right)^3 u\left(\frac{1+u}{(1-u)^3} + \sum_{n=1}^{\infty} q^n \left(\frac{1+q^n u}{(1-q^n u)^3} + \frac{q^n + u}{(q^n - u)^3}\right)\right) .$$

Note that the formula for $\wp'(z)$ in the first printing of [Sil] is incorrect.

As usual, we must do reductions of τ and z before applying the crude formulas, and this gives the following algorithm.

Algorithm 7.4.5 ($\wp(z)$ and $\wp'(z)$). Given ω_1 and ω_2 generating a lattice L, and $z \in \mathbb{C}$, this algorithm computes $\wp(z)$ and $\wp'(z)$.

1. [Initialize and reduce] If $\mathrm{Im}(\omega_2/\omega_1) < 0$, exchange ω_1 and ω_2. Then set $\tau \leftarrow \omega_2/\omega_1$. Using Algorithm 7.4.2, find a matrix $A = \begin{pmatrix} a & b \\ c & d \end{pmatrix} \in \mathrm{SL}_2(\mathbb{Z})$ such that $A\tau$ is in the the fundamental domain $\mathcal{F}$. Finally, set $\tau \leftarrow A\tau$ and $\omega_1 \leftarrow c\omega_2 + d\omega_1$.

2. [Reduce z] Set $z \leftarrow z/\omega_1$, $n \leftarrow \lfloor \mathrm{Im}(z)/\mathrm{Im}(\tau) \rfloor$, $z \leftarrow z - n\tau$ and $z \leftarrow z - \lfloor \mathrm{Re}(z) \rfloor$.

3. [Compute] If $z = 0$, output a message saying that $z \in L$. Otherwise compute $\wp(z)$ and $\wp'(z)$ using the formulas given in Proposition 7.4.4 (with $u = e^{2i\pi z}$ since we have already divided z by ω_1) and terminate the algorithm.

Remark. For the above computations it is more efficient to use the formulas that link elliptic functions with the σ function, since the latter are theta series and so can be computed efficiently. For reasonable accuracy however (say less than 100 decimal digits) the above formulas suffice.

We now consider the inverse problems. Given g_2 and g_3 defining a Weierstraß equation, we want to compute a basis ω_1 and ω_2 of the corresponding lattice.

First, recall the definition of the Arithmetic-Geometric Mean (AGM) of two numbers.

Definition 7.4.6. *Let a and b be two positive real numbers. The Arithmetic-Geometric mean of a and b, denoted by $\mathrm{AGM}(a, b)$ is defined as the common limit of the two sequences a_n and b_n defined by $a_0 = a$, $b_0 = b$, $a_{n+1} = (a_n + b_n)/2$ and $b_{n+1} = \sqrt{a_n b_n}$.*

It is an easy exercise to show that these sequences converge and that they have a common limit $\mathrm{AGM}(a, b)$ (see Exercise 10). It can also be proved quite easily that

$$\frac{\pi}{2\,\mathrm{AGM}(a, b)} = \int_0^{\pi/2} \frac{dt}{\sqrt{a^2 \cos^2 t + b^2 \sin^2 t}}$$

(see Exercise 11) and this can easily be transformed into an elliptic integral, which explains the relevance of the AGM to our problems. For many more details on the AGM, I refer to the marvelous book of Borwein and Borwein [Bor-Bor].

Apart from their relevance to elliptic integrals, the fundamental property of the AGM sequences a_n and b_n is that they converge quadratically, i.e. the number of significant decimals approximately *doubles* with each iteration (see Exercise 10). For example, there exists AGM-related methods for computing π to high precision (see again [Bor-Bor]), and since $2^{20} > 10^6$ only 20 iterations are needed to compute 1000000 decimals of π!

The AGM can also be considered when a and b are not positive real numbers but are arbitrary complex numbers. Here the situation is more complicated, but can be summarized as follows. At each stage of the iteration, we must choose some square root of $a_n b_n$. Assume that for n sufficiently large the same branch of the square root is taken (for example the principal branch, but it can be any other branch). Then the sequences again converge quadratically to the same limit, but this limit of course now depends on the choices made for the square roots. In addition, the set of values of $\pi / \mathrm{AGM}(a, b)$ (where now $\mathrm{AGM}(a, b)$ has infinitely many values) together with 0 form a *lattice* L in $\mathbb{C}$. The precise link with elliptic curves is as follows. Let e_1, e_2, e_3 be the three complex roots of the polynomial $4x^3 - g_2 x - g_3$ such that $y^2 = 4x^3 - g_2 x - g_3$ defines an elliptic curve E. Then, when the AGM runs through all its possible determinations $\pi / \mathrm{AGM}(\sqrt{e_1 - e_3}, \sqrt{e_1 - e_2})$ gives all the lattice points (except 0) of the lattice L such that $E \simeq \mathbb{C}/L$.

We however will usually use the AGM over the positive real numbers, where it is single-valued, since the elliptic curves that we will mainly consider are defined over $\mathbf{R}$, and even over $\mathbf{Q}$. In this case, the following algorithm gives a basis of the period lattice L. Since our curves will usually be given by a generalized Weierstraß equation $y^2 + a_1 xy + a_3 y = x^3 + a_2 x^2 + a_4 x + a_6$ instead of the simpler equation $Y^2 = 4X^3 - g_2 X - g_3$, we give the algorithm in that context.

Algorithm 7.4.7 (Periods of an Elliptic Curve over $\mathbf{R}$). Given real numbers $a_1, \ldots, a_6$, this algorithm computes the basis (ω_1, ω_2) of the period lattice of E such that ω_1 is a positive real number and ω_2/ω_1 has positive imaginary part and a real part equal to 0 or $-1/2$.

1. [Initialize] Using Formulas (7.1), compute b_2, b_4, b_6 and Δ, and if $\Delta < 0$ go to step 3.

2. [Disconnected case] Let e_1, e_2 and e_3 be the three real roots of the polynomial $4x^3 + b_2 x^2 + 2b_4 x + b_6 = 0$ with $e_1 > e_2 > e_3$. Set $\omega_1 \leftarrow \pi/\operatorname{AGM}(\sqrt{e_1 - e_3}, \sqrt{e_1 - e_2})$, $\omega_2 \leftarrow i\pi/\operatorname{AGM}(\sqrt{e_1 - e_3}, \sqrt{e_2 - e_3})$ and terminate the algorithm.

3. [Connected case] Let e_1 be the unique real root of $4x^3 + b_2 x^2 + 2b_4 x + b_6 = 0$. Set $a \leftarrow 3e_1 + b_2/4$ and $b \leftarrow \sqrt{3e_1^2 + (b_2/2)e_1 + b_4/2}$. Then set $\omega_1 \leftarrow 2\pi/\operatorname{AGM}(2\sqrt{b}, \sqrt{2b + a})$, $\omega_2 \leftarrow -\omega_1/2 + i\pi/\operatorname{AGM}(2\sqrt{b}, \sqrt{2b - a})$ and terminate the algorithm.

Note that the "real period" Ω occurring in the Birch and Swinnerton-Dyer conjecture 7.3.9 is $2\omega_1$ when $\Delta > 0$, and ω_1 otherwise, and that ω_2/ω_1 is not necessarily in the standard fundamental domain for $\mathcal{H}/\operatorname{SL}_2(\mathbf{Z})$.

Finally, we need an algorithm to compute the functional inverse of the $\wp$ function.

The Weierstraß parametrization $(\wp(z) : \wp'(z) : 1)$ can be seen as an exponential morphism from the universal covering $\mathbf{C}$ of $E(\mathbf{C})$. It can be considered as the composition of three maps:

$$\mathbf{C} \to \mathbf{C}^* \to \mathbf{C}^*/q^{\mathbf{Z}} \to E(\mathbf{C})$$

$$z \mapsto u = e^{2i\pi z/\omega_1} \mapsto u \mod q^{\mathbf{Z}} \mapsto (\wp(z), \wp'(z)) \ ,$$

the last one being an isomorphism. Its functional inverse, which we can naturally call the elliptic logarithm, is thus a multi-valued function. In fact, Algorithm 7.4.7 can be extended so as to find the inverse image of a given point. Since square roots occur, this give rise to the same indeterminacy as before, i.e. the point z is defined only up to addition of a point of the period lattice L. As in the previous algorithm, taking the positive square root in the real case gives directly the unique u such that $|q| < |u| \le 1$. We will therefore only give the description for a real point.

Algorithm 7.4.8 (Elliptic Logarithm). Given real numbers $a_1, \ldots, a_6$ defining a generalized Weierstraß equation for an elliptic curve E and a point $P = (x, y)$ on $E(\mathbb{R})$, this algorithm computes the unique complex number z such that $\wp(z) = x + b_2/12$ and $\wp'(z) = 2y + a_1 x + a_3$, where $\wp$ is the Weierstraß function corresponding to the period lattice of E, and which satisfies the following additional conditions. Either z is real and $0 \leq z < \omega_1$, or $\Delta > 0$, $z - \omega_2/2$ is real and satisfies $0 \leq z - \omega_2/2 < \omega_1$.

1. [Initialize] Using Formulas (7.1), compute b_2, b_4, b_6 and Δ. If $\Delta < 0$ go to step 6.

2. [Disconnected case] Let e_1, e_2 and e_3 be the three real roots of the polynomial $4x^3 + b_2 x^2 + 2b_4 x + b_6 = 0$ with $e_1 > e_2 > e_3$. Set $a \leftarrow \sqrt{e_1 - e_3}$ and $b \leftarrow \sqrt{e_1 - e_2}$. If $x < e_1$ set $f \leftarrow 1$, $\lambda \leftarrow y/(x - e_3)$ and $x \leftarrow \lambda^2 + a_1 \lambda - a_2 - x - e_3$, otherwise set $f \leftarrow 0$. Finally, set $c \leftarrow \sqrt{x - e_3}$.

3. [Loop] Repeat $(a, b, c) \leftarrow ((a + b)/2, \sqrt{ab}, (c + \sqrt{c^2 + b^2 - a^2})/2)$ until the difference $a - b$ is sufficiently small.

4. [Connected component] If $f = 0$ and $2y + a_1 x + a_3 < 0$ or $f = 1$ and $2y + a_1 x + a_3 \geq 0$ set $z \leftarrow \arcsin(a/c)/a$. Otherwise set $z \leftarrow (\pi - \arcsin(a/c))/a$. If $f = 0$ output z and terminate the algorithm.

5. [Other component] Compute $\omega_2 \leftarrow i\pi / \mathrm{AGM}(\sqrt{e_1 - e_3}, \sqrt{e_2 - e_3})$ as in Algorithm 7.4.7 (unless of course this has already been done). Output $z + \omega_2/2$ and terminate the algorithm.

6. [Connected case] Let e_1 be the unique real root of $4x^3 + b_2 x^2 + 2b_4 x + b_6 = 0$. Set $\beta \leftarrow \sqrt{3e_1^2 + (b_2/2)e_1 + b_4/2}$, $\alpha \leftarrow 3e_1 + b_2/4$, $a \leftarrow 2\sqrt{\beta}$, $b \leftarrow \sqrt{\alpha + 2\beta}$ and $c \leftarrow (x - e_1 + \beta)/\sqrt{x - e_1}$.

7. [Loop] Repeat $(a, b, c) \leftarrow ((a + b)/2, \sqrt{ab}, (c + \sqrt{c^2 + b^2 - a^2})/2)$ until the difference $a - b$ is sufficiently small.

8. [Terminate] If $(2y + a_1 x + a_3)((x - e_1)^2 - \beta^2) < 0$, set $z \leftarrow \arcsin(a/c)/a$ otherwise set $z \leftarrow (\pi - \arcsin(a/c))/a$. If $2y + a_1 x + a_3 > 0$, set $z \leftarrow z + \pi/a$. Output z and terminate the algorithm.

Note that we could have avoided the extra AGM in step 5, but this would have necessitated using the complex AGM and arcsin. Hence, it is simpler to proceed as above. In addition, in practice ω_2 will have already been computed previously and so there is not really any extra AGM to compute.

7.4.2 Algorithm for Reducing a General Cubic

The problem that we want to solve here is the following. Given a general non-singular irreducible projective plane cubic over an arbitrary field K, say

$$s_1 U^3 + s_2 U^2 V + s_3 U V^2 + s_4 V^3$$
$$+ (s_5 U^2 + s_6 U V + s_7 V^2)W + (s_8 U + s_9 V)W^2 + s_{10} W^3 ,$$

where $(U : V : W)$ are the projective coordinates, and a K-rational point $P_0 = (u_0 : v_0 : w_0)$ on the curve, find a birational transformation which transforms this into a generalized Weierstraß equation.

We will explain how to do this in the generic situation (i.e. assuming that no expression vanishes, that our points are in general position, etc ...), and then give the algorithm in general. We also assume for simplicity that our field is of characteristic different from 2.

We first make a couple of reductions. Since the curve is non-singular, its partial derivatives with respect to U and V cannot vanish simultaneously on the curve. Hence, by exchanging if necessary U and V, we may assume that it is the derivative with respect to V at P_0 which is different from zero. Consider now the tangent at P_0 to the curve. This tangent will then have a (rational) slope λ, and intersects the curve in a unique third point which we will call $P_1 = (u_1 : v_1 : w_1)$. After making the change of coordinates $(U', V') = (U - u_1, V - v_1)$ we may assume that P_1 has coordinates $(0 : 0 : 1)$, i.e. is at the origin, or in other words that the new value of s_{10} is equal to zero. We now have the following theorem (for simplicity we state everything with affine coordinates, but the conversion to projective coordinates is easy to make).

Theorem 7.4.9. *We keep the above notations and reductions. Call $c_j(U, V)$ the coefficient of degree W^{3-j} in the equation of the curve (so that c_j is a homogeneous polynomial of degree j), and set*

$$d(U, V) = c_2(U, V)^2 - 4c_1(U, V)c_3(U, V) \ .$$

Furthermore, if λ is the slope of the tangent at P_0 as defined above, set

$$d(U, \lambda U + 1) = AU^4 + BU^3 + CU^2 + DU + E \ .$$

Then

(1) *We have $A = 0$ and $B \neq 0$.*
(2) *The transformation*

$$X = \frac{BU}{V - \lambda U}$$

$$Y = \frac{B}{(V - \lambda U)^2} \left(2c_3(U, V) + c_2(U, V)\right)$$

is a birational transformation whose inverse is given by

$$U = X \frac{BY - c_2(X, \lambda X + B)}{2c_3(X, \lambda X + B)}$$

$$V = (\lambda X + B) \frac{BY - c_2(X, \lambda X + B)}{2c_3(X, \lambda X + B)} \ .$$

(3) *This birational map transforms the equation of the curve into the Weierstraß equation*

$$Y^2 = X^3 + CX^2 + BDX + B^2 E \ .$$

Proof. The line $V = \lambda U$ is the new equation of the tangent at P_0 that we started from. This means that it is tangent to the curve. Solving for U, one has the trivial solution $U = 0$ corresponding to the point P_1, and the two other roots must be equal. In other words we must have $d(1, \lambda) = 0$, since this is the discriminant of the quadratic equation. Since clearly $A = d(1, \lambda)$, this shows that $A = 0$.

Now solving for the double root, we see that the coordinates of P_0 (in the new coordinate system of course) are $(\alpha, \lambda\alpha)$, where we set

$$\alpha = -\frac{c_2(1, \lambda)}{2c_3(1, \lambda)} .$$

Now I claim that we have the equalities

$$B = \frac{\partial d}{\partial V}(1, \lambda) = -4c_3(1, \lambda)\frac{\partial f}{\partial V}(\alpha, \lambda\alpha) ,$$

where $f(U, V) = 0$ is the (affine) equation of the curve. Assuming this for a moment, this last partial derivative is the partial derivative of f with respect to V at the point P_0, hence is different from zero by the first reduction made above. Furthermore, $c_3(1, \lambda) \neq 0$ also since otherwise P_0 would be at infinity and we have assumed (for the moment) that P_0 is in general position. This shows that $B \neq 0$ and hence the first part of the theorem. To prove my claim, note that the first equality is trivial. For the second, let us temporarily abbreviate $c_j(1, \lambda)$ to c_j and $\frac{\partial c_j}{\partial V}(1, \lambda)$ to c_j'. Then by homogeneity, one sees immediately that

$$\frac{\partial f}{\partial V}(\alpha, \lambda\alpha) = \frac{c_2^2 c_3' - 2c_2 c_3 c_2' + 4c_1' c_3^2}{4c_3^2} .$$

We know that $A = c_2^2 - 4c_1 c_3 = 0$ (and this can be checked once again explicitly if desired). Therefore we can replace c_2^2 by $4c_1 c_3$, thus giving

$$\frac{\partial f}{\partial V}(\alpha, \lambda\alpha) = \frac{4c_1 c_3' + 4c_1' c_3 - 2c_2 c_2'}{4c_3}$$

and the claim follows by differentiating the formula $d = c_2^2 - 4c_1 c_3$.

By simple replacement, one sees immediately that, since $B \neq 0$, the maps $(U, V) \rightarrow (X, Y)$ and $(X, Y) \rightarrow (U, V)$ are inverse to one another, hence the second part is clear.

For the last part, we simply replace U and V by their expressions in terms of X and Y. We can multiply by $c_3(X, \lambda X + B)$ (which is not identically zero), and we can also simplify the resulting equation by $BY - c_2(X, \lambda X + B)$ since B is different from zero and the curve is irreducible (why?). After expanding and simplifying we obtain the equation

$$B^2 Y^2 = d(X, \lambda X + B) \ .$$

Now since $d(U, V)$ is a homogeneous polynomial of degree 4, one sees immediately that

$$d(X, \lambda X + B) = B^2 X^3 + C B^2 X^2 + D B^3 X + E B^4 \ ,$$

thus finishing the proof of the theorem.   □

It is now easy to generalize this theorem to the case where the point P_0 is not in general position, and this leads to the following algorithm, which we give in projective coordinates.

Algorithm 7.4.10 (Reduction of a General Cubic). Let K be a field of characteristic different from 2, and let $f(U, V, W) = 0$ be the equation of a general cubic, where

$$f(U, V, W) = s_1 U^3 + s_2 U^2 V + s_3 U V^2 + s_4 V^3$$
$$+ (s_5 U^2 + s_6 U V + s_7 V^2) W + (s_8 U + s_9 V) W^2 + s_{10} W^3 \ .$$

Finally, let $P_0 = (u_0 : v_0 : w_0)$ be a point on the cubic, i.e. such that $f(u_0, v_0, w_0) = 0$. This algorithm, either outputs a message saying that the curve is singular or reducible, or else gives a Weierstraß equation for the curve and a pair of inverse birational maps which transform one equation into the other. We will call $(X : Y : T)$ the new projective coordinates, and continue to call s_i the coefficients of the transformed equation g during the algorithm.

1. [Initialize] Set $(m_1, m_2, m_3) \leftarrow (U, V, W)$, $(n_1, n_2, n_3) \leftarrow (X, Y, T)$ and $g \leftarrow f$. (Here $(m_1 : m_2 : m_3)(U, V, W)$ and $(n_1 : n_2 : n_3)(X, Y, T)$ will be the pair of inverse birational maps. The assignments given in this algorithm for these maps and for g are formal, i.e. we assign polynomials or rational functions, not values. In addition, it is understood that the modifications of g imply the modifications of the coefficients s_i.)

2. [Send P_0 to $(0 : 0 : 1)$] If $w_0 \neq 0$, set $(m_1, m_2, m_3) \leftarrow (w_0 m_1 - u_0 m_3, w_0 m_2 - v_0 m_3, w_0 m_3)$, $(n_1, n_2, n_3) \leftarrow (w_0 n_1 + u_0 n_3, w_0 n_2 + v_0 n_3, w_0 n_3)$, $g \leftarrow g(w_0 U + u_0 W, w_0 V + v_0 W, w_0 W)$ and go to step 3. Otherwise, if $u_0 \neq 0$, set $(m_1, m_2, m_3) \leftarrow (u_0 m_3, u_0 m_2 - v_0 m_1, u_0 m_1)$, $(n_1, n_2, n_3) \leftarrow (u_0 n_3, u_0 n_2 + v_0 n_3, u_0 n_1)$, $g \leftarrow g(u_0 W, u_0 V + v_0 W, u_0 U)$ and go to step 3. Finally, if $w_0 = u_0 = 0$ (hence $v_0 \neq 0$), exchange m_2 and m_3, n_2 and n_3, and set $g \leftarrow g(U, W, V)$.

3. [Exchange U and V?] (Here $s_{10} = 0$). If $s_8 = s_9 = 0$, output a message saying that the curve is singular at P_0 and terminate the algorithm. Otherwise, if $s_9 = 0$, exchange m_1 and m_2, n_1 and n_2, and set $g \leftarrow g(V, U, W)$.

4. [Send P_1 to $(0 : 0 : 1)$] (Here $s_9 \neq 0$.) Set $\lambda \leftarrow (-s_8/s_9)$, $c_2 \leftarrow s_7 \lambda^2 + s_6 \lambda + s_5$, $c_3 \leftarrow s_4 \lambda^3 + s_3 \lambda^2 + s_2 \lambda + s_1$. Then, if $c_3 \neq 0$, set $(m_1, m_2, m_3) \leftarrow (c_3 m_1 + c_2 m_3, c_3 m_2 + \lambda c_2 m_3, c_3 m_3)$, $(n_1, n_2, n_3) \leftarrow (c_3 n_1 - c_2 n_3, c_3 n_2 - \lambda c_2 n_3, c_3 n_3)$,

$g \leftarrow g(c_3U - c_2W, c_3V - \lambda c_2W, c_3W)$ and go to step 5. Otherwise, if $c_2 = 0$ output a message saying that the curve is reducible and terminate the algorithm. Finally, if $c_3 = 0$ and $c_2 \neq 0$, set $(m_1, m_2, m_3) \leftarrow (m_3, m_2 - \lambda m_1, m_1)$, $(n_1, n_2, n_3) \leftarrow (n_3, n_2 + \lambda n_3, n_1)$ and $g \leftarrow g(W, V + \lambda W, U)$, then set $\lambda \leftarrow 0$.

5. [Apply theorem] (Here we are finally in the situation of the theorem.) Let as in the theorem $c_j(U, V)$ be the coefficient of W^{3-j} in $g(U, V, W)$, and $d(U, V) \leftarrow c_2(U, V)^2 - 4c_1(U, V)c_3(U, V)$. Compute B, C, D and E such that $d(U, \lambda U + 1) = BU^3 + CU^2 + DU + E$. Then set

$$(m_1, m_2, m_3) \leftarrow (Bm_1(m_2 - \lambda m_1)m_3,$$
$$B(2c_3(m_1, m_2) + c_2(m_1, m_2)m_3), (m_2 - \lambda m_1)^2 m_3) ,$$

$$(n_1, n_2, n_3) \leftarrow (n_1(Bn_2n_3 - c_2(n_1, \lambda n_1 + Bn_3)),$$
$$(\lambda n_1 + Bn_3)(Bn_2n_3 - c_2(n_1, \lambda n_1 + Bn_3)), 2c_3(n_1, \lambda n_1 + Bn_3)) .$$

Output the maps $(X, Y, T) \leftarrow (m_1, m_2, m_3)$ and $(U, V, W) \leftarrow (n_1, n_2, n_3)$, the projective Weierstraß equation

$$Y^2T = X^3 + CX^2T + DBXT^2 + EB^2T^3$$

and terminate the algorithm.

7.4.3 Algorithms for Elliptic Curves over $\mathbb{F}_p$

The only algorithms which we will need here are algorithms which count the number of points of an elliptic curve over $\mathbb{F}_p$, or equivalently the numbers a_p such that $|E(\mathbb{F}_p)| = p + 1 - a_p$. We first describe the naïve algorithm which expresses a_p as a sum of Legendre symbols, then give a much faster algorithm using Shanks's baby-step giant-step method and a trick of Mestre.

Counting the number of points over $\mathbb{F}_2$ or $\mathbb{F}_3$ is trivial, so we assume that $p \geq 5$. In particular, we may simplify the Weierstraß equation, i.e. assume that $a_1 = a_2 = a_3 = 0$, so the equation of the curve is of the form $y^2 = x^3 + ax + b$. The curve has one point at infinity $(0 : 1 : 0)$, and then for every $x \in \mathbb{F}_p$, there are $1 + \left(\frac{x^3 + ax + b}{p}\right)$ values of y. Hence we have $N_p = p + 1 + \sum_{x \in \mathbb{F}_p}\left(\frac{x^3 + ax + b}{p}\right)$, thus giving the formula

$$a_p = -\sum_{x \in \mathbb{F}_p}\left(\frac{x^3 + ax + b}{p}\right) .$$

This formula gives a $O(p^{1+o(1)})$ time algorithm for computing a_p, and this is reasonable when p does not exceed 10000, say.

However we can use Shanks's baby step-giant step method to obtain a much better algorithm. By Hasse's theorem, we know that $p + 1 - 2\sqrt{p} <$

$N_p < p + 1 + 2\sqrt{p}$, hence we can apply Algorithm 5.4.1 with $C = p + 1 - 2\sqrt{p}$ and $B = p + 1 + 2\sqrt{p}$. This will give an algorithm which runs in time $(B - C)^{1/2+o(1)} = p^{1/4+o(1)}$, and so will be much faster for large p. Now the reader will recall that one problem with Shanks's method is that if our group is not cyclic, or if we do not start with a generator of the group, we need to do some extra work which is not so easy to implement. There is a nice trick due, I believe to Mestre, which tells us how to do this extra work in a very simple manner.

If one considers all the curves over $\mathbb{F}_p$ defined by $y^2 = x^3 + ad^2x + bd^3$ with $d \neq 0$, then there are exactly two isomorphism classes of such curves: those for which $\left(\frac{d}{p}\right) = 1$ are all isomorphic to the initial curve correspond to $d = 1$, and those for which $\left(\frac{d}{p}\right) = -1$ are also all isomorphic, but to another curve. Call E' one of these other curves. Then one has the following proposition.

Proposition 7.4.11. *Let*

$$E(\mathbb{F}_p) \simeq \mathbb{Z}/d_1\mathbb{Z} \times \mathbb{Z}/d_2\mathbb{Z} \quad and \quad E'(\mathbb{F}_p) \simeq \mathbb{Z}/d_1'\mathbb{Z} \times \mathbb{Z}/d_2'\mathbb{Z}$$

be the Abelian group structures of $E(\mathbb{F}_p)$ and $E'(\mathbb{F}_p)$ respectively, with $d_1 \mid d_2$ and $d_1' \mid d_2'$ (see Proposition 7.1.9). Then for $p > 457$ we have

$$\max(d_2, d_2') > 4\sqrt{p} \; .$$

This proposition shows that on at least one of the two curves E or E' there will be points of order greater than $4\sqrt{p}$, hence according to Hasse's theorem, sufficiently large so as to obtain the cardinality of $E(\mathbb{F}_p)$ (or of $E'(\mathbb{F}_p)$) immediately using Shanks's baby-step giant-step method. In addition, since each value of x gives either two points on one of the curves and none on the other, or one on each, it is clear that if $|E(\mathbb{F}_p)| = p + 1 - a_p$, we have $|E'(\mathbb{F}_p)| = p + 1 + a_p$, so computing one value gives immediately the other one.

This leads to the following algorithm.

Algorithm 7.4.12 (Shanks-Mestre). Given an elliptic curve E over $\mathbb{F}_p$ with $p > 457$ by a Weierstraß equation $y^2 = x^3 + ax + b$, this algorithm computes the a_p such that $|E(\mathbb{F}_p)| = p + 1 - a_p$.

1. [Initialize] Set $x \leftarrow -1$, $A \leftarrow 0$, $B \leftarrow 1$, $k_1 = 0$.

2. [Get next point] (Here we have $|E(\mathbb{F}_p)| \equiv A \pmod{B}$). Repeat $x \leftarrow x + 1$, $d \leftarrow x^3 + ax + b$, $k \leftarrow \left(\frac{d}{p}\right)$ until $k \neq 0$ and $k \neq k_1$. Set $k_1 \leftarrow k$. Finally, if $k_1 = -1$ set $A_1 \leftarrow 2p + 2 - A \bmod B$ else set $A_1 \leftarrow A$.

3. [Find multiple of the order of a point] Let m be the smallest integer such that $m > p + 1 - 2\sqrt{p}$ and $m \equiv A_1 \pmod{B}$. Using Shanks's baby-step giant-step strategy, find an integer n such that $m \leq n < p + 1 + 2\sqrt{p}$, $n \equiv m \pmod{B}$

and such that $n \cdot (xd, d^2) = 0$ on the curve $Y^2 = X^3 + ad^2X + bd^3$ (note that this will be isomorphic to the curve E or E' according to the sign of k_1).

4. [Find order] Factor n, and deduce from this the exact order h of the point (xd, d^2).

5. [Finished?] Using for instance the Chinese remainder algorithm, find the smallest integer h' which is a multiple of h and such that $h' \equiv A_1 \pmod{B}$. If $h' < 4\sqrt{p}$ set $B \leftarrow LCM(B, h)$, then $A \leftarrow h' \bmod B$ if $k_1 = 1$, $A \leftarrow 2p + 2 - h' \bmod B$ if $k_1 = -1$, and go to step 2.

6. [Compute a_p] Let N be the unique multiple of h' such that $p + 1 - 2\sqrt{p} < N < p + 1 + 2\sqrt{p}$. Output $a_p = p + 1 - k_1 N$ and terminate the algorithm.

The running time of this algorithm is $O(p^{1/4+\epsilon})$ for any $\epsilon > 0$, but it is much easier to implement than the algorithm for class numbers because of the simpler group structure. It should be used instead of the algorithm using Legendre symbols as soon as p is greater than 457. Note that one can prove that 457 is best possible, but it is easy to modify slightly the algorithm so that it works for much lower values of p.

Note also that, as in the case of class groups of quadratic fields, we can use the fact that the inverse of a point is trivial to compute, and hence enlarge by a factor $\sqrt{2}$ the size of the giant steps. In other words, in step 3 the size of the giant steps should be taken equal to the integer part of $\sqrt{2\sqrt{p}/B}$.

Another algorithm for computing a_p has been discovered by R. Schoof ([Scho]). What is remarkable about it is that it is a *polynomial time* algorithm, more precisely it runs in time $O(\ln^8 p)$. The initial version did not seem to be very useful in practice, but a lot of progress has been done since.

Schoof's idea, which we will not explain in detail here, is to use the *division polynomials* for the Weierstraß $\wp$ function, i.e. polynomials which express $\wp(nz)$ and $\wp'(nz)$ in terms of $\wp(z)$ and $\wp'(z)$ for integer n (in fact a prime number n). This gives *congruences* for the a_p, and using the Chinese remainder theorem we can glue together these congruences to compute the a_p.

An interesting blend of the baby-step giant-step algorithm and Schoof's algorithm is to compute Schoof-type congruences for a_p modulo a few primes ℓ. If for example we find the congruences modulo 2, 3 and 5, we can divide the search interval by 30 in the algorithm above, and hence this allows the treatment of larger primes.

The main practical problem with Schoof's idea is that the equations giving the division polynomials are of degree $(n^2 - 1)/2$, and this becomes very difficult to handle as soon as n is a little large.

Recently N. Elkies has been able to show that for approximately one half of the primes n, this degree can be reduced to $n + 1$, which is much more manageable. J.-M. Couveignes has also shown how to use n which are powers of small primes and not only primes.

Combining all these ideas, Morain and Lercier (Internet announcement) have been able to deal with a 500-digit prime, which is the current record at the time of this writing.

7.5 Algorithms for Elliptic Curves over $\mathbb{Q}$

7.5.1 Tate's algorithm

Given an elliptic curve E defined over $\mathbb{Q}$, using Algorithm 7.4.10 we can assume that E is given by a generalized Weierstraß equation $y^2 + a_1xy + a_3y = x^3 + a_2x^2 + a_4x + a_6$ with coefficients in $\mathbb{Q}$. We would first like to find a global *minimal* Weierstraß equation of E (see [Sil], [LN476] and Algorithm 7.5.3 for the precise definitions). This will be a canonical way of representing the curve E since this equation exists and is unique. (As already remarked, it is essential at this point that we work over $\mathbb{Q}$ and not over an arbitrary number field.) Note that this is a major difference with the case of equations defining number fields, where no really canonical equation for the field can be found, but only partial approaches such as the pseudo-canonical polynomial given by Algorithm 4.4.12. In addition, it is necessary to know this minimal equation for several other algorithms.

Two elliptic curves with different parameters may be isomorphic over $\mathbb{Q}$. Such an isomorphism must be given by transformations $x = u^2x' + r$, $y = u^3y' + su^2x' + t$, where $u \in \mathbb{Q}^*$, $r, s, t \in \mathbb{Q}$. We obtain a new model for the same elliptic curve. Using the same quantities as those used in Formulas (7.1), the parameters of the new model are given by

$$
\begin{aligned}
&ua_1' = a_1 + 2s, \quad u^2a_2' = a_2 - sa_1 + 3r - s^2 \\
&u^3a_3' = a_3 + ra_1 + 2t \\
&u^4a_4' = a_4 - sa_3 + 2ra_2 - (t + rs)a_1 + 3r^2 - 2st \\
&u^6a_6' = a_6 + ra_4 + r^2a_2 + r^3 - ta_3 - t^2 - rta_1 \\
&u^2b_2' = b_2 + 12r, \quad u^4b_4' = b_4 + rb_2 + 6r^2 \\
&u^6b_6' = b_6 + 2rb_4 + r^2b_2 + 4r^3 \\
&u^8b_8' = b_8 + 3rb_6 + 3r^2b_4 + r^3b_2 + 3r^4 \\
&u^4c_4' = c_4, \quad u^6c_6' = c_6, \quad u^{12}\Delta' = \Delta, \quad j' = j, \quad u^{-1}\omega' = \omega \ .
\end{aligned}
\tag{7.2}
$$

Using these formulas, we may now assume that the coefficients of the equations are integers. We will make this assumption from now on. We first want to find a model for E which is minimal with respect to a given prime p, and we also want to know the type of the fiber at p of the elliptic pencil defined by E over $\mathbb{Z}$ (see [Sil], [LN476]). The possible types are described by symbols known as Kodaira types. They are $I_0, I_\nu, II, III, IV, I_0^*, I_\nu^*, II^*, III^*, IV^*$, where ν

is a positive integer. We need also to compute the coefficient c_p which appears in the formulation of the Birch and Swinnerton-Dyer Conjecture 7.3.9, that is, the index in $E(\mathbb{Q}_p)$ of the group $E^0(\mathbb{Q}_p)$ of points which do not reduce to the singular point.

The following algorithm is due to Tate (cf [LN476]). We specialize his description to the case of rational integers. The situation is a bit simpler when the prime p is greater than 3, so let us start with that case.

Algorithm 7.5.1 (Reduction of an Elliptic Curve Modulo p). Given integers $a_1, \ldots, a_6$ and a prime $p > 3$, this algorithm determines the Kodaira symbol associated with the curve modulo p. In addition, it computes the exponent f of p in the arithmetic conductor of the curve, the index $c = [E(\mathbb{Q}_p) : E^0(\mathbb{Q}_p)]$ and integers u, r, s, t such that $a_1', \ldots, a_6'$ linked to $a_1, \ldots, a_6$ via Formulas (7.2) give a model with the smallest possible power of p in its discriminant.

1. [Initialize] Compute c_4, c_6, Δ and j using Formulas (7.1). If $v_p(j) < 0$ set $k \leftarrow v_p(\Delta) + v_p(j)$ else set $k \leftarrow v_p(\Delta)$.

2. [Minimal?] If $k < 12$ set $u \leftarrow 1$, $r \leftarrow 0$, $s \leftarrow 0$, and $t \leftarrow 0$. Otherwise, set $u \leftarrow p^{\lfloor k/12 \rfloor}$; if a_1 is odd then set $s \leftarrow (u - a_1)/2$ else set $s \leftarrow -a_1/2$. Set $a_2' \leftarrow a_2 - sa_1 - s^2$. Set $r \leftarrow -a_2'/3, (u^2 - a_2')/3$ or $(-u^2 - a_2')/3$ depending on a_2' being congruent to 0, 1 or -1 modulo 3. Set $a_3' \leftarrow a_3 + ra_1$. If a_3' is odd, then set $t \leftarrow (u^3 - a_3')/2$ else set $t \leftarrow -a_3'/2$. Finally, set $k \leftarrow k \bmod 12$, $\Delta \leftarrow \Delta/u^{12}$, $c_4 \leftarrow c_4/u^4$ and $c_6 \leftarrow c_6/u^6$.

3. [Non-integral invariant] If $v_p(j) < 0$, then set $\nu \leftarrow -v_p(j)$. k must be equal to 0 or 6. If $k = 0$, set $f \leftarrow 1$, and set $c \leftarrow 1$ if $\left(\frac{-c_6}{p}\right) = 1$ or $c \leftarrow \gcd(2, \nu)$ if $\left(\frac{-c_6}{p}\right) = -1$, then output Kodaira type I_ν. If $k = 6$ set $f \leftarrow 2$, and set $c \leftarrow 3 + \left(\frac{\Delta c_6 p^{-9-\nu}}{p}\right)$ if ν is odd, $c \leftarrow 3 + \left(\frac{\Delta p^{-6-\nu}}{p}\right)$ if ν is even, then output Kodaira type I_ν^*. In any case, output f, c, u, r, s, t and terminate the algorithm.

4. [Integral invariant] If $k = 0$ then set $f \leftarrow 0$ else set $f \leftarrow 2$. The possible values for k are 0, 2, 3, 4, 6, 8, 9 and 10. Set $c \leftarrow 1, 1, 2, 2 + \left(\frac{-6c_6 p^{-2}}{p}\right), 1 +$ the number of roots of $4X^3 - 3c_4 p^{-2}X - c_6 p^{-3}$ in $\mathbb{Z}/p\mathbb{Z}$, $2 + \left(\frac{-6c_6 p^{-4}}{p}\right), 2, 1$ respectively. Output respectively the Kodaira types $I_0, II, III, IV, I_0^*, IV^*, III^*, II^*$. In any case, output f, c, u, r, s, t and terminate the algorithm.

When $p = 2$ or $p = 3$, the algorithm is much more complicated.

Algorithm 7.5.2 (Reduction of an Elliptic Curve Modulo 2 or 3). Given integers $a_1, \ldots, a_6$ and $p = 2$ or 3, this algorithm determines the Kodaira symbol associated with the curve modulo p. In addition, it computes the exponent f of p in the arithmetic conductor of the curve, the index $c = [E(\mathbb{Q}_p) : E^0(\mathbb{Q}_p)]$ and integers u, r, s, t such that $a_1', \ldots, a_6'$ linked to $a_1, \ldots, a_6$ via Formulas (7.2) give a model with the smallest possible power of p in its discriminant. To simplify the presentation, we use a variable T which will hold the Kodaira type, coded in any way one likes.

1. [Initialize] Set $u \leftarrow 1$, $r \leftarrow 0$, $s \leftarrow 0$, and $t \leftarrow 0$. Compute Δ and j using Formulas (7.1). Set $\nu \leftarrow v_p(\Delta)$.

2. [Type I_0] If $\nu = 0$ then set $f \leftarrow 0$, $c \leftarrow 1$, $T \leftarrow I_0$ and go to step 22.

3. [Type I_ν] If $p \nmid b_2 = a_1^2 + 4a_2$ then set $f \leftarrow 1$, and set $c \leftarrow \nu$ if $X^2 + a_1 X - a_2$ has a root in $\mathbb{Z}/p\mathbb{Z}$, set $c \leftarrow \gcd(2, \nu)$ otherwise, then set $T \leftarrow I_\nu$ and go to step 22.

4. [Change Equation] If $p = 2$, then set $r_1 \leftarrow a_4 \bmod 2$, $s_1 \leftarrow (r_1 + a_2) \bmod 2$ and $t_1 \leftarrow (a_6 + r_1(a_4 + s_1)) \bmod 2$, otherwise compute b_6 using Formulas (7.1) and set $r_1 \leftarrow -b_6 \bmod 3$, $s_1 \leftarrow a_1 \bmod 3$ and $t_1 \leftarrow (a_3 + r_1 a_1) \bmod 3$. Use Formulas (7.2) with the parameters $1, r_1, s_1, t_1$ to compute $a_1', \ldots, a_6'$, then set $a_1 \leftarrow a_1'$, $a_2 \leftarrow a_2'$, $\ldots$, $a_6 \leftarrow a_6'$, $r \leftarrow r + u^2 r_1$, $s \leftarrow s + u s_1$ and $t \leftarrow t + u^3 t_1 + u^2 s r_1$.

5. [Type II] If $p^2 \nmid a_6$, then set $f \leftarrow \nu$, $c \leftarrow 1$, $T \leftarrow II$ and go to step 22.

6. [Type III] Compute b_8 using Formulas (7.1). If $p^3 \nmid b_8$, then set $f \leftarrow \nu - 1$, $c \leftarrow 2$, $T \leftarrow III$ and go to step 22.

7. [Type IV] Compute b_6 using Formulas (7.1). If $p^3 \nmid b_6$, then set $f \leftarrow \nu - 2$ and set $c \leftarrow 3$ if $X^2 + a_3/pX - a_6/p^2$ has a root in $\mathbb{Z}/p\mathbb{Z}$, set $c \leftarrow 1$ otherwise, then set $T \leftarrow IV$ and go to step 22.

8. [Change Equation] If $p^3 \nmid a_6$ do the following. If $p = 2$, then set $k \leftarrow 2$, otherwise set $k \leftarrow a_3 \bmod 9$. Use Formulas (7.2) with parameters $1, 0, 0, k$ to compute $a_1', \ldots, a_6'$, then set $a_1 \leftarrow a_1'$, $a_2 \leftarrow a_2'$, $\ldots$, $a_6 \leftarrow a_6'$ and finally set $t \leftarrow t + u^3 k$.

9. [Type I_0^*] (At this point, we have $p \mid a_2, p^2 \mid a_4 and p^3 \mid a_6$.) Set $P \leftarrow X^3 + a_2/pX^2 + a_4/p^2 X + a^6/p^3$. If P has distinct roots modulo p, then set $f \leftarrow \nu - 4$, set $c \leftarrow 1+$ the number of roots of P in $\mathbb{Z}/p\mathbb{Z}$, $T \leftarrow I_0^*$ and go to step 22.

10. [Change Equation] Let a be the multiple root of the polynomial P modulo p. If $a \neq 0$, then use Formulas (7.2) with parameters $1, ap, 0, 0$ to compute $a_1', \ldots, a_6'$, then set $a_1 \leftarrow a_1'$, $a_2 \leftarrow a_2'$, $\ldots$, $a_6 \leftarrow a_6'$, $r \leftarrow r + u^2 ap$ and $t \leftarrow t + u^2 sap$. If a is a double root, then go to step 16.

11. [Type IV^*] (Here $p^2 \mid a_3, p^4 \mid a_6$.) Set $P \leftarrow X^2 + a_3/p^2 X + a_6/p^4$. If P has a double root in $\mathbb{Z}/p\mathbb{Z}$, then let a be that root. Otherwise set $f \leftarrow \nu - 6$, set $c \leftarrow 3$ if P splits over $\mathbb{Z}/p\mathbb{Z}$ and $c \leftarrow 1$ otherwise, set $T \leftarrow IV^*$ and go to step 22.

12. [Change Equation] If $a \neq 0$ then use Formulas (7.2) with parameters $1, 0, 0, ap^2$ to compute $a_1', \ldots, a_6'$, then set $a_1 \leftarrow a_1'$, $a_2 \leftarrow a_2'$, $\ldots$, $a_6 \leftarrow a_6'$ and $t \leftarrow t + u^3 ap^2$.

13. [Type III^*] If $p^4 \nmid a_4$, then set $f \leftarrow \nu - 7$, $c \leftarrow 2$, $T \leftarrow III^*$ and go to step 22.

14. [Type II^*] If $p^6 \nmid a_6$, then set $f \leftarrow \nu - 8$, $c \leftarrow 1$ $T \leftarrow II^*$ and go to step 22.

15. [Non-minimal equation] Use Formulas (7.2) with parameters $p, 0, 0, 0$ to compute $a'_1, \ldots, a'_6$, then set $a_1 \leftarrow a'_1$, $a_2 \leftarrow a'_2$, $\ldots$, $a_6 \leftarrow a'_6$, $u \leftarrow pu$, $v \leftarrow v - 12$ and go to step 2.

16. [Initialize Loop] Set $f \leftarrow v - 5$, $v \leftarrow 1$, $q \leftarrow p^2$.

17. [Type I^*_v, day in] Set $P \leftarrow X^2 + a_3/qX - a_6/q^2$. If P has distinct roots modulo p, then set $c \leftarrow 4$ if these roots are in $\mathbb{Z}/p\mathbb{Z}$, set $c \leftarrow 2$ otherwise, then set $T \leftarrow I^*_v$ and go to step 22.

18. [Change Equation] Let a be the double root of P modulo p. If $a \neq 0$, use Formulas (7.2) with parameters $1, 0, 0, aq$ to compute $a'_1, \ldots, a'_6$, then set $a_1 \leftarrow a'_1$, $a_2 \leftarrow a'_2$, $\ldots$, $a_6 \leftarrow a'_6$ and $t \leftarrow t + u^3 aq$.

19. [Type I^*_v, day out] Set $v \leftarrow v + 1$ and $P \leftarrow a_2/pX^2 + a_4/(pq)X + a_6/(pq^2)$. If P has distinct roots modulo p, then set $c \leftarrow 4$ if these roots are in $\mathbb{Z}/p\mathbb{Z}$, set $c \leftarrow 2$ otherwise, then set $T \leftarrow I^*_v$ and go to step 22.

20. [Change Equation] Let a be the double root of P modulo p. If $a \neq 0$, use Formulas (7.2) with parameters $1, aq, 0, 0$ to compute $a'_1, \ldots, a'_6$, then set $a_1 \leftarrow a'_1$, $a_2 \leftarrow a'_2$, $\ldots$, $a_6 \leftarrow a'_6$, $r \leftarrow r + u^2 aq$ and $t \leftarrow t + u^2 saq$.

21. [Loop] Set $v \leftarrow v + 1$, $q \leftarrow p \cdot q$ and go to step 17.

22. [Common termination] Output the Kodaira type T, the numbers f, c, u, r, s, t and terminate the algorithm.

Let us turn now to the global counterpart of this process: what is the best equation for an elliptic curve defined over $\mathbb{Q}$?.

Algorithm 7.5.3 (Global Reduction of an Elliptic Curve). Given $a_1, \ldots, a_6 \in \mathbb{Z}$, this algorithm computes the arithmetic conductor N of the curve and integers u, r, s, t such that $a'_1, \ldots, a'_6$ linked to $a_1, \ldots, a_6$ via Formulas (7.2) give a model with the smallest possible discriminant (in absolute value) and such that $a'_1, a'_3 \in \{0, 1\}$ and $a'_2 \in \{0, \pm 1\}$.

1. [Initialize] Set $N \leftarrow 1$, $u \leftarrow 1$, $r \leftarrow 0$, $s \leftarrow 0$ and $t \leftarrow 0$. Compute $D \leftarrow |\Delta|$ using Formulas (7.1).

2. [Finished ?] If $D = 1$, then output N, u, r, s, t and terminate the algorithm.

3. [Local Reduction] Find a prime divisor p of D. Then use Algorithm 7.5.1 or 7.5.2 to compute the quantities f_p, u_p, r_p, s_p (the quantity c_p may be discarded if it is not wanted for other purposes). Set $N \leftarrow Np^{f_p}$. If $u_p \neq 1$, set $u \leftarrow uu_p$, $r \leftarrow r + u^2 r_p$, $s \leftarrow s + us_p$ and $t \leftarrow t + u^3 t_p + u^2 sr_p$. Finally, set $D \leftarrow D/p$ until $p \nmid D$, then go to step 2.

Note that if only the minimal Weierstraß equation of the curve is desired, and not all the local data as well, we can use a simpler algorithm due to Laska (see [Las] and Section 3.2 of [Cre] for a version due to Kraus and Connell).

7.5.2 Computing rational points

We now turn to the problem of trying to determine the group $E(\mathbb{Q})$ of rational points on E. As already mentioned, this is a difficult problem for which no algorithm exists unless we assume some of the standard conjectures.

On the other hand, the determination of the torsion subgroup $E(\mathbb{Q})_{\text{tors}}$ is easy. (This is the elliptic curve analog of computing the subgroup of roots of unity in a number field, see Algorithms 4.9.9 and 4.9.10.)

By considering the formal group associated with the elliptic curve, one can prove (see [Sil]) that torsion points of composite order in any number field have integral coordinates in any Weierstraß model with integral coefficients. Moreover, there are bounds on the denominators of the coordinates of torsion points of order p^n where p is a prime. Over $\mathbb{Q}$, these bounds tell us that only the points of order 2 may have non-integral coordinates in a generalized Weierstraß model, and in that case the denominator of the x-coordinate is at most 4. Using the fact that if P is a torsion point, then $2P$ is also one, one obtains the following theorem, due to Nagell and Lutz (see [Sil]).

Theorem 7.5.4 (Nagell-Lutz). *If $P = (x, y)$ is a rational point of finite order $n > 2$ on the elliptic curve $y^2 = x^3 + Ax + B$, where A and B are integers, then x and y are integers and y^2 divides the discriminant $-(4A^3 + 27B^2)$.*

This result, together with Mazur's Theorem 7.1.11 gives us the following algorithm.

Algorithm 7.5.5 (Rational Torsion Points). Given integers $a_1, \ldots, a_6$, this algorithm lists the rational torsion points on the corresponding elliptic curve E.

1. [2-Division Points] Using Formulas (7.1), compute b_2, b_4, b_6, b_8 and Δ. Output the origin of the curve $((0 : 1 : 0)$ in projective coordinates). Set $P \leftarrow 4X^3 + b_2X^2 + 2b_4X + b_6$. For each rational root α of P, output the point $(\alpha, -(a_1\alpha + a_3)/2)$.

2. [Initialize Loop] Set $n \leftarrow 4 \prod_{p|\Delta} p^{\lfloor v_p(\Delta)/2 \rfloor}$, the largest integer whose square divides 16Δ. Form the list $\mathcal{L}$ of all positive divisors of n.

3. [Loop on $2y + a_1x + a_3$] If $\mathcal{L}$ is empty, terminate the algorithm. Otherwise, let d be the smallest element of $\mathcal{L}$, and remove d from $\mathcal{L}$. For each rational root α of $P - d^2$ execute step 4, then go to step 3.

4. [Check if torsion] Set $P_1 \leftarrow (\alpha, (d - a_1\alpha - a_3)/2)$. Compute the points $2P_1$, $3P_1$, $4P_1$, $5P_1$ and $6P_1$, and let $x_2, \ldots, x_6$ be their x-coordinates. If one of these points is the origin of the curve, or if one of the x_i is equal to the x-coordinate of a point found in step 1, or if $x_2 = x_3$ or $x_3 = x_4$ or $x_4 = x_5$, then output the two points P_1 and $P_2 \leftarrow (\alpha, -(d + a_1\alpha + a_3)/2)$.

Indeed, from Mazur's Theorem 7.1.11, it is clear that P_1 will be a torsion point if and only if kP_1 is a point of order dividing 2 for $k \leq 6$ or if $kP_1 =$

$-(k+1)P_1$ for $k \leq 4$, and since opposite points have equal x-coordinates in a Weierstraß model, we deduce the test for torsion used in step 4.

Note that to obtain the torsion *subgroup* from this algorithm is very easy: if the polynomial P of step 1 has three rational roots, the torsion subgroup is isomorphic to $(\mathbb{Z}/2\mathbb{Z}) \times (\mathbb{Z}/(N/2)\mathbb{Z})$ otherwise it is isomorphic to $\mathbb{Z}/N\mathbb{Z}$, where N is the total number of torsion points output by the algorithm.

The last algorithm that we will see in this section is an algorithm to compute the canonical height of a rational point.

The Weil height of a point $P = (\dfrac{a}{e^2}, \dfrac{b}{e^3})$ on an elliptic curve E is defined to be $h(P) = \ln|e|$. It is known that the limit

$$\hat{h}(P) = \lim_{n \to \infty} \frac{h(2^n P)}{2^{2n}}$$

exists and defines a positive definite quadratic form on $\mathbb{R} \otimes E(\mathbb{Q})$, known as the canonical height function on $E(\mathbb{Q})$. The existence of this limit means that when a rational point with large denominator is multiplied by some integer m for the group law on the curve, the number of digits of its denominator is multiplied by m^2.

The symmetric bilinear form $\langle P, Q \rangle = \hat{h}(P+Q) - \hat{h}(P) - \hat{h}(Q)$ is called the canonical height pairing and is used to compute the regulator in the Birch and Swinnerton-Dyer Conjecture 7.3.9. The canonical height has properties analogous to those of the logarithmic embedding for number fields (Theorem 4.9.7). More precisely, $\hat{h}(P) = 0$ if and only if P is a point of finite order. More generally if $P_1, \ldots, P_r$ are points on E, then $\det(\langle P_i, P_j \rangle) = 0$ if and only if there exists a linear combination of the points (for the group law of E) which is a point of finite order. Hence this determinant is called the (elliptic) *regulator* of the points P_i.

If $P_1, \ldots, P_r$ form a *basis* of the torsion-free part of $E(\mathbb{Q})$, the regulator $R(E/\mathbb{Q})$ which enters in the Birch and Swinnerton-Dyer conjecture is the elliptic regulator of the points P_i.

The height function $\hat{h}(P)$ has a very interesting structure (see [Sil]). We will only note here that it can be expressed as a sum of local functions, one for each prime number p and one for the "Archimedean prime" ∞. To compute the contribution of a prime p we use an algorithm due in this form to Silverman (see [Sil2]). We will always assume that the elliptic curve is given by a global minimal equation, obtained for example by Algorithm 7.5.3.

Algorithm 7.5.6 (Finite part of the height). Given $a_1, \ldots, a_6 \in \mathbb{Z}$ the coefficients of the global minimal equation of an elliptic curve E and the coordinates (x, y) of a rational point P on E, this algorithm computes the contribution of the finite primes to the canonical height $\hat{h}(P)$.

1. [Initialize] Using Formulas (7.1), compute b_2, b_4, b_6, b_8, c_4, and Δ. Set $z \leftarrow (1/2)\ln(\text{denominator of } x)$, $A \leftarrow \text{numerator of } 3x^2 + 2a_2x + a_4 - a_1y$,

$B \leftarrow$ numerator of $2y + a_1x + a_3$, $C \leftarrow$ numerator of $3x^4 + b_2x^3 + 3b_4x^2 + 3b_6x + b_8$ and $D \leftarrow \gcd(A, B)$.

2. [Loop on p] If $D = 1$, output z and terminate the algorithm. Otherwise, choose a prime divisor p of D and set $D \leftarrow D/p$ until $p \nmid D$.

3. [Add local contribution] If $p \nmid c_4$, then set $N \leftarrow v_p(\Delta)$, $n \leftarrow \min(v_p(B), N/2)$ and $z \leftarrow z - (n(N - n)/(2N)) \ln p$. Otherwise, if $v_p(C) \geq 3v_p(B)$ set $z \leftarrow z - (v_p(B)/3) \ln p$ else set $z \leftarrow z - (v_p(C)/8) \ln p$. Go to step 2.

The Archimedean contribution has a more interesting history from the computational point of view. Initially, it was defined using logarithms of σ functions on the curve, but such objects are not easy to compute by hand or with a hand-held calculator. Tate then discovered a very nice way to compute it using a simple series. Silverman's paper [Sil2] also contains an improvement to that method. However, that series converges only geometrically (the n–th term is bounded by a constant times 4^{-n}). The original definition, while more cumbersome, has a faster rate of convergence by using q-expansions, so it should be preferred for high-precision calculations.

Algorithm 7.5.7 (Height Contribution at ∞).   Given $a_1, \ldots, a_6 \in \mathbb{R}$ and the coordinates (x, y) of a point P on $E(\mathbb{R})$, this algorithm computes the Archimedean contribution of the canonical height of P.

1. [Initialize] Using Formulas (7.1), compute b_2, b_4, b_6 and Δ. Using Algorithm 7.4.7, compute ω_1 and ω_2. Using Algorithm 7.4.8, compute the elliptic logarithm z of the point P. Set $\lambda \leftarrow 2\pi/\omega_1$, $t \leftarrow \lambda \operatorname{Re}(z)$ and $q \leftarrow e^{2i\pi\omega_2/\omega_1}$. (Note that q is a real number and $|q| < 1$.)

2. [Compute theta function] Set

$$\theta \leftarrow \sum_{n=0}^{\infty} \sin((2n + 1)t)(-1)^n q^{n(n+1)/2}$$

(stopping the sum when $q^{n(n+1)/2}$ becomes sufficiently small).

3. [Terminate] Output

$$\frac{1}{32} \ln \left| \frac{\Delta}{q} \right| + \frac{1}{8} \ln \left(\frac{x^3 + (b_2/4)x^2 + (b_4/2)x + b_6/4}{\lambda} \right) - \frac{1}{4} \ln |\theta|$$

and terminate the algorithm.

The canonical height $\hat{h}(P)$ is the sum of the two contributions coming from Algorithms 7.5.6 and 7.5.7.

7.5.3 Algorithms for computing the L-function

As we have seen, according to the Birch and Swinnerton-Dyer conjecture, most of the interesting arithmetical invariants of an elliptic curve E are grouped together in the behavior of $L(E, s)$ around the point $s = 1$, in a manner similar to the case of number fields. In this section, we would like to explain how to compute this L function at $s = 1$, assuming of course that E is a modular elliptic curve. The result is analogous to Propositions 5.3.14 and 5.6.11 but is in fact simpler since it (apparently) does not involve any higher transcendental functions.

Proposition 7.5.8. *Let E be a modular elliptic curve, let N be the conductor of E, let $L(E, s) = \sum_{n \geq 1} a_n n^{-s}$ be the L-series of E and finally let $\varepsilon = \pm 1$ be the sign in the functional equation for $L(E, s)$. Then if A is any positive real number, we have*

$$L(E, 1) = \sum_{n=1}^{\infty} \frac{a_n}{n} \left(e^{-2\pi n A / \sqrt{N}} + \varepsilon e^{-2\pi n / (A \sqrt{N})} \right)$$

and in particular

$$L(E, 1) = (1 + \varepsilon) \sum_{n=1}^{\infty} \frac{a_n}{n} e^{-2\pi n / \sqrt{N}} \ .$$

As in the case of quadratic fields, we have given the general formula involving a real parameter A, but here the purpose is different. In the case of quadratic fields, it gave the possibility of checking the correctness of the computation of certain higher transcendental functions. Here, its use is very different: since the expression must be independent of A, it gives an indirect but quite efficient way to compute the sign ε (and also the conductor N for that matter), which otherwise is not so easy to compute (although there exist algorithms for doing so which are rather tedious). Indeed, we compute the right hand side of the formula giving $L(E, 1)$ for two different values of A, say $A = 1$ and $A = 1.1$ (A should be close to 1 for optimal speed), and the results must agree. Only one of the two possible choices for ε will give results which agree. Hence the above proposition enables us, not only to compute $L(E, 1)$ to great accuracy (the series converges exponentially) but also to determine the sign of the functional equation. Also note that the a_p are computed using Algorithm 7.4.12 or simply as a sum of Legendre symbols, and the a_n are computed using the relations $a_1 = 1$, $a_{mn} = a_m a_n$ if m and n are coprime, and $a_{p^k} = a_p a_{p^{k-1}} - p a_{p^{k-2}}$ for $k \geq 2$.

This is not the whole story. Assume that we discover in this way that $\varepsilon = -1$. Then $L(E, 1) = 0$ for trivial antisymmetry reasons, but the Birch and Swinnerton-Dyer conjecture tells us that the interesting quantity to compute is now the derivative $L'(E, 1)$ of $L(E, s)$ at $s = 1$. In that case we have the following proposition which now involves higher transcendental functions.

Proposition 7.5.9. *Let E be a modular elliptic curve, let N be the conductor of E, and let $L(E, s) = \sum_{n \geq 1} a_n n^{-s}$ be the L-series of E. Assume that the sign ϵ of the functional equation for $L(E, s)$ is equal to -1 (hence trivially $L(E, 1) = 0$). Then*

$$L'(E, 1) = 2 \sum_{n=1}^{\infty} \frac{a_n}{n} E_1 \left(\frac{2\pi n}{\sqrt{N}} \right)$$

where E_1 is the exponential integral function already used in Proposition 5.6.11.

In the case where $L(E, s)$ vanishes to order greater than 1 around $s = 1$, there exist similar formulas for $L^{(r)}(E, 1)$ using functions generalizing the function $E_1(x)$. We refer to [BGZ] for details. If we assume the Birch and Swinnerton-Dyer conjecture, these formulas allow us to compute the rank of the curve E as the exact order of vanishing of $L(E, s)$ around $s = 1$. Note that although the convergence of the series which are obtained is exponential, we need at least $O(\sqrt{N})$ terms before the partial sums start to become significantly close to the result, hence the limit of this method, as in the case of quadratic fields, is for N around 10^{10}. In particular, if we want to estimate the rank of elliptic curves having a much larger conductor, other methods must be used (still dependent on all standard conjectures). We refer to [Mes2] for details.

7.6 Algorithms for Elliptic Curves with Complex Multiplication

7.6.1 Computing the Complex Values of $j(\tau)$

We first describe an efficient way to compute the numerical value of the function $j(\tau)$ for $\tau \in \mathcal{H}$.

Note first that, as in most algorithms of this sort, it is worthwhile to have τ with the largest possible imaginary part, hence to use $j(\tau) = j(\gamma(\tau))$ for any $\gamma \in \mathrm{SL}_2(\mathbb{Z})$. For this, we use Algorithm 7.4.2.

After this preliminary step, there are numerous formulas available to us for computing $j(\tau)$, as is the case for all modular forms or functions. We could for example use Algorithm 7.4.3 for computing g_2 and g_3. It would also be possible to use formulas based on the use of the arithmetic-geometric mean which are quadratically convergent. This would be especially useful for high precision computations of $j(\tau)$.

We will use an intermediate approach which I believe is best suited for practical needs. It is based on the following formulas.

Set as usual $q = e^{2i\pi\tau}$, and

$$\Delta(\tau) = q \left(1 + \sum_{n \geq 1} (-1)^n \left(q^{n(3n-1)/2} + q^{n(3n+1)/2} \right) \right)^{24}.$$

This expression should be computed as written. Note that the convergence is considerably better than that of an ordinary power series since the exponents grow quadratically. It is a well known theorem on modular forms that

$$g_2^3 - 27g_3^2 = \left(\frac{2\pi}{\omega_2} \right)^{12} \Delta .$$

Now the formula that we will use for computing $j(\tau)$ is

$$j(\tau) = \frac{(256f(\tau) + 1)^3}{f(\tau)} \quad \text{where} \quad f(\tau) = \frac{\Delta(2\tau)}{\Delta(\tau)}$$

(note that changing τ into 2τ changes q into q^2).

7.6.2 Computing the Hilbert Class Polynomials

Our second goal is to compute the equation of degree $h(D)$ satisfied by $j(\tau)$, which we will call the *Hilbert class polynomial* for the discriminant D. For this we directly apply Theorem 7.2.14. This leads to the following algorithm, which is closely modeled on Algorithm 5.3.5.

Algorithm 7.6.1 (Hilbert Class Polynomial). Given a negative discriminant D, this algorithm computes the monic polynomial of degree $h(D)$ in $\mathbb{Z}[X]$ of which $j((D + \sqrt{D})/2)$ is a root. We make use of a polynomial variable P.

1. [Initialize] Set $P \leftarrow 1$, $b \leftarrow D \bmod 2$ and $B \leftarrow \left\lfloor \sqrt{|D|/3} \right\rfloor$.

2. [Initialize a] Set $t \leftarrow (b^2 - D)/4$ and $a \leftarrow \max(b, 1)$.

3. [Test] If $a \nmid t$ go to step 4. Otherwise compute $j \leftarrow j((-b + \sqrt{D})/(2a))$ using the above formulas. Now if $a = b$ or $a^2 = t$ or $b = 0$ set $P \leftarrow P \cdot (X - j)$, else set $P \leftarrow P \cdot (X^2 - 2\operatorname{Re}(j)X + |j|^2)$.

4. [Loop on a] Set $a \leftarrow a + 1$. If $a^2 \leq t$, go to step 3.

5. [Loop on b] Set $b \leftarrow b + 2$. If $b \leq B$ go to step 2, otherwise round the coefficients of P to the nearest integer, output P and terminate the algorithm.

An important remark must be made, otherwise this algorithm would not make much sense. The final coefficients of P (known to be integers) must be computed within an error of 0.5 at most. For this, we need to make some a priori estimate on the size of the coefficients of P. In practice, we look at the constant term, which is usually not far from being the largest. This term is equal to the product of the values $j((-b + \sqrt{D})/(2a))$ over all reduced forms (a, b, c), and the modulus of this is approximately equal to $e^{\pi \sqrt{|D|}/(2a)}$ hence the modulus of the constant term is relatively close to 10^k, where

$$k = \frac{\pi \sqrt{|D|}}{\ln(10)} \sum \frac{1}{a} ,$$

the sum running over all reduced forms (a, b, c) of discriminant D.

Hence in step 3, the computation of the j-values should be done with at least $k + 10$ significant digits, 10 being an empirical constant which is sufficient in practice. Note that the value of $\sum 1/a$ is not known in advance, so it should be computed independently (by again applying a variant of Algorithm 5.3.5), since this will in any case take a negligible proportion of the time spent.

7.6.3 Computing Weber Class Polynomials

One of the main applications of computing the Hilbert class polynomials is to explicitly generate the *Hilbert class field* of $K = \mathbb{Q}(\sqrt{D})$ when D is a negative fundamental discriminant. As already mentioned, the coefficients of these polynomials will be very large, and it is desirable to make them smaller. One method is to use the POLRED Algorithm 4.4.11. An essentially equivalent method is given in [Kal-Yui]. A better method is to start by using some extra algebraic information.

We give an example. Set

$$\eta(\tau) = e^{2i\pi\tau/24} \left(1 + \sum_{n \geq 1} (-1)^n \left(q^{n(3n-1)/2} + q^{n(3n+1)/2} \right) \right)$$

(this is the 24-th root of the function $\Delta(\tau)$ defined above, and is called *Dedekind's eta-function*). Define

$$f_1(\tau) = \frac{\eta(\tau/2)}{\eta(\tau)} .$$

Then if $D \equiv \pm 8 \pmod{32}$ and $3 \nmid D$, if we set

$$u = f_1(\sqrt{D})^2 \sqrt{2} ,$$

we can use u instead of j for generating the class field. Indeed, one can show that $K(j) = K(u)$, that u is an algebraic integer (of degree equal to $h(D)$), and what is more important, that the coefficients of the minimal monic polynomial of u (which we will call the *Weber class polynomial* for D) have approximately 12 times fewer digits than those of the Hilbert class polynomials.

Note that one can easily recover j from u if needed. For example, in our special case above we have

$$j = \frac{(256 - u^{12})^3}{u^{24}}.$$

This takes care only of certain congruence classes for D, but most can be treated in a similar manner. We refer the interested reader to [Atk-Mor] or to [Kal-Yui] for complete details.

The algorithm for computing the Weber class polynomials is essentially identical to the one for Hilbert class polynomials: we replace j by u, and furthermore use a much lower precision for the computation of u. For example, in the case $D \equiv \pm 8 \pmod{32}$ and $3 \nmid D$, we can take approximately one twelfth of the number of digits that were needed for the Hilbert class polynomials.

7.7 Exercises for Chapter 7

1. (J. Cremona) Given c_4 and c_6 computed by Formulas (7.1), we would like to recover the b_i and a_i, where we assume that the a_i are in $\mathbf{Z}$. Show that the following procedure is valid. Let b_2 be the unique integer such that $-5 \le b_2 \le 6$ and $b_2 \equiv 0$, $-c_6$, $-4c_6$ or $3c_6$ modulo 12 if $c_4 \equiv 0$, ± 1, ± 2 or 3 modulo 6 respectively. Then set $b_4 = (b_2^2 - c_4)/24$, $b_6 = (-b_2^3 + 36b_2b_4 - c_6)/216$. Finally set $a_1 = b_2 \bmod 2 \in \{0, 1\}$, $a_2 = (b_2 - a_1)/4 \in \{-1, 0, 1\}$, $a_3 = b_6 \bmod 2 \in \{0, 1\}$, $a_4 = (b_4 - a_1 a_3)/2$ and $a_6 = (b_6 - a_3)/4$.

2. Let E be an elliptic curve with complex multiplication by the complex quadratic order of discriminant D. Show that if p is a prime such that $\left(\frac{D}{p}\right) = -1$, then $|E(\mathbf{Z}/p\mathbf{Z})| = p + 1$.

3. Using the result of Exercise 2, show that the only torsion points on the elliptic curve $y^2 = x^3 - n^2 x$ (which has complex multiplication by $\mathbf{Z}[i]$) are the 4 points of order 1 or 2. (Hint: use Dirichlet's theorem on the infinitude of primes in arithmetic progressions.)

4. Show that the elliptic curve $y^2 = 4x^3 - 30x - 28$ has complex multiplication by $\mathbf{Z}[\sqrt{-2}]$ and give explicitly the action of multiplication by $\sqrt{-2}$ on a point (x, y).

5. Given an elliptic curve defined over $\mathbf{Q}$ by a generalized Weierstraß equation, write an algorithm which determines whether this curve has complex multiplication, and if this is the case, gives the complex quadratic order $\mathrm{End}(E)$. (This exercise requires some additional knowledge about elliptic curves.)

6. Using Algorithm 7.4.10, find a Weierstraß equation for the elliptic curve E given by the projective equation

$$x^3 + y^3 = dt^3$$

with $(1 : -1 : 0)$ as given rational point.

7. Given the point $(2 : 1 : 1)$ on the elliptic curve whose projective equation is $x^3 + y^3 = 9t^3$, find another rational point with positive coordinates (apart from the point $(1 : 2 : 1)$ of course). It may be useful to use the result of Exercise 6.

8. Given an elliptic curve E by a general Weierstraß equation $y^2 + a_1 xy + a_3 y = x^3 + a_2 x^2 + a_4 x + a_6$ and a complex number z, give the formulas generalizing those of Proposition 7.4.4 for the coordinates (x, y) on $E(\mathbb{C})$ corresponding to z considered as an element of $\mathbb{C}/L$ where L is the lattice associated to E.

9. (J.-F. Mestre) Let r_1, r_2, r_3 and r_4 be distinct rational numbers and let t be a parameter (which we will also take to be a rational number). Consider the polynomial of degree 12

$$P(X) = \prod_{1 \leq i,j \leq 4,\ i \neq j} (X - (r_i + tr_j)) \ .$$

a) By considering the Laurent series expansion of $P^{1/3}$ show that for any monic polynomial P of degree 12 there exists a unique polynomial $g \in \mathbb{Q}[X]$ such that $\deg(P(X) - g^3(X)) \leq 7$, and show that in our special case we have in fact $\deg(P(X) - g^3(X)) \leq 6$.

b) Show that there exists $q(X) \in \mathbb{Q}[X]$ and $r(X) \in \mathbb{Q}[X]$ such that $P(X) = g^3(X) + q(X)g(X) + r(X)$ with $\deg(q) \leq 2$ and $\deg(r) \leq 3$.

c) Deduce from this that the equation $Y^3 + q(X)Y + r(X) = 0$ is the equation of a cubic with rational coefficients, and that the 12 points $(r_i + tr_j, g(r_i + tr_j))_{i \neq j}$ are 12 (not necessarily distinct) rational points on this cubic.

d) Give explicit values of the r_i and t such that the cubic is non-singular, the 12 points above are distinct and in fact linearly independent for the group law on the cubic.

e) Using Algorithm 7.4.10, find a Weierstraß equation corresponding to the cubic, and give explicitly an elliptic curve defined over $\mathbb{Q}$ whose rank is at least equal to 11 as well as 11 independent points on the elliptic curve (note that we have to "lose" a point in order to obtain an elliptic curve). To answer the last two questions of this exercise, the reader is strongly advised to use a package such as those described in Appendix A. In [Nag] it is shown how to refine this construction in order to have infinite families of elliptic curves of rank 13 instead of 11.

10. Prove that the AGM of two positive real numbers exists, i.e. that the two sequences a_n and b_n given in the text both converge and to the same limit. Show also that the convergence is quadratic.

11. The goal of this exercise is to prove the formula giving $\mathrm{AGM}(a, b)$ in terms of an elliptic integral.

a) Set

$$I(a, b) = \int_0^{\pi/2} \frac{dt}{\sqrt{a^2 \cos^2 t + b^2 \sin^2 t}} \ .$$

By making the change of variable $\sin t = 2a \sin u/((a + b) + (a - b) \sin^2 u)$ show that $I(a, b) = I((a + b)/2, \sqrt{ab})$.

b) Deduce from this the formula $I(a, b) = \pi/(2\,\mathrm{AGM}(a, b))$ given in the text.

c) By making the change of variable $x = a + (b - a) \sin^2 t$, express $I(a, b)$ as an elliptic integral.

Chapter 8

Factoring in the Dark Ages

I owe this title to a talk given by Hendrik Lenstra at MSRI Berkeley in the spring of 1990.

8.1 Factoring and Primality Testing

Since Fermat, it is known that the problem of decomposing a positive integer N into the product of its prime factors splits in fact in three subproblems. The first problem is to decide quickly whether N is composite or probably prime. Such tests, giving a correct answer when N is composite, but no real answer when N is prime, will be called *compositeness tests* (and certainly not primality tests). We will study them in Section 8.2. The second problem is, if one is almost sure that N is prime, to prove that it is indeed prime. Methods used before 1980 to do this will be studied in Section 8.3. Modern methods are the subject matter of Chapter 9. The third problem is that once one knows that N is composite, to factor N. Methods used before the 1960's (i.e. in the dark ages) will be studied starting at Section 8.4. Modern methods are the subject matter of Chapter 10.

Note that factoring/primality testing is usually a recursive process. Given a composite number N, a factoring method will not in general give the complete factorization of N, but only a non-trivial factor d, i.e. such that $1 < d < N$. One then starts working on the two pieces d and N/d. Finding a non-trivial divisor d of N will be called *splitting* N, or even sometimes by abuse of language, factoring N.

Before going to the next section, it should be mentioned that the most naïve method of trial division (which simultaneously does factoring and primality testing) deserves a paragraph. Indeed, in most factoring methods, it usually never hurts to trial divide up to a certain bound to remove small factors. Now we want to divide N by primes up to the square root of N. For this, we may or may not have at our disposal a sufficiently large table of primes. If this is not the case, it is clear that we can divide N by numbers d in given congruence classes, for example 1 and 5 modulo 6, or 1, 7, 11, 13, 17, 19, 23, 29 modulo 30. We will then make unnecessary divisions (by composite numbers), but the result will still be correct. Hence we may for instance use the following algorithm.

Algorithm 8.1.1 (Trial Division). We assume given a table of prime numbers $p[1] = 2$, $p[2] = 3$, ... , $p[k]$, with $k > 3$, an array $t \leftarrow [6, 4, 2, 4, 2, 4, 6, 2]$, and an index j such that if $p[k]$ mod 30 is equal to 1, 7, 11, 13, 17, 19, 23 or 29 then j is set equal to equal to 0, 1, 2, 3, 4, 5, 6 or 7 respectively. Finally, we give ourselves an upper bound B such that $B \geq p[k]$, essentially to avoid spending too much time.

Then given a positive integer N, this algorithm tries to factor (or split N), and if it fails, N will be free of prime factors less than or equal to B.

1. [Initialize] If $N \leq 5$, output the factorization $1 = 1$, $2 = 2$, $3 = 3$, $4 = 2^2$, $5 = 5$ corresponding to the value of N, and terminate the algorithm. Otherwise, set $i \leftarrow -1$, $m \leftarrow 0$, $l \leftarrow \lfloor \sqrt{N} \rfloor$.

2. [Next prime] Set $m \leftarrow m + 1$. If $m > k$ set $i \leftarrow j - 1$ and go to step 5, otherwise set $d \leftarrow p[m]$.

3. [Trial divide] Set $r \leftarrow N$ mod d. If $r = 0$, then output d as a non-trivial divisor of N and terminate the algorithm (or set $N \leftarrow N/d$, $l \leftarrow \lfloor \sqrt{N} \rfloor$ and repeat step 3 if we want to continue finding factors of N).

4. [Prime?] If $d \geq l$, then if $N > 1$ output a message saying that the remaining N is prime and terminate the algorithm. Otherwise, if $i < 0$ go to step 2.

5. [Next divisor] Set $i \leftarrow i + 1$ mod 8, $d \leftarrow d + t[i]$. If $d > B$, then output a message saying that the remaining prime divisors of N are greater than B, otherwise go to step 3.

Note that we have $i = -1$ as long as we are using our prime number table, $i \geq 0$ if not.

This test should not be used for factoring completely, except when N is very small (say $N < 10^8$) since better methods are available for that purpose. On the other hand, it is definitely useful for removing small factors.

Implementation Remark. I suggest using a table of primes up to 500000, if you can spare the memory (this represents 41538 prime numbers). Trial division up to this limit usually never takes more than a few seconds on modern computers. Furthermore, only the difference of the primes (or even half of these differences) should be stored and not the primes themselves, since $p[k] - p[k-1]$ can be held in one byte instead of four when $p[k] \leq 436273009$, and $(p[k] - p[k-1])/2$ can be held in one byte if $p[k] \leq 304599508537$ (see [Bre3]).

Also, I suggest not doing any more divisions after exhausting the table of primes since there are better methods to remove small prime factors. Finally, note that it is not really necessary to compute $l \leftarrow \lfloor \sqrt{N} \rfloor$ in the initialization step, since the test $d \geq l$ in step 4 can be replaced by the test $q \leq l$, where q is the Euclidean quotient of N by d usually computed simultaneously with the remainder in step 3.

8.2 Compositeness Tests

The first thing to do after trial dividing a number N up to a certain bound, is to check whether N (or what remains of the unfactored part) is probably prime or composite. The possibility of doing this easily is due to Fermat's theorem $a^{p-1} \equiv 1 \pmod{p}$ when p is a prime not dividing a. Fermat's theorem in itself would not be sufficient however, even for getting a probable answer.

The second reason Fermat's theorem is useful is that $a^{p-1} \bmod p$ can be computed quickly using the powering algorithms of Section 1.2. This is in contrast with for instance Wilson's theorem stating that $(p-1)! \equiv -1 \pmod{p}$ if and only if p is prime. Although superficially more attractive than Fermat's theorem since it gives a necessary and sufficient condition for primality, and not only a necessary one, it is totally useless because nobody knows how to compute $(p-1)! \bmod p$ in a reasonable amount of time.

The third reason for the usefulness of Fermat's theorem is that although it gives only a necessary condition for primality, exceptions (i.e. composite numbers which satisfy the theorem) are rare. They exist, however. For example the number $N = 561 = 3 \cdot 11 \cdot 17$ is such that $a^{N-1} \equiv 1 \pmod{N}$ as soon as $(a, N) = 1$. Such numbers are called Carmichael numbers. It has just recently been proved by Alford, Granville and Pomerance ([AGP]) that there are infinitely many Carmichael numbers and even that up to x their number is at least $C \cdot x^{2/7}$ for some positive constant C.

It is not difficult to strengthen Fermat's theorem. If p is an odd prime and p does not divide a, then $a^{(p-1)/2} \equiv \pm 1 \pmod{p}$ (more precisely it is congruent to the Legendre symbol $\left(\frac{a}{p}\right)$, see Section 1.4.2). This is stronger than Fermat, and for example eliminates 561. It does not however eliminate all counterexamples, since for instance $N = 1729$ satisfies $a^{(N-1)/2} \equiv 1 \pmod{N}$ for all a coprime to N.

The first test which is really useful is due to Solovay and Strassen ([Sol-Str]). It is based on the fact that if we require not only $a^{(N-1)/2} \equiv \pm 1 \pmod{N}$ but $a^{(N-1)/2} \equiv \left(\frac{a}{N}\right) \pmod{N}$, where $\left(\frac{a}{N}\right)$ is the Jacobi-Kronecker symbol, then this will be satisfied by at most $N/2$ values of a when N is not a prime. This gives rise to the first compositeness test, which is probabilistic in nature: for 50 (say) randomly chosen values of a, test whether the congruence is satisfied. If it is not for any value of a, then N is composite. If it is for all 50 values, then we say that N is probably prime, with probability of error less than $2^{-50} \approx 10^{-15}$, lower in general than the probability of a hardware error.

This test has been superseded by a test due to Miller and Rabin ([Mil], [Rab]), which has two advantages. First, it does not require any Jacobi symbol computation, and second the number of a which will satisfy the test will be at most $N/4$ instead of $N/2$, hence fewer trials have to be made to ensure a given probability. In addition, one can prove that if a satisfies the Rabin-Miller test, then it will also satisfy the Solovay-Strassen test, so the Miller-Rabin test completely supersedes the Solovay-Strassen test.

Definition 8.2.1. *Let N be an odd positive integer, and a be an integer. Write $N - 1 = 2^t q$ with q odd. We say that N is a strong pseudo-prime in base a if either $a^q \equiv 1 \pmod{N}$, or if there exists an e such that $0 \leq e < t$ and $a^{2^e q} \equiv -1 \pmod{N}$.*

If p is an odd prime, it is easy to see that p is a strong pseudo-prime in any base not divisible by p (see Exercise 1). Conversely, one can prove (see for example [Knu2]) that if p is not prime, there exist less than $p/4$ bases a such that $1 < a < p$ for which p is a strong pseudo-prime in base a. This leads to the following algorithm.

Algorithm 8.2.2 (Rabin-Miller).   Given an odd integer $N \geq 3$, this algorithm determines with high probability if N is composite. If it fails, it will output a message saying that N is probably prime.

1. [Initialize] Set $q \leftarrow N - 1$, $t \leftarrow 0$, and while q is even set $q \leftarrow q/2$ and $t \leftarrow t+1$ (now $N - 1 = 2^t q$ with q odd). Then set $c \leftarrow 20$.

2. [Choose new a] Using a random number generator, choose randomly an a such that $1 < a < N$. Then set $e \leftarrow 0$, $b \leftarrow a^q \bmod N$. If $b = 1$, go to step 4.

3. [Squarings] While $b \not\equiv \pm 1 \pmod{N}$ and $e \leq t - 2$ set $b \leftarrow b^2 \bmod N$ and $e \leftarrow e + 1$. If $b \neq N - 1$ output a message saying that N is composite and terminate the algorithm.

4. [Repeat test] Set $c \leftarrow c - 1$. If $c > 0$ go to step 2, otherwise output a message saying that N is probably prime.

The running time of this algorithm is essentially the same as that of the powering algorithm which is used, i.e. in principle $O(\ln^3 N)$. Note however that we can reasonably restrict ourselves to single precision values of a (which will not be random any more, but it probably does not matter), and in that case if we use the left-right Algorithms (1.2.2 to 1.2.4), the time drops to $O(\ln^2 N)$. Hence, it is essentially as fast as one could hope for.

This algorithm is the workhorse of compositeness tests, and belongs in almost any number theory program. Note once again that it will prove the compositeness of essentially all numbers, but it will never prove their primality. In fact, by purely theoretical means, it is usually possible to construct composite numbers which pass the Rabin-Miller test for any given reasonably small finite set of bases a ([Arn]). For example, the composite number

$$1195068768795265792518361315725116351898245581$$

$$= 2444451644843139244746 1 \cdot 48889032896862784894921$$

is a strong pseudo-prime to bases 2, 3, 5, 7, 11, 13, 17, 19, 23, 29 and 31 and several others.

There is a variation on this test due to Miller which is as follows. If one assumes the Generalized Riemann Hypothesis, then one can prove that if N

is not prime, there exists an $a < C \ln^2 N$ such that N will not be a strong pseudo-prime in base a, C being an explicit constant. Hence this gives a non-probabilistic primality and compositeness test, but since it is based on an unproven hypothesis, it cannot be used for the moment. Note that the situation is completely different in factoring algorithms. There, we can use any kinds of unproven hypotheses or crystal balls for that matter, since once the algorithm (or pseudo-algorithm) finishes, one can immediately check whether we have indeed obtained a factor of our number N, without worrying about the manner in which it was obtained. Primality testing however requires rigorous mathematical proofs.

Note also that even if one uses the best known values of the constant C, for our typical range of values of N (say up to 10^{500}), the modern methods explained in Chapter 9 are in practice faster.

8.3 Primality Tests

We now consider the practical problem of rigorously *proving* that a number N is prime. Of course, we will try to do this only after N has successfully passed the Rabin-Miller test, so that we are morally certain that N is indeed prime.

8.3.1 The Pocklington-Lehmer $N - 1$ Test

We need a sort of converse to Fermat's theorem. One such converse was found by Pocklington, and improved by Lehmer. It is based on the following result.

Proposition 8.3.1. Let N be a positive integer, and let p be a prime divisor of $N - 1$. Assume that we can find an integer a_p such that $a_p^{N-1} \equiv 1 \pmod{N}$ and $(a_p^{(N-1)/p} - 1, N) = 1$. Then if d is any divisor of N, we have $d \equiv 1 \pmod{p^{\alpha_p}}$, where p^{α_p} is the largest power of p which divides $N - 1$.

Proof. It is clearly enough to prove the result for all prime divisors of N, since any divisor is a product of prime divisors. Now if d is a prime divisor of N, we have $a_p^{d-1} \equiv 1 \pmod{d}$, since a_p is coprime to N (why?) hence to d. On the other hand, since $(a_p^{(N-1)/p} - 1, N) = 1$, we have $a_p^{(N-1)/p} \not\equiv 1 \pmod{d}$. If e is the exact order of a_p modulo d (i.e. the smallest positive exponent such that $a_p^e \equiv 1 \pmod{d}$), this means that $e \mid d - 1$, $e \nmid (N - 1)/p$ but $e \mid N - 1$, hence $p^{\alpha_p} \mid e \mid d - 1$ showing that $d \equiv 1 \pmod{p^{\alpha_p}}$. $\qquad\square$

Corollary 8.3.2. *Assume that we can write $N - 1 = F \cdot U$ where $(F, U) = 1$, F is completely factored, and $F > \sqrt{N}$. Then, if for each prime p dividing F we can find an a_p satisfying the conditions of Proposition 8.3.1, N is prime. Conversely, if N is prime, for any prime p dividing $N - 1$, one can find a_p satisfying the conditions of Proposition 8.3.1.*

Proof. If the hypotheses of this corollary are satisfied, it follows immediately from Proposition 8.3.1 that all divisors of N are congruent to 1 mod F. Since $F > \sqrt{N}$, this means that N has no prime divisor less than its square root, hence N is prime.

Conversely, when N is prime, if we take for a_p a *primitive root* modulo N, i.e. a generator of the multiplicative group $(\mathbb{Z}/N\mathbb{Z})^*$, it is clear that the conditions of the proposition are satisfied since the order of a_p is exactly equal to $N - 1$. $\square$

This corollary gives us our first true primality test. Its main drawback is that we need to be able to factor $N - 1$ sufficiently, and this is in general very difficult. It is however quite useful for numbers having special forms where $N - 1$ factors easily, for example the Fermat numbers $2^{2^k} + 1$ (see Exercise 9).

The condition $F > \sqrt{N}$ of the corollary can be weakened if we make an extra test:

Proposition 8.3.3. *Assume that we can write $N - 1 = F \cdot U$ where $(F, U) = 1$, F is completely factored, all the prime divisors of U are greater than B, and $B \cdot F \geq \sqrt{N}$. Then if for each prime p dividing F we can find an a_p satisfying the conditions of Proposition 8.3.1, and if in addition we can find a_U such that $a_U^{N-1} \equiv 1 \pmod{N}$ and $(a_U^F - 1, N) = 1$, then N is prime. Conversely, if N is prime, such a_p and a_U can always be found.*

Proof. We follow closely the proof of Proposition 8.3.1. Let d be any prime divisor of N. Proposition 8.3.1 tells us that $d \equiv 1 \pmod{F}$. If e is the exact order of a_U modulo d, then $e \mid d - 1$, $e \mid N - 1$ and $e \nmid F = (N - 1)/U$. Now one cannot have $(e, U) = 1$, otherwise from $e \mid N - 1 = FU$ one would get $e \mid F$, contrary to the hypothesis. Hence $(e, U) > 1$, and since U has all its prime factors greater than B, $(e, U) > B$. Finally, since $(F, U) = 1$, from $d \equiv 1 \pmod{e}$ and $d \equiv 1 \pmod{F}$ we obtain $d \equiv 1 \pmod{(e, U) \cdot F}$ hence $d > B \cdot F \geq \sqrt{N}$, showing that N has no prime divisor less than or equal to its square root, hence that N is prime. $\square$

Note that the condition that U has all its prime factors greater than B is very natural in practice since the factorization $N - 1 = F \cdot U$ is often obtained by trial division.

8.3.2 Briefly, Other Tests

Several important generalizations of this test exist. First, working in the multiplicative group of the field $\mathbb{F}_{N^2}$ instead of $\mathbb{F}_N$, one obtains a test which uses the factorization of $N + 1$ instead of $N - 1$. This gives as a special case the Lucas-Lehmer test for Mersenne numbers $N = 2^p - 1$. In addition, since $\mathbb{F}_N$ is a subfield of $\mathbb{F}_{N^2}$, it is reasonable to expect that one can combine the information coming from the two tests, and this is indeed the case. One can also use higher degree finite fields ($\mathbb{F}_{N^3}$, $\mathbb{F}_{N^4}$ and $\mathbb{F}_{N^6}$) which correspond to using in addition the completely factored part of $N^2 + N + 1$, $N^2 + 1$ and $N^2 - N + 1$ respectively. These numbers are already much larger, however, and do not always give much extra information. Other finite fields give even larger numbers. One last improvement is that, as in Proposition 8.3.3 one can use the upper bound used in doing the trial divisions to find the factors of $N - 1$, $N + 1$, etc ... For details, I refer to [BLS], [Sel-Wun] or [Wil-Jud].

8.4 Lehman's Method

We now turn our attention to factoring methods. The spirit here will be quite different. For example, we do not need to be completely rigorous since if we find a number which may be factor of N, it will always be trivial to check if it is or not. It will however be useful to have some understanding of the asymptotic behavior of the algorithm.

Although several methods were introduced to improve trial division (which is, we recall, a $O(N^{1/2+\epsilon})$ algorithm), the first method which has a running time which could be proved to be substantially lower was introduced by Lehman (see [Leh1]). Its execution time is at worst $O(N^{1/3+\epsilon})$, and it is indeed faster than trial division already for reasonably small values of N. The algorithm is as follows.

Algorithm 8.4.1 (Lehman). Given an integer $N \geq 3$, this algorithm finds a non-trivial factor of N if N is not prime, or shows that N is prime.

1. [Trial division] Set $B \leftarrow \lfloor N^{1/3} \rfloor$. Trial divide N up to the bound B using Algorithm 8.1.1. If any non-trivial factor is found, output it and terminate the algorithm. Otherwise set $k \leftarrow 0$.

2. [Loop on k] Set $k \leftarrow k + 1$. If $k > B$, output the fact that N is prime and terminate the algorithm. Otherwise, set $r = 1$ and $m = 2$ if k is even, $r = k + N$ and $m = 4$ if k is odd.

3. [Loop on a] For all integers a such that $4kN \leq a^2 \leq 4kN + B^2$ and $a \equiv r$ (mod m) do as follows. Set $c \leftarrow a^2 - 4kN$. Using Algorithm 1.7.3, test whether c is a square. If it is, let $c = b^2$, output $\gcd(a+b, N)$ (which will be a non-trivial divisor of N) and terminate the algorithm. Otherwise, use the next value of a if any. If all possible values of a have been tested, go to step 2.

Proof (D. Zagier). We only give a sketch, leaving the details as an exercise to the reader.

If no factors are found during step 1, this means that all the prime factors of N are greater than $N^{1/3}$ hence N has at most two prime factors.

Assume first that N is prime. Then the test in step 3 can never succeed. Indeed, if $a^2 - 4kN = b^2$ then $N \mid a^2 - b^2$ hence $N \mid (a - b)$ or $N \mid (a + b)$ so $a + b \geq N$, but this is impossible since the given inequalities on k and a imply that $a < 2N^{2/3} + 1$ and $b < N^{1/3}$ so $N \leq 13$. An easy check shows that for $3 \leq N \leq 13$, N prime, the test in step 3 does not succeed.

Assume now that N is composite, so that $N = pq$ with p and q not necessarily distinct primes, where we may assume that $p \leq q$. Consider the convergents u_n/v_n of the continued fraction expansion of q/p. Let n be the unique index such that $u_n v_n < N^{1/3} < u_{n+1} v_{n+1}$ (which exists since $pq > N^{1/3}$). Using the elementary properties of continued fractions, if we set $k = u_n v_n$ and $a = pv_n + qu_n$, it is easily checked that the conditions of step 3 are met, thus proving the validity of the algorithm.   □

For each value of k there are at most $1/2(\sqrt{4kN + N^{2/3}} - \sqrt{4kN}) \approx N^{1/6}k^{-1/2}/8$ values of a, and since $\sum_{k \leq x} k^{-1/2} \approx 2x^{1/2}$, the running time of the algorithm is indeed $O(N^{1/3+\epsilon})$ as claimed.

We refer to [Leh1] for ways of fine tuning this algorithm, which is now only of historical interest.

8.5 Pollard's ρ Method

8.5.1 Outline of the Method

The idea behind this method is the following. Let $f(X)$ be a polynomial with integer coefficients. We define a sequence by taking any initial x_0, and setting $x_{k+1} = f(x_k) \bmod N$. If p is a (unknown) prime divisor of N, then the sequence $y_k = x_k \bmod p$ satisfies the same recursion. Now if $f(X)$ is chosen suitably, it is not unreasonable to assume that this sequence will behave like the sequence of iterates of a *random* map from $\mathbb{Z}/p\mathbb{Z}$ into itself. Such a sequence must of course be ultimately periodic, and a mathematical analysis shows that it is reasonable to expect that the period and preperiod will have length $O(\sqrt{p})$. Now if $y_{k+t} = y_k$, this means that $x_{k+t} \equiv x_k \pmod{p}$, hence that $(x_{k+t} - x_k, N) > 1$. Now this GCD will rarely be equal to N itself, hence we obtain in this way, maybe not p, but a non-trivial factor of N, so N is split and we can look at the pieces. The number of necessary steps will be $O(\sqrt{p}) = O(N^{1/4})$, and the total time in bit operations will be $O(N^{1/4} \ln^2 N)$.

Of course, we have just given a rough outline of the method. It is clear however that it will be efficient since the basic operations are simple, and furthermore that its running time depends mostly on the size of the smallest prime factor of N, not on the size of N itself, hence it can replace trial division or Lehman's method to cast out small factors. In fact, it is still used along with more powerful methods for that purpose. Finally, notice that, at least in a primitive form, it is very easy to implement.

We must now solve a few related problems:

(1) How does one find the periodicity relation $y_{k+t} = y_k$?
(2) How does one choose f and x_0?
(3) What is the expected average running time, assuming f is a random map?

I would like to point out immediately that although it is believed that the polynomials that we give below behave like random maps, this is not at all proved, and in fact the exact mathematical statement to prove needs to be made more precise.

8.5.2 Methods for Detecting Periodicity

From now on, we consider a sequence $y_{k+1} = f(y_k)$ from a finite set E into itself. Such a sequence will be ultimately periodic, i.e. there exists M and $T > 0$ such that for $k \geq M$, $y_{k+T} = y_k$ but $y_{M-1+T} \neq y_{M-1}$. The number M will be called the preperiod, and T (chosen as small as possible) will be the period. If the iterates are drawn on a piece of paper starting at the bottom and ending in a circle the figure that one obtains has the shape of the Greek letter ρ, whence the name of the method.

We would like to find a reasonably efficient method for finding k and $t > 0$ such that $y_{k+t} = y_k$ (we do not need to compute M and T). The initial method suggested by Pollard and Floyd is to compute simultaneously with the sequence y_k the sequence z_k defined by $z_0 = y_0$, $z_{k+1} = f(f(z_k))$. Clearly $z_k = y_{2k}$, and if k is any multiple of T which is larger than M, we must have $z_k = y_{2k} = y_k$, hence our problem is solved. This leads to a simple-minded but nonetheless efficient version of Pollard's ρ method. Unfortunately we need three function evaluations per step, and this may seem too many.

An improvement due to Brent is the following. Let $l(m)$ be the largest power of 2 less than or equal to m, i.e.

$$l(m) = 2^{\lfloor \lg m \rfloor} ,$$

so that in particular $l(m) \leq m < 2l(m)$. Then I claim that there exists an m such that $y_m = y_{l(m)-1}$. Indeed, if one chooses

$$m = 2^{\lceil \lg \max(M+1,T) \rceil} + T - 1 ,$$

we clearly have $l(m) = 2^{\lceil \lg \max(M+1,T) \rceil}$ hence $l(m) - 1 \geq M$ and $m - (l(m) - 1) = T$, thus proving our claim.

If instead of computing an extra sequence z_k we compute only the sequence y_k and keep y_{2^e-1} each time we hit a power of two minus one, for every m such that $2^e \leq m < 2^{e+1}$ it will be enough to compare y_m with y_{2^e-1} (note that at any time there is only one value of y to be kept).

Hence Brent's method at first seems definitely superior. It can however be shown that the number of comparisons needed before finding an equality $y_m = y_{l(m)-1}$ will be on average almost double that of the initial Pollard-Floyd method. In practice this means that the methods are comparable, the lower number of function evaluations being compensated by the increased number of comparisons which are needed.

However a modification of Brent's method gives results which are generally better than the above two methods. It is based on the following proposition.

Proposition 8.5.1.

(1) *There exists an m such that*

$$y_m = y_{l(m)-1} \quad and \quad \frac{3}{2}l(m) \leq m < 2l(m) \ .$$

(2) *the least such m is $m_0 = 3$ if $M = 0$ and $T = 1$ (i.e. if $y_1 = y_0$), and otherwise is given by*

$$m_0 = 2^{\lceil \lg \max(M+1,T) \rceil} + T \left\lceil \frac{l(M)+1}{T} \right\rceil - 1 \ ,$$

where we set $l(0) = 0$.

Proof. Set $e = \lceil \lg \max(M+1,T) \rceil$. We claim that, as in Brent's original method, we still have $l(m_0) = 2^e$. Clearly, $2^e \leq m_0$, so we must prove that $m_0 < 2^{e+1}$ or equivalently that

$$T \left\lceil \frac{l(M)+1}{T} \right\rceil \leq 2^e \ .$$

We consider two cases. First, if $T \leq l(M)$, then

$$T \left\lceil \frac{l(M)+1}{T} \right\rceil \leq l(M) + T \leq 2l(M) = 2^{\lceil \lg(M+1) \rceil} \leq 2^e \ ,$$

since $\lfloor \lg M \rfloor + 1 = \lceil \lg(M+1) \rceil$. On the other hand, if $T \geq l(M) + 1$, then $\left\lceil \frac{l(M)+1}{T} \right\rceil = 1$, and we clearly have $T \leq 2^e$.

Now that our claim is proved, since $m_0 \geq M$ and $m_0 - (l(m_0) - 1)$ is a multiple of T we indeed have $y_m = y_{l(m)-1}$ for $m = m_0$. To finish proving the first part of the proposition, we must show that $\frac{3}{2}l(m_0) \leq m_0$ (the other inequality being trivial), or equivalently, keeping our notations above, that

$$T \left\lceil \frac{l(M)+1}{T} \right\rceil - 1 \geq 2^{e-1} \ .$$

Now clearly the left hand side is greater than or equal to $T - 1$, and on the other hand $2^{\lceil \lg T \rceil - 1} \leq 2^{\lg T} - 1 = T - 1$. Furthermore, the left hand side is also greater than or equal to $l(M) = 2^{\lfloor \lg M \rfloor}$, but one sees easily that $2^{\lceil \lg(M+1) \rceil - 1} = 2^{\lfloor \lg M \rfloor}$, thus showing the first part of the proposition. The proof of the second part (that is, the claim that m_0 is indeed the smallest) is similar (i.e. not illuminating) and is left to the reader.   □

Using this proposition, we can decrease the number of comparisons in Brent's method since it will not be necessary to do anything (apart from a function evaluation) while m is between 2^e and $\frac{3}{2}2^e$.

8.5.3 Brent's Modified Algorithm

We temporarily return to our problem of factoring N. We must first explain how to choose f and x_0. The choice of x_0 seems to be quite irrelevant for the efficiency of the method. On the other hand, one must choose f carefully. In order to minimize the number of operations, we will want to take for f a polynomial of small degree. It is intuitively clear (and easy to prove) that linear polynomials f will not be random and hence give bad results. The quadratic polynomials on the other hand seem in practice to work pretty well, as long as we avoid special cases. The fastest to compute are the polynomials of the form $f(x) = x^2 + c$. Possible choices for c are $c = 1$ or $c = -1$. On the other hand $c = 0$ should, of course, be avoided. We must also avoid $c = -2$ since the recursion $x_{k+1} = x_k^2 - 2$ becomes trivial if one sets $x_k = u_k + 1/u_k$.

As already explained in Section 8.5.1, the "comparisons" $y_{k+t} = y_k$ are done by computing $(x_{k+t} - x_k, N)$. Now, even though we have studied efficient methods for GCD computation, such a computation is slow compared to a simple multiplication. Hence, instead of computing the GCD's each time, we batch them up by groups of 20 (say) by multiplying modulo N, and then do a single GCD instead of 20. If the result is equal to 1 (as will unfortunately usually be the case) then all the GCD's were equal to 1. If on the other hand it is non-trivial, we can backtrack if necessary.

The results and discussion above lead to the following algorithm.

Algorithm 8.5.2 (Pollard ρ). Given a composite integer N, this algorithm tries to find a non-trivial factor of N.

1. [Initialize] Set $y \leftarrow 2$, $x \leftarrow 2$, $x_1 \leftarrow 2$, $k \leftarrow 1$, $l \leftarrow 1$, $P \leftarrow 1$, $c \leftarrow 0$.

2. [Accumulate product] Set $x \leftarrow x^2 + 1 \bmod N$, $P \leftarrow P \cdot (x_1 - x) \bmod N$ and $c \leftarrow c + 1$. (We now have $m = 2l - k$, $l = l(m)$, $x = x_m$, $x_1 = x_{l(m)-1}$.) If $c = 20$, compute $g \leftarrow (P, N)$, then if $g > 1$ go to step 4 else set $y \leftarrow x$ and $c \leftarrow 0$.

3. [Advance] Set $k \leftarrow k - 1$. If $k \neq 0$ go to step 2. Otherwise, compute $g \leftarrow (P, N)$. If $g > 1$ go to step 4 else set $x_1 \leftarrow x$, $k \leftarrow l$, $l \leftarrow 2l$, then repeat k times $x \leftarrow x^2 + 1 \bmod N$, then set $y \leftarrow x$, $c \leftarrow 0$ and go to step 2.

4. [Backtrack] (Here we know that a factor of N has been found, maybe equal to N). Repeat $y \leftarrow y^2 + 1 \bmod N$, $g \leftarrow (x_1 - y, N)$ until $g > 1$ (this must occur). If $g < N$ output g, otherwise output a message saying that the algorithm fails. Terminate the algorithm.

Note that the algorithm may fail (indicating that the period modulo the different prime factors of N is essentially the same). In that case, do *not* start with another value of x_0, but rather with another polynomial, for example $x^2 - 1$ or $x^2 + 3$.

This algorithm has been further improved by P. Montgomery ([Mon2]) and R. Brent ([Bre2]).

8.5.4 Analysis of the Algorithm

As has already been said, it is not known how to analyze the above algorithms without assuming that f is a random map. Hence the analysis that we give is in fact an analysis of the iterates of a random map from a finite set E of cardinality p into itself. We also point out that some of the arguments given here are not rigorous but can be made so. We have given very few detailed analysis of algorithms in this book, but we make an exception here because the mathematics involved are quite pretty and the proofs short.

Call $P(M,T)$ the probability that a sequence of iterates y_m has preperiod M and period T. Then $y_0, \ldots, y_{M+T-1}$ are all distinct, and $y_{M+T} = y_M$. Hence we obtain

$$P(M,T) = \frac{1}{p} \prod_{1 \le k < M+T} \left(1 - \frac{k}{p}\right) .$$

Now we will want to compute the asymptotic behavior as $p \to \infty$ of the average of certain functions over all maps f, i.e. of sums of the form

$$S = \sum_{M,T} P(M,T)g(M,T) .$$

Now if we set $M = \mu\sqrt{p}$ and $T = \lambda\sqrt{p}$, we have

$$\ln(p \cdot P(M,T)) = \sum_{k<(\lambda+\mu)\sqrt{p}} \ln\left(1 - \frac{k}{p}\right) = \sum_{k<(\lambda+\mu)\sqrt{p}} \left(-\frac{k}{p} + O\left(\frac{k^2}{p^2}\right)\right)$$

$$= -\frac{(\lambda+\mu)^2}{2} + O\left(\frac{\lambda+\mu}{\sqrt{p}}\right) + O\left(\frac{(\lambda+\mu)^3}{\sqrt{p}}\right) .$$

Hence the limiting distribution of $P(M,L)dM\,dL$ is

$$\frac{1}{p}e^{-(\lambda+\mu)^2/2}\sqrt{p}\,d\mu\sqrt{p}\,d\lambda = e^{-(\lambda+\mu)^2/2}d\mu\,d\lambda ,$$

so our sum S is asymptotic to

$$\int_0^\infty \int_0^\infty g(\mu\sqrt{p}, \lambda\sqrt{p})e^{-(\lambda+\mu)^2/2}d\mu\,d\lambda. \qquad (*)$$

As a first application, let us compute the asymptotic behavior of the average of the period T.

Proposition 8.5.3. *As $p \to \infty$, the average of T is asymptotic to*

$$\sqrt{\frac{\pi p}{8}} .$$

Proof. Using $(*)$, we see that the average of T is asymptotic to

$$\sqrt{p} \int_0^\infty \int_0^\infty y e^{-(x+y)^2/2} \, dx \, dy \ .$$

By symmetry, this is equal to one half of the integral with $x + y$ instead of y, and this is easily computed and gives the proposition. □

Now we need to obtain the average of the other quantities entering into the expression for m_0 given in Proposition 8.5.1. Note that

$$T \left\lceil \frac{l(M) + 1}{T} \right\rceil = T \left\lfloor \frac{l(M)}{T} \right\rfloor + T \ .$$

We then have

Proposition 8.5.4. *As $p \to \infty$, the average of $T \left\lfloor \frac{l(M)}{T} \right\rfloor$ is asymptotic to*

$$\left(\frac{\ln \pi - \gamma}{2 \ln 2} \right) \sqrt{\frac{\pi p}{8}}$$

where $\gamma = 0.57721\ldots$ is Euler's constant.

Proof. The proof is rather long, so we only sketch the main steps. Using $(*)$, the average of the quantity that we want to compute is asymptotic to

$$S = \int_0^\infty \int_0^\infty y\sqrt{p} \left\lfloor \frac{2^{\lfloor \lg(x\sqrt{p}) \rfloor}}{y\sqrt{p}} \right\rfloor e^{-(x+y)^2/2} \, dx \, dy \ .$$

By splitting up the integral into pieces where the floor is constant, it is then a simple matter to show that

$$S = \sqrt{p} \sum_{n=1}^\infty \int_0^\infty y F \left(\frac{1}{\sqrt{p}} 2^{\lceil \lg(ny\sqrt{p}) \rceil} + y \right) \, dy \ ,$$

where $F(y) = \int_y^\infty e^{-t^2/2} \, dt$. Now we assume that if we replace $\lceil \lg(ny\sqrt{p}) \rceil$ by $\lg(ny\sqrt{p}) + u$, where u is a uniformly distributed variable between 0 and 1, then S will be replaced by a quantity which is asymptotic to S (this step can be rigorously justified), i.e.

$$S \sim \sqrt{p} \sum_{n=1}^\infty \int_0^1 du \int_0^\infty y F(2^u ny + y) \, dy \ .$$

Now using standard methods like integration by parts and power series expansions, we find

$$S \sim \sqrt{\frac{\pi p}{8}} \frac{G(1) - G(1/2)}{\ln 2} \ ,$$

where

$$G(x) = \sum_{k=2}^{\infty} (-1)^k \frac{k-1}{k} \zeta(k) x^k$$

and $\zeta(s)$ is the Riemann zeta function. Now from the Taylor series expansion of the logarithm of the gamma function near $x = 1$, we immediately see that

$$G(x) = x \frac{\Gamma'(x+1)}{\Gamma(x+1)} - \ln \Gamma(x+1) \ ,$$

and using the special values of the gamma function and its derivative, we obtain Proposition 8.5.3.   □

In a similar way (also by using the trick with the variable u), we can prove:

Proposition 8.5.5. *As $p \to \infty$, the average of*

$$2^{\lceil \lg \max(M+1,T) \rceil}$$

is asymptotic to

$$\frac{3}{2 \ln 2} \sqrt{\frac{\pi p}{8}} \ .$$

Combining these three propositions, we obtain the following theorem.

Theorem 8.5.6. *As $p \to \infty$, the average number of function evaluations in Algorithm 8.5.2 is asymptotic to*

$$FE = \left(\frac{3 + \ln 4\pi - \gamma}{2 \ln 2} \right) \sqrt{\frac{\pi p}{8}} \approx 3.1225 \sqrt{p} \ ,$$

and the number of multiplications mod N (i.e. implicitly of GCD's) is asymptotic to

$$MM = \left(\frac{\ln 4\pi - \gamma}{2 \ln 2} \right) \sqrt{\frac{\pi p}{8}} \approx 0.8832 \sqrt{p} \ .$$

This terminates our analysis of the Pollard ρ algorithm. As an exercise, the reader can work out the asymptotics for the unmodified Brent method and for the Pollard-Floyd method of detecting periodicity.

8.6 Shanks's Class Group Method

Another $O(N^{1/4+\epsilon})$ method (and even $O(N^{1/5+\epsilon})$ if one assumes the GRH) is due to Shanks. It is a simple by-product of the computation of the class number of an imaginary quadratic field (see Section 5.4). Indeed, let $D = -N$ if $N \equiv 3 \pmod 4$, $D = -4N$ otherwise. If h is the class number of $\mathbb{Q}(\sqrt{D})$ and if N is composite, then it is known since Gauss that h must be even (this is the start of the theory of genera into which we will not go). Hence, there must be an element of order exactly equal to 2 in the class group. Such an element will be called an ambiguous element, or in terms of binary quadratic forms, a form whose square is equivalent to the unit form will be called an ambiguous form.

Clearly, (a, b, c) is ambiguous if and only if it is equivalent to its inverse $(a, -b, c)$, and if the form is reduced this means that we have three cases.

(1) Either $b = 0$, hence $D = -4ac$, so $N = ac$.
(2) Or $a = b$, hence $D = b(b - 4c)$, hence $N = (b/2)(2c - b/2)$ if b is even, $N = b(4c - b)$ if b is odd.
(3) Or finally $a = c$, hence $D = (b - 2a)(b + 2a)$ hence $N = (b/2 + a)(a - b/2)$ if b is even, $N = (2a - b)(b + 2a)$ if b is odd.

We see that each ambiguous form gives a factorization of N (and this is a one-to-one correspondence).

Hence, Shanks's factoring method is roughly as follows: after having computed the class number h, look for an ambiguous form. Such a form will give a factorization of N (which may be trivial). There must exist a form which gives a non-trivial factorization however, and in practice it is obtained very quickly.

There remains the problem of finding ambiguous forms. But this is easy and standard. Write $h = 2^t q$ with q odd. Take a form f at random (for example one of the prime forms f_p used in Algorithm 5.4.10) and compute $g = f^q$. Then g is in the 2-Sylow subgroup of the class group, and if g is not the unit form, there exists an exponent m such that $0 \le m < t$ and such that g^{2^m} is an ambiguous form. This is identical in group-theoretic terms to the idea behind the Rabin-Miller compositeness test (Section 8.2 above).

We leave to the reader the details of the algorithm which can be found in Shanks's paper [Sha1], as well as remarks on what should be done when the trivial factorization is found too often.

8.7 Shanks's SQUFOF

Still another $O(N^{1/4+\epsilon})$ method, also due to Shanks, is the SQUFOF (SQUare FOrm Factorization) method. This method is very simple to implement and also has the big advantage of working exclusively with numbers which are at most $2\sqrt{N}$, hence essentially half of the digits of N. Therefore it is eminently practical and fast when one wants to factor numbers less than 10^{19}, even on a pocket calculator. This method is based upon the infrastructure of real quadratic fields which we discussed in Section 5.8, although little of that appears in the algorithm itself.

Let D be a positive discriminant chosen to be a small multiple of the number N that we want to factor (for example we could take $D = N$ if $N \equiv 1$ (mod 4), $D = 4N$ otherwise). Without loss of generality, we may assume that if $D \equiv 0$ (mod 4), then $D/4 \equiv 2$ or 3 (mod 4), since otherwise we may replace D by $D/4$, and furthermore we may assume that D/N is squarefree, up to a possible factor of 4.

As in Shanks's class group method seen in the preceding section, we are going to look for ambiguous forms of discriminant D. Since here D is positive, we must be careful with the definitions. Recall from Chapter 5 that we have defined composition of quadratic forms only modulo the action of Γ_∞. We will say that a form is ambiguous if its square is *equal* to the identity modulo the action of Γ_∞, and not simply equivalent to it. In other words, the square of $f = (a, b, c)$ as given by Definition 5.4.6 must be of the form $(1, b', c')$. Clearly this is equivalent to $a \mid b$. Hence, a will be a factor of D, so once again ambiguous forms give us factorizations of D. The notion of ambiguous form must not be confused with the weaker notion of form belonging to an ambiguous cycle (see Section 5.7) which simply means that its square is equivalent to the identity modulo the action of $\mathrm{PSL}_2(\mathbb{Z})$ and not only of Γ_∞, i.e. belongs to the principal cycle.

Now let $g = (a, b, c)$ be a reduced quadratic form of discriminant D such that $a \mid c$. We note that since g is reduced hence primitive, we must have $\gcd(a, b) = 1$. Using Definition 5.4.6, one obtains immediately that

$$g^2 = (a^2, b, c/a) \ ,$$

this form being of course not necessarily reduced. This suggests the following idea.

We start from the identity form and use the ρ reduction operator used at length in Chapter 5 to proceed along the principal cycle, and we look for a form $f = (A, B, C)$ such that A is a square (such a form will be called a *square form*). We will see in a moment how plausible it is to believe that we can find such a form. Assume for the moment that we have found one, and set $A = a^2$ and $g = (a, B, aC)$.

Now g may not be primitive. In that case let p be a prime dividing the coefficients of g. Then if $p = 2$ we have $4 \mid A$ and $2 \mid B$. Hence, $D \equiv B^2 \equiv$

0 or 4 (mod 16), contradicting $D/4 \equiv 2$ or 3 (mod 4) when $4 \mid D$. If $p > 2$, then $p^2 \mid D$ hence since D/N or $D/(4N)$ is squarefree, we have $p^2 \mid N$. Although this case is rare in practice, it could occur, so we must compute $\gcd(a, B)$, and if this is not equal to 1 it gives a non-trivial factor of N (in fact its square divides N), and we can start the factorization after removing this factor.

Therefore we may assume that g is primitive. It is then clear from the definition that $g^2 = f$, whence the name "square form" given to f.

Now we start from $g^{-1} = (a, -B, aC)$ (which may not be reduced) and proceed along its cycle by applying the ρ operator. Since g^2 lies on the principal cycle, the reduced forms equivalent to g^{-1} will be on an ambiguous cycle.

Now we have the following proposition.

Proposition 8.7.1. *Keeping the above notations, there exists an ambiguous form g_1 on the cycle of g^{-1} at exactly half the distance (measured with the δ function introduced in Chapter 5) of f from the unit form.*

Proof. We prove this in the language of ideals, using the correspondence between classes of forms modulo Γ_∞ and classes of ideals modulo multiplication by $\mathbb{Q}^*$ given in Section 5.2.

Let $\mathfrak{a}$ be a representative of the ideal class (modulo $\mathbb{Q}^*$) corresponding to the quadratic form $g = (a, B, aC)$. Then by assumption, $\mathfrak{a}^2 = \gamma \mathbb{Z}_K$ for some $\gamma \in K$ which is of positive norm since $A = a^2 > 0$, and hence, in particular, $\mathcal{N}(\gamma) = \mathcal{N}(\mathfrak{a})^2$. Set

$$\beta = \gamma + \mathcal{N}(\mathfrak{a}) \quad \text{and} \quad \mathfrak{b} = \beta^{-1}\mathfrak{a} \ .$$

(Note that if desired, we can choose $a > 0$ and $\mathfrak{a}$ to be the unique primitive integral ideal corresponding to g, and then $\mathcal{N}(\mathfrak{a}) = a$.)

If, as usual, σ denotes real conjugation in K, we have chosen β such that

$$\frac{\sigma(\beta)}{\beta} = \frac{\mathcal{N}(\mathfrak{a})}{\gamma} = \frac{\sigma(\gamma)}{\mathcal{N}(\mathfrak{a})} \ .$$

Although it is trivial to give β explicitly, the knowledgeable reader will recognize that the existence of such a β is guaranteed by Hilbert's Theorem 90.

Now I claim that the quadratic form corresponding to $\mathfrak{b}$ is the ambiguous form that we are looking for. First, using the equations given above, we have

$$\mathfrak{b}^2 = \beta^{-2}\mathfrak{a}^2 = \frac{\gamma}{\beta^2}\mathbb{Z}_K = \frac{\mathcal{N}(\mathfrak{a})}{\mathcal{N}(\beta)}\mathbb{Z}_K$$

so the ideal $\mathfrak{b}^2$ is indeed equivalent up to multiplication by an element of $\mathbb{Q}^*$ to the unit ideal, so if g_1 is the quadratic form corresponding to $\mathfrak{b}$, it is ambiguous.

Second, we clearly have $\gamma/\sigma(\gamma) = (\beta/\sigma(\beta))^2$ hence

$$\delta(g_1, g) = \frac{1}{2} \ln \left| \frac{\beta}{\sigma(\beta)} \right| = \frac{1}{4} \ln \left| \frac{\gamma}{\sigma(\gamma)} \right| = \frac{1}{2} \delta(1, f)$$

thus proving the proposition.   □

Using this proposition, we see that with approximately half the number of applications of the ρ operator that were necessary to go from the identity to f, we go back from g^{-1} to an ambiguous form. In fact, since we know the exact distance that we have to go, we could use a form of the powering algorithm to make this last step much faster.

Now there are two problems with this idea. First, some ambiguous forms will correspond to trivial factorizations of N. Second, we have no guarantee that we will find square forms other than the identity. This will for instance be the case when the principal cycle is very short.

For the first problem, we could simply go on along the principal cycle if a trivial factorization is found. This would however not be satisfactory since for each square form that we encounter which may correspond to a trivial factorization, we would have to go back half the distance starting from g^{-1} before noticing this.

A good solution proposed by Shanks is as follows. Assume for the moment that $D = N$ or $D = 4N$. We obtain trivial factorizations of N exactly when the ambiguous cycle on which g^{-1} lies is the principal cycle itself. Hence, $f = g^2$ will be a square form which is equal to the square of a form on the principal cycle. Since all the forms considered are reduced, this can happen only if $g = (a, b, c)$ with $a^2 < \sqrt{D}$, hence $|a| < D^{1/4}$, which is quite a rare occurrence. When such an a occurs, we store $|a|$ in a list of dubious numbers, which Shanks calls the *queue*. Note that the condition $|a| < D^{1/4}$ is a necessary, but in general not a sufficient condition for the form g to be on the principal cycle, hence we may be discarding some useful numbers. In practice, this has little importance.

Now when a square form (A, B, C) with $A = a^2$ is found, we check whether a is in the queue. If it is, we ignore it. Otherwise, we are certain that the corresponding square root g is not in the principal cycle. (Note that the distance of the identity to $f = g^2$ is equal to twice the distance of the identity to g. This means that if g was in the principal cycle, we would have encountered it *before* encountering f.) Hence, we get a non-trivial factorization of D. This may of course give the spurious factors occurring in D/N, in which case one must go on. In fact, one can in this case modify the queue so that these factorizations are also avoided.

The second problem is more basic: what guarantee do we have that we can find a square form different from the identity in the principal cycle? For example, when the length of the cycle is short, there are none. This is the case, for example, for numbers N of the form $N = a^2 + 4$ for a odd, where the length of the cycle is equal to 1.

There are two different and complementary answers to this question. First, a heuristic analysis of the algorithm shows that the average number of reduction steps necessary to obtain a useful square form is $O(N^{1/4})$ (no ϵ here). This is much shorter than the usual length of the period which is in general of the order of $O(N^{1/2})$, so we can reasonably hope to obtain a square form before hitting the end of the principal cycle.

Second, to avoid problems with the length of the period, it may be worthwhile to work simultaneously with two discriminants D which are multiples of N, for example N and $5N$ when $N \equiv 1 \pmod 4$, $3N$ and $4N$ when $N \equiv 3 \pmod 4$. It is highly unlikely that both discriminants will have short periods. In addition, although the average number of reduction steps needed is on the order of $N^{1/4}$, experiments show that there is a very large dispersion around the mean, some numbers being factored much more easily than others. This implies that by running simultaneously two discriminants, one may hope to gain a substantial factor on average, which would compensate for the fact that twice as much work must be done.

We now give the basic algorithm, i.e. using only $D = N$ if $N \equiv 1 \pmod 4$, $D = 4N$ otherwise, and not using the fact than once g is found we can go back much faster by keeping track of distances.

Algorithm 8.7.2 (Shanks's SQUFOF). Given an odd integer N, this algorithm tries to find a non-trivial factor of N.

1. [Is N prime?] Using Algorithm 8.2.2, check whether N is a probable prime. If it is, output a message to that effect and terminate the algorithm.

2. [Is N square?] Using Algorithm 1.7.3, test whether N is a square. If it is, let n be its square root (also given by the algorithm), output n and terminate the algorithm.

3. [Initializations] If $N \equiv 1 \pmod 4$, let $D \leftarrow N$, $d \leftarrow \lfloor \sqrt{D} \rfloor$, $b \leftarrow 2\lfloor (d - 1)/2 \rfloor + 1$. Otherwise, let $D \leftarrow 4N$, $d \leftarrow \lfloor \sqrt{D} \rfloor$, $b \leftarrow 2\lfloor d/2 \rfloor$. Then set $f \leftarrow (1, b, (b^2 - D)/4)$, $Q \leftarrow \emptyset$ (Q is going to be our queue), $i \leftarrow 0$, $L \leftarrow \lfloor \sqrt{d} \rfloor$.

4. [Apply rho] Let $f = (A, B, C) \leftarrow \rho(f)$, where ρ is given by Definition 5.6.4, and set $i \leftarrow i + 1$. If i is odd, go to step 7.

5. [Squareform?] Using Algorithm 1.7.3, test whether A is a square. If it is, let a be the (positive) square root of A (which is also output by Algorithm 1.7.3) and if $a \notin Q$ go to step 8.

6. [Short period?] If $A = 1$, output a message saying that the algorithm ran through the i elements of the principal cycle without finding a non-trivial squareform, and terminate the algorithm.

7. [Fill queue and cycle] If $|A| \leq L$, set $Q \leftarrow Q \cup \{|A|\}$. Go to step 4.

8. [Initialize back-cycle] (Here we have found a non-trivial square form). Let $s \leftarrow \gcd(a, B, D)$. If $s > 1$, output s^2 as a factor of N and terminate the algorithm (or start again with N replaced by N/s^2). Otherwise, set $g \leftarrow (a, -B, aC)$. Apply ρ to g until g is reduced, and write $g = (a, b, c)$.

9. [Back-cycle] Let $b_1 \leftarrow b$ and $g = (a, b, c) \leftarrow \rho(g)$. If $b_1 \neq b$ go to step 9. Otherwise, output $|a|$ if a is odd, $|a/2|$ if a is even, and terminate the algorithm.

Some remarks are in order. First, it is essential that N be a composite number, otherwise the queue will fill up indefinitely without the algorithm finding a square form. Also, N must not be a square, otherwise we do not have a quadratic field to work with. This is the reason why steps 1 and 2 have been explicitly included.

Second, once these cases out of the way, experiment shows that the queue stays small. A storage capacity of 50 is certainly more than sufficient.

Third, during the back-cycle part of the algorithm, we need to test whether we hit upon our ambiguous form. To do this, we could use the necessary and sufficient condition that $a \mid b$. It is however a simple exercise (see Exercise 12) to show that this is equivalent to the condition $b_1 = b$ used in step 9.

Several improvements are possible to this basic algorithm, including those mentioned earlier. For example, the queue could be used to shorten the back-cycle length, starting at hg^{-1} instead of g^{-1}, where h is the form corresponding to the last element put in the queue. We will not dwell on this here.

One of the main reasons why SQUFOF is attractive is that it works exclusively with reduced quadratic forms (a, b, c) of discriminant at most a small multiple of N, hence such that a, b and c are of the order of $N^{1/2}$. This implies that the basic operations in SQUFOF are much faster than in the other factoring algorithms where operations on numbers of size N or N^2 must be performed. Of course, this is only a constant factor, but in practice it is very significant. Furthermore, the algorithm is extremely simple, so it can easily be implemented even on a 10-digit pocket calculator, and one can then factor numbers having up to 19 or 20 digits without any multi-precision arithmetic.

Unfortunately, SQUFOF is not sensitive to the size of the small prime factors of N, hence contrary to Pollard's rho method, cannot be used to cast out small primes. So if N has more than 25 digits, say, SQUFOF becomes completely useless, while Pollard rho still retains its value (although it is superseded by ECM for larger numbers, see Chapter 10).

8.8 The $p - 1$-method

The last factoring method which we will study in this chapter is a little special for two reasons. First, it is not a general purpose factoring method, but a way to find quickly prime factors of N that may be very large, but which possess certain properties. Second, the idea behind the method has successfully been used in some of the most successful modern factoring method like the elliptic curve method (see Section 10.3). Hence it is important to understand this method at least as an introduction to Chapter 10.

8.8.1 The First Stage

We need a definition.

Definition 8.8.1. *Let B be a positive integer. A positive integer n will be said to be B-smooth if all the prime divisors of n are less than or equal to B. We will say that n is B-powersmooth if all prime powers dividing n are less than or equal to B.*

These notions of smoothness are quite natural in factoring methods, and we will see that they become essential in the modern methods. The idea behind the $p-1$ method is the following. Let p be a prime dividing the number N that we want to split (p is of course a priori unknown). Let $a > 1$ be an integer (which we can assume coprime to N by computing a GCD, otherwise N will have split). Then by Fermat's theorem, $a^{p-1} \equiv 1 \pmod{p}$. Now assume that $p-1$ is B-powersmooth for a certain B which is not too large. Then by definition $p-1$ divides the least common multiple of the numbers from 1 to B, which we will denote by $\mathrm{lcm}[1..B]$. Hence, $a^{\mathrm{lcm}[1..B]} \equiv 1 \pmod{p}$, which implies that

$$(a^{\mathrm{lcm}[1..B]} - 1, N) > 1 .$$

As in the Pollard ρ method, if this is tested for increasing values of B, it is highly improbable that this GCD will be equal to N, hence we will have found a non-trivial divisor of N. This leads to the following algorithm, which in this form is due to Pollard.

Algorithm 8.8.2 ($p-1$ First Stage). Let N be a composite number, and B be an a priori chosen bound. This algorithm will try to find a non-trivial factor of N, and has a chance of succeeding only when there exists a prime factor p of N such that $p-1$ is B-powersmooth. We assume that we have precomputed a table $p[1], \ldots, p[k]$ of all the primes up to B.

1. [Initialize] Set $x \leftarrow 2$, $y \leftarrow x$, $c \leftarrow 0$, $i \leftarrow 0$, and $j \leftarrow i$.
2. [Next prime] Set $i \leftarrow i+1$. If $i > k$, compute $g \leftarrow (x-1, N)$. If $g = 1$ output a message saying that the algorithm has not succeeded in splitting N, and terminate, else set $i \leftarrow j$, $x \leftarrow y$ and go to step 5. Otherwise (i.e. if $i \leq k$), set $q \leftarrow p[i]$, $q_1 \leftarrow q$, $l \leftarrow \lfloor B/q \rfloor$.
3. [Compute power] While $q_1 \leq l$, set $q_1 \leftarrow q \cdot q_1$. Then, set $x \leftarrow x^{q_1} \bmod N$, $c \leftarrow c+1$ and if $c < 20$ go to step 2.
4. [Compute GCD] Set $g \leftarrow (x-1, N)$. If $g = 1$, set $c \leftarrow 0$, $j \leftarrow i$, $y \leftarrow x$ and go to step 2. Otherwise, set $i \leftarrow j$ and $x \leftarrow y$.
5. [Backtrack] Set $i \leftarrow i+1$, $q \leftarrow p[i]$ and $q_1 \leftarrow q$.
6. [Finished?] Set $x \leftarrow x^q \bmod N$, $g \leftarrow (x-1, N)$. If $g = 1$, set $q_1 \leftarrow q \cdot q_1$ and if $q_1 \leq B$, go to step 6, else go to step 5. Otherwise (i.e. if $g > 1$), if $g < N$ output g and terminate the algorithm. Finally, if $g = N$ (a rare occurrence), output that the algorithm has failed and terminate.

Note that this algorithm may fail for two completely different reasons. The first one, by far the most common, occurs in step 2, and comes because N does not have any prime divisor p such that $p-1$ is B-powersmooth. In fact, it proves this. The second reason why it may fail occurs in step 6, but this is extremely rare. This would mean that all the prime p divisors of N are found simultaneously. If this is the case, then this means that there certainly exists a p dividing N which is B-powersmooth. Hence, it may be worthwhile to try the algorithm with a different initial value of x, for example $x \leftarrow 3$ instead of $x \leftarrow 2$.

Even in this simple form, the behavior of the $p-1$ algorithm is quite impressive. Of course, it does not pretend to be a complete factoring algorithm (in fact when $N = (2p+1)(2q+1)$ where p, q, $2p+1$ and $2q+1$ are primes with p and q about the same size, the running time of the algorithm will in general be $O(N^{1/2+\epsilon})$ if we want to factor N completely, no better than trial division). On the other hand, it may succeed in finding very large factors of N, since it is not the size of the prime factors of N which influence the running time but rather the smoothness of the prime factors minus 1.

The size of B depends essentially on the time that one is willing to spend. It is also however also strongly conditioned by the existence of a second stage to the algorithm as we shall see presently. Usual values of B which are used are, say, between 10^5 and 10^6.

8.8.2 The Second Stage

Now an important practical improvement to the $p-1$ algorithm (which one also uses in the modern methods using similar ideas) is the following. It may be too much to ask that there should exist a prime divisor p of N such that $p-1$ is B-powersmooth. It is more reasonable to ask that $p-1$ should be completely factored by trial division up to B. But this means that $p-1 = fq$, where f is B-smooth, and q is a prime which may be much larger than B (but not than B^2). For our purposes, we will slightly strengthen this condition and assume that N has a prime factor p such that $p-1 = fq$ where f is B_1-powersmooth and q is a prime such that $B_1 < q \leq B_2$, where B_1 is our old B, and B_2 is a much larger constant. We must show how are we going to find such a p. Of course, $p-1$ is B_2-powersmooth so we could use the $p-1$ algorithm with B_1 replaced by B_2. This is however unrealistic since B_2 is much larger than B_1.

Now we have as usual

$$(a^{q \operatorname{lcm}[1..B_1]} - 1, N) > 1$$

and we will proceed as follows. At the end of the first stage (i.e. of Algorithm 8.8.2 above), we will have computed $b \leftarrow a^{\operatorname{lcm}[1..B_1]} \bmod N$. We store a table of the difference of primes from B_1 to B_2. Now these differences are small, and there will not be many of them. So we can quickly compute b^d for all possible differences d, and obtain all the b^q by *multiplying* successively an initial power of b by these precomputed b^d. Hence, for each prime, we replace a powering operation by a simple multiplication, which is of course *much* faster, and this is why we can go much further. This leads to the following algorithm.

Algorithm 8.8.3 ($p-1$ with Stage 2). Let N be a composite number, and B_1 and B_2 be a priori chosen bounds. This algorithm will try to find a non-trivial factor of N, and has a chance of succeeding only when there exists a prime factor p of N such that $p-1$ is equal to a B_1-powersmooth number times a prime less than or equal to B_2. We assume that we have precomputed a table $p[1], \ldots ,p[k_1]$ of all the primes up to B_1 and a table $d[1], \ldots ,d[k_2]$ of the differences of the primes from B_1 to B_2, with $d[1] = p[k_1 + 1] - p[k_1]$, etc $\ldots$

1. [First stage] Using $B = B_1$, try to split N using Algorithm 8.8.2 (i.e. the first stage. If this succeeds, terminate the algorithm. Otherwise, we will have obtained a number x at the end of Algorithm 8.8.2, and we set $b \leftarrow x$, $c \leftarrow 0$, $P \leftarrow 1$, $i \leftarrow 0$, $j \leftarrow i$ and $y \leftarrow x$.

2. [Precomputations] For all values of the differences $d[i]$ (which are small and few in number), precompute and store $b^{d[i]}$. Set $x \leftarrow x^{p[k_1]}$.

3. [Advance] Set $i \leftarrow i+1$, $x \leftarrow x \cdot b^{d[i]}$ (using the precomputed value of $b^{d[i]}$), $P \leftarrow P \cdot (x-1)$, $c \leftarrow c+1$. If $i \geq k_2$, go to step 6. Otherwise, if $c < 20$, go to step 3.

4. [Compute GCD] Set $g \leftarrow (P, N)$. If $g = 1$, set $c \leftarrow 0$, $j \leftarrow i$, $y \leftarrow x$ and go to step 3.

5. [Backtrack] Set $i \leftarrow j$, $x \leftarrow y$. Then repeat $x \leftarrow x \cdot b^{d[i]}$, $i \leftarrow i+1$, $g \leftarrow (x-1, N)$ until $g > 1$ (this must occur). If $g < N$ output g and terminate the algorithm. Otherwise (i.e. if $g = N$, a rare occurrence), output that the algorithm has failed (or try again using $x \leftarrow 3$ instead of $x \leftarrow 2$ in the first step of Algorithm 8.8.2), and terminate.

6. [Failed?] Set $g \leftarrow (P, N)$. If $g = 1$, output that the algorithm has failed and terminate. Otherwise go to step 5.

In this form, the $p-1$ algorithm is much more efficient than using the first stage alone. Typical values which could be used are $B_1 = 2 \cdot 10^6$, $B_2 = 10^8$. See also [Mon2] and [Bre2] for further improvements.

8.8.3 Other Algorithms of the Same Type

The main drawback of the $p-1$ algorithm is that there is no reason for N to have a prime divisor p such that $p-1$ is smooth. As with the primality tests (see Section 8.3.2), we can also detect the primes p such that $p+1$ is smooth, or also $p^2 + p + 1$, $p^2 + 1$, $p^2 - p + 1$ (although since these numbers are much larger, their probability of being smooth for a given bound B is much smaller). We leave as an exercise for the reader (Exercise 13) to write an algorithm when $p+1$ is B-powersmooth.

We see that the number of available groups which give numbers of reason-
able size (here $\mathbb{F}_p^*$ and $\mathbb{F}_{p^2}^*/\mathbb{F}_p^*$, which give $p-1$ and $p+1$ respectively) is very
small (2) and this limits the usefulness of the method. The idea of the elliptic
curve method (ECM) is to use the group of points of an elliptic curve over
$\mathbb{F}_p$, which also has approximately p elements by Hasse's Theorem 7.1.8, and
this will lead to a much better algorithm since we will have at our disposal a
large number of groups of small size instead of only two. See Section 10.3 for
details.

8.9 Exercises for Chapter 8

1. Show that an odd prime number p is a strong pseudo-prime in any base not
    divisible by p.

2. If N is the 46 digit composite number due to Arnault given in the text as an
    example of a strong pseudoprime to all prime bases $a \leq 31$, compute explicitly
    $a^{(N-1)/4} \bmod N$ for these a and show that -1 has at least 5 different square
    roots modulo N (showing clearly N that is not prime even without knowing its
    explicit factorization). From this remark, deduce a strengthening of the Rabin-
    Miller test which would not be passed for example by Arnault's number.

3. Show that if N is any odd integer, the congruence

    $$a^{N-1} \equiv -1 \pmod{N}$$

    is impossible. More generally, show that

    $$a^k \equiv -1 \pmod{N}$$

    implies that

    $$v_2(k) \leq v_2\left(\frac{N-1}{2}\right) .$$

    The following four exercises are due to H. W. Lenstra.

4. Show that there are only a finite number of integers N such that for all $a \in \mathbb{Z}$
    we have

    $$a^{N+1} \equiv a \pmod{N} ,$$

    and give the complete list.

5. Let N be a positive integer such that $2^N \equiv 1 \pmod{N}$. Show that $N = 1$.

6. Let a be a positive integer such that $a^4 + 4^a$ is a prime number. Show that $a = 1$.

7. Show that there exists infinitely many n for which at least one of $2^{2^n} + 1$ or
    $6^{2^n} + 1$ is composite.

8. Denote by F_k the k-th Fermat number, i.e. $F_k = 2^{2^k} + 1$.
    a) Show manually that F_k is prime for $0 \leq k \leq 4$ but that $641 \mid F_5$.
    b) Let $h > 1$ be an integer such that $h \equiv 1 \pmod{F_0 F_1 F_2 F_3 F_4}$. If $h2^n + 1$
    is prime show that $32 \mid n$.
    c) Conclude that there exists an a such that if

$$h \equiv a \pmod{F_0 F_1 F_2 F_3 F_4 F_5}$$

and $h > 1$, then for all n, $h2^n + 1$ is composite.

9. Let $N = 2^{2^k} + 1$ be a Fermat number. Prove that in this case Proposition 8.3.1 can be made more precise as follows: N is prime if and only if $3^{(N-1)/2} \equiv -1 \pmod{N}$ (use the quadratic reciprocity law).

10. Using implicitly the finite field $\mathbb{F}_{N^2}$, write a primality testing algorithm in the case where $N + 1$ is completely factored, using a proposition similar to 8.3.1.

11. Using the algorithm developed in Exercise 10, show that the Mersenne number $N = 2^p - 1$ is prime if and only p is prime and (for $p \neq 2$) if the sequence defined by $u_0 = 4$ and $u_{k+1} = u_k^2 - 2 \bmod N$ satisfies $u_{p-2} = 0$ (this is called the Lucas-Lehmer test).

12. Let $g = (a, b, c)$ and $g_1 = (a_1, b_1, c_1) = \rho(g)$ be reduced forms with positive discriminant. Show that g_1 is an ambiguous form if and only if $b = b_1$.

13. The $p-1$-algorithm is based on the properties of the finite field $\mathbb{F}_p$. Using instead the field $\mathbb{F}_{p^2}$, develop a $p + 1$-factoring algorithm for use when a prime factor p of N is such that $p + 1$ is B-powersmooth for some reasonable bound B.

14. Let N be a number to be factored. Assume that after one of the factoring algorithms seen in this chapter we have found a number a such that $d = \gcd(N, a)$ satisfies $1 < d < N$ hence gives a non-trivial divisor of N. Write an algorithm which extracts as much information as possible from this divisor d, i.e. which finds N_1 and N_2 such that $N = N_1 N_2$, $\gcd(N_1, N_2) = 1$ and $d \mid N_1$.

Chapter 9

Modern Primality Tests

In Section 8.3, we studied various primality tests, essentially the $N - 1$ test, and saw that they require knowing the factorization of $N - 1$ (or $N + 1, \ldots$), which are large numbers. Even though only partial factorizations are needed, the tests of Section 8.3 become impractical as soon as N has more than 100 digits, say. A breakthrough was made in 1980 by Adleman, Pomerance and Rumely, that enabled testing the primality of much larger numbers. The APR test was further simplified and improved by H. W. Lenstra and the author, and the resulting APRCL test was implemented in 1981 by A. K. Lenstra and the author, with the help of D. Winter. It is now possible to prove the primality of numbers with 1000 decimal digits in a not too unreasonable amount of time. The running time of this algorithm is $O((\ln N)^{C \ln \ln \ln N})$ for a suitable constant C. This is almost a polynomial time algorithm since for all practical purposes the function $\ln \ln \ln N$ acts like a constant. (Note that the practical version of the algorithm is probabilistic, but that there exists a non-probabilistic but less practical version.)

We will describe the algorithm in Section 9.1, without giving all the the implementation tricks. The reader will find a detailed description of this algorithm and its implementation in [Coh-Len2], [Coh-Len3] and [Bos-Hul].

In 1986, another primality testing algorithm was invented, first for theoretical purposes by Goldwasser and Kilian, and then considerably modified so as to obtain a practical algorithm by Atkin. This algorithm has been implemented by Atkin and Morain, and is also practical for numbers having up to 1000 digits. The expected running time of this algorithm is $O(\ln^6 N)$, hence is polynomial time, but this is only on average since for some numbers the running time could be much larger. A totally non-practical version using a higher dimensional analog of this test has been given by Adleman and Huang, and they can prove that their test is polynomial time. In other words, they prove the following theorem ([Adl-Hua]).

Theorem 9.1. *There exists a probabilistic polynomial time algorithm which can prove or disprove that a given number N is prime.*

Their proof is pretty but very complex, and this theorem is one of the major achievements of theoretical algorithmic number theory.

We will describe Atkin's practical primality test in Section 9.2, and we refer to [Atk-Mor] and to [Mor2] for implementation details.

9.1 The Jacobi Sum Test

The idea of the APRCL method is to test Fermat-type congruences in higher degree number fields, and more precisely in certain well chosen cyclotomic fields. We need a few results about group rings in this context.

9.1.1 Group Rings of Cyclotomic Extensions

Recall first the following definitions and results about cyclotomic fields (see [Was]).

Definition 9.1.1. *If n is a positive integer, the n-th cyclotomic field is the number field $\mathbb{Q}(\zeta_n)$, where ζ_n is a primitive n-th root of unity, for example $\zeta_n = e^{2i\pi/n}$.*

Proposition 9.1.2. *Let $K = \mathbb{Q}(\zeta_n)$ be the n-th cyclotomic field.*

(1) *The extension $K/\mathbb{Q}$ is a Galois extension, with Abelian Galois group given by*

$$G = \mathrm{Gal}(K/\mathbb{Q}) = \{\sigma_a, \ (a,n) = 1, \ \text{where } \sigma_a(\zeta_n) = \zeta_n^a\} \ .$$

In particular, the degree of $K/\mathbb{Q}$ is $\phi(n)$, where ϕ is Euler's phi function.
(2) *The ring of integers of K is $\mathbb{Z}_K = \mathbb{Z}[\zeta_n]$.*

We now come to the definition of a group ring. We could of course bypass this definition, but the notations would become very cumbersome.

Definition 9.1.3. *Let G be any finite group. The group ring $\mathbb{Z}[G]$ is the set of maps (not necessarily homomorphisms) from G to $\mathbb{Z}$ with the following two operations. If f_1 and f_2 are in $\mathbb{Z}[G]$, we naturally define*

$$(f_1 + f_2)(\sigma) = f_1(\sigma) + f_2(\sigma)$$

for all $\sigma \in G$. The multiplication law is more subtle, and is defined by

$$f_1 \cdot f_2(\sigma) = \sum_{\tau \in G} f_1(\tau) f_2(\tau^{-1}\sigma) \ .$$

The name group ring is justified by the easily checked fact that the above operations do give a ring structure to $\mathbb{Z}[G]$. If for $f \in \mathbb{Z}[G]$, we set formally

$$f = \sum_{\sigma \in G} f(\sigma)[\sigma] \ ,$$

where $[\sigma]$ is just a notation, then it is easy to see that addition and multiplication become natural $\mathbb{Z}$-algebra laws, if we set, as is natural, $[\sigma_1] \cdot [\sigma_2] = [\sigma_1 \sigma_2]$. This is the notation which we will use. Note also that although we have only defined group rings $\mathbb{Z}[G]$ for finite groups G, it is easy to extend this to infinite groups by requiring that all but a finite number of images of the maps be equal to 0 (in order to have finite sums).

We can consider $\mathbb{Z}$ as a subring of $\mathbb{Z}[G]$ by identifying n with $n[1]$, where 1 is the unit element of G, and we will use this identification from now on.

We now specialize to the situation where $G = \mathrm{Gal}(K/\mathbb{Q})$ for a number field K Galois over $\mathbb{Q}$, and in particular to the case where K is a cyclotomic field. By definition, the group G acts on K, and also on all objects naturally associated to K: the unit group, the class group, etc ... One can extend this action of G in a natural way to an action of $\mathbb{Z}[G]$ in the following way. If $f \in \mathbb{Z}[G]$ and $x \in K$, then we set

$$f(x) = \prod_{\sigma \in G} \sigma(x)^{f(\sigma)} .$$

In the expanded form where we write $f = \sum_{\sigma \in G} n_\sigma[\sigma]$, one sees immediately that this corresponds to a *multiplicative* extension of the action of G, and suggests using the notation x^f instead of $f(x)$ so that

$$x^f = \prod_{\sigma \in G} \sigma(x)^{n_\sigma} .$$

Indeed, it is easy to check the following properties (x, x_1 and x_2 are in K and f, f_1 and f_2 are in $\mathbb{Z}[G]$):

(1) $x^{f_1 + f_2} = x^{f_1} \cdot x^{f_2}$.
(2) $x^{f_1 \cdot f_2} = (x^{f_1})^{f_2} = (x^{f_2})^{f_1}$.
(3) $(x_1 + x_2)^f = x_1^f + x_2^f$
(4) $(x_1 x_2)^f = x_1^f x_2^f$

We now fix a prime number p and an integer k, and consider the n-th cyclotomic field K, where $n = p^k$. Let G be its Galois group, which is the set of all σ_a for $a \in (\mathbb{Z}/n\mathbb{Z})^*$ by Proposition 9.1.2. Since it is Abelian, the group ring $\mathbb{Z}[G]$ is a commutative ring. Set

$$\mathfrak{p} = \{f \in \mathbb{Z}[G] \;/\; \zeta_p^f = 1\} ,$$

where $\zeta_p = e^{2i\pi/p}$ is a primitive p^{th} root (not p^k) of unity. Then one checks immediately that $\mathfrak{p}$ is an ideal of $\mathbb{Z}[G]$. In fact, if $f = \sum_{a \in (\mathbb{Z}/n\mathbb{Z})^*} n_a[\sigma_a]$, then $f \in \mathfrak{p}$ if and only if $\sum_{a \in (\mathbb{Z}/n\mathbb{Z})^*} a n_a \equiv 0 \pmod{p}$. This shows that the number of cosets of $\mathbb{Z}[G]$ modulo $\mathfrak{p}$ is equal to p (the number of different incongruent sums $\sum a n_a$ modulo p), hence that $\mathfrak{p}$ is in fact a *prime* ideal of degree one (i.e. of norm equal to p). Clearly, it is generated over $\mathbb{Z}$ by p (i.e. $p[1]$) and all the $a - [\sigma_a]$.

9.1.2 Characters, Gauss Sums and Jacobi Sums

Recall that a *character* (more precisely a Dirichlet character) χ modulo q is a group homomorphism from $(\mathbb{Z}/q\mathbb{Z})^*$ to $\mathbb{C}^*$ for some integer q. This can be naturally extended to a multiplicative map from $(\mathbb{Z}/q\mathbb{Z})$ to $\mathbb{C}$ by setting $\chi(x) = 0$ if $x \notin (\mathbb{Z}/q\mathbb{Z})^*$. It can then be lifted to a map from $\mathbb{Z}$ to $\mathbb{C}$, which by abuse of notation we will still denote by χ. The set of characters modulo q forms a group, and for instance using Section 1.4.1 one can easily show that this group is (non-canonically) isomorphic to $(\mathbb{Z}/q\mathbb{Z})^*$, and in particular has $\phi(q)$ elements. The unit element of this group is the character χ_0 such that $\chi_0(x) = 1$ if $(x, q) = 1$ and 0 otherwise.

Proposition 9.1.4. *Let χ be a character different from χ_0. Then*

$$\sum_{x \in (\mathbb{Z}/q\mathbb{Z})^*} \chi(x) = 0 .$$

Dually, if $x \not\equiv 1 \pmod{q}$, then

$$\sum_{\chi} \chi(x) = 0 ,$$

where the sum is over all characters modulo q.

Proof. Since $\chi \neq \chi_0$, there exists a number a coprime to q such that $\chi(a) \neq 1$. Set $S = \sum_x \chi(x)$. Since χ is multiplicative we have $\chi(a)S = \sum_x \chi(ax)$. Since a is coprime to q and hence invertible modulo q, the map $x \mapsto ax$ is a bijection of $(\mathbb{Z}/q\mathbb{Z})^*$ onto itself. It follows that $\chi(a)S = \sum_y \chi(y) = S$, and since $\chi(a) \neq 1$, this shows that $S = 0$ as claimed. The second part of the proposition is proved in the same way using the existence of a character χ_1 such that $\chi_1(x) \neq 1$ when $x \not\equiv 1 \pmod{q}$. $\square$

The *order* of a character χ is the smallest positive n such that $\chi(a)^n = 1$ for all integers a prime to q, in other words it is the order of χ considered as an element of the group of characters modulo q.

Definition 9.1.5.

(1) *Let χ be a character modulo q. The* Gauss sum $\tau(\chi)$ *is defined by*

$$\tau(\chi) = \sum_{x \in (\mathbb{Z}/q\mathbb{Z})^*} \chi(x)\zeta_q^x ,$$

where as usual $\zeta_q = e^{2i\pi/q}$.

(2) *Let χ_1 and χ_2 be two characters modulo q. The* Jacobi sum $j(\chi_1, \chi_2)$ *is defined by*

$$j(\chi_1, \chi_2) = \sum_{x \in (\mathbb{Z}/q\mathbb{Z})^*} \chi_1(x)\chi_2(1 - x) .$$

Note that since we have extended characters by 0, we can replace $(\mathbb{Z}/q\mathbb{Z})^*$ by $\mathbb{Z}/q\mathbb{Z}$, and also that in the definition of Jacobi sums, one could exclude $x = 1$ which contributes 0 to the sum.

From the definitions, it is clear that if χ is a character modulo q of order n (hence $n \mid \phi(q)$), then

$$\tau(\chi) \in \mathbb{Z}[\zeta_n, \zeta_q] ,$$

while if χ_1 and χ_2 are two characters modulo q of order dividing n, then

$$j(\chi_1, \chi_2) \in \mathbb{Z}[\zeta_n] .$$

This will in general be a much simpler ring than $\mathbb{Z}[\zeta_n, \zeta_q]$, and this observation will be important in the test.

The basic results about Gauss sums and Jacobi sums that we will need are summarized in the following proposition. Note that we assume that q is a prime, which makes things a little simpler.

Proposition 9.1.6.

(1) *Let $\chi \neq \chi_0$ be a character modulo a prime q. Then*

$$\tau(\chi)\tau(\bar{\chi}) = \chi(-1)q \quad and \quad |\tau(\chi)| = \sqrt{q} .$$

(2) *Let χ_1 and χ_2 be two characters modulo q such that $\chi_1\chi_2 \neq \chi_0$. Then*

$$j(\chi_1, \chi_2) = \frac{\tau(\chi_1)\tau(\chi_2)}{\tau(\chi_1\chi_2)} .$$

Proof. To simplify notations, except if explicitly stated otherwise, the summations will always be over $(\mathbb{Z}/q\mathbb{Z})^*$, and we abbreviate ζ_q to ζ. We have:

$$\tau(\chi)\tau(\bar{\chi}) = \sum_x \chi(x)\zeta^x \sum_y \bar{\chi}(y)\zeta^y = \sum_t \chi(t) \sum_y \chi(y)\bar{\chi}(y)\zeta^{y(1+t)} ,$$

by setting $x = ty$. Since $\chi(y)\bar{\chi}(y) = 1$, the inner sum is simply a sum of powers of ζ, and since q is prime, is a geometric series whose sum is equal to -1 if $1 + t \neq 0$ and to $q - 1$ otherwise. Hence, our product is equal to

$$-\sum_{t \neq -1} \chi(t) + (q-1)\chi(-1) = q\chi(-1) - \sum_t \chi(t) = q\chi(-1)$$

by Proposition 9.1.4. Finally, note that

$$\overline{\tau(\chi)} = \sum_x \bar{\chi}(x)\zeta^{-x} = \sum_x \bar{\chi}(-x)\zeta^x = \chi(-1)\tau(\bar{\chi}) ,$$

and the first part of the proposition is proved.

The second part is proved analogously. We have

$$\tau(\chi_1)\tau(\chi_2) = \sum_x \sum_y \chi_1(x)\chi_2(y)\zeta^{x+y} = \sum_t \sum_y \chi_1(t)\chi_1\chi_2(y)\zeta^{y(1+t)}$$

by setting $x = ty$. Now by setting $x = ay$ it is clear that for any $\chi \neq \chi_0$ we have

$$\sum_y \chi(y)\zeta^{ay} = \begin{cases} 0 & \text{if } a \equiv 0 \pmod{q} \\ \bar{\chi}(a)\tau(\chi) & \text{otherwise.} \end{cases}$$

Hence, since $\chi_1\chi_2 \neq \chi_0$, we have

$$\tau(\chi_1)\tau(\chi_2) = \tau(\chi_1\chi_2)\sum_{t\neq-1}\chi_1(t)\overline{\chi_1\chi_2}(1+t) = \tau(\chi_1\chi_2)\sum_u \chi_1(u)\chi_2(1-u)$$

if we set $u = t/(1+t)$ which sends bijectively $(\mathbb{Z}/q\mathbb{Z}) \setminus \{0,-1\}$ onto $(\mathbb{Z}/q\mathbb{Z}) \setminus \{0,1\}$, proving the identity. $\square$

9.1.3 The Basic Test

We now come back to our basic purpose, i.e. testing the primality of a number N. It is assumed that N has already passed the Rabin-Miller test 8.2.2, so that it is highly improbable that N is composite. The aim is to *prove* that N is prime.

In this section, we fix a prime p and a character χ of order p^k modulo a prime q (hence with $p^k \mid (q-1)$). We can of course assume that N is prime to p and q. We set for simplicity $n = p^k$, and denote by $\langle\zeta_n\rangle$ the group of n-th roots of unity, which is generated by ζ_n. We shall use a modified version of Fermat's theorem as follows.

Proposition 9.1.7. *Let $\beta \in \mathbb{Z}[G]$. Then if N is prime, there exists $\eta(\chi) \in \langle\zeta_n\rangle$ such that*

$$\tau(\chi)^{\beta(N-\sigma_N)} \equiv \eta(\chi)^{-\beta N} \pmod{N}, \qquad (*_\beta)$$

where in fact $\eta(\chi) = \chi(N)$.

Note that we consider $\mathbb{Z}[G]$ as acting not only on $\mathbb{Q}(\zeta_n)$ but also on $\mathbb{Q}(\zeta_n, \zeta_q)$, the action being trivial on ζ_q. Note also that the congruences modulo N are in fact modulo $N\mathbb{Z}[\zeta_n, \zeta_q]$.

Proof. We know that in characteristic N, $(\sum a_k)^N = \sum a_k^N$ since the binomial coefficients $\binom{N}{i}$ are divisible by N if $0 < i < N$. Hence,

$$\tau(\chi)^N \equiv \sum_x \chi(x)^N \zeta_q^{Nx} \equiv \sum_x \chi(N^{-1}x)^N \zeta_q^x \equiv \chi(N)^{-N}\tau(\chi^N) \pmod{N}$$

and the proposition follows since $\tau(\chi^N) = \tau(\chi)^{\sigma_N}$ by definition of σ_N. Note that $\tau(\chi)$ is also coprime to N since by Proposition 9.1.6, $\tau(\chi)\overline{\tau(\chi)} = q$ is coprime to N. $\qquad\square$

This proposition is a generalization of Fermat's theorem since one checks immediately that if we take $n = p = 2$ and $\beta = 1$, the proposition is equivalent to the statement $q^{(N-1)/2} \equiv \pm 1 \pmod{N}$. What we are now going to prove is in essence that if, conversely, condition $(*_\beta)$ is satisfied for a number of characters χ (with different p^k and q), then we can easily finish the proof that N is prime. First, we prove the following

Lemma 9.1.8. *Let N be any integer, and assume that $(*_\beta)$ is satisfied. Then*

(1) *For all $i > 0$*

$$\tau(\chi)^{\beta(N^i - \sigma_{N^i})} \equiv \eta(\chi)^{-\beta i N^i} \pmod{N} .$$

(2)

$$\tau(\chi)^{\beta\left(N^{(p-1)p^{k-1}} - 1\right)} \equiv \eta(\chi)^{\beta p^{k-1}} \pmod{N} .$$

(3) *If r is prime and coprime to p and q then*

$$\tau(\chi)^{\left(r^{(p-1)p^{k-1}} - 1\right)} \equiv \chi(r)^{p^{k-1}} \pmod{r} .$$

Proof. Assertion (1) follows from $(*_\beta)$ by induction on i using the identity

$$N^{i+1} - \sigma_{N^{i+1}} = N^i(N - \sigma_N) + \sigma_N(N^i - \sigma_{N^i})$$

and $\eta(\chi)^{\sigma_N} = \eta(\chi)^N$ since $\eta(\chi) \in \langle \zeta_n \rangle$. For (2) we apply the first assertion to $i = (p-1)p^{k-1}$ and use Euler's Theorem 1.4.2 which tells us that

$$N^{(p-1)p^{k-1}} \equiv 1 \pmod{p^k} .$$

The last assertion follows immediately since Proposition 9.1.7 tells us that $(*_\beta)$ is satisfied for a prime number r with $\beta = 1$ and $\eta(\chi) = \chi(r)$. $\qquad\square$

We now introduce a condition which will be crucial to all our future work. We will show that this condition is a consequence of $(*_\beta)$ conditions for suitable characters χ. This means that it will have a similar nature to the Fermat tests, but it is much more convenient to isolate it from the rest of the tests.

Definition 9.1.9. *We say that condition $\mathcal{L}_p$ is satisfied (with respect to N of course) if for all prime divisors r of N and all integers $a > 0$ we can find an integer $l_p(r, a)$ such that*

$$r^{p-1} \equiv N^{(p-1)l_p(r,a)} \pmod{p^a} .$$

Note that if N is prime this condition is trivially satisfied with $l_p(r,a) = 1$. We will see later that this condition is not as difficult as it looks and that it can easily be checked. For the moment, let us see what consequences we can deduce from it. Note first that if $l_p(r,a)$ exists for all primes r dividing N, it exists by additivity for every divisor r of N.

Note also that condition $\mathcal{L}_p$ is more nicely stated in p-adic terms, but we will stay with the present definition. One consequence of this fact which we will use (and prove later) is the following result.

Lemma 9.1.10. Let $u = v_p\left(N^{p-1} - 1\right)$ if $p \geq 3$, $u = v_2\left(N^2 - 1\right)$ if $p = 2$. Then for $a \geq b \geq u$ we have

$$l_p(r,a) \equiv l_p(r,b) \pmod{p^{b-u}} .$$

The main consequence of condition $\mathcal{L}_p$ which we need is the following.

Proposition 9.1.11. *Assume that condition $\mathcal{L}_p$ is satisfied.*

(1) *If χ satisfies $(*_\beta)$ for some $\beta \notin \mathfrak{p}$, then for all sufficiently large a and all $r \mid N$ we have*

$$\chi(r) = \chi(N)^{l_p(r,a)} \quad \text{and} \quad \eta(\chi) = \chi(N) .$$

(2) *If ψ is a character modulo a power of p and of order a power of p, then we also have*

$$\psi(r) = \psi(N)^{l_p(r,a)}$$

for sufficiently large a.

Proof. Set for simplicity $x = \tau(\chi)^\beta$. From the first part of Lemma 9.1.8 we have

$$x^{N^{(p-1)p^k} - 1} \equiv 1 \pmod{N} .$$

Set $N^{(p-1)p^k} - 1 = p^e N_1$ with $p \nmid N_1$. Set $\ell = l_p(r, \max(e, k+1))$. Then again using the first part of Proposition 9.1.8 we have

$$x^{N^{(p-1)\ell}} \equiv \eta(\chi)^{-\beta(p-1)\ell N^{(p-1)\ell}} \tau\left(\chi^{N^{(p-1)\ell}}\right)^\beta \pmod{N}$$

$$\equiv \eta(\chi)^{-\beta(p-1)\ell r^{p-1}} \tau\left(\chi^{r^{p-1}}\right)^\beta \pmod{N}$$

since $\eta(\chi)$ and χ are of order dividing p^k. If r is a prime divisor of N, we have by Proposition 9.1.7

$$x^{r^{p-1}} \equiv \chi(r)^{-\beta(p-1)r^{p-1}} \tau \left(\chi^{r^{p-1}} \right)^{\beta} \pmod{r}$$

hence, since $\tau \left(\chi^{r^{p-1}} \right)^{\beta}$ is invertible modulo r by Proposition 9.1.6, we obtain finally

$$x^{\left(N^{(p-1)\ell} - r^{p-1} \right)} \equiv \zeta^{\beta(p-1)r^{p-1}} \pmod{r} \quad \text{with } \zeta = \chi(r)\eta(\chi)^{-\ell} .$$

Now from our choice of ℓ, we have $N^{(p-1)\ell} \equiv r^{p-1} \pmod{p^e}$, hence

$$N_1 \left(N^{(p-1)\ell} - r^{p-1} \right) \equiv 0 \pmod{N^{(p-1)p^k} - 1} .$$

So if we combine this with our preceding congruences we obtain

$$x^{N_1 \left(N^{(p-1)\ell} - r^{p-1} \right)} \equiv 1 \equiv \zeta^{N_1 \beta(p-1)r^{p-1}} \pmod{r} .$$

Now we trivially have $N_1 \beta(p-1)r^{p-1} \notin \mathfrak{p}$ since $\mathfrak{p}$ is a prime ideal and none of the factors belong to $\mathfrak{p}$. Since ζ is a p^k-th root of unity, the definition of $\mathfrak{p}$ implies that it must be equal to 1, i.e. that

$$\chi(r) = \eta(\chi)^{\ell} = \eta(\chi)^{l_p(r,a)}$$

for a sufficiently large, and for all prime r dividing N. By additivity of l_p (i.e. $l_p(rr', a) = l_p(r, a) + l_p(r', a)$) it immediately follows that this is true for all divisors r of N, not only prime ones. In particular, it is true for $r = N$ and since we can take $l_p(N, a) = 1$ we have $\chi(N) = \eta(\chi)$ and the first part of the proposition is proved.

For the second part, if ψ is of order p^{k_1} modulo p^{k_2} then if we take $\ell = l_p(r, \max(k_1, k_2))$ it is clear that $\psi \left(r^{p-1} \right) = \psi \left(N^{p-1} \right)^{\ell}$ and since $p-1$ is coprime to the order of ψ we immediately get the second part of the proposition. Note that we have implicitly used Lemma 9.1.10 in the proof of both parts. $\square$

From this result, we obtain the following theorem which is very close to our final goal of proving N to be prime.

Theorem 9.1.12. *Let t be an even integer, let*

$$e(t) = 2 \prod_{\substack{q \text{ prime} \\ (q-1)|t}} q^{v_q(t)+1}$$

and assume that $(N, te(t)) = 1$. For each pair of prime numbers (p, q) such that $(q-1) \mid t$ and $p^k \| (q-1)$, let $\chi_{p,q}$ be a character modulo q of order p^k (for example $\chi_{p,q} \left(g_q^a \right) = \zeta_{p^k}^a$ if g_q is a primitive root modulo q). Assume that

(1) *For each pair (p, q) as above the character $\chi = \chi_{p,q}$ satisfies condition $(*_\beta)$ for some $\beta \notin \mathfrak{p}$ (but of course depending on p and q).*
(2) *For all primes $p \mid t$, condition $\mathcal{L}_p$ is satisfied.*

Then for every divisor r of N there exists an integer i such that $0 \le i < t$ satisfying

$$r \equiv N^i \pmod{e(t)} .$$

Proof. From Proposition 9.1.11 and Lemma 9.1.10, there exists a sufficiently large a such that $\chi(r) = \chi(N)^{l_p(r,a)}$ for every a and every $\chi = \chi_{p,q}$. By the Chinese remainder Theorem 1.3.9, we can find $l(r)$ defined modulo t such that $l(r) \equiv l_p(r,a) \pmod{p^{v_p(t)}}$ for all primes p dividing t, hence since $p^k \mid (q-1) \mid t$, for all p and q as above we have

$$\chi_{p,q}(r) = \chi_{p,q}\left(N^{l(r)}\right) .$$

Now I claim that $\chi_q = \prod_{p|(q-1)} \chi_{p,q}$ is a character of order exactly $q-1$. Indeed, if χ_0 is the trivial character modulo q, then $\chi_q^a = \chi_0$ implies that for every $p^k \| (q-1)$,

$$\chi_{p,q}^{a(q-1)/p^k} = \chi_0 ,$$

hence since $\chi_{p,q}$ is of order a power of p, hence prime to $(q-1)/p^k$, that $\chi_{p,q}^a = \chi_0$. This shows that $p^k \mid a$ since $\chi_{p,q}$ is of order exactly equal to p^k. Since this is true for every $p \mid q-1$, we have $(q-1) \mid a$, thus proving our assertion.

Hence, χ_q is a generator of the group of characters modulo q, and this implies that for any character χ_1 modulo q we have $\chi_1(r) = \chi_1\left(N^{l(r)}\right)$.

Now let χ be a character modulo $q^{v_q(t)+1+\delta}$ where $\delta = 0$ if $q > 2$, $\delta = 1$ if $q = 2$. We can write $\chi = \chi_1\chi_2$, where χ_1 is a character modulo q and χ_2 modulo $q^{v_q(t)+1+\delta}$ of order $q^{v_q(t)+1+\delta-(1+\delta)} = q^{v_q(t)}$ (this follows from Theorem 1.4.1). Hence, if $q \nmid t$, $\chi = \chi_1$ so $\chi(r) = \chi\left(N^{l(r)}\right)$. On the other hand, if $q \mid t$, then by assumption, condition $\mathcal{L}_q$ is satisfied. Hence, by Proposition 9.1.11 (2) we have

$$\chi_2(r) = \chi_2(N)^{l(r)} = \chi_2\left(N^{l(r)}\right)$$

since χ_2 is of order $q^{v_q(t)}$ and $l(r) \equiv l_q(r,a) \pmod{q^{v_q(t)}}$ for a sufficiently large. Therefore for every χ modulo $e(t)$ this equality is true, and this proves that

$$r \equiv N^{l(r)} \pmod{e(t)} .$$

Finally note that for every prime q such that $(q-1) \mid t$ we have

$$N^{(q-1)q^{v_q(t)}} \equiv 1 \pmod{q^{v_q(t)+1+\delta}} .$$

Hence, $N^t \equiv 1 \pmod{e(t)}$, so we may reduce the exponent $l(r)$ modulo t, thus proving the theorem. $\square$

Corollary 9.1.13. *We keep all the notations and assumptions of the theorem. Set $r_i = N^i \bmod e(t)$, so that $0 < r_i < e(t)$. If $e(t) > \sqrt{N}$ and if for every i such that $0 < i < t$ we have $r_i = 1$ or $r_i = N$ or $r_i \nmid N$, then N is prime.*

Proof. If N was not prime, there would exist a prime divisor r of N such that $1 < r \leq \sqrt{N} < e(t)$, and by the theorem there would exist $i < t$ such that $r \equiv N^i \pmod{e(t)}$ hence $r = r_i$, contradiction. $\quad\Box$

9.1.4 Checking Condition $\mathcal{L}_p$

We must now see how to check condition $\mathcal{L}_p$, and incidentally prove Lemma 9.1.10. We have the following result:

Lemma 9.1.14.

(1) *If $p \geq 3$, condition $\mathcal{L}_p$ is equivalent to the inequality*

$$v_p\left(r^{p-1} - 1\right) \geq v_p\left(N^{p-1} - 1\right) .$$

(2) *For $p = 2$, condition $\mathcal{L}_2$ is equivalent to the inequality*

$$\max(v_2(r - 1), v_2(r - N)) \geq v_2\left(N^2 - 1\right) .$$

Proof. That condition $\mathcal{L}_p$ implies the above inequalities is trivial and left to the reader. Conversely, assume they are satisfied, and consider first the case $p \geq 3$. Set $u = v_p\left(N^{p-1} - 1\right)$. Then it is easy to prove by induction on $a \geq 0$ that there exist integers x_i for $0 \leq i < l$ satisfying $0 \leq x_i < p$ and such that if we set $l_p(r, a + u) = \sum_{0 \leq i < l} x_i p^i$, we will have

$$r^{p-1} \equiv N^{(p-1)l_p(r,a+u)} \pmod{p^{a+u}} .$$

A similar induction works for $p = 2$ with $u = v_2\left(N^2 - 1\right)$ and $a + u$ replaced by $a + u - 1$. This proves both the above lemma and Lemma 9.1.10 since the x_i are independent of a. $\quad\Box$

Corollary 9.1.15. *If $p \geq 3$ and $N^{p-1} \not\equiv 1 \pmod{p^2}$, then condition $\mathcal{L}_p$ is satisfied.*

This is clear, since in this case $v_p\left(N^{p-1} - 1\right) = 1$. $\quad\Box$

This result is already useful for testing $\mathcal{L}_p$, but it is not a systematic way of doing so. Before giving a more systematic result, we need another lemma.

Lemma 9.1.16. *Let a and b be positive integers, and let x be in $\mathbb{Z}[\zeta_{p^k}, \zeta_q]$. Assume that for an integer r coprime to p we have the congruences*

$$x^a \equiv \eta_a \pmod{r} \qquad and \qquad x^b \equiv \eta_b \pmod{r} ,$$

where η_a and η_b are primitive roots of unity of order p^{l_a} and p^{l_b} respectively, where l_a and l_b are less than or equal to k.

Assume, in addition, that $l_a \geq l_b$ and $l_a \geq 1$. Then:

$$v_p(b) - v_p(a) = l_a - l_b \quad if \quad l_b > 0,$$
$$v_p(b) - v_p(a) \geq l_a \quad if \quad l_b = 0 .$$

Proof. Write $a = p^{v_p(a)}m$, $b = p^{v_p(b)}n$ so $p \nmid mn$. If we had $v_p(a) > v_p(b)$, then, computing x^{an} in two different ways ($an = p^{v_p(a)-v_p(b)}bm$) we would obtain

$$\eta_a^n = \eta_b^{mp^{v_p(a)-v_p(b)}}$$

so $l_a < l_b$, contrary to our assumption. Hence, $v_p(b) \geq v_p(a)$, and we can now similarly compute x^{mb} in two different ways, giving

$$\eta_b^m = \eta_a^{np^{v_p(b)-v_p(a)}} .$$

This immediately implies the lemma. Note that a congruence between roots of unity of order a power of p is in fact an equality since p is coprime to r. $\square$

The main result which allows us to test condition $\mathcal{L}_p$ is the following:

Proposition 9.1.17. *Assume that we can find a character χ modulo q, of order p^k and a $\beta \notin \mathfrak{p}$, for which $(*_\beta)$ is satisfied with $\eta(\chi)$ a primitive p^k-th root of unity. Then, if one of the following supplementary conditions is true, condition $\mathcal{L}_p$ is satisfied:*

(1) *If $p \geq 3$;*
(2) *If $p = 2$, $k = 1$ and $N \equiv 1 \pmod 4$;*
(3) *If $p = 2$, $k \geq 2$ and $q^{(N-1)/2} \equiv -1 \pmod N$.*

Proof. Assume that $p \geq 3$. By Lemma 9.1.8, if r is a prime divisor of N and if we set $x = \tau(\chi)^\beta$, then we have

$$x^{N^{(p-1)p^{k-1}}-1} \equiv \eta(\chi)^{\beta p^{k-1}} \pmod r$$

and

$$x^{r^{(p-1)p^{k-1}}-1} \equiv \chi(r)^{\beta p^{k-1}} \pmod r .$$

Since $\beta \notin \mathfrak{p}$, $\eta(\chi)^{\beta p^{k-1}}$ is a primitive p-th root of unity. From Lemma 9.1.16, we deduce that

$$v_p\left(r^{(p-1)p^{k-1}} - 1\right) - v_p\left(N^{(p-1)p^{k-1}} - 1\right) \geq 0 .$$

But, since $p \geq 3$ for any integer m we have

$$v_p\left(m^{(p-1)p^{k-1}} - 1\right) = k - 1 + v_p\left(m^{p-1} - 1\right) ,$$

hence

$$v_p\left(r^{p-1} - 1\right) \geq v_p\left(N^{p-1} - 1\right)$$

and this proves the theorem in this case by Lemma 9.1.14.

The proof of the two other cases is similar and left to the reader (see Exercise 5). □

It is easy to show that if N is prime, one can always find a χ satisfying the hypotheses of Proposition 9.1.17. In practice, such a χ, if not already found among the χ which are used to test ($*_\beta$), will be found after a few trials at most. Strictly speaking, however, this part of the algorithm makes it probabilistic, but in a weak sense. A non-probabilistic, but less practical version also exists (see [APR]).

9.1.5 The Use of Jacobi Sums

It is clear that we now have an asymptotically fast primality testing algorithm. In this form, however, it is far from being practical. The main reason is as follows: we essentially have to test a number of conditions of the form ($*_\beta$) for certain β's and characters. This number is not that large, for example if N has less than 100 decimal digits, less than 80 tests will usually be necessary. The main problem lies in the computation of $\tau(\chi)^{\beta(N-\sigma_N)} \bmod N$. One needs to work in the ring $\mathbb{Z}[\zeta_{p^k}, \zeta_q]$, and this will be hopelessly slow (to take again the case of $N < 10^{100}$, we can take $t = 5040$, hence p^k will be very small, more precisely $p^k \leq 16$, but q will be much larger, the largest value being $q = 2521$). We must therefore find a better way to test these conditions. The reader may have wondered why we have carried along the element $\beta \in \mathbb{Z}[G]$, which up to now was not necessary. Now, however we are going to make a specific choice for β, and it will not be $\beta = 1$. We have the following proposition.

Proposition 9.1.18. *Let χ be a character modulo q of order p^k, and let a and b be integers such that $p \nmid ab(a+b)$. Denote by E be the set of integers x such that $1 \leq x < p^k$ and $p \nmid x$. Finally, let*

$$\alpha = \sum_{x \in E} \left\lfloor \frac{Nx}{p^k} \right\rfloor \sigma_x^{-1}$$

and

$$\beta = -\sum_{x \in E} \left(\left\lfloor \frac{xa}{p^k} \right\rfloor + \left\lfloor \frac{xb}{p^k} \right\rfloor - \left\lfloor \frac{x(a+b)}{p^k} \right\rfloor\right) \sigma_x^{-1} .$$

Then, we have

$$\tau(\chi)^{\beta(N-\sigma_N)} = j(\chi^a, \chi^b)^\alpha .$$

Proof. Set

$$\Theta = \sum_{x \in E} x \sigma_x^{-1} \in \mathbb{Z}[G] \ .$$

An easy computation shows that for any integer r not divisible by p we have

$$\Theta(\sigma_r - r) = -p^k \sum_{x \in E} \left\lfloor \frac{rx}{p^k} \right\rfloor \sigma_x^{-1} \ .$$

Using this formula for $r = N$, a, b and $a + b$ (which are all coprime to p) we obtain

$$\Theta(N - \sigma_N) = p^k \alpha$$

and

$$\Theta(\sigma_a + \sigma_b - \sigma_{a+b}) = \Theta(\sigma_a - a + \sigma_b - b - (\sigma_{a+b} - (a + b))) = p^k \beta \ ,$$

hence

$$\beta(N - \sigma_N) = \alpha(\sigma_a + \sigma_b - \sigma_{a+b}) \ .$$

Now it follows from Proposition 9.1.6 that

$$j(\chi^a, \chi^b) = \tau(\chi)^{\sigma_a + \sigma_b - \sigma_{a+b}} \ ,$$

and our proposition follows. $\square$

One sees from this proposition that if we can find suitable values of a and b, we can replace taking powers of $\tau(\chi)$, which are in a large ring, by powers of a Jacobi sum, which are in the much smaller ring $\mathbb{Z}[\zeta_{p^k}]$. This is the basic observation needed to make this test practical.

However this is not enough. First, note that the condition $p \nmid ab(a + b)$ excludes immediately the case $p = 2$, which will, as usual, have to be treated separately. Hence, we first assume that $p \geq 3$. Recall that to get anything useful from $(*_\beta)$ we must have $\beta \notin \mathfrak{p}$. This is easily dealt with by the following lemma.

Lemma 9.1.19. *With the notations of the above proposition, a necessary and sufficient condition for $\beta \notin \mathfrak{p}$ is that*

$$a^p + b^p \not\equiv (a + b)^p \pmod{p^2} \ .$$

Proof. If we set

$$K = -\sum_{x \in E} \left(\left\lfloor \frac{xa}{p^k} \right\rfloor + \left\lfloor \frac{xb}{p^k} \right\rfloor - \left\lfloor \frac{x(a + b)}{p^k} \right\rfloor \right) x^{-1}$$

where x^{-1} is an inverse of x modulo p^k, it is clear from the definition of $\mathfrak{p}$ that $\beta \notin \mathfrak{p}$ is equivalent to $p \nmid K$. Now by computing the product of ax for $x \in E$ in two different ways, it is easy to show that if $p \nmid a$

$$\sum_{x \in E} \left\lfloor \frac{xa}{p^k} \right\rfloor x^{-1} \equiv a \frac{a^{(p-1)p^{k-1}} - 1}{p^k} \pmod{p^k} \qquad \text{(A)}$$

(see Exercise 1). The lemma follows immediately from this identity and the congruence

$$\frac{a^{(p-1)p^{k-1}} - 1}{p^k} \equiv \frac{a^{p-1} - 1}{p} \pmod{p}$$

(see Exercise 2). $\qquad \qquad \square$

From this we obtain the following.

Proposition 9.1.20. *If $3 \le p < 6 \cdot 10^9$ and $p \ne 1093, 3511$, we can take $a = b = 1$. In other words, if we take*

$$\beta = \sum_{p^k/2 < x < p^k, p \nmid x} \sigma_x^{-1}$$

then $\beta \notin \mathfrak{p}$ and condition $(*_\beta)$ is equivalent to the congruence

$$j(\chi, \chi)^\alpha \equiv \eta(\chi)^{-cN} \pmod{N} \ ,$$

where as before

$$\alpha = \sum_{x \in E} \left\lfloor \frac{Nx}{p^k} \right\rfloor \sigma_x^{-1}$$

and

$$c = 2 \frac{2^{(p-1)p^{k-1}} - 1}{p^k} \ .$$

Proof. By the preceding lemma, we can take $a = b = 1$ if we have $2^p \not\equiv 2 \pmod{p^2}$. This congruence is exactly the Wieferich congruence which occurs for the first case of Fermat's last theorem and has been tested extensively (see [Leh2]). One knows that the only solutions for $p < 6 \cdot 10^9$ are $p = 1093$ and $p = 3511$. The proposition now follows from Proposition 9.1.18 and formula (A) for $a = 2$. $\qquad \square$

Note that the restriction on p in the above proposition is completely irrelevant in practice. Even if we were capable one day of using this test to prove the primality of numbers having 10^9 decimal digits, we would never need primes as large as 1093. This means that we have solved the practical problem of testing $(*_\beta)$ for $p \ge 3$.

The case $p = 2$ is a little more complicated, since we cannot use the above method. Let us first assume that $k \geq 3$. We must now consider the *triple Jacobi sum* defined by

$$j_3(\chi_1, \chi_2, \chi_3) = \sum_{x+y+z=1} \chi_1(x)\chi_2(y)\chi_3(z) \; ,$$

where the variables x, y and z range over $\mathbb{F}_q$. A similar proof to the proof of Proposition 9.1.6 shows that if $\chi_1\chi_2\chi_3$ is not the trivial character, then

$$j_3(\chi_1, \chi_2, \chi_3) = \frac{\tau(\chi_1)\tau(\chi_2)\tau(\chi_3)}{\tau(\chi_1\chi_2\chi_3)}$$

and in particular,

$$j_3(\chi, \chi, \chi) = \tau(\chi)^{3-\sigma_3} \; .$$

Now what we want is an analog of Proposition 9.1.18. This can be easily obtained for one half of the values of N as follows.

Proposition 9.1.21. *Let χ be a character modulo q of order 2^k with $k \geq 3$. Denote by E be the set of integers x such that $1 \leq x < 2^k$ and x congruent to 1 or 3 modulo 8. Finally, let*

$$\alpha = \sum_{x \in E} \left\lfloor \frac{Nx}{2^k} \right\rfloor \sigma_x^{-1}$$

and

$$\beta = \sum_{x \in E} \left\lfloor \frac{3x}{2^k} \right\rfloor \sigma_x^{-1} \; .$$

Then, if N is congruent to 1 or 3 modulo 8, we have

$$\tau(\chi)^{\beta(N-\sigma_N)} = j_3(\chi, \chi, \chi)^{\alpha} \; .$$

Furthermore, $\beta \notin \mathfrak{p}$.

Proof. The proof is essentially the same as that of Proposition 9.1.18, using $\Theta = \sum_{x \in E} x\sigma_x^{-1}$. The condition on N is necessary since $\Theta(\sigma_r - r)$ does not take any special form if r is not congruent to 1 or 3 modulo 8. The restriction to these congruences classes is also mandatory since $(\mathbb{Z}/2^k\mathbb{Z})^*$ is not cyclic but has cyclic subgroups of index 2. (We could also have taken for E those x congruent to 1 or 5 modulo 8, but that would have required the use of quintuple Jacobi sums). $\qquad \square$

When N is congruent to 5 or 7 modulo 8, we use the following trick: $-N$ will be congruent to 1 or 3 modulo 8, hence $\Theta(\sigma_{-N} + N)$ will have a nice form. But on the other hand, it is immediate to transform condition $(*_\beta)$ into a condition involving $\sigma_{-N} + N$:

$$\tau(\chi)^{\sigma_{-N}+N} = \tau(\chi)^{N-\sigma_N}\tau(\chi^N)\tau(\chi^{-N})$$

and by Proposition 9.1.6 we have

$$\tau(\chi^N)\tau(\chi^{-N}) = \chi(-1)q = -q \ ,$$

the last equality coming from $\chi(-1) = (-1)^{(q-1)/2^k} = -1$. This enables us to give a proposition analogous to Proposition 9.1.21 for N congruent to 5 or 7 modulo 8.

Proposition 9.1.22. *Let χ be a character modulo q of order 2^k with $k \geq 3$. Denote by E be the set of integers x such that $1 \leq x < 2^k$ and x congruent to 1 or 3 modulo 8. Finally, let*

$$\alpha_1 = \sum_{x \in E} \left(\left\lfloor \frac{Nx}{2^k} \right\rfloor + 1 \right) \sigma_x^{-1}$$

and

$$\beta = \sum_{x \in E} \left\lfloor \frac{3x}{2^k} \right\rfloor \sigma_x^{-1} \ .$$

Then, if N is congruent to 5 or 7 modulo 8, we have

$$\tau(\chi)^{\beta(N-\sigma_N)} = j_3(\chi, \chi, \chi)^{\alpha_1}(-q)^{-\beta} \ .$$

Furthermore, $\beta \notin \mathfrak{p}$.

The proof of this proposition follows immediately from what we have said before and is left to the reader.   □

Corollary 9.1.23. *Let χ and E be as in the proposition. Set $\delta_N = 0$ if N is congruent to 1 or 3 modulo 8, $\delta_N = 1$ if N is congruent to 5 or 7 modulo 8. We may replace condition $(*_\beta)$ by the following condition:*

$$(j(\chi, \chi)j(\chi, \chi^2))^\alpha \, j^{2\delta_n} \left(\chi^{2^{k-3}}, \chi^{3 \cdot 2^{k-3}} \right) \equiv (-1)^{\delta_N}\eta(\chi)^{-cN} \pmod{N} \ ,$$

where

$$\alpha = \sum_{x \in E} \left\lfloor \frac{xN}{2^k} \right\rfloor \sigma_x^{-1}$$

and

$$c = 3\frac{3^{2^{k-2}} - 1}{2^k} \ .$$

Proof. Note first that using the formulas linking triple Jacobi sums with Gauss sums, and the analogous formula for ordinary Jacobi sums (Proposition 9.1.6), we have

$$j_3(\chi,\chi,\chi) = j(\chi,\chi)j(\chi,\chi^2)$$

and this is the most efficient way to compute j_3.

Now if N is congruent to 1 or 3 modulo 8, the result follows immediately from Proposition 9.1.21 and formula (A) for $a = 3$.

Assume now that N is congruent to 5 or 7 modulo 8. From Proposition 9.1.22, formula (A) and the identity

$$\sum_{x\in E}\left\lfloor\frac{3x}{2^k}\right\rfloor = 2^{k-2}-1 \ ,$$

we obtain

$$j_3(\chi,\chi,\chi)^{\alpha_1} \equiv \eta(\chi)^{-cN}(-q)^d$$

with $d = 2^{k-2}-1$. It is clear that the corollary will follow from this formula and the following lemma:

Lemma 9.1.24. *Set $\gamma = \sum_{x\in E}\sigma_x^{-1}$ and $d = 2^{k-2}-1$. We have the identity:*

$$j_3(\chi,\chi,\chi)^\gamma = q^d j^2\left(\chi^{2^{k-3}},\chi^{3\cdot 2^{k-3}}\right) \ .$$

Proof. Using the formula expressing triple Jacobi sums in terms of Gauss sums, we have

$$j_3(\chi,\chi,\chi) = \prod_{x\in E}\tau^2\left(\chi^x\right) \ .$$

Now we have the following theorem, due to Hasse and Davenport (see for example [Was] and [Ire-Ros]).

Theorem 9.1.25 (Hasse-Davenport). *Let ψ be any character and χ_1 a character of order exactly equal to m. We have the identity*

$$\prod_{0\le x<m}\tau\left(\psi\chi_1^x\right) = -\tau\left(\psi^m\right)\psi^{-m}(m)\prod_{0\le x<m}\tau\left(\chi_1^x\right) \ .$$

Applying this identity to $\psi = \chi^a$, $\chi_1 = \chi^{2^{k-l}}$, one easily shows by induction on l that

$$\prod_{0\le n<2^l}\tau^2\left(\chi^{a+n2^{k-l}}\right) = q^{2^l-1}\tau^2\left(\chi^{2^l a}\right)\chi(2)^{-al2^{l+1}} \ .$$

If we now take $l = k - 3$ and multiply the identities for $a = 1$ and $a = 3$, we easily obtain the lemma by using Proposition 9.1.6, thus proving our corollary.

$\square$

Note that one can give a direct proof of Lemma 9.1.24 without explicitly using the Hasse-Davenport theorem (see Exercise 3).

We have assumed that $k \geq 3$. What remains is the easy case of $k \leq 2$. Here we have the following proposition, whose proof is an immediate consequence of Proposition 9.1.6.

Proposition 9.1.26. *For $p = 2$ and $k = 1$, condition $(*_1)$ is equivalent to the congruence*

$$(-q)^{(N-1)/2} \equiv \eta(\chi) \pmod{N} .$$

For $p = 2$ and $k = 2$, condition $(*_1)$ is equivalent to the congruence

$$j(\chi, \chi)^{(N-1)/2} q^{(N-1)/4} \equiv \eta(\chi)^{-1} \pmod{N}$$

if $N \equiv 1 \pmod{4}$, and to the congruence

$$j(\chi, \chi)^{(N+1)/2} q^{(N-3)/4} \equiv -\eta(\chi) \pmod{N}$$

if $N \equiv 3 \pmod{4}$.

This ends our transformation of condition $(*_\beta)$ into conditions involving only the ring $\mathbb{Z}[\zeta_{p^k}]$.

9.1.6 Detailed Description of the Algorithm

We can now give a detailed and complete description of the Jacobi sum primality test.

Algorithm 9.1.27 (Precomputations). Let B be an upper bound on the numbers that we want to test for primality using the Jacobi sum test. This algorithm makes a number of necessary precomputations which do not depend on N but only on B.

1. [Find t] Using a table of $e(t)$, find a t such that $e^2(t) > B$.

2. [Compute Jacobi sums] For every prime q dividing $e(t)$ with $q \geq 3$, do as follows.

(1) Using Algorithm 1.4.4, compute a primitive root g_q modulo q, and a table of the function $f(x)$ defined for $1 \leq x \leq q - 2$ by $1 - g_q^x = g_q^{f(x)}$ and $1 \leq f(x) \leq q - 2$.

(2) For every prime p dividing $q - 1$, let $k = v_p(q-1)$ and let $\chi_{p,q}$ be the character defined by $\chi_{p,q}(g_q^x) = \zeta_{p^k}^x$.

(3) If $p \geq 3$ or $p = 2$ and $k = 2$, compute

$$J(p,q) = j(\chi_{p,q}, \chi_{p,q}) = \sum_{1 \leq x \leq q-2} \zeta_{p^k}^{x+f(x)} .$$

If $p = 2$ and $k \geq 3$, compute $J(2,q)$ as above,

$$j(\chi_{2,q}^2, \chi_{2,q}) = \sum_{1 \leq x \leq q-2} \zeta_{2^k}^{2x+f(x)} ,$$

$$J_3(q) = j_3(\chi_{2,q}, \chi_{2,q}, \chi_{2,q}) = J(2,q)j(\chi_{2,q}^2, \chi_{2,q})$$

and

$$J_2(q) = j^2 \left(\chi_{2,q}^{2^{k-3}}, \chi_{2,q}^{3 \cdot 2^{k-3}} \right) = \left(\sum_{1 \leq x \leq q-2} \zeta_8^{3x+f(x)} \right)^2 .$$

Note that it is very easy to build once and for all a table of $e(t)$. For example, $e(5040) \approx 1.532 \cdot 10^{52}$ hence $t = 5040$ can be used for numbers having up to 104 decimal digits, $e(720720) \approx 2.599 \cdot 10^{237}$, for numbers having up to 474 decimal digits (see however the remarks at the end of this section).

The Jacobi sum primality testing algorithm is then as follows.

Algorithm 9.1.28 (Jacobi Sum Primality Test). Let N be a positive integer. We assume that N is a strong pseudo-prime in 20 randomly chosen bases (so that N is almost certainly prime). We also assume that $N \leq B$ and that the precomputations described in the preceding algorithm have been made. This algorithm decides (rigorously!) whether N is prime or not.

1. [Check GCD] If $(te(t), N) > 1$, then N is composite and terminate the algorithm.

2. [Initialize] For every prime $p \mid t$, set $l_p \leftarrow 1$ if $p \geq 3$ and $N^{p-1} \not\equiv 1 \pmod{p^2}$, $l_p \leftarrow 0$ otherwise.

3. [Loop on characters] For each pair (p,q) of primes such that $p^k \| (q-1) \mid t$, execute step 4a if $p \geq 3$, step 4b if $p = 2$ and $k \geq 3$, step 4c if $p = 2$ and $k = 2$, step 4d if $p = 2$ and $k = 1$. Then go to step 5.

4a.[Check $(*_\beta)$ for $p \geq 3$] Let E be the set of integers between 0 and p^k which are not divisible by p. Set $\Theta \leftarrow \sum_{x \in E} x \sigma_x^{-1}$, $r \leftarrow N \bmod p^k$, $\alpha \leftarrow \sum_{x \in E} \left\lfloor \frac{rx}{p^k} \right\rfloor \sigma_x^{-1}$, and compute $s_1 \leftarrow J(p,q)^\Theta \bmod N$, $s_2 \leftarrow s_1^{\lfloor N/p^k \rfloor} \bmod N$, and finally $S(p,q) = s_2 J(p,q)^\alpha \bmod N$.

If there does not exist a p^k-th root of unity η such that $S(p,q) \equiv \eta \pmod N$, then N is composite and terminate the algorithm. If η exists and if it is a primitive p^k-th root of unity, set $l_p \leftarrow 1$.

4b.[Check $(*_\beta)$ for $p = 2$ and $k \geq 3$] Let E be the set of integers between 0 and 2^k which are congruent to 1 or 3 modulo 8. Set $\Theta \leftarrow \sum_{x \in E} x\sigma_x^{-1}$, $r \leftarrow N \bmod 2^k$, $\alpha \leftarrow \sum_{x \in E} \left\lfloor \frac{rx}{2^k} \right\rfloor \sigma_x^{-1}$, and compute $s_1 \leftarrow J_3(q)^\Theta \bmod N$, $s_2 \leftarrow s_1^{\lfloor N/p^k \rfloor} \bmod N$, and finally $S(2, q) = s_2 J_3(q)^\alpha J_2(q)^{\delta_N}$, where $\delta_N = 0$ if $r \in E$ (i.e. if N if congruent to 1 or 3 modulo 8), $\delta_N = 1$ otherwise.

    If there does not exist a 2^k-th root of unity η such that $S(2, q) \equiv \eta$ $(\bmod\ N)$, then N is composite and terminate the algorithm. If η exists and is a primitive 2^k-th root of unity, and if in addition $q^{(N-1)/2} \equiv -1 \pmod{N}$, set $l_2 \leftarrow 1$.

4c.[Check $(*_\beta)$ for $p = 2$ and $k = 2$] Set $s_1 \leftarrow J(2, q)^2 \cdot q \bmod N$, $s_2 \leftarrow s_1^{\lfloor N/4 \rfloor} \bmod N$, and finally $S(2, q) \leftarrow s_2$ if $N \equiv 1 \pmod 4$, $S(2, q) \leftarrow s_2 J(2, q)^2$ if $N \equiv 3 \pmod 4$.

    If there does not exist a fourth root of unity η such that $S(2, q) \equiv \eta$ $(\bmod\ N)$, then N is composite and terminate the algorithm. If η exists and is a primitive fourth root of unity (i.e. $\eta = \pm i$), and if in addition $q^{(N-1)/2} \equiv -1$ $(\bmod\ N)$, set $l_2 \leftarrow 1$.

4d.[Check $(*_\beta)$ for $p = 2$ and $k = 1$] Compute $S(2, q) \leftarrow (-q)^{(N-1)/2} \bmod N$. If $S(2, q) \not\equiv \pm 1 \pmod{N}$, then N is composite and terminate the algorithm. If $S(2, q) \equiv -1 \pmod{N}$ and $N \equiv 1 \pmod 4$, set $l_2 \leftarrow 1$.

5. [Check conditions $\mathcal{L}_p$] For every $p \mid t$ such that $l_p = 0$, do as follows. Choose random primes q such that $q \nmid e(t)$, $q \equiv 1 \pmod p$, $(q, N) = 1$, execute step 4a, 4b, 4c, 4d according to the value of the pair (p, q). To do this, we will have to compute a number of new Jacobi sums, since these will not have been precomputed, and we do this as explained in the precomputation algorithm.

    If after a reasonable number of attempts, some l_p is still equal to 0, then output a message saying that the test has failed (this is highly improbable).

6. For $i = 1, \ldots, t - 1$, compute (by induction of course, not by the binary powering algorithm) $r_i \leftarrow N^i \bmod e(t)$. If for some i, r_i is a non-trivial divisor of N, then N is composite and terminate the algorithm. Otherwise (i.e. if for every i either $r_i \nmid N$ or $r_i = 1$ or $r_i = N$), output the message that N is prime and terminate the algorithm.

9.1.7 Discussion

The above algorithm works already quite well both in theory and in practice. Pomerance and Odlyzko have shown that the running time of the Jacobi sum algorithm is

$$O((\ln N)^{C \ln \ln \ln N})$$

for some constant C. Hence this is almost (but not quite) a polynomial time algorithm. Many improvements are however still possible.

    For example, it is not difficult to combine the Jacobi sum test with the information gained from the Pocklington $N - 1$ and $N + 1$ tests (Proposition

8.3.1). One can go even further and combine the test with the so-called Galois theory test. This has been done by Bosma and van der Hulst (see [Bos-Hul]).

Note also that the part of the algorithm which is the most time-critical is the computation of $s_2 \leftarrow s_1^{\lfloor N/p^k \rfloor}$. To do this, we of course use the fastest powering algorithms possible, in practice the 2^k-left to right Algorithm 1.2.4. But we must also do multiplications in the rings $\mathbb{Z}[\zeta_{p^k}]$ which is of dimension $n = \phi(p^k) = (p-1)p^{k-1}$ over $\mathbb{Z}$. A priori such a multiplication would require n^2 multiplications in $\mathbb{Z}$. Using the same tricks as explained in Section 3.1.2, it is possible to substantially decrease the number of necessary multiplications. Furthermore, special squaring routines must also be written. All this is explained in complete detail in [Coh-Len2] and [Coh-Len3].

Another important improvement uses an algorithm due to H. W. Lenstra (see [Len2]) for finding in polynomial time factors of N which are in a given residue class modulo s when $s > N^{1/3}$. This can be applied here, and allows us to replace the condition $e^2(t) > B$ of the precomputations by $e^3(t) > B$. This gives a substantial saving in time since one can choose a much smaller value of t. We give the algorithm here, and refer to [Len2] for its proof.

Algorithm 9.1.29 (Divisors in Residue Classes). Let r, s, N be integers such that $0 \le r < s < N$, $(r,s) = 1$ and $s > \sqrt[3]{N}$. This algorithm determines all the divisors d of N such that $d \equiv r \pmod{s}$.

1. [Initialization] Using Euclid's extended Algorithm 1.3.6 compute u and v such that $ur + vs = 1$. Set $r' \leftarrow uN \bmod s$ (hence $0 \le r' < s$), $a_0 \leftarrow s$, $b_0 \leftarrow 0$, $c_0 \leftarrow 0$, $a_1 \leftarrow ur' \bmod s$, $b_1 \leftarrow 1$, $c_1 \leftarrow u(N - rr')/s \bmod s$ and $j \leftarrow 1$. Finally, if $a_1 = 0$ set $a_1 = s$ (so $0 < a_1 \le s$).

2. [Compute c] If j is even let $c \leftarrow c_j$. Otherwise, let $c \leftarrow c_j + s\lfloor(N + s^2(a_j b_j - c_j))/s^3\rfloor$ and if $c < 2a_j b_j$ go to step 6.

3. [Solve quadratic equation] If $(cs + a_j r + b_j r')^2 - 4a_j b_j N$ is not the square of an integer, go to step 5. Otherwise, let t_1 and t_2 be the two (integral) solutions of the quadratic equation $T^2 - (cs + a_j r + b_j r')T + a_j b_j N = 0$.

4. [Divisor found?] If $a_j \mid t_1$, $b_j \mid t_2$, $t_1/a_j \equiv r \pmod{s}$ and $t_2/b_j \equiv r' \pmod{s}$, then output t_1/a_j as a divisor of N congruent to r modulo s.

5. [Other value of c] If j is even and $c > 0$, set $c \leftarrow c - s$ and go to step 3.

6. [Next j] If $a_j = 0$, terminate the algorithm. Otherwise, set $j \leftarrow j + 1$, and $q_j \leftarrow \lfloor a_{j-2}/a_{j-1}\rfloor$ if j is even, $q_j \leftarrow \lfloor(a_{j-2} - 1)/a_{j-1}\rfloor$ if j is odd. Finally, set $a_j \leftarrow a_{j-2} - q_j a_{j-1}$, $b_j \leftarrow b_{j-2} - q_j b_{j-1}$, $c_j \leftarrow c_{j-2} - q_j c_{j-1}$ and go to step 2.

Remarks.

(1) [Len2] also shows that under the conditions of this algorithm, there exist at most 11 divisors of N congruent to r modulo s.

(2) In step 4, t_2/b_j is a divisor of N congruent to r' modulo s. Since in the case of the Jacobi sum test $r = N^i \bmod s$ and so $r' = N^{1-i} \bmod s$, Lenstra's

algorithm allows us to test simultaneously two residue classes modulo s, reducing the time spent in step 6 of Algorithm 9.1.28.

9.2 The Elliptic Curve Test

We now come to the other modern primality test, based on the use of elliptic curves over finite fields. Here, instead of looking for suitably strong generalizations of Fermat's theorem in cyclotomic fields, or equivalently instead of implicitly using the multiplicative group of $\mathbb{F}_{N^d}$, we will use the group of points of elliptic curves over $\mathbb{F}_N$ itself.

Now recall that when we start using a primality test, we are already morally certain that our number N is prime, since it has passed the Rabin-Miller pseudo-primality test. Hence, we can work as if N was prime, for example by assuming that any non-zero element modulo N is invertible. In the unlikely event that some non-zero non-invertible element appears, we can immediately stop the algorithm since we know not only that N is composite, but even an explicit prime factor by taking a GCD with N.

We will consider an "elliptic curve over $\mathbb{Z}/N\mathbb{Z}$". What this means is that we consider a Weierstraß equation

$$ y^2 = x^3 + ax + b\ ,\qquad a,b \in \mathbb{Z}/N\mathbb{Z}\ ,\quad (4a^3 + 27b^2) \in (\mathbb{Z}/N\mathbb{Z})^*\ . $$

(It is not necessary to consider a completely general Weierstraß equation since we may of course assume that $(N,6) = 1$.)

We then add points on this curve *as if* N was prime. Since the group law involves only addition/subtraction/multiplication/division in $\mathbb{Z}/N\mathbb{Z}$, the only phenomenon which may happen if N is not prime is that some division is impossible, and in that case as already mentioned, we know that N is composite and we stop whatever algorithm we are executing.

Hence, from now on, we implicitly assume that all operations take place without any problems.

9.2.1 The Goldwasser-Kilian Test

The basic proposition which will enable us to prove that N is prime is the following analog of Pocklington's Theorem 8.3.1.

Proposition 9.2.1. *Let N be an integer coprime to 6 and different from 1. and E be an elliptic curve modulo N.*

Assume that we know an integer m and a point $P \in E(\mathbb{Z}/N\mathbb{Z})$ satisfying the following conditions.

(1) *There exists a prime divisor q of m such that*

$$q > \left(\sqrt[4]{N} + 1 \right)^2 .$$

(2) $m \cdot P = O_E = (0 : 1 : 0)$.

(3) $(m/q) \cdot P = (x : y : t)$ with $t \in (\mathbb{Z}/N\mathbb{Z})^*$.

Then N is prime. (As above, it is assumed that all the computations are possible.)

Proof. Let p be a prime divisor of N. By reduction modulo p, we know that in the group $E(\mathbb{Z}/p\mathbb{Z})$, the image of P has order a divisor of m, but not a divisor of m/q since $t \in (\mathbb{Z}/N\mathbb{Z})^*$. Since q is a prime, this means that q divides the order of the image of P in $E(\mathbb{Z}/p\mathbb{Z})$, and in particular $q \leq |E(\mathbb{Z}/p\mathbb{Z})|$. By Hasse's Theorem 7.1.8, we thus have

$$q < (\sqrt{p} + 1)^2 .$$

Assume that N was not prime. We can then choose for p the smallest prime divisor of N which will be less than or equal to $\sqrt{N}$. Hence we obtain $q < (\sqrt[4]{N} + 1)^2$, contradicting the hypothesis on the size of q and thus proving the proposition. $\qquad \square$

For this proposition to be of any use, we must explain three things. First, how one chooses the elliptic curve, second how one finds P, and finally how one chooses m. Recall that for all these tasks, we may as well assume that N is prime, since this only helps us in making a choice. Only the above proposition will give us a *proof* that N is prime.

The only non-trivial choice is that of the integer m. First, we have:

Proposition 9.2.2. *Let N be a prime coprime to 6, E an elliptic curve modulo N and let*

$$m = |E(\mathbb{Z}/N\mathbb{Z})| .$$

If m has a prime divisor q satisfying

$$q > (\sqrt[4]{N} + 1)^2 ,$$

then there exists a point $P \in E(\mathbb{Z}/N\mathbb{Z})$ such that

$$m \cdot P = O_E \quad and \quad (m/q) \cdot P = (x : y : t) \quad with \quad t \in (\mathbb{Z}/N\mathbb{Z})^* .$$

Proof. First note that any point P will satisfy $m \cdot P = O_E$. Second, since N is assumed here to be prime, $t \in (\mathbb{Z}/N\mathbb{Z})^*$ means $t \neq 0$ hence the second condition is $(m/q) \cdot P \neq O_E$.

Set $G = E(\mathbb{Z}/N\mathbb{Z})$ and assume by contradiction that for every $P \in G$ we have $(m/q) \cdot P = O_E$. This means that the order of any P is a divisor of m/q, hence that the *exponent* of the Abelian group G divides m/q. (Recall that the exponent of an Abelian group is the LCM of the orders of the elements of the group.)

Now, by Theorem 7.1.9, we know that G is the product of at most two cyclic groups, i.e. that

$$G \simeq \mathbb{Z}/d_1\mathbb{Z} \times \mathbb{Z}/d_2\mathbb{Z} \quad \text{with} \quad d_2 \mid d_1$$

(and $d_2 = 1$ if G is cyclic). Hence the exponent of G is equal to d_1, while the cardinality of G is equal to $d_1 d_2 \le d_1^2$. Thus we obtain

$$m = |G| \le d_1^2 \le (m/q)^2 \;,$$

hence $q^2 \le m$. Using our hypothesis on the size of q and Hasse's bound 7.1.8 on m, we obtain

$$(\sqrt[4]{N} + 1)^2 < \sqrt{N} + 1 \;,$$

and this is clearly a contradiction, thus proving the proposition. $\square$

We now know that Proposition 9.2.1 can in principle be applied to prove the primality of N, by choosing $m = |E(\mathbb{Z}/N\mathbb{Z})|$, where this cardinality is computed as if N was prime. But that is precisely the main question: how is this computed? We could of course use the baby-step giant-step Algorithm 7.4.12, but this is a $O(N^{1/4})$ algorithm, hence totally unsuitable.

The idea of Goldwasser and Kilian ([Gol-Kil]) is to make use of the remarkable algorithm of Schoof already mentioned in Section 7.4.3 ([Scho]), which computes $m = |E(\mathbb{Z}/N\mathbb{Z})|$ in time $O(\ln^8 N)$. Of course, this algorithm may fail since it is not absolutely certain that N is prime, but if it fails, we will know that N is composite.

Once m has been computed, we trial divide m by small primes, hoping that the unfactored part will be a large strong pseudo-prime. In fact, Goldwasser and Kilian's aim was purely theoretical, and in that case one looks for m equal to twice a strong pseudo-prime. If this is the case, and q is the large pseudo-prime that remains (large meaning larger than $(\sqrt[4]{N} + 1)^2$ of course), we temporarily assume that q is prime, and look at random for a point P so as to satisfy the hypothesis of Proposition 9.2.1. This will be possible (and in fact quite easy) by Proposition 9.2.2.

If such a P is found, there remains the task of proving that our strong pseudo-prime q is prime. For this, we apply the algorithm recursively. Indeed, since $q \le m/2 \le (N + 2\sqrt{N} + 1)/2$, the size of N will decrease by a factor which is at least approximately equal to 2 at each iteration, hence the number of recursive uses of the algorithm will be $O(\ln N)$. We stop using this algorithm as soon as N becomes small enough so that other algorithms (even trial division!) may be used.

The algorithm may be formally stated as follows.

Algorithm 9.2.3 (Goldwasser-Kilian). Let N be a positive integer different from 1 and coprime to 6. This algorithm will try to prove that N is prime. If N is not a prime, the algorithm may detect it, or it may run indefinitely (hence we must absolutely use the Rabin-Miller test before entering this algorithm).

1. [Initialize] Set $i \leftarrow 0$ and $N_i \leftarrow N$.

2. [Is N_i small?] If $N_i < 2^{30}$, trial divide N_i by the primes up to 2^{15}. If N_i is not prime go to step 9.

3. [Choose a random curve] Choose a and b at random in $\mathbb{Z}/N_i\mathbb{Z}$, and check that $4a^3 + 27b^2 \in (\mathbb{Z}/N_i\mathbb{Z})^*$. Let E be the elliptic curve whose affine Weierstraß equation is $y^2 = x^3 + ax + b$.

4. [Use Schoof] Using Schoof's algorithm, compute $m \leftarrow |E(\mathbb{Z}/N_i\mathbb{Z})|$. If Schoof's algorithm fails go to step 9.

5. [Is m OK?] Check whether $m = 2q$ where q passes the Rabin-Miller test 8.2.2 (or more generally, trial divide m up to a small bound, and check that the remaining factor q passes the Rabin-Miller test and is larger than $(\sqrt[4]{N_i}+1)^2$). If this is not the case, go to step 3.

6. [Find P] Choose at random $x \in \mathbb{Z}/N_i\mathbb{Z}$ until the Legendre-Jacobi symbol $\left(\frac{x^3+ax+b}{N_i}\right)$ is equal to 0 or 1 (this will occur after a few trials at most). Then using Algorithm 1.5.1, compute $y \in \mathbb{Z}/N_i\mathbb{Z}$ such that $y^2 = x^3 + ax + b$ (again, if this algorithm fails, go to step 9).

7. [Check P] Compute $P_1 \leftarrow m \cdot P$ and $P_2 \leftarrow (m/q) \cdot P$. If during the computations some division was impossible, go to step 9. Otherwise, check that $P_1 = O_E$, i.e. that $P_1 = (0 : 1 : 0)$ in projective coordinates. If $P_1 \neq O_E$, go to step 9. Finally, if $P_2 = O_E$, go to step 6.

8. [Recurse] Set $i \leftarrow i + 1$, $N_i \leftarrow q$ and go to step 2.

9. [Backtrack] (We are here when N_i is not prime, which is a very unlikely occurrence.) If $i = 0$, output a message saying that N is composite and terminate the algorithm. Otherwise, set $i \leftarrow i - 1$ and go to step 3.

Some remarks are in order. As stated in the algorithm, if N is not prime, the algorithm may run indefinitely and so should perhaps not be called an "algorithm" in our sense. Note however that it will never give a false answer. But even if N is prime, the algorithm is probabilistic in nature since we need to find an elliptic curve whose number of points has a special property, and in addition a certain point P on that curve. It can be shown that under reasonable hypotheses on the distribution of primes in short intervals, the expected running time of the algorithm is $O(\ln^{12} N)$, hence is polynomial in $\ln N$. Therefore it is asymptotically faster than the Jacobi sum test. Note however that the Goldwasser-Kilian test is not meant to be practical.

The sequence of primes $N_0 = N, N_1, \ldots N_i, \ldots$ together with the elliptic curves E_i, the points P_i and the cardinality m_i obtained in the algorithm is called a *primality certificate*. The reason for this is clear: although it may have been difficult to find E_i, P_i or m_i, once they are given, to check that the conditions of Proposition 9.2.1 are satisfied (with $q = N_{i+1}$) is very easy, so anybody can prove to his or her satisfaction the primality of N using much less work than executing the algorithm. This is quite different from the Jacobi sum test where to check that the result given by the algorithm is correct, there is little that one can do but use a different implementation and run the algorithm again.

To finish this (sub)section, note that, as stated in the beginning of this chapter, an important theoretical advance has been made by Adleman and Huang.

Their idea is to use, in addition to elliptic curves, Jacobians of curves of genus 2, and a similar algorithm to the one above. Although their algorithm is also not practical, the important point is that they obtain a probabilistic primality testing algorithm which runs in polynomial time, in other words they prove Theorem 9.1. Note that the Goldwasser-Kilian test is not of this kind since only the expected running time is polynomial, but the worst case may not be.

9.2.2 Atkin's Test

Using the same basic idea, i.e. Proposition 9.2.1, Atkin has succeeded in finding a practical version of the elliptic curve test. It involves a number of new ideas. This version has been implemented by Atkin and by Morain, and has been able to prove the primality of *titanic numbers*, i.e. numbers having more than 1000 decimal digits. The Jacobi sum test could of course do the same, but time comparisons have not been done, although it seems that at least up to 800 digits the Jacobi sum test is slightly faster. Of course, since asymptotically Atkin's test is polynomial while the Jacobi sum test is not, the former must win for N sufficiently large.

The main (if not the sole) practical stumbling block in the algorithm of Goldwasser-Kilian is the computation of $m = |E(\mathbb{Z}/N\mathbb{Z})|$ using Schoof's algorithm. Although progress has been made in the direction of making Schoof's algorithm practical, for example by Atkin and Elkies, Atkin has found a much better idea.

Instead of taking random elliptic curves, we choose instead elliptic curves with complex multiplication by an order in a quadratic number field $K = \mathbb{Q}(\sqrt{D})$ where N splits as a product of two *elements*. This will enable us to use Theorem 7.2.15 which (if N is prime) gives us immediately the cardinality of $E(\mathbb{Z}/N\mathbb{Z})$.

The test proceeds as follows. As always we can work as if N was prime. We must first find a negative discriminant D such that N splits in the order of discriminant D as a product of two elements. This is achieved by using Cornacchia's Algorithm 1.5.3. Indeed, Cornacchia's algorithm gives us, if it exists, a solution to the equation $x^2 + dy^2 = 4p$, where $d = -D$, hence $\pi\bar{\pi} = p$, with

$$\pi = \frac{x + y\sqrt{D}}{2} .$$

Once such a D is found, using Theorem 7.2.15 we obtain that, if N is prime,

$$m = |E(\mathbb{Z}/N\mathbb{Z})| = N + 1 - \pi - \bar{\pi} = N + 1 - x$$

with the above notations, if E is an elliptic curve with complex multiplication by the order of discriminant D. We now check whether m satisfies the condition which will enable us to apply Proposition 9.2.1, i.e. that m is not prime, but its largest prime factor is larger than $(\sqrt[4]{N} + 1)^2$. Since we are describing a practical algorithm, this is done much more seriously than in Goldwasser-Kilian's test: we trial divide m up to a much higher bound, and then we can also use Pollard ρ and $p - 1$ to factor m.

If m is not suitable, we still have at least another chance. Recall from Section 5.3 that we denote by $w(D)$ the number of roots of unity in the quadratic order of discriminant D, hence $w(D) = 2$ if $D < -4$, $w(-4) = 4$ and $w(-3) = 6$.

Then it can be shown that there exist exactly $w(D)$ isomorphism classes of elliptic curves modulo N with complex multiplication by the quadratic order of discriminant D. These correspond to the factorizations $N = (\zeta\pi)(\overline{\zeta\pi})$ where ζ runs over all $w(D)$-th roots of unity (in particular $\zeta = \pm 1$ if $D < -4$).

Hence we can compute $w(D)$ different values of m in this way and hope that at least one of them is suitable. If none are, we go on to another discriminant.

Therefore, let us assume that we have found a suitable value for m corresponding to a certain discriminant D. It remains to find explicitly the equations of elliptic curves modulo N with complex multiplication by the order of discriminant D.

Now since N splits in the order of discriminant D, we have $w(D) \mid N - 1$ and there exist $(N - 1)/2$ values of $g \in \mathbb{Z}/N\mathbb{Z}$ $((N - 1)/3$ if $D = -3)$ such that $g^{(N-1)/p} \neq 1$ for each prime $p \mid w(D)$. Choose one of these values of g.

If $D = -4$ (resp. $D = -3$), then the four (resp. six) isomorphism classes of elliptic curves with complex multiplication by the order of discriminant -4 are given by the affine equations

$$y^2 = x^3 - g^k x \quad \text{for} \quad 0 \le k \le 3$$

(resp.

$$y^2 = x^3 - g^k \quad \text{for} \quad 0 \le k \le 5) .$$

If D is not equal to -3 or -4, we set

$$c = j/(j - 1728) , \quad \text{where} \quad j = j\left(\frac{D + \sqrt{D}}{2}\right)$$

is the j-invariant which corresponds to the order of discriminant D. Then the two isomorphism classes of elliptic curves with complex multiplication by the order of discriminant D can be given by the affine equations

$$y^2 = x^3 - 3cg^{2k}x + 2cg^{3k} \quad \text{for} \quad k = 0 \text{ or } 1 .$$

Note that $j = j((D + \sqrt{D})/2)$ has been defined in Section 7.2.1 as a complex number, and not as an element of $\mathbb{Z}/N\mathbb{Z}$. Hence we must make sense of the above definition.

Recall that according to Theorem 7.2.14, j is an algebraic integer of degree exactly equal to $h(D)$. Furthermore, it can easily be shown that our hypothesis that N splits into a product of two elements is equivalent (if N is prime) to the fact that the minimal monic polynomial T of j in $\mathbb{Z}[X]$ splits completely modulo N as a product of linear factors. Since the roots of T in $\mathbb{C}$ are the conjugates of $j((D + \sqrt{D})/2)$, any one will define by the above equations the isomorphism classes of elliptic curves with complex multiplication by the order of discriminant D, hence we define j as being any of the $h(D)$ roots of $T(X)$ in $\mathbb{Z}/N\mathbb{Z}$.

Once the elliptic curve has been found, the rest of the algorithm proceeds as in the Goldwasser-Kilian algorithm, i.e. we must find a point P on the curve satisfying the required properties, etc ...

There are, however, two remarks to be made. First, we have $w(D)$ elliptic curves modulo N at our disposal, but a priori only one corresponds to a suitable value of m, and it is not clear which one. For $D = -3$ and $D = -4$, it is easy to give a recipe that will tell us which elliptic curve to choose. For $D < -4$, such a recipe is more difficult to find, and we then simply compute $m \cdot P$ for our suitable m and a random P on one of the two curves. If this is not equal to the identity, we are on the wrong curve. If it is equal to the identity, this does not prove that we are on the right curve, but if P has really been chosen randomly, we can probably still use the curve to satisfy the hypotheses of Proposition 9.2.1.

The second remark is much more important. To obtain the equation of the curve, it is necessary to obtain the value of j modulo N. This clearly is more difficult if the class number $h(D)$ is large. Hence, we start by considering discriminants whose class number is as small as possible. So we start by looking at the 13 quadratic orders with class number 1, then class number 2, etc ...

But now a new difficulty appears. The coefficients in the minimal polynomial T of j become large when the class number grows. Of course, they will afterwards be reduced modulo N, but to compute them we will need to use high precision computations of the values of $j(\tau)$ for every quadratic irrational τ corresponding to a reduced quadratic form of discriminant D. Since this computation is independent of N, it could be done only once and the results stored, but the coefficients are so large that even for a moderately sized table we would need an enormous amount of storage.

Several methods are available to avoid this. First, one can use the notion of *genus field* to reduce the computations to a combination of relative computations of smaller degree. Second, we can use *Weber functions*, which are meromorphic functions closely related to the function $j(\tau)$ and which have analogous arithmetic properties. In the best cases, these functions reduce the number of digits of the coefficients of the minimal polynomial T by a factor 24 (see Section 7.6.3).

All these tricks and many more, and the detailed implementation procedures, are described completely in [Atk-Mor] and in Morain's thesis [Mor2]. Here, we will simply give a formal presentation of Atkin's algorithm without any attempt at efficiency.

Algorithm 9.2.4 (Atkin).   Given an integer N coprime to 6 and different from 1, this algorithm tries to prove that N is prime. It is assumed that N is already known to be a strong pseudo-prime in the sense of the Rabin-Miller test 8.2.2. We assume that we have a list of negative discriminants D_n ($n \geq 1$) ordered by increasing computational complexity (for example as a first approximation by increasing class number).

1.   [Initialize] Set $i \leftarrow 0$, $n \leftarrow 0$ and $N_i \leftarrow N$.

2.   [Is N_i small?] If $N_i < 2^{30}$, trial divide N_i by the primes up to 2^{15}. If N_i is not prime go to step 14.

3.   [Choose next discriminant] Let $n \leftarrow n + 1$ and $D \leftarrow D_n$. If $\left(\frac{D}{N}\right) \neq 1$, go to step 3. Otherwise, use Cornacchia's Algorithm 1.5.3 to find a solution, if it exists, of the equation $x^2 + |D|y^2 = 4N$. If no solution exists, go to step 3.

4.   [Factor m]' For $m = N + 1 + x$, $m = N + 1 - x$ (and in addition for $m = N + 1 + 2y$, $m = N + 1 - 2y$ if $D = -4$, or $m = N + 1 + (x + 3y)$, $m = N + 1 - (x + 3y)$, $m = N + 1 + (x - 3y)$, $m = N + 1 - (x - 3y)$ if $D = -3$), factor m using trial division (up to 1000000, say), then Pollard ρ and $p - 1$. It is worthwhile to spend *some* time factoring m here.

5.   [Does a suitable m exist?] If, using the preceding step, for at least one value of m we can find a q dividing m which passes the Rabin-Miller test 8.2.2 and is larger than $(\sqrt[4]{N_i} + 1)^2$, then go to step 6, otherwise go to step 3.

6.   [Compute elliptic curve] If $D = -4$, set $a \leftarrow -1$ and $b \leftarrow 0$. If $D = -3$, set $a \leftarrow 0$, $b \leftarrow -1$. Otherwise, using Algorithm 7.6.1, compute the minimal polynomial $T \in \mathbb{Z}[X]$ of $j((D + \sqrt{D})/2)$. Then reduce T modulo N_i and let j be one of the roots of $\overline{T} = T \bmod N_i$ obtained by using Algorithm 1.6.1 (note that we know that $\overline{T} \mid X^{N_i} - X$ so the computation of $A(X)$ in step 1 of that algorithm is not necessary, we can simply set $A \leftarrow \overline{T}$). Then set $c \leftarrow j/(j - 1728) \bmod N_i$, $a \leftarrow -3c \bmod N_i$, $b \leftarrow 2c \bmod N_i$.

7.   [Find g] By making several random choices of g, find g such that g is a quadratic non-residue modulo N_i and in addition if $D = -3$, $g^{(N_i - 1)/3} \not\equiv 1$ (mod N_i).

8. [Find P] Choose at random $x \in \mathbb{Z}/N_i\mathbb{Z}$ until the Legendre-Jacobi symbol $\left(\frac{x^3+ax+b}{N_i}\right)$ is equal to 0 or 1 (this will occur after a few trials at most). Then using Algorithm 1.5.1, compute $y \in \mathbb{Z}/N_i\mathbb{Z}$ such that $y^2 = x^3 + ax + b$. (If this algorithm fails, go to step 14, but see also Exercise 6.) Finally, set $k \leftarrow 0$.

9. [Find right curve] Compute $P_2 \leftarrow (m/q) \cdot P$ and $P_1 \leftarrow q \cdot P_2$ on the curve whose affine equation is $y^2 = x^3 + ax + b$. If during the computations some division was impossible, go to step 14. If $P_1 = (0 : 1 : 0)$ go to step 12.

10. Set $k \leftarrow k + 1$. If $k \geq w(D)$ go to step 14, else if $D < -4$ set $a \leftarrow ag^2$, $b \leftarrow bg^3$, if $D = -4$ set $a \leftarrow ag$, if $D = -3$ set $b \leftarrow bg$ and go to step 8.

11. [Find a new P] Choose at random $x \in \mathbb{Z}/N_i\mathbb{Z}$ until the Legendre-Jacobi symbol $\left(\frac{x^3+ax+b}{N_i}\right)$ is equal to 0 or 1 (this will occur after a few trials at most). Then using Algorithm 1.5.1, compute $y \in \mathbb{Z}/N_i\mathbb{Z}$ such that $y^2 = x^3 + ax + b$ (if this algorithm fails, go to step 14). If $P_1 \neq (0 : 1 : 0)$ go to step 10.

12. [Check P] If $P_2 = O_E$, go to step 11.

13. [Recurse] Set $i \leftarrow i + 1$, $N_i \leftarrow q$ and go to step 2.

14. [Backtrack] (We are here when N_i is not prime, which is unlikely.) If $i = 0$, output a message saying that N is composite and terminate the algorithm. Otherwise, set $i \leftarrow i - 1$ and go to step 3.

Most remarks that we have made about the Goldwasser-Kilian algorithm are still valid here. In particular, this algorithm is probabilistic, but its expected running time is polynomial in $\ln N$. More important, it is practical, and as already mentioned, it has been used to prove the primality of numbers having more than 1000 decimal digits, by using weeks of workstation time.

Also, as for the Goldwasser-Kilian test, it gives a certificate of primality for the number N, hence the primality of N can be re-checked much faster.

9.3 Exercises for Chapter 9

1.   a) Let p be a prime, E the set of integers x such that $1 \leq x < p^k$ and $p \nmid x$, and a an integer such that $p \nmid a$. By computing the product of ax for $x \in E$ in two different ways, show that we have

$$\sum_{x \in E} \left\lfloor \frac{xa}{p^k} \right\rfloor x^{-1} \equiv a \frac{a^{(p-1)p^{k-1}} - 1}{p^k} \pmod{p^k} .$$

     b) Generalize this result, replacing p^k by an arbitrary integer m and the condition $p \nmid a$ by $(a, m) = 1$.

2. Show that if p is an odd prime and $p \nmid a$, we have

$$\frac{a^{(p-1)p^{k-1}} - 1}{p^k} \equiv \frac{a^{p-1} - 1}{p} \pmod{p} .$$

3. Prove Lemma 9.1.24 without explicitly using the Hasse-Davenport relations.

4. (Wolstenholme's theorem)
    a) Let p be a prime, and set

    $$\sum_{1 \leq x \leq p-1} \frac{1}{x} = \frac{A_p}{B_p}$$

    where A_p and B_p are coprime integers. By first adding together the terms for x
    and for $p - x$, show that $p^2 \mid A_p$ (note that $p \mid A_p$ is immediate).
    b) As in Exercise 1, generalize to arbitrary integers m, replacing $\sum_{1 \leq x \leq p-1}$
    by $\sum_{1 \leq x \leq m, (x,m)=1}$.

5. Let $a \in \mathbb{Z}$ and assume that $a^{(N-1)/2} \equiv -1 \pmod{N}$.
    a) Show that for every $r \mid N$ we have $v_2(r - 1) \geq v_2(N - 1)$.
    b) Show that equality holds if and only if $\left(\frac{a}{r}\right) = -1$, and in particular that
    $\left(\frac{a}{N}\right) = -1$.
    c) If $N \equiv 1 \pmod 4$ show that condition $\mathcal{L}_2$ is satisfied.
    d) If $N \equiv 3 \pmod 8$ and $a = 2$ show that condition $\mathcal{L}_2$ is satisfied.

6. Show how to avoid the search in step 8 of Algorithm 9.2.4 by setting $d \leftarrow$
    $x^3 + ax + b$ for some x and modifying the equation of the curve as in step 3 of
    Algorithm 7.4.12.

7. Let χ be a character modulo q, where q is not necessarily prime. We will say that
    χ is *primitive* if for all divisors d of q such that $d < q$, there exists an x such that
    $x \equiv 1 \pmod d$ and $\chi(x) \neq 1$. Set $\zeta = e^{2i\pi/q}$, and $\psi(a) = \sum_{x \in (\mathbb{Z}/q\mathbb{Z})^*} \chi(x)\zeta^{ax}$.
    a) Let a be such that $d = (a, q) = 1$. Show that $\psi(a) = \overline{\chi}(a)\tau(\chi)$.
    b) Assume that χ is a primitive character and that $d = (a, q) > 1$. Show that
    there exists a $u \in (\mathbb{Z}/q\mathbb{Z})^*$ such that $au = d$. Deduce from this that $\psi(a) = 0$,
    and hence that the formula $\psi(a) = \overline{\chi}(a)\tau(\chi)$ is still valid.

8. Let χ be a primitive character modulo $q > 1$, as defined in the preceding
    exercise, and set $S(x) = \sum_{n \leq x} \chi(n)$.
    a) Using the preceding exercise, give an explicit formula for $\tau(\overline{\chi})S(x)$.
    b) Deduce that

    $$\sqrt{q}\,|S(x)| \leq \sum_{1 \leq m < q,\ m \neq q/2} \frac{1}{\sin \frac{\pi m}{q}}.$$

    c) Show finally the *Polya-Vinogradov inequality*

    $$|S(x)| = \left| \sum_{1 \leq n \leq x} \chi(n) \right| \leq \sqrt{q} \log q.$$

Chapter 10

Modern Factoring Methods

The aim of this chapter is to give an overview of the fastest factoring methods known today. This could be the object of a book in itself, hence it is unreasonable to be as detailed here as we have been in the preceding chapters. In particular, most methods will not be written down as formal algorithms as we have done before. We hope however that we will have given sufficient information so that the reader may understand the methods and be able to implement them, at least in unoptimized form. The reader who wants to implement these methods in a more optimized form is urged to read the abundant literature after reading this chapter, before doing so.

10.1 The Continued Fraction Method

We will start this survey of modern factoring methods by the continued fraction factoring algorithm (CFRAC). Although superseded by better methods, it is important for two reasons. First, because it was historically the first algorithm which is asymptotically of sub-exponential running time (although this is only a heuristic estimate and was only realized later), and also because in the late 60's and 70's it was the main factoring method in use. The second reason is that it shares a number of properties with more recent factoring methods: it finds a large number of congruences modulo N, and the last step consists in Gaussian elimination over the field $\mathbb{Z}/2\mathbb{Z}$. Since the ideas underlying it are fairly simple, it is also a natural beginning.

The main idea of CFRAC, as well as the quadratic sieve algorithm (Section 10.4) or the number field sieve (Section 10.5), is to find integers x and y such that
$$x^2 \equiv y^2 \pmod{N}, \qquad x \not\equiv \pm y \pmod{N}.$$
Since $x^2 - y^2 = (x-y)(x+y)$, it is clear that the $\gcd(N, x+y)$ will be a non-trivial factor of N.

Now finding randomly such integers x and y is a hopeless task. The trick, common to the three factoring methods mentioned above, is to find instead congruences of the form
$$x_k^2 \equiv (-1)^{e_{0k}} p_1^{e_{1k}} p_2^{e_{2k}} \cdots p_m^{e_{mk}} \pmod{N}$$

where the p_i are "small" prime numbers. If we find sufficiently many such congruences, by Gaussian elimination over $\mathbb{Z}/2\mathbb{Z}$ we may hope to find a relation of the form

$$\sum_{1 \leq k \leq n} \epsilon_k(e_{0k}, \cdots, e_{mk}) \equiv (0, \cdots, 0) \pmod 2$$

where $\epsilon = 0$ or 1, and then if

$$x = \prod_{1 \leq k \leq n} x_k^{\epsilon_k} \ , \quad y = (-1)^{v_0} p_1^{v_1} \cdots p_m^{v_m}$$

where $\sum_k \epsilon_k(e_{0k}, \cdots, e_{mk}) = 2(v_0, \cdots, v_m)$, it is clear that we have $x^2 \equiv y^2$ (mod N). This splits N if, in addition $x \not\equiv \pm y$ (mod N), condition which will usually be satisfied.

The set of primes p_i (for $1 \leq i \leq m$) which are chosen to find the congruences is called the *factor base*. We will see in each of the factoring methods how to choose it in an optimal manner. These methods differ mainly in the way they generate the congruences.

The CFRAC method, stemming from ideas of Legendre, Kraitchik, Lehmer and Powers, and developed for computer use by Brillhart and Morrison ([Bri-Mor]), consists in trying to find *small* values of t such that $x^2 \equiv t$ (mod N) has a solution. In that case, since t is small, it has a reasonably good chance of being a product of the primes of our factor base, thus giving one of the sought for congruences.

Now if t is small and $x^2 \equiv t$ (mod N), we can write $x^2 = t + kd^2N$ for some k and d, hence $(x/d)^2 - kN = t/d^2$ will be small. In other words, the rational number x/d is a good approximation to the quadratic number $\sqrt{kN}$. Now it is well known (and easy, see [H-W]) that continued fraction expansions of real numbers give good (and in a certain sense the best) rational approximations. This is the basic idea behind CFRAC. We compute the continued fraction expansion of $\sqrt{kN}$ for a number of values of k. This gives us good rational approximations P/Q, say, and we then try to factor the corresponding integer $t = P^2 - Q^2kN$ (which will be not too large) on our factor base. If we succeed, we will have a new congruence.

Now from Section 5.7, we know that it is easy to compute the continued fraction expansion of a quadratic number, using no real approximations, but only rather simple integer arithmetic. Note that although we know that the expansion will be ultimately periodic (in fact periodic after one term in the case of $\sqrt{kN}$), this is completely irrelevant for us since, except for very special numbers, we will never compute the expansion on a whole period or even a half period. The main point which I stress again is that the expansion can be computed *simply*, in contrast with more general numbers.

The formulas of Sections 5.6 and 5.7, adapted to our situation, are as follows. Let $\tau = (-U + \sqrt{D})/2V$ be a quadratic number in the interval $[0, 1[$ with $4V \mid U^2 - D$ and $V > 0$ (hence $|U| < \sqrt{D}$). We have

$$1/\tau = \frac{2V(U + \sqrt{D})}{D - U^2} = \frac{U + \sqrt{D}}{2V'}$$

where $V' = (D - U^2)/(4V)$ is a positive integer. Hence, if we set

$$a = \left\lfloor \frac{1}{\tau} \right\rfloor = \left\lfloor \frac{U + \sqrt{D}}{2V'} \right\rfloor \, ,$$

then

$$\frac{-U + \sqrt{D}}{2V} = \frac{1}{a + \dfrac{-U' + \sqrt{D}}{2V'}} = \frac{1}{a + \tau'}$$

with $U' = U - 2aV'$. Clearly $\tau' \in [0, 1[$, and since $4VV' = D - U^2 \equiv D - U'^2$ (mod $4V'$) the conditions on (U, V) are also satisfied for (U', V') hence the process can continue. Thus we obtain the continued fraction expansion of our initial τ.

Note we have simply repeated the proof of Proposition 5.6.6 (2) that if a quadratic form $f = (V, U, (U^2 - D)/(4V))$ is reduced, then $\rho(f)$ is also reduced. In addition, Proposition 5.6.3 tells us that we will always have U and V less than $\sqrt{D}$ if we start with a reduced form. This will be the case for the form corresponding to the quadratic number $\tau = \sqrt{D} - \lfloor \sqrt{D} \rfloor$. If we denote by a_n (resp U_n, V_n, τ_n), the different quantities a, U, V and τ occurring in the above process, we have, with the usual notation of continued fractions

$$\sqrt{D} = [a_0, a_1, a_2, \cdots, a_n + \tau_n]$$

where we have set $a_0 = \lfloor \sqrt{D} \rfloor$. Hence, if we set

$$[a_0, a_1, \cdots, a_n] = \frac{P_n}{Q_n} \, ,$$

we have the usual recursions

$$(P_{n+1}, Q_{n+1}) = a_{n+1}(P_n, Q_n) + (P_{n-1}, Q_{n-1})$$

with $(P_{-1}, Q_{-1}) = (1, 0)$, $(P_0, Q_0) = (a_0, 1)$.

Returning to our factoring process, we apply this continued fraction algorithm to $D = kN$ for squarefree values of k such that $kN \equiv 0$ or 1 (mod 4). Then P_n/Q_n will be a good rational approximation to $\sqrt{kN}$, hence $t = P_n^2 - Q_n^2 kN$ will not be too large (more precisely $|t| < 2\sqrt{kN}$, see Proposition 5.7.3), and we can try to factor it on our factor base. For every success, we obtain a congruence

$$x^2 \equiv (-1)^{e_0} p_1^{e_1} p_2^{e_2} \cdots p_m^{e_m} \pmod{N}$$

as above, and as already explained, once we have obtained at least $m + 2$ such congruences then by Gaussian elimination over $\mathbb{Z}/2\mathbb{Z}$ we can obtain a congruence $x^2 \equiv y^2 \pmod{N}$, and hence (usually) a non-trivial splitting of N.

Remarks.

(1) For a prime p to be useful in our factor base we must have $\left(\frac{kN}{p}\right) = 0$ or 1. Indeed, if $p \mid P_n^2 - Q_n^2 kN$, we cannot have $p \mid Q_n$ otherwise P_n and Q_n would not be coprime. Hence kN is congruent to a square modulo p, which is equivalent to my claim.

(2) An important improvement to the method of factoring on a fixed factor base is to use the so-called *large prime variation* which is a follows. A large number of residues will not quite factor completely on our factor base, but will give congruences of the form $x^2 \equiv Fp \pmod{N}$ where F does factor completely and p is a large prime number not in the factor base. A single such relation is of course useless. But if we have two with the same large prime p, say $x_1^2 \equiv F_1 p \pmod{N}$ and $x_2^2 \equiv F_2 p \pmod{N}$, we will have $(x_1 x_2/p)^2 \equiv F_1 F_2 \pmod{N}$ which is a useful relation.

Now since p is large (typically more than 10^5), it could be expected that getting the same p twice is very rare. That this is not true is an instance of the well known "birthday paradox". What it says in our case is that if k numbers are picked at random among integers less than some bound B, then if $k > B^{1/2}$ (approximately) there will be a probability larger than $1/2$ that two of the numbers picked will be equal (see Exercise 5). Hence this large prime variation will give us quite a lot of extra relations essentially for free.

(3) Another important improvement to CFRAC is the so-called *early abort strategy*. It is based on the following idea. Most of the time is being spent in the factorization of the residues (this is why methods using sieves such as MPQS or NFS are so much faster). Instead of trying to factor completely on our factor base, we can decide that if after a number of primes have been tried the unfactored portion is too large, then we should abort the factoring procedure and generate the next residue. With a suitable choice of parameters, this gives a considerable improvement.

(4) Finally, note that the final Gaussian elimination over $\mathbb{Z}/2\mathbb{Z}$ is a non-trivial task since the matrices involved can be huge. These matrices are however very sparse, hence special techniques apply. See for example the "intelligent Gaussian elimination" method used by LaMacchia and Odlyzko ([LaM-Odl]), as well as [Cop1], [Cop2].

10.2 The Class Group Method

10.2.1 Sketch of the Method

The continued fraction method, as well as the more recent quadratic sieve (Section 10.4) or number field sieve (Section 10.5) have sub-exponential running time, which make them quite efficient, but require also sub-exponential space.

The class group method due to Schnorr and Lenstra was the first sub-exponential method which required a negligible amount of space, say polynomial space. The other prominent method having this characteristic is the elliptic curve method (see Section 10.3).

Note that we name this method after Schnorr and Lenstra since they published it ([Schn-Len]), but essentially the same method was independently discovered and implemented by Atkin and Rickert, who nicknamed it SPAR (Shanks, Pollard, Atkin, Rickert).

The idea of the method is as follows. We have seen in Section 8.6 that the determination of the 2-Sylow subgroup of the class group of the quadratic field $\mathbb{Q}(\sqrt{-N})$ is equivalent to knowing all the factorizations of N. In a manner analogous to the continued fraction method, we consider the class numbers $h(-kN)$ of $\mathbb{Q}(\sqrt{-kN})$ for several values of k. Then, if $h(-kN)$ is smooth, we will be able to apply the $p-1$ method, replacing the group $\mathbb{F}_p^*$ by the class group of $\mathbb{Q}(\sqrt{-kN})$. As for the $p-1$ method, this will enable us to compute the (unknown) order of a group, the only difference being that from the order of $\mathbb{F}_p^*$ we split N by computing a GCD with N, while in our case we will split N by using ambiguous forms.

Since we will use $p-1$-type methods, we need to specify the bounds B_1 (for the first stage), and B_2 (for the second stage). Since we have a large number of groups at our disposal, we will be able to create a method which will be a systematic factoring method by choosing B_1 and B_2 appropriately, since we can hope that $h(-kN)$ will be smooth for a value of k which is not too large.

To choose these values appropriately, we need a fundamental theorem about smooth numbers. The upper bound was first proved by de Bruijn ([de-Bru]), and the complete result by Canfield, Erdős and Pomerance ([CEP]). It is as follows.

Theorem 10.2.1 (Canfield, Erdős, Pomerance). *Let*

$$\psi(x,y) = |\{n \leq x, n \text{ is } y\text{-smooth }\}| \ .$$

Then if we set $u = \ln x / \ln y$, we have

$$\psi(x,y) = xu^{-u(1+o(1))}$$

uniformly for $x \to \infty$ if $(\ln x)^\epsilon < u < (\ln x)^{1-\epsilon}$ for a fixed $\epsilon \in (0,1)$.

In particular, if we set

$$L(x) = e^{\sqrt{\ln x \ln \ln x}} \, ,$$

then

$$\psi(x, L(x)^a) = xL(x)^{-1/2a+o(1)} \, .$$

Now heuristic methods (see Section 5.10 and [Coh-Len1]) seem to indicate that class numbers are not only as smooth, but even slightly smoother than average. Furthermore, it is not difficult to see that there is little quantitative difference between B-smoothness and B-powersmoothness. Hence, it is not unreasonable to apply Theorem 10.2.1 to estimate the behavior of powersmoothness of class numbers. In addition, the class number $h(-N)$ is $O(N^{1/2+\epsilon})$ (for example $h(-N) < \frac{1}{\pi}\sqrt{N}\ln N$, see Exercise 27 of Chapter 5).

Hence, if we take $x = \sqrt{N}$ and $B = L(x)^a$, we expect that the probability that a given class number of size around x is B-powersmooth should be at least $L(x)^{-1/2a+o(1)}$, hence the expected number of values of k which we will have to try before hitting a B-powersmooth number should be approximately $L(x)^{1/2a+o(1)}$. (Note that the class number $h(-kN)$ is still $O(N^{1/2+\epsilon})$ for such values of k.) Hence, ignoring step 2 of the $p-1$ algorithm (which in any case influences only on the O constant, not the exponents), the expected running time with this choice of B is $O(L(x)^{a+1/2a+o(1)})$, and this is minimal for $a = 1/\sqrt{2}$. Since $L(x)^{1/\sqrt{2}} \approx L(N)^{1/2}$, we see that the optimal choice of B is approximately $L(N)^{1/2}$, and the expected running time is $L(N)^{1+o(1)}$. Note also that the storage is negligible.

10.2.2 The Schnorr-Lenstra Factoring Method

We now give the algorithm. Note that contrary to the $p-1$ method, we do not need to do any backtracking since if x is an ambiguous form which is not the unit form (i.e. is of order exactly equal to 2), so is x^r for any odd number r).

Algorithm 10.2.2 (Schnorr-Lenstra). Let N be a composite number. This algorithm will attempt to split N. We assume that we have precomputed a table $p[1], \ldots, p[k]$ of all the primes up to $L(N)^{1/2}$.

1. [Initialize] Set $B \leftarrow \lfloor L(N)^{1/2} \rfloor$, $K \leftarrow 1$, $e \leftarrow \lfloor \lg B \rfloor$.

2. [Initialize for K] Let $D = -KN$ if $KN \equiv 3 \pmod 4$, $D = -4KN$ otherwise.

3. [Choose form] Let f_p be a random primeform of discriminant D (see Algorithm 5.4.10). Set $x \leftarrow f_p$, $c \leftarrow 0$ and $i \leftarrow 1$.

4. [Next prime] Set $i \leftarrow i+1$. If $i > k$, set $K \leftarrow K+1$ and go to step 2. Otherwise, set $q \leftarrow p[i]$, $q_1 \leftarrow q$, $l \leftarrow \lfloor B/q \rfloor$.

5. [Compute power] While $q_1 \leq l$, set $q_1 \leftarrow q \cdot q_1$. Then, set $x \leftarrow x^{q_1}$ (powering in the class group), $c \leftarrow c + 1$ and if $c < 20$ go to step 4.

6. [Success?] Set $e_1 \leftarrow 0$, and while x is not an ambiguous form and $e_1 < e$ set $x \leftarrow x^2$ and $e_1 \leftarrow e_1 + 1$. Now if x is not an ambiguous form, set $c \leftarrow 0$, and go to step 4.

7. [Finished?] (Here x is an ambiguous form.) Find the factorization of KN corresponding to x. If this does not split N (for example if x is the unit form), go to step 3. Otherwise, output a non-trivial factor of N and terminate the algorithm.

Note that if in step 7 we obtain an ambiguous form which does not succeed in splitting N, this very probably still means that the K used is such that $h(-KN)$ is B-powersmooth. Therefore we must keep this value of K and try another random form in the group, but we should not change the group anymore. Note also that the first prime tried in step 4 is $p[2] = 3$, and *not* $p[1] = 2$.

To give a numerical example of the numbers involved, for $N = 10^{60}$, which is about the maximum size of numbers which one can factor in a reasonable amount of time with this method, we have $B \approx 178905$, and since we need the primes only up to B, this is quite reasonable. In fact, it is better to take a lower value of $B_1 = B$, and use the second stage of the $p - 1$ method with quite a larger value for B_2. This reduces the expected running time of the algorithm, but the optimal values to take are implementation dependent. We leave as an exercise for the reader the incorporation of step 2 of the $p - 1$ method into this algorithm, using these remarks (see Exercise 2).

As in all algorithms using class groups of quadratic fields, the basic operation in this algorithm is composition of quadratic forms. Even with the use of optimized methods like NUDUPL and NUCOMP (Algorithms 5.4.8 and 5.4.9), this is still a slow operation. Hence, although this method is quite attractive because of its running time, which is as good as all the other modern factoring algorithms with the exception of the number field sieve, and although it uses little storage, to the author's knowledge it has never been used intensively in factoring projects. Indeed, the elliptic curve method for instance has the same characteristics as the present one as far as speed and storage are concerned, but the group operations on elliptic curves can be done faster than in class groups, especially when (as will be the case), several curves have to be dealt with simultaneously (see Section 10.3).

Also note that it has been proved by Lenstra and Pomerance that for composite numbers of a special form the running time of this algorithm is very poor (i.e. exponential time).

10.3 The Elliptic Curve Method

10.3.1 Sketch of the Method

We now come to another method which also uses ideas from the $p-1$-method, but uses the group of points of an elliptic curve over $\mathbb{Z}/p\mathbb{Z}$ instead of the group $(\mathbb{Z}/p\mathbb{Z})^*$. This method, due to H. W. Lenstra, is one of the three main methods in use today, together with the quadratic sieve (see Section 10.4) and the number field sieve (see Section 10.5). In addition it possesses a number of properties which make it useful even if it is only used in conjunction with other algorithms. Like the class group method, it requires little storage and has a similar expected running time. Unique among modern factoring algorithms however, it is sensitive to the size of the prime divisors. In other words, its running time depends on the size of the smallest prime divisor p of N, and not on N itself. Hence, it can be profitably used to remove "small" factors, after having used trial division and the Pollard ρ method 8.5.2. Without too much trouble, it can find prime factors having 10 to 20 decimal digits. On the other hand, it very rarely finds prime factors having more than 30 decimal digits. This means that if N is equal to a product of two roughly equal prime numbers having no special properties, the elliptic curve method will not be able to factor N if it has more than, say, 70 decimal digits. In this case, one should use the quadratic sieve or the number field sieve.

We now describe the algorithm. As in the class group algorithm, for simplicity we give only the version which uses stage 1 of the $p-1$-method, the extension to stage 2 being straightforward.

Recall that the group law on an elliptic curve of the form $y^2 = x^3 + ax + b$ is given by formulas which generically involve the expression $(y_2 - y_1)/(x_2 - x_1)$. This makes perfect sense in a field (when $x_2 \neq x_1$), but if we decide to work in $\mathbb{Z}/N\mathbb{Z}$, this will not always make sense since $x_2 - x_1$ will not always be invertible when $x_2 \neq x_1$. But this is exactly the point: if $x_2 - x_1$ is not invertible in $\mathbb{Z}/N\mathbb{Z}$ with $x_2 \neq x_1$, this means that $(x_2 - x_1, N)$ is a non-trivial divisor of N, and this is what we want. Hence we are going to work on an elliptic curve modulo N (whatever that is, we will define it in Section 10.3.2), and work as if N is prime. Everything will work out as long as every non-zero number modulo N that we encounter is invertible. As soon as it does not work out, we have found a non-trivial factorization of N. At this point, the reader may wonder what elliptic curves have to do with all this. We could just as well choose numbers x at random modulo N and compute (x, N), hoping to find a non-trivial divisor of N. It is easy to see that this would be a $O(N^{1/2+\epsilon})$ algorithm, totally unsuitable. But if N has a prime divisor p such that our elliptic curve E has a smooth number of points modulo p, the $p-1$-method will discover this fact, i.e. find a power of a point giving the unit element of the curve modulo p. This means that we will have some x_1 and x_2 such that $x_1 \equiv x_2 \pmod{p}$, hence $(x_2 - x_1, N) > 1$, and as with all these methods, this is in fact equal to a non-trivial divisor of N. This means it is reasonable to expect that something will break down, which is what we hope in this case.

Before turning to the detailed description of the algorithm, it is instructive to compare the different methods using the $p-1$-idea. For this discussion, we assume that we obtain exactly the prime p which is at the basis of the method. Let B be the stage 1 bound, $M = \mathrm{lcm}[1..B]$, and let G be the underlying group and a an element of G.

(1) In the $p-1$ method itself (or its variants like the $p+1$ method), $G = \mathbb{F}_p^*$ (or $G = \mathbb{F}_{p^2}^*$), and we obtain p directly as $\gcd(a^M - 1, N)$.

(2) In the class group method, $G = Cl(\mathbb{Q}(\sqrt{-KN}))$ for a suitable K, and we obtain p indirectly through the correspondence between a factorization $KN = p \cdot KN/p$ and some ambiguous forms x in G, which is obtained as $a^{M/2^t}$ for a suitable value of t.

(3) In the elliptic curve method, $G = E(\mathbb{F}_p)$ and we obtain p indirectly because of the impossibility of computing a^M modulo N (that is, we encountered a non-invertible element).

We see that the reasons why we obtain the factorization of N are quite diverse. The running time is essentially governed by the abundance of smooth numbers, i.e. by the theorem of Canfield, Erdős and Pomerance, and so it is not surprising that the running time of the elliptic curve method will be similar to that of the class group method, with the important difference of being sensitive to the size of p.

10.3.2 Elliptic Curves Modulo N

Before giving the details of the method, it is useful to give some idea of projective geometry over $\mathbb{Z}/N\mathbb{Z}$ when N is not a prime. When N is a prime, the projective line over $\mathbb{Z}/N\mathbb{Z}$ can simply be considered as the set $\mathbb{Z}/N\mathbb{Z}$ to which is added a single "point at infinity", hence has $N+1$ elements. When N is not a prime, the situation is more complicated.

Definition 10.3.1. *We define projective n-space over $\mathbb{Z}/N\mathbb{Z}$ as follows. Let $E = \{(x_0, x_1, \ldots, x_n) \in (\mathbb{Z}/N\mathbb{Z})^{n+1}, \gcd(x_0, x_1, \ldots, x_n, N) = 1\}$. If $\mathcal{R}$ is the relation on E defined by multiplication by an invertible element of $\mathbb{Z}/N\mathbb{Z}$, then $\mathcal{R}$ is an equivalence relation, and we define*

$$\mathbb{P}_n(\mathbb{Z}/N\mathbb{Z}) = E/\mathcal{R} \ ,$$

i.e. the set of equivalence classes of E modulo the relation $\mathcal{R}$.

We will denote by $(x_0 : x_1 : \cdots : x_n)$ the equivalence class in $\mathbb{P}_n(\mathbb{Z}/N\mathbb{Z})$ of $(x_0, x_1, \ldots, x_n)$.

Remarks.

(1) Note that even though the x_i are in $\mathbb{Z}/N\mathbb{Z}$, it makes sense to take their GCD together with N by taking any representatives in $\mathbb{Z}$ and then computing the GCD.

(2) We recover the usual definition of projective n-space over a field when N is prime.

(3) The set $(\mathbb{Z}/N\mathbb{Z})^n$ can be naturally embedded into $\mathbb{P}_n(\mathbb{Z}/N\mathbb{Z})$ by sending $(x_0, x_1, \ldots, x_{n-1})$ to $(x_0 : x_1 : \cdots : x_{n-1} : 1)$. This subset of $\mathbb{P}_n(\mathbb{Z}/N\mathbb{Z})$ will be called for our purposes *its* affine subspace, and denoted $\mathbb{P}_n^{\text{Aff}}(\mathbb{Z}/N\mathbb{Z})$, although it is not canonically defined.

(4) If p is a prime divisor of N (or in fact any divisor), there exists a natural map from $\mathbb{P}_n(\mathbb{Z}/N\mathbb{Z})$ to $\mathbb{P}_n(\mathbb{Z}/p\mathbb{Z})$ induced by reducing projective coordinates modulo p. Then P belongs to $\mathbb{P}_n^{\text{Aff}}(\mathbb{Z}/N\mathbb{Z})$ if and only if the reduction of P modulo every prime divisor p of N belongs to $\mathbb{P}_n^{\text{Aff}}(\mathbb{Z}/p\mathbb{Z})$.

(5) When N is a prime, we have a natural decomposition $\mathbb{P}_n(\mathbb{Z}/N\mathbb{Z}) = \mathbb{P}_n^{\text{Aff}}(\mathbb{Z}/N\mathbb{Z}) \cup \mathbb{P}_{n-1}(\mathbb{Z}/N\mathbb{Z})$, by identifying $(x_0 : x_1 : \cdots : x_{n-1})$ with $(x_0 : x_1 : \cdots : x_{n-1} : 0)$. In the general case, this is no longer true. We can still make the above identification of $\mathbb{P}_{n-1}$ with a subspace of $\mathbb{P}_n$. (It is easy to check that it is compatible with the equivalence relation defining the projective spaces.) There is however a third subset which enters, made up of points $P = (x_0 : x_1 : \cdots : x_n)$ such that x_n is neither invertible nor equal to 0 modulo N, i.e. such that (x_n, N) is a non-trivial divisor of N. We will call this set the *special subset*, and denote it by $\mathbb{P}_n^s(\mathbb{Z}/N\mathbb{Z})$. For any subset E of $\mathbb{P}_n(\mathbb{Z}/N\mathbb{Z})$ we will denote by E^{Aff}, E_{n-1} and E^s the intersection of E with $\mathbb{P}_n^{\text{Aff}}$, $\mathbb{P}_{n-1}$ and $\mathbb{P}_n^s$ respectively. Hence, we have the disjoint union

$$E = E^{\text{Aff}} \cup E_{n-1} \cup E^s \ .$$

Let us give an example. The projective line over $\mathbb{Z}/6\mathbb{Z}$ has 12 elements, which are $(0 : 1)$, $(1 : 1)$, $(2 : 1)$, $(3 : 1)$, $(4 : 1)$, $(5 : 1)$, $(1 : 2)$, $(3 : 2)$, $(5 : 2)$, $(1 : 3)$, $(2 : 3)$ and $(1 : 0)$ (denoting by the numbers 0 to 5 the elements of $\mathbb{Z}/6\mathbb{Z}$). The first 6 elements make up the affine subspace, and the last element $(1 : 0)$ corresponds to the usual point at infinity, i.e. to $\mathbb{P}_0$. The other 5 elements are the special points.

It is clear that finding an element in the special subset of $\mathbb{P}_n(\mathbb{Z}/N\mathbb{Z})$ will immediately factor N, hence the special points are the ones which are interesting for factoring.

We leave as an exercise for the reader to show that

$$|\mathbb{P}_n(\mathbb{Z}/N\mathbb{Z})| = N^n \prod_{p|N} \left(1 + \frac{1}{p} + \cdots + \frac{1}{p^n}\right) \ ,$$

and in particular

$$|\mathbb{P}_1(\mathbb{Z}/N\mathbb{Z})| = N \prod_{p|N} \left(1 + \frac{1}{p}\right)$$

(see Exercise 6).

Definition 10.3.2. *Let N be a positive integer coprime to 6. We define an elliptic curve E over $\mathbb{Z}/N\mathbb{Z}$ as a projective equation of the form*

$$y^2 t = x^3 + axt^2 + bt^3$$

where $(x : y : t)$ are the projective coordinates, and a and b are elements of $\mathbb{Z}/N\mathbb{Z}$ such that $4a^3 + 27b^2$ is invertible modulo N.

As usual, by abuse of notation we shall use affine equations and affine coordinates even though it is understood that we work in the projective plane.

Now if N is a prime, the above definition is indeed the definition of an elliptic curve over the field $\mathbb{F}_N$. When N is not a prime the reduction maps modulo the prime divisors p of N clearly send $E(\mathbb{Z}/N\mathbb{Z})$ into $E(\mathbb{Z}/p\mathbb{Z})$. (Note that the condition that $4a^3 + 27b^2$ is invertible modulo N ensures that the reduced curves will all be elliptic curves.) Hence, as with any other set we can write

$$E(\mathbb{Z}/N\mathbb{Z}) = E^{\text{Aff}} \cup E_1 \cup E^s ,$$

and E^s is the set of points $(x : y : t)$ such that t is neither invertible nor equal to 0 modulo N. This means, in particular, that the reduction of $(x : y : t)$ modulo p will not always be in the affine part modulo p.

Warning. Note that if the reduction of $(x : y : t)$ modulo every prime divisor p of N is the point at infinity, this does *not* imply that t is equal to 0 modulo N. What it means is that t is divisible by all the primes dividing N, and this implies $t = 0 \pmod{N}$ only if N is squarefree.

Now we can use the addition laws given by Proposition 7.1.7 to try and define a group law on $E(\mathbb{Z}/N\mathbb{Z})$. They will of course not work as written, since even if $x_1 \neq x_2$, $x_1 - x_2$ may not be invertible modulo N. There are two ways around this. The first one, which we will not use, is to define the law on the projective coordinates. This can be done, and involves essentially looking at 9 different cases (see [Bos]). We then obtain a true group law, and on the affine part it is clear that the reduction maps modulo p are compatible with the group laws.

The second way is to stay ignorant of the existence of a complete group law. After all, we only want to factor N. Hence we use the formulas of Proposition 7.1.7 as written. If we start with two points in the affine part, their sum P will either be in the affine part, or of the form $(x : y : 0)$ (i.e. belong to E_1), or finally in the special part. If P is in the special part, we immediately split N since (t, N) is a non-trivial factor of N. If $P = (x : y : 0)$, then note that since $P \in E(\mathbb{Z}/N\mathbb{Z})$ we have $x^3 \equiv 0 \pmod{N}$. Then either $x \equiv 0 \pmod{N}$, corresponding to the non-special point at infinity of E, or (x, N) is a non-trivial divisor of N, and again we will have succeeded in splitting N.

10.3.3 The ECM Factoring Method of Lenstra

Before giving the algorithm in detail, we must still settle a few points. First, we must explain how to choose the elliptic curves, and how to choose the stage 1 bound B.

As for the choice of elliptic curves, one can simply choose $y^2 = x^3 + ax + 1$ which has the point $(0 : 1 : 1)$ on it, and a is small. For the stage 1 bound, since the number of points of E modulo p is around p by Hasse's theorem, one expects $E(\mathbb{Z}/p\mathbb{Z})$ to be $L(p)^a$-powersmooth with probability $L(p)^{-1/(2a)+o(1)}$ by the Canfield-Erdős-Pomerance theorem, hence if we take $B = L(p)^a$ we expect to try $L(p)^{1/(2a)+o(1)}$ curves before getting a smooth order, giving as total amount of work $L(p)^{a+1/(2a)+o(1)}$ group operations on the curve. This is minimal for $a = 1/\sqrt{2}$, giving a running time of $L(p)^{\sqrt{2}+o(1)}$ group operations.

Since, when N is composite, there exists a $p \mid N$ with $p \leq \sqrt{N}$, this gives the announced running time of $L(N)^{1+o(1)}$. But of course what is especially interesting is that the running time depends on the size of the smallest prime factor of N, hence the ECM can be used in a manner similar to trial division. In particular, contrary to the class group method, the choice of B should be done not with respect to the size of N, but, as in the original $p - 1$ method, with respect to the amount of time that one is willing to spend, more precisely to the approximate size of the prime p one is willing to look for.

For example, if we want to limit our search to primes less than 10^{20}, one can take $B = 12000$ since this is close to the value of $L(10^{20})^{1/\sqrt{2}}$, and we expect to search through 12000 curves before successfully splitting N. Of course, in actual practice the numbers will be slightly different since we will also use stage 2. The algorithm is then as follows.

Algorithm 10.3.3 (Lenstra's ECM). Let N be a composite integer coprime to 6, and B be a bound chosen as explained above. This algorithm will attempt to split N. We assume that we have precomputed a table, $p[1], \ldots, p[k]$ of all the primes up to B.

1. [Initialize curves] Set $a \leftarrow 0$ and let E be the curve $y^2 t = x^3 + axt^2 + t^3$.

2. [Initialize] Set $x \leftarrow (0 : 1 : 1)$, $i \leftarrow 0$.

3. [Next prime] Set $i \leftarrow i+1$. If $i > k$, set $a \leftarrow a+1$ and go to step 2. Otherwise, set $q \leftarrow p[i]$, $q_1 \leftarrow q$, $l \leftarrow \lfloor B/q \rfloor$.

4. [Compute power] While $q_1 \leq l$, set $q_1 \leftarrow q \cdot q_1$. Then, try to compute $x \leftarrow q_1 \cdot x$ (on the curve E) using the law given by Proposition 7.1.7. If the computation never lands in the set of special points or the $n - 1$ part of E (i.e. if one does not hit a non-invertible element t modulo N), go to step 3.

5. [Finished?] (Here the computation has failed, which is what we want.) Let t be the non-invertible element. Set $g \leftarrow (t, N)$ (which will not be equal to 1). If $g < N$, output g and terminate the algorithm. Otherwise, set $a \leftarrow a+1$ and go to step 2.

Note that when $g = N$ in step 5, this means that our curve has a smooth order modulo p, hence, as with the class group algorithm, we should keep the same curve and try another point. Finding another point may however not be easy since N is not prime, so there is no easy way to compute a square root modulo N (this is in fact essentially equivalent to factoring N, see Exercise

1). Therefore we have no other choice but to try again. As usual, this is an exceedingly rare occurrence, and so in practice it does not matter.

10.3.4 Practical Considerations

The ECM algorithm as given above in particular involves one division modulo N per operation on the elliptic curve, and this needs approximately the same time as computing a GCD with N. Thus we are in a similar situation to the Schnorr-Lenstra Algorithm 10.2.2 where the underlying group is a class group and the group operation is composition of quadratic forms, which also involves computing one, and sometimes two GCD's. Hence, outside from the property that ECM usually gives small factors faster, it seems that the practical running time should be slowed down for the same reason, i.e. the relative slowness of the group operation.

In the case of the ECM method however, many improvements are possible which do not apply to the class group method. The main point to notice is that here all the GCD's (or extended GCD's) are with the *same* number N. Hence, we can try grouping all these extended GCD's by working with several curves in parallel. That this can easily be done was first noticed by P. Montgomery. We describe his trick as an algorithm.

Algorithm 10.3.4 (Parallel Inverse Modulo N). Given a positive integer N and k integers $a_1, \ldots, a_k$ which are not divisible by N, this algorithm either outputs a non-trivial factor of N or outputs the inverses $b_1, \ldots, b_k$ of the a_i modulo N.

1. [Initialize] Set $c_1 \leftarrow a_1$ and for $i = 2, \ldots, k$ set $c_i \leftarrow c_{i-1} \cdot a_i \bmod N$.
2. [Apply Euclid] Using one of Euclid's extended algorithms of Section 1.3, compute (u, v, d) such that $u c_k + vN = d$ and $d = (c_k, N)$. If $d = 1$ go to step 3. Otherwise, if $d = N$, then set $d \leftarrow (a_i, N)$ for $i = 1, \ldots, k$ until $d > 1$ (this will happen). Output d as a non-trivial factor of N and terminate the algorithm.
3. [Compute inverses] For $i = k, k - 1, \ldots i = 2$ do the following. Output $b_i \leftarrow u c_{i-1} \bmod N$, and set $u \leftarrow u a_i \bmod N$. Finally, output $b_1 \leftarrow u$ and terminate the algorithm.

Proof. We clearly have $c_i = a_1 \cdots a_i \bmod N$, hence at the beginning of step 3 we have $u = (a_1 \cdots a_i)^{-1} \bmod N$, showing that the algorithm is valid.   □

Let us see the improvements that this algorithm brings. The naïve method would have required k extended Euclid to do the job. The present algorithm needs only 1 extended Euclid, plus $3k - 3$ multiplications modulo N. Hence, it is superior as soon as 1 extended Euclid is slower than 3 multiplications modulo N, and this is almost always the case.

Now recall from Chapter 7 that the computation of the sum of two points on an elliptic curve $y^2 = x^3 + ax + b$ requires the computation of $m = (y_2 -$

$y_1)(x_2 - x_1)^{-1}$ if the points are distinct, $m = (3x_1^2 + a)(2y_1)^{-1}$ if the points coincide, plus 2 multiplications modulo N and a few additions or subtractions. Since the addition/subtraction times are small compared to multiplication modulo N, we see that by using Montgomery's trick on a large number C of curves, the actual time taken for a group operation on the curve in the context of the ECM method is $6 + T/C$ multiplications modulo N when the points are distinct, or $7 + T/C$ when they are equal, where T is the ratio between the time of an extended GCD with N and the time of a multiplication modulo N. (Incidentally, note that in every other semi-group that we have encountered, including $\mathbb{Z}$, $\mathbb{R}$, $\mathbb{Z}[X]$ or even class groups, squaring is always faster than general multiplication. In the case of elliptic curves, it is the opposite.) If we take C large enough (say $C = 50$) this gives numbers which are not much larger than 6 (resp. 7), and this is quite reasonable.

Another way to speed up group computations on elliptic curves modulo N is to use projective coordinates instead of affine ones. The big advantage is then that no divisions modulo N are required at all. Unfortunately, since we must now keep track of three coordinates instead of two, the total number of operations increases, and the best that one can do is 12 multiplications modulo N when the points are distinct, 13 when they are equal (see Exercise 3). Thanks to Montgomery's trick, this is worse than the affine method when we work on many curves simultaneously.

By using other parametrizations of elliptic curves than the Weierstraß model $y^2 = x^3 + ax + b$, one can reduce the number 12 to 9 (see [Chu] and Exercise 4), but this still does not beat the $6 + T/C$ above when C is large. Hence, in practice I suggest using affine coordinates on the Weierstraß equation and Montgomery's trick.

Finally, as for the class group method, it is necessary to include a stage 2 into the algorithm, as for the $p - 1$ method. The details are left to the reader (see [Mon2], [Bre2]).

As a final remark in this section, we note than one can try to use other algebraic groups than elliptic curves, for example Abelian varieties. D. and G. Chudnovsky have explored this (see [Chu]), but since the group law requires a lot more operations modulo N, this does not seem to be useful in practice.

10.4 The Multiple Polynomial Quadratic Sieve

We now describe the quadratic sieve factoring algorithm which, together with the elliptic curve method, is the most powerful general factoring method in use at this time (1994). (The number field sieve has been successfully applied to numbers of a special form, the most famous being the ninth Fermat number $2^{2^9} + 1 = 2^{512} + 1$, a 155 digit number, but for general numbers, the quadratic sieve is still more powerful in the feasible range.) This method is due to C. Pomerance , although some of the ideas were already in Kraitchik.

10.4.1 The Basic Quadratic Sieve Algorithm

As in the continued fraction method CFRAC explained in Section 10.1, we look for many congruences of the type

$$x_k^2 \equiv (-1)^{e_{0k}} p_1^{e_{1k}} p_2^{e_{2k}} \cdots p_m^{e_{mk}} \pmod{N}$$

where the p_i are "small" prime numbers, and if we have enough, a Gaussian stage will give us a non-trivial congruence $x^2 \equiv y^2 \pmod{N}$ and hence a factorization of N. The big difference with CFRAC is the way in which the congruences are generated. In CFRAC, we tried to keep $x^2 \bmod N$ as small as possible so that it would have the greatest possible chance of factoring on our factor base of p_i. We of course assume that N is not divisible by any element of the factor base.

Here we still want the $x^2 \bmod N$ to be not too large but we allow residues larger than $\sqrt{N}$ (although still $O(N^{1/2+\epsilon})$). The simplest way to do this is to consider the polynomial

$$Q(a) = \left(\left\lfloor \sqrt{N} \right\rfloor + a\right)^2 - N \ .$$

It is clear that $Q(a) \equiv x^2 \pmod{N}$ for $x = \left\lfloor \sqrt{N} \right\rfloor + a$ and as long as $a = O(N^\epsilon)$, we will have $Q(a) = O(N^{1/2+\epsilon})$.

Although this is a simpler and more general way to generate small squares modulo N than CFRAC, it is not yet that interesting. The crucial point, from which part of the name of the method derives, is that contrary to CFRAC we do not need to (painfully) factor all these $x^2 \bmod N$ over the factor base. (In fact, most of them do not factor so this would represent a waste of time.) Here, since $Q(a)$ is a polynomial with integer coefficients, we can use a *sieve*. Let us see how this works. Assume that for some number m we know that $m \mid Q(a)$. Then, for every integer k, $m \mid Q(a + km)$ automatically. To find an a (if it exists) such that $m \mid Q(a)$ is of course very easy since we solve $x^2 \equiv N \pmod{m}$ using the algorithm of Exercise 30 of Chapter 1, and take $a = x - \left\lfloor \sqrt{N} \right\rfloor \bmod m$.

Since we are going to sieve, without loss of generality we can restrict to sieving with prime powers $m = p^k$. If p is an odd prime, then $x^2 \equiv N \pmod{p^k}$ has a solution (in fact two) if and only if $\left(\frac{N}{p}\right) = 1$, so we include only those primes in our factor base (this was also the case in the CFRAC algorithm) and we compute explicitly the two possible values of $a \pmod{p^k}$ such that $p^k \mid Q(a)$, say a_{p^k} and b_{p^k}. If $p = 2$ and $k \geq 3$, then $x^2 \equiv N \pmod{2^k}$ has a solution (in fact four) if and only if $N \equiv 1 \pmod 8$ and we again compute them explicitly. Finally, if $p = 2$ and $k = 2$, we take $x = 1$ if $N \equiv 1 \pmod 4$ (otherwise a does not exist) and if $p = 2$ and $k = 1$ we take $x = 1$.

Now for a in a very long interval (the sieving interval), we compute very crudely $\ln |Q(a)|$. (As we will see, an absolute error of 1 for instance is enough, hence we certainly will *not* use the internal floating point log but some ad hoc program.) We then store this in an array indexed by a. For every prime p in our factor base, and more generally for small prime powers when p is small (a good rule of thumb is to keep all possible p^k less than a certain bound), we subtract a crude approximation to $\ln p$ to every element of the array which is congruent to a_{p^k} or to b_{p^k} modulo p^k (this is the sieving part). When all the primes of the factor base have been removed in this way, it is clear that a $Q(a)$ will factor on our factor base if and only if what remains at index a in our array is close to 0 (if the logs were exact, it would be exactly zero). In fact, if $Q(a)$ does not factor completely, then the corresponding array element will be at least equal to $\ln B$ (where B is the least prime which we have not included in our factor base), and since this is much larger than 1 this explains why we can take very crude approximations to logs.

It can be shown on heuristic grounds, again using the theorem of Canfield, Erdős and Pomerance, that using suitable sieving intervals and factor bases, the running time is of the form $O(L(N)^{1+o(1)})$. Although this is comparable to the class group or ECM methods, note that the basic operation in the quadratic sieve is a single precision subtraction, and it is difficult to have a faster basic operation than that! As a consequence, for practical ranges (say up to 100 decimal digits) the quadratic sieve runs faster than the other methods that we have seen, although as already explained, ECM may be lucky if N has a relatively small prime divisor.

The method that we have just briefly explained is the basic quadratic sieve (QS). Many improvements are possible. The two remarks made at the end of Section 10.1 also apply here. First, only primes p such that $p = 2$ or $\left(\frac{N}{p}\right) = 1$ need to be taken in the prime base (or more generally $\left(\frac{kN}{p}\right) = 0$ or 1 if a multiplier is used). Second, the large prime variation is just as useful here as before. (This is also the case for the number field sieve, and more generally for any algorithm which uses in some way factor bases, for example McCurley or Buchmann's sub-exponential algorithms for class group and regulator computation.)

10.4.2 The Multiple Polynomial Quadratic Sieve

There is however a specific improvement to the quadratic sieve which explains the first two words of the complete name of the method (MPQS). The polynomial $Q(a)$ introduced above is nice, but unfortunately it stands all alone, hence the values of $Q(a)$ increase faster than we would like. The idea of the Multiple Polynomial Quadratic Sieve is to use several polynomials Q so that the size of $Q(a)$ can be kept as small as possible. The following idea is due to P. Montgomery.

We will take quadratic polynomials of the form $Q(x) = Ax^2 + 2Bx + C$ with $A > 0$, $B^2 - AC > 0$ and such that $N \mid B^2 - AC$. This gives congruences just as nicely as before since

$$AQ(x) = (Ax + B)^2 - (B^2 - AC) \equiv (Ax + B)^2 \pmod{N} .$$

In addition, we want the values of $Q(x)$ to be as small as possible on the sieving interval. If we want to sieve on an interval of length $2M$, it is therefore natural to center the interval at the minimum of the function Q, i.e. to sieve in the interval

$$I = [-B/A - M, -B/A + M] .$$

Then, for $x \in I$, we have $Q(-B/A) \leq Q(x) \leq Q(-B/A + M)$. Therefore to minimize the absolute value of $Q(x)$ we ask that $Q(-B/A) \approx -Q(-B/A+M)$, which is equivalent to $A^2 M^2 \approx 2(B^2 - AC)$ i.e. to

$$A \approx \frac{\sqrt{2(B^2 - AC)}}{M}$$

and we will have

$$\max_{x \in I} |Q(x)| \approx \frac{B^2 - AC}{A} \approx M\sqrt{(B^2 - AC)/2} .$$

Since we want this to be as small as possible, but still have $N \mid B^2 - AC$, we will choose A, B and C such that $B^2 - AC = N$ itself, and the maximum of $|Q(x)|$ will then be approximately equal to $M\sqrt{N/2}$.

This is of the same order of magnitude (in fact even slightly smaller) than the size of the values of our initial polynomial $Q(x)$, but now we have the added freedom to change polynomials as soon as the size of the residues become too large for our taste.

To summarize, we first choose an appropriate sieving length M. Then we choose A close to $\sqrt{2N}/M$ such that A is prime and $\left(\frac{N}{A}\right) = 1$. Using Algorithm 1.5.1 we find B such that $B^2 \equiv N \pmod{A}$ and finally we set $C = (B^2 - N)/A$.

Now as in the ordinary quadratic sieve, we must compute for each prime power p^k in our factor base the values $a_{p^k}(Q)$ and $b_{p^k}(Q)$ with which we will initialize our sieve. These are simply the roots mod p^k of $Q(a) = 0$. Hence, since the discriminant of Q has been chosen equal to N, they are equal to $(-B + a_{p^k})/A$ and $(-B + b_{p^k})/A$, where a_{p^k} and b_{p^k} denote the square roots of N modulo p^k which should be computed once and for all. The division by A (which is the only time-consuming part of the operation) is understood modulo p^k.

As for the basic quadratic sieve, heuristically the expected running time of MPQS is $O(L(N)^{1+o(1)})$, as for the class group method and ECM. However, as already mentioned above, the basic operation being so simple, MPQS is much faster than these other methods on numbers which are difficult to factor (numbers equal to a product of two primes having the same order of magnitude).

10.4.3 Improvements to the MPQS Algorithm

The detailed aspects of the implementation of the MPQS algorithm, such as
the choice of the sieving intervals, the size of the factor base and criteria to
switch from one polynomial to the next, are too technical to be given here. We
refer the interested reader to [Sil1] which contains all the necessary information
for a well tuned implementation of this algorithm.

A number of improvements can however be mentioned. We have already
discussed above the large prime variation. Other improvements are as follows.

(1) One improvement is the double large prime variation. This means that we
    allow the unfactored part of the residues to be equal not only to a single
    prime, but also to a product of two primes of reasonable size. This idea
    is a natural one, but it is then more difficult to keep track of the true
    relations that are obtained, and A. Lenstra and M. Manasse have found a
    clever way of doing this. I refer to [LLMP] for details.

(2) A second improvement is the small prime variation which is as follows.
    During the sieving process, the small primes or prime powers take a very
    long time to process since about $1/p$ numbers are divisible by p. In ad-
    dition, their contribution to the logarithms is the smallest. So we do not
    sieve at all with prime powers less than 100, say. This makes it necessary
    keep numbers whose residual logarithm is further away from zero than
    usual, but practice shows that it makes little difference. The main thing
    is to avoid missing any numbers which factor, at the expense of having a
    few extra which do not.

(3) A third improvement is the self-initialization procedure. This is as follows.
    We could try changing polynomials extremely often, since this would be
    the best chance that the residues stay small, hence factor. Unfortunately,
    as we have mentioned above, each time the polynomial is changed we
    must "reinitialize" our sieve, i.e. recompute starting values $a_{p^k}(Q)$ and
    $b_{p^k}(Q)$ for each p^k in our factor base. Although all the polynomials have
    the same discriminant N and the square roots have been precomputed (so
    no additional square root computations are involved), the time-consuming
    part is to invert the leading coefficient A modulo each element of the factor
    base. This prevents us from changing polynomial too often since otherwise
    this would dominate the running time.

    The self-initialization procedure deals with this problem by choosing
    A not to be a prime, but a product of a few (say 10) distinct medium-
    sized primes p such that $\left(\frac{N}{p}\right) = 1$. The number of possible values for B
    (hence the number of polynomials with leading term A) is equal to the
    number of solutions of $B^2 \equiv N \pmod{A}$, and this is equal to 2^{t-1} is t is
    the number of prime factors of A (see Exercise 30 of Chapter 1). Hence
    this procedure essentially divides by 2^{t-1} most of the work which must be
    done in initializing the sieve.

10.5 The Number Field Sieve

10.5.1 Introduction

We now come to the most recent and potentially the most powerful known factoring method, the number field sieve (NFS). For complete details I refer to [Len-Len2]. The basic idea is the same as in the quadratic sieve: by a sieving process we look for congruences modulo N by working over a factor base, and then we do a Gaussian elimination over $\mathbb{Z}/2\mathbb{Z}$ to obtain a congruence of squares, hence hopefully a factorization of N.

Before describing in detail the method, we will comment on its performance. Prior to the advent of the NFS, all modern factoring methods had an expected running time of at best $O(e^{\sqrt{\ln N \ln \ln N}(1+o(1))})$. Because of the theorem of Canfield, Erdős and Pomerance, some people believed that this could not be improved, except maybe for the $(1+o(1))$. The invention by Pollard of the NFS has now changed this belief, since under reasonable heuristic assumptions, one can show that the expected running time of the NFS is

$$O\left(e^{(\ln N)^{1/3}(\ln \ln N)^{2/3}(C+o(1))}\right)$$

for a small constant C (an admissible value is $C = (64/9)^{1/3}$ and this has been slightly lowered by Coppersmith). This is asymptotically considerably better than what existed before. Unfortunately, the practical situation is less simple. First, for a number N having no special form, it seems that the practical cutoff point with, say, the MPQS method, is for quite large numbers, maybe around 130 digits, and these numbers are in any case much too large to be factored by present methods. On the other hand, for numbers having a special form, for example Mersenne numbers $2^p - 1$ or Fermat numbers $2^{2^k} + 1$, NFS can be considerably simplified (one can in fact decrease the constant C to $C = (32/9)^{1/3}$), and stays practical for values of N up to 120 digits. In fact, using a system of distributed e-mail computing, and the equivalent of years of CPU time on small workstations, A. K. Lenstra and Manasse succeeded in 1990 in factoring the ninth Fermat number $F_9 = 2^{512} + 1$, which is a number of 155 decimal digits. The factors have respectively 7, 49 and 99 digits and the 7-digit factor was of course already known. Note that the knowledge of this 7-digit factor does not help NFS at all in this case.

The idea of the number field sieve is as follows. We choose a number field $K = \mathbb{Q}(\theta)$ for some algebraic integer θ, let $T(X) \in \mathbb{Z}[X]$ be the minimal monic polynomial of θ, and let d be the degree of K. Assume that we know an integer m such that $T(m) = kN$ for a small integer k. Then we can define a ring homomorphism ϕ from $\mathbb{Z}[\theta]$ to $\mathbb{Z}/N\mathbb{Z}$, by setting

$$\phi(\theta) = m \bmod N .$$

This homomorphism can be extended to $\mathbb{Z}_K$ in the following way. Let $f = [\mathbb{Z}_K : \mathbb{Z}[\theta]]$ be the index of $\mathbb{Z}[\theta]$ in $\mathbb{Z}_K$. We may assume that $(f, N) = 1$

otherwise we have found a non-trivial factor of N. Hence f is invertible modulo N, and if $u \in \mathbb{Z}$ is an inverse of f modulo N, for all $\alpha \in \mathbb{Z}_K$ we can set $\phi(\alpha) = u\phi(f\alpha)$ since $f\alpha \in \mathbb{Z}[\theta]$.

We can use ϕ as follows. To take the simplest example, if we can find integers a and b such that $a + bm$ is a square (in $\mathbb{Z}$), and also such that $a + b\theta$ is a square (in $\mathbb{Z}_K$), then we may have factored N: write $a + bm = x^2$, and $a + b\theta = \beta^2$. Since ϕ is a ring homomorphism, $\phi(a + b\theta) = a + bm \equiv y^2$ (mod N) where we have set y (mod N) $= \phi(\beta)$, hence $x^2 \equiv y^2$ (mod N), so $(x - y, N)$ may be a non-trivial divisor of N. Of course, in practice it will be impossible to obtain such integers a and b directly, but we can use techniques similar to those which we used in the continued fraction or in the quadratic sieve method, i.e. factor bases. Here however the situation is more complicated. We can take a factor base consisting of primes less than a given bound for the $a + bm$ numbers. But for the $a + b\theta$, we must take prime *ideals* of $\mathbb{Z}_K$. In general, if K is a number field with large discriminant, this will be quite painful. This is the basic distinction between the general number field sieve and the special one: if we can take for K a simple number field (i.e. one for which we know everything: units, class number, generators of small prime ideals, etc ...) then we are in the special case.

We will start by describing the simplest case of NFS, which can be applied only to quite special numbers, and in the following section we will explain what must be done to treat numbers of a general form.

10.5.2 Description of the Special NFS when $h(K) = 1$

In this section we not only assume that K is a simple number field in the sense explained above, but in addition that $\mathbb{Z}_K$ has class number equal to 1 (we will see in the next section what must be done if this condition is not satisfied).

Let $\alpha \in \mathbb{Z}_K$ and write

$$\alpha \mathbb{Z}_K = \prod_i \mathfrak{p}_i^{v_i} \;,$$

where we assume that for all i, $v_i > 0$. We will say that α is B-smooth if $\mathcal{N}_{K/\mathbb{Q}}(\alpha)$ is B-smooth, or in other words if all the primes below $\mathfrak{p}_i$ are less than or equal to B. Since $\mathbb{Z}_K$ has class number equal to 1, we can write

$$\alpha = \prod_{u \in U} u^{\lambda_u} \prod_{g \in G} g^{\mu_g} \;,$$

where U is a generating set of the group of units of K (i.e. a system of fundamental units plus a generator of the subgroup of roots of unity in K), and G is a set of $\mathbb{Z}_K$-generators for the prime ideals $\mathfrak{p}$ above a prime $p \leq B$ (since the ideals $\mathfrak{p}$ are principal).

If a lift of $\phi(\alpha)$ to $\mathbb{Z}$ is also B-smooth (in practice we always take the lift in $[-N/2, N/2]$) then we have

$$\phi(\alpha) \equiv \prod_{p \le B} p^{v_p}$$

hence the congruence

$$\prod_{u \in U} \phi(u)^{\lambda_u} \prod_{g \in G} \phi(g)^{\mu_g} \equiv \prod_{p \le B} p^{v_p} \pmod{N} .$$

If $\mathcal{P}$ is the set of primes less than or equal to B, then as in the quadratic sieve and similar algorithms, if we succeed in finding more than $|U| + |G| + |\mathcal{P}|$ such congruences, we can factor N by doing Gaussian elimination over $\mathbf{Z}/2\mathbf{Z}$.

By definition an HNF basis of $\mathbf{Z}_K$ is of the form $(1, (u\theta + v)/w, \dots)$. Replacing, if necessary, θ by $(u\theta + v)/w$, without loss of generality we may assume that there exists an HNF basis of $\mathbf{Z}_K$ of the form $(\omega_1, \omega_2, \omega_3, \dots, \omega_d)$ where $\omega_1 = 1$, $\omega_2 = \theta$ and ω_i is of degree exactly equal to $i - 1$ in θ. We will say in this case that θ is *primitive*.

This being done, we will in practice choose α to be of the form $a + b\theta$ with a and b in $\mathbf{Z}$ and coprime. We have the following lemma.

Lemma 10.5.1. *If a and b are coprime integers, then any prime ideal $\mathfrak{p}$ which divides $a + b\theta$, either divides the index $f = [\mathbf{Z}_K : \mathbf{Z}[\theta]]$ or is of degree 1.*

Proof. Let p be the prime number below $\mathfrak{p}$. Then $p \nmid b$ otherwise $a \in \mathfrak{p} \cap \mathbf{Z}$ hence $p \mid a$, contradicting a and b being coprime. Now assume that $p \nmid f$, and let b^{-1} be an inverse of b modulo p and u be an inverse of f modulo p. We have $\theta \equiv -ab^{-1} \pmod{\mathfrak{p}}$. Hence, if $x \in \mathbf{Z}_K$, $fx \in \mathbf{Z}[\theta]$ so there exists a polynomial $P \in \mathbf{Z}[X]$ such that $x \equiv uP(-ab^{-1}) \pmod{\mathfrak{p}}$ so any element of $\mathbf{Z}_K$ is congruent to a rational integer modulo $\mathfrak{p}$, hence to an element of the set $\{0, 1, \dots, p-1\}$, thus proving the lemma. $\square$

Let $d = \deg(T)$ be the degree of the number field K. By Theorem 4.8.13, prime ideals of degree 1 dividing a prime number p not dividing the index correspond to linear factors of $T(X)$ modulo p, i.e. to roots of $T(X)$ in $\mathbf{F}_p$. These can be found very simply by using Algorithm 1.6.1.

For any root $c_p \in \{0, 1, \dots, p-1\}$ of $T(X)$ modulo p, we thus have the corresponding prime ideal of degree 1 above p generated over $\mathbf{Z}_K$ by $(p, \theta - c_p)$. Now when we factor numbers α of the form $a + b\theta$ with $(a, b) = 1$, we will need to know the $\mathfrak{p}$-adic valuation of α for all prime ideals $\mathfrak{p}$ such that $\alpha \in \mathfrak{p}$. But clearly, if p does not divide f, then $\alpha \in \mathfrak{p}$ if and only if $p \mid a + bc_p$, and if this is the case then α does not belong to any other prime above p since the c_p are distinct. Hence, if $p \mid a + bc_p$, the $\mathfrak{p}$-adic valuation of α (with $\mathfrak{p} = (p, \theta - c_p)$) is equal to the p-adic valuation of $\mathcal{N}(\alpha)$ which is simple to compute.

For $p \mid f$, we can use an HNF basis of $\mathfrak{p}$ with respect to θ, where we may assume that θ is primitive. This basis will then be of the form $(p, -c_p + y\theta, \gamma_2, \dots, \gamma_{d-1})$ where c_p and y are integers with $y \mid p$ and the

γ_i are polynomials of degree exactly i in θ (not necessarily with integral coefficients). It is clear that $a + b\theta \in \mathfrak{p}$ if and only if $y \mid b$ and $a \equiv -bc_p/y \pmod{p}$. But $p \mid b$ is impossible since as before it would imply $p \mid a$ hence a and b would not be coprime. It follows that we must have $y = 1$. Hence, $\alpha \in \mathfrak{p}$ if and only if $p \mid a + bc_p$. Furthermore, $\theta - c_p \in \mathfrak{p}$ implies clearly that $T(c_p) \equiv 0 \pmod{p}$, i.e. that c_p is a root of T modulo p. The condition is therefore exactly the same as in the case $p \nmid f$. Note however that now there may be several prime ideals $\mathfrak{p}$ with the same value of c_p, so in that case the $\mathfrak{p}$-adic valuation of α should be computed using for example Algorithm 4.8.17. (Since this will be done only when we know that α and $\phi(\alpha)$ are B-smooth, it does not matter in practice that Algorithm 4.8.17 takes longer than the computation of $v_p(\mathcal{N}(\alpha))$.)

Thus, we will compute once and for all the roots c_p of the polynomial $T(X)$ modulo each prime $p \leq B$, and the constants $\beta_{\mathfrak{p}}$ (β in the notation of Algorithm 4.8.17) necessary to apply directly step 3 of Algorithm 4.8.17 for each prime ideal $\mathfrak{p}$ dividing the index. It is then easy to factor $\alpha = a + b\theta$ into prime *ideals* as explained above. Note that in the present situation, it is not necessary to split completely the polynomial $T(X)$ modulo p using one of the methods explained in Chapter 3, but only to find its roots modulo p, and in that case Algorithm 1.6.1 is much faster.

We must however do more, that is we need to factor α into prime *elements* and units. This is more delicate.

First, we will need to find explicit generators of the prime ideals in our factor base (recall that we have assumed that $\mathbb{Z}_K = \mathbb{Z}[\theta]$ is a PID). This can be done by computing norms of a large number of elements of $\mathbb{Z}_K$ which can be expressed as polynomials in θ with small coefficients, and combining the norms to get the desired prime numbers. This operation is quite time consuming, and can be transformed into a probabilistic algorithm, for which we refer to [LLMP]. This part is the essential difference with the general NFS since in the general case it will be impossible in practice to find generators of principal ideals. (The fact that $\mathbb{Z}_K$ is not a PID in general also introduces difficulties, but which are less important.)

Second, we also need generators for the group of units. This can be done during the search for generators of prime ideals. We find in this way a generating system for the units, and the use of the complex logarithmic embedding allows us to extract a multiplicative basis for the units as in Algorithm 6.5.9.

Choosing a factor base limit B, we will take as factor base for the numbers $a + bm$ the primes p such that $p \leq B$, and as factor base for the numbers $a + b\theta$ we will take a system G of non-associate prime *elements* of $\mathbb{Z}_K$ whose norm is either equal to $\pm p$, where p is a prime such that $p \leq B$ and $p \nmid f$, or equal to $\pm p^k$ for some k if $p \leq B$ and $p \mid f$, plus a generating system of the group of units of $\mathbb{Z}_K$.

We have seen that $\alpha \in \mathfrak{p}$ if and only if $p \mid a + bc_p$ which is a linear congruence for a and b. Hence, we can sieve using essentially the same sieving procedure as the one that we have described for the quadratic sieve.

1) By sieving on small primes, eliminate pairs (a, b) divisible by a small prime. (We will therefore keep a few pairs with $(a, b) > 1$, but this will not slow down the procedure in any significant way.)

2) Initialize the entries in the sieving interval to a crude approximation to $\ln(a + mb)$.

3) First sieve: for every $p^k \leq B$, subtract $\ln p$ from the entries where $p^k \mid a + mb$ by sieving modulo p, p^2, $\ldots$

4) Set a flag on all the entries which are still large (i.e. which are not B-smooth), and initialize the other entries with $\mathcal{N}(a + b\theta)$.

5) Second sieve: for every pair (p, c_p), subtract $\ln p$ from the unflagged entries for which $p \mid a + bc_p$. Note that we cannot sieve modulo p^2, $\ldots$

6) For each entry which is smaller than $2 \ln B$ (say), check whether the corresponding $\mathcal{N}(a + b\theta)$ is indeed smooth and in that case compute the complete factorization of $a + b\theta$ on $G \cup U$. Note that since we have not sieved with powers of prime ideals, we must check some entries which are larger than $\ln B$.

In practice, the factorization of $a + b\theta$ is obtained as follows. Since $\mathcal{N}(a + b\theta)$ is smooth we know that $\mathcal{N}(a + b\theta) = \prod_{p \leq B} p^{v_p}$. We can obtain the element relations as follows. If only one prime ideal $\mathfrak{p}$ above p corresponds to a given c_p (this is always true if $p \nmid f$), then if we let d be the degree of $\mathfrak{p}$ (1 if $p \nmid f$), the $\mathfrak{p}$-adic valuation of $a + b\theta$ is v_p/d, and the $\mathfrak{p}'$-adic valuation is zero for every other prime ideal above p. If several prime ideals correspond to the same c_p (this is possible only in the case $p \mid f$), then we use Algorithm 4.8.17 to compute the $\mathfrak{p}$-adic valuations. As already mentioned, this will be done quite rarely and does not really increase the running time which is mainly spent in the sieving process. Using the set G of explicit generators of our prime ideals, we thus obtain a decomposition

$$a + b\theta = u \prod_{g \in G} g^{\mu_g}$$

where u is a unit. If $(u_1, \ldots, u_r)$ is a system of fundamental units of K and ζ is a generator of the group of roots of unity in K, we now want to write

$$u = \zeta^{n_0} \prod_{i=1}^{r} u_i^{n_i} .$$

To achieve this, we can use the logarithmic embedding L (see Definition 4.9.6) and compute $L(a + b\theta) - \sum_{g \in G} \mu_g L(g)$. This will lie in the hyperplane $\sum x_i = 0$ of $\mathbb{R}^{r_1 + r_2}$, and by Dirichlet's theorem, the $L(u_i)$ form a basis of this hyperplane, hence we can find the n_i for $i \geq 1$ by solving a linear system (over $\mathbb{R}$, but we know that the solution is integral). Finally, n_0 can be obtained by comparing arguments of complex numbers (or even more simply by comparing signs if everything is real, which can be assumed if d is odd).

10.5.3 Description of the Special NFS when $h(K) > 1$

In this section, we briefly explain what modifications should be made to the above method in the case $h(K) > 1$, hence when $\mathbb{Z}_K$ is not a PID.

In this case we do not try to find generators of the prime ideals, but we look as before for algebraic integers (not necessarily of the form $a + b\theta$) with small coordinates in an integral basis, having a very smooth norm. More precisely, let $\mathfrak{p}_1, \mathfrak{p}_2, \ldots$ be the prime ideals of norm less than or equal to B ordered by increasing norm. We first look for an algebraic integer a_1 whose decomposition gives $a_1\mathbb{Z}_K = \mathfrak{p}_1^{k_{1,1}}$ where $k_{1,1}$ is minimal and hence is equal to the order of $\mathfrak{p}_1$ in $Cl(K)$. Then we look for another algebraic integer a_2 such that $a_2\mathbb{Z}_K = \mathfrak{p}_1^{k_{1,2}}\mathfrak{p}_2^{k_{2,2}}$ where $k_{2,2}$ is minimal and hence is equal to the order of $\mathfrak{p}_2$ in $Cl(K)/ < \mathfrak{p}_1 >$. We may also assume that $k_{1,2} < k_{1,1}$. We proceed in this way for each $\mathfrak{p}_i$ of norm less than or equal to B, and thus we have constructed an upper triangular matrix M whose rows correspond to the prime ideals and whose columns correspond to the numbers a_i. With high probability we have $h(K) = \prod_i k_{i,i}$, but it does not matter if this is not the case.

We can now replace the set G of generators of the $\mathfrak{p}_i$ which was used in the case $h(K) = 1$ by the set of numbers a_i in the following way.

Assume that α is B-smooth and that $\alpha\mathbb{Z}_K = \prod_i \mathfrak{p}_i^{v_i}$. Let V is the column vector whose components are the v_i. It is clear that $\alpha\mathbb{Z}_K = \prod_j a_j^{\mu_j}\mathbb{Z}_K$ where the μ_j are the components of the vector $M^{-1}V$ which are integers by construction of the matrix M. Hence $\alpha = u \prod_j a_j^{\mu_j}$ where u is a unit, and we can proceed as before. Note that since M is an upper triangular matrix it is easy to compute $M^{-1}V$ by induction.

An Example of the Special NFS. Assume that N is of the form $r^e - s$, where r and s are small. Choose a suitable degree d ($d = 5$ is optimal for numbers having 70 digits or more), and set $k = \left\lceil \dfrac{e}{d} \right\rceil$. Consider the polynomial

$$T(X) = X^d - sr^{kd-e} .$$

Since $0 \leq kd - e < d$ and s and r are small, so is sr^{kd-e}. If we choose $m = r^k$, it is clear that $T(m) = r^{kd-e}N$ is a small multiple of N. If T is an irreducible polynomial, we will work in the number field K of degree d defined by T. (If T is reducible, which almost never happens, we usually obtain a nontrivial factorization of N from a non-trivial factorization of T.) Since typically $d = 5$, and sr^{kd-e} is small, K is a simple field, i.e. it will not be difficult to find generators for ideals of small norm, the class number and a generating system for the group of units.

As mentioned above, the first success of the special NFS was obtained by [LLMP] with the ninth Fermat number $N = 2^{512} + 1$ which is of the above form. They chose $d = 5$, hence $k = 103$ and $T(X) = X^5 + 8$, thus $K = \mathbb{Q}(2^{1/5})$ which happens to be a field with class number equal to 1.

10.5.4 Description of the General NFS

The initial ideas of the general NFS are due to Buhler and Pomerance (see [BLP]). We do not assume anymore that K is a simple field. Hence it is out of the question to compute explicit generators for prime ideals of small norm, a system of fundamental units, etc ... Hence, we must work with ideals (and not with algebraic numbers) as long as possible.

So we proceed as before, but instead of keeping relations between elements (which is not possible anymore), we keep relations between the prime ideals themselves. As usual in our factor base we take the prime ideals of degree 1 whose norm is less than or equal to B and the prime ideals of norm less than or equal to B which divide the index f; since the index may not be easy to compute, we can use instead the prime ideals above primes $p \leq B$ such that p^2 divides the discriminant of the polynomial T).

After the usual Gaussian elimination step over $\mathbb{Z}/2\mathbb{Z}$, we will obtain algebraic numbers of the form

$$y = \prod (a + b\theta)^{\varepsilon_{a,b}}$$

where without loss of generality we may assume that $\varepsilon_{a,b} = 0$ or 1, such that

$$\phi(y) = \prod_{p \leq B} p^{v_p} \quad \text{(i.e. } \phi(y) \text{ is } B\text{-smooth), and}$$

$$y\mathbb{Z}_K = \prod_{\mathfrak{p}} \mathfrak{p}^{2v_{\mathfrak{p}}} ,$$

this last product being over the prime ideals of our factor base. Although the principal ideal $y\mathbb{Z}_K$ is equal to the square of an ideal, this does not imply that it is equal to the square of a *principal* ideal. Fortunately, this difficulty can easily be overcome by using a trick due to L. Adleman (see [Adl]).

Let us say that a non-zero algebraic number $y \in K$ is *singular* if $y\mathbb{Z}_K$ is the square of a fractional ideal. Let S be the multiplicative group of singular numbers. If $U(K)$ is the group of units of K, it is easy to check that we have an exact sequence

$$1 \longrightarrow U(K)/U(K)^2 \longrightarrow S/K^{*2} \longrightarrow Cl(K)[2] \longrightarrow 1 ,$$

where for any Abelian group G, $G[2]$ is the subgroup of elements of G whose square is equal to the identity (see Exercise 9). This exact sequence can be considered as an exact sequence of vector spaces over $\mathbb{F}_2 = \mathbb{Z}/2\mathbb{Z}$. Furthermore, using Dirichlet's Theorem 4.9.5 and the parity of the number $w(K)$ of roots of unity in K, it is clear that

$$\dim_{\mathbb{F}_2} U(K)/U(K)^2 = r_1 + r_2 .$$

For any finite Abelian group G, the exact sequence

$$1 \longrightarrow G[2] \longrightarrow G \longrightarrow G \longrightarrow G/G^2 \longrightarrow 1 \ ,$$

where the map from G to G is squaring, shows that $|G[2]| = |G/G^2|$ hence

$$\dim_{\mathbf{F}_2} G[2] = rk_2(G) \ ,$$

where the 2-rank $rk_2(G)$ of G is by definition equal to $\dim_{\mathbf{F}_2} G/G^2$ (and also to the number of even factors in the decomposition of G into a direct product of cyclic factors). Putting all this together, we obtain

$$\dim_{\mathbf{F}_2}(S/K^{*2}) = r_1 + r_2 + rk_2(Cl(K)) \ .$$

Hence, if we obtain more than $e = r_1 + r_2 + rk_2(Cl(K))$ singular numbers which are algebraic integers, a suitable multiplicative combination with coefficients 0 or 1 will give an element of $\mathbb{Z}_K \cap K^{*2}$, i.e. a square of $\mathbb{Z}_K$, as in the special NFS, hence a true relation of the form we are looking for. Since e is very small, this simply means that instead of stopping at the first singular integer that we find, we wait till we have at least $e + 1$ more relations than the cardinality of our factor base. Note that it is *not* necessary (and in practice not possible) to compute $rk_2(Cl(K))$. Any guess is sufficient, since afterwards we will have to check that we indeed obtain a square with a suitable combination, and if we do not obtain a square, this simply means that our guess is not large enough.

To find a suitable combination, following Adleman we proceed as follows. Choose a number r of prime ideals $\mathfrak{p}$ which do not belong to our factor base. A reasonable choice is $r = 3e$, where e can (and must) be replaced by a suitable upper bound. For example, we can choose for $\mathfrak{p}$ ideals of degree 1 above primes which are larger than B. Then $\mathfrak{p} = (p, \theta - c_p)$. We could also choose prime ideals of degree larger than 1 above primes (not dividing the index) less than B.

Whatever choice is made, the idea is then to compute a generalized Legendre symbol $\left(\frac{a+b\theta}{\mathfrak{p}}\right)$ (see Exercise 19 of Chapter 4) for every $a + b\theta$ which is kept after the sieving process. Hence each relation will be stored as a vector over $\mathbb{Z}/2\mathbb{Z}$ with $|E| + |\mathcal{P}| + r$ components, where E is the set of prime ideals in our factor base. As soon as we have more relations than components, by Gaussian elimination over $\mathbb{Z}/2\mathbb{Z}$ we can find an algebraic number x which is a singular integer and which is a quadratic residue modulo our r extra primes $\mathfrak{p}$. It follows that x is quite likely a square.

Assuming this to be the case, one of the most difficult problems of the general number field sieve, which is not yet satisfactorily solved at the time of this writing, is the problem of finding an algorithm to compute a square root y of x. Note that in practice x will be a product of thousands of $a + b\theta$, hence will be an algebraic number with coefficients (as polynomials in θ, say) having several hundred thousand decimal digits. Although feasible in principle, it does not seem that the explicit computation of x as a polynomial in θ will be of much help because of the size of the coefficients involved. Similarly for any other practical representation of x, for example by its minimal polynomial.

Let us forget this practical difficulty for the moment. We would like an algorithm which, given an algebraic integer x of degree d, either finds $y \in \mathbb{Q}[x]$ such that $y^2 = x$, or says that such a y does not exist. A simple-minded algorithm to achieve this is as follows.

Algorithm 10.5.2 (Square Root in a Number Field).   Given an algebraic integer x by its minimal monic polynomial $A(X) \in \mathbb{Z}[X]$ of degree d, this algorithm finds a y such that $y^2 = x$ and $y \in \mathbb{Q}[x]$, or says that such a y does not exist. (If x is given in some other way than by its minimal polynomial, compute the minimal polynomial first.) We let $K = \mathbb{Q}[x]$.

1. [Factor $A(X^2)$] Factor the polynomial $A(X^2)$ in $\mathbb{Z}[X]$. If $A(X^2)$ is irreducible, then y does not exist and terminate the algorithm. Otherwise, let $A(X^2) = \pm S(X)S(-X)$ for some monic polynomial $S \in \mathbb{Z}[X]$ of degree d be the factorization of $A(X^2)$ (it is necessarily of this form with S irreducible, see Exercise 10).

2. [Reduce to degree 1] Let $S(X) = (X^2 - x)Q(X) + R(X)$ be the Euclidean division of $S(X)$ by $X^2 - x$ in $K[X]$.

3. [Output result] Write $R(X) = aX + b$ with a and b in K and $a \neq 0$. Output $y \leftarrow -b/a$ and terminate the algorithm.

The proof of the validity of this algorithm is easy and left to the reader (Exercise 10).

Unfortunately, in our case, simply computing the polynomial $A(X)$ is already not easy, and factoring $A(X^2)$ will be even more difficult (although it will be a polynomial of degree 10 for example, but with coefficients having several hundred thousand digits). So a new idea is needed at this point. For example, H. W. Lenstra has suggested looking for y of the form $y = \prod(a+b\theta)$, the product being over coprime pairs (a,b) such that $a + b\theta$ is smooth, but not necessarily $a + bm$. This has the advantage that many more pairs (a, b) are available, and also leads to a linear system over $\mathbb{Z}/2\mathbb{Z}$. Future work will tell whether this method or similar ones are sufficiently practical.

10.5.5 Miscellaneous Improvements to the Number Field Sieve

Several improvements have been suggested to improve the (theoretical as well as practical) performance of NFS. Most of the work has been done on the general NFS, since the special NFS seems to be in a satisfactory state. We mention only two, since lots of work is being done on this subject.

The most important choice in the general NFS is the choice of the number field K, i.e. of the polynomial $T \in \mathbb{Z}[X]$ such that $T(m) = kN$ for some small integer k. Choosing a fixed degree d (as already mentioned, $d = 5$ is optimal for numbers having more than 60 or 70 digits), we choose $m = \lfloor N^{1/d} \rfloor$. If $N = m^d + a_{d-1}m^{d-1} + \cdots + a_0$ is the base m expansion of N (with $0 \leq a_i < m$), we can choose

$$T(X) = X^d + a_{d-1}X^{d-1} + \cdots + a_0 .$$

It is however not necessary to take the base m expansion of N in the strictest sense, since any base m expansion of N whose coefficients are at most of the order of m is suitable. In addition, we can choose to expand some small multiple kN of N instead of N itself. This gives us additional freedom.

Another idea is to use $m = \lceil N^{1/(d+1)} \rceil$ instead of $\lfloor N^{1/d} \rfloor$. The base m expansion of N is then of the form $N = a_d m^d + a_{d-1} m^{d-1} + \cdots + a_0$ with a_d not necessarily equal to 1, but still less than m. We take as before

$$T(X) = a_d X^d + a_{d-1}X^{d-1} + \cdots + a_0 ,$$

and if θ is a root of T, then θ is not an algebraic integer if $a_d > 1$. We can now use Exercise 15 of Chapter 4 which tells us that $a_d\theta$, $a_d\theta^2 + a_{d-1}\theta$, $\ldots$ are algebraic integers. The map ϕ is defined as usual by $\phi(\theta) = m$ and extended to polynomials in θ with integer coefficients. In particular, if a and b are integers, $a_d(a + b\theta)$ is an algebraic integer and

$$\phi(a_d(a + b\theta)) = a_d(a + mb)$$

is always divisible by a_d. Also,

$$\mathcal{N}(a_d(a + b\theta)) = (-1)^d a_d^{d-1} b^d T(-a/b)$$

with $b^d T(-a/b) \in \mathbb{Z}$. We then proceed as before, but using numbers of the form $a_d(a + b\theta)$ with a and b coprime, instead of simply $a + b\theta$.

To get rid of a_d in the final relations, it is not necessary to include the prime factors of a_d in the factor base, but simply to take an even number of factors in each relation.

A second type of improvement, studied by D. Coppersmith, is to use *several* number fields K. This leads to an improvement of the constant in the exponent of the running time of NFS, but its practicality has not yet been tested. The idea is a little similar to the use of several polynomials in MPQS.

10.6 Exercises for Chapter 10

1. Show that the problem of computing a square root modulo an arbitrary integer N is probabilistically polynomial time equivalent to the problem of factoring N in the following sense. If we have an algorithm for one of the problems, then we can solve the other in probabilistic polynomial time.

2. Generalize Algorithm 10.2.2 by incorporating a second stage in the manner of Algorithm 8.8.3.

3. Show how to write the addition law on an elliptic curve modulo N given by a Weierstraß equation using projective coordinates, using 12 multiplications modulo N, or 13 for the double of a point.

4. By using a Fermat parametrization of an elliptic curve, i.e. a projective equation of the form $x^3 + ay^3 = bt^3$, show how to compute the addition law using only 9 multiplications modulo N, or 10 for the double of a point.

5. Let B and k be large integers, and let $a_1, \ldots a_k$ be a randomly chosen sequence of integers less than B. Give an estimate of the average number of pairs (i, j) such that $a_i = a_j$. You may assume that $k > B^{1/2}$.

6. Let n be fixed, and set $f(N) = |\mathbb{P}_n(\mathbb{Z}/N\mathbb{Z})|$.
   a) Show that $f(N) = g(N)/\phi(N)$ where $\phi(N)$ is the Euler ϕ function, and $g(N)$ is the number of $n + 1$-uples $(x_0, \ldots, x_n) \in (\mathbb{Z}/N\mathbb{Z})^{n+1}$ such that $\gcd(x_0, \ldots, x_n, N) = 1$.
   b) Show that $\sum_{d|N} g(d) = N^{n+1}$.
   c) Using the Möbius inversion formula (see [H-W] Section 16.4), prove the formula for $f(N)$ given in the text.

7. In the multiple polynomial version of the quadratic sieve factoring algorithm, we have $aQ(x) \equiv y^2 \pmod{N}$ for some N, and not $Q(x)$ itself. Then why do we take into account in the explanation the maximum of $|Q(x)|$ and not of $|aQ(x)|$?

8. Let $\mathfrak{p} = (p, \theta - c_p)$ be a prime ideal of degree 1 in $\mathbb{Z}_K$, where $K = \mathbb{Q}(\theta)$. If $x = a + b\theta \in \mathbb{Z}_K$, show that $\left(\frac{x}{\mathfrak{p}}\right) = \left(\frac{a+bc_p}{p}\right)$, where $\left(\frac{x}{\mathfrak{p}}\right)$ is defined in Exercise 19 of Chapter 4.

9. Prove that, as claimed in the text, if S is the group of singular numbers, the following sequence is exact:

$$1 \longrightarrow U(K)/U(K)^2 \longrightarrow S/K^{*2} \longrightarrow Cl(K)[2] \longrightarrow 1 \ ,$$

where $Cl(K)[2]$ is the subgroup of elements of $Cl(K)$ whose square is equal to the identity.

10. Let $A(X)$ be an irreducible monic polynomial in $\mathbb{Z}[X]$.
    a) Show that either $A(X^2)$ is irreducible in $\mathbb{Z}[X]$, or there exists an irreducible monic polynomial $S \in \mathbb{Z}[X]$ such that $A(X^2) = \pm S(X)S(-X)$.
    b) Prove the validity of Algorithm 10.5.2.

11. For any finite Abelian group G and $n \geq 1$ show that

$$G[n] \simeq G/G^n$$

(although this isomorphism is not canonical in general).

Appendix A

Packages for Number Theory

There exist several computer packages which can profitably be used for number-theoretic computations. In this appendix, I will briefly describe the advantages and disadvantages of some of these systems.

Most general-purpose symbolic algebra packages have been written primarily for applied mathematicians, engineers and physicists, and are not always well suited for number theory. These packages roughly fall into two categories. In the first category one finds computer algebra systems developed in the 1970's, of which the main representatives are Macsyma and Reduce. Because of their maturity, these systems have been extensively tested and have probably less bugs than more recent systems. In addition they are very often mathematically more robust. In the second category, I include more recent packages developed in the 1980's of which the most common are Mathematica, by Wolfram Research, Inc., Maple, by the University of Waterloo, Canada, and more recently Axiom, developed by IBM and commercialized by NAG. These second-generation systems being more recent have more bugs and have been less tested. They are also often more prone to mathematical errors. On the other hand they have been aggressively commercialized and as a consequence have become more popular. However, the older systems have also been improved, and in particular recently Macsyma was greatly improved in terms of speed, user friendliness and efficiency and now compares very favorably to more recent packages. Mathematica has a very nice user interface, and its plotting capabilities, for example on the Macintosh, are superb. Maple is faster and often simpler to use, and has my preference. Axiom is a monster (in the same sense that ADA is a monster as a programming language). It certainly has a large potential for developing powerful applications, but I do not believe that there is the need for such power (which is usually obtained at the expense of speed) for everyday (number-theoretic) problems.

Some other packages were specially designed for small machines like Personal Computers (PC's). One of these is Derive, which is issued from μ-Math, and requires only half a megabyte of main memory. Derive even runs on some pocket computers! Another system, the Calculus Calculator (CC), is a symbolic manipulator with three-dimensional graphics and matrix operations which also runs on PC's. A third system, Numbers, is a shareware calculator for number theory that runs on PC's. It is designed to compute number theoretic functions for positive integers up to 150 decimal digits (modular arithmetic, primality testing, continued and Farey fractions, Fibonacci and Lucas numbers, encryption and decryption).

In addition to commercial packages, free software systems (which are not complete symbolic packages) also exist. One is Ubasic, written by Y. Kida, which is a math-oriented high-precision Basic for PC's (see the review in the Notices of the AMS of March 1991). Its extensions to Basic allow it to handle integers and reals of several thousand digits, as well as fractions, complex numbers and polynomials in one variable. Many number-theoretic functions are included in Ubasic, including the factoring algorithm MPQS. Since the package is written in assembly language, Ubasic is very fast.

Another package, closer to a symbolic package, is Pari, written by the author and collaborators (see the review in the Notices of the AMS of October 1991). This package can be used on Unix workstations, Macintosh, Amiga, PC's, etc. Its kernel is also written in assembler, so it is also very fast. Furthermore, it has been specially tailored for number-theoretic computations. In addition, it provides tools which are rarely or never found in other symbolic packages such as the direct handling of concrete mathematical objects, for example p-adic numbers, algebraic numbers and finite fields, etc ... It also gives mathematically more correct results than many packages on fundamental operations (e.g. subtraction of two real numbers which are approximately equal).

Source is included in the package so it is easy to correct, improve and expand. Essentially all of the algorithms described in the present book have been implemented in Pari, so I advise the reader to obtain a copy of it.

Apart from those general computer algebra systems, some special-purpose systems exist: GAP, Kant, Magma, Simath. The Magma system is designed to support fast computations in algebra (groups, modules, rings, polynomial rings over various kinds of coefficient domains), number theory and finite geometry. It includes general machinery for classical number theory (for example the ECM program of A.K. Lenstra), finite fields and cyclotomic fields and facilities for computing in a general algebraic number field. It will eventually include a MPQS factoring algorithm, a Jacobi sum-type primality test and a general purpose elliptic curve calculator. According to the developers, it should eventually include "just about all of the algorithms of this book". GAP (Groups, Algorithms and Programming) is specially designed for computations in group theory. It includes some facilities for doing elementary number theory, in particular to calculate with arbitrary length integers and rational numbers, cyclotomic fields and their subfields, and finite fields. It has functions for integer factorization (based on elliptic curves), for primality testing, and for some elementary functions from number theory and combinatorics. Its programming language is Maple-like. Kant (Komputational Algebraic Number Theory) is a subroutine package for algorithms from the geometry of numbers and algebraic number theory, which will be included in Magma. Simath, developed at the university of Saarbrucken, is another system for number-theoretic computations which is quite fast and has a nice user interface called simcalc.

In addition to specific packages, handling of multi-precision numbers or more general types can be easily achieved with several languages, Lisp, C and C++. For Lisp, the INRIA implementation LeLisp (which is not public domain) contains a package written in assembler to handle large numbers, and hence is very fast. The GNU Calc system is an advanced desk calculator for GNU Emacs, written in Emacs Lisp. An excellent public domain C++ compiler can be obtained from the Free Software Foundation, and its library allows to use multi-precision numbers or other types. The library is however written in C++ hence is *slow*, so it is strongly advised to write a library in assembler for number-theoretic uses. Another multi-precision system written in C is the desk calculator (Calc) of Hans-J. Boehm for Unix workstations. Its particularity is to handle "constructive" real numbers, that is to remember the best known approximation to a number already computed. For PC's, Timothy C. Frenz has developed an "infinite" precision calculator, also named Calc.

Finally, a few free packages exist which have been specifically written for handling multi-precision integers as part of a C library in an efficient way. In addition to Pari mentioned above, there is the Bignum package of DEC PRL (which is essentially the package used in LeLisp as mentioned above) which can be obtained by sending an e-mail message to librarian@decprl.dec.com, and the GNU multi-precision package Gmp which can be obtained by anonymous ftp from prep.ai.mit.edu, the standard place where one can ftp all the GNU software.

Conclusions.

My personal advice (which is certainly not objective) is the following. If you are on an IBM-PC 286, you do not have much choice. Obtain Ubasic, Derive or the Calculus Calculator. On an IBM-PC 386 or more, Maple, Macsyma, Mathcad (see Maple below) and Pari are also available. If you are on a MacII or on a Unix workstation then, if you really need all the power of a symbolic package, buy either Maple or Mathematica, my preference going to Maple. If you want a system that is already specialized for number theoretic computations, then buy Magma. In any case, as a complement to this package, obtain Pari.

Where to obtain these packages.

You can order Maple at the following address: Waterloo Maple Software, 160 Columbia St. W., Waterloo, Ontario, Canada N2L 3L3, phone (519) 747-2373, fax (519) 747-5284, e-mail wmsi@daisy.waterloo.edu. Maple has been ported to many different machines and it is highly probable that it has been ported to the machine that you want. There is also a system named Mathcad that uses some parts of Maple for its symbolic manipulations; Mathcad runs under Microsoft Windows and is published by MathSoft Inc., 201 Broadway, Cambridge, Massachussets, USA, 02139 Phone: (617) 577-1017.

You can order Mathematica from Wolfram Research, Inc. at the following address: Wolfram Research, 100 Trade Center Drive, Champaign, IL 61820, phone 800-441-Math, fax 217-398-0747, e-mail info@wri.com. Mathematica has also been ported to quite a number of machines, and in addition you can use a friendly "front-end" like the Macintosh II linked to a more powerful computer (including supercomputers) which will do the actual computations.

Macsyma exists in two flavors : the commercial versions (Macsyma, AL-JABR, ParaMacs) are licensed from MIT, the non-commercial versions (Vaxima, Maxima, and DOE-Macsyma) officially come from the American Department of Energy (DOE). All these versions are derived from the Macsyma developed by the Mathlab Group at MIT. The commercial version runs on PC 386, Symbolics computers, VMS machines and most Unix workstations; the address to order it is: Macsyma Inc., 20 Academy Street, Suite 201, Arlington MA 02174-6436, phone (617) 646-4550 or 1-800-MACSYMA (free from the U.S.), fax (617) 646-3161, e-mail info-macsyma@macsyma.com. Vaxima is available from the Energy Science and Technology Software Center (ESTSC), P.O. Box 1020, Oak Ridge, Tennessee 37831, phone (615) 576-2606. Maxima is a Common Lisp version maintained by William Schelter (e-mail wfs@math.utexas.edu) at Texas University. Although it is a non-commercial version, one must get a license from the Energy Science and Technology Software Center (see above) to use it. For more information, get the file README.MAXIMA by anonymous ftp on rascal.ics.utexas.edu. Para-Macs, is available from Leo Harten, Paradigm Associates, Inc., 29 Putnam Avenue, Suite 6, Cambridge, MA 02139, phone (617) 492-6079, fax (617) 876-8186, e-mail lph@paradigm.com. ALJABR is available from Fort Pond Research, 15 Fort Pond Road, Acton, MA 01720, phone 508-263-9692, e-mail aljabr@fpr.com. It runs on Macintosh, Sun and SGI computers.

There are many distributors of Reduce, depending on the machine and version of Lisp that is used. The main one is Herbert Melenk, Konrad-Zuse-Zentrum für Informationstechnik Berlin (ZIB), Heilbronner Str. 10, D 1000 Berlin 31, Germany, phone 30-89604-195, fax 30-89604-125, e-mail melenk@sc.zib-berlin.de. You will get detailed informations if you send an electronic message with send info-package as subject to reduce-netlib@rand.org.

Axiom on IBM RS/6000 is distributed by NAG: contact the Numerical Algorithms Group Ltd., Wilkinson House, Jordan Hill Rd., Oxford, UK OX2 8DR, phone (0)-865-511245, e-mail nagttt@vax.oxford.ac.uk. A Sparc version is also available.

Derive is available from Soft Warehouse, Inc., 3615 Harding Avenue, Suite 505, Honolulu, Hawaii 96816, USA, phone (808) 734-5801, fax (808) 735-1105.

You can obtain Ubasic by anonymous ftp at shape.mps.ohio-state.edu or wuarchive.wustl.edu. Or you can write directly to Kida at the following address: Prof. Yuji Kida, Department of Mathematics, Rikkyo University, Nishi-Ikebukuro 3, Tokyo 171, JAPAN, e-mail kida@rkmath.rikkyo.ac.jp.

The Calculus Calculator (CC) is developed by David Meredith, Department of Mathematics, San Francisco State University, 1600 Holloway Avenue, San Francisco, CA 94132, phone (415) 338-2199. Version 3 (CC3) is published with a 200 page manual by Prentice Hall, phone (201) 767-5937. Version 4 (CC4) is available by anonymous ftp from wuarchive.wustl.edu.

You can order Magma from The Secretary, Computational Algebra Group, Pure Mathematics, University of Sydney, NSW 2006, Australia, phone (2) 692-3338, fax (2) 692-4534, e-mail magma@maths.su.oz.au. It runs on Sun, HP, Apollo, VAX/VMS, Convex and various IBM machines.

GAP is available free of charge through ftp from Aachen: the ordinary mail address is Lehrstuhl D für Mathematik, RWTH Aachen, Templergraben 64, D-5100 Aachen, Germany. For technical questions, contact Martin Schoenert (e-mail martin@math.rwth-aachen.de), and for more general questions, contact Prof. Joachim Neubüser (e-mail neubueser@math.rwth-aachen.de).

There are two versions of Kant: Kant V1 is written in Ansi-Fortran 77, while Kant V2 is built on the Magma Platform and written in Ansi-C. These two versions are available from the KANT Group: e-mail to pohst@math.tu-berlin.de or daberkow@math.tu-berlin.de. You can get the system by anonymous ftp from ftp.math.tu-berlin.de, directory /pub/algebra/Kant. Note that Kant V2 is now also part of the Magma package.

You can obtain Simath by anonymous ftp from ftp.math.uni-sb.de.

Numbers is developed by Ivo Düntsch, Moorlandstr. 59, W-4500 Osnabrück, phone (541) 189-106, fax (541) 969-2470, e-mail duentsch@dosuni1.bitnet. You can get the system by anonymous ftp from dione.rz.uni-osnabrueck.de.

You can obtain Gmp (as well as all software from the Free Software Foundation) by anonymous ftp on prep.ai.mit.edu.

The three multi-precision systems named Calc can all be obtained by anonymous ftp: the GNU calculator (written and maintained by Dave Gillespie, e-mail daveg@csvax.cs.caltech.edu, 256-80 Caltech, Pasadena, CA 91125) from csvax.cs.caltech.edu, the calculator of Hans-J. Boehm from arisia.xerox.com and the calculator of Timothy C. Frenz (5361 Amalfi Drive, Clay, NY 13041) from the site wuarchive.wustl.edu.

Finally, you can obtain Pari by anonymous ftp from the sites megrez.ceremab.u-bordeaux.fr, ftp.inria.fr and math.ucla.edu.

Internet addresses and numbers for ftp

arisia.xerox.com	13.1.64.94	Boehm-Calc
csvax.cs.caltech.edu	131.215.131.131	GNU Calc
dione.rz.uni-osnabrueck.de	131.173.128.15	Numbers
ftp.math.tu-berlin.de	130.149.12.72	Kant
ftp.math.uni-sb.de	134.96.32.23	Simath
math.ucla.edu	128.97.4.254	Pari
megrez.ceremab.u-bordeaux.fr	147.210.16.17	Pari
prep.ai.mit.edu	18.71.0.38	Gmp
rascal.ics.utexas.edu	128.83.138.20	Maxima
shape.mps.ohio-state.edu	128.146.110.30	Ubasic
wuarchive.wustl.edu	128.252.135.4	Most packages

Appendix B

Some Useful Tables

In this appendix, we give five short tables which may be useful as basic data on which to work in algebraic number fields and on elliptic curves. The first two tables deal with quadratic fields and can be found in many places.

The third and fourth table give the corresponding tables for complex and totally real cubic fields respectively, and have been produced by M. Olivier using the method explained in Section 6.4.1 and the KANT package (see Appendix A).

The fifth table is a short table of elliptic curves extracted from [LN476] and [Cre].

I give here a list of references to the main tables that I am aware of. Not included are tables which have been superseded, and also papers containing only a few of the smallest number fields.

For quadratic fields see [Buel] and [Ten-Wil].

For cubic fields see [Enn-Tur1], [Enn-Tur2], [Gras], [Ang], [Sha-Wil] and [Ten-Wil].

For quartic fields see [Ford3], [Buc-Ford] and [BFP].

For quintic fields see [Diaz] and [SPD].

For sextic fields see [Oli3], [Oli4], [Oli5] and [Oli6].

Finally, for an extensive table of elliptic curves see Cremona's book [Cre].

B.1 Table of Class Numbers of Complex Quadratic Fields

Recall that the group of units of complex quadratic fields is equal to ± 1 except when the discriminant is equal to -3 or -4 in which case it is equal to the group of sixth or fourth roots of unity respectively.

The following table list triples $(d, h(d), H(-d))$ where d is negative and congruent to 0 or 1 modulo 4, $h(d)$ is the class number of the quadratic order of discriminant d, and $H(-d)$ is the Hurwitz class number of discriminant d (see Definition 5.3.6). Note that $h(d) = H(-d)$ if and only if d is a fundamental discriminant, that $H(-d)$ has a denominator equal to 2 (resp. 3) if and only if d is of the form $-4f^2$ (resp. $-3f^2$) and otherwise is an integer.

$(-3,1,1/3)$	$(-4,1,1/2)$	$(-7,1,1)$	$(-8,1,1)$
$(-11,1,1)$	$(-12,1,4/3)$	$(-15,2,2)$	$(-16,1,3/2)$
$(-19,1,1)$	$(-20,2,2)$	$(-23,3,3)$	$(-24,2,2)$
$(-27,1,4/3)$	$(-28,1,2)$	$(-31,3,3)$	$(-32,2,3)$
$(-35,2,2)$	$(-36,2,5/2)$	$(-39,4,4)$	$(-40,2,2)$
$(-43,1,1)$	$(-44,3,4)$	$(-47,5,5)$	$(-48,2,10/3)$
$(-51,2,2)$	$(-52,2,2)$	$(-55,4,4)$	$(-56,4,4)$
$(-59,3,3)$	$(-60,2,4)$	$(-63,4,5)$	$(-64,2,7/2)$
$(-67,1,1)$	$(-68,4,4)$	$(-71,7,7)$	$(-72,2,3)$
$(-75,2,7/3)$	$(-76,3,4)$	$(-79,5,5)$	$(-80,4,6)$
$(-83,3,3)$	$(-84,4,4)$	$(-87,6,6)$	$(-88,2,2)$
$(-91,2,2)$	$(-92,3,6)$	$(-95,8,8)$	$(-96,4,6)$
$(-99,2,3)$	$(-100,2,5/2)$	$(-103,5,5)$	$(-104,6,6)$
$(-107,3,3)$	$(-108,3,16/3)$	$(-111,8,8)$	$(-112,2,4)$
$(-115,2,2)$	$(-116,6,6)$	$(-119,10,10)$	$(-120,4,4)$
$(-123,2,2)$	$(-124,3,6)$	$(-127,5,5)$	$(-128,4,7)$
$(-131,5,5)$	$(-132,4,4)$	$(-135,6,8)$	$(-136,4,4)$
$(-139,3,3)$	$(-140,6,8)$	$(-143,10,10)$	$(-144,4,15/2)$
$(-147,2,7/3)$	$(-148,2,2)$	$(-151,7,7)$	$(-152,6,6)$
$(-155,4,4)$	$(-156,4,8)$	$(-159,10,10)$	$(-160,4,6)$
$(-163,1,1)$	$(-164,8,8)$	$(-167,11,11)$	$(-168,4,4)$
$(-171,4,5)$	$(-172,3,4)$	$(-175,6,7)$	$(-176,6,10)$
$(-179,5,5)$	$(-180,4,6)$	$(-183,8,8)$	$(-184,4,4)$
$(-187,2,2)$	$(-188,5,10)$	$(-191,13,13)$	$(-192,4,22/3)$
$(-195,4,4)$	$(-196,4,9/2)$	$(-199,9,9)$	$(-200,6,7)$
$(-203,4,4)$	$(-204,6,8)$	$(-207,6,9)$	$(-208,4,6)$
$(-211,3,3)$	$(-212,6,6)$	$(-215,14,14)$	$(-216,6,8)$
$(-219,4,4)$	$(-220,4,8)$	$(-223,7,7)$	$(-224,8,12)$
$(-227,5,5)$	$(-228,4,4)$	$(-231,12,12)$	$(-232,2,2)$
$(-235,2,2)$	$(-236,9,12)$	$(-239,15,15)$	$(-240,4,8)$
$(-243,3,13/3)$	$(-244,6,6)$	$(-247,6,6)$	$(-248,8,8)$
$(-251,7,7)$	$(-252,4,10)$	$(-255,12,12)$	$(-256,4,15/2)$
$(-259,4,4)$	$(-260,8,8)$	$(-263,13,13)$	$(-264,8,8)$
$(-267,2,2)$	$(-268,3,4)$	$(-271,11,11)$	$(-272,8,12)$
$(-275,4,5)$	$(-276,8,8)$	$(-279,12,15)$	$(-280,4,4)$
$(-283,3,3)$	$(-284,7,14)$	$(-287,14,14)$	$(-288,4,9)$
$(-291,4,4)$	$(-292,4,4)$	$(-295,8,8)$	$(-296,10,10)$
$(-299,8,8)$	$(-300,6,28/3)$	$(-303,10,10)$	$(-304,6,10)$
$(-307,3,3)$	$(-308,8,8)$	$(-311,19,19)$	$(-312,4,4)$
$(-315,4,6)$	$(-316,5,10)$	$(-319,10,10)$	$(-320,8,14)$
$(-323,4,4)$	$(-324,6,17/2)$	$(-327,12,12)$	$(-328,4,4)$
$(-331,3,3)$	$(-332,9,12)$	$(-335,18,18)$	$(-336,8,12)$
$(-339,6,6)$	$(-340,4,4)$	$(-343,7,8)$	$(-344,10,10)$
$(-347,5,5)$	$(-348,6,12)$	$(-351,12,16)$	$(-352,4,6)$

(−355,4,4)	(−356,12,12)	(−359,19,19)	(−360,8,10)
(−363,4,13/3)	(−364,6,8)	(−367,9,9)	(−368,6,12)
(−371,8,8)	(−372,4,4)	(−375,10,12)	(−376,8,8)
(−379,3,3)	(−380,8,16)	(−383,17,17)	(−384,8,14)
(−387,4,5)	(−388,4,4)	(−391,14,14)	(−392,8,9)
(−395,8,8)	(−396,6,12)	(−399,16,16)	(−400,4,15/2)
(−403,2,2)	(−404,14,14)	(−407,16,16)	(−408,4,4)
(−411,6,6)	(−412,5,10)	(−415,10,10)	(−416,12,18)
(−419,9,9)	(−420,8,8)	(−423,10,15)	(−424,6,6)
(−427,2,2)	(−428,9,12)	(−431,21,21)	(−432,6,40/3)
(−435,4,4)	(−436,6,6)	(−439,15,15)	(−440,12,12)
(−443,5,5)	(−444,8,16)	(−447,14,14)	(−448,4,8)
(−451,6,6)	(−452,8,8)	(−455,20,20)	(−456,8,8)
(−459,6,8)	(−460,6,8)	(−463,7,7)	(−464,12,18)
(−467,7,7)	(−468,8,10)	(−471,16,16)	(−472,6,6)
(−475,4,5)	(−476,10,20)	(−479,25,25)	(−480,8,12)
(−483,4,4)	(−484,6,13/2)	(−487,7,7)	(−488,10,10)
(−491,9,9)	(−492,6,8)	(−495,16,20)	(−496,6,12)
(−499,3,3)	(−500,10,12)	(−503,21,21)	(−504,8,12)

B.2 Table of Class Numbers and Units of Real Quadratic Fields

In the following table of real quadratic fields K we list the following data from left to right: the discriminant $d = d(K)$, the class number $h = h(K)$, the regulator $R = R(K)$, the norm of the fundamental unit and finally the fundamental unit itself given as a pair of coordinates (a, b) on the canonical integral basis $(1, \omega)$ where $\omega = (1 + \sqrt{d})/2$ if $d \equiv 1 \pmod 4$, $\omega = \sqrt{d}/2$ if $d \equiv 0 \pmod 4$.

d	h	R	$\mathcal{N}(\epsilon)$	ϵ
5	1	0.4812	−1	(0,1)
8	1	0.8814	−1	(1,1)
12	1	1.317	1	(2,1)
13	1	1.195	−1	(1,1)
17	1	2.095	−1	(3,2)
21	1	1.567	1	(2,1)
24	1	2.292	1	(5,2)
28	1	2.769	1	(8,3)
29	1	1.647	−1	(2,1)
33	1	3.828	1	(19,8)
37	1	2.492	−1	(5,2)

40	2	1.818	−1	(3,1)
41	1	4.159	−1	(27,10)
44	1	2.993	1	(10,3)
53	1	1.966	−1	(3,1)
56	1	3.400	1	(15,4)
57	1	5.710	1	(131,40)
60	2	2.063	1	(4,1)
61	1	3.664	−1	(17,5)
65	2	2.776	−1	(7,2)
69	1	3.217	1	(11,3)
73	1	7.667	−1	(943,250)
76	1	5.829	1	(170,39)
77	1	2.185	1	(4,1)
85	2	2.209	−1	(4,1)
88	1	5.976	1	(197,42)
89	1	6.908	−1	(447,106)
92	1	3.871	1	(24,5)
93	1	3.366	1	(13,3)
97	1	9.324	−1	(5035,1138)
101	1	2.998	−1	(9,2)
104	2	2.312	−1	(5,1)
105	2	4.407	1	(37,8)
109	1	5.565	−1	(118,25)
113	1	7.347	−1	(703,146)
120	2	3.089	1	(11,2)
124	1	8.020	1	(1520,273)
129	1	10.43	1	(15371,2968)
133	1	5.153	1	(79,15)
136	2	4.248	1	(35,6)
137	1	8.157	−1	(1595,298)
140	2	2.478	1	(6,1)
141	1	5.247	1	(87,16)
145	4	3.180	−1	(11,2)
149	1	4.111	−1	(28,5)
152	1	4.304	1	(37,6)
156	2	3.912	1	(25,4)
157	1	5.361	−1	(98,17)
161	1	10.07	1	(10847,1856)
165	2	2.559	1	(6,1)
168	2	3.257	1	(13,2)
172	1	8.849	1	(3482,531)
173	1	2.571	−1	(6,1)
177	1	11.73	1	(57731,9384)
181	1	7.174	−1	(604,97)
184	1	10.79	1	(24335,3588)

185	2	4.913	−1	(63,10)
188	1	4.564	1	(48,7)
193	1	15.08	−1	(1637147,253970)
197	1	3.333	−1	(13,2)
201	1	13.85	1	(478763,72664)
204	2	4.605	1	(50,7)
205	2	3.761	1	(20,3)
209	1	11.44	1	(43331,6440)
213	1	4.290	1	(34,5)
217	1	15.86	1	(3583111,521904)
220	2	5.182	1	(89,12)
221	2	2.704	1	(7,1)
229	3	2.712	−1	(7,1)
232	2	5.288	−1	(99,13)
233	1	10.74	−1	(21639,3034)
236	1	6.966	1	(530,69)
237	1	4.344	1	(36,5)
241	1	18.77	−1	(66436843,9148450)
248	1	4.836	1	(63,8)
249	1	16.66	1	(8011739,1084152)
253	1	7.529	1	(872,117)
257	3	3.467	−1	(15,2)
264	2	4.867	1	(65,8)
265	2	9.405	−1	(5699,746)
268	1	11.49	1	(48842,5967)
269	1	5.100	−1	(77,10)
273	2	7.282	1	(683,88)
277	1	7.868	−1	(1228,157)
280	2	6.219	1	(251,30)
281	1	14.57	−1	(1000087,126890)
284	1	8.848	1	(3480,413)
285	2	2.830	1	(8,1)
293	1	2.837	−1	(8,1)
296	2	4.454	−1	(43,5)
301	1	10.03	1	(10717,1311)
305	2	6.886	1	(461,56)
309	1	8.526	1	(2379,287)
312	2	4.663	1	(53,6)
313	1	19.35	−1	(119691683,14341370)
316	3	5.075	1	(80,9)
317	1	4.489	−1	(42,5)
321	3	6.064	1	(203,24)
328	4	2.893	−1	(9,1)
329	1	15.37	1	(2245399,262032)
332	1	5.100	1	(82,9)

337	1	21.43	−1	(960491695,110671282)
341	1	5.624	1	(131,15)
344	1	9.943	1	(10405,1122)
345	2	9.512	1	(6397,728)
348	2	4.025	1	(28,3)
349	1	9.821	−1	(8717,986)
353	1	11.87	−1	(67471,7586)
357	2	2.942	1	(9,1)
364	2	8.055	1	(1574,165)
365	2	2.947	−1	(9,1)
373	1	9.234	−1	(4853,530)
376	1	15.27	1	(2143295,221064)
377	2	6.144	1	(221,24)
380	2	4.357	1	(39,4)
381	1	7.616	1	(963,104)
385	2	12.16	1	(90947,9768)
389	1	7.849	−1	(1217,130)
393	1	18.35	1	(44094699,4684888)
397	1	8.145	−1	(1637,173)
401	5	3.690	−1	(19,2)
408	2	5.308	1	(101,10)
409	1	26.13	−1	(106387620283,11068353370)
412	1	13.03	1	(227528,22419)
413	1	4.111	1	(29,3)
417	1	18.96	1	(81144379,8356536)
421	1	13.01	−1	(211627,21685)
424	2	8.988	−1	(4005,389)
428	1	7.562	1	(962,93)
429	2	4.977	1	(69,7)
433	1	23.39	−1	(6883177307,694966754)
437	1	3.042	1	(10,1)
440	2	3.737	1	(21,2)
444	2	6.380	1	(295,28)
445	4	3.047	−1	(10,1)
449	1	19.75	−1	(180529627,17883410)
453	1	5.004	1	(71,7)
456	2	7.626	1	(1025,96)
457	1	25.50	−1	(56325840235,5528222698)
460	2	7.720	1	(1126,105)
461	1	5.900	−1	(174,17)
465	2	10.37	1	(15135,1472)
469	3	4.174	1	(31,3)
472	1	13.33	1	(306917,28254)
473	3	5.159	1	(83,8)
476	2	5.481	1	(120,11)

481	2	14.47	−1	(920179,87922)
485	2	3.785	−1	(21,2)
488	2	3.093	−1	(11,1)
489	1	23.44	1	(7249279379,686701192)
492	2	5.497	1	(122,11)
493	2	4.710	−1	(53,5)
497	1	14.69	1	(1147975,107824)

B.3 Table of Class Numbers and Units of Complex Cubic Fields

Any number field can be defined as $K = \mathbb{Q}[\alpha]$ where α is a primitive algebraic integer (see Section 10.5.2), and we will denote by $A(X)$ the minimal monic polynomial of α. We will choose A so that the index $f = [\mathbb{Z}_K : \mathbb{Z}[\alpha]]$ is as small as possible and with small coefficients (hence A will not always be the pseudo-canonical polynomial given by Algorithm 4.4.12). The choice of the particular polynomials A which we will give is therefore not at all canonical.

Let now K be a cubic field. Since we have chosen α primitive, there exists an integral basis of the form $(1, \alpha, \beta)$. Furthermore any cubic field has at least one real embedding hence the set of roots of unity is always equal to ± 1. On the other hand complex cubic fields have unit rank equal to 1, while real cubic fields have unit rank equal to 2. Since the norm of -1 is equal to -1, there is no such thing as the sign of the norm of fundamental units.

The following is a table of the first hundred complex cubic fields. For each field K we give the following data from left to right: the discriminant $d = d(K)$, the index $f = [\mathbb{Z}_K : \mathbb{Z}[\alpha]]$, the polynomial A, the third element β of an integral basis $(1, \alpha, \beta)$, the class number $h = h(K)$, the regulator $R = R(K)$ and the fundamental unit ϵ expressed on the integral basis (for example $(2, 3, 1)$ means $2 + 3\alpha + \beta$). Since the signature of K is equal to $(1, 1)$, the Galois group of the Galois closure of K is always equal to the symmetric group S_3.

d	f	A	β	h	R	ϵ
-23	1	$X^3 + X^2 - 1$	α^2	1	0.2812	(0,1,1)
-31	1	$X^3 - X^2 - 1$	α^2	1	0.3822	(0,1,0)
-44	1	$X^3 - X^2 - X - 1$	α^2	1	0.6094	(0,1,0)
-59	1	$X^3 + 2X - 1$	α^2	1	0.7910	(2,0,1)
-76	1	$X^3 - 2X - 2$	α^2	1	1.019	(1,1,0)
-83	1	$X^3 - X^2 + X - 2$	α^2	1	1.041	(1,0,1)
-87	1	$X^3 + X^2 + 2X - 1$	α^2	1	0.9348	(2,1,1)
-104	1	$X^3 - X - 2$	α^2	1	1.576	(1,1,1)
-107	1	$X^3 - X^2 + 3X - 2$	α^2	1	1.256	(3,0,1)
-108	1	$X^3 - 2$	α^2	1	1.347	(1,1,1)

-116	1	$X^3 - X^2 - 2$	α^2	1	1.718	(1,1,1)
-135	1	$X^3 + 3X - 1$	α^2	1	1.133	(3,0,1)
-139	1	$X^3 + X^2 + X - 2$	α^2	1	1.664	(3,2,1)
-140	1	$X^3 + 2X - 2$	α^2	1	1.474	(3,1,1)
-152	1	$X^3 - X^2 - 2X - 2$	α^2	1	2.131	(-1,-1,-1)
-172	1	$X^3 + X^2 - X - 3$	α^2	1	1.882	(-2,-2,-1)
-175	1	$X^3 - X^2 + 2X - 3$	α^2	1	1.289	(2,0,1)
-199	1	$X^3 - X^2 + 4X - 1$	α^2	1	1.337	(4,-1,1)
-200	1	$X^3 + X^2 + 2X - 2$	α^2	1	2.604	(9,5,3)
-204	1	$X^3 - X^2 + X - 3$	α^2	1	2.355	(4,1,2)
-211	1	$X^3 - 2X - 3$	α^2	1	2.238	(-2,-2,-1)
-212	1	$X^3 - X^2 + 4X - 2$	α^2	1	2.713	(-15,2,-4)
-216	1	$X^3 + 3X - 2$	α^2	1	3.024	(-17,-3,-5)
-231	1	$X^3 + X^2 - 3$	α^2	1	1.745	(2,2,1)
-239	1	$X^3 - X - 3$	α^2	1	2.097	(2,2,1)
-243	1	$X^3 - 3$	α^2	1	2.525	(4,3,2)
-244	1	$X^3 + X^2 - 4X - 6$	α^2	1	3.303	(5,6,2)
-247	1	$X^3 + X - 3$	α^2	1	1.545	(2,1,1)
-255	1	$X^3 - X^2 - 3$	α^2	1	1.993	(-2,-1,-1)
-268	1	$X^3 + X^2 - 3X - 5$	α^2	1	2.521	(3,3,1)
-283	1	$X^3 + 4X - 1$	α^2	2	1.401	(4,0,1)
-300	1	$X^3 - X^2 - 3X - 3$	α^2	1	3.149	(2,3,2)
-307	1	$X^3 + X^2 + 3X - 2$	α^2	1	2.958	(-15,-6,-4)
-324	1	$X^3 - 3X - 4$	α^2	1	4.048	(-9,-11,-5)
-327	1	$X^3 - X^2 - 2X - 3$	α^2	1	2.199	(1,1,1)
-331	1	$X^3 - X^2 + 3X - 4$	α^2	2	1.503	(3,0,1)
-335	1	$X^3 + X^2 + 4X - 1$	α^2	1	1.456	(4,1,1)
-339	1	$X^3 + X^2 - X - 4$	α^2	1	3.546	(11,10,4)
-351	1	$X^3 + 3X - 3$	α^2	1	1.702	(-4,-1,-1)
-356	2	$X^3 - X^2 + 4X - 8$	$(\alpha + \alpha^2)/2$	1	3.755	(-25,2,-10)
-364	1	$X^3 + 4X - 2$	α^2	1	2.936	(17,2,4)
-367	1	$X^3 + X^2 + 2X - 3$	α^2	1	1.856	(4,2,1)
-379	1	$X^3 - X^2 + X - 4$	α^2	1	3.273	(9,3,4)
-411	1	$X^3 - X^2 + 5X - 2$	α^2	1	4.029	(57,-7,12)
-419	1	$X^3 - 4X - 5$	α^2	1	3.345	(-4,-5,-2)
-424	2	$X^3 - 2X - 8$	$\alpha^2/2$	1	4.859	(31,21,18)
-431	2	$X^3 - X - 8$	$(\alpha + \alpha^2)/2$	1	6.155	(133,42,72)
-436	1	$X^3 + X - 4$	α^2	1	4.948	(-61,-29,-21)
-439	1	$X^3 + X^2 - 2X - 5$	α^2	1	2.430	(3,3,1)
-440	2	$X^3 + 2X - 8$	$\alpha^2/2$	1	4.534	(-43,-15,-18)
-451	1	$X^3 + X^2 - 5X - 8$	α^2	1	3.576	(-7,-7,-2)
-459	1	$X^3 - 6X - 7$	α^2	1	3.669	(-5,-6,-2)
-460	1	$X^3 - X^2 + 5X - 3$	α^2	1	3.671	(38,-3,8)
-472	1	$X^3 - 5X - 6$	α^2	1	5.380	(29,35,13)
-484	1	$X^3 + X^2 + 4X - 2$	α^2	1	5.303	(171,53,37)
-491	1	$X^3 + X^2 + X - 4$	α^2	2	1.891	(3,2,1)
-492	1	$X^3 + X^2 + 3X - 3$	α^2	1	4.421	(59,24,14)
-499	1	$X^3 + 4X - 3$	α^2	1	3.874	(-40,-6,-9)
-503	2	$X^3 - X^2 - 2X - 8$	$(\alpha + \alpha^2)/2$	1	7.027	(-211,-56,-146)
-515	1	$X^3 - X^2 - X - 4$	α^2	1	3.646	(-7,-5,-4)
-516	2	$X^3 - X^2 - 4X - 8$	$(\alpha + \alpha^2)/2$	1	6.385	(-81,-35,-63)
-519	1	$X^3 + X^2 - 4X - 7$	α^2	1	2.681	(3,3,1)
-524	1	$X^3 - X^2 + 3X - 5$	α^2	1	3.422	(18,2,5)
-527	1	$X^3 + 5X - 1$	α^2	1	1.617	(5,0,1)

-543	1	$X^3 - X^2 + 2X - 5$	α^2	1	3.013	(-9,-2,-3)
-547	1	$X^3 - X^2 - 3X - 4$	α^2	1	4.367	(9,10,6)
-563	1	$X^3 - X^2 + 5X - 4$	α^2	2	1.737	(5,0,1)
-567	1	$X^3 - 3X - 5$	α^2	1	2.464	(-2,-2,-1)
-588	1	$X^3 + X^2 + 5X - 1$	α^2	3	1.654	(5,1,1)
-620	1	$X^3 - X^2 - 5X - 5$	α^2	1	3.553	(3,4,2)
-628	2	$X^3 + X^2 - 3X - 11$	$(1+\alpha^2)/2$	1	6.494	(-138,-123,-74)
-643	1	$X^3 - 2X - 5$	α^2	2	2.359	(2,2,1)
-648	2	$X^3 - 3X - 10$	$(\alpha+\alpha^2)/2$	3	2.234	(2,1,1)
-652	1	$X^3 - 8X - 10$	α^2	1	4.320	(-11,0,1)
-655	1	$X^3 + X^2 - 5$	α^2	1	2.906	(-7,-5,-2)
-671	1	$X^3 - X - 5$	α^2	1	2.345	(-3,-2,-1)
-675	1	$X^3 - 5$	α^2	1	4.812	(-41,-24,-14)
-676	2	$X^3 + X^2 - 4X - 12$	$(\alpha+\alpha^2)/2$	3	2.186	(2,1,1)
-679	1	$X^3 + X - 5$	α^2	1	3.443	(13,6,4)
-680	1	$X^3 + X^2 - 6X - 10$	α^2	1	6.071	(-79,-77,-21)
-687	1	$X^3 + X^2 + 4X - 3$	α^2	1	3.455	(-25,-8,-5)
-695	1	$X^3 - X^2 - 5$	α^2	1	2.151	(2,1,1)
-696	1	$X^3 + X^2 - 2X - 6$	α^2	1	7.810	(-673,-589,-207)
-707	1	$X^3 + 2X - 5$	α^2	1	4.187	(34,12,9)
-716	1	$X^3 - 4X - 6$	α^2	1	6.405	(-95,-101,-40)
-728	1	$X^3 - X^2 + 6X - 2$	α^2	1	6.052	(-433,49,-75)
-731	1	$X^3 + X^2 + 3X - 4$	α^2	2	2.013	(-5,-2,-1)
-743	1	$X^3 + 5X - 3$	α^2	1	4.556	(-85,-9,-16)
-744	1	$X^3 - X^2 - 6X - 6$	α^2	1	8.294	(-347,-451,-193)
-748	1	$X^3 + X^2 + X - 5$	α^2	1	4.532	(-43,-25,-11)
-751	1	$X^3 - X^2 + 6X - 1$	α^2	2	1.768	(6,-1,1)
-755	1	$X^3 + X^2 + 5X - 2$	α^2	1	4.904	(121,30,22)
-756	2	$X^3 + 9X - 2$	$(\alpha+\alpha^2)/2$	1	7.107	(1208,-104,267)
-759	1	$X^3 - X^2 + 6X - 3$	α^2	1	3.137	(23,-2,4)
-771	1	$X^3 - X^2 + 3X - 6$	α^2	1	6.140	(-251,-36,-65)
-780	1	$X^3 - X^2 - X - 5$	α^2	1	6.159	(94,59,44)
-804	1	$X^3 - X^2 + 4X - 6$	α^2	1	8.571	(-3499,-270,-784)
-808	1	$X^3 - X^2 + 2X - 6$	α^2	1	7.625	(-875,-201,-259)
-812	1	$X^3 - X^2 - 7X - 7$	α^2	1	3.844	(4,5,2)
-815	1	$X^3 - 7X - 9$	α^2	1	5.064	(20,22,7)

B.4 Table of Class Numbers and Units of Totally Real Cubic Fields

The following is a table of the first hundred totally real cubic fields. We give the following data from left to right: the discriminant $d(K)$, the index $[\mathbb{Z}_K : \mathbb{Z}[\alpha]]$, the polynomial $A(X)$, the third element β of an integral basis $(1, \alpha, \beta)$, the class number $h(K)$, the regulator $R(K)$ and a pair of fundamental units ϵ_1 and ϵ_2 expressed on the integral basis $(1, \alpha, \beta)$. The Galois group of the Galois closure of K is equal to S_3 except for the fields whose discriminant is marked with an asterisk, which are cyclic cubic fields, i.e. with Galois group equal to C_3.

d	f	A	β	h	R	ϵ_1	ϵ_2
49*	1	$X^3 + X^2 - 2X - 1$	α^2	1	0.5255	(-1,1,1)	(2,0,-1)
81*	1	$X^3 - 3X - 1$	α^2	1	0.8493	(2,1,-1)	(0,-1,0)
148	1	$X^3 + X^2 - 3X - 1$	α^2	1	1.662	(0,1,0)	(2,0,-1)
169*	1	$X^3 - X^2 - 4X - 1$	α^2	1	1.365	(2,2,-1)	(0,-1,0)
229	1	$X^3 - 4X - 1$	α^2	1	2.355	(0,1,0)	(2,1,0)
257	1	$X^3 - 5X - 3$	α^2	1	1.975	(4,1,-1)	(5,1,-1)
316	1	$X^3 + X^2 - 4X - 2$	α^2	1	3.913	(-3,1,1)	(-5,1,1)
321	1	$X^3 + X^2 - 4X - 1$	α^2	1	2.569	(0,-1,0)	(-1,2,1)
361*	1	$X^3 + X^2 - 6X - 7$	α^2	1	1.952	(4,1,-1)	(5,0,-1)
404	1	$X^3 - X^2 - 5X - 1$	α^2	1	3.760	(0,-1,0)	(1,-1,-1)
469	1	$X^3 + X^2 - 5X - 4$	α^2	1	3.853	(-1,-1,0)	(-1,2,1)
473	1	$X^3 - 5X - 1$	α^2	1	2.843	(0,-1,0)	(-2,-1,0)
564	1	$X^3 + X^2 - 5X - 3$	α^2	1	5.403	(-2,1,0)	(-1,-1,1)
568	1	$X^3 - X^2 - 6X - 2$	α^2	1	6.087	(-5,-1,1)	(-7,-4,2)
621	1	$X^3 - 6X - 3$	α^2	1	5.400	(-2,-1,0)	(1,2,0)
697	1	$X^3 - X^2 - 8X - 5$	α^2	1	2.712	(6,2,-1)	(7,2,-1)
733	1	$X^3 + X^2 - 7X - 8$	α^2	1	5.309	(1,1,0)	(-5,-2,0)
756	1	$X^3 - 6X - 2$	α^2	1	5.692	(5,0,-1)	(11,1,-2)
761	1	$X^3 - X^2 - 6X - 1$	α^2	1	3.526	(0,1,0)	(2,1,0)
785	1	$X^3 + X^2 - 6X - 5$	α^2	1	4.098	(1,1,0)	(-4,1,1)
788	1	$X^3 - X^2 - 7X - 3$	α^2	1	5.987	(2,1,0)	(-1,-2,0)
837	1	$X^3 - 6X - 1$	α^2	1	6.801	(0,-1,0)	(-3,-6,-2)
892	1	$X^3 + X^2 - 8X - 10$	α^2	1	8.323	(3,1,-1)	(1,3,1)
940	1	$X^3 - 7X - 4$	α^2	1	8.908	(-11,-2,2)	(-3,1,1)
961*	2	$X^3 + X^2 - 10X - 8$	$(\alpha^2 + \alpha)/2$	1	12.20	(-1,2,2)	(3,4,-2)
985	1	$X^3 + X^2 - 6X - 1$	α^2	1	3.724	(0,1,0)	(-2,1,0)
993	1	$X^3 + X^2 - 6X - 3$	α^2	1	5.555	(5,-1,-1)	(5,0,-1)
1016	1	$X^3 + X^2 - 6X - 2$	α^2	1	10.13	(7,-1,-1)	(-11,-1,1)
1076	1	$X^3 - 8X - 6$	α^2	1	6.932	(1,1,0)	(-7,-3,0)
1101	1	$X^3 + X^2 - 9X - 12$	α^2	1	9.184	(5,2,-1)	(-7,-4,2)
1129	1	$X^3 - 7X - 3$	α^2	1	6.728	(-8,0,1)	(1,2,-1)
1229	1	$X^3 + X^2 - 7X - 6$	α^2	1	8.232	(-1,-1,0)	(11,15,4)
1257	1	$X^3 + X^2 - 8X - 9$	α^2	1	6.197	(-1,-1,0)	(2,-2,-1)
1300	1	$X^3 - 10X - 10$	α^2	1	6.550	(-1,-1,0)	(-1,2,1)
1304	2	$X^3 - X^2 - 11X - 1$	$(\alpha^2 + 1)/2$	1	11.93	(0,-1,0)	(-5,14,10)
1345	1	$X^3 - 7X - 1$	α^2	1	4.923	(0,1,0)	(2,2,-1)
1369*	1	$X^3 - X^2 - 12X - 11$	α^2	1	3.126	(6,3,-1)	(9,2,-1)
1373	1	$X^3 - 8X - 5$	α^2	1	9.423	(-6,0,1)	(-13,-2,2)
1384	1	$X^3 + X^2 - 10X - 14$	α^2	1	10.38	(-3,-2,0)	(-5,1,1)
1396	1	$X^3 + X^2 - 7X - 5$	α^2	1	8.146	(-8,0,1)	(-9,1,1)
1425	1	$X^3 - X^2 - 8X - 3$	α^2	1	6.676	(-2,-1,0)	(1,2,-1)
1436	1	$X^3 - 11X - 12$	α^2	1	12.70	(5,2,0)	(-11,-6,2)
1489	1	$X^3 + X^2 - 12X - 19$	α^2	1	3.361	(10,1,-1)	(11,1,-1)
1492	1	$X^3 - X^2 - 9X - 5$	α^2	1	7.646	(-2,-1,0)	(-1,-1,1)
1509	1	$X^3 + X^2 - 7X - 4$	α^2	1	11.30	(3,1,0)	(-3,-6,-1)
1524	1	$X^3 + X^2 - 7X - 1$	α^2	1	10.45	(0,1,0)	(-6,-11,6)
1556	1	$X^3 + X^2 - 9X - 11$	α^2	1	8.376	(8,0,-1)	(19,0,-2)
1573	1	$X^3 + X^2 - 7X - 2$	α^2	1	8.445	(-3,-1,0)	(1,4,1)
1593	1	$X^3 - 9X - 7$	α^2	1	6.331	(1,1,0)	(5,2,0)
1620	1	$X^3 - 12X - 14$	α^2	1	10.17	(9,1,-1)	(5,5,1)
1708	1	$X^3 - X^2 - 8X - 2$	α^2	1	12.87	(7,1,-1)	(-29,-9,5)
1765	1	$X^3 + X^2 - 11X - 16$	α^2	1	9.445	(-3,-1,0)	(-7,-6,-1)

1772	2	$X^3 - 14X - 12$	$\alpha^2/2$	1	15.37	(-1,-1,0)	(-23,-36,-18)
1825	1	$X^3 + X^2 - 8X - 7$	α^2	1	4.488	(1,1,0)	(3,1,0)
1849*	2	$X^3 - X^2 - 14X - 8$	$(\alpha^2 + \alpha)/2$	1	18.92	(-9,2,0)	(-17,-4,2)
1901	1	$X^3 - X^2 - 9X - 4$	α^2	1	10.66	(-1,-2,0)	(-5,0,1)
1929	1	$X^3 + X^2 - 10X - 13$	α^2	1	8.218	(3,1,0)	(5,5,1)
1937	1	$X^3 - X^2 - 8X - 1$	α^2	1	6.542	(0,-1,0)	(-3,1,1)
1940	1	$X^3 - 8X - 2$	α^2	1	11.09	(3,-1,0)	(39,1,-5)
1944	1	$X^3 - 9X - 6$	α^2	1	15.60	(1,3,-1)	(-1,0,2)
1957	1	$X^3 + X^2 - 9X - 10$	α^2	2	4.551	(1,1,0)	(3,1,0)
2021	1	$X^3 - 8X - 1$	α^2	1	11.52	(0,-1,0)	(-1,-28,-10)
2024	1	$X^3 - X^2 - 10X - 6$	α^2	1	15.77	(5,6,-2)	(-11,-9,3)
2057	1	$X^3 - 11X - 11$	α^2	1	6.782	(1,1,0)	(-1,-3,-1)
2089	2	$X^3 - 13X - 4$	$(\alpha^2 + \alpha)/2$	1	20.76	(-1,-4,2)	(-15,4,0)
2101	1	$X^3 - X^2 - 11X - 8$	α^2	1	8.543	(-1,-1,0)	(15,2,-2)
2177	1	$X^3 + X^2 - 8X - 5$	α^2	1	7.518	(-3,-1,0)	(17,-1,-2)
2213	1	$X^3 - X^2 - 13X - 12$	α^2	1	12.68	(-1,-1,0)	(-1,9,4)
2228	1	$X^3 - 14X - 18$	α^2	1	11.09	(-7,-3,1)	(-41,-16,6)
2233	1	$X^3 + X^2 - 8X - 1$	α^2	1	5.523	(0,1,0)	(-1,3,1)
2241	1	$X^3 - 9X - 5$	α^2	1	8.264	(-4,-2,1)	(-2,-3,1)
2292	2	$X^3 + X^2 - 13X - 1$	$(\alpha^2 + 1)/2$	1	14.36	(0,1,0)	(-4,36,17)
2296	1	$X^3 - X^2 - 14X - 14$	α^2	1	14.27	(13,3,-1)	(-5,-4,0)
2300	1	$X^3 + X^2 - 8X - 2$	α^2	1	18.12	(5,-2,0)	(73,-7,-9)
2349	1	$X^3 - 12X - 13$	α^2	1	11.92	(-4,-2,1)	(15,4,-2)
2429	1	$X^3 - X^2 - 15X - 16$	α^2	1	13.28	(-11,-2,1)	(85,16,-7)
2505	1	$X^3 - X^2 - 10X - 5$	α^2	1	10.68	(-2,-3,1)	(7,6,-2)
2557	1	$X^3 - X^2 - 9X - 2$	α^2	1	10.72	(-1,2,1)	(1,4,-1)
2589	2	$X^3 + X^2 - 14X - 12$	$(\alpha^2 + \alpha)/2$	1	16.29	(-5,-1,1)	(31,38,-20)
2597	1	$X^3 + X^2 - 9X - 8$	α^2	3	4.796	(1,1,0)	(-3,-1,0)
2636	1	$X^3 - X^2 - 16X - 18$	α^2	1	18.38	(-5,-2,0)	(25,13,-3)
2673	1	$X^3 - 9X - 3$	α^2	1	7.760	(10,0,-1)	(-8,0,1)
2677	1	$X^3 - 10X - 7$	α^2	1	11.16	(-12,0,1)	(2,2,-1)
2700	1	$X^3 - 15X - 20$	α^2	1	20.37	(1,-1,-1)	(-59,-22,8)
2708	1	$X^3 - X^2 - 11X - 7$	α^2	1	12.95	(6,7,-2)	(9,6,-2)
2713	1	$X^3 - 13X - 15$	α^2	1	12.34	(-13,-2,1)	(-17,-4,2)
2777	1	$X^3 + X^2 - 14X - 23$	α^2	2	3.949	(-2,-1,0)	(-3,-1,0)
2804	1	$X^3 - X^2 - 9X - 1$	α^2	1	15.24	(0,-1,0)	(10,56,21)
2808	1	$X^3 - 9X - 2$	α^2	1	20.31	(-1,-9,3)	(-1,-4,2)
2836	1	$X^3 + X^2 - 9X - 7$	α^2	1	9.692	(10,0,-1)	(-17,0,2)
2857	1	$X^3 + X^2 - 10X - 11$	α^2	1	4.870	(-1,-1,0)	(-3,-1,0)
2917	1	$X^3 + X^2 - 13X - 20$	α^2	1	11.93	(3,1,0)	(13,6,-1)
2920	2	$X^3 + X^2 - 16X - 20$	$(\alpha^2 + \alpha)/2$	1	17.94	(-9,-8,4)	(-4,-3,1)
2941	1	$X^3 - X^2 - 17X - 20$	α^2	1	13.72	(3,2,0)	(-17,-4,1)
2981	1	$X^3 + X^2 - 11X - 14$	α^2	1	14.63	(3,1,0)	(15,10,-1)
2993	1	$X^3 + X^2 - 12X - 17$	α^2	1	7.514	(-3,-1,0)	(3,2,0)
3021	1	$X^3 + X^2 - 9X - 6$	α^2	1	17.40	(-5,-4,2)	(5,9,2)
3028	1	$X^3 - 10X - 6$	α^2	1	20.35	(-1,-1,1)	(5,13,4)
3124	2	$X^3 - 16X - 12$	$\alpha^2/2$	1	19.56	(-5,-1,1)	(115,121,-68)
3132	2	$X^3 - 18X - 20$	$\alpha^2/2$	1	22.49	(7,2,0)	(7,7,2)

B.5 Table of Elliptic Curves

In the table below we give a table of all modular elliptic curves defined over $\mathbb{Q}$ with conductor N less than or equal to 44 (up to isomorphism). Recall that according to the Taniyama-Weil Conjecture 7.3.8, all elliptic curves defined over $\mathbb{Q}$ are modular.

To every elliptic curve is attached quite a large set of invariants. We refer to [Cre] for details and a complete table. In the following table, we only give the minimal Weierstraß equation of the curve, its rank and its torsion subgroup. The rank is always equal to 0 except in the two cases $N = 37$ (curve A1) and $N = 43$ for which it is equal to 1, and in these two cases a generator of the group $E(\mathbb{Q})$ is the point with coordinates $(0,0)$. The canonical height of this point, computed using Algorithms 7.5.6 and 7.5.7 is equal to $0.0255557041\ldots$ for $N = 37$ and to $0.0314082535\ldots$ for $N = 43$.

The Kodaira types and the constants c_p can be found by using Tate's Algorithm 7.5.1. The coefficients a_p of the L-series can be computed using Algorithm 7.4.12 or simply by adding Legendre symbols if p is small. The periods can be computed using Algorithm 7.4.7. In the limit of the present table the Tate-Shafarevitch group III is always trivial.

We follow the notations of [Cre]. We give from left to right: the conductor N of the curve E, an identifying label of the curve among those having the same conductor. This label is of the form letter-number. The letter (A or B) denotes the isogeny class, and the number is the ordinal number of the curve in its isogeny class. Curves numbered 1 are the strong Weil curves (see [Sil]). The next 5 columns contain the coefficients a_1, a_2, a_3, a_4 and a_6. The last two columns contain the rank r and the torsion subgroup T of $E(\mathbb{Q})$ expressed as t if $T \simeq \mathbb{Z}/t\mathbb{Z}$ and as $t_1 \times t_2$ if $T \simeq \mathbb{Z}/t_1\mathbb{Z} \times \mathbb{Z}/t_2\mathbb{Z}$.

N		a_1	a_2	a_3	a_4	a_6	r	T
11	A1	0	−1	1	−10	−20	0	5
11	A2	0	−1	1	−7820	−263580	0	1
11	A3	0	−1	1	0	0	0	5
14	A1	1	0	1	4	−6	0	6
14	A2	1	0	1	−36	−70	0	6
14	A3	1	0	1	−171	−874	0	2
14	A4	1	0	1	−1	0	0	6
14	A5	1	0	1	−2731	−55146	0	2
14	A6	1	0	1	−11	12	0	6
15	A1	1	1	1	−10	−10	0	2 × 4
15	A2	1	1	1	−135	−660	0	2 × 2
15	A3	1	1	1	−5	2	0	2 × 4
15	A4	1	1	1	35	−28	0	8
15	A5	1	1	1	−2160	−39540	0	2

15	A6	1	1	1	−110	−880	0	2
15	A7	1	1	1	−80	242	0	4
15	A8	1	1	1	0	0	0	4
17	A1	1	−1	1	−1	−14	0	4
17	A2	1	−1	1	−6	−4	0	2 × 2
17	A3	1	−1	1	−91	−310	0	2
17	A4	1	−1	1	−1	0	0	4
19	A1	0	1	1	−9	−15	0	3
19	A2	0	1	1	−769	−8470	0	1
19	A3	0	1	1	−1	0	0	3
20	A1	0	1	0	4	4	0	6
20	A2	0	1	0	−1	0	0	6
20	A3	0	1	0	−36	−140	0	2
20	A4	0	1	0	−41	−116	0	2
21	A1	1	0	0	−4	−1	0	2 × 4
21	A2	1	0	0	−49	−136	0	2 × 2
21	A3	1	0	0	−39	90	0	8
21	A4	1	0	0	1	0	0	4
21	A5	1	0	0	−784	−8515	0	2
21	A6	1	0	0	−34	−217	0	2
24	A1	0	−1	0	−4	4	0	2 × 4
24	A2	0	−1	0	−24	−36	0	2 × 2
24	A3	0	−1	0	−64	220	0	4
24	A4	0	−1	0	1	0	0	4
24	A5	0	−1	0	−384	−2772	0	2
24	A6	0	−1	0	16	−180	0	2
26	A1	1	0	1	−5	−8	0	3
26	A2	1	0	1	−460	−3830	0	1
26	A3	1	0	1	0	0	0	3
26	B1	1	−1	1	−3	3	0	7
26	B2	1	−1	1	−213	−1257	0	1
27	A1	0	0	1	0	−7	0	3
27	A2	0	0	1	−270	−1708	0	1
27	A3	0	0	1	0	0	0	3
27	A4	0	0	1	−30	63	0	3
30	A1	1	0	1	1	2	0	6
30	A2	1	0	1	−19	26	0	2 × 6
30	A3	1	0	1	−14	−64	0	2
30	A4	1	0	1	−69	−194	0	6
30	A5	1	0	1	−289	1862	0	6
30	A6	1	0	1	−334	−2368	0	2 × 2
30	A7	1	0	1	−5334	−150368	0	2
30	A8	1	0	1	−454	−544	0	2
32	A1	0	0	0	4	0	0	4
32	A2	0	0	0	−1	0	0	2 × 2

32	A3	0	0	0	−11	−14	0	2
32	A4	0	0	0	−11	14	0	4
33	A1	1	1	0	−11	0	0	2 × 2
33	A2	1	1	0	−6	−9	0	2
33	A3	1	1	0	−146	621	0	4
33	A4	1	1	0	44	55	0	2
34	A1	1	0	0	−3	1	0	6
34	A2	1	0	0	−43	105	0	6
34	A3	1	0	0	−103	−411	0	2
34	A4	1	0	0	−113	−329	0	2
35	A1	0	1	1	9	1	0	3
35	A2	0	1	1	−131	−650	0	1
35	A3	0	1	1	−1	0	0	3
36	A1	0	0	0	0	1	0	6
36	A2	0	0	0	−15	22	0	6
36	A3	0	0	0	0	−27	0	2
36	A4	0	0	0	−135	−594	0	2
37	A1	0	0	1	−1	0	1	1
37	B1	0	1	1	−23	−50	0	3
37	B2	0	1	1	−1873	−31833	0	1
37	B3	0	1	1	−3	1	0	3
38	A1	1	0	1	9	90	0	3
38	A2	1	0	1	−86	−2456	0	1
38	A3	1	0	1	−16	22	0	3
38	B1	1	1	1	0	1	0	5
38	B2	1	1	1	−70	−279	0	1
39	A1	1	1	0	−4	−5	0	2 × 2
39	A2	1	1	0	−69	−252	0	2
39	A3	1	1	0	−19	22	0	4
39	A4	1	1	0	1	0	0	2
40	A1	0	0	0	−7	−6	0	2 × 2
40	A2	0	0	0	−107	−426	0	2
40	A3	0	0	0	−2	1	0	4
40	A4	0	0	0	13	−34	0	4
42	A1	1	1	1	−4	5	0	8
42	A2	1	1	1	−84	261	0	2 × 4
42	A3	1	1	1	−104	101	0	2 × 2
42	A4	1	1	1	−1344	18405	0	4
42	A5	1	1	1	−914	−10915	0	2
42	A6	1	1	1	386	1277	0	2
43	A1	0	1	1	0	0	1	1
44	A1	0	1	0	3	−1	0	3
44	A2	0	1	0	−77	−289	0	1

Bibliography

Essential Introductory Books.

[Bo-Sh] Z.I. Borevitch and I.R. Shafarevitch, *Number Theory*, Academic Press, New York, 1966.

A classic must which gives a fairly advanced introduction to algebraic number theory, with applications for example to Fermat's last theorem. Contains numerous exercises.

[GCL] K. Geddes, S. Czapor and G. Labahn, *Algorithms for Computer Algebra*, Kluwer Academic Publishers, Boston, Dordrecht, London, 1992.

This book contains a very detailed description of the basic algorithms used for handling fundamental mathematical objects such as polynomials, power series, rational functions, as well as more sophisticated algorithms such as polynomial factorization, Gröbner bases computation and symbolic integration. The algorithms are those which have been implemented in the Maple system (see Appendix A). This is required reading for anyone wanting to understand the inner workings of a computer algebra system.

[H-W] G.H. Hardy and E.M. Wright, *An Introduction to the Theory of Numbers*, (5-th ed.), Oxford University Press, Oxford, 1979.

This is another classic must for a beginning introduction to number theory. The presentation is very clear and simple, and the book contains all basic essential material. Avoid reading parts like the "elementary" proof of the prime number theorem. Proofs based on complex function theory, while requiring deeper concepts, are much more enlightening.

[Ire-Ros] K. Ireland and M. Rosen, *A Classical Introduction to Modern Number Theory*, (2nd ed.), Graduate texts in Math. **84**, Springer-Verlag, New York, 1982.

A remarkable introductory book on the more analytic and computational parts of algebraic number theory, with numerous concrete examples and exercises. This book can be read profitably jointly with the present book for a deeper understanding of several subjects such as Gauss and Jacobi sums and related identities (used in Chapter 9), quadratic and cyclotomic fields (Chapters 5 and 9), zeta functions of varieties (Chapter 7), etc .... A must.

[Knu1] D.E. Knuth, *The Art of Computer Programming, Vol. 1: Fundamental Algorithms*, (2nd ed.), Addison-Wesley, Reading, Mass., 1973.

This is the first volume of the "bible" of computer science. Although not specifically targeted to number theory, this volume introduces a large number of fundamental concepts and techniques (mathematical or otherwise) which are of constant use to anyone implementing algorithms. The style of writing is crystal clear, and I have copied the style of presentation of algorithms from Knuth. A must.

[Knu2] D.E. Knuth, *The Art of Computer Programming, Vol. 2: Seminumerical Algorithms*, (2nd ed.), Addison-Wesley, Reading, Mass., 1981.

This is the second volume of the "bible" of computer science. Essentially all the contents of chapter 4 of Knuth's book is basic to computational number theory, and as stated in the preface, some parts of chapters 1 and 3 of the present book have been merely adapted from Knuth. The section on factoring and primality testing is of course outdated. The book contains also a huge number of fascinating exercises, with solutions. An absolute must.

[Knu3] D.E. Knuth, *The Art of Computer Programming, Vol. 3: Sorting and Searching*, Addison-Wesley, Reading, Mass., 1973.

This is the third volume of the "bible" of computer science. The description of searching and sorting methods (in particular heapsort and quicksort) as well as hashing techniques can be used for number-theoretic applications.

[Lang1] S. Lang, *Algebra*, (2nd ed.), Addison-Wesley, Reading, Mass., 1984.

This book is quite abstract in nature and in fact contains little concrete examples. On the other hand one can find the statements and proofs of most of the basic algebraic results needed in number theory.

[Mar] D.A. Marcus, *Number Fields*, Springer-Verlag, New York, 1977.

An excellent textbook on algebraic number theory with numerous very concrete examples, not far from the spirit of this book, although much less algorithmic in nature.

[Rie] H. Riesel, *Prime Numbers and Computer Methods for Factorization*, Birkhäuser, Boston, 1985.

An excellent elementary text on prime number theory and algorithms for primality testing and factoring. As in the present book the algorithms are ready to implement, and in fact implementations of many of them are given in Pascal. The subject matter of the algorithmic part overlaps in a large part with chapters 8 to 10 of this book.

[Sam] P. Samuel, *Théorie algébrique des nombres*, Hermann, Paris, 1971.

Another excellent textbook on algebraic number theory. Gives the basic proofs and results in a very nice and concise manner.

[Ser] J.-P. Serre, *A Course in Arithmetic*, Springer-Verlag, New York, 1973.

A very nice little book which contains an introduction to some basic number-theoretic objects such as $\mathbf{Z}/n\mathbf{Z}$, finite fields, quadratic forms, modular forms, etc . . . . A must, although further reading is necessary in almost all cases. The original was published in French in 1970.

Other Books and Volumes.

[AHU] A. Aho, J. Hopcroft and J. Ullman, *The Design and Analysis of Computer Algorithms*, Addison-Wesley, Reading, Mass., 1974.

This book discusses many issues related to basic computer algorithms and their complexity. In particular, it discusses in detail the notion of NP-complete problems, and has chapters on integer and polynomial arithmetic, on the LUP decomposition of matrices and on the Fast Fourier transform.

[Bac-Sha] E. Bach and J. Shallit, *Algorithmic Number Theory, Vol. 1: Efficient Algorithms*, MIT Press, Cambridge, Mass, 1996.

Studies in detail the complexity of number-theoretic algorithms.

[Bor-Bor] J. Borwein and P. Borwein, *Pi and the AGM*, Canadian Math. Soc. Series, John Wiley and Sons, New York, 1987.

A marvelous book containing a wealth of formulas in the style of Ramanujan, including formulas coming from complex multiplication for computing π to great accuracy.

[Bue] D. Buell, *Binary Quadratic Forms: Classical Theory and Modern Computations*, Springer-Verlag, New York, 1990.

A nice and easy to read book on the theory of binary quadratic forms, which expands on some of the subjects treated in Chapter 5.

[Cas] J. Cassels, *Lectures on elliptic curves*, Cambridge Univ. Press, 1991.

An excellent small introductory book to the subject of elliptic curves containing a wealth of deeper subjects not so easily accessible otherwise. The viewpoint is different from Silverman's, and hence is a highly recommended complementary reading.

[Cas-Frö] J. Cassels and A. Fröhlich, *Algebraic number theory*, Academic Press, London and New York, 1967.

This book has been one of the main reference books for a generation of algebraic number theorists and is still the standard book to read before more sophisticated books like Shimura's.

[Cohn] H. Cohn, *A Classical Introduction to Algebraic Numbers and Class Fields*, Universitext, Springer-Verlag, New York, 1978.

A highly recommended concrete introduction to algebraic number theory and class field theory, with a large number of detailed examples.

[Con-Slo] J. Conway and N. Sloane, *Sphere Packings, Lattices and Groups*, Grundlehren der math. Wiss. 290, Springer-Verlag, New York, 1988.

The bible on lattices and sphere packings. Everything you ever wanted to know and much more, including a large number of tables. An irreplaceable tool for research in the Geometry of Numbers.

[Cox] D. Cox, *Primes of the Form $x^2 + ny^2$. Fermat, Class Field Theory and Complex Multiplication*, John Wiley and Sons, New York, 1989.

This is an excellent book on class field theory and complex multiplication. It is written in a very concrete manner with many examples and exercises, and I recommend it highly.

[Cre] J. Cremona, *Algorithms for Modular Elliptic Curves*, Cambridge Univ. Press, 1992.

An extension of [LN476] to conductors less than 1000, and much more information. Also many algorithms related to elliptic curves are listed, most of which are not given in this book. A must on the subject.

[Dah-Bjö] G. Dahlquist and A. Björk (translated by N. Anderson), *Numerical Methods*, Prentice Hall, Englewood Cliffs, N.J., 1974.

A basic reference book on numerical algorithms, especially for linear algebra.

[Del-Fad] B.N. Delone and D.K. Fadeev, *The Theory of Irrationalities of the Third Degree*, Trans. Math. Mon. 10, A.M.S., Providence, R.I., 1964.

Although quite old, this book contains a wealth of theoretical and algorithmic information on cubic fields.

[Gol-Van] G. Golub and C. Van Loan, *Matrix Computations*, (2nd ed.), Johns Hopkins Univ. Press, Baltimore and London, 1989.

An excellent comprehensive introduction to basic techniques of numerical analysis used in linear algebra.

[Hus] D. Husemoller, *Elliptic Curves*, Graduate texts in Math. 111, Springer-Verlag, New York, 1987.

Simpler than Silverman's book, this gives a good introduction to elliptic curves.

[Kap] I. Kaplansky, *Commutative Rings*, Allyn and Bacon, Boston, 1970.

A very nicely written little book on abstract algebra.

[Kob] N. Koblitz, *Introduction to Elliptic Curves and Modular Forms*, Graduate texts in Math. 97, Springer-Verlag, New York, 1984.

This nice book gives the necessary tools for obtaining the complete solution of the congruent number problem modulo a weak form of the Birch-Swinnerton Dyer conjecture. In passing, a lot of very concrete material on elliptic curves and modular forms is covered.

[Lang2] S. Lang, *Algebraic Number Theory*, Addison-Wesley, Reading, Mass., 1970.

An advanced abstract introduction to the subject.

[Lang3] S. Lang, *Elliptic Functions*, Addison Wesley, Reading, Mass., 1973.

A nice introductory book on elliptic functions and elliptic curves.

[Lang4] S. Lang, *Introduction to Modular Forms*, Springer-Verlag, Berlin, Heidelberg, New York, 1976.

A nice introductory book on modular forms.

[LN476] B. Birch and W. Kuyk (eds.), *Modular Forms in one Variable IV*, LN in Math. **476**, Springer-Verlag, Berlin, Heidelberg, 1975.

A fundamental book of tables and algorithms on elliptic curves, containing in particular a detailed description of all elliptic curves of conductor less than or equal to 200. A must on the subject.

[MCC] H.W. Lenstra and R. Tijdeman (eds.), *Computational Methods in Number Theory*, Math. Centre tracts **154/155**, Math. Centrum Amsterdam, 1982.

A very nice two volume collection on computational number theory, covering many different topics.

[Nau-Qui] P. Naudin and C. Quitté, *Algorithmique Algébrique*, Masson, Paris, 1992.

A very nice and leisurely introduction to computational algebra (in French) with many detailed algorithms and a complete chapter devoted to the use of the Fast Fourier Transform in computer algebra.

[Ogg] A. Ogg, *Modular Forms and Dirichlet Series*, Benjamin, 1969.

A nice little introductory book on modular forms, containing in particular a detailed proof of Weil's Theorem 7.3.7.

[PPWZ] A. Pethő, M. Pohst, H. Williams and H.G. Zimmer (eds.), *Computational Number Theory*, Walter de Gruyter, 1991.

Similar to [MCC] but very up to date and more oriented towards algebraic number theory. Contains very important contributions which are referenced separately here.

[Poh] M. Pohst (ed.), *Algorithmic Methods in Algebra and Number Theory*, Academic Press, 1987.

A special volume of the Journal of Symbolic Computation devoted to computational number theory, and containing a number of important individual contributions which are referenced separately here.

[Poh-Zas] M. Pohst and H. Zassenhaus, *Algorithmic Algebraic Number Theory*, Cambridge Univ. Press, 1989.

The reference book on algorithmic algebraic number theory. Contains detailed descriptions of numerous algorithms for solving the fundamental tasks of algebraic number theory in the general number field case. The notation is sometimes heavy, and direct computer implementation of the algorithms is not always easy, but the wealth of information is considerable. A must for further reading on the subject.

[Poh5] M. Pohst, *Computational Algebraic Number Theory*, DMV Seminar **21**, Birkhäuser, Boston, 1993.

Writeup of a course given by the author in 1990. This can be considered as an update to parts of [Poh-Zas].

[PFTV] W. Press, B. Flannery, S. Teukolsky and W. Vetterling, *Numerical Recipes in C*, (2nd ed.), Cambridge University Press, Cambridge, 1988.

The algorithms presented in this book are essentially unrelated to number theory, but this is a basic reference book for implementing algorithms in numerical analysis, and in particular for number theory, polynomial root finding and linear algebra over $\mathbb{R}$. A must for implementing numerical analysis-related algorithms.

[Sha] H. Williams (ed.), Math. Comp **48(January)** (1987).

A special volume of Mathematics of Computation dedicated to D. Shanks. Contains a large number of important individual contributions.

[Shi] G. Shimura, *Introduction to the Arithmetic Theory of Automorphic Functions*, Iwanami Shoten and Princeton Univ. Press, Princeton, 1971.

This book is one of the great classics of advanced number theory, in particular about class fields, elliptic curves and modular forms. It contains a great wealth of information, and even though it is quite old, it is still essentially up to date and still a basic reference book. Beware however that the mathematical sophistication is high. A must for people wanting to know more about class fields, complex multiplication and modular forms at a high level.

[Sil] J. Silverman, *The Arithmetic of Elliptic Curves*, Graduate texts in Math. **106**, Springer-Verlag, New York, 1986.

This excellent book has now become *the* reference book on elliptic curves, and a large part is of very advanced level. It is excellently written, contains numerous exercises and is a great pleasure to study. A must for further study of elliptic curves.

[Sil3] J. Silverman, *Advanced Topics in the Arithmetic of Elliptic Curves*, Graduate texts in Math. **151**, Springer-Verlag, New York, 1994.

The long awaited sequel to [Sil].

[Was] L. Washington, *Introduction to Cyclotomic Fields*, Graduate Texts in Math. **83**, Springer-Verlag, New York, 1982.

An excellent advanced introduction to algebraic number theory, with many concrete examples.

[W-W] E. Whittaker and G. Watson, *A Course of Modern Analysis*, (4th ed.), Cambridge Univ. Press, 1927.

Still the reference book on practical use of complex analysis. The chapters on elliptic functions and theta functions are of special interest to number theorists.

[Zag] D. Zagier, *The Analytic Theory of Modular Forms*, in preparation.

A thorough introduction to the analytic theory of modular forms, including a number of advanced topics. Very clear exposition. A must on the subject (when it comes out).

[Zim] H. Zimmer, *Computational Problems, Methods and Results in Algebraic Number Theory*, LN in Math. **262**, Springer-Verlag, Berlin, Heidelberg, 1972.

A very thorough list of commented bibliographic references on computational number theory prior to 1971.

Papers and other references

[Adl] L. Adleman, *Factoring numbers using singular integers*, Proc. 18th Annual ACM Symp. on Theory of Computing (1991), 64–71.

[Adl-Hua] L. Adleman and M. Huang, *Primality testing and Abelian varieties over finite fields*, LN in Math **1512**, Springer-Verlag, Berlin, Heidelberg, 1992.

[APR] L. Adleman, C. Pomerance and R. Rumely, *On distinguishing prime numbers from composite numbers*, Ann. of Math. **117** (1983), 173–206.

[AGP] R. Alford, A. Granville and C. Pomerance, *There are infinitely many Carmichael numbers*, Ann. of Math. **139** (1994), 703–722.

[Ang] I. Angell, *A table of complex cubic fields*, Bull. London Math. Soc. **5** (1973), 37–38.

[Arn] F. Arnault, *The Rabin-Miller primality test: composite numbers which pass it*, Math. Comp. **64** (1995), 335–361.

[ARW] S. Arno, M. Robinson and F. Wheeler, *Imaginary quadratic fields with small odd class number* (to appear).

[Atk1] O. Atkin, *Composition of binary quadratic forms*, manuscript (1990).

[Atk2] O. Atkin, *The number of points on an elliptic curve modulo a prime*, manuscript (1991).

[Atk-Mor] O. Atkin and F. Morain, *Elliptic curves and primality proving*, Math. Comp. **61** (1993), 29–68.

[Ayo] R. Ayoub, *An Introduction to the Analytic Theory of Numbers*, Mathematical surveys **10**, A.M.S., 1963.

[Bach] E. Bach, *Explicit bounds for primality testing and related problems*, Math. Comp. **55** (1990), 355–380.

[Bar] E. Bareiss, *Sylvester's identity and multistep integer-preserving Gaussian elimination*, Math. Comp. **22** (1968), 565–578.

[BeMaOl] A.-M. Bergé, J. Martinet and M. Olivier, *The computation of sextic fields with a quadratic subfield*, Math. Comp. **54** (1990), 869–884.

[Ber] E. Berlekamp, *Factoring polynomials over large finite fields*, Math. Comp. **24** (1970), 713–735.

[Bir-SwD] B. Birch and H.P.F. Swinnerton-Dyer, *Notes on elliptic curves I*, J. Reine Angew. Math. **212** (1963), 7–25; *II*, ibid. **218** (1965), 79–108.

[BFHT] A. Borodin, R. Fagin, J. Hopcroft and M. Tompa, *Decreasing the nesting depth of expressions involving square roots*, J. Symb. Comp. **1** (1985), 169–188.

[Bos] W. Bosma, *Primality testing using elliptic curves*, Report 85-12, Math. Instituut, Univ. of Amsterdam (1985).

[Bos-Hul] W. Bosma and M.-P. van der Hulst, *Primality proving with cyclotomy*, thesis, Univ. of Amsterdam, 1990.

[Bra] G. Bradley, *Algorithms for Hermite and Smith normal form matrices and linear Diophantine equations*, Math. Comp. **25** (1971), 897–907.

[Brau] R. Brauer, *On the Zeta-function of algebraic number fields I*, Amer. J. Math. **69** (1947), 243–250; *II*, ibid. **72** (1950), 739–746.

[Bre1] R.P. Brent, *Some integer factorization algorithms using elliptic curves*, in Proc. 9th Australian Computer science conference (1985).

[Bre2] R.P. Brent, *An improved Monte-Carlo factorization algorithm*, BIT **20** (1980), 176–184.

[Bre3] R.P. Brent, *The first occurence of large gaps between successive primes*, Math. Comp. **27** (1973), 959–963.

[BLSTW] J. Brillhart, D.H. Lehmer, J. Selfridge, B. Tuckerman and S. Wagstaff, *Factorizations of $b^n \pm 1$, $b = 2, 3, 5, 6, 7, 10, 11, 12$, up to high powers*, Contemporary Mathematics **22**, A.M.S., Providence, R.I., 1983.

[Bri-Mor] J. Brillhart and M. Morrison, *A method of factoring and the factorization of F_7*, Math. Comp. **29** (1975), 183–205.

[BLS] J. Brillhart, D.H. Lehmer and J. Selfridge, *New primality criteria and factorizations of $2^m \pm 1$*, Math. Comp. **29** (1975), 620–647.

[BCS] S. Brlek, P. Castéran and R. Strandh, *On addition schemes*, TAPSOFT 1991, LN in Comp. Sci. **494**, 1991, pp. 379–393.

[deBru] N. G. de Bruijn, *The asymptotic behavior of a function occurring in the theory of primes*, J. Indian Math. Soc. (N. S.) **15** (1951), 25–32.

[Buc1] J. Buchmann, *A generalization of Voronoi's unit algorithm I and II*, J. Number Theory **20** (1985), 177–209.

[Buc2] J. Buchmann, *On the computation of units and class numbers by a generalization of Lagrange's algorithm*, J. Number Theory **26** (1987), 8-30.

[Buc3] J. Buchmann, *On the period length of the generalized Lagrange algorithm*, J. Number Theory **26** (1987), 31–37.

[Buc4] J. Buchmann, *Zur Komplexität der Berechnung von Einheiten und Klassenzahlen algebraischer Zahlkörper*, Habilitationsschrift, University of Düsseldorf, 1988.

[Buc-Dül] J. Buchmann and S. Düllmann, *A probabilistic class group and regulator algorithm and its implementation*, in [PPWZ], 1991, pp. 53–72.

[Buc-Ford] J. Buchmann and D. Ford, *On the computation of totally real quartic fields of small discriminant*, Math. Comp. **52** (1989), 161-174.

[BFP] J. Buchmann, D. Ford and M. Pohst, *Enumeration of quartic fields of small discriminant*, Math. Comp. **61** (1993), 873–879.

[Buc-Len] J. Buchmann and H.W. Lenstra, *Computing maximal orders and factoring over $\mathbb{Z}_p$*, preprint.

[Buc-Len2] J. Buchmann and H.W. Lenstra, *Approximating rings of integers in number fields*, J. Th. des Nombres Bordeaux (Série 2) **6** (1994), 221–260.

[Buc-Pet] J. Buchmann and A. Pethő, *On the computation of independent units in number fields by Dirichlet's method*, Math. Comp. **52** (1989), 149–159.

[Buc-Poh-Sch] J. Buchmann, M. Pohst and J. Graf von Schmettow, *On the computation of unit groups and class groups of totally real quartic fields*, Math. Comp. **53** (1989), 387–397.

[Buc-Thi-Wil] J. Buchmann, C. Thiel and H. Williams, *Short representation of quadratic integers*, Computational Algebra and Number Theory, Mathematics and its Applications, Kluwer, Dordrecht, 1995, pp. 159–185.

[Buc-Wil] J. Buchmann and H. Williams, *On principal ideal testing in algebraic number fields*, J. Symb. Comp. **4** (1987), 11–19.

[Buel] D. Buell, *The expectation of success using a Monte-Carlo factoring method—some statistics on quadratic class numbers*, Math. Comp. **43** (1984), 313–327.

[BGZ] J. Buhler, B. Gross and D. Zagier, *On the conjecture of Birch and Swinnerton-Dyer for an elliptic curve of rank 3*, Math. Comp. **44** (1985), 473–481.

[BLP] J. Buhler, H. W. Lenstra and C. Pomerance, *Factoring integers with the number field sieve*, [Len-Len2], 1993, pp. 50–94.

[But-McKay] G. Butler and J. McKay, *The transitive groups of degree up to eleven*, Comm. in Algebra **11** (1983), 863–911.

[CEP] E.R. Canfield, P. Erdös and C. Pomerance, *On a problem of Oppenheim concerning "Factorisatio Numerorum"*, J. Number Theory **17** (1983), 1–28.

[Can-Zas] D. Cantor and H. Zassenhaus, *A new algorithm for factoring polynomials over finite fields*, Math. Comp. **36** (1981), 587–592.

[Car] H. Carayol, *Sur les représentations l-adiques associées aux formes modulaires de Hilbert*, Ann. Sci. E.N.S. **19** (1986), 409–468.

[Chu] D. and G. Chudnovsky, *Sequences of numbers generated by addition in formal groups and new primality and factorization tests*, Adv. in Appl. Math. **7** (1986), 187–237.

[Coa-Wil] J. Coates and A. Wiles, *On the conjecture of Birch and Swinnerton-Dyer*, Invent. Math. **39** (1977), 223–251.

[Coh1] H. Cohen, *Variations sur un thème de Siegel et Hecke*, Acta Arith. **30** (1976), 63–93.

[Coh2] H. Cohen, *Formes modulaires à une et deux variables*, Thesis, Univ. de Bordeaux I, 1976.

[Coh3] P. Cohen, *On the coefficients of the transformation polynomials for the elliptic modular function*, Math. Proc. Cambridge Phil. Soc. **95** (1984), 389–402.

[Coh-Diaz] H. Cohen and F. Diaz y Diaz, *A polynomial reduction algorithm*, Sem. Th. Nombres Bordeaux (Série 2) **3** (1991), 351–360.

[CohDiOl] H. Cohen, F. Diaz y Diaz and M. Olivier, *Calculs de nombres de classes et de régulateurs de corps quadratiques en temps sous-exponentiel*, Séminaire de Théorie des Nombres Paris 1990–91 (1993), 35–46.

[Coh-Len1] H. Cohen and H.W. Lenstra, *Heuristics on class groups of number fields*, Number Theory, Noordwijkerhout 1983, LN in Math. **1068**, Springer-Verlag, 1984, pp. 33–62.

[Coh-Len2] H. Cohen and H.W. Lenstra, *Primality testing and Jacobi sums*, Math. Comp. **42** (1984), 297–330.

[Coh-Len3] H. Cohen and A.K. Lenstra, *Implementation of a new primality test*, Math. Comp. **48** (1987), 103–121.

[Coh-Mar1] H. Cohen and J. Martinet, *Class groups of number fields: numerical heuristics*, Math. Comp. **48** (1987), 123–137.

[Coh-Mar2] H. Cohen and J. Martinet, *Etude heuristique des groupes de classes des corps de nombres*, J. Reine Angew. Math. **404** (1990), 39–76.

[Coh-Mar3] H. Cohen and J. Martinet, *Heuristics on class groups: some good primes are not too good*, Math. Comp. **63** (1994), 329–334.

[Col] G. Collins, *The calculation of multivariate polynomial resultants*, JACM **18** (1971), 515–532.

[Cop1] D. Coppersmith, *Solving linear equations over* GF(2), RC **16997**, IBM Research, T.J. Watson research center (1991).

[Cop2] D. Coppersmith, *Solving homogeneous linear equations over* GF(2) *via block Wiedemann algorithm*, Math. Comp. **62** (1994), 333–350.

[Del] P. Deligne, *La conjecture de Weil I*, Publ. Math. IHES **43** (1974), 273–307.

[Deu] Deuring, *Die Klassenkörper der komplexen Multiplication*, Enzyklopädie der mathematischen Wissenschaften **12** (Book 10, Part II), Teubner, Stuttgart, 1958.

[Diaz] F. Diaz y Diaz, *A table of totally real quintic number fields*, Math. Comp. **56** (1991), 801–808 and S1–S12.

[DKT] P. Domich, R. Kannan and L. Trotter, *Hermite normal form computation using modulo determinant arithmetic*, Math. Oper. Research **12** (1987), 50–59.

[Duk] W. Duke, *Hyperbolic distribution functions and half-integral weight Maass forms*, Invent. Math. **92** (1988), 73–90.

[Duv] D. Duval, *Diverses questions relatives au calcul formel avec des nombres algébriques*, Thesis, Univ. of Grenoble, 1987.

[Eic1] Y. Eichenlaub, *Méthodes de calcul des groupes de Galois sur* Q, Mémoire DEA, 1990.

[Eic2] M. Eichler, *On the class number of imaginary quadratic fields and the sums of divisors of natural numbers*, J. Indian Math. Soc. **19** (1955), 153–180.

[Enn-Tur1] V. Ennola and R. Turunen, *On totally real cubic fields*, Math. Comp. **44** (1985), 495–518.

[Enn-Tur2] V. Ennola and R. Turunen, *On cyclic cubic fields*, Math. Comp. **45** (1985), 585–589.

[Fal] G. Faltings, *Endlichkeitssätze für abelsche Varietäten über Zahlkörpern*, Invent. Math. **73** (1983), 349–366.

[Fer1] S. Fermigier, *Un exemple de courbe elliptique définie sur* Q *de rang* $\geq$ 19, C.R. Acad. Sci. Paris **315** (1992), 719–722.

[Fer2] S. Fermigier, in preparation.

[Fin-Poh] U. Fincke and M. Pohst, *Improved methods for calculating vectors of short length in a lattice, including a complexity analysis*, Math. Comp. **44** (1985), 463–471.

[Ford1] D. Ford, *On the computation of the maximal order in a Dedekind domain*, Thesis, Ohio State Univ., 1978.

[Ford2] D. Ford, *The construction of maximal orders over a Dedekind domain*, J. Symb. Comp. **4** (1987), 69–75.

[Ford3] D. Ford, *Enumeration of totally complex quartic fields of small discriminant*, in [PPWZ], 1991, pp. 129–138.

[Fri] E. Friedman, *Analytic formulas for the regulator of a number field*, Invent. math. **98** (1989), 599–622.

[Gir] K. Girstmair, *On invariant polynomials and their application in field theory*, Math. Comp. **48** (1987), 781–797.

[Gol] D. Goldfeld, *The class number of quadratic fields and the conjectures of Birch and Swinnerton-Dyer*, Ann. Sc. Norm. Super. Pisa **3** (1976), 623–663.

[Gol-Kil] S. Goldwasser and J. Kilian, *Almost all primes can be quickly certified*, Proc. 18th Annual ACM Symp. on Theory of Computing (1986), 316–329.

[Gras] M.-N. Gras, *Méthodes et algorithmes pour le calcul numérique du nombre de classes et des unités des extensions cubiques cycliques de* Q, J. Reine Angew. Math. **277** (1975), 89–116.

[Gro-Zag1] B. Gross and D. Zagier, *On singular moduli*, J. Reine Angew. Math. **355** (1985), 191–220.

[Gro-Zag2] B. Gross and D. Zagier, *Heegner points and derivatives of L-series*, Invent. Math. **84** (1986), 225–320.

[GKZ] B. Gross, W. Kohnen and D. Zagier, *Heegner points and derivatives of L-series II*, Math. Ann. **278** (1987), 497–562.

[Haf-McCur1] J. Hafner and K. McCurley, *A rigorous subexponential algorithm for computation of class groups*, Journal American Math. Soc. **2** (1989), 837–850.

[Haf-McCur2] J. Hafner and K. McCurley, *Asymptotically fast triangularization of matrices over rings*, SIAM J. Comput. **20** (1991), 1068–1083.

[Has] H. Hasse, *Arithmetische Theorie der kubischen Zahlkörper auf klassenkörpertheoretischer Grundlage*, Math. Zeit. **31** (1930), 565–582; *Math. Abhandlungen*, Walter de Gruyter, 1975, pp. 423–440.

[HJLS] J. Hastad, B. Just, J.C. Lagarias and C.P. Schnorr, *Polynomial time algorithms for finding integer relations among real numbers*, Siam J. Comput. **18** (1989), 859–881.

[Her] O. Hermann, *Über die Berechnung der Fouriercoefficienten der Funktion $j(\tau)$*, J. Reine Angew. Math. **274/275** (1975), 187–195.

[Hül] A. Hülpke, in preparation.

[Hun] J. Hunter, *The minimum discriminants of quintic fields*, Proc. Glasgow Math. Ass. **3** (1957), 57–67.

[Kal-Yui] E. Kaltofen and N. Yui, *Explicit construction of the Hilbert class fields of imaginary quadratic fields by integer lattice reduction*, New York Number Theory Seminar 1989–1990, Springer-Verlag, 1991, pp. 150–202.

[Kam] S. Kamienny, *Torsion points on elliptic curves and q-coefficients of modular forms*, Invent. Math. **109** (1992), 221–229.

[Kan-Bac] R. Kannan and A. Bachem, *Polynomial algorithms for computing the Smith and Hermite normal form of an integer matrix*, Siam J. Comput. **8** (1979), 499–507.

[Kol1] V.A. Kolyvagin, *Finiteness of $E(\mathbb{Q})$ and $\text{Ш}(E/\mathbb{Q})$ for a subclass of Weil curves*, Izv. Akad. Nauk. SSSR **52** (1988), 522–540.

[Kol2] V.A. Kolyvagin, *Euler systems*, Progress in Math. **87**, Grothendieck Festschrift II, Birkhäuser, Boston, 1991, pp. 435–483.

[LaM] B. LaMacchia, *Basis reduction algorithms and subset sum problems*, Thesis, MIT Artificial Intelligence Lab., 1991.

[LaM-Odl] B. LaMacchia and A.M. Odlyzko, *Solving large sparse linear systems over finite fields*, Advances in cryptology: Crypto 90, A. Menezes and S. Vanstone (eds.), LN in Comp. Sci. **537**, Springer-Verlag, 1991, pp. 109–133.

[Las] M. Laska, *An algorithm for finding a minimal Weierstraß equation for an elliptic curve*, Math. Comp. **38** (1982), 257–260.

[Leh1] S. Lehman, *Factoring large integers*, Math. Comp. **28** (1974), 637–646.

[Leh2] D.H. Lehmer, *On Fermat's quotient, base two*, Math. Comp. **36** (1981), 289–290.

[Len1] H.W. Lenstra, *On the computation of regulators and class numbers of quadratic fields*, Lond. Math. Soc. Lect. Note Ser. **56** (1982), 123–150.

[Len2] H.W. Lenstra, *Divisors in residue classes*, Math. Comp. **42** (1984), 331-334.

[Len3] H.W. Lenstra, *Factoring integers with elliptic curves*, Ann. of Math. **126** (1987), 649–673.

[Len4] A.K. Lenstra, *Polynomial time algorithms for the factorization of polynomials*, dissertation, Univ. of Amsterdam, 1984.

[Len-Len1] A.K. Lenstra and H.W. Lenstra, *Algorithms in number theory*, Handbook of theoretical computer science, J. Van Leeuwen, A. Mayer, M. Nivat, M. Patterson and D. Perrin (eds.), Elsevier, Amsterdam, 1990, pp. 673–715.

[Len-Len2] A.K. Lenstra and H.W. Lenstra (eds.), *The development of the nmber field sieve*, LN in Math. **1554**, Springer-Verlag, Berlin, Heidelberg, New-York, 1993.

[LLL] A.K. Lenstra, H.W. Lenstra and L. Lovász, *Factoring polynomials with rational coefficients*, Math. Ann. **261** (1982), 515–534.

[LLMP] A.K. Lenstra, H.W. Lenstra, M.S. Manasse and J.M. Pollard, *The Number Field Sieve*, in [Len-Len2], 1993, pp. 11–42.

[Llo-Quer] P. Llorente and J. Quer, *On the 3-Sylow subgroup of the class group of quadratic fields*, Math. Comp. **50** (1988), 321–333.

[Mah] K. Mahler, *On a class of non-linear functional equations connected with modular functions*, J. Austral. Math. Soc. **22A** (1976), 65–118.

[Mart] J. Martinet, *Méthodes géométriques dans la recherche des petits discriminants*, Progress in Math **59**, 1985, pp. 147–179.

[Maz] B. Mazur, *Rational isogenies of prime degree*, Invent. Math. **44** (1978), 129–162.

[McCur] K. McCurley, *Cryptographic key distribution and computation in class groups*, Proceedings of NATO ASI Number Theory and applications, Kluwer Academic Publishers, 1989, pp. 459–479.

[Mer] L. Merel, *Bornes pour la torsion des courbes elliptiques sur les corps de nombres*, Invent. Math. **124** (1996), 437–449.

[Mes1] J.-F. Mestre, *Construction d'une courbe elliptique de rang ≥ 12*, C.R. Acad. Sci. Paris **295** (1982), 643–644.

[Mes2] J.-F. Mestre, *Formules explicites et minorations de conducteurs de variétés algébriques*, Compositio Math. **58** (1986), 209–232.

[Mes3] J.-F. Mestre, *Courbes elliptiques de rang ≥ 12 sur $\mathbb{Q}(t)$*, C.R. Acad. Sci. Paris (1991), 171–174.

[Mes4] J.-F. Mestre, *Un exemple de courbe elliptique sur $\mathbb{Q}$ de rang ≥ 15*, C.R. Acad. Sci. Paris **314** (1992), 453–455.

[Mes5] J.-F. Mestre, private communication.

[Mig] M. Mignotte, *An inequality about factors of polynomials*, Math. Comp. **28** (1974), 1153–1157.

[Mil] G. Miller, *Riemann's hypothesis and tests for primality*, J. Comput. and System Sc. **13** (1976), 300–317.

[Mol-Wil] R. Mollin and H. Williams, *Computation of the class number of a real quadratic field*, Utilitas Math. **41** (1992), 259–308.

[Mon-Nar] J. Montes and E. Nart, *On a theorem of Ore*, Journal of Algebra **146** (1992), 318–339.

[Mon1] P. Montgomery, *Modular multiplication without trial division*, Math. Comp. **44** (1985), 519–521.

[Mon2] P. Montgomery, *Speeding the Pollard and elliptic curve methods of factorization*, Math. Comp. **48** (1987), 243–264.

[Mor1] F. Morain, *Résolution d'équations de petit degré modulo de grands nombres premiers*, Rapport de recherche INRIA **1085** (1989).

[Mor2] F. Morain, *Courbes elliptiques et tests de primalité*, Thesis, Univ. Claude Bernard, Lyon, 1990.

[Mor-Nic] F. Morain and J.-L. Nicolas, *On Cornacchia's algorithm for solving the Diophantine equation $u^2 + dv^2 = m$* (to appear).

[Nag] K. Nagao, *An example of elliptic curve over $\mathbb{Q}(T)$ with rank ≥ 13*, Proc. Japan Acad. **70** (1994), 152–153.

[Nag-Kou] K. Nagao and T. Kouya, *An example of elliptic curve over $\mathbb{Q}$ with rank ≥ 21*, Proc. Japan Acad. **70** (1994), 104–105.

[Nic] J.-L. Nicolas, *Etre ou ne pas être un carré*, Dopo le Parole, (a collection of not always serious papers for A. K. Lenstra's doctorate), Amsterdam, 1984.

[Odl] A.M. Odlyzko, *Bounds for discriminants and related estimates for class numbers, regulators and zeros of zeta functions: a survey of recent results*, Sem. Th. des Nombres Bordeaux (Série 2) **2** (1991), 117–141.

[Oes] J. Oesterlé, *Nombre de classes des corps quadratiques imaginaires*, in Séminaire Bourbaki 1983–84, Astérisque **121–122**, Soc. Math. de France, 1985, pp. 309–323.

[Oli1] M. Olivier, *Corps sextiques primitifs*, Ann. Institut Fourier **40** (1990), 757–767.

[Oli2] M. Olivier, *The computation of sextic fields with a cubic subfield and no quadratic subfield*, Math. Comp. **58** (1992), 419–432.

[Oli3] M. Olivier, *Tables de corps sextiques contenant un sous-corps quadratique (I)*, Sém. Th. des Nombres Bordeaux (Série 2) **1** (1989), 205–250.

[Oli4] M. Olivier, *Corps sextiques contenant un corps quadratique (II)*, Sém. Th. des Nombres Bordeaux (Série 2) **2** (1990), 49–102.

[Oli5] M. Olivier, *Corps sextiques contenant un corps cubique (III)*, Sém. Th. des Nombres Bordeaux (Série 2) **3** (1991), 201–245.

[Oli6] M. Olivier, *Corps sextiques primitifs (IV)*, Sém. Th. des Nombres Bordeaux (Série 2) **3** (1991), 381–404.

[Ore] Ö. Ore, *Newtonsche Polygone in der Theorie der algebraischen Körper*, Math. Ann. **99** (1928), 84–117.

[Poh1] M. Pohst, *On the computation of number fields of small discriminants including the minimum discriminants of sixth degree fields*, J. Number Theory **14** (1982), 99–117.

[Poh2] M. Pohst, *A modification of the LLL-algorithm*, J. Symb. Comp. **4** (1987), 123–128.

[Poh3] M. Pohst, *On computing isomorphisms of equation orders*, Math. Comp. **48** (1987), 309–314.

[Poh4] M. Pohst, *A note on index divisors*, in [PPWZ], 1991, pp. 173–182.

[Poh-Wei-Zas] M. Pohst, P. Weiler and H. Zassenhaus, *On effective computation of fundamental units I and II*, Math. Comp. **38** (1982), 275–329.

[Poh-Zas1] M. Pohst and H. Zassenhaus, *Über die Berechnung von Klassenzahlen und Klassengruppen algebraische Zahlkörper*, J. Reine Angew. Math. **361** (1985), 50-72.

[Pol1] J. Pollard, *Theorems on factorization and primality testing*, Proc. Cambridge Phil. Soc. **76** (1974), 521–528.

[Pol2] J. Pollard, *A Monte-Carlo method for factorization*, BIT **15** (1975), 331–334.

[Pom] C. Pomerance, *Analysis and comparison of some integer factoring algorithms*, in [MCC], 1983, pp. 89–139.

[Quer] J. Quer, *Corps quadratiques de 3-rang 6 et courbes elliptiques de rang 12*, C.R. Acad. Sci. Paris **305** (1987), 1215–1218.

[Rab] M. Rabin, *Probabilistic algorithms for testing primality*, J. Number Theory **12** (1980), 128–138.

[Rib] K. Ribet, *On modular representations of $\mathrm{Gal}(\overline{\mathbb{Q}}/\mathbb{Q})$ arising from modular forms*, Invent. Math. **100** (1990), 431–476.

[Rub] K. Rubin, *Tate-Shafarevitch groups and L-functions of elliptic curves with complex multiplication*, Invent. Math. **93** (1987), 527–560.

[von Schm1] J. Graf v. Schmettow, *Beiträge zur Klassengruppenberechnung*, Dissertation, Univ. Düsseldorf, 1991.

[von Schm2] J. Graf v. Schmettow, *KANT – a tool for computations in algebraic number fields*, in [PPWZ], 1991, pp. 321–330.

[Schn] C.P. Schnorr, *A more efficient algorithm for lattice basis reduction*, J. Algorithms **9** (1988), 47–62.

[Schn-Euch] C.P. Schnorr and M. Euchner, *Lattice basis reduction: Improved practical algorithms and solving subset sum problems*, Proc. of the FCT 1991, LN in Comp. Sci. **529**, Springer-Verlag, Berlin, Heidelberg, 1991, pp. 68–85.

[Schn-Len] C.P. Schnorr and H.W Lenstra, *A Monte-Carlo factoring algorithm with linear storage*, Math. Comp. **43** (1984), 289–312.

[Schön] A. Schönhage, *Probabilistic computation of integer polynomial GCD*, J. Algorithms **9** (1988), 365–371.

[Scho] R. Schoof, *Elliptic curves over finite fields and the computation of square roots mod p*, Math. Comp. **43** (1985), 483–494.

[Scho2] R. Schoof, *Counting points of elliptic curves over finite fields*, J. Th. des Nombres Bordeaux (Série 2) **7** (1995), 219–254.

[SPD] A. Schwarz, M. Pohst and F. Diaz y Diaz, *A table of quintic number fields*, Math. Comp. **63** (1994), 361–376.

[Sel-Wun] J. Selfridge and M. Wunderlich, *An efficient algorithm for testing large numbers for primality*, Proc. Fourth Manitoba Conf. Numer. Math. (1974), 109–120.

[Ser1] J.-P. Serre, *Sur les représentations modulaires de degré 2 de $\mathrm{Gal}(\overline{\mathbb{Q}}/\mathbb{Q})$*, Duke Math. J. **54** (1987), 179–230.

[Sey1] M. Seysen, *A probabilistic factorization algorithm with quadratic forms of negative discriminants*, Math. Comp. **48** (1987), 757-780.

[Sey2] M. Seysen, *Simultaneous reduction of a lattice basis and its reciprocal basis*, Combinatorica **13** (1993), 363-376.

[Sha1] D. Shanks, *Class number, a theory of factorization, and genera*, Proc. Symp. in Pure Maths. **20**, A.M.S., Providence, R.I., 1969, pp. 415-440.

[Sha2] D. Shanks, *On Gauss and composition I and II*, Number theory and applications, R. Mollin (ed.), Kluwer Academic Publishers, 1989, pp. 163-204.

[Sha3] D. Shanks, *The infrastructure of a real quadratic field and its applications*, Proc. 1972 Number theory conference, Boulder (1972), 217-224.

[Sha4] D. Shanks, *Incredible identities*, Fibon. Quart. **12** (1974).

[Sha-Wil] D. Shanks and H. Williams, *A note on class number one in pure cubic fields*, Math. Comp. **33** (1979), 1317-1320.

[Shi1] G. Shimura, *On the zeta-function of an Abelian variety with complex multiplication*, Ann. of Math. **94** (1971), 504-533.

[Shi2] G. Shimura, *On elliptic curves with complex multiplication as factors of the Jacobians of modular function fields*, Nagoya Math. J. **43** (1971), 199-208.

[Sie] C.L. Siegel, *Über die Classenzahl quadratischer Zahlkörper*, Acta Arith. **1** (1935), 83-86.

[Sil1] R. Silverman, *The multiple polynomial quadratic sieve*, Math. Comp. **48** (1987), 329-340.

[Sil2] J. Silverman, *Computing heights on elliptic curves*, Math. Comp. **51** (1988), 339-358.

[Soi] L. Soicher, *The computation of Galois groups*, Thesis, Concordia Univ., Montreal, 1981.

[Soi-McKay] L. Soicher and J. McKay, *Computing Galois groups over the rationals*, J. Number Theory **20** (1985), 273-281.

[Sol-Str] R. Solovay and V. Strassen, *A fast Monte-Carlo test for primality*, SIAM J. Comput. **6** (1977), 84-85;erratum ibid. **7** (1978), p. 118.

[Star] H. Stark, *Class numbers of complex quadratic fields*, in Modular Functions of one variable I, LN in Math. **320**, Springer-Verlag, Berlin, Heidelberg, 1973, pp. 153-174.

[Stau] R.P. Stauduhar, *The determination of Galois groups*, Math. Comp. **27** (1973), 981-996.

[Tay-Wil] R. Taylor and A. Wiles, *Ring-theoretic properties of certain Hecke algebras*, Ann. of Math. **141** (1995), 553-572.

[Ten-Wil] M. Tennenhouse and H. Williams, *A note on class number one in certain real quadratic and pure cubic fields*, Math. Comp. **46** (1986), 333-336.

[Tra] B. Trager, *Algebraic factoring and rational function integration*, Proceedings of SYMSAC '76 (1976), 219-226.

[Val] B. Vallée, *Une approche géométrique des algorithmes de réduction en petite dimension*, Thesis, Univ. of Caen, 1986.

[Wag] C. Wagner, *Class number 5, 6 and 7*, Math. Comp. **65** (1996), 785-800.

[de Weg] de Weger B., *Algorithms for Diophantine equations*, Dissertation, Centrum voor Wiskunde en Informatica, Amsterdam, 1988.

[Weil] A. Weil, *Number of solutions of equations in finite fields*, Bull. A.M.S. **55** (1949), 497-508.

[Wie] D. Wiedemann, *Solving sparse linear equations over finite fields*, IEEE Trans. Information Theory **32** (1986), 54-62.

[Wiles] A. Wiles, *Modular elliptic curves and Fermat's last theorem*, Ann. of Math. **141** (1995), 443-551.

[Wil-Jud] H. Williams and J. Judd, *Some algorithms for prime testing using generalized Lehmer functions*, Math. Comp. **30** (1976), 157-172 and 867-886.

[Wil-Zar] H. Williams and C. Zarnke, *Some algorithms for solving a cubic congruence modulo p*, Utilitas Math. **6** (1974), 285-306.

[Zag1] D. Zagier, *On the values at negative integers of the zeta-function of a real quadratic field*, Ens. Math. **22** (1976), 55-95.

[Zag2] D. Zagier, *Modular forms whose Fourier coefficients involve zeta-functions of quadratic fields*, Modular functions of one variable VI, LN in Math. **627**, Springer-Verlag, Berlin, Heidelberg, New-York, 1977, pp. 105–169.

[Zag3] D. Zagier, *Large integral points on elliptic curves*, Math. Comp. **48** (1987), 425–436.

[Zim1] R. Zimmert, *Ideale kleiner Norm in Idealklassen und eine Regulatorabschätzung*, Invent. Math. **62** (1981), 367–380.

[Zip] R. Zippel, *Simplification of expressions involving radicals*, J. Symb. Comp. **1** (1985), 189–210.

Index

Errata et Addenda to the Third and Fourth Corrected Printings of
A Course in Computational Algebraic Number Theory
by Henri Cohen
(20001127 version)

Warning. The errata presented here are of course to be taken into account for the first and second printing, but the page and line numbering given here corresponds to the third and fourth printings and is quite different from that of the preceding printings.

Graduate Texts in Mathematics 138, Springer-Verlag, 1993,
Third, Corrected Printing 1996, XX + 545 pages.
ISBN 3-540-55640-0 Springer-Verlag Berlin Heidelberg New York
ISBN 0-387-55640-0 Springer-Verlag New York Berlin Heidelberg

p. VI   at Shanks, add the footnote "Daniel Shanks died on September 6, 1996"

p. VI   middle and p. VII line 11, instead of "Francois Dress" read "François Dress"

p. VI   line -1, instead of "Jean-Francois Mestre, Francois Morain" read "Jean-François Mestre, François Morain"

p. 11   just before "Quite a different way" insert the following long text

"Perhaps surprisingly, we can easily improve on Algorithm 1.2.4 by using a flexible window of size at least k bits, instead of using a window of fixed size k. Indeed, it is easy to see that any positive integer N can be written in a unique way as

$$N = 2^{t_0}(a_0 + 2^{t_1}(a_1 + \cdots + 2^{t_e}a_e))$$

where $t_i \geq k$ for $i \geq 1$ and the a_i are odd integers such that $1 \leq a_i \leq 2^k - 1$ (in Algorithm 1.2.4 we took $t_0 = 0$, $t_i = k$ for $i \geq 1$, and $0 \leq a_i \leq 2^k - 1$ odd or even).

As before, we can precompute $g^3, g^5, \ldots, g^{2^k-1}$ and then compute g^N by successive squarings and multiplications by g^{a_i}. To find the a_i and t_i, we use the following immediate sub-algorithm.

Sub-Algorithm 1.2.4.1 (Flexible Base 2^k Digits). Given a positive integer N and $k \geq 1$, this sub-algorithm computes the unique integers t_i and a_i defined above. We use $[N]_{b,a}$ to denote the integer obtained by extracting bits a through b (inclusive) of N, where bit 0 is the least significant bit.

1. [Compute t_0] Let $t_0 \leftarrow v_2(N)$, $e \leftarrow 0$ and $s \leftarrow t_0$.

2. [Compute a_e] Let $a_e \leftarrow [N]_{s+k-1,s}$.

3. [Compute t_e] Set $m \leftarrow [N]_{\infty,s+k}$. If $m = 0$, terminate the sub-algorithm. Otherwise, set $e \leftarrow e + 1$, $t_e \leftarrow v_2(m) + k$, $s \leftarrow s + t_e$ and go to step 2.

The flexible window algorithm is then as follows.

Algorithm 1.2.4.2 (Flexible Left-Right Base 2^k). Given $g \in G$ and $n \in \mathbb{Z}$, this algorithm computes g^n in G. We write 1 for the unit element of G.

1. [Initialize] If $n = 0$, output 1 and terminate. If $n < 0$ set $N \leftarrow -n$ and $z \leftarrow g^{-1}$. Otherwise, set $N \leftarrow n$ and $z \leftarrow g$.

2. [Compute the a_i and t_i] Using the above sub-algorithm, compute a_i, t_i and e such that $N = 2^{t_0}(a_0 + 2^{t_1}(a_1 + \cdots + 2^{t_e}a_e))$ and set $f \leftarrow e$.

3. [Precomputations] Compute and store $z^3, z^5, \ldots, z^{2^k-1}$.

4. [Loop] If $f = e$ set $y \leftarrow z^{a_f}$ otherwise set $y \leftarrow z^{a_f} \cdot y$. Then repeat t_f times $y \leftarrow y \cdot y$.

5. [Finished?] If $f = 0$, output y and terminate the algorithm. Otherwise, set $f \leftarrow f - 1$ and go to step 4.

We have used above the word "surprisingly" to describe the behavior of this algorithm. Indeed, it is not a priori clear why it should be any better than Algorithm 1.2.4. An easy analysis shows, however, that the average number of multiplications which are not squarings is now of the order of $2^{k-1} + \lg|n|/(k+1)$ (instead of $2^{k-1} + \lg|n|/k$ in Algorithm 1.2.4), see Exercise 33. The optimal value of k is the smallest integer satisfying the inequality $\lg|n| \leq (k+1)(k+2)2^{k-1}$.

In the above example where n has 100 decimal digits, the flexible base 2^5 algorithm takes on average $(3/4)332 + 16 + 332/6 \approx 320$ multiplications, another 3% improvement. In fact, using a simple modification, in certain cases we can still easily improve (very slightly) on Algorithm 1.2.4.2, see Exercise 34."

p. 11 line -11, instead of "the 2^k algorithm" read "the flexible 2^k algorithm"

p. 17 in Algorithm 1.3.7, remove the initializations "$A \leftarrow 1$, $B \leftarrow 0$, $C \leftarrow 0$, $D \leftarrow 1$" from step 1 and put them instead at the end of step 2

p. 45 add the following exercises.

"33. Show that, as claimed in the text, the average number of multiplications which are not squarings in the flexible left-right base 2^k algorithm is approximately $2^{k-1} + \lg|n|/(k+1)$, and that the optimal value of k is the smallest integer such that $\lg|n| \leq (k+1)(k+2)2^{k-1}$.

34. Consider the following modification to Algorithm 1.2.4.2. We choose some odd number L such that $2^{k-1} < L < 2^k$ and precompute only $z, z^3, \ldots, z^L$. Show that one can write any integer N in a unique way as $N = 2^{t_0}(a_0 + 2^{t_1}(a_1 + \cdots + 2^{t_e}a_e))$ with a_i odd, $a_i \leq L$, and $t_i \geq k - 1$ for $i \geq 1$, but $t_i = k - 1$ only if $a_i > L - 2^{k-1}$. Analyze the resulting algorithm and show that, in certain cases, it is slightly faster than Algorithm 1.2.4.2."

p. 52 line -1, instead of "column" read "column, with $k + 1 \leq i \leq n$"

p. 69 step 4 of Algorithm 2.4.5, instead of "set $k \leftarrow k + 1$ and go to step 5" read "set $k \leftarrow k + 1$, and if $l > 1$ and $i = l$ set $l \leftarrow l - 1$, then go to step 5"

p. 72 line 4 of step 4 of Algorithm 2.4.8, instead of "$W_j \leftarrow W_j - qW_i$" read "$W_j \leftarrow W_j - qW_i \bmod R$"

p. 73 line -3, instead of "last $n - r + 1$" read "last $n - r$"

p. 129 line 4, instead of "$p = 2$" read "$p > 2$"

p. 129 line -20, instead of "$U \circ T$" read "$U \circ T \bmod A$"

p. 156 line -17, instead of "$A_1(b) < 0$ when" read "$A_1(b) < 0$ if and only if"

p. 157 line -8, instead of "Proposition 4.8.6" read "Theorem 4.8.6"

p. 159 line -5, instead of "$r_{i,k}$" read "$r_{k,i}$"

p. 159 line -4, instead of "$(r_{0,k}, r_{1,k}, \ldots, r_{n-1,k}, 1)$" read "$(r_{k,0}, r_{k,1}, \ldots, r_{k,n-1}, 1)$" and instead of "$r_{i,k}$" read "$r_{k,i}$"

p. 159 line -3, instead of "$r_{i,0}$" read "$r_{0,i}$"

p. 160 line 10, instead of "$r_{i,j}$" read "$r_{k,i}$"

p. 161   middle, instead of "This will in practice be considered as a $r_1 + 2r_2 = n$-uplet of real numbers. Now operations" read "Operations"

p. 168   line 8 and 9, instead of "p is an odd prime" read "p is a prime"

p. 176   lines 12 to 15, replace the four lines of the end of the proof starting with "If we set γ..." by "It follows that the vector of the $(P(\beta_i))$ and of the $\alpha_{\phi(i)}$ are both solutions of the linear system $\sum_{1 \le i \le n} v_i \beta_i^h = s_h$, and since the β_i are distinct this system has a unique solution, so the vectors are equal, thus proving the proposition."

p. 179   line -3 and p. 180 line -11, instead of "a^{m-1}" read "a^{n-1}"

p. 184   line -8, instead of "so $M \subset M'$" read "so M annihilates IH/IJ hence $M \subset M'$"

p. 193   line 11, add "Note that this is simply the proof of the Chinese remainder theorem for ideals."

p. 195   line 2 of Algorithm 4.7.10, instead of "$\mathbb{Z}_K$-generators" read "$\mathbb{Z}$-generators"

p. 195   line 1 of step 3 of Algorithm 4.7.10, instead of "$2 \le i \le m$" read "$2 \le i \le k$"

p. 195   line 2 of step 4 of Algorithm 4.7.10, instead of "$j + 1 \le i \le m$" read "$j + 1 \le i \le k$"

p. 200   line 14, instead of "$e_i = f_i$" read "$d_i = e_i$"

p. 201   line -5, instead of "Then $y \notin xR$ and $y\mathfrak{p} \subset xR$, hence $a = y/x$" read "Since $y\mathfrak{p} \subset xR$, the element $a = y/x$"

p. 202   line -10, instead of " $\displaystyle\sum_{1 \le \le n}$ " read " $\displaystyle\sum_{1 \le i \le n}$ "

p. 204   line 3 of step 3, instead of "0" read "0"

p. 204   line 1 of step 5, instead of "If $p \nmid A_{n,n}$," read "Using Algorithm 2.4.8, replace A by its HNF. Then, if $p \nmid A_{n,n}$,"

p. 206   line -7, instead of "determinant $d(K)$" read "discriminant $d(K)$"

p. 211   line -13 and -12, instead of "where $\|x\|$ denotes the absolute value of x when x is real and the square of the modulus of x when x is complex" read "where $\|\sigma(x)\| = |\sigma(x)|$ if σ is a real embedding and $\|\sigma(x)\| = |\sigma(x)|^2$ if σ is a complex embedding"

p. 216   line 1, instead of "$\dfrac{1}{6}$" read "$\dfrac{1}{60}$"

p. 217   line -4, instead of "$A \leftarrow 8b - 3a^2$" read "$A \leftarrow 3a^2 - 8b$" and line -2, instead of "$(r_1, r_2) = (0, 4)$ iff $D > 0$ and $AB < 0$" read "$(r_1, r_2) = (0, 2)$ iff $D > 0$ and either $A \le 0$ or $B \le 0$"

p. 224   line -2, replace 4 times small parentheses by larger ones

p. 227   line 1, instead of "$(-b + \sqrt{D})/2a$" read "$(-b + \sqrt{D})/(2a)$"

p. 234   line -6, instead of "$H(0) = -1/12$" read "$H(N) = -1/12$"

p. 237   line -13, instead of "Let D be a negative fundamental discriminant" read "Let D be a negative discriminant (not necessarily fundamental)"

p. 237   line -10 and -9, instead of "entire function satisfying" read "entire function. If in addition D is a fundamental discriminant, this function satisfies the functional equation"

p. 240   line 4, instead of "time" read "average time"

p. 246   line 7, instead of "$I_i = a_i\mathbb{Z} + \tau_i\mathbb{Z}$" read "$I_i = a_i(\mathbb{Z} + \tau_i\mathbb{Z})$"

p. 246   line 9, instead of "$\tau_3 = ua_1\tau_2 + va_2\tau_1 + w\tau_1\tau_2$" read "$\tau_3 = (d/d_0)(u\tau_2 + v\tau_1 + w\tau_1\tau_2)$"

p. 248 line 2 of step 3 of Algorithm 5.4.8, instead of "$c_2 = c_2 + gd_1$" read "$c_2 \leftarrow c_2 + gd_1$"

p. 249 line 2 of step 6 of Algorithm 5.4.9, instead of "$c_3 \leftarrow v_3 d + gd_1$" read "$c_3 \leftarrow v_3 f + gd_1$"

p. 250 line -16, instead of "guess that $h(D)$" read "guess that, for $D < -4$, $h(D)$"

p. 252 line 7, instead of "[McCur-Haf]" read "[Haf-McCur1]"

p. 262 line -5, instead of "reduced form" read "quadratic form"

p. 262 line -3, after (1) insert "If (a, b, c) is reduced, then"

p. 262 line -2, instead of "More precisely" read "More precisely, if (a, b, c) is reduced"

p. 276 line -7, add a white square at the end of the line

p. 280 line -6, instead of "positive norm." read "positive norm. By abuse of notation, we will again denote by $\delta(f, g)$ the unique representative belonging to the interval $[0, R^+[$, and similarly for the distance between ideals."

p. 282 line -16, instead of "very small." read "very small. More precisely, it can be proved (see [Len1]) that $\delta(f, \rho^2(f)) > \ln 2$, hence the number of reduction steps is at most $4 \ln(D)/\ln 2$."

p. 289 line -19, instead of "$\delta(1, f) = (eL \ln 2 + \ln R)/2$" read "$\delta(f_0, f) = (eL \ln 2 + \ln R)/2$ for some fixed form f_0 equivalent to f"

p. 290 line 20, instead of "$\delta(1, f)$" read "$\delta(\prod_{p \leq P} f_p^{e_p}, f)$"

p. 290 line -6, instead of "$g = \prod_{p \leq P} f_p^{v_p}$" read "$g = \prod_{p \leq P} f_p^{\varepsilon_p v_p}$"

p. 290 line -1, instead of "$f_{p_i}^{a_{i,j}}$" read "$f_{p_i}^{-a_{i,j}}$"

p. 291 line 9, instead of "$a_{n+1,j} =$" read "$a_{n+1,j} \equiv$"

p. 304 lines -12 and -11, instead of "again by the binomial theorem" read "using this time the multinomial theorem instead of the binomial theorem"

p. 321 step 9, instead of "s" read "f" (3 times)

p. 322 step 14, instead of "r" read "s" (7 times) and instead of "d" read "r" (3 times)

p. 340 line 3, instead of "$(\overline{f}, \overline{gh}) = 1$" read "$(\overline{f}, \overline{g}, \overline{h}) = 1$"

p. 343 line 3 of Corollary 6.4.12, instead of "$\dfrac{-3v \pm u}{6v}$" read "$\dfrac{-3v \mp u}{6v}$"

p. 359 line 4 of step 5, instead of "the matrix is not of maximal rank" read "one of the matrices H or C is not of maximal rank"

p. 368 and following major correction (oversight in all the previous printings): exchange in most places "ω_1" and "ω_2", except p. 415 where the "ω_2" is correct. In particular, the canonical basis (ω_1, ω_2) for a real elliptic curve is now such that ω_2 is real and ω_1 is in the upper half plane. Specifically, the corrections are p. 368, p. 370, p. 378, twice p. 395, five times page 396, twice p. 398 and p. 412 replace "ω_2/ω_1" by "ω_1/ω_2", twice p. 395 and twice p. 396 replace "$2\pi/\omega_1$" by "$2\pi/\omega_2$", twice p. 396 replace "$c\omega_2 + d\omega_1$" by "$c\omega_1 + d\omega_2$", twice p. 396 and p. 398 replace z/ω_1 by z/ω_2, three times p. 396, six times p. 398, twice p. 399 and p. 412 replace an isolated "ω_1" by "ω_2", twice p. 398 and five times p. 399 (but *not* p. 415) replace an isolated "ω_2" by "ω_1". Although not mathematically necessary, it is then more aesthetic to replace everywhere "$\mathbb{Z} + \mathbb{Z}\tau$" by "$\mathbb{Z}\tau + \mathbb{Z}$".

p. 392 line -16, instead of "$n \geq 2$" read "$n \leq 2$"

p. 395 line -1, add the following: "**Warning.** The condition $m \geq 1$ in step 3 should in practice be implemented as $m > 1 - \varepsilon$ for some small $\varepsilon > 0$ depending on the current accuracy. If this precaution is not taken the algorithm may loop indefinitely, and the cost is simply that the final τ may land very close to but not exactly in the standard fundamental domain, and this has absolutely no consequence for practical computations."

p. 407   line 2 of step 3, instead of "set $c \leftarrow 1$" read "set $c \leftarrow \nu$"

p. 408   line 1 of step 9, instead of "$a_4 and p^3$" read "a_4 and p^3"

p. 408   line 2 of step 9, instead of "a^6" read "a_6"

p. 408   line 1 of step 11, instead of "$X^2 + a_3/p^2 X + a_6/p^4$" read "$X^2 + a_3/p^2 X - a_6/p^4$"

p. 408   line -1, instead of "$c \leftarrow 1\ T \leftarrow II^*$" read "$c \leftarrow 1,\ T \leftarrow II^*$"

p. 416   line -4, instead of "$f_1(\sqrt{D})$" read "$f_1(\sqrt{D/4})$"

p. 417   line 4 of Exercise 1, instead of "and $b_2 \equiv 0,...$ respectively" read "and $b_2 \equiv -c_6 \pmod{12}$"

p. 425   line 3 of Section 8.4, instead of "may be factor" read "may be a factor"

p. 432   line 8, instead of "Proposition 8.5.3" read "Proposition 8.5.4"

p. 435   line -3, instead of "corresponding to $\mathfrak{b}$" read "corresponding to $\mathfrak{b}^{-1}$"

p. 436   line 1, instead of "$\delta(g_1, g) =$" read "$\delta(g_1, g^{-1}) =$"

p. 440   line 19, instead of "It is also however also" read "It is however also"

p. 440   line -10, instead of "We must show how are we going" read "We must explain how we are going"

p. 452   line -6, instead of "$\max(e, k + 1))$" read "$\max(e, k + u))$ where u is as in Lemma 9.1.10"

p. 452   line -5, instead of "Proposition 9.1.8" read "Lemma 9.1.8"

p. 453   middle, after "prime r dividing N" insert "(by Lemma 9.1.10 and our choice of ℓ)"

p. 454   line -16, instead of "of order" read "of order dividing"

p. 462   line 3 of the proof of Lemma 9.1.24, instead of "$j_3(\chi, \chi, \chi) =$" read "$j_3(\chi, \chi, \chi)^\gamma =$"

p. 474   step 4 of Algorithm 9.2.4, instead of "$(x+3y)$" read "$(x+3y)/2$" (twice) and instead of "$(x - 3y)$" read "$(x - 3y)/2$" (twice)

p. 476   line 3 of Exercise 7. instead of "$\chi(x) \neq 1$" read "$\chi(x) \neq 0$ and 1"

p. 476   Exercise 7, add the following question.

" c) Show that if χ is a primitive character modulo q which is not necessarily a prime, we still have $|\tau(\chi)| = \sqrt{q}$."

p. 478   line 5, instead of "$\varepsilon = 0$ or 1" read "$\varepsilon_k = 0$ or 1"

p. 480   line 2 of the second remark, instead of "a follows" read "as follows"

p. 482   lines 4, 14, 16, 19, instead of "$1/2a$" read "$1/(2a)$"

p. 487   middle, instead of "$t = 0 \pmod{N}$" read "$t \equiv 0 \pmod{N}$"

p. 490   line -12, instead of "we note than one can" read "we note that one can"

p. 490   line -1, instead of "Pomerance ," read "Pomerance,"

p. 494   line -4, instead of "is t is the" read "if t is the"

p. 499   line 9, instead of "$\mathcal{N}(a + b\theta)$" read "$\ln(\mathcal{N}(a + b\theta))$"

p. 500   line 19, instead of "Let V is the column" read "Let V be the column"

p. 528   before [Lang1], add the following: "One can find at the URL
        http://www-cs-faculty.stanford.edu/~knuth/index.html
nearly 350 pages of corrections and additions to [Knu1], [Knu2] and [Knu3], absolutely necessary for those having the older editions of Knuth's books. This has been incorporated in a new 3 volume set which came out in 1996."

p. 535   in [Len-Len2], instead of "nmber field" read "number field"

p. 538   instead of "[**de Weg**] de Weger B.," read "[**deWeg**] B. de Weger,"

Graduate Texts in Mathematics

(continued from page ii)